AF572372

Sir Thomas Kirke Rose

Thomas Kirke Rose was born August 14, 1865. He entered the Royal School of Mines in 1883 and graduated in 1887 with the A.R.S.M. in Metallurgy. From 1888 to 1889 he worked as chemist and assayer for the Colorado Gold and Silver Extraction Co. in Denver, Colorado. In 1890 he returned to London and joined the staff of the Royal Mint as Assistant Assayer. In 1891 he took the degree B.Sc. at London University and D.Sc. in 1895.

In 1894, at the age of 29, he published the first edition of his *Metallurgy of Gold.* The book went through seven editions over the course of 43 years. The early editions were edited by the renowned Professor Roberts-Austin. Roberts-Austin was the Chief Chemist and Assayer at the Mint and appears to have been an important influence in Rose's early career. The book soon became a standard reference in the industry and established Rose as a leading expert in the field. When Roberts-Austin died in 1902, Rose succeeded him as Chief Chemist and Assayer of the Mint. He held that position until he retired in 1926.

Dr. Rose's career roughly coincides with a golden age in gold metallurgical research. British gold coinage also peaked during his career, reaching 36,500,000 pieces in 1912. He actively participated in numerous professional organizations, including the Institution of Mining and Metallurgy, The Institute of Metals, The Royal Institute of Chemistry, and The Chemical Society. He also continued to write extensively throughout his career. (See List of Rose's publications.) He was knighted in 1914.

Dr. Rose was consulted by the Transvaal Chamber of Mines concerning the design of the Rand Refinery. The Rand Refinery was completed in December, 1921 and is the world's largest gold refinery.

Dr. Rose was known as an extremely energetic and humorous person. His leisure activities included mountain and rock climbing. His favorite climbing areas were Wales and Switzerland. In 1914, at the age of 49, he was caught in Switzerland at the outbreak of World War I. He was one of many Englishmen who had to find their way home by devious routes. In his later years he also played on representative teams of chess and bridge.

Sir Thomas Kirke Rose died on May 10th, 1953 at Hindhead, Surrey at the age of 87.

List of Publications by Sir Thomas Kirke Rose

The Detection Of Gold In Dilute Solutions, Chemical News, 66, 1892, p. 271.

The Volatilisation Of Metallic Gold, Chem. Soc. J., 63, 1893, pp. 714-724.

Limits Of Accuracy Attained In Gold Bullion Assay, Chem. Soc. J., 63, 1893, pp. 700-713.

Some Physical Properties Of The Chlorides Of Gold, Chem. Soc. J., 67, 1895, pp. 905-906.

The Dissociation Of Chloride Of Gold, Chem. Soc. J., 67, 1895, pp. 881-904.

Note on Liquation (Segregation) in Crystalline Standard Gold, Chem. Soc. J., 67, 1895, pp. 552-556.

The Klondike Placers, Nature, 56, 1897, pp. 615-616.

The Cause Of Loss Incurred In Roasting Gold Ores Containing Tellurium, Brit. Ass. Rep., 1897, pp. 623-624.

Electrical Percipitation Of Gold On Amalgamated Plates, Inst. Min. Met. Trans., 8, 1899-1901, pp. 369-378.

Note On Volhard's Method For The Assay Of Silver Bullion, Chem. Soc. J., 77, 1900, PT. 1, pp. 232-233.

On Certain Properties Of The Alloys Of The Gold-Copper Series, Royal Soc. Proc., 67, 1901, pp. 105-112 (With Sir Roberts-Austin).

Alloys, Royal Soc. Of Arts–London J., 49, 1901.

Certain Properties Of The Gold-Silver Series, Royal Soc. Lon. Proc., 71, 1902, pp. 161-163 (With Sir Roberts-Austin).

Properties Of The Alloys Of Silver And Cadmium, Royal Soc. Lon. Proc., 74, 1904, pp. 218-230.

Refining Gold Bullion And Cyanide Precipitates With Oxygen Gas, Inst. Min. Met. Proc., 14, 1905, pp. 377-422.

Cupellation And Parting In Ore Assaying, J. Chem. Met. Min. Soc. South Africa, 5, 1905, pp. 165-168.

Alloys Of Gold And Tellurium, Inst. Min. Met.–London, 17, PT.1, 1908, pp. 285-289.

The Precious Metals, Comprising Gold, Silver And Platinum, Book, 295 Pages, 1909, Van Nostrand C., New York and London.

Distribution Of Gold Produced On The Rand, J. Chem. Met. Min. Soc. South Africa, 9, 1909, pp. 265-270.

The Annealing Of Coinage Alloys, Inst. Metals–London, 8, 1912, pp. 86-125, Also: J. Soc. Chem. Ind., 31, 1912, p. 989.

The Annealing Of Gold, Inst. Metals–London, 10, 1913, pp. 150-174.

Estimation Of Zinc In Coinage Bronze, Soc. Chem. Ind. J., 33, 1914, pp. 170-172.

Electrolytic Refining Of Gold, Inst. Min. Met.–London, 24, 1915, President's Address, pp. XXXV-LV, Also: J. Soc. Chem. Ind., 30, 1911, p. 216; Abstract Of Paper Read At 7th Int'l. Congr. Appl. Chem., 1909, Sect. IIIA, 15.

Loss Of Gold During Melting, Inst. Min. Met.–London, 28, 1919, pp. 135-148.

Treatment Of Tin Ores By Volatilization Of Tin As Chlorides, Report Of The Tin And Tungsten Research Board, 1922, H.M. Stationery Office, London.

Report Of The Tin And Tungsten Research Board By T.K. Rose et al, 1922, 100 pages, H.M. Stationery Office, London.

Working Of Nickel For Coinage, J. Inst. Metals, 32, 1924, pp. 271-282. (With J.H. Watson)

Density Of Rhodium, J. Inst. Metals, 33, 1925, pp. 109-114.

About The Seven Editions

Rose's *Metallurgy of Gold* went through seven editions as follows:

1st Edition	1894
2nd Edition	1896
3rd Edition	1898
4th Edition	1902
5th Edition	1906
6th Edition	1915
7th Edition	1937

The 7th Edition was co-authored with W.A.C. Newman, Senior Technical Assistant at the Royal Mint.

The book evolved considerably over the years. To a large extent this evolution reflected the changing state of the gold industry and what was considered important at the time.

Later editions were stronger in the areas of alloys, occurrence, cyanide, crushing and grinding, and flotation. At the same time the later editions place less emphasis on placer, chlorination, refining, and to some extent, assaying. The last edition dropped the extensive bibliography. The bibliography from the 5th edition has been added to this reprint of the 7th edition.

Anyone doing serious research should consider referring to earlier editions. The following matrix shows the number of pages devoted to each subject in the various editions. Page numbers are approximate and the 7th edition has a slightly larger page size.

Dave Schneller

CHAPTERS	Edition						
	1	2	3	4	5	6	7
General:							
Physical and Chemical Properties of Gold and Its Alloys	15 pg.	18 pg.	19 pg.	18 pg.	20 pg.	23 pg.	20 pg.
Alloys of Gold						34 pg.	46 pg.
The Chemistry of the Compounds of Gold	12 pg.	14 pg.	14 pg.	14 pg.	14 pg.	15 pg.	12 pg.
The Mode of Occurrence and Distribution of Gold	11 pg.	7 pg.	7 pg.	8 pg.	8 pg.	23 pg.	23 pg.
Placer:							
Placer Mining—Shallow Deposits	20 pg.	19 pg.	19 pg.	22 pg.	24 pg.	29 pg.	
Deep Placer Deposits	25 pg.	25 pg.	26 pg.	23 pg.	21 pg.	15 pg.	
Treatment of Placer Deposits							27 pg.
Stamp Batteries and Amalgamation:							
Quartz Crushing in the Stamp Battery	33 pg.	33 pg.	39 pg.	39 pg.	34 pg.	33 pg.	34 pg.
Amalgamation in the Stamp Battery	24 pg.	25 pg.	25 pg.	26 pg.	29 pg.	33 pg.	19 pg.
Other Forms of Crushing and Amalgamating Machinery	18 pg.	17 pg.	19 pg.	17 pg.	13 pg.	16 pg.	
Gravity Concentration In Stamp Mills	20 pg.	23 pg.	23 pg.	23 pg.	16 pg.	17 pg.	28 pg.
Stamp Battery Practice In Particular Localities	17 pg.	24 pg.	24 pg.	24 pg.			
Chlorination:							
Chlorination: The Preparation of Ore for Treatment	33 pg.	33 pg.	36 pg.	37 pg.			
Chlorination: The Vat Process	12 pg.	12 pg.	12 pg.	13 pg.	7 pg.		
Chlorination: The Barrel Process	21 pg.	21 pg.	21 pg.	21 pg.	19 pg.		
Chlorination: Practice in Particular Mills	20 pg.	21 pg.	22 pg.	18 pg.			
Chlorination: The Plattner Process					17 pg.		
Chlorination						20 pg.	

CHAPTERS	Edition 1	2	3	4	5	6	7
Cyanide:							
The Cyanide Process	24 pg.	26 pg.	34 pg.	61 pg.	62 pg.	45 pg.	75 pg.
The Chemistry of the Cyanide Process	15 pg.	19 pg.	25 pg.	30 pg.	26 pg.	24 pg.	32 pg.
Cyanide Process—Special Methods and Examples of Practice						28 pg.	45 pg.
Smelting and Roasting:							
Pyritic Smelting	3 pg.	3 pg.	3 pg.				
Roasting					27 pg.	18 pg.	9 pg.
Crushing and Grinding:							
Dry Crushing					10 pg.	8 pg.	5 pg.
Intermediate and Fine Grinding							27 pg.
Fine Grinding						23 pg.	
Regrinding					11 pg.		
Flotation:							
Flotation							37 pg.
Refining:							
Refining and Parting of Gold Bullion	40 pg.	50 pg.	53 pg.	54 pg.	58 pg.	67 pg.	
Melting and Refining of Gold Bullion							35 pg.
Assaying:							
The Assay of Gold Ores	24 pg.	23 pg.	24 pg.	25 pg.	35 pg.	34 pg.	24 pg.
Ore Testing							6 pg.
The Assay of Gold Bullion	32 pg.	30 pg.	30 pg.	30 pg.	28 pg.	30 pg.	18 pg.
Other:							
Economic Considerations		8 pg.	9 pg.	9 pg.			
Statistics of Gold Production					6 pg.	7 pg.	
Bibliography		11 pg.	11 pg.	11 pg.	12 pg.	10 pg.	

Notice Concerning Safety, First Aid, and Pollution Control

Due to the enormous changes in these fields over the past fifty years, you should not rely on the information presented in this book but should consult current sources on these subjects.

Reduction and Cyanide Plants at Modderfontein East.

THE

METALLURGY OF GOLD

BY

SIR THOMAS KIRKE ROSE, D.SC.,

ASSOCIATE OF THE ROYAL SCHOOL OF MINES; LATE PRESIDENT OF THE INSTITUTION OF MINING AND METALLURGY; HON. MEMBER OF THE CHEMICAL, METALLURGICAL, AND MINING SOCIETY OF SOUTH AFRICA; LATE MEMBER OF COUNCIL OF THE INSTITUTE OF METALS; FELLOW OF THE CHEMICAL SOCIETY; LATE CHEMIST AND ASSAYER OF THE ROYAL MINT.

AND

W. A. C. NEWMAN, B.SC., F.I.C.,

ASSOCIATE OF THE ROYAL SCHOOL OF MINES; ASSOCIATE OF THE ROYAL COLLEGE OF SCIENCE; DIPLOMA OF THE IMPERIAL COLLEGE OF SCIENCE AND TECHNOLOGY; MEMBER OF THE INSTITUTION OF MINING AND METALLURGY; MEMBER OF THE INSTITUTE OF METALS; MEMBER OF THE ELECTRODEPOSITORS' TECHNICAL SOCIETY; SENIOR TECHNICAL ASSISTANT, ROYAL MINT.

Seventh Edition, Revised Throughout and Re-set

WITH FRONTISPIECE AND 238 OTHER ILLUSTRATIONS (INCLUDING 5 PLATES)

Published By
Met-Chem Research, Inc.
from the 7th edition
originally published by
Charles Griffin & Company Ltd.

ISBN 0-931913-05-5

1st Edition	1894
2nd Edition	1896
3rd Edition	1898
4th Edition	1902
5th Edition	1906
6th Edition	1915
7th Edition	1937
Out of Print	1945

7th Edition Reprinted 1986 by
Met-Chem Research Inc.
P.O. Box 3014 Highmar Station
Boulder, Colorado 80307

PREFACE TO SEVENTH EDITION.

The last edition of this book has been out of print for some considerable time and is now over 20 years old, so that the preparation of a new edition presented a serious problem. The greater part of the book has always consisted of a description of the current methods of extracting gold from its ores, but since 1915 so much change has taken place in this direction that very little of the older practice remains. Accordingly, we decided to discard almost the whole of the old text and to write what is practically a new book. Even in the chapters on Refining and Assaying—old arts which have stood the test of time—this new edition is more concise and has little in common with the earlier one. Obsolete methods of ore treatment have been left out, except for brief historical summaries, so that whole chapters, *e.g.*, that on Chlorination, have disappeared, while new ones on Flotation and Ore Testing have been added. The publishers have courageously faced their heavy task, sharing our hope that metallurgists will show at least the same amount of confidence in the new book as in the old one.

Nevertheless, the structure of the book remains much the same, for the reason that the principles on which the work is founded have not been changed. These are fully explained in the preface of the sixth edition. They are, briefly, that the principles underlying practice are of more importance than the details of the machines employed, and also that full information on the properties of gold, its compounds and its ores is needed for the use of metallurgists desirous of making progress in the older methods and of breaking new ground wherever possible.

One aspect of gold extraction takes into account the impetus given by the increase in value of gold in terms of commodities owing to Government action. Besides the consequent lowering of the grade of payable ores, there have been renewed efforts in prospecting and in developing small properties. The needs of those working in small detached mills have therefore been carefully kept in view.

We are indebted to many individuals, Institutions and Firms for information, drawings and blocks dealing with special subjects, and we have endeavoured to make full acknowledgement in the text. Our especial thanks are due to Mr. Philip Rabone, A.R.S.M., D.I.C., for kindly assistance in reading the proofs of the chapter on Flotation, and to Mr. Bowyer of the Perth Mint for information on Chlorine Refining.

T. K. ROSE.

W. A. C. NEWMAN.

November, 1936.

PREFACE TO THE SIXTH EDITION.

FOLLOWING an old definition, which can hardly be bettered, a book on the metallurgy of gold should give an account of the extraction of gold from its ores and of adapting it for use. The book would then be useful to students and to metallurgists engaged in their profession. The most important function, however, which a book on metallurgy has to fulfil is to help those who are taking part in attempts to improve the existing practice. Progress in metallurgy depends on the capacity of metallurgists to apply their knowledge of physical science and engineering to the problems presented to them, and in this they are aided by a full understanding of the causes of the phenomena which they observe.

Hence, in addition to making an attempt to give, in as few words as possible, a complete picture of present practice, my aim in this edition has been to give full and accurate information as to the properties of gold, its alloys and compounds, and of the bearings of these on the work to be done. This has involved the rewriting of much of the work and a great expansion of certain sections corresponding with the rapid advance of science. References to the sources of information are added in every case, to enable the reader to consult the original memoirs if he desires to do so.

A summary of the present general position in the working of placers, ore dressing, stamp milling and the cyanide process is given, with references to the lengthy treatises expressly devoted to each of these subjects. Some account is included of the treatment of gold ores in particular mills or districts, but this depends in great measure on local conditions and is best dealt with in separate text-books, such as the excellent one on *Rand Metallurgical Practice.* The chapters on the refining of gold and on assaying will, I trust, be found of value. Much care has, in particular, been devoted to the discussion of the electrolytic refining processes. Throughout the volume more attention is paid to the principles underlying practice than to the details of the machines employed, although descriptions are given of a number of machines which are typical of their class.

In the preparation of a portion of this edition I have been aided by my colleague, Mr. W. A. C. Newman, A.R.S.M., B.Sc., whose help has been specially useful in passing the work through the press. I am also indebted to many correspondents whose kind assistance has, I believe, been acknowledged in every case in the text, and last, but not least, to my publishers, who, with their usual thoroughness, have reset the work throughout on a larger page so as to compass in handy form the greatly extended text.

T. K. ROSE.

CONTENTS.

CHAPTER I.

THE PHYSICAL AND CHEMICAL PROPERTIES OF GOLD.

CHAPTER II.

ALLOYS OF GOLD.

CHAPTER III.

CHEMISTRY OF THE COMPOUNDS OF GOLD.

CHAPTER IV.

MODE OF OCCURRENCE AND DISTRIBUTION OF GOLD.

CHAPTER V.

TREATMENT OF PLACER DEPOSITS.

CHAPTER VI.

PRIMARY ORE CRUSHING.

CHAPTER VII.

AMALGAMATION.

CHAPTER VIII.

INTERMEDIATE AND FINE GRINDING.

CHAPTER IX.

GRAVITY CONCENTRATION IN GOLD MILLS.

CHAPTER X.

FLOTATION.

CHAPTER XI.

DRY CRUSHING.

CHAPTER XII.

ROASTING.

CHAPTER XIII.

THE CYANIDE PROCESS. CHEMICAL REACTIONS.

CHAPTER XIII.—*Continued.*

CHAPTER XIV.

THE CYANIDE PROCESS. GENERAL METHODS.

CHAPTER XV.

THE CYANIDE PROCESS. SPECIAL METHODS.

CHAPTER XVI.

EXAMPLES OF PRACTICE.

CHAPTER XVII.

THE MELTING AND REFINING OF GOLD BULLION.

CHAPTER XVIII.

THE ASSAY OF GOLD ORES.

CHAPTER XIX.

THE ASSAY OF GOLD BULLION.

CHAPTER XX.

ORE TESTING.

POSITION OF PLATES.

THE METALLURGY OF GOLD.

CHAPTER I.

THE PHYSICAL AND CHEMICAL PROPERTIES OF GOLD.

Introduction.—From very early times the ancients were attracted by the beautiful colour, the brilliant lustre, and the indestructibility of gold.

Prof. Gowland points out [1] that on account of its wide distribution in the sands and gravels of rivers, and its distinctive appearance, it must have been the first metal to attract the attention of prehistoric man in most regions of the world. He also observes, however, that it could not have been used even for ornaments until the art of melting had been invented, and this could hardly have happened until man had passed the Stone Age culture and entered the Bronze Age.

No objects, he says, consisting of gold have been found with undoubted Stone Age remains. The earliest mining and metallurgical operations of which traces remain were those carried on in Egypt in dealing with the ores of gold. "The ancient mines are scattered over Upper Egypt, Nubia, and the Sudan," and consist of shallow pits in detritus, and trenches and shafts in hard rocks. The ore was broken by stone hammers, ground in stone mills or querns, and treated on inclined stone tables, on which the particles of rock were washed away from the gold. Shallow earthen dishes were used for the final washings, and the residual gold was melted with purifying fluxes in crucibles and cast into ingots. Remains of all the implements have been found, but their exact age is doubtful.[2]

Among the pictorial rock carvings of Upper Egypt there are several illustrations of the gold-extraction processes mentioned above. The earliest indications appear to be certain inscriptions on monuments of the Fourth Dynasty (4000 B.C.), depicting gold washing.[3] Certain stelæ of the Twelfth Dynasty (2400 B.C.) in the British Museum refer to gold washing in the Sudan, and one of them appears to indicate the working of gold ore as distinguished from alluvial.[4]

In somewhat later times, in the collection of alluvial gold, the sands were washed down over smooth sloping rocks by means of running water, and the particles of gold, sinking to the bottom of the stream, were entangled and caught in the hair of raw hides spread on the rocks—thus anticipating the modern blanket practice. Among the hides used were sheepskins, and hence originated the form of the legend of the Golden Fleece. Stripped of its heroic dress, this legend merely describes a successful piratical expedition

---

[1] *J. Anthrop. Inst.*, 1912, **42**, 252-262. [2] Gowland, *loc. cit.*
[3] Wilkinson, "*The Ancient Egyptians*" (London, 1874), vol. ii., p. 137.
[4] Hoover, "*Translation of Agricola*" (London, 1912), p. 279.

about 1200 B.C. to win gold, which was being laboriously obtained from streams with the help of sheepskins or goatskins by the inhabitants of what is now Armenia.

For many years, when auriferous sands were washed, the aid was also invoked of what Baron Born called in 1786 the "elective affinity" of mercury for gold when mixed with impurities. The ease with which gold-amalgam can be collected, in spite of its being less dense than gold itself, is due to the fact that it is wetted by mercury.

It is also of interest to remember that the earliest dawn of the science of chemistry was heralded by the study of the properties of gold, and by the efforts which were made to invest other matters with these properties. From the fourth to the fifteenth century, chemistry, which was first called "chemia" (χημεια), and then "alchemy," was defined as the art of transmuting base metals into gold and silver, almost all the labours of philosophers being intended to aid directly or indirectly in solving this problem. At the end of this period, while Paracelsus was giving to chemistry a new aim—that of investigating the composition of drugs, and their effect on the human body—Agricola was reducing to order the numerous empirical facts which together made up the art of metallurgy, and although alchemy died hard, its era of usefulness may be said to have ended here. Gold has doubtless been the cause of many of the wars and marauding expeditions from which the world has suffered, but on the other hand it has been instrumental, in a far greater degree than most other commodities, in promoting the growth of civilisation, the efforts of the alchemists having laid the foundations of the science of chemistry, and those of the gold-seekers having resulted in the discovery of new countries, and in the spread of knowledge of all kinds.[1]

Colour.—The lustre and fine colour of gold have given rise to most of the words which are used to denote it in different languages. The word "*gold*" is probably connected with the Sanscrit word "*jvalita*," which is derived from the verb "*jval*," to shine. It is the only metal which has a yellow colour when in mass and in a state of purity. Impurities greatly modify this colour, small quantities of silver lowering the tint, while copper raises it. In a finely divided state, when prepared by volatilisation or precipitation, gold assumes various colours, such as deep violet, ruby and reddish-purple, the tint varying to brownish-purple and thence to dark brown and black. This purple colour is not due to the formation of a coloured oxide of gold. No oxygen can be obtained from it, and it probably consists of metallic gold. Similar colours are seen in Purple of Cassius, and in Roberts-Austen's purple alloy of aluminium and gold, the colour in each case being probably due to a particular form of finely divided gold. "Faraday's gold," a ruby-coloured liquid prepared by the action of phosphorus dissolved in carbon bisulphide or in ether, or of formaldehyde on a cold dilute solution of chloride of gold, is a solution of colloidal metallic gold in water.[2] Faraday used a solution containing 0·6 of a grain of gold in a quart in preparing ruby gold. For further details as to colloidal gold, see Chap. III. Finely divided gold gives a faint blue tinge to light transmitted through the liquid in which it is suspended. Very thin films of gold are translucent, and appear green by transmitted light, while remaining yellow by reflected light. On heating, the green

---

[1] See T. A. Rickard, "The Early Use of Metals," *J. Inst. Metals*, 1930, **43**, 297.

[2] Faraday, *Phil. Trans.*, 1857, p. 145; Zsigmondy, *Liebig's Annalen*, 1898, **301**, 29, 361.

colour changes to some shade between ruby-red and violet, or disappears entirely owing to the breaking up of the continuous film into a network of metal through which white light passes. The green colour is restored by burnishing.[1] Bragg [2] has shown from X-ray data that in fine films of unannealed gold the crystals are oriented with their faces parallel to the sheet. Molten gold is green, and its vapour is greenish-yellow.

Malleability and Ductility.—Malleability and ductility are possessed by gold at all temperatures to a far higher degree than by any other metal. A single grain of gold can be drawn out into a wire over 500 feet long, and leaves of not more than $\frac{1}{300000}$ of an inch in thickness can be obtained by beating. Faraday has shown that the thickness of these leaves may be still further reduced by floating them in a dilute solution of potassium cyanide, by which they are partly dissolved. Annealing is advantageous during the cold working of pure gold, but the temperature required is low, so that the goldbeater's skin is uninjured.[3]

Additions of even more than 1 per cent. of iron, up to 1 per cent. of tin or 0·1 per cent. of aluminium, have only a slight effect on the rolling properties, whereas less than 0·1 per cent. bismuth, tellurium or lead renders the gold brittle, owing to the distribution of Bi, $AuTe_3$ or Au_2Pb between the grains. Tellurium and lead are most harmful, the maximum quantities permissible being 0·01 and 0·005 per cent. respectively.[4]

Hardness.—Gold is softer than silver and harder than tin. Its hardness, according to Auerbach, is 2·5 to 3·0, and according to Rydberg 2·5, in the scale in which the diamond is 10 and talc 1.[5] The hardness of pure gold, however, like that of other metals, varies with its physical condition, as follows :—[6]

	Ludwik's Cone Machine.	Shore's Scleroscope, Magnifier Hammer.
Cast, ...	22	4·5
Hammered or rolled, . .	60-65	30-35
Annealed, ...	25	6

The scales are not the same. In the Ludwik scale, lead is 4·5, and quenched steel containing 0·9 per cent. carbon about 260. In the scleroscope scale, lead is 2 and steel about 175. Edwards gives the Brinell hardness of cast and slowly cooled gold as 33·01.[7] Auerbach cites the absolute hardness as 91 kg./mm.^{2} (diamond 2500, talc 5).

Surface Tension.—On stretching electrolytically-prepared gold foil between two rods and then applying heat, Tammann and Boehme [8] found that shrinkage becomes noticeable at 200° C. with foil 0·2 μ thick and at 600° C. with foil 0·8 μ thick. Shrinkage ceases after about 15 minutes' heating at any one temperature. From the shrinkage determination the surface tension was calculated to be as follows :—1·23 grms./cm. at 700° C., 1·18 grms./cm. at 850° C. (and by extrapolation 1·12 grms./cm. at 1063° C.). Smith [9]

[1] Faraday, *loc. cit.* ; G. T. Beilby, *Proc. Roy. Soc.*, 1903, **72**, 226 ; T. Turner, *Proc. Roy. Soc.*, 1908, **81A**, 301.
[2] *Nature*, 1924, **113**, 639.
[3] A history of the goldbeater's art is given in *Helios*, 1925, **31**, 449.
[4] Nowack, *Zeit. Metallkunde*, 1927, **19**, 238.
[5] "*Landolt-Börnstein's Tabellen*," 1923, p. 91.
[6] T. K. Rose, *J. Inst. Metals*, 1912, **8**, 86 ; 1913, **10**, 150.
[7] *Met. Ind.*, 1921, **17**, 221.
[8] *Ann. Physik*, 1932, **12**, 820.
[9] *J. Inst. Metals*, 1914, **12**, 168 ; 1917, **17**, 65.

gives σ, the surface tension, as 1,018 dynes per cm. and the capillary constant a^2 as 11·29 sq. mm.

Tenacity.—According to Roberts-Austen,[1] pure gold when cast breaks with a load of 7 tons per square inch and an elongation of 30·8 per cent. on a test piece 3 inches long. In "*Landolt's Tabellen*,"[2] the elastic limit of hard-drawn gold wire is given as 14 kilos. per square mm. (8·9 tons per square inch), and its tensile strength as 27 kilos. per square mm. (17·1 tons per square inch). The tensile strength of annealed gold is 10 kilos. per square mm. (6·3 tons per square inch).

Miss C. F. Elam[3] has investigated the tensile strength of large gold crystals. The inclinations of the slip planes tend to reach the limiting value previously observed in aluminium, viz. 61° 52′. Under increasing extension gold is initially stronger than silver but finally weaker.

Young's modulus for gold in the annealed state is given by Jacquerod and Mügeli[4] as 8060 kg./mm.^{2} and the temperature coefficient between 0° and 100° C. as $3{\cdot}988 \times 10^{-4}$. For drawn wires Wertheim[5] gives the value of 81·31 kg./mm.^{2} Mallock obtained 1·32 as the ratio between Young's moduli at 0° K. and at 273° K.

Specific Gravity.—The specific gravity of gold is about 19·3. When cast it is liable to contain cavities, by which the density is diminished, and after compression the density is again higher, although it is not supposed that the true specific gravity can be increased in that way. Roberts-Austen and Rigg[6] gave cast gold as 19·2945, after compression 19·3203, at 0°/4° C. The specific gravity of rolled sheet gold at 0°/4° C. is 19·2965,[7] that of soft annealed wire at 20°/4° C. is 19·26, and of hard wire 19·25.[8] When crystallised from solution the specific gravity is 19·431.[8] Kahlbaum[9] gives the specific gravity of vacuum-distilled gold as (*a*) cast, 20°/4° C.—18·884, (*b*) compressed under 10,000 atmospheres, 20°/4° C.—19·2685. Henry Louis has shown[10] that the specific gravity of unannealed "parted" gold (*i.e.* the residue left after boiling silver-gold alloys in nitric acid) is 20·3, its density being lowered by the process of annealing. Lowry and Parker[11] give the density of filings as 19·217. The density as derived from X-ray data is given as $19{\cdot}210 \pm 0{\cdot}057$.[12]

Taking the density of pure gold at 19·3, then

1 c.c. of pure gold weighs 19·3 grammes, or 0·6205 oz. troy.
1 cubic inch weighs 316·25 grammes, or 10·168 ozs. troy.
1 cubic foot weighs 546·485 kg., or 17569·9 ozs. troy.

The volume of 1 kilogramme of gold is 51·81 c.c., or 3·162 cubic inches.
,, 100 ozs. troy is 161·16 c.c., or 9·835 cubic inches.
,, 1 ton avoirdupois is 1·86 cubic feet.

---

[1] *Phil. Trans.*, 1888, **179**, 339.
[2] *Op. cit.*, p. 87.
[3] *Proc. Roy. Soc.*, 1926, **112A**, 289.
[4] *Helv. Physica Acta*, 1931, **4**, 3.
[5] *Proc. Roy. Soc.*, 1919, **95A**, 429.
[6] *Seventh Ann. Report of the Royal Mint*, 1876, p. 44.
[7] T. K. Rose, *J. Inst. Metals*, 1912, **8**, 111.
[8] "*Landolt-Börnstein's Tabellen*," 1923, p. 287.
[9] *Zeitsch. anorg. Chem.*, 1902, **29**, 177.
[10] *Trans. Am. Inst. of Mng. Eng.*, Chicago Meeting, 1893.
[11] *Journ. Chem. Soc.*, 1915, **107**, 1005.
[12] W. P. Davey, *Phys. Rev.*, 1924, **23**, 292.

According to Quincke the specific gravity of liquid gold at 1,120° C. is 18·38.

The specific volume of liquid gold is a linear function of the temperature (Jouniaux).

Cohesion.—On heating, gold can be welded below the point of fusion like iron, and finely divided gold agglomerates on heating without being subjected to pressure. Pressure alone is also sufficient to make gold dust cohere, while a true flow of the particles of gold can be induced in the case of the pure metal and some of its alloys.

Specific Heat.—The specific heat of gold is 0·0297 between − 188° and + 20° C., 0·03120 at 18° C., and 0·03141 at 100° C. (Richards and Jackson).[1] The average between 0° and 100° C. is 0·0316 (Violle). It is 0·0345 at 900° C., and 0·0352 at 1,020° C. (Violle).

Fusibility.—Gold fuses, after passing through a pasty stage, at a clear cherry-red heat, just below the fusing point of copper and much above that of silver. The metal expands considerably on fusing, and contracts again on solidifying. Endo [2] has stated that the contraction on solidification amounts to 5·17 per cent. The freezing point was given by Berthelot as 1,064°,[3] by Day and Sosman as 1,062·4°,[4] by Jacquerod and Perrot as 1,067·2°,[5] by Roberts-Austen and Rose [6] and by the Reichsanstalt (1916) as 1,063° C. Gold is monatomic below its melting point but as the temperature rises the molecular weight increases rapidly.[7]

Ruff and Konschak [8] found the boiling point of gold to be 2,677° C.

Viscosity.—Usually metals become less viscous at low temperatures, but there is a slower rate of decrease in the case of gold between 0° and − 100° C.[9]

Latent Heat.—The latent heat of fusion of gold has been given as 16·3,[10] and as 15·73 [11] The normal lowering of the freezing point for 1 atom of impurity in 100 atoms of gold is 10·6° C.[10]

Magnetism.—Gold is diamagnetic, its specific magnetism being 3·47 (Becquerel), if that of iron is taken as 100. Hanriot says that yellow gold is more diamagnetic than brown. Hanriot and Raoult [12] give the magnetic susceptibility of pure gold as not less than $-0{\cdot}234 \times 10^{-6}$. According to Honda [13] the value is $-0{\cdot}15 \times 10^{-6}$ mass units between 18° and 1,064° C.; there is a discontinuity at the melting point.

Conductivity and Expansion.—Its electrical conductivity is $45{\cdot}5 \times 10^4$ at 0°, that of silver being $68{\cdot}12 \times 10^4$, and that of copper $64{\cdot}06 \times 10^4$ (Dewar and Fleming). Northrup [14] found the electrical resistance of gold at 20° C. to be 2·316 microhms per c.c. and at 500° C. 6·62. It probably vanishes

---

1 "*Landolt-Börnstein's Tabellen*," 1923, p. 1244.
2 *J. Inst. Met.*, 1923, **30**, 132.
3 *Compt. rend.*, 1898, **126**, 473.
4 *Ann. Phys.*, 1901, **4**, [iv.], 99.
5 "*Landolt-Börnstein's Tab.*," 1923, p. 318.
6 *Proc. Roy. Soc.*, 1900, **67**, 105.
7 Jouniaux, *Bull. Soc. chim.*, 1924, [iv.], **35**, 463.
8 *Z. Elektrochem.*, 1926, **32**, 515.
9 Mallock, *Proc. Roy. Inst.*, 1921, **23**, 377.
10 Roberts-Austen, *Proc. Roy. Soc.*, 1891, **49**, 352.
11 Umino, *Sci. Rep. Tôhoku Imp. Univ.*, 1926, **15**, 597.
12 *Bull. Soc. chim.*, 1911, **9**, 1052; *Compt. rend.*, 1911, **153**, 182.
13 *Ann. Physik*, 1910, **32**, 1027.
14 *Journ. Frank. Inst.*, 1914, **177**, 287.

at the absolute zero. The change in the resistance of molten gold with rise in temperature is linear, and of solid gold almost linear. The temperature coefficient of electrical resistance over the range 0°-34° C. is 0·00391 $\pm$ 0·00001. At current densities of the order of 5×10^6 amps. per sq. cm. on leaf gold Bridgman [1] discovered a variation of about 1 per cent. from Ohm's law. The coefficient of thermal conductivity, K, is 0·7003 at 18°, that of silver being 1·006 (Jaegar and Diesselhorst). The coefficient of linear expansion of gold is 0·0000144 between 0° and 100° C. (Fizeau). The coefficient between 0° and 520° C.[2] is $(14{\cdot}157t + 0{\cdot}002150t^2)\ 10^{-6}$.

Atomic Weight and Volume.—Its atomic weight is 197·2, compared with oxygen = 16·00. The atomic volume of gold is 10·2.

Emissivity.—Molten gold has a stronger power of emission in red light and a weaker power of emission in blue light than solid gold.

Spectrum.—Fig. 1 shows arc spectra of gold purified as far as possible for use in check assays in the valuation of gold bullion.[3] The spectra were photographed with a quartz spectrograph, using poles 5 mm. $\times$ 1 to 1·5 mm. in section, with a direct current of 2·3 to 2·5 amperes. Under these conditions the arc is not very steady and the restarting necessary to complete the exposure is probably responsible for the prominence, and variation, of the air bands in the photographs.

The "proof" gold samples all assayed from 999·97 to 1000·00 fine, in comparison with the purest gold obtainable. The majority of the spectra show the persistent lines of silver, copper and calcium. Some comparison spectra are shown of "proof" gold to which small proportions of silver and copper have been added. An addition of 0·001 per cent. of either metal produces the characteristic lines in the spectra considerably more strongly than in any of the samples of "proof" gold.

The calcium lines are remarkably strong in some spectra, but no corresponding difference can be detected in the gold assay to 0·01 per 1,000; so that the strong lines referred to appear to be due to less than 0·01 per 1,000 of calcium. No means appear to be available of introducing, or estimating by analysis, a comparably small proportion of the metal.

There are at least 50 lines recorded in the spectrum between wavelengths 5656·5 and 8928·5 A.

It has been stated that about 0·0001 per cent. of gold may be detected in minerals by examination of the arc and spark spectra, but confirmation is needed. The most persistent line in the arc spectrum is that of wavelength 2428·1 A. For further information see Chapter XIX. ("Assay by Spectroscope").

Lattice Constant.—Davey [4] found this to be 4·076 $\pm$ 0·004 A.

Absorption of X-Rays.—The critical absorption wavelength of X-rays [5] is $1{\cdot}532 \times 10^{-8}$ cm.

Volatility of Gold.—When gold is melted and allowed to solidify there is always some loss of the metal and the natural assumption is that part has been volatilised. Purple stains consisting of finely divided gold are formed on the edges or lid of the containing vessel or in the flue of the

---

[1] *Proc. Amer. Acad. Arts Sci.*, 1922, 57, 131.
[2] *Phys. Zeit.*, 1916, 17, 29.
[3] The photographs are published by the courtesy of Dr. S. W. Smith, Chief Assayer, Royal Mint. The description is abstracted from one supplied by J. Phelps, M.A.
[4] *Phys. Rev.*, 1924, 23, 292.
[5] *Proc. Nat. Acad. Sci.*, 1920, [vi.], 607.

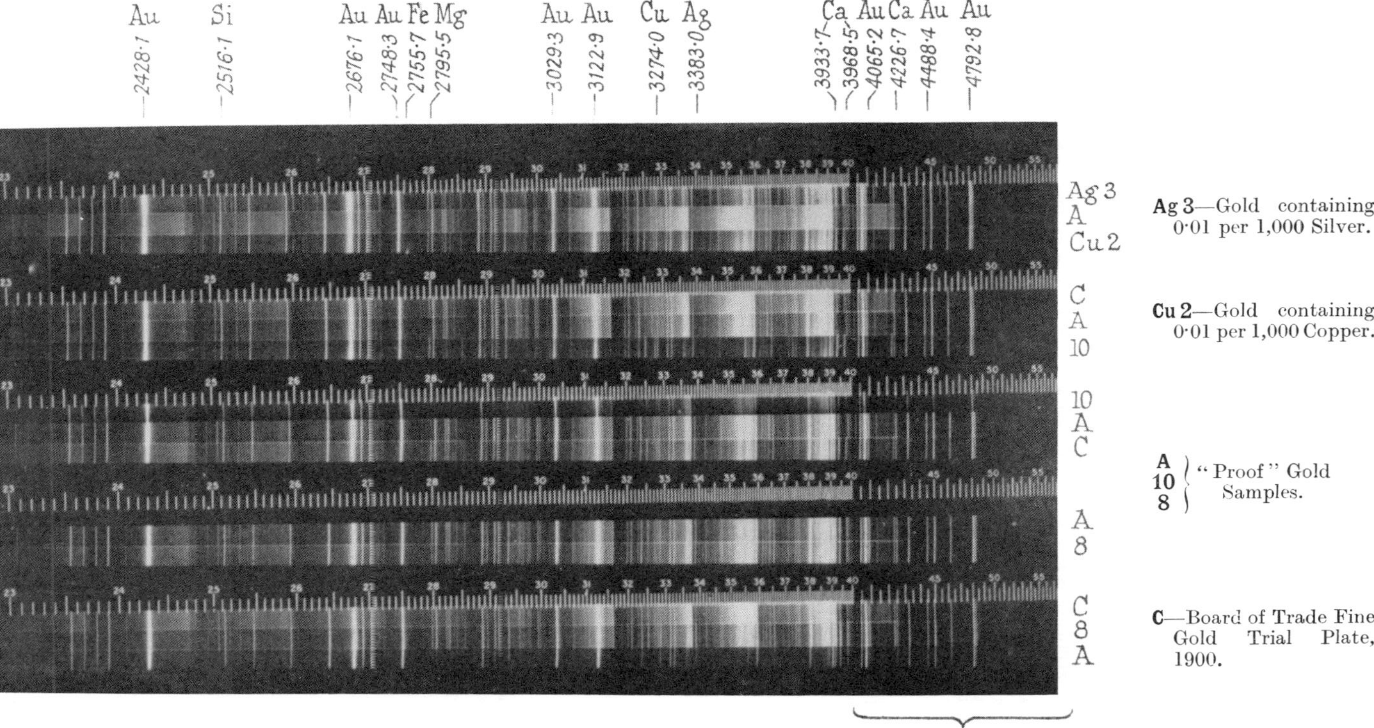

Ag 3—Gold containing 0·01 per 1,000 Silver.

Cu 2—Gold containing 0·01 per 1,000 Copper.

A, **10**, **8** } "Proof" Gold Samples.

C—Board of Trade Fine Gold Trial Plate, 1900.

Fig. 1.—Gold Arc Spectra.

furnace and are regarded as condensations of the volatilised portion. A discharge of high-tension electricity through a fine gold wire stretched on paper converts it into a purple streak of metal which has been volatilised and then condensed.

It was found by Krafft and Bergfeld [1] that gold begins to volatilise at 1,070° C. *in vacuo* in a quartz vessel and boils at 1,800° C. under the same conditions. Richards [2] estimated from these results that the boiling point of the metal at atmospheric pressure is 2,530° C. Moissan [3] found that gold can readily be distilled when heated in an electric arc furnace. With a current of 300 amperes at 70 volts acting for 6 minutes, 59 grams remained out of 107 grams placed in a crucible. Copious fumes of a greenish-yellow colour were evolved. The gold was condensed as small, regular, yellow spheres, as brilliant yellow cubical crystals, as deep yellow leafy crystals, as filaments, or as a purple powder. In the distillation of gold-copper or gold-tin alloys the residual ingot was richer in gold than the original alloy.

Mostowitsch and Pletneff [4] found that fine gold melted and heated in a stream of gas for periods of 30 minutes up to 3 hours suffered no loss at temperatures up to 1,400° C. when the gas used was O_2, N_2, CO_2 or CO. With hydrogen there was no loss up to 1,100° C., but from 1,200° C. upwards there were losses which increased with temperature and time, and the silica boat was coloured red, which did not happen with the other gases.

The losses in hydrogen were :—

Temp., ° C.	Time, mins.	Loss, per cent.
1,250	25	0·055
1,300	25	0·090
1,350	25	0·105
1,400	25	0·250

There is apparently a volatile compound of gold and hydrogen, perhaps Au_2H_2, but it is unstable and breaks up.

Crystallisation of Gold.—Gold crystallises in the cubic system, occurring frequently in nature in the form of cubes, octahedra, rhombic dodecahedra, and trapezohedra. X-ray examination by Bragg's method has shown that gold has a face-centred cubic lattice [5] with sides 4·08 A. Cubes and octahedra are often elongated, giving rise to rod-shaped crystals, and plates are also not uncommon. Twinning is frequent, giving rise to dendritic groups, tessellated surfaces and various complicated forms. Some of these present the appearance of hexagonal pyramids or monoclinic prisms, and have even been described as such. Cleavage is never exhibited. Single detached crystals are comparatively rare, and the crystals are usually attached end to end, forming strings, and branching, arborescent, or moss-like masses, which are composed of microscopic crystals, usually octahedra.[6]

---

[1] *Ber.*, 1903, **36**, 1670 ; 1905, **38**, 254.
[2] "*Metallurgical Calculations*" (1908), Part iii, 588.
[3] *Compt. rend.*, 1893, **116**, 1429 ; 1905, **141**, 977.
[4] *Met. and Chem. Eng. N.Y.*, Feb. 1, 1917, **16**, 153.
[5] Vegard, *Phil. Mag.*, 1916, [vi.], **32**, 141 ; Sherrer, *Phys. Zeit.*, 1918, **19**, 23 ; Gross, *Umschau.*, 1920, **34**, 51.
[6] See E. S. Dana, "On the Crystallisation of Gold," *Amer. J. Sci.*, 1886, **32**, 132, where other references are also given.

These forms occur frequently in quartz veins, but the single crystals, which are usually of larger size—viz., from $\frac{1}{4}$ to $1\frac{1}{2}$ inches in diameter—are mainly found in drift deposits. They are rarely perfect or of brilliant lustre, although such crystals were found at the Princeton Gold Mine, Mariposa County, California, but occur more frequently with rounded angles, raised edges, and cavernous faces, which are often marked with parallel striations, and possess little or no lustre (Fig. 2). The octahedra found in California are usually flattened parallel to two opposite faces, or elongated, or otherwise distorted. Still more frequently they are only partially developed, as in Figs. 3 and 4. Crystals of greater complexity, containing many modifying faces, occur chiefly in Siberia, Transylvania and Brazil. The most common

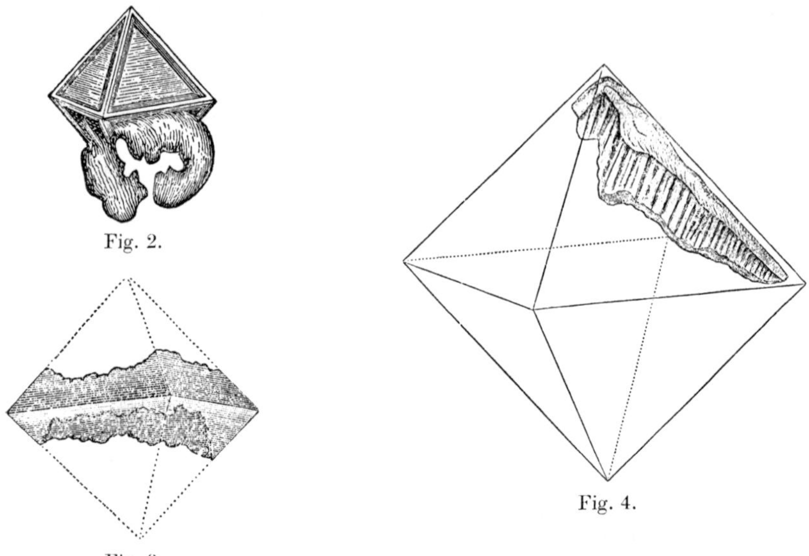

Fig. 2.

Fig. 3.

Fig. 4.

Figs. 2, 3, and 4.—Octahedral Crystals of Native Gold from California.

forms occurring naturally in Australia are the octahedron and the rhombic dodecahedron.[1]

Liversedge found [2] that, in the majority of cases, the gold embedded in massive quartz is remarkably free from any traces of crystalline form, and the larger the fragments of gold, the less crystalline form do they present. Widmanstätten figures were occasionally obtained by polishing and etching. Liversedge stated that all well-shaped crystals of gold appear to have been formed in what are now cavities, usually left by the removal of iron pyrite, or else in very soft matrices like iron oxides, clay, calcite, and serpentine. Crystallised gold is not usually met with in the quartz of the reef itself, but in the upper portions of the ferruginous and argillaceous casing of the reef and in the detritus near its outcrop.

---

[1] For a full account of the crystalline forms of native gold, see a Paper by W. P. Blake in "*Precious Metals of the U.S.A.*," 1884, p. 573.

[2] *J. Roy. Soc. of New South Wales*, 1893, **27**, 299; *Chem. News*, 1894, **49**. 162.

On the other hand, polished and etched sections of nuggets [1] always show marked crystalline structure closely resembling that of a fused mass of gold. This is shown in Fig. 5, a section of a nugget from Coolgardie, Western Australia, which weighed 9·94 ozs., and consisted of gold 890, silver 105. Fig. 6 shows the outside of the same nugget. There are no traces of the concentric structure which might have been expected if the nugget had been built up of successive coatings round a nucleus.

Carpenter and Tamura [2] confirmed the observations of Liversedge, including the appearance of twin structure in a nugget from Australia. They conclude that this type of crystal is a secondary formation due to straining of the primary deposit, the effect of which is determined by the

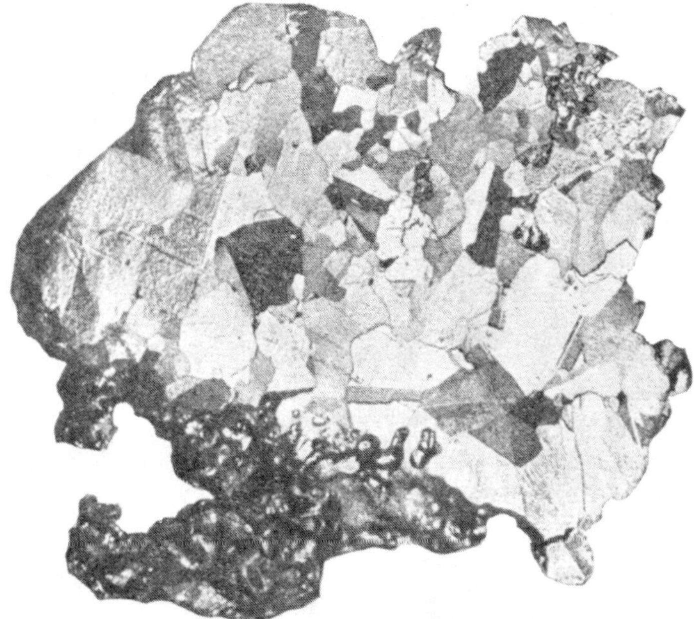

Fig. 5.—Gold Nugget, Internal Structure. × 1½ diam.

temperature and duration of time over which rearrangement of the atoms has been possible. Movements of the earth's crust would cause sufficient stress to explain the structures that were found.

Fisher,[3] on the other hand, found three varieties of zonal structure in several specimens obtained from the Molobe field. The normal occurrence consists of light and dark concentric bands around regular-shaped nuclei. The bands themselves are vividly coloured, due probably to interference effects of a thin film of silver chloride formed during etching. Contemporaneous twinning was observed in some of the crystals, *i.e.* twinning brought about during the crystal growth. Occasionally, especially in the

---

[1] Liversedge, *J. Chem. Soc.*, 1897, 71, 1125; Figs. 5 and 6 are reproduced from the Journal with permission. See also Oebbeke and Von Schwarz, *Zeit. Krist.*, 1924, 59, 62.

[2] *Trans. Inst. Min. Met.*, 1927-28, 37, 365.

[3] *Trans. Inst. Min. Met.*, 1935, 44.

granular zonal structures, twinning due to physical straining was detected. In this case the gold is assumed to have been derived mechanically from denuded gold quartz veins. The greater purity of alluvial deposits is

Fig. 6.—The same, Outside View. × 1½ diam.

attributed to electrolytic corrosion, resulting in the removal of silver and the redeposition of the gold.

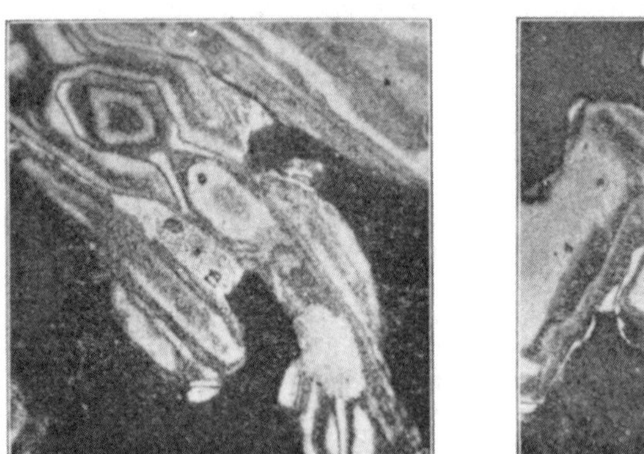

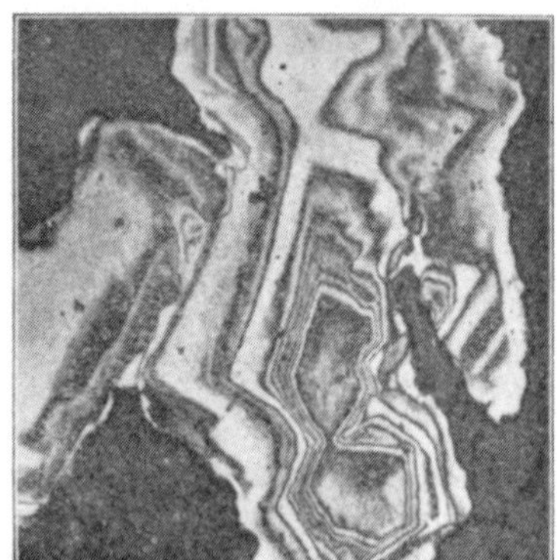

[Reproduced by permission of the Institution of Mining and Metallurgy.

Fig. 7. Fig. 8.

Figs. 7 and 8 show two specimens, having zonal structure. They were etched with aqua regia.

From a thermodynamic point of view native metals having an irregular atomic arrangement are not in a stable condition. There is a constant tendency for them to change to a stable condition involving the rearrangement of atoms, and resulting in an internal structure similar to that of manufactured metals which have been recrystallised by annealing.

Artificial crystals can be obtained in several ways, but with some difficulty. Feathery crystalline plates are precipitated in the electrolysis of a solution of the chlorides of gold and ammonium. Formaldehyde in the presence of hydrochloric or nitric acid precipitates crystalline gold from solutions of gold chloride or bromide.[1] Crystals belonging to the cubic system are formed by the precipitation of gold from its solution as chloride by means of ether, phosphorus in ether, oxalic acid, ferrous sulphate, etc. When copper pyrite, mispickel, blende, etc., are used as precipitants, however, minute prisms, beautifully sharp and well-defined, are sometimes obtained.[2] The prisms are often grouped in six-rayed stars, or six-sided plates.

By keeping an amalgam containing 5 per cent. of gold at a temperature

Fig. 9.—Gold Crystals.

of 80° C. for some days, and then digesting it at 30° C. with dilute nitric acid. bright crystals of gold can be obtained. These crystals are prismatic needles, and are said by Chester to be regular hexagonal prisms with pyramidal terminations. Adam states that oxalic acid in precipitating gold from solutions of $AuCl_3$ sometimes gives octahedra.[3] He also found that if mercury containing 1 per cent. of gold is warmed for two days, octahedra of gold are formed. After 7 or 8 days the crystals are larger. If the mercury contains 2 per cent. of gold, hexagonal prisms result. With a gold content between 1 and 2 per cent. a mixture of the two kinds of crystals is produced. The crystals are left behind when the mercury is dissolved in nitric acid, but still contain from 5 to 25 per cent. of mercury. The hexagonal crystals are pseudomorphs of a gold-mercury alloy.

The gold-mercury system has been examined by Braley and Schneider,[4]

---

[1] Awerkieff, *J. Chem. Soc.*, 1903, **84**, [ii.], 218, 603.
[2] Liversedge, *Chem. News*, 1894, **49**, 172.
[3] *J. Chem. Met. Mng. Soc. S.A.*, 1924, **24**, 258.
[4] *J. Am. Chem. Soc.*, 1921, **43**, 740.

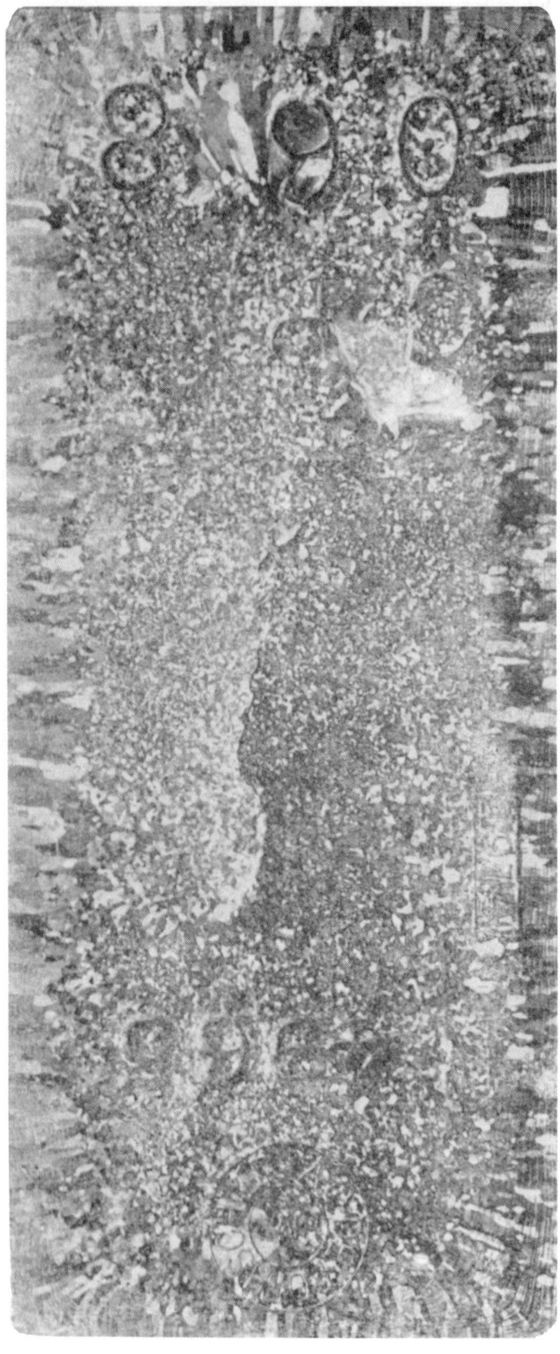

Fig. 10.—Cast Gold, Large Ingot (8 inches × $3\frac{1}{2}$ inches), Etched. (Upper surface, slightly reduced.)

who found that three compounds are formed, one of which, Au_2Hg, freezes at 490° C. In the Percy collection in the South Kensington Museum are some gold crystals found in the mercury troughs in an amalgamation mill. The mercury has been removed by nitric acid, and the gold crystals are well-defined combinations of the octahedron, rhombic dodecahedron, and cube.

When gold is cooled from fusion very slowly, the metal first solidified, owing to contraction during solidification, forms fern-like structures in relief on the surface, showing the rectangular arrangement of the axes. Sometimes faces and, less frequently, angles of octahedra are visible on the surface of the metal. The purer the gold the more likely these crystals are to be observable. The presence of small quantities of copper reduces the size of the crystals. On pouring a partly solidified mass of pure gold from a

Fig. 11. The same Ingot as Fig. 10. (Lower surface × 3·2.)

crucible, Roberts-Austen obtained a shell consisting of an aggregate of well-formed crystals, apparently octahedra.[1] These crystals are preserved at the Royal Mint, London (see Fig. 9, which is about half-size).

On solidification from fusion, whether quickly or slowly, gold, like other metals, sets in crystal grains or allotriomorphic crystals, consisting of irregular polygons of considerable size, which are larger as the rate of cooling is slower. These crystals may be seen without magnification by lightly etching the gold ingot for 30 to 60 minutes with aqua regia diluted with an equal bulk of water at the ordinary temperature. Photographs of cast pure gold etched in this way are shown in Figs. 10 and 11.

Fig. 10 is a photograph of the upper surface of a gold ingot, 999·9 fine by assay, and weighing 400 ozs. It was cast in an open mould and occupied

[1] "*Encyclopædia Britannica*" (9th edition), article "Gold."

a considerable time in cooling. Crystals bounded by straight lines were accordingly formed, the slow cooling producing the same effect as annealing. In small ingots which are cooled more quickly, smaller crystals are formed

Fig. 12.—Gold, Small Ingot before Annealing (Etched). × 100.

Fig. 13.—The Same, after Annealing. × 100.

with irregular boundaries, as is seen in Fig. 12. On annealing, the irregular boundaries give place to straight lines, as in Fig 13.

Bannister has shown [1] that when gold is solidifying, as for instance in the form of a cupellation bead, the solidification proceeds radially from centres. At the interference of solidification from two centres there is a straight line, from three centres three lines at angles of 120° to one another, and from four centres lines at right-angles. The method of solidification, *i.e.* with or without "flashing," changes the nature of the crystalline surface. When "flashing" occurs the crystalline surface shows an irregular polygonal structure.

On rolling, the original crystals are distorted by elongation, with the laminated effect shown in Fig. 14. Annealing causes the metal to recrystallise,

Fig. 14.—Fine Gold (Rolled and Etched). × 3.

each large lamina breaking up into a number of small crystals, which appear and rapidly increase in number in particular laminæ, while others remain unaltered [2] (Fig. 15). The first appearance of recrystallisation, which is simultaneous with softening, has been noted at 80° C. after 100 hours, and takes place in a few seconds at 200° C.[3] In course of time, or with rise of temperature, the new crystals increase in size, and obliterate the boundaries of the original crystal grains. Twinned crystals make their appearance, giving a characteristic banded structure, shown as certain narrow parallel-sided strips in Fig. 16. Parravano and Agostini [4] have found that melting

[1] *J. Inst. Met.*, 1929, **42**, 141.
[2] T. K. Rose, *J. Inst. Met.*, 1913, **10**, 162.
[3] See also Jeffries and Archer, *Met. Chem. Eng.*, 1922, **26**, 343.
[4] *Rend. della Reale Accad. Linc., Roma*, 1921, **30**, 481.

in nitrogen or hydrogen raises the annealing temperature, while melting in carbon dioxide lowers it. The temperature of annealing of gold containing

Fig. 15.—Gold (Rolled, Incompletely Annealed, and Etched). × 4.

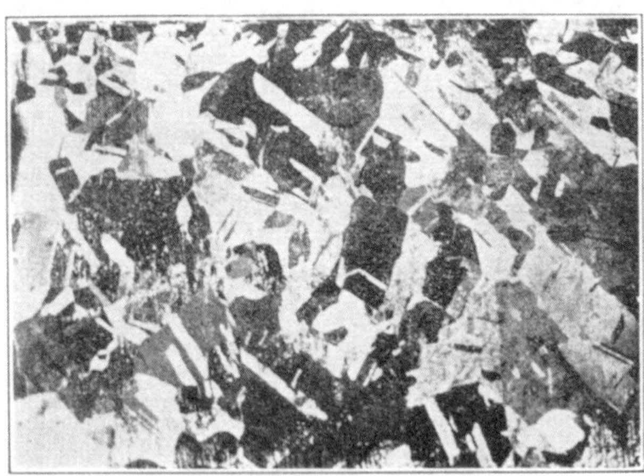

Fig. 16.—Fine Gold (Heated at 800° C. for 1 Hour). × 7.

0·002 per cent. of hydrogen is 300° C., and that of gold containing 0·05 per cent. of copper is 250° C.[1]

Absorption of Gases.—Gold is not attacked by hot or cold air in either the moist or dry condition, but absorbs furnace gases in varying amounts.

---

[1] Rose, *loc. cit.*

When the metal is heated to 1,215° C. in air 0·25 part of gas per 1,000 is absorbed, and when heated in coal gas 7·8 parts per 1,000. Trowbridge found that molten gold can absorb 37 to 46 times its volume of hydrogen. It can also take up 33 to 48 times its volume of oxygen.

Dissolution of Gold.—Gold is readily soluble in aqua regia, or in any other mixture producing nascent chlorine. Almost any chloride, bromide, or iodide will dissolve gold in the presence of an oxidising agent. The action is much more rapid if heat is applied, or if the gold is alloyed with one of the base metals. The solution of gold in aqua regia takes place according to the equation—

$$Au + HNO_3 + 4HCl = 2H_2O + NO + HAuCl_4$$

and these proportions are most economical for employment.[1] They correspond to one part by weight of nitric acid of specific gravity 1·42 to four parts by weight of hydrochloric acid of specific gravity 1·2, or by volume, 1 part to 4·7 parts. The presence of silver in the gold retards the process, a scale of insoluble chloride of silver being formed over the metal, and the action may eventually be completely stopped if the percentage of silver present is large. Gold is also dissolved by liquids containing chlorine and bromine or a mixture producing bromine. The action is much slower than that with aqua regia, and subject to the same difficulties if silver is present; heat assists the dissolution. Iodine dissolves gold only if it is nascent, or if dissolved in iodides or in ether or alcohol. Gold dissolves in hydrochloric acid in the presence of organic substances—*e.g.* methyl or ethyl alcohol, chloroform, glycerol, etc. The action is accelerated by heat.[2] Metallic gold does not dissolve in strong sulphuric acid unless a little nitric acid is added, when a yellow liquid is formed, which, when diluted with water, deposits the metal as a violet or brown powder. The mixture of nitric and sulphuric acids is a more rapid solvent for gold than nitric acid alone.

When gold is heated with concentrated selenic acid, H_2SeO_4, it dissolves with liberation of selenium dioxide and formation of a reddish-yellow solution of auric selenate, $Au_2(SeO_4)_3$; the action begins at 230° C., but proceeds more readily towards 300° C.[3]

Gold is soluble in ferric chloride,[4] and in cupric chloride. In a re-examination of these observed facts, Stokes[5] found that the dissolution takes place readily at 200° C., thus—

$$Au + 3FeCl_3 \rightleftarrows AuCl_3 + 3FeCl_2$$
$$Au + 3CuCl_2 \rightleftarrows AuCl_3 + 3CuCl$$

Equilibrium is reached after a time, and no further action takes place, except on the addition of more ferric or cupric chloride or a rise in the temperature. A further addition of ferrous or cuprous chloride or a fall in the temperature causes some gold to be reprecipitated with the formation of ferric or cupric chloride.

M'Ilhiney, however, found[6] that ferric chloride acts as an efficient carrier of chlorine in the presence of hydrochloric acid and oxygen. Stokes

[1] Priwoznik, *J. Chem. Soc.*, 1911, **100**, [ii.], 484.
[2] Awerkieff, *J. Chem. Soc.*, 1908, **94**, [ii.], 859.
[3] Lenher, *J. Amer. Chem Soc.*, 1902, **24**, 354.
[4] Napier, *Phil. Mag.*, 1844, [ii.], **24**, 370; Schild, *Berg. und Hütt. Zeit.*, 1888, **47**, 251.
[5] "*A Treatise on Metamorphism*," by Van Hise. *U.S. Geol. Survey*, 1904, **47**, 1090.
[6] *Amer. J. Sci.*, 1896, **2**, 293.

also observed [1] that gold is not appreciably dissolved in ferric sulphate unless chlorides are present at the same time, thus furnishing ferric chloride. These observations are of importance in considering the action of descending solutions in the belt of weathering in auriferous lodes.

Gold is somewhat readily soluble in a 10 per cent. solution of sodium carbonate, and also in an 8 per cent. solution of sodium carbonate containing excess of carbonic acid and sodium silicate. It is easily soluble in sodium sulphide and in sodium sulphydrate.

Some other haloid compounds only attack gold in the presence of ether, in which solvent even hydriodic acid has a slight effect. Iodic acid has also been mentioned as a solvent for gold, but its action is very slight, much less, for example, than that of concentrated hydrochloric acid under similar conditions. A mixture of iodic and sulphuric acids dissolves gold when heated to 300° C. (Prat, also Victor Lenher). The effect of nitric and nitrous acids and of mixtures of them is described in Chapter XIX. Alkaline sulphides attack gold slowly in the cold, and more rapidly if heated, producing sulphide of gold which is subsequently dissolved. Gold is also soluble in the thiosulphates of calcium, sodium, potassium, and magnesium, in the presence of an oxidising agent. According to H. A. White,[2] the solubility of gold in sodium thiosulphate is greatly accelerated by ferric chloride and some other oxidising agents. He also found that ammonium thiocyanate alone would not dissolve gold, but that potassium thiocyanate in the presence of ferric chloride dissolved gold far more rapidly than was the case with thiosulphates. Fresh solutions of ferric thiocyanate dissolved gold slowly.

Spring has shown [3] that gold is soluble in hydrochloric acid if heated with it to 150° C. in a closed tube, and is subsequently reduced by the liberated hydrogen and deposited as microscopic crystals on the side of the tube. T. K. Rose found that boiling concentrated hydrochloric acid dissolves gold and maintains it in solution. Berthelot [4] found that the action takes place slowly in the cold in the presence of light and air, but not in the dark. Gold is also soluble in solutions of ferric and stannic salts in presence of hydrochloric acid, and is still more freely dissolved by a solution containing $CuCl_2$ and HCl, especially on heating.[5] Victor Lenher showed [6] that in the presence of sulphuric acid many oxidising substances, such as telluric acid, manganese dioxide, lead dioxide, red lead, chromium trioxide, and nickel oxide, cause gold to pass into solution. In some cases phosphoric acid may be substituted for sulphuric acid. When gold is used as an anode, it is oxidised, and if the electrolyte is strong sulphuric acid, phosphoric acid or caustic soda or potash, part of the oxide is dissolved. C. Lossen [7] has pointed out that if a solution of potassium bromide is electrolysed, the resulting alkaline solution containing hypobromite and bromate of potassium, is capable of dissolving gold. Gold is dissolved by aqueous solutions of simple cyanides in presence of an oxidising agent. This action is fully discussed in Chap. XIII. Sulphocyanides, ferrocyanides, and some other double cyanides also dissolve gold, both at ordinary temperatures and on heating, but the

---

[1] Van Hise, *loc. cit.*
[2] *J. Chem. Met. Mng. Soc. S.A.*, 1905, **6**, 109.
[3] *Zeitsch. anorg. Chem.*, 1893, **1**, 240.
[4] *J. Chem. Soc.*, 1904, **86**, [ii.], 569.
[5] M'Caughey, *J. Amer. Chem. Soc.*, 1909, **31**, 1261.
[6] *Eng. and Min. J.*, 1904, **77**, 963 ; *J. Amer. Chem. Soc.*, 1904, **26**, 550.
[7] *Ber.*, 1894, **27**, 2726.

action is very slow, even in the presence of oxidising agents. Beutel has shown[1] that when potassium ferrocyanide is used, potassium aurocyanide is formed, and the resulting ferro-ions are oxidised by the air, giving ferric hydroxide. The solution formed is alkaline, and the reaction is probably represented as follows :—

$$3Au + K_4Fe(CN)_6 + 2H_2O + O_2 = 3KAu(CN)_2 + Fe(OH)_3 + KOH$$

Moir found[2] that finely divided gold dissolves slowly in acid (hydrochloric or sulphuric) solutions of thiocarbamide, and that the action is rapid in the presence of oxidising agents such as ferric chloride or hydrogen peroxide, especially if heated to 50° C. The gold is very slowly reprecipitated by ferrous sulphate or stannous chloride.

Allotropic Forms of Gold.—Little is known of these. The marked influence of traces of other metals on the properties of gold has already been touched on ; from this and from the variations in colour and other properties the existence of several allotropic modifications of gold might be inferred.

Wilm[3] states that if gold is dissolved in dilute sodium amalgam under water, the aqueous liquid becomes dark violet, and when this is acidulated with hydrochloric acid, a black precipitate of pure gold is obtained. The black gold differs from the ordinary modifications in its extreme lightness ; moreover, it is soluble in alkaline solutions, and does not amalgamate with mercury or with sodium amalgam. When heated, it yields the ordinary modifications as a violet red powder. This form of gold appears, from Wilm's account, to resemble the black precipitate obtained on digesting certain aluminium-gold alloys with hydrochloric acid and that obtained by the action of water on potassium-gold alloys.

Julius Thomsen[4] stated that different allotropic modifications of gold are obtained by the reducing action of sulphurous acid on various solutions of gold compounds. The supposed allotropic forms obtained by the reduction of (*a*) neutral auric chloride, (*b*) auric bromide, (*c*) aurous chloride, bromide, or iodide have been designated respectively as gold, gold α, and gold β. Thus gold α gave out 3·2 calories in passing into gold. Van Heteren,[5] however, found that the potential differences between these samples of gold were no greater than between two samples of the first allotropic form, and consequently the forms were identical.

Hanriot[6] studied the properties of brown gold, obtained by dissolving away the silver from silver-gold alloys with nitric acid, and concluded that it is a spongy mass formed of a mixture of ordinary or α gold, with a new modification, β gold. He leaves it as an open question whether the differences are due merely to the nature of the crystals, or whether they are sufficient to warrant the use of the name allotropic modification. The β variety has not been prepared in a pure state. It is always mixed with α gold, and also with traces of silver, copper, lead, iron, etc., as revealed by the spectroscope. It has a lower coefficient of diamagnetism than α gold, and is transformed into α gold at temperatures above 300° C. Between 300° and 650° C. brown

[1] *J. Chem. Soc.*, 1910, 98, [i.], 723.
[2] *J. Chem. Soc.*, 1906, 89, 1345.
[3] *Zeitsch. anorg. Chem.*, 1893, **4**, 325.
[4] "*Thermochemische Untersuchungen,*" vol. iii, p. 398.
[5] *J. Chem. Soc.*, 1905, 88, [ii.], 260.
[6] *Bull. Soc. chim.*, 4th series, 1911, 9, 139, 339, and 1052.

gold is metastable, and tends to change into ordinary gold with a contraction of 40 per cent. in the length of thin plates. This change is hastened by touching it with a plate of the α variety.

It has also been shown by Hanriot and Raoult[1] that brown gold is more soluble in nitric acid and in hydrochloric acid than yellow gold. They also found that brown gold is freely soluble in a hot hydrochloric acid solution of auric chloride, and that on cooling beautiful crystals of β gold are formed, consisting of a mixture of tetrahedra and rhombic dodecahedra. These crystals are very soluble in auric chloride.

Louis had previously found[2] that brown gold has a density of 20·3, and that it expands on being transformed by heat into the yellow variety.

The evidence in favour of the existence of amorphous or hard gold in cold-worked or polished specimens is similar to that in the case of other metals.[3]

---

[1] *Compt. rend.*, 1912, **155**, 1085.

[2] *Trans. Amer. Inst. Min. Eng.*, Chicago Meeting, 1893, **22**, 117.

[3] See G. T. Beilby, *Proc. Roy. Soc.*, 1907, **A**, **79**, 463 ; *J. Inst. Metals*, 1911, **6**, 5.

CHAPTER II.

ALLOYS OF GOLD.

Introduction.—Gold can be made to alloy with almost all other metals, but many of the bodies thus formed are of little or no importance to metallurgists. The binary alloys of gold with one other metal have been studied in a number of instances, but little systematic study has been devoted to gold alloys containing three or more metals, although the metallurgist has to deal chiefly with these. In the following pages the binary alloys are described successively.

Alloys of gold are generally prepared by melting their constituents together, but they may in certain cases be prepared by simultaneous precipitation from solution, as was shown by Mylius and Fromm in 1892.[1] Thus a gold-zinc alloy, containing equal weights of the two metals, and approximating in composition to $AuZn_3$, was obtained in the form of black, spongy flocks, by adding a solution of gold sulphate to water in which a zinc plate was placed. The gold-zinc slime obtained in the cyanide process (*q.v.*) may be compared with this. Gold-cadmium, similarly obtained, is a lead-grey crystalline precipitate, having the composition $AuCd_3$. If the gold-zinc alloy is shaken with a solution containing a cadmium salt, the gold-cadmium alloy and a zinc salt are obtained. Similarly, a copper plate, acting on solutions containing gold, yields a black, spongy compound of gold and copper ; and gold-lead and gold-tin alloys in the form of black slimes are also readily prepared.

The diffusion of gold into other metals, both liquid and solid, was investigated by Roberts-Austen.[2] The rates of diffusion in liquid metals are of the same order as those of soluble salts in water, but the diffusion in solids is very slow. If gold is placed at the base of a cylinder of solid lead 70 mm. high and the temperature kept at 251° C. some is found to have reached the top in thirty days. The rate of diffusion is still measurable at 100°, but is almost inappreciable at ordinary temperatures.

Fraenkel and Houben [3] diffused silver into gold in the solid state by heating a cylinder of the latter, into which a plug of the former had been driven, at 870° C. The mean value for the coefficient of diffusion over a surface area was 0·000037 cm.^{2} per day.

Gold and Aluminium.—These alloys were investigated by Roberts-Austen,[4] and by Heycock and Neville.[5] Several true compounds of the two metals were proved to exist, the most remarkable and stable being Au_2Al, a hard white alloy with a freezing point of only 622° C., and $AuAl_2$, Roberts-Austen's

---

[1] *J. Chem. Soc.*, 1894, 66, [ii.], 236 ; *Ber.*, 1894, **27**, 630.
[2] *Proc. Roy. Soc.*, 1896, **59**, 281 ; Bakerian Lecture, *Phil. Trans.*, 1896, **A**, **187**, 383.
[3] *Zeit. anorg. Chem.*, 1921, **116**, 1.
[4] *Proc. Roy. Soc.*, 1892, **50**, 367.
[5] *Phil. Trans.*, 1900, **A**, **194**, 201 ; *Proc. Roy. Soc.*, 1914, **90**, 560.

beautiful purple alloy, which melts at about the fusing point of pure gold. This compound, which contains 21·5 per cent. by weight of aluminium, is always formed when mixtures of gold and aluminium are fused, and allowed to cool, provided that not more than about 90 per cent. of gold is present. The purple alloy is seen in patches on a white ground containing the excess of aluminium or of white compounds of gold and aluminium.

The solidus [1] descends from the freezing point of gold to a point corresponding to 13 atomic per cent. aluminium at 543° C. Thence a horizontal branch leads to a sloping branch from 18·8 to 21·5 per cent. aluminium, where the eutectic line starts at 527° C.

The structure of the α solid solution is normal and homogeneous. The range of the β solid solution narrows down until a eutectoid point is reached at 19 atomic per cent. aluminium and 424° C. This component is obtained only in quenched specimens, and normally shows a striated structure when etched. There is some similarity between the constituents α, β and δ (Al_3Au_8) and those in the copper-tin series.

A remarkable recalescence, starting below 400° C. and finishing at 470° C. or thereabouts, and not usually occurring until the eutectoid has formed, is observed in the alloys containing about 20 atomic per cent. aluminium. It is probably due to a reaction between α and δ constituents, resulting in the formation of a new constituent γ (? $AlAu_4$). The recalescence usually does not occur spontaneously above 410° C. but may be instigated up to 515° C. by touching the alloys with a cold steel wire. In some instances the rise in temperature is sufficient to melt part of the alloy.

In etching these alloys aqua regia is used. The α constituent is bright, β and γ dark and δ dark in the presence of α but bright in the presence of β or γ.

Gold and Antimony.—Gold is very easily dissolved by molten antimony. A piece of gold wire held in the heated material can be readily seen to be going into solution, forming radial stream lines.

A small amount of antimony lowers the melting point of gold considerably, and causes a long pasty stage. Molten gold absorbs the vapour of antimony and becomes brittle. Roberts-Austen [2] found that an addition of 0·2 per cent. of antimony to gold reduced its tensile strength from 7·0 to 6·0 tons per square inch. T. K. Rose [3] found that about 1·0 per cent. of antimony makes gold brittle, but that the brittleness is removed by annealing.

The alloys are not malleable. They are all formed with contraction. They are hard, pale yellow or grey, and not easily attacked by acids, except aqua regia. Cosmo Newbery has shown that when the alloys are finely divided and ground with mercury they are slowly decomposed, yielding gold amalgam and black metallic antimony, mixed with antimonide of mercury.

From Vogel's results [4] it appears that the liquidus curve consists of three branches, along which three different series of crystals separate. The two series containing most gold do not appear to consist of solid solutions. The third series of crystals corresponds to the formation of the compound $AuSb_2$, at 460° C., and 55 per cent. Sb, the point of maximum thermal effect in the reaction. At the same point there is a break in the liquidus curve. Vogel

---

[1] Heycock and Neville, *Phil. Trans.*, 1914, **A**, **214**, 267.

[2] *Phil. Trans.*, 1888, **A**, **198**, 339.

[3] T. K. Rose, *33rd Annual Report of the Royal Mint*, 1902, p. 73.

[4] *Zeitsch. anorg. Chem.*, 1906, **50**, 145-157; *J. Chem. Soc.*, 1906, **90**, [ii.], 679; Friedrich, *Metallurgie*, 1908, **5**, 603.

records that these alloys are strongly susceptible to undercooling, and that only by inoculation can solidification be promoted. In this respect they resemble the tellurium-gold alloys.

The compound $AuSb_2$ is of the same colour as antimony, is extremely brittle, and considerably harder than the individual components. There is a well-marked eutectic at 360° C. and 24 per cent. Sb.

Grigoriew [1] has confirmed these observations by electrical conductivity measurements. Neal Almin and Westgren [2] have also substantiated them by X-ray methods, especially with regard to the compound $AuSb_2$. At temperatures higher than 360° C. gold retains 1 atomic per cent. of antimony in solid solution.

Gold and Arsenic.—These resemble the gold-antimony alloys. They are readily fusible, hard, brittle, pale yellow or grey alloys, formed with contraction and decomposed by mercury like the antimony alloys. They are soluble with difficulty in aqua regia. The vapour of arsenic, acting on a suspended plate of gold at a red heat, forms a readily fusible alloy which drips off the gold, so that the residue of the gold plate is malleable and free from arsenic.

The alloys containing from 0 to 11 per cent. of arsenic were examined thermally by Schleicher.[3] The freezing point curve indicates a eutectic point at about 25 per cent. of arsenic, the melting point being 665° C. The eutectic is still observable when only a small proportion of arsenic is present. Arsenic is evolved suddenly at the eutectic temperature on melting.

Gold and Bismuth.—Bismuth is an element which, when present in even the smallest amounts, renders ordinary commercial gold very brittle and unworkable. Of all metals, it has the greatest effect in reducing the ductility of gold. Thus one of us found [4] that fine gold containing 0·25 per 1,000 of bismuth is fragile, breaks under the hammer, and cannot be rolled. The same proportion of bismuth in standard gold (gold 916·6, copper 83·3) also causes brittleness, and differs from lead in the fact that the brittleness is not removed by annealing. Molten gold absorbs the vapours of bismuth and becomes brittle. All the alloys exhibit a strong tendency to segregation,[5] so that when solid they are not uniform in composition.

It was observed by Arnold and Jefferson [6] that in gold containing 0·2 per cent. of bismuth, when somewhat quickly cooled from fusion, the structure consists of irregular grains, forming a series of cells cemented together with what is probably the gold-bismuth eutectic. The whole thus presents the appearance of a network. Slowly cooled specimens are similar in appearance, but the grains are larger. When broken by working, the fracture takes place along the cell walls, and any detached grain is perfectly malleable, probably consisting of pure gold, and can easily be beaten out into gold leaf. Similar effects were noted when lead or tellurium was substituted for bismuth.

---

[1] *Ann. Inst. Platine*, 1929, [i.], 45.

[2] *Z. physikal. Chem.*, 1931, **B**, **14**, 81. See also Oftedal, *Z. physikal. Chem.*, 1928, **A**, **135**, 291.

[3] *Internat. Zeitsch. Metallographie*, 1914, **6**, 18.

[4] T. K. Rose, 33*rd Annual Report of the Royal Mint*, 1902, pp. 72-73.

[5] The term "segregation" (sometimes called "liquation") is used in this chapter to denote the partial separation of the constituents of an alloy during solidification in such a way as to render the alloy non-uniform in composition. (See Smith, *Trans. Inst. Min. Met.*, 1926, **35**, 248.)

[6] *Engineering*, 1896, **61**, 176; also Andrews, *ibid.*, 1898, **66**, 541.

Osmond and Roberts-Austen [1] found that in quickly cooled ingots containing 0·2 per cent. of bismuth the thickness of the joints between the crystals was 2·5 μ (0·0025 mm.). On annealing at 200° to 250° C., the large grains were subdivided into small polyhedral grains.

The thermal equilibrium diagram was constructed by Vogel.[2] The liquidus curve has two branches, which meet in a eutectic point at 82 per cent. bismuth, and at a temperature of 240° C. The eutectic horizontal reaches to pure bismuth on the one side, and to gold containing less than 0·025 per cent. bismuth on the other side. The action of gravity causes the bottom of a culôt of any of the alloys to be enriched in gold. Vogel states that in these alloys the primary crystals of gold have a zoned structure, being yellow in the interior (pure gold) and white and poorer in gold on the exterior (probably nearly pure eutectic).

The structure of the eutectic alloy is quite altered by quenching from 400° C. The white element is replaced by a mixture of what appear to be yellow and black components arranged in lamellæ. The gold-bismuth alloys are all brittle, but are not harder than their constituents, and, with the exception of the gold-rich alloys, have the white colour of bismuth. The gold-rich alloys are pale yellow or greyish.

A peculiar effect has been observed on the surface of bismuth-rich alloys. After cooling, glittering spherical nodules of metal appear, which have been attributed to the well-known property of bismuth to expand on solidification. These globular masses have a composition approximating always to 18 per cent. gold and 82 per cent. bismuth, which is that of the eutectic. This phenomenon may be compared with the effect of heating gold-tellurium alloys or native tellurides, or the " sweating " of certain base metal alloys.

Gold and Cadmium.—The complete equilibrium diagram was constructed by Vogel in 1906,[3] and Soldau [4] also investigated the gold-cadmium series. The latter concludes that there are only two compounds, AuCd and $AuCd_3$. AuCd alone gives a maximum on the freezing point curve. It takes up Cd over a wide range. Electrical conductivity and Brinell hardness curves are characteristic.

Durrant [5] differs from Soldau in the construction of the equilibrium diagram in that he obtains a maximum point at 28·6 atomic per cent. gold and 500° C., whence the liquidus falls to a eutectic point at 30 atomic per cent. and 496° C. A horizontal at 496° C. represents a eutectic level. At 39·4 atomic per cent. gold and 540° C. a peritectic line meets the liquidus. A new solid solution area has also been detected. This solution undergoes two polymorphic changes—at 500° C. and 375° C. The latter appears to be analogous to the change in the β phase in brass at 460° C. Durrant obtained no evidence of the compound $AuCd_3$ at the liquidus, but suggests that Au_2Cd_5 and Au_2Cd_3 exist and are both dissociated at high temperatures.

Ölander [6] has published a complete phase diagram of the gold-cadmium series, and postulates the occurrence of three new transition points.

Alloys of gold and cadmium can be " parted " with nitric acid, and are dissolved by aqua regia. They may be also parted with dilute sulphuric acid.

---

[1] *Phil. Trans.*, 1896, **A**, **187**, 417-432.
[2] *Zeitsch. anorg. Chem.*, 1906, **50**, 145-157.
[3] *Zeitsch. anorg. Chem.*, 1906, **48**, 333-346.
[4] *Internat. Zeitsch. Metallographie*, 1915, **7**, 3. *Chem. Abs.*, 1925, **19**, 2440.
[5] *Journ. Inst. Metals*, 1929, **41**, 139.
[6] *J. Amer. Chem. Soc.*, 1932, **54**, 3819.

Gold and Chromium.—Vogel [1] has studied the equilibrium diagram for these two metals. The alloys find no use in the arts, owing to their brittleness, which is enhanced by impurities such as nitrogen, oxygen or carbon. These are absorbed principally during melting.

In the solid state chromium takes up 10 per cent. by weight of gold and gold 7 per cent. by weight of chromium into solid solution.[2]

Gold and Cobalt.—Wahl [3] studied the thermal, magnetic, and microscopic properties of these alloys. The experimental difficulties were great, owing to the ease with which cobalt oxide is formed on heating. This influences the results by depressing the true freezing points. The difficulty was overcome by heating the alloys in an atmosphere of nitrogen in porcelain tubes. The cobalt first attacks the porcelain, but the oxide present being thus removed, subsequent fusing does not cause any further attack.

Pure cobalt melts at 1,480° C., but as much as 216° C. of supercooling has been observed. The cobalt-gold alloys exhibit a similar tendency to pass their melting point before becoming solid. Wahl obtained a rather unusual form of liquidus curve with many turning points by plotting atomic percentages against temperatures. Speaking generally, there are two branches to the liquidus curve, meeting at a eutectic at 997° C., and corresponding to a composition of about 90·1 per cent. of gold; but the accuracy of these figures is open to doubt, as few experiments were made. The saturated solid solutions (ends of the horizontal eutectic line) contain 3·5 and 94 per cent. gold respectively.

All the alloys were found to be magnetic, the magnetisability decreasing, at first rapidly and afterwards more slowly, with increase in the gold content. The magnetisability also decreases with decrease in temperature, being three times as great at the eutectic temperature as when cold.

The transition point of cobalt has been determined by magnetic means to be about 1,140° C., and is found to remain practically unaltered by the addition of gold. α-Cobalt comes into direct contact with the melt at the lower temperatures. Crystals which are rich in cobalt are cubic above 1,140° and isomorphous with β-cobalt.

Wahl, in his paper, describes the appearance of six-pointed stars in the eutectic, which he considers may be hexagonal cobalt crystals, but similar hexagonal crystals have been observed in the undoubtedly cubic metals, gold and copper.

The addition of cobalt to gold makes it brittle in a greater degree than the addition of nickel. Standard gold becomes brittle if one-sixteenth part of cobalt is melted with it.

Gold and Copper.[4]—These metals are miscible in all proportions when molten and form solid alloys of sensibly uniform composition. The equilibrium curve is shown in Fig. 17 and includes the modifications below 500° C. which have been suggested by Haughton and Payne (*vide infra*).

In 1916, Kurnakow, Zemczuzny and Zasedatelev re-examined the copper-gold series, especially in the solid state.[5] They concluded that the α solution

---

[1] *Umschau.*, 1924, **28**, 297.

[2] Vogel and E. Trilling, *Z. anorg. Chem.*, 1923, **129**, 276.

[3] *Zeitsch. anorg. Chem.*, 1910, **66**, 60-72.

[4] Roberts-Austen and Rose, *Proc. Roy. Soc.*, 1900, **67**, 105; Kurnakow and Zemczuzny, *Zeitsch. anorg. Chem.*, 1907, **54**, 149; *J. Russ. Phys. and Chem. Soc.*, 1907. **39**, 211.

[5] *J. Inst. Met.*, 1916, **15**, 305.

TABLE I.—Gold and Copper (Kurnakow and Zemczuzny).

	Percentage by Weight.		Atomic Percentage. Au	Solidification Temperature, °C.	
	Au	Cu		Beginning.	End.
1	100	...	100	1,063°	...
2	98·33	1·67	95	1,034°	1,011°
3	94·42	5·58	85	978°	949°
4	90·29	9·71	75	934°	916°
5	86·0	14·0	65·93	890°	887°
6	82·0	18·0	59·5	884°	883°
7	80·0	20·0	56·34	886°	884°
8	69·90	30·1	42·94	900°	894°
9	50·87	49·13	25·00	942°	920°
10	30	70	12·55	1,018°	980°
11	14·02	85·97	5·00	1,056°	1,030°
12	...	100	...	1,084°	...

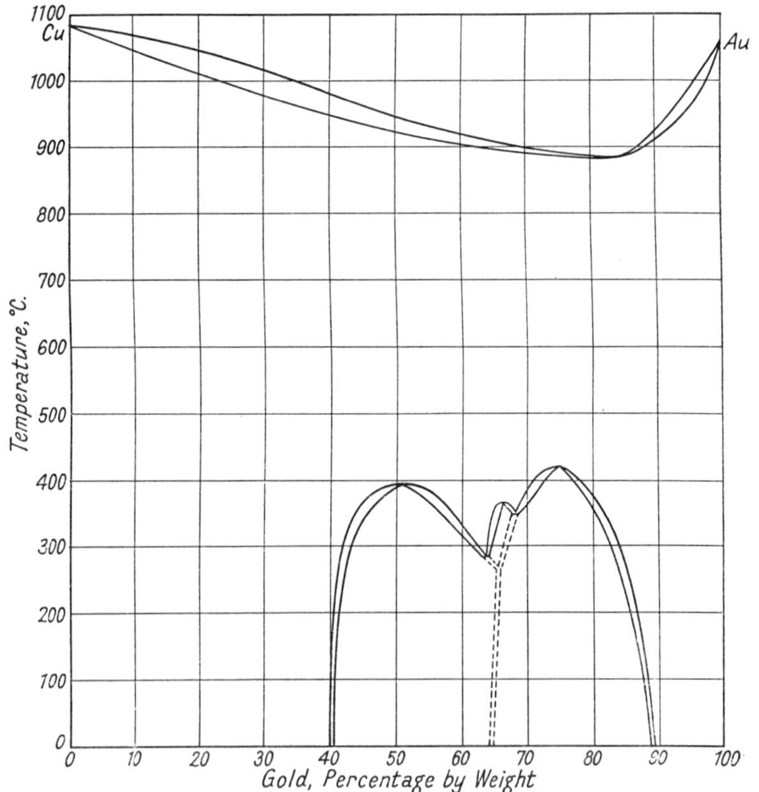

Fig. 17.—Thermal Equilibrium Curve of Gold and Copper.

of copper in gold exhibits transition points at 371° C. (50 atomic per cent. Au) and 367° C. (25 atomic per cent. Au). Below these points gradual decomposition occurs with the formation of the compounds CuAu and Cu_3Au. The α solution is comparatively soft and plastic. CuAu is hard and brittle and hence cannot be rolled or drawn. In an alloy, for example, containing 75·62 per cent. by weight of gold, quenching and annealing have effects opposite to those with normal alloys. Annealing after quenching gives a harder and more brittle product than that obtained after quenching alone—a phenomenon akin to age-hardening.[1] It was found that the electrical conductivity of the quenched alloys containing 40 to 60 atomic per cent. Au decreased by 5 per cent. in an hour and a half and could therefore be used for defining the starting point of the dispersion of the compound CuAu. The conductivity curves of the annealed alloys show two sharp maxima corresponding to the formation of the compounds Cu_3Au and CuAu, but at the same time confirm the fact that quenching and annealing do not affect the properties of the homogeneous α solid solution. Careful Brinell hardness measurements substantiate these observations.

Owing to the formation of compounds on cooling, gold-copper alloys show anomalies in their expansion curves according to their previous heat-treatment.[2]

Haughton and Payne [3] have studied the transformations in the solid state of the gold-copper series of alloys, following the observation by Kurnakow and his collaborators of one such effect at 370° C. in the alloys most nearly corresponding to the compounds AuCu and $AuCu_3$. Johanssen and Linde showed that at all temperatures above the transformation point the structure is an irregular cubic one, but that on cooling it merges into the regular tetragonal in the region of the compound AuCu and the regular cubic around $AuCu_3$. Borelius, Johannsen and Linde, by means of electrical resistivity and X-ray measurements, found maxima at 400° C. on cooling and at 430° C. on heating in the case of the alloy containing 50 atomic per cent. gold, and maxima at 382° C. and 390° C. on cooling and heating respectively an alloy containing 25 atomic per cent. gold. The temperatures observed by Kurnakow were 367° and 371° C. respectively, and these, as well as the hysteresis originally found by Borelius, were confirmed by Haughton. Broadly speaking it was shown that, though the system is a complex one, the high-temperature phase is stable above 405° C., and the low-temperature phase below 390° C. Between these temperatures either phase can be stable. Minute changes in composition, and especially the presence of oxygen, vitally affect the transformation points. In the case of the $AuCu_3$ group of compounds the hysteresis is small and can readily be overcome. Haughton and Payne have detected a third maximum on the transformation curve at 40 atomic per cent. of gold, corresponding to the compound Au_2Cu_3.

Kurnakow and Ageew [4] have examined this series of alloys by electrical

---

[1] Age (or precipitation) hardening may be defined as that hardness which is acquired by an alloy after quenching it from a high temperature and then allowing it to stand for some time, or heating it to a comparatively low temperature for a much shorter period. The increase in hardness following this treatment is attributed to the dispersion in a fine state of a hard constituent (in this case AuCu.)

[2] Portevin and Durand, *Compt. rend.*, 1921, **172**, 325.

[3] *J. Inst. Met.*, 1931, **46**, 457.

[4] *J. Inst. Met.*, 1931, **46**, 481.

resistance methods and by means of the dilatometer. They conclude that the solid solution of gold and copper which exists at high temperatures undergoes a transformation at about 425°-450° C., passing into the chemical compounds AuCu and $AuCu_3$. The formation of both compounds can be retarded by rapid chilling, but when they are established the change is accompanied by a marked decrease in volume.

Grube and his collaborators [1] measured the electrical resistance and thermal expansion of alloys containing 17 to 85 per cent. gold for temperatures between 20° and 460° C. They showed that three new phases are produced by decomposition of the α solid solution—β up to 34 per cent. gold, γ from 38 to 62 per cent. gold, and δ above 64·5 per cent. gold. Between β and γ, and between γ and δ, are heterogeneous regions in which the surrounding phases are in equilibrium. The β solid solution shows an oriented distribution of the compound $AuCu_3$, and the γ solid solution that of the AuCu compound.

Le Blanc and Wehner,[2] from thermoelectric power and X-ray investigations, have found the third compound in the copper-gold series—Au_2Cu_3 (see also Haughton and Payne)—stable below 400° C. At 250° C. all the compounds undergo a transformation not detectable by X-rays.

Thermodynamic calculations indicate that the liquid and solid solutions in the copper-gold system are derived from monatomic molecules of the two metals.[3]

The freezing point of the British coinage alloy, containing gold 916·6, copper 83·3 parts, per 1,000, is 949° C., according to the curve in Fig. 17, or 951° according to Roberts-Austen and Rose. The freezing point of the alloy with gold 900, copper 100 parts per 1,000, is 931° by the curve. The alloy with the lowest freezing point contains gold 820, copper 180, solidifying at 884° without a pasty stage. This alloy is the only brittle member of the series, breaking with conchoidal fracture, and, being homogeneous, resembles a eutectic, but it is doubtful if it can be regarded as one. Portevin and Durand [4] describe the fracture as columnar. The other alloys are malleable and ductile, breaking with rough granular fracture after being nicked and bent backwards and forwards. The presence of only 0·04 per cent. of silicon in standard gold alloy renders it unfit for rolling. According to Nowack [5] the effects of small quantities of impurities in gold-copper alloys are as follows :—

Iron.—Forms a solid solution with gold up to about 15 per cent. iron by weight. Alloys containing up to 10 per cent. iron can be worked without fracturing.

Tin.—Forms a solid solution when not more than 5 per cent. tin is present. Above this figure compounds are formed. An alloy with 1 per cent. tin shows a heterogeneous structure even after annealing, though its working qualities are fair. 10 per cent. tin gives a brittle alloy.

Bismuth.—With 0·1 per cent. bismuth the bars broke easily on rolling. An alloy containing 0·01 per cent. could be rolled further, but was appreciably more brittle than fine gold containing 0·01 per cent. bismuth.

Antimony.—Alloys containing up to 0·1 per cent. antimony rolled fairly

[1] *Z. anorg. Chem.*, 1931, **201**, 41.
[2] *Ann. Physik*, 1932, **14**, 481.
[3] Jeffery, *Trans. Far. Soc.*, 1932, **28**, 705.
[4] *Rev. de Met.*, 1919, **16**, 149.
[5] *Zeit. Metallkunde*, 1927, **19**, 238.

well, but with 1·0 per cent. the presence of a compound made the alloy quite unworkable.

Aluminium.—3 per cent. of aluminium is the maximum permissible limit if good working qualities are to be assured.

Tellurium.—0·1 per cent. of tellurium makes an alloy unworkable. 0·01 per cent. or less is harmless.

Lead.—An alloy with 0·06 per cent. lead is unworkable, and according to T. K. Rose, 0·01 per cent. lead makes standard gold unfit for coinage.

In the solidification of alloys rich in gold, cores are formed of solid particles containing a higher percentage of gold than the part remaining liquid, and successive layers solidify round the cores, each containing less gold than the previous one, until the last portions to be solidified form a network of copper-rich material. The grains vary in size with the velocity of cooling, being largest in slowly-cooled ingots. By long continued annealing the alloys would probably become completely uniform in composition, but with an increase in the size of the grains. After annealing the grains are nearly regular polygons. Figs. 12 and 13 (page 14) resemble standard gold before and after annealing.

In the solidification of alloys poor in gold, the cores are copper-rich, the succeeding layers containing more and more gold. For microscopic study, the alloys may be etched by aqua regia, or by ferric chloride in a hydrochloric acid solution.

According to D. M. Liddell [1] gold and silver dissolved in large amounts of copper tend to segregate, accumulating towards the bottom of a molten charge. His results on a slowly-cooled culôt were as follows:—

TABLE II.

	Silver, Oz. per Ton.	Gold, Oz. per Ton.
Top, . . .	53·14	5·86
Middle—Edge, .	53·08	5·82
Centre, .	59·92	6·18
Bottom, . .	69·58	6·72

With gold and copper only, in a quickly-cooled culôt, the results were not so marked, and showed no definite law. Van Heteren and Haagen Smit [2] found, however, that gold tends to concentrate in the inner portions of castings containing more than 82·6 per cent. gold, and in the outer portion in alloys of lower fineness, which is contrary to Liddell's results. In the former case the greatest segregation is said to occur in the alloy of 916·6 fineness.

Segregation takes place [3] in standard gold made brittle by small quantities of lead or bismuth, the presence of 0·2 per cent. of either of these metals causing the centre of a sphere 3 inches in diameter to be enriched in gold to the extent of about one part per thousand. This is due to the fact that a eutectic alloy of gold and lead remains molten after the remainder

[1] *Eng. and Mng. J.*, 1910, **90**, 418.
[2] *Met. und Erz*, 1923, **20**, 183.
[3] T. K. Rose, *J. Chem. Soc.*, 1895, **67**, 552.

of the mass is solidified, and is consequently driven towards the centre of the sphere, and that this fusible alloy contains much gold but very little copper. Jefferson and Arnold,[1] and also Andrews,[2] by micrographic studies, found that brittleness in standard gold containing a little lead, bismuth, tellurium, or certain other impurities is due to the presence of films composed of eutectic alloys separating the crystals of gold from each other, but Osmond and Roberts-Austen [3] were unable to observe these films in quickly cooled ingots.

Segregation in bars of standard gold (24″ × 1½″ × ½″) has been examined by S. W. Smith.[4] It is found that those portions in contact with the chill face of the mould are richer in copper. The degree of segregation,

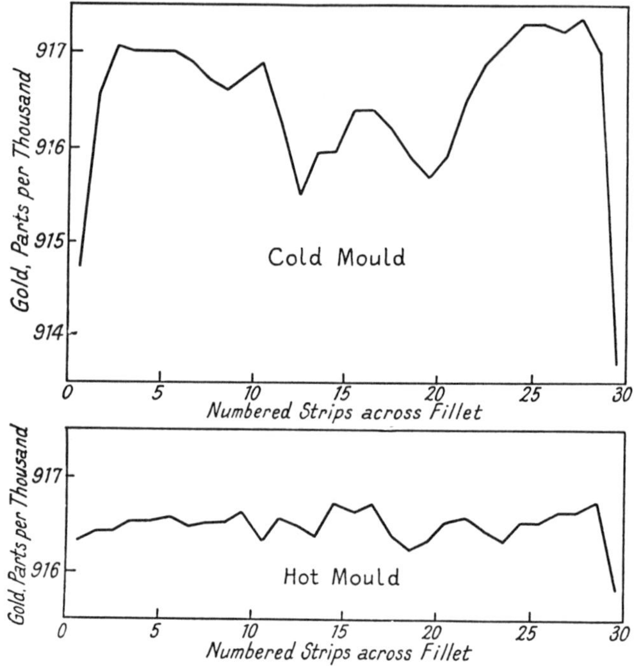

Fig. 18.—" Liquation " in Standard Gold (Smith).

however, depends on the temperature of casting and the temperature of the moulds (see Fig. 18). A corresponding gold enrichment occurs within a very short distance of the face of the bar. This surface variation in fineness is often reflected in the uneven rolling of the edges of the strips produced from the bars. From the curves it will be noticed that when the moulds are warm the extent of the segregation is much less.

The alloys of gold and copper are less malleable, harder, and more elastic than gold, and possess a reddish tint. Those with less than 12 per cent. of copper are easily worked; when more than this is present they are more

---

[1] *Engineering*, 1896, 61, 176.
[2] *Ibid.*, 1898, 66, 541.
[3] *Phil. Trans.*, 1896, **A**, 187, 417.
[4] 61*st Annual Report of Royal Mint*, 1930, p. 55.

difficult to work owing to their hardness. The hardening effect of cold work is shown in the following table :—

TABLE III.

Carat.	Gold, per cent.	Brinell Hardness.	
		Annealed.	After 60 per cent. Redn.
22	91·7	66	155
20	83·3	109	197
18	75·0	109	202
16	66·7	102	200
14	58·5	83	185

Gold-copper alloys containing 50 atomic per cent. of gold may be age-hardened by quenching after annealing at a comparatively low temperature and then re-annealing at 250° C. for ½ hour or at 200° C. for eight hours.[1] According to Sachs,[2] who employed radiographic means of determination,

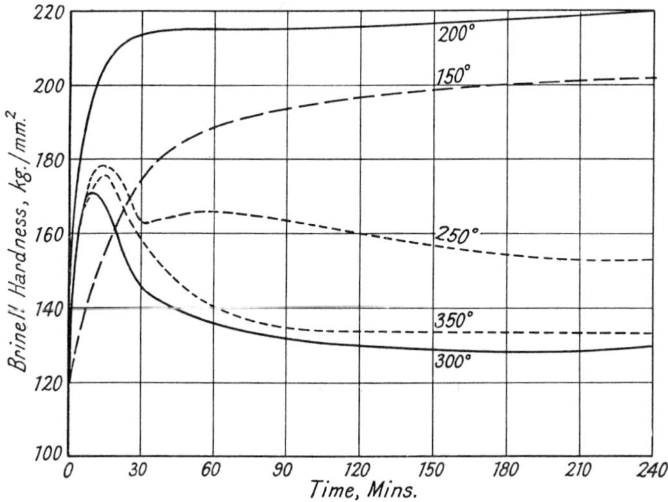

Fig. 19.—Brinell Hardness of Aged Copper-Gold Alloys (Nowack).

the maximum age-hardening is obtained after 8 hours at 200° C. or ½ hour at 350° C. It is probable, however, that these results were confused by other factors. In the case of gold alloys containing 25 atomic per cent. of gold, where mixed crystals are present at high temperatures and the compound $AuCu_3$ at lower temperatures, no evidence of age-hardening has been discovered. This is contrary to expectations. Nowack suggests that this is because the two phases do not possess the same type of lattice.

Fig. 19 shows the variations in hardness of a gold alloy containing 50

[1] Nowack, *Z. Metallk.*, 1930, **22**, 101.
[2] *Z. Metallk.*, 1925, **17**, 88.

atomic per cent. gold which has been quenched from 450° C. and then annealed at various temperatures for different periods of time. It will be noticed that annealing at temperatures up to 200° C. results first in a comparatively rapid increase and subsequently a more gradual increase in hardness, while annealing at higher temperatures causes a maximum increase in hardness after some 15 minutes, followed by a rapid decrease to a hardness comparable with that produced on straight annealing.

Since no change of volume occurs when gold-copper alloys are formed, their densities may be calculated from those of the constituent metals. The densities of gold-copper alloys as cast are given in the following table :—[1]

TABLE IV.

Gold.	Copper.	Specific Gravity at 15° C.
Per 1,000	Per 1,000	
1,000	0	19·26
917	83	17·35
833	167	15·86
750	250	14·74
583	417	12·69
250	750	10·035
0	1,000	8·7

The densities of worked gold-copper alloys with large percentages of gold are given by Roberts-Austen as follows,[2] at 0° C. compared with water at 4° C., the metal being in the form of discs compressed in a coinage press :—

TABLE V.

Gold.	Density.	Gold.	Density.
Per 1,000		Per 1,000	
1,000	19·30	923	17·57
980	18·84	916·6	17·48
969	18·58		(Broch)
959	18·36	900	17·17
948	18·12	880	16·80
938	17·93	861	16·48
932	17·79		

Later determinations by T. K. Rose of hard-rolled standard gold (gold 916·6, copper 83·3) gave a density of 17·45 $\pm$ 0·005 at 0°C. compared with water at 4° C. The density of a gold coin is sometimes as high as 17·48, owing to the presence of silver.

Many of the alloys have been used for coinage at various times. The Greeks and Romans, after electrum had fallen into disuse, employed the purest gold they could procure—viz., that from 990 to 997 fine. Under the Roman Emperors, however, copper was intentionally added, and in the two centuries preceding the fall of Rome very base alloys were used,

[1] Hoitsema, *Zeitsch. anorg. Chem.*, 1904, **41**, 65.
[2] *7th Annual Report of the Royal Mint*, 1876, p. 44.

some containing only 2 per cent. of gold or even less.[1] In the middle ages these base alloys were discarded, and the "byzant" of Constantinople and the "florin" of Florence were both nearly pure gold, while the first gold coins struck by the nations of Western Europe were also intended to be absolutely fine. The standard 916·6 or $\frac{11}{12}$ (*i.e.* 916·6 parts of gold in 1,000) was adopted by England in the year 1526, the standard of 994·8, which had been introduced in 1343, being finally abandoned in 1637. The 900 standard was introduced in France in 1794, and subsequently adopted in other countries. The Austrian ducat has a fineness of 986$\frac{1}{9}$, and that of Holland a fineness of 983, but these coins are used only for trade purposes in Asia and Africa. The 900 and 916·6 standards keep their colour fairly well, and resist wear better than richer alloys. The 900 alloy is harder and wears better than the 916·6 alloy, but the difference is not great. The alloys used in coinage formerly contained from two to twelve parts of silver per 1,000 in addition to the gold, but new coins now generally contain only one or two parts of silver per 1,000.

Gold-copper alloys tarnish on exposure to air owing to oxidation of the copper, and blacken on heating in air from the same cause. This oxidised coating may be removed and the colour of fine gold (not that of the original alloy) produced by plunging the metal into dilute acids or alkaline solutions, the operation being technically known as "blanching." The colour of some alloys may be improved without previous oxidation by dissolving out some copper by acids, a film of pure gold being thus left on the outside which can be burnished. French jewellers use a hot solution of two parts of nitrate of potash, one part of alum, and one of common salt for this purpose.

Nitric or sulphuric acid dissolves out the copper from gold-copper alloys under conditions similar to those under which it removes silver from silver-gold alloys (see Chapter XIX). If the copper falls below 6·5 per cent, the alloy is not attacked by these acids (Pearce). Keller has shown [2] that when copper containing a small proportion of gold is dissolved in nitric acid, part of the gold is dissolved, and that more gold is dissolved from slowly cooled than from quickly cooled alloys. Aqua regia dissolves all the alloys completely.

Tammann [3] has examined the electrolytic potentials of gold-copper alloys in various electrolytes, and finds that the limit of action is usually at $\frac{1}{8}$ mole gold. When the quantity of gold is less than this, some 14-point cubes containing only copper occur.

Gold and Iron.—In the fused state the metals are miscible in all proportions, and form an interrupted series of isomorphous mixtures on solidification.[4] The break in the formation of these occurs at 28 to 63 per cent. of gold and becomes greater at lower temperatures, owing to the transformation of the iron into another allotropic form. The break is confirmed by electrical conductivity measurements.[5] The minimum point on the liquidus curve (1,040° C.) is at Au 95, Fe 5. From this point, the liquidus curve rises steadily as the proportion of iron increases up to pure iron (1,535° C.).

The transition temperature of iron is unaffected by the addition of gold.

---

[1] Roberts-Austen, Cantor Lecture, *J. Soc. of Arts*, Aug., 1884.
[2] *Trans. Amer. Inst. Min. Eng.*, 1912, 43, 582.
[3] *Zeitsch. anorg. Chem.*, 1921, 118, 48.
[4] Isaac and Tammann, *Zeitsch. anorg. Chem.*, 1907, 53, 281-297.
[5] Guertler and Schulzer, *Zeitsch. physikal. Chem.*, 1923, 104, 90.

The hardness of the alloy containing 10 per cent. of gold is rather greater than that of pure iron, but beyond this point the hardness gradually diminishes, and an alloy containing 70 per cent. gold is considerably softer than iron. The alloys generally are hard, of high tenacity, and both malleable and ductile, so long as the proportion of iron does not exceed 80 per cent. All the alloys form with expansion and are hardened by quenching. The main difficulty in preparing the alloys is due to the high melting point of iron, but iron is gradually taken up by molten gold at a temperature of 1,100° to 1,200°, and more rapidly at higher temperatures. If cast iron, melting at 1,130°, is used, this difficulty is avoided. It is obvious that gold and its alloys when molten should not be stirred with an iron rod.

Alloys containing up to 20 per cent. iron may be age-hardened by annealing at 450° C. after quenching from temperatures just above the solidus. The maximum hardness of 200 Brinell (for the 15 per cent. alloy) is obtained after 3 hours.[1]

The alloys with 8 to 10 per cent. of iron are pale yellow, very ductile, and take a beautiful polish. Those with 15 to 20 per cent. of iron are used in France for jewellery under the name *or gris*. They are yellowish-grey, and, although hard, are not difficult to work. The alloy with 25 per cent. iron (*or bleu*) is also used in jewellery. Alloys with 75 to 80 per cent. of iron are silver-white in colour, and strongly magnetic.

Gold-iron alloys are not easily purified by cupellation, but the iron can be removed by sulphuric acid at 100° C., if its proportion is not too small. Nitric or hydrochloric acid may also be used. The action is slow. All the alloys dissolve completely in aqua regia.

Gold and Lead.—The liquidus curve has four distinct branches[2] (see Fig. 20). There are three eutectic points ; at A, 215° C. (gold 14·8 per cent.), B, 254° C. (gold 27 per cent.), and C, 418° C. (gold 57 per cent.). Three series of isomorphous mixtures separate out in the whole range from gold to lead, and a consideration of the time required for the crystallisation of the eutectic shows (according to Tammann) the existence of two definite compounds of the two metals—namely, $AuPb_2$ (32 per cent. gold) and Au_2Pb (65·5 per cent. gold). The existence of these compounds has been confirmed by microscopic examination. $AuPb_2$ undergoes a polymorphic transformation at 211° C., and the conversion from the α to the β variety is reversible. This alloy forms long white needle-shaped crystals with rounded contours, while the crystals of Au_2Pb are larger, rhombic, and well-formed, thus being easily distinguishable from the other constituents.

In alloys containing 45 per cent. Pb three structural elements are observable—viz., (1) primary crystals of gold occurring in irregular form ; (2) the compound Au_2Pb, which is almost completely surrounded by gold crystals ; and (3) a eutectic. Alloys with 45 to 72 per cent. lead do not show primary gold crystals, but large crystals of Au_2Pb are plainly visible in the section. These crystals are surrounded by a eutectic composed of the two compounds, Au_2Pb and $AuPb_2$. From 72 to 85 per cent. lead primary $AuPb_2$ crystals separate, and afterwards merely solid solutions of gold in pure lead.

The alloys, which are all brittle, range in colour from pale yellow through greenish-yellow to bluish-white. All are formed with expansion.

---

[1] Nowack, *Z. Metallk.*, 1930, **22**, 97.
[2] Vogel, *Zeitsch. anorg. Chem.*, 1905, **45**, 11-23.

Considerable attention has been paid to the influence of small proportions of lead, bismuth, and tellurium on the chemical and physical properties of

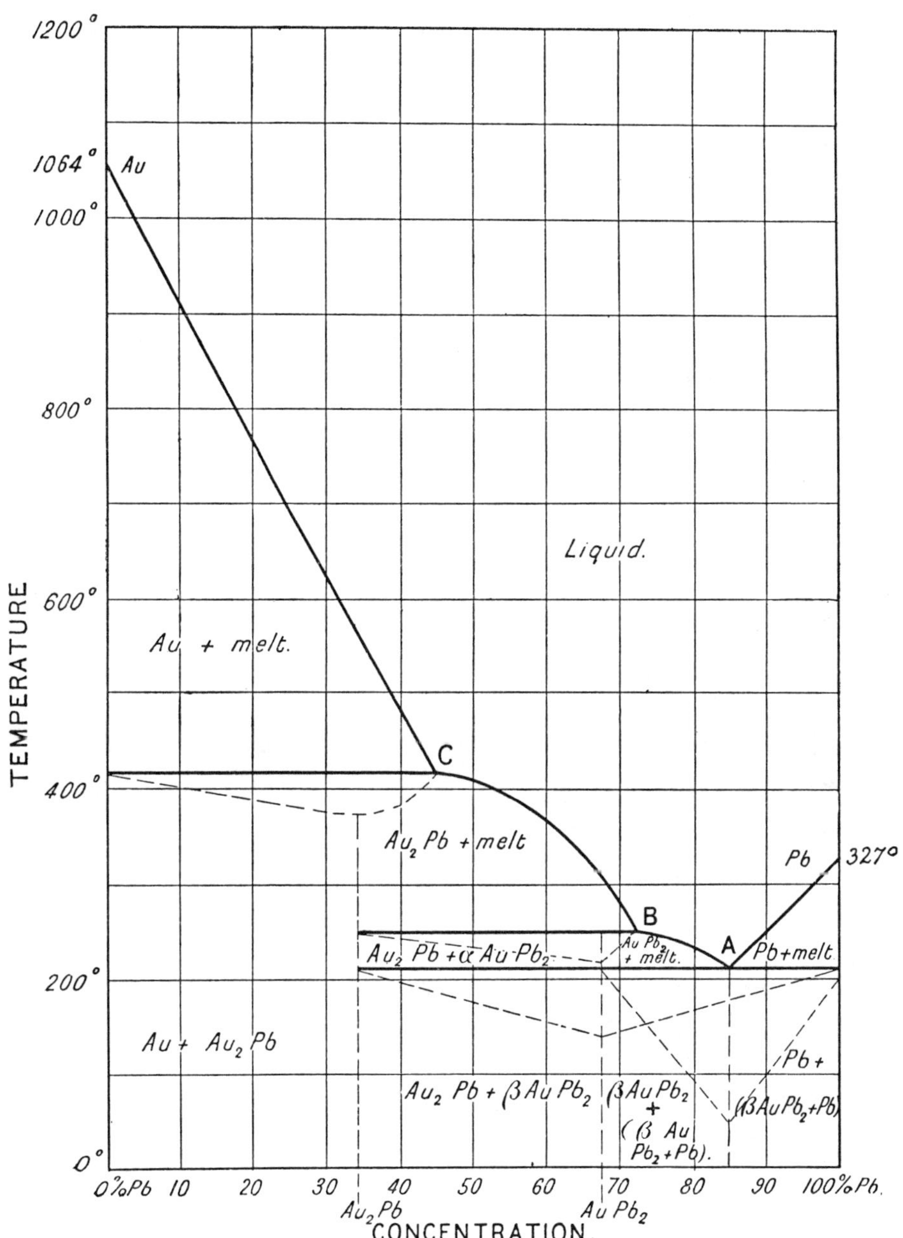

Fig. 20.—Thermal Equilibrium Curve of Gold and Lead.

gold.[1] It is known that small quantities of lead increase the brittleness of gold considerably, and particularly is this the case with coinage alloys, 0·02 per cent. of lead being sufficient to cause cold-shortness.[2] In pure gold 0·15 per cent. of lead produces slight brittleness. Even with the very small percentages of lead, certain heterogeneous constituents (see Gold-bismuth alloys) appear in the structure, to which is attributed the rapid decrease in ductility and tensile strength of these alloys. Nowack [3] states that a 10 per cent. copper-gold alloy and gold each containing 0·06 per cent. lead are unrollable, but that an addition of 0·005 per cent. lead is not harmful. The lead, he surmises, lies in the grain boundaries, probably as the compound Au_2Pb.

The vapour of lead, like that of bismuth, is absorbed by gold, making it brittle.[4]

The impurities in lead bars, with the exception of silver and gold, collect nearer the top than the bottom when the bars are slowly cooled, the differentiation being probably due to the differences in the specific gravities of the various constituents. Hofman [5] quotes figures to show that the precious metals concentrate more on the bottom when the bars are rapidly cooled.

In the Pattinson process for refining lead containing gold, the eutectic alloy containing 14·8 per cent. of gold remains as the mother liquor, which cannot be enriched beyond this point.

Gold and Magnesium.—Gold forms solid solutions with magnesium up to about 5 per cent. of magnesium.[6] The eutectic between pure gold and the point of formation of the compound AuMg at 1,150° C. occurs at 827° C., corresponding to a concentration of 5·7 per cent. of magnesium. The eutectic range is from 5 to 5·7 per cent. of the base metal. The eutectic between the maxima on the melting point curve corresponding to $AuMg_2$ and $AuMg_3$ occurs at 770° C. and 70 atomic per cent. of magnesium.

Magnesium-gold alloys containing up to 18 per cent. of magnesium retain the yellow colour, but all others are silver-grey. Alloys with 33 to 66 per cent. of the base metal are brittle, with a maximum hardness of 5, and have a glassy fracture.

From the practical point of view alloys of gold and magnesium are of little use.

Gold and Manganese.—Parravano [7] has found that there is a compound AuMn, which is shown by a maximum on the freezing point curve at 1,225° C. A minimum occurs at 990° C., 10 per cent. Mn, and a eutectic at 1,080° C., 46 per cent. Mn, with a horizontal branch, showing a region of partial miscibility, between 50 and 57·5 per cent. Mn.

Gold and Mercury—Amalgams.—A piece of gold rubbed with mercury is immediately penetrated by it and becomes exceedingly brittle. The ductility is not always restored when the mercury is removed by distillation,

---

[1] Hatchett, *Phil. Trans.*, 1803, p. 43; Roberts-Austen, *Phil. Trans.*, 1888, **A**, **179**, 339; *ibid.*, 1896, **A**, **187**, 417; Heycock and Neville, *J. Chem. Soc.*, 1892, **61**, 909; Andrews, *Engineering*, 1898, **66**, 541; Jefferson and Arnold, *ibid.*, 1896, **61**, 176; T. K. Rose, 33*rd Ann. Report of the Royal Mint*, 1902, p. 73.

[2] Rose, *loc. cit.*

[3] *Z. anorg. Chem.*, 1926, **154**, 395.

[4] Hatchett, *loc. cit.*

[5] "*Metallurgy of Lead*" (5th edition), p. 347.

[6] Vogel, *Zeitsch. anorg. Chem.*, 1909, **63**, 109, 183; Urasow, *ibid.*, 1909, **64**, 375, 396: Vogel and Urasow, *ibid.*, 1910, **67**, 442.

[7] *Gazz. chim. ital.*, 1915, **45**, [i], 293.

a crystalline structure being often induced, perhaps because part of the mercury is retained by the gold. A particle of gold "wetted" by mercury at once loses its colour. A solid amalgam is formed, but is not readily dissolved in an excess of mercury.

The amalgams of gold were studied early in the last century, but the earlier workers devoted themselves mainly to qualitative work. Henry [1] obtained an amalgam in which he showed the solubility of gold in mercury to be 0·14 per cent., and also demonstrated the formation of a liquid phase in the series of alloys. Kasantseff [2] isolated crystals at various temperatures by passing the liquid amalgam through a glass tube of 0·15 to 0·40 mm. diameter, and found the following figures for the solubility :—

Temperature ° C.	Percentage solubility.
0	0·110 per cent. gold.
20	0·126 ,,
100	0·650 ,,

These results are in agreement with those cited by Henry (0·14 per cent.) and by Gouy (0·13 per cent.).

Tammann has studied a small portion of the liquidus curve corresponding to the addition of minute quantities of gold to mercury, and finds that the melting point of the latter is raised.

Percentage addition of gold.	Rise in M.P. of mercury.
0·006	0·1°
0·012	0·1°
0·025	0·2°

The curve has a steep ascent, with which the sparing solubility of gold in mercury is in agreement.

Crookewitt [3] obtained a crystalline residue containing 32·75 per cent. gold by filtration of the amalgam through chamois leather. This composition corresponds to the formula $AuHg_2$. Henry repeated this experiment, and obtained a residue with 86 per cent. of gold by weight. Knaffl [4] noted that the quantity of crystals in the pulpy amalgam appreciably increases with a fall in temperature, and in more recent years Fay and North (1901), by centrifuging the amalgam, succeeded in isolating a residue corresponding to the composition Au_4Hg.

Wilm [5] dissolved gold in sodium amalgam, and obtained results which were somewhat different from those of previous workers. The solution of gold in the sodium amalgam was boiled repeatedly with nitric acid, and crystals remained which contained 9·71, 11·45, 9·67, and 5·45 per cent. by weight of mercury. Henry [6] had also obtained a similar residue, containing 11·26 per cent. of mercury, which would correspond to the formula Au_8Hg. In alloys having higher gold content than the above the attack by nitric acid is considerably reduced, which tends to prove that in alloys containing

---

[1] *Phil. Mag.*, 1855, [iv.], **9**, 468.
[2] *Bull. Soc. chim.*, 1878, **30**, [ii.], 20.
[3] *J. prakt. Chem.*, 1848, **45**, 87 ; *Ann. Chim. Pharm.*, 1848, **68**, 289.
[4] *Dingl. Poly. J.*, 1863, **168**, 882.
[5] *Zeitsch. anorg. Chem.*, 1893, [iv.], 325.
[6] *Phil. Mag.*, 1855, **9**, [iv.], 468.

up to about 10 per cent. of mercury the latter enters into a state of solid solution.

The maximum amount of mercury, however, which can be absorbed by solid gold is much greater. Thus a native amalgam of gold, Au_2Hg_3, is found in small yellowish crystals of specific gravity 15·47 in the native mercury of Mariposa, in California, containing gold 39·02 to 41·63 per cent., mercury 60·98 to 58·37 per cent. An amalgam of gold and silver occurring in white soft grains at Choco, New Granada, contains 38·39 per cent. gold, 4·21 per cent. silver, and 57·40 per cent. mercury.[1] By treating 1 part of gold with 50 parts of boiling mercury, keeping it heated for some days, allowing to cool, and carefully squeezing, Louis[2] obtained a hard alloy of gold and mercury, crystallising in silver-white, long delicate interlacing needles, and consisting of 41·43 per cent. of gold and 58·57 per cent. of mercury, thus corresponding exactly to the composition of native amalgam, but having a different crystalline form.

Parravano[3] found that in liquid amalgam poorer in gold, the compound Au_3Hg separates on cooling from temperatures above 100° C. This then decomposes into gold and Au_2Hg_3 below the temperature named. Solid amalgams containing over 90 per cent. gold are solid solutions of mercury in the latter.

According to Plaksin[4] two compounds, $AuHg_2$ and Au_2Hg, exist. The first is stable up to 306° C. and the second up to 420° C. Above the latter temperature a solid solution of 16 per cent. mercury in gold is present.

Anderson[5] records that a break occurs in the solubility curve at about 310° C. He prefers to consider the gold-rich compound as Au_2Hg.

It has been shown by MacPhail Smith,[6] in studying the phenomena of diffusion, that gold does not dissolve in mercury in the form of free atoms, but as molecules of the type Au_4Hg, and that although intermediate crystals have not been obtained in the series, nevertheless compounds of Hg and Au do appear to be formed in the liquid phase.

White[7] states as a result of tests that the force exerted between a 2 cm. square of mill gold and a free surface of mercury was equal to 13 grammes weight; that on dry amalgamated copper it was 105 grammes, and on a wet copper plate 160 grammes, increasing, on standing for 2 hours, to 200 grammes.

In gold-mill practice the solid amalgam filtered off from the mercury by squeezing it through chamois leather usually contains about 1 part of gold to 2 parts of mercury, but it is doubtful whether this is a homogeneous mixture. The composition of a saturated solid solution of mercury in gold may, therefore, be regarded as unknown, with a presumption in favour of the view that mercury is present to the extent of not less than 60 per cent. This saturated solid solution is "wetted" by fresh mercury, but if more mercury is added the excess can be removed by filtering under pressure. Pasty amalgam is left in the filter bag, and becomes "drier" and less pasty

---

[1] Rammelsberg, "*Mineralchemie*," p. 10; Watt, "*Dictionary of Chemistry*," 1864, **2**, 927.

[2] "*Handbook of Gold Milling*," 1894, p. 82.

[3] *Gazz. chim. ital.*, 1918, **48**, 123.

[4] *Proc. Russ. Phys. Chem. Soc.*, 1927, **33**, 88; *Ann. Inst. anal. Phys. Chim.* (*Leningrad*), 1928, **4**, 336.

[5] *J. Phys. Chem.*, 1932, **36**, 2145.

[6] *Ann. Phys.*, 1908, **25**, [iv.], 252.

[7] "*Rand Metallurgical Practice*," 1926, **1**, 545.

the more the pressure is increased, until the whole of the liquid phase has been removed. The amount of pressure, however, does not alter the composition, either of the liquid or the solid phase.

The amalgam recovered in mills may be regarded as a collection of little nuggets of gold, coated and partly saturated with mercury. The amalgam sinks to the bottom of the mercury. Coarse particles of gold are not saturated with mercury to their centres, although the outside layers of the particles consist of a saturated solution. The saturation of the interior, depending on the diffusion of mercury through solid amalgam, would probably not be complete for many days. The finer (*i.e.* smaller) the particles of gold, the more nearly the interior approaches to saturation. It follows that coarse gold gives rich amalgam, and fine gold poor amalgam. The limits are pure gold on the one hand and a saturated solution of mercury in gold on the other, neither of which exists in practice. The practical limits are amalgams containing about 50 per cent. and 25 per cent. of gold respectively. These gold amalgams usually contain impurities in the shape of amalgams of silver and of base metals, as well as non-metallic substances.

When amalgams are gradually heated, the mercury is distilled off by degrees, the action soon ceasing if the temperature is allowed to become stationary, and distillation recommencing if it is again raised. At 440° C. (somewhat below a red heat), an amalgam containing about three parts of gold to one of mercury is obtained, and at a bright red heat almost all the mercury is expelled, and if the heating has not been pushed too rapidly the vapours contain but little gold. According to Louis,[1] when properly heated the gold retains 1 to $1\frac{1}{2}$ per cent. of mercury, and the mercury vapours contain 0·005 per 1,000 of gold.

The application of heat causes the crystals of gold amalgam to lose their contained mercury, but without change of their shape. To some extent they retain their surface lustre. From the crystal structure it would appear that after heating, a porous mass remains, in which the interstitial spaces are the places from which the easily volatilised component—*i.e.* the mercury—has been removed. Similar phenomena of residual porous structure have been observed in the case of other amalgams, such as those of chromium and manganese.

Gold and Nickel.—Fraenkel and Stern [2] have shown that the gold-nickel series consists of a continuous series of solid solutions. Their equilibrium diagram is shown in Fig. 21. Grube [3] considers, on the other hand, that below 850° C. the homogeneous solid solution phase with 5 to 85 per cent. nickel decomposes into a mixture of two saturated solid solutions—the compositions of which are dependent on the ageing temperature of the alloys after quenching from the higher range.

Homogeneous nickel-gold alloys vary in colour from yellow at the gold end to a pleasing platinum white with 50 atomic per cent. nickel. Heat-treatment tends to cause the colours to change.

Alloys whose nickel content is greater than that of the minimum melting-point alloy are more readily attacked by nitric acid when they are slowly cooled than when they are quenched.

The transformation point of nickel-gold alloys lies very near the

---

[1] *Loc. cit.* [2] *Z. anorg. Chem.*, 1926, **151**, 105.

[3] *Z. physikal. Chem.*, 1931, 187. See also Heike and Kessner, *J. Inst. Met.*, 1929, **42**, 453; E. M. Ware, *Trans. Amer. Inst. Min. Met.*, 1928, T.P. 147, and *Z. anorg. Chem.*, 1929, **178**, 272.

transformation temperature of pure nickel (350-365°)—*i.e.* the temperature of transformation of pure nickel from the α to the β variety is not appreciably altered by the addition of pure gold.

Similar results to the above have been obtained from a study of the magnetic properties of the alloys. At the transformation temperature of pure nickel, passing from a higher to a lower temperature, there is formed a material of very slight permeability from one of much greater permeability.

Kournakoff [1] and Achnazaroff have shown that the rate of cooling has a marked effect on the hardness of the minimum melting-point alloy of

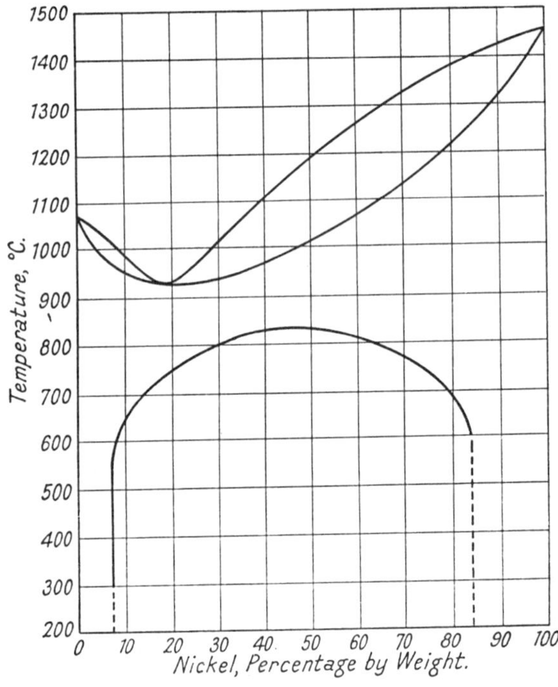

Fig. 21.—Thermal Equilibrium Diagram of Gold and Nickel (*Fraenkel and Stern*).

the gold-nickel series (75 per cent. Au). When quenched the hardness is 235 (Brinell); when slowly cooled 150.

As the nickel content of gold-nickel alloys increases, their hardness on quenching from 860° C. increases up to a maximum of 360 Brinell for an alloy containing 70 atomic per cent. nickel. A hardness of 200 is never exceeded by slowly-cooled alloys. By quenching the alloys containing up to 30 per cent. nickel from a high temperature and then annealing for 40 minutes at 450° C., the hardness may be raised to 350-360 Brinell. Annealing at higher temperatures must only be carried out for short periods if any increase in hardness is to be effected.[2] Wise [3] also found that the alloys

[1] *Zeit. anorg. allgem. Chem.*, 1922, **125**, 185.
[2] Nowack, *Z. Metallk.*, 1930, **22**, 98.
[3] *Amer. Inst. Min. Met. Eng.*, Tech. Pub. No. 147 1928

containing the higher percentages of nickel reach their maximum hardness by ageing at 400°-500° C., but that the higher copper alloys need only to be annealed at temperatures between 200° and 400° C.

An alloy containing 50 atomic per cent. gold shows a face-centred cubic lattice, $a = 3{\cdot}85$A, when quenched from 860° C. After annealing at 455° C. the structure shows two face-centred lattices corresponding with solid solutions containing 14·8 atomic per cent. of nickel and 3·7 atomic per cent. of gold respectively. The alloy has a platinum colour.

A hot solution of potassium sulphide has been recommended for etching polished specimens of these alloys.

White Gold.—The original white gold alloys were made up to simulate platinum when the latter metal was at a very high price. Their utility in jewellery has been so marked that they have continued to be made in increasing quantities. The first alloys tried contained gold, nickel and zinc. Palladium, platinum and nickel have since been used as alloying elements. Silver is rarely used as its whitening effect on gold is not sufficiently marked. Copper in small amounts improves the behaviour of these alloys. Zinc decreases the tendency to gas porosity. Sulphur is deleterious, especially when nickel is a main constituent, as little as 0·01 per cent. being the maximum that may be permitted without causing serious cracking of the bars during rolling. Alloys containing 80 per cent. gold together with palladium or nickel are completely white.

The mechanical properties of the white gold alloys (particularly the gold-nickel-zinc alloys) are less favourable in manufacture than those of the yellow and red gold alloys. There are considerable difficulties in the treatment of the various groups, especially in the melting, casting and heat-treatment. In the gold-nickel-zinc mixtures the variation in properties does not closely conform to changes in composition, and but few individual mixtures have suitable characteristics. The gold-palladium mixtures are, however, more amenable, easier to work, not so hard on the tool and more readily enamelled. The high price of palladium has so far, however, restricted its use except for expensive jewellery. (Wise[1] gives a list embracing a few of the commercially adaptable gold alloys whose mechanical properties are satisfactory.)

The alloys with lowest melting point in the gold-nickel-zinc series lie within the range 15 to 20 per cent. nickel and 10 per cent. zinc. Their melting points are about 940°-980° C., but considerably higher temperatures are requisite during casting. In the manufacture Fröhlich[2] advocates melting the nickel first, using an excess of boric acid to prevent oxidation. The gold is then added gradually and the melt allowed to cool to about 1,100° C., when it is deoxidised by a little magnesium wire. Next the zinc is added, then a little more magnesium, and the metal poured. Cupro-manganese has also been suggested as a suitable deoxidant. Pure metals only should be used. The nickel should contain only traces of, if any, sulphur, carbon, oxygen and iron.

It has been found advisable, particularly if wholly virgin metals have been used in the first make-up, to melt the alloy a second time, after first rolling it and cutting it up into pieces.

---

[1] *Amer. Inst. Min. Met. Eng.*, 1928, Tech. Pub. 147.

[2] *Met. Ind.*, 1932, **41**, 28. See also E. A. Smith, "*Working in Precious Metals*," N.A.G. Press London, 1933, pp. 342, 352, 373.

In treating scrap white gold alloy, it is common practice first to boil it with 10 per cent. sulphuric acid in order to remove most of the base impurities, of which tin is the worst; copper is deleterious only in so far as it tends to discolour the white gold during any subsequent pickling operations.

The alloys, especially those rich in gold, are sensitive to any abrupt changes of temperature during heat-treatment. Air-cooling usually is satisfactory, but quenching may lead to softness. The annealing temperatures should be kept as low as possible (750°-800° C.), and cooling should never be rapid and certainly not achieved by quenching.

White gold alloys are especially susceptible to "season cracking,"[1] the danger from which, however, can be removed by a low-temperature anneal. The high zinc alloys give most trouble in this direction.

A good 14-carat white alloy[2] contains 58·35 per cent. gold, 16 per cent. copper, 17 per cent. nickel and 8·65 per cent. zinc. It has a tensile strength of roughly 50 tons per square inch, a hardness of 200 Brinell after annealing, and of 360 after hard rolling. The 18-carat alloy, with 3·5 per cent. copper, 16·5 per cent. nickel and 5 per cent. zinc has a tensile strength of approximately 45 tons per square inch. For enamelling purposes an 18-carat alloy containing palladium and a little silver is recommended. The colour is not quite so white as the nickeliferous alloy, but the metal is much softer, is easily annealed and takes enamel more satisfactorily.

Gold and Palladium.[3]—Native gold alloys containing roughly 85 to 90 per cent. gold and 8 to 9 per cent. palladium have been found in Brazil.

Gold containing 10 to 20 per cent. palladium is nearly white, and is hard and ductile. The 50 per cent. alloy is iron grey in colour, and is not ductile.

Gold and palladium form a continuous series of isomorphous mixtures with cube face-centred lattice. The equilibrium curve is a simple one concave to the axis denoting composition, and having only a small interval between the solidus and the liquidus.

The melting point of palladium is lowered 7° by the addition of 10 per cent. of gold, and that of gold is raised 207° by 10 per cent. palladium. The hardness increases with the palladium content up to 70 per cent. of that metal, but beyond this point it decreases. The electrical conductivity and the temperature coefficient sink to minimum values at 55 per cent. of gold, while the tensile strength has a maximum at 73 per cent. of gold. The thermoelectric power, as measured against platinum, has the greatest value for an alloy containing 60 per cent. of gold.

The capacity of palladium for occluding hydrogen is lowered by the addition of gold. Graham[4] showed that a 50 per cent. alloy exhibits an appreciable diminution in the power to absorb hydrogen, as the following figures show:—

	Volume of hydrogen absorbed (in times own volume).
Pure palladium, ...	956·3
50 per cent. alloy, ...	459·9

[1] "Season cracking" is the term applied to the fissures which form in some hard-worked alloys, especially those that have been spun or drawn. It may be prevented by subjecting the pieces to a low-temperature anneal—sufficient to release the internal stresses, without seriously impairing the mechanical properties produced by working.

[2] Nowack, *Metallwirtschaft*, 1928, **7**, 465.

[3] Gubel, *Zeitsch. anorg. Chem.*, 1910, **69**, 38; Ruer, *ibid.*, 1906, **51**, 391-6.

[4] *Proc. Roy. Soc.*, 1869, **17**, 500.

Berry,[1] by using a cell in which the cathode was composed of a palladium-gold alloy and the anode of pure platinum, obtained results showing that 1 gramme of pure palladium absorbs 70 c.c. of hydrogen under normal conditions, and that this value decreases in a straight line to zero for gold alloys containing less than 25 per cent. of palladium.

French patent BF607428, 1926, describes an alloy containing Au 61·5 per cent., Pd 28·5 per cent., Pt 10 per cent., which is suitable for laboratory apparatus, electrical instruments and resistances. It melts between 1,600° and 1,700° C. and has a density of 17.

Gold and Platinum.—Additions of platinum to pure gold raise the melting point of the latter, and form malleable alloys. Up to 10 or 15 per cent. of platinum the yellow colour of gold persists, though in a diminishing degree, but alloys with a greater proportion of platinum assume the white colour of

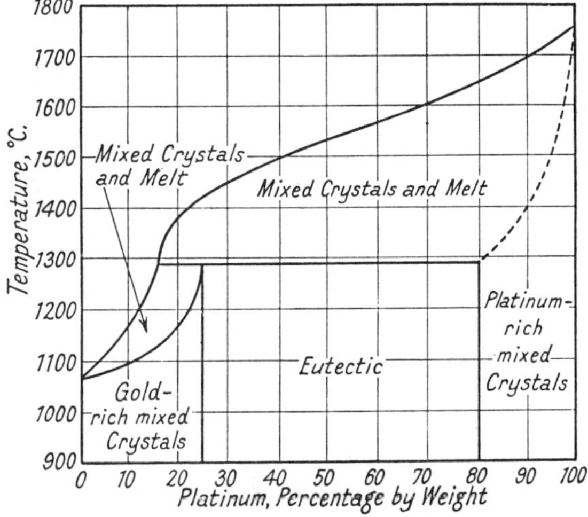

Fig. 22.—Gold-Platinum Thermal Equilibrium Diagram (*Grigoriew*).

the latter element.[2] An alloy of 75 per cent. platinum and 25 per cent. gold resembles grey cast iron in its hardness and brittleness.

Grigoriew,[3] using the purest gold and platinum, agrees with Dörinckel [4] that up to 25 per cent. platinum the alloys are homogeneous solid solutions, while above that concentration they are mechanical mixtures of the solid solution of platinum in gold, with a maximum concentration of 25 per cent. platinum, and of gold in platinum, with a maximum of 80 per cent. platinum. The transition point is sharply marked. Alloys containing more than 80 per cent. platinum are homogeneous. Complete solution of platinum in gold is slow, even after prolonged annealing.

It is probable that the eutectic limit in the diagram (see Fig. 22) extends too far towards the platinum-rich side.

[1] *J. Chem. Soc.*, 1911, 99, 463-6.
[2] Grigoriew, *Z. anorg. Chem.*, 1929, 178, 97.
[3] *Ann. Inst. Platine*, 1928, [6], 184. *Zeitsch. anorg. Chem.*, 1929, 178, 97.
[4] *Zeitsch. anorg. Chem.*, 1907, 54, 345.

Johannsen and Linde[1] show that tempering above 500° C., followed by quenching, renders gold-platinum alloys ductile. They contest Grigoriew's results and state that above 1,150° C. the metals form a homogeneous series of mixed crystals, Dörinckel's liquidus and solidus curves being confirmed. Below 1,150° C. a gap of immiscibility broadens with fall in temperature and extends from 25 to 92 per cent. platinum at 800° C. Tempering of alloys containing not more than 25 per cent. platinum at low temperatures (550° C.) after quenching from above 1,150° C., causes age-hardening, which is enhanced by the addition of small quantities of iron and zinc. Intermediate phases with a regular distribution of the atoms in the lattice appear to be formed during ageing and can exist only below 600° C. The nature of the various phases is still very obscure.

Fig. 23 shows the effect of ageing at 550° C. of the 20 per cent. and 25 per cent. platinum alloys after quenching from 1,300° C. In the case of the 25 per cent. alloy the increase in hardness is at first rapid and then more gradual.

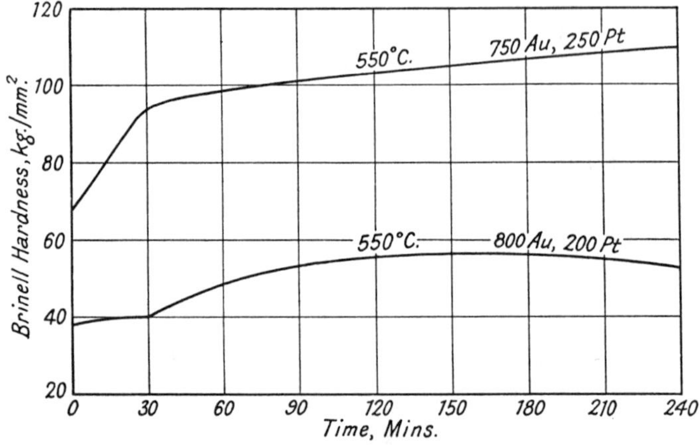

Fig. 23.—Brinell Hardness of Aged Gold-Platinum Alloys (*Grigoriew*).

With only 20 per cent. platinum the initial hardening is less marked, afterwards increasing and then decreasing as the time of heating is extended.

In triple or quadruple gold alloys which comprise (*a*) gold or a gold alloy consisting of homogeneous mixed crystals, (*b*) two other components which form an intermediate compound and which are only sparingly soluble in (*a*), thus forming a quasibinary system, the compound is more soluble in the gold at higher than at lower temperatures, thus satisfying the main condition for age-hardening. Examples of such alloys are :—

Gold.	Ni	Si	Al	Pd	Pt
99·5	0·4	0·1	...	...	...
79·0	1·0	0·25	...	19·75	...
88·0	1·6	0·4	...	...	10·00
97·9	2·0	...	0·1	...	...

[1] *Ann. Physik*, 1930, 5, 762. See also Steuzel and Weerts, *J. Inst. Met.*, 1932, 50, 86

With regard to the addition of small quantities of iron to the alloys, age-hardening occurs when 2 per cent. platinum and 0·2 per cent. iron are present. A maximum hardness is obtained with 6 per cent. platinum and 0·2 per cent. iron. The iron addition accelerates the ageing process without increasing the hardness.[1]

It has been shown by Edward Matthey [2] that in ingots containing 5 to 20 per cent. of platinum segregation occurs, and the platinum is concentrated in the interior.

Platinum cannot be separated from gold by the ordinary methods of cupellation and parting. Both metals are dissolved by aqua regia, however.

Gold-platinum alloys are used in dentistry in the form of wire, and for some varieties of filling. Benedicks suggests an alloy of gold with 7 per cent. platinum for contact points for potential regulators with a relatively high current loading.

Gold and Selenium.—Gold dissolves in selenium only to a very slight extent, forming a very unstable combination.

Gold and Silver.—Gold and silver are miscible in all proportions, forming an uninterrupted series of solid solutions with properties intermediate between those of the two metals. The curve of equilibrium has been traced out by several observers [3] (see Fig. 24).

Jänecke's table of solidification points is as follows :—

TABLE VI.

Composition.	Beginning of Solidification. °C.	End of Solidification. °C.
Gold, 100,	1,064°	...
„ 94, silver, 6,	1,060°	1,053°
„ 88½, „ 11½,	1,054°	1,042°
„ 81½, „ 18½,	1,047°	1,035°
„ 73½, „ 26½,	1,036°	1,026°
„ 64½, „ 35½,	1,031°	1,018°
„ 54½, „ 45½,	1,015°	1,003°
„ 43½, „ 56½,	1,007°	997°
„ 32, „ 68,	993°	982°
„ 18, „ 82,	977°	970°
Silver, 100,	961·5°	...

The first additions of gold to silver raise its melting point.

The alloys are homogeneous, malleable, soft, and ductile. They are nearly uniform in composition, but there is a slight tendency for gold to

[1] Goedecke, *Fest.* 3, 50 *j*, "*Bestehen der Platinschmelz G. m. b. H. Hanau,*" 1931 p. 100.

[2] *Phil. Trans.*, 1892, **A**, 183, 629.

[3] Roberts-Austen and Rose, *Proc. Roy. Soc.*, 1902, 71, 161 ; Raydt, *Zeitsch. anorg. Chem.*, 1912, 75, 58 ; Jänecke, *Metallurgie*, 1911, 8, 597 ; Heycock and Neville *Phil. Trans.*, 1897, **A**, 189, 69.

concentrate in the interior of the ingot, the greatest divergence occurring in the alloy containing 36·1 per cent. gold. The colour of gold is sensibly diminished by the addition of very small quantities of silver, and, on increasing the proportion of the latter, the colour changes to greenish-yellow (when from 20 to 40 per cent. of silver is present), to a faint yellowish-white (when 50 per cent. of silver is present), and to white with a scarcely perceptible yellow tinge (when 60 per cent. of silver is present). The yellow colour finally disappears when some proportion between 60 and 70 per cent. of silver is present. Tabulated results for many physical properties of this series of alloys are given by Broniewski and Wesolowski.[1]

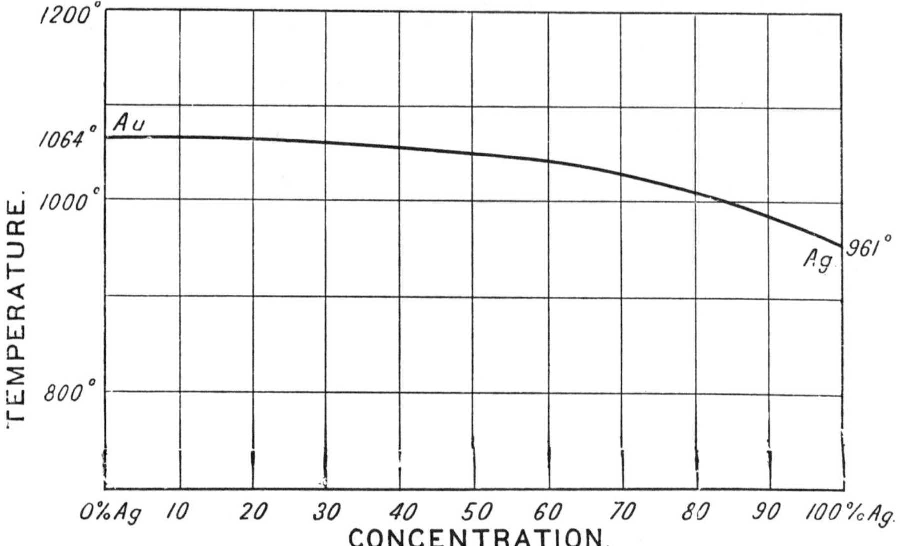

Fig. 24.—Thermal Equilibrium Curve of Gold and Silver.

Alloys of gold and silver were much used for coinage before the methods of parting became well known and inexpensive.

Electrum includes pale yellow alloys with from 15 to 35 per cent. of silver. It occurs native, and was much used for ornaments and coins by the Greeks and Romans, and by the nations which acquired their arts. It was the metal used for the earliest known coins, which were made by Gyges in Lydia about B.C. 727. Rods of electrum, containing gold 651 parts, silver 334 parts per 1,000, were used as money in Asia Minor at an earlier period.[2] The use of silver in the gold-copper coinage alloys was not discontinued until comparatively recently, all English guineas and the Australian sovereigns manufactured at Sydney up to the year 1871 containing some of it.

---

[1] *Compt. rend.*, 1932, **194**, 2047.

[2] Schliemann's "*Ilios*," p. 496, London, 1880. See Roberts-Austen, *J. Soc. Arts*, Aug., 1884.

The hardening effect of cold work, *e.g.* rolling and drawing, is exemplified by the figures in the following table (Carter) :—

TABLE VII.

Carat.	Gold, per cent.	Brinell Hardness.	
		Annealed.	After 60 per cent. Reduction.
22	91·67	30	74
20	83·33	31	84
18	75	32	93
16	66·67	35	90
14	58·5	34	97

Both nitric and sulphuric acids attack silver-gold alloys, almost completely dissolving out the silver if it is present in amounts variously stated as at least 60 to 70 per cent., while, if the proportion falls below 60 per cent., some of the silver is left undissolved with the gold. The solution process follows the equation $p = \varphi\left(\frac{t}{a}\right)$ where p is the amount of silver dissolved in the time t and φ and a are constants depending on the character of the alloy. The nature of pre-treatment has no effect on the solution, but thick plates are attacked more vigorously than thin ones. Hydrochloric acid scarcely attacks these alloys, and the action of aqua regia is soon arrested if the proportion of silver is considerable. The alloys may be dissolved by a mixture of nitric acid and a concentrated solution of common salt.

The densities of the alloys of silver and gold are as in Table VIII.[1]

TABLE VIII.—Densities of Gold-Silver Alloys.

Gold.	Silver.	Specific Gravity at 15° C.
Per 1,000	Per 1,000	
1,000	0	19·26
917	83	18·08
879	121	17·54
843	157	16·96
750	250	16·03
667	333	15·07
500	500	13·60
333	667	13·00
250	750	11·78
167	833	11·28
0	1,000	10·45

Gold, Silver and Copper.—These alloys have found extensive use in jewellery and general goldsmith's work. One of the most important features

[1] Hoitsema, *Zeitsch. anorg. Chem.*, 1904, **41**, 66. These results agree closely with those obtained by Matthiessen, *Proc. Roy. Soc.*, 1859, **10**, 12.

is the range of colours which can be obtained, varying from various shades of red in those alloys containing large proportions of copper, to pale yellow and green in those having large percentages of silver.

Generally speaking, those triple alloys are hardest which contain silver and copper in equal proportions. The hardening effect of cold work, *e.g.*

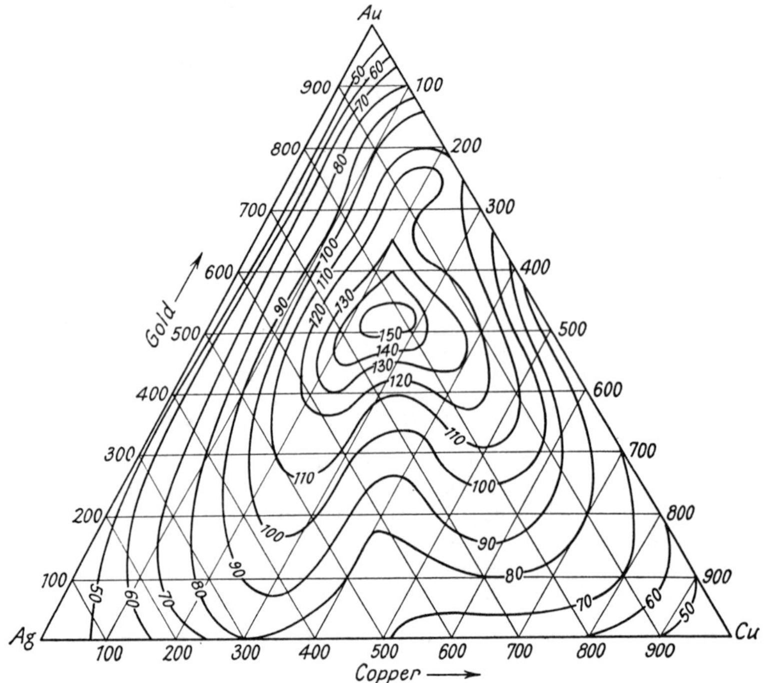

Fig. 25.—Brinell Hardness of Gold-Silver-Copper Alloys—Annealed and Quenched from above 400° C. (*Sterner Rainer*).

rolling and drawing, on such alloys, is exemplified in the following table (E. A. Smith):—

TABLE IX.

Carat.	Gold, per cent. (Silver and Copper—equal parts).	Hardness.	
		Annealed.	After 60 per cent. Reduction in Thickness.
22	91·67	57	123
20	83·33	88	167
18	75·00	105	182
16	66·67	115	195
14	58·3	136	186
12	50·00	150	200
8	33·33	103	171

Most of the alloys exhibit a high degree of malleability and will suffer a reduction in thickness of 50 per cent. or more without annealing. Ductility, which is dependent on tenacity, and to a lesser degree on hardness, is also pronounced ; it is the determining factor in spinning and stamping. The following Erichsen figures, which give a measure of ductility, are quoted by Carter[1] :—

Carat.	Au per cent.	Ag per cent.	Cu per cent.	Depth of Cup (mms.)	
				Slowly Cooled.	Quenched.
18	75	25	...	7·3	13·0
18	75	12·5	12·5	3·6	9·5
18	75	...	25	...	...
14	58·3	41·7	...	11·4	11·3
14	58·3	21·0	21·0	6·9	8·9
14	58·3	...	41·7	11·3	11·4

Fig. 25 shows the Brinell hardness diagram of the gold-silver-copper alloys. It will be observed that the hardest alloy is that containing gold 520, silver 220, and copper 260 parts per 1,000. Sterner Rainer states that the alloys in which the copper and silver are near to the eutectic proportion (280 : 720) and those in which the gold and copper are in the ratio 3 : 2 exhibit pronounced hardness.

The electrical conductivity of gold-silver-copper alloys has been studied by Schulzer.[2]

The equilibrium diagram has been worked out by Jänecke [3] and by Sterner Rainer.[4] The curve given by the latter is shown in Fig. 26.

According to Sterner Rainer the range of the compound AuCu in the triple system Au-Ag-Cu, lies on a surface bounded by two lines originating from the silver corner of the diagram. All alloys in this series having gold contents of 83·3, 75, 66·7 and 58·5 per cent. and copper content between 10 and 30 per cent. exhibit pronounced age-hardening. The approach to definite carat alloys is noteworthy.

Age-hardened gold alloys find a use in orthodonty where continual working on a small part tends to leave it under irregular stress. They are also employed for wires which have to be soldered. Their use has also brought a new technique into dentistry.

In general, gold alloys may be satisfactorily annealed by heating to 650°-700° C. for a few minutes. There is risk of burning the metal if higher temperatures are employed, and great care should be taken that the heating is uniform in order to avoid internal strains. The number of anneals is dependent on the composition of the alloy and the degree of work to which it is subjected. During heating oxidation may be prevented by surrounding the objects with a neutral (*e.g.* steam or organic gas such as butane) or reducing (*e.g.* hydrogen, coal gas or cracked ammonia) atmosphere, or by applying a paste of boric acid and ochre. In this manner not only is the formation of

[1] *Amer. Inst. Min. Met. Eng.*, 1928, Tech. Pub. 86.
[2] *Helios*, 1925, **31**, 457, 465.
[3] *Metallurgie*, 1911, 8, 597.
[4] *Zeitsch. Metallkunde*, 1926, **18**, 143.

the hard oxide scale prevented, but also the interpenetration of the oxide between the grains is avoided.

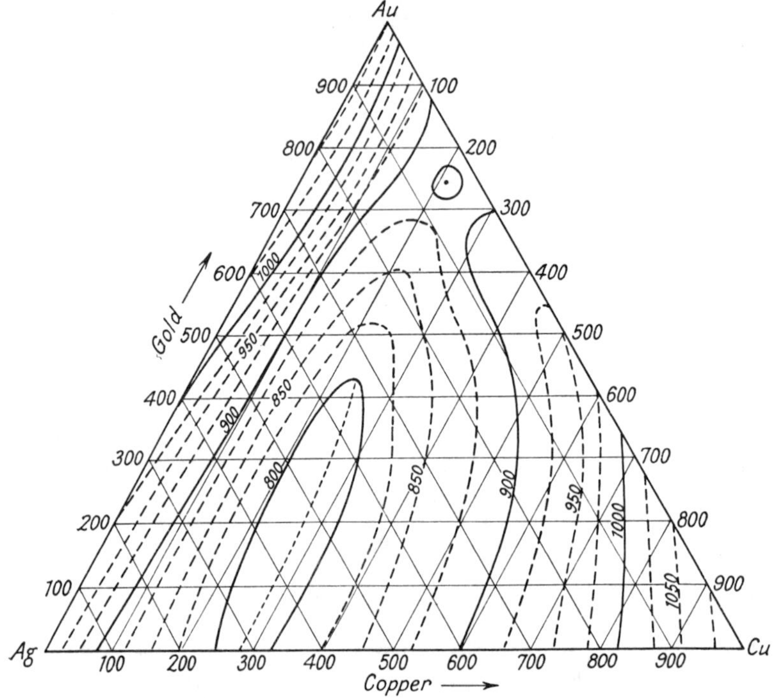

Fig. 26.—Melting Points of the Gold-Silver-Copper Alloys (*Sterner Rainer*).

Gold and Tellurium.—The thermal curve for the equilibrium between these elements [1] (see Fig. 27) shows a maximum at the point A corresponding to a concentration of 44 per cent. of gold ($AuTe_2$), and a temperature of about 470° C. (Pellini and Quercigh). It will be noticed that this is the formula assigned to the natural mineral calaverite. It appears that the compound $AuTe_2$, whether natural or artificial, can only be formed on solidification from fusion. No other compound exists. The curve exhibits two eutectics, which occur at 39 per cent., B (Fig. 27), and 79 per cent., C, of tellurium respectively.

The solubility of tellurium in gold crystals is very small and consequently its effects on the mechanical properties of gold are great. Minute quantities (0·025 per cent.) cause extreme brittleness in gold, rendering the commercial working of the latter impossible.

Gold containing small quantities of tellurium is remarkable for the fact that on annealing at a low red or black heat it becomes more brittle than before annealing. Bismuth-gold is unaltered on annealing, but lead-gold becomes less brittle, the lead appearing to pass into solution at the annealing temperature.[2]

[1] T. K. Rose, *Trans. Inst. Mng. and Met.*, 1907-8, **17**, 285.
[2] T. K. Rose, 33*rd Ann. Report of the Royal Mint* (1902), p. 73.

On heating in air, most of the tellurium in alloys burns to oxide, leaving moderately pure gold, which still remains brittle however from the presence of some tellurium.

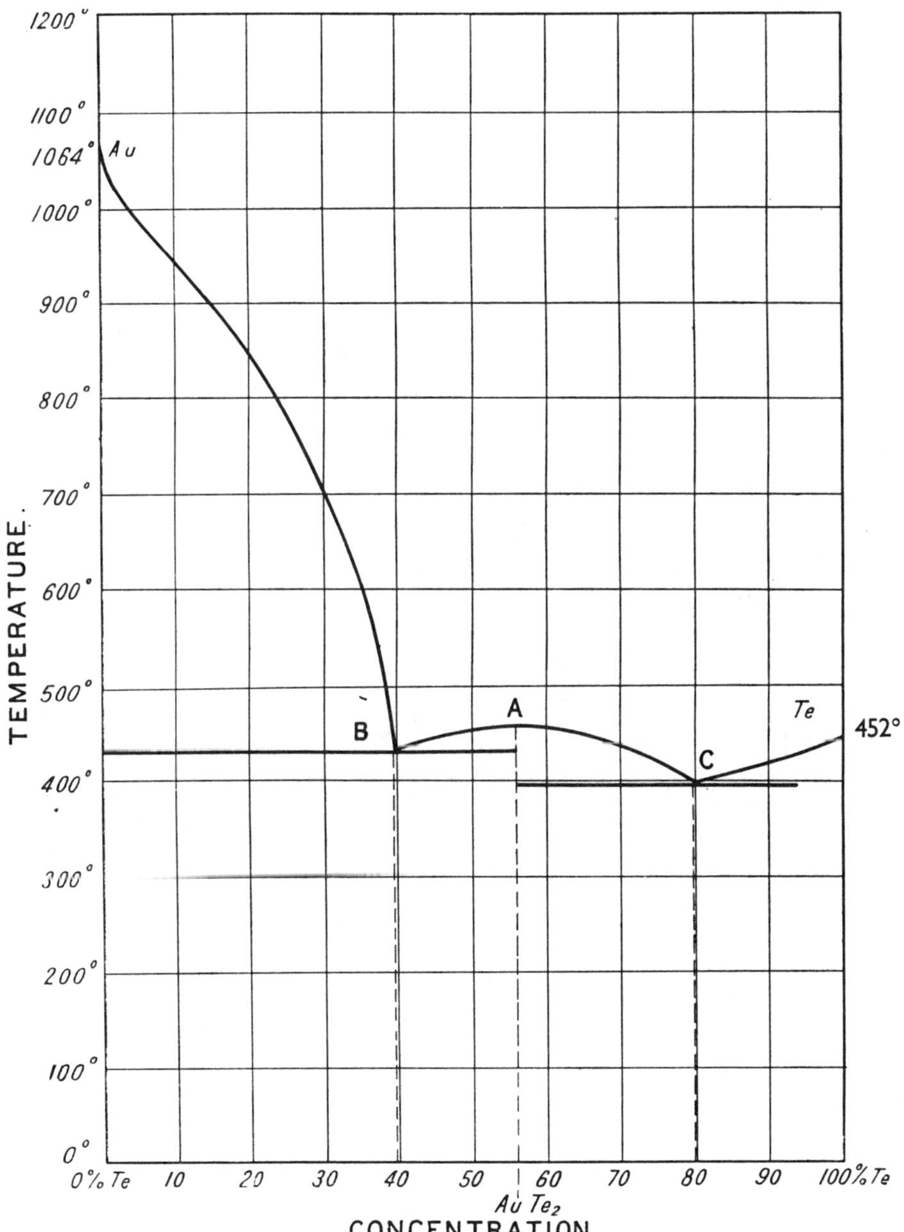

Fig. 27—Thermal Equilibrium Curve of Gold and Tellurium.

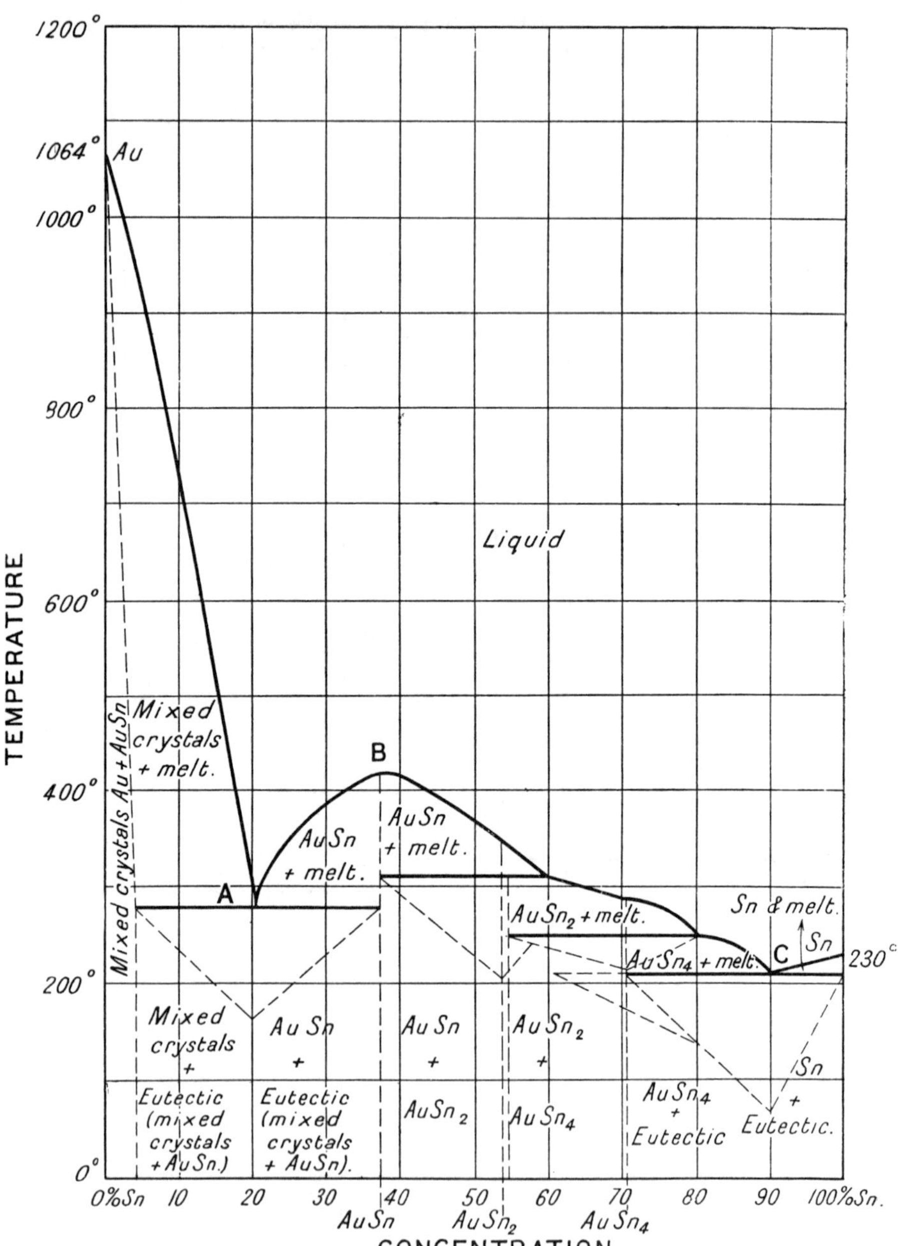

Fig. 28.—Thermal Equilibrium Curve of Gold and Tin.

Gold and Thallium.—Small quantities of thallium act like lead, causing extreme brittleness in gold, especially on heating. The thermal curve of equilibrium [1] is composed of two branches in the liquidus meeting at a single eutectic at 131° C. (27 per cent. Au). The yellow crystals of the gold show zoned colourings from the centre to the periphery of the individual crystals, whence it may be inferred that possibly quite appreciable amounts of thallium may be absorbed by gold on reaching the equilibrium state (*i.e.* on annealing the brittleness should disappear).

The thallium-gold eutectic melts at a lower temperature than any other alloy of gold except amalgam. A very small quantity of thallium causes a long pasty stage in the solidification of gold, terminating at 131° C.

Gold and Tin.—The curve of thermal equilibrium is given in Fig. 28.[2] From the melting point of pure gold the liquidus falls very sharply to a eutectic at 280° C. (A, Fig. 28), and 20 per cent. tin. In this range there separate primary crystals of gold, surrounded by mixed crystals of gold and tin, together with small portions of eutectic. The curve rises again to a maximum, B, at 418° C. and 37·63 per cent. tin, denoting a compound with the formula AuSn. Another eutectic appears at 217° C. and 90 per cent. tin, the point C. In addition to the compound AuSn, two others, $AuSn_2$ and $AuSn_4$, have been detected, and are formed on perieutectic horizontals at the two breaks at 308° C. and 258° C. respectively.

The compound AuSn is formed with slight dilatation of the volume of the reacting mixture. It has a metallic, silver-grey appearance, and is hard but brittle. Its electrical conductivity is greater than that of all other gold-tin alloys, except those with 95 per cent. or more of gold, and it is very resistant to acids.

Pure gold takes up 5 per cent. by weight of tin in solid solution. Small quantities of tin are therefore not harmful in their effects upon the mechanical properties of gold.

Under the usual cooling conditions, when 5 to 35 per cent. tin is present, crystals of tin dissolved in gold are yellow and rich in gold in the interior, while on the outside they are white and rich in tin, the whole crystals being enclosed within much eutectic. By long-continued heating at 800° C. the eutectic structure disappears, and the zoned portions in the crystals are replaced by a more homogeneous structure.

Stenbeck [3] and Westgren confirm Vogel's results by X-ray analysis, and also assert that a new β phase exists in alloys with 12 to 16 atomic per cent. tin.

With only a few per cent. of tin alloyed with it, gold loses its yellow colour, and assumes a grey appearance. Most of the alloys with comparatively large proportions of tin are very brittle, so that tin-gold alloys are unsuitable for commercial purposes, in spite of the fact that such materials have a high resistance to chemical attack.

Gold and Zinc.—The alloys of gold with zinc are of interest in connection with the Parkes process for the separation of precious metals from lead. They closely resemble the brasses in constitution.

According to Vogel [4] the thermal curve has a eutectic point at 15 per cent. zinc, and a maximum at 25 per cent. corresponding to a compound AuZn. Also there probably exist two other compounds, Au_3Zn_5 and $AuZn_3$.

The alloys with less than 14 per cent. of zinc are pale yellow, and about

[1] Levin, *Zeitsch. anorg. Chem.*, 1905, 45, 31-38.
[2] Vogel, *Zeitsch. anorg. Chem.*, 1905, 46, 60-75.
[3] *Zeitsch. physikal. Chem.*, 1931, **B** 14, 91.
[4] *Zeitsch. anorg. Chem.*, 1906, 48, 319-332.

as hard as gold. They increase gradually in brittleness as the percentage of zinc increases. The first additions of zinc rapidly reduce the melting point of gold, giving rise to a long pasty stage during solidification. As the proportion of zinc increases from 14 to 25 per cent. the colour changes gradually from pale yellow to a reddish-lilac tint, the colour of the compound AuZn (*cf.* AgZn and AgCd), which forms lustrous crystals and is brittle.

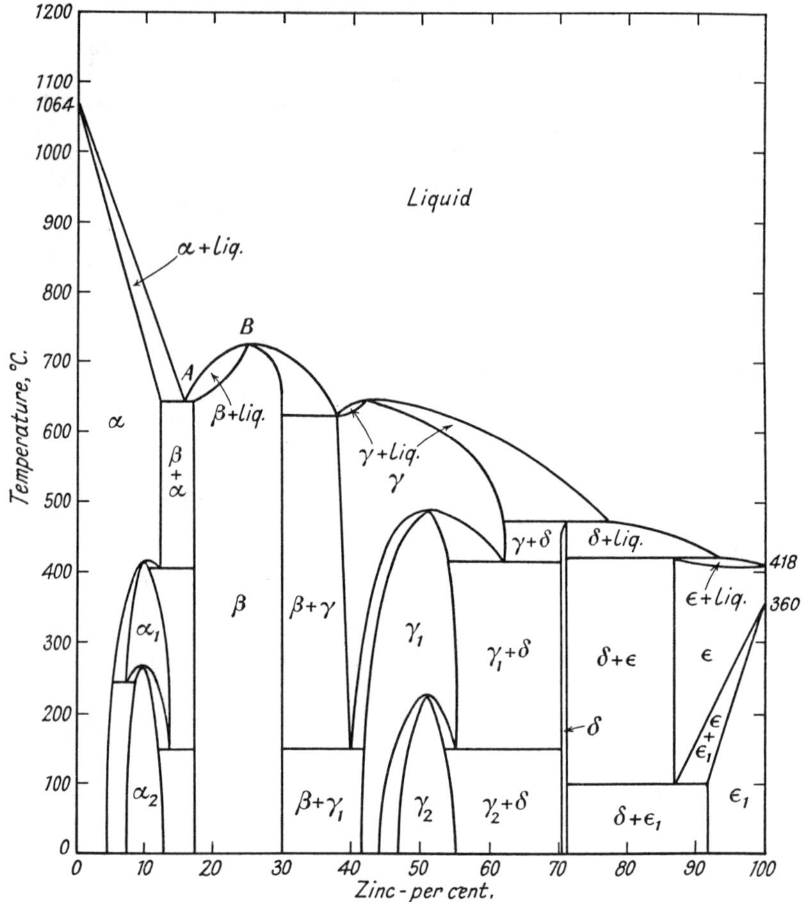

Fig. 29.—Thermal Equilibrium Curve of Gold and Zinc (*Soldau*).

The alloys containing from 25 to 30 per cent. of zinc consist of lilac-coloured polygonal crystals of AuZn set in a white matrix.

Soldau [1] has re-examined the alloys of the gold-zinc series, using methods involving the measurement of electrical conductivity at high temperatures, thermal analysis and microscopic analysis. The presence of the compounds AuZn, Au_3Zn and $AuZn_3$ is affirmed, but Vogel's Au_3Zn_5 is denied.

[1] *J. Inst. Met.*, 1923, **30**, 351; 1926, **36**, 454; *Zeit. anorg. Chem.*, 1925, **141**, 325; *Chem. Abst.*, 1927, **21**, 3802. See also Westgren and Phragmen, *Phil. Mag.*, 1925, [vi.], **50**, 311.

Soldau's diagram is given in Fig. 29, and indicates the phases which exist in the various temperature ranges. The point A represents the formation of a eutectic containing 16 per cent. of zinc and the point B that of the compound AuZn, containing 25 per cent. of zinc. A number of changes in constitution occurs in the solid alloys as they cool down to atmospheric temperature.

Etching of Gold Alloys.—In many instances a solution of aqua regia may be employed, the action being delayed, if necessary, by dilution. Electrolytic etching, using an electrolyte composed of concentrated hydrochloric acid containing a little ferric chloride, may also be adopted; the current density should be high (0·25 amp. per sq. cm.) and any contact of the gold with base metal whilst in the solution should be avoided. A mixture containing equal parts of 10 per cent. solutions of ammonium persulphate and potassium cyanide is also recommended; this mixture is only effective, however, for a short time after preparation.

Gold Solders.—These solders consist of alloys having fusion points lower than the gold objects with which they are to be used. They usually contain gold, silver, copper, zinc and cadmium in proportions varying according to the colour of the alloy required. The following are typical analyses (Sterner Rainer):—

TABLE X.

Percentage Composition.						Freezing Point.			Tensile Strength. Kg/mm².		Hardness Number.	Colour.
Au	Ag	Cu	Zn	Cd		Liquidus.	Arrest.	Solidus.	Joint.	Solder.		
20	40	28	2·2	9·8	Repairing,	756	727	724	46	45·7	167	Yellowish-white.
39	34·5	23	3	0·5	Hard,	779	771	767	..	52·1	271	Pale yellow.
50	25	16	1·6	7·4	Dental,	752	..	741	..	55·3	252	Golden yellow.
53	23·2	22·5	0·4	0·9		800	791	781	52·4	53·0	265	,,
58	10·4	17·6	2·5	11·5	14 ct.,	740	731	722	..	54·2	238	Reddish-yellow.
66·6	8·2	14·0	2·0	9·2	16 ct.,	777	769	737	..	55·6	204	Pale golden yell.
75	6·2	10·4	1·5	6·9	18 ct.,	810	797	791	45·9	47·7	204	,,
83·3	4	5	1·4	6·3	20 ct.,	850	843	832	..	29·1	147	Golden-yellow.
87·5	3	4·5	0·9	4·1	21 ct.,	923	912	900	..	23·1	119	,,

The preparation of solders with such low freezing points is only possible by introduction of Zn and Cd in approximately eutectic proportions. In Germany the following solder for 22 carat gold is used:—91·6-95·8 Au, 3-4 Cu, 1·2-4·1 per cent. of an alloy containing Ag 45, Cd 45, Zn 10 per cent. Other typical German solders are:—

TABLE XI.

	Au	Ag	Cu	Zn	Cd
Soft, . .	20·5	45·0	32·5	2·0	...
Hard, Coloured,	42·3	40·0	16·4	1·3	...
Hard, . .	50·6	24·2	23·0	2·2	...
,, . .	50·0	20·0	20·0	...	10
,, . .	55·0	15·0	18·0	...	12
14 ct., . .	58·3	11·5	18·6	...	11·6
18 ct., . .	75·0	3	10·0	...	12·0
Enamelling, .	59·2	22·8	18·0	...	...
,, .	80·4	19·6	...	...	...

Gold Alloys for Jewellery.—A great number of alloys containing gold together with a variety of other metals is in use at the present time for goldsmiths' work. Copper and silver, however, are the chief alloying metals. Reference is made elsewhere to the production of white gold (p. 41) as a substitute for platinum.

Until the 14th century practically pure gold was used for ornamental work in this country. In 1300 an alloy containing $19\frac{1}{5}$ parts of gold and $4\frac{4}{5}$ parts of alloy, equivalent to a fineness of 800/1000, was, however, adopted as standard. The 18-carat alloy (*i.e.* $\frac{18}{24}$ gold) was introduced in 1477, the 22-carat alloy in 1573 and the 15-, 12- and 9-carat standards in 1854. In 1932 the 15- and 12-carat alloys were abolished as legal standards and in their place a new standard containing 585 parts of gold per 1,000 has been adopted, this closely approximating to the true 14-carat gold (583·3 parts of gold per 1,000).

9-*Carat Gold Alloys.*—Owing to their comparative cheapness, combined with durability, these alloys are used for the greatest amount of jewellery. They offer much scope for variations in colour by changing the proportions of the alloying metals. The latter are not confined to silver and copper, but include zinc, cadmium, nickel and occasionally iron. Alloys of this quality melt at comparatively low temperatures. The one with the highest melting point contains gold and copper only, fusing at 1,000° C.; that with the lowest melting point, 790° C., contains 22·0 per cent. silver and 40·5 per cent. copper. The binary gold-silver alloy melts at 985° C. In the preparation of these alloys a small quantity of zinc is frequently added in the form of a brass containing 34 per cent. zinc. The zinc acts as a deoxidiser and assists in producing sound metal. The brass—or "Compo"—should be made from pure materials and care should be taken that by its addition the balance between copper and silver is not disturbed. The hardness, tensile strength and elongation (ductility) of certain 9-carat alloys in the annealed state are given in the following table (Smith):—

TABLE XIa.

Composition			Brinell Hardness.	Tensile Strength, tons/sq. in.	Elongation, per cent.
Au	Ag	Cu			
37·5	62·5	0	34	12·7	39
37·5	60·5	2	60	19·1	35
37·5	55·0	7·5	80	25·3	33
37·5	53·0	9·5	85	28·5	30
37·5	43·0	19·5	120	31·7	25
37·5	31·25	31·25	110	34·8	25
37·5	28·0	34·5	115	33·0	25
37·5	13·5	49·0	100	31·7	32
37·5	6·0	56·5	80	28·5	40
37·5	0	62·5	70	27·3	37

14-*Carat Gold Alloys.*—The 14-carat alloys are suitable for all stamping, drawing and pressing operations. They should be made from carefully selected pure metals, and melted and cast under controlled oxygen-free conditions.

With annealed alloys maximum hardness occurs in the alloy containing approx. equal proportions of silver and copper. With further additions of copper hardness decreases until the composition of the simple gold-copper alloy is attained. With hard rolled alloys the maximum hardness occurs in an alloy containing silver and copper in the ratio of 1 to 1·22. Annealing at too high a temperature (say above 650° C.), for too long a time or in an oxidising atmosphere will impair the working qualities of these alloys.

The tensile strength increases as copper is substituted for silver to a maximum at 27·7 per cent. copper. The ductility is about the same as that of copper, but a little lower than that of brass used for stamping.

Although most of the alloys may be made in a malleable condition if due precaution is exercised, zinc is often added as a deoxidiser and degasifier, especially to the alloys containing roughly equal proportions of silver and copper. Carter [1] gives the following figures for the hardness of these alloys in the states as shown :—

TABLE XII.

Silver, per cent.	Copper, per cent.	Hardness (Brinell).							
		As Cast.	Hard Rolled.	Cooled Slowly after Annealing at			Quenched after Annealing at		
				500° C.	600° C.	700° C.	500° C.	600° C.	700° C.
41·67	0	53	115	112	64	59	110	62	56
34·97	6·70	118	210	218	140	128	195	110	111
33·33	8·34	144	240	240	166	168	200	131	128
31·25	10·42	170	278	278	224	242	245	166	143
29·67	12·00	217	270	285	250	258	251	180	160
27·77	13·90	242	286	293	272	270	270	200	174
22·92	18·75	250	322	280	217	250	215	203	184
20·83	20·84	250	320	269	206	251	243	217	181
18·75	22·92	290	335	285	215	263	251	220	172
16·00	25·67	255	322	292	231	237	254	206	190
12·00	29·67	206	302	255	214	210	235	198	178
10·42	31·25	183	286	264	174	167	237	176	145
8·34	33·33	172	282	242	170	179	201	168	166
6·70	34·97	154	276	242	160	160	186	166	150
0	41·67	113	268	174	130	126	166	126	124

It will be noticed that in every case quenching after annealing above 500° C. results in a greater degree of softness, which is attributed to the fact that the hard compound of gold and copper has had no opportunity to form. On rolling, those alloys which contain silver or copper in excess harden considerably, but where these two metals are about equal in amount there is not so much hardening. When the hard worked material is heated to a temperature somewhat below that at which softening begins, there is an initial increase in hardness and then a fall. Erichsen tests show that with the intermediate alloys ductility is improved by quenching from 750° C. The silver-rich alloys have the lower electrical resistance and the lower electromotive force, measured against platinum.

[1] *Amer. Inst. Min. Met. Eng.*, 1928, T.P. 86. See also E. A. Smith, *Met. Industry*, 1932, **41**, 28.

These alloys are of interest especially in the range of green golds (5 to 12 per cent. copper) and the range of red and yellow golds (25 to 41 per cent. copper). Except for tarnishing their properties are excellent, but their melting points are on the low side.

TABLE XIII.—FUSIBILITY OF 14-CARAT GOLD ALLOYS (STERNER-RAINER).

Gold 58·5 per cent.		Melting point, °C.
Silver, per cent.	Copper, per cent.	
35·6	5·9	960
31·0	10·5	900
27·75	13·75	874
20·75	20·75	845
13·8	27·7	867
10·4	31·1	886
6·0	35·5	908

18-*Carat Gold Alloys.*—This standard holds a premier position on account of its colour, its retention of the physical and mechanical properties of pure gold, and its suitability for dentures. Colour variations may be observed with changing composition in this series of alloys as in the case of the 14-carat series. The 18-carat alloys are more easily cast gas-free than the 14-carats, but zinc is frequently added as a deoxidant. The alloy containing no silver does not roll as well as the other alloys ; that containing 15 per cent. silver is pale coloured.

TABLE XIV.—HARDNESS OF 18-CARAT ALLOYS (CARTER).

Silver, per cent.	Copper, per cent.	Hardness (Brinell), 2 mm. ball, 120 kilos.							
		As Cast.	Hard-rolled.	Cooled Slowly after Annealing at			Quenched after Annealing at		
				500° C.	650° C.	800° C.	500° C.	650° C.	800° C.
25·0	0	45	113	70	54	52	85	54	52
22·0	3·0	83	166	98	85	82	96	83	82
20·0	5·0	103	188	130	106	94	155	104	98
19·0	6·0	117	201	151	121	103	170	114	104
17·0	8·0	139	220	152	130	112	194	128	112
14·0	11·0	158	234	176	144	135	174	139	130
13·0	12·0	170	245	179	144	134	174	142	130
12·5	12·5	166	234	182	152	144	198	152	138
12·0	13·0	156	265	181	148	141	204	148	137
11·0	14·0	166	265	201	156	145	194	150	140
8·0	17·0	174	267	201	161	156	204	156	141
6·0	19·0	192	282	190	181	165	196	160	148
5·0	20·0	224	302	190	196	167	198	167	156
3·0	22·0	245	302	276	186	162	186	167	160
0	25·0	265	321	347	330	236	186	176	167

The softening effect of quenching, it will be seen, does not become generally operative until a temperature of about 650° C. is reached, and it is relatively greater in the alloys containing most copper.

Most of the gold-silver-copper alloys respond to age-hardening processes and strengths up to 50 tons per square inch can be developed in certain alloys by quenching from about 700° C. and then tempering at about 325° C. for half an hour.

TABLE XV.—Range of Colour in 18-Carat Gold Alloys.[1]

Silver, per cent.	Copper, per cent.	Colour.
25·0	0	Green.
21·4	3·6	Pale greenish-yellow.
16·7	8·3	Pale yellow.
12·5	12·5	Bright yellow.
8·3	16·7	Pale red.
3·6	21·4	Orange.
0	25·0	Red.

The ductility of these alloys is a property which enters into many commercial operations, *e.g.* drawing, stamping, etc. Erichsen figures, measured by Carter on samples 3¼ inches square and 0·04 inch thick, annealed at 750° C., are given in the following table :—

TABLE XVI.

Silver, per cent.	Copper, per cent.	Depth of Cup, mm.	
		Cooled Slowly.	Quenched.
25	0	7·3	13·0
22	3	6·5	11·7
17	8	3·8	11·0
12·5	12·5	3·6	9·5
8	17	2·9	8·8
3	22	2·6	5·8
0	25	...	...

Experience generally has shown that the best annealing temperature for 18-carat alloys is about 650° C. for 15 minutes. A higher temperature, whilst probably giving greater softness, will at the same time increase the crystal size inconveniently.

22-*Carat Gold Alloys.*—These are commonly known as standard gold. They constituted an early legal standard for coinage, but were not legalised for jewellery and gold wares until nearly 50 years later. Table XVIa shows the mechanical properties of some annealed and hard-rolled alloys compared with those of fine gold.

[1] E. A. Smith, *Met. Ind.*, 1931, **39**, 123.

TABLE XVIa.

Composition.			Brinell Hardness.		Tensile Strength, tons/sq. in.		Elongation, per cent.	
Au	Ag	Cu	Annealed.	Hard-rolled.	Annealed.	Hard-rolled.	Annealed.	Hard-rolled.
100	0	0	25	58	7·5	14·0	45·0	4·2
91·7	8·3	0	30	74	10·2	18·3	40·5	3·2
91·7	6·2	2·1	48	105	16·4	28·7	38·3	2·4
91·7	4·1	4·2	57	123	18·8	32·6	34·5	1·9
91·7	2·0	6·3	64	140	20·4	34·4	34·9	2·1
91·7	0	8·3	66	155	24·2	39·5	40·6	2·1

Rolled Gold.—Rolled gold [1] is the product obtained when bars of alloyed gold and base metal (usually brass or "gilding metal") are sweated or soldered together and then rolled. The industry dates back to the end of the eighteenth century. The essential features in the process are cleanliness of the bars, absolute contact between the bars along their whole length in order to avoid blistering at later stages, careful regulation of the temperature during the critical operation of sweating, and prevention of oxidation by coating the composite ingot with a paste of boric acid. During rolling the gold and base metal suffer similar amounts of reduction, so that their ultimate thicknesses are in the same ratio as their original ones. This ratio is very variable and may range from one of gold to twenty or more of base, depending on the quality and price to be catered for. The actual thickness of the gold may reach 0·003 inch. Solid and hollow wires and tubing are also being made in like fashion. Finishing, in all cases, is done in highly polished rolls. Usually rolled gold is of 18-, 14- or 10-carat quality. Even the cheaper qualities, where the gold covering is very thin, have considerably greater lasting properties than electro-deposited or "washed gold" faces.

In the "Fancy Jewellery (Standard Trade Descriptions) Bulletin, 1928," the following definitions are given :—

Gold Front.—Sheet of gold of a standard not lower than 9-carat, which is made in a separate portion and afterwards attached to a silver or base metal back so that the separate sheet of gold can be removed from the back of silver or base metal.

Rolled Gold.—Sheet of gold of a standard not lower than 9-carat, sweated or soldered to a sheet or bar of silver or base metal, the whole being rolled or drawn down together in such a way that on annealing and pickling a surface of gold will be shown.

Gold Filled.—Sheet of gold of a standard not lower than 9-carat, sweated or soldered to each side of a bar or ingot of silver or base metal, the whole being rolled or drawn down together in such a way that on annealing and pickling a surface of gold will be shown on each side.

Gold Shell.—A casing of gold of a standard not lower than 9-carat on silver or base metal, the casing being of such a thickness that when the base is dissolved in acid the gold shell is left intact.

Gold Cased.—Silver or base metal having thereon an electro-deposit of

[1] E. A. Smith, *J. Inst. Met.*, 1930, 44, 175.

gold of such density that neither nitric acid of 1·40 specific gravity nor lunar caustic applied for two minutes will discolour or stain it.

Gilt.—Silver or base metal having gold deposited thereon by a chemical or electro-deposition process.

Fine Gilt or Mercurial Gilt.—A covering of gold dissolved in mercury from which the mercury has been burnt out by firing.

Gold Plating.—This is often practised in order to simulate solid gold ware and to combine wearing with aesthetic properties. There are three methods in use.

(1) *Electro-gilding*—for the production of thick deposits. Suitable bath compositions are—

(*a*) [1]

Gold,	$\frac{1}{2}$ oz. Tr.
Total KCN,	3 ozs. Av.
Water,	1 gallon.

(*b*) [2]

Gold,	5 dwt.
NaCN,	2 ozs.
Na_3PO_4,	1 oz.
Water,	1 gallon.

Using (*b*) solution with low current density the deposit at 20° C. is pale in colour; at medium current density and 55° C. it is bright yellow.

The baths are prepared by dissolving the gold in aqua regia, evaporating off excess acid, redissolving the residue and re-evaporating carefully to remove last traces of acid. The gold may then be precipitated with ammonia,[3] filtered off, washed and the precipitate dissolved in sodium cyanide solution. The solutions should be heated during use. Various colours of gold may be obtained, according to the temperature and the current density employed.

(2) *Immersion.*—2 grams of gold are made into damp fulminate as in (1) and then added to NaCN, 40 grams, Na_2CO_3, 50 grams, water, 1 litre. The solution is kept hot and the work simply immersed in it.

A bath containing 24 ozs. potassium ferrocyanide, 12 ozs. sodium carbonate, 0·25 oz. sodium hydroxide, 0·125 oz. ferric chloride and 3 dwt. of gold per gallon is also suggested.[4]

(3) *Mercurial Gilding.*—The article is rubbed over with a wire brush that has been dipped in mercuric nitrate, and then with a similar brush which has first been passed over the surface of some " dry " gold amalgam.

The operation may be repeated several times, and the mercury is finally expelled by heating to about 420°-450° C. To improve the colour the work is passed through a paste of alum, nitre and salt, dried and heated to fuse the salts, which may then be washed off.

Melting of Gold Alloys.—*Furnaces.*—These may be constructed as unit furnaces or as banks when large quantities have to be dealt with. In the former case an iron or steel casing of circular or square section is lined with firebrick. Within the casing is a second lining of firebricks about 4½ inches thick, separated from the first by a thinner layer of firebrick or by a packing of an insulating material such as magnesia, kieselguhr or asbestos. When

---

[1] Field and Weill, " *Electro-plating*," Pitman, 1930.

[2] Sizelove, *Rev. Am. Electro. Soc.*, 1931, **18**, 45; *Met. Ind.* (N.Y.), 1931, **29**, 205.

[3] *Note.*—Fulminate of gold is precipitated. It is explosive when dry.

[4] Sizelove, *loc. cit.*

the furnaces are built in banks the whole structure may be of firebrick, supported on the outside by girders and iron strapping. In laying the bricks they should be dipped in a slurry of finely ground refractory cement and pressed into position, thus ensuring that the joints are of minimum thickness. In some instances the inner lining, instead of being made of firebrick, is formed of a high-grade refractory cement rammed behind wooden formers whose outside dimension is the internal one of the finished furnace. The formers are removed when the ramming is finished.

When a furnace is newly built it should be dried-out carefully, and the temperature gradually increased to full heat during about 24 hours. This obviates spalling. A further protection may be given by applying a wash consisting of a mixture of carborundum cement and 15 per cent. of its weight of fireclay to the interior face. The wash may, with advantage, be repeated, drying-out thoroughly between each application.

Except in the case of electric furnaces, a flue is necessary in order to carry away the products of combustion and convey them to the stack. The dimensions of the flue are important and should bear such a relation to the volume of the furnace chamber that neither throttling of the gases occurs, causing too high a pressure within the chamber, nor too free a flow of heat escaping up the stack. For a furnace 19 inches in diameter, and 28 inches deep, a flue of 4 inches square section is found satisfactory. In addition the height of the stack should not be so great that too much draught is induced. Alternatively a damper may be placed in the flue to regulate the draught and thereby the volume of gases drawn through.

As the heating of the crucible in gas- and oil-fired furnaces is mainly by radiation from the furnace walls, the narrower the gap between the two the greater the efficiency. This gap should only be sufficiently large to permit the lowering of tongs for lifting the crucible. In coke furnaces the gap should be wide enough to ensure a sufficient supply of fuel.

Coke, gas and oil are the principal fuels used. Coke should be of a good furnace variety containing sufficient volatile matter to give ready ignition. Too much sulphur is detrimental. The fuel rests on ordinary firebars and is built up round the crucible. Air under pressure may be admitted under the firebars in order to increase the intensity of the heat. The flames from gas and oil burners are admitted tangentially into the furnace chamber. They should not strike the crucible but pass between it and the wall and circulate upwards towards the flue outlet. In some instances the hot gases pass into a second or preheating chamber between the furnace and the flue, where the temperature of a new charge can be raised by the waste heat. Under ideal conditions the gas or oil should suffer complete combustion within a very short distance of entry. Air under pressure is usually admitted with the fuel to burners having special mixing chambers. The combustion space for oil is always greater than that for gas. In some instances low-pressure air and high-pressure gas are used ; in others both air and gas are under pressure, while recently a system has been introduced in which primary air and gas are premixed in a special compressor in proportions below those which would give rise to an explosion and then passed into the furnace, inducing secondary air for complete combustion.

The crucibles containing the metal are made of a graphite-fireclay mixture and stand on a stool of the same composition but covered with fine coke dust to prevent sticking. During the earlier stages, a ring or muffle of the same diameter as the crucible is placed on top of the latter to prevent spillage when charging and to prevent too free an access of the injurious furnace gases to the metal. The muffle may be surmounted by a lid. Both lid and muffle are removed before pouring. Crucibles suitable for coke furnaces are not equally suitable for those fired by gas. The maximum charge per crucible is about 3,000 ozs.

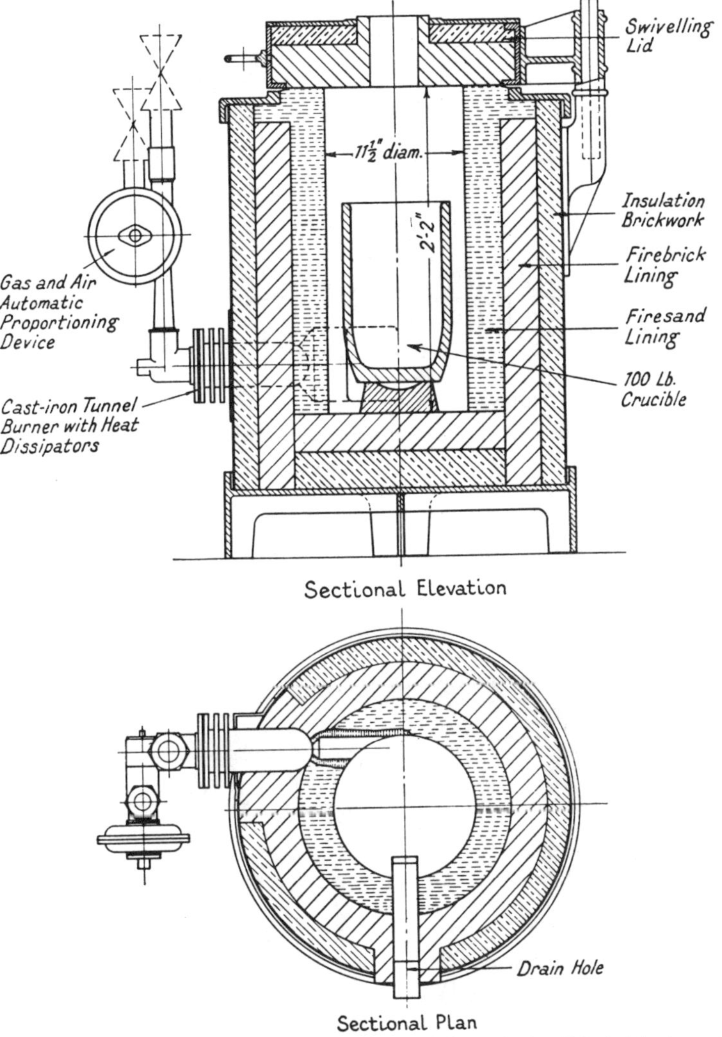

[*Reproduced by permission of the Gas Light & Coke Co., Ltd.*

Fig. 30.—Unit Stationary 100 lb. Gas Fired Crucible Furnace.

Crucibles containing appreciable quantities of silicon carbide are not to be recommended owing to absorption of silicon which embrittles gold alloys.

Figs. 30, 31 show stationary and tilting types of gas-fired furnaces. Fig. 32 is a photograph of the external appearance of the tilting furnace.

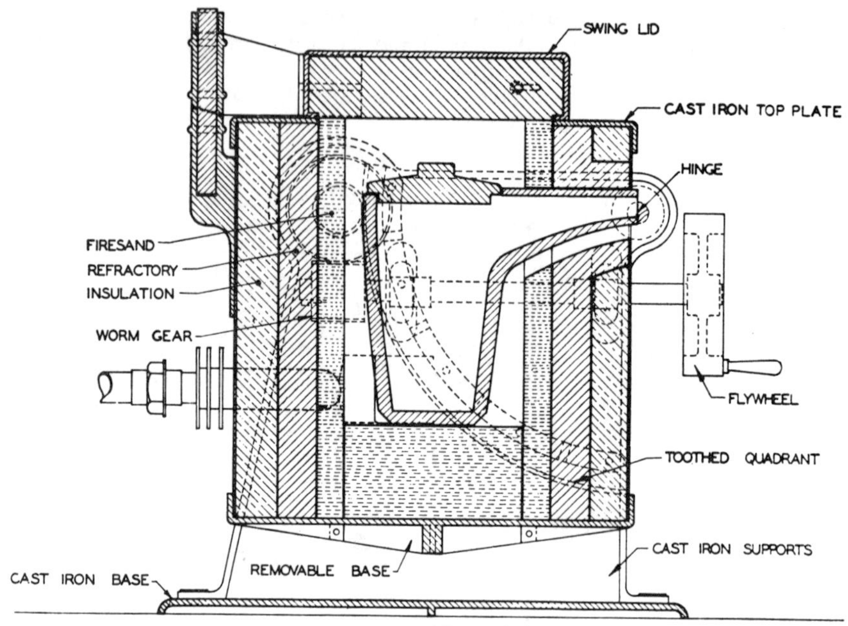

[*By permission of the Gas Light and Coke Co., Ltd.*

Fig. 31.—G.L.C. Tilting Crucible Furnace.

[*By permission of the Gas Light and Coke Co., Ltd.*

Fig. 32.—Tilting Crucible Furnace.

It will be noticed that the mouth is in line with the tilting trunnions. An automatic gas-air proportioning device is also shown (Fig. 30). This ensures that the correct mixture for complete combustion is always employed.

Fig. 33 shows a cross-section of a bank of gas-fired furnaces as used at the Royal Mint. They are heated by single burners, each with a blast of air. The gas used is ordinary town gas at a pressure of 2½ inches (water gauge). Air at 3 lbs. per square inch pressure is mixed with it in a special chamber, which is part of the feeding tube to the burner. Valves control the supplies of air and gas.

Hocking[1] states that 88·5 per cent. increased output was obtained at the Royal Mint when gas furnaces replaced coke furnaces for melting gold alloys. The gas consumption was 350 cubic feet per cwt.

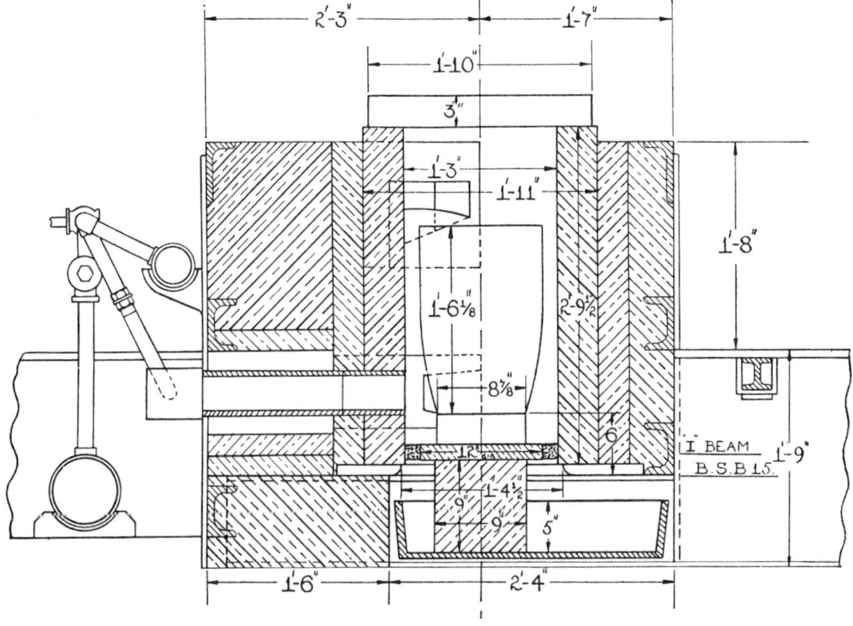

Fig. 33.—200 lb. Stationary Gas Fired Furnace for Melting Bullion.

Tilting furnaces with reverberatory hearths having a capacity up to 30,000 ozs. have also been used.

An electric tilting furnace of the Baily type was installed at the Pretoria Mint in 1927.[2] It had a capacity of 20,000 ozs. of standard gold. The hearth was made of carborundum mixture with a clay bond, and was bowl-shaped. Granulated graphite contained in a trough attached to the interior of the furnace walls carried the current and the heat from the white-hot resistor was radiated to the roof, and then reflected on to the charge. It was found that the hearth absorbed considerable amounts of gold which had ultimately to be recovered, and there was danger of silicon absorption from the carborundum. The heating of crucibles, containing the alloys, on the hearth

[1] *J. Inst. Met.*, 1917, 17, 162.
[2] 64*th Ann. Report Royal Mint*, 1933, p. 139.

of this furnace was successful, and the loss amounted to only 0·03 part per thousand.

Melting.—The charge is usually placed in the warm crucible, covered with wood charcoal, and heated until it is melted. Practice is not uniform as to when the various ingredients should be added. It is contended that the lowest melting point materials should be added last, when they will be readily absorbed. On the other hand, if added first they quickly form a pool of molten metal into which the remainder may sink, thus facilitating melting, the alloying process under these conditions proceeding rapidly. Volatile constituents are best added in the form of an alloy whose composition can be determined, as it is usually the case that the loss by volatilisation of an alloyed metal is less than that of the pure material. In some instances, however, zinc may be added to the crucible first, covered with a layer of charcoal, followed by finely cut pieces of a second metal or alloy and finally the remainder of the charge. The addition of zinc at the end causes losses by volatilisation, although a little may serve as deoxidiser. The base metals used for alloying should be of the purest varieties.

Loss of Gold in Melting.—It is the common experience in Mints that the gross loss in melting gold-copper coinage alloys is usually from 0·2 to 0·25 part per 1,000 and, after taking account of the amounts recovered from condensation chambers, flues, etc., and from the ground-up crucibles, ashes, floor-sweepings and furnace bricks, the net loss amounts to 0·1 to 0·15 part per 1,000. The net loss at the Royal Mint was an annual average of £9,714 or 0·147 part per 1,000 for the five years 1909-1913.[1] The loss there was somewhat less when using gas-fired furnaces with charges of 2,750 oz. than in coke-fired furnaces taking charges of 1,200 oz., but the difference was not great.

The loss in melting rough gold at gold mills cannot be determined precisely owing to the difficulty of valuing the gold in the charge.

The net loss in melting is ascribed to flue loss. In the case of an electric resistance furnace at Pretoria the loss in crucible melting was only 0·03 part per 1,000, flue loss being eliminated.

The flue loss is not due to true volatilisation, which is inappreciable below 1,400° C.,[2] a temperature never attained in molten gold in industrial furnaces. It appears to be due to spirting,[3] caused by slight changes in the composition of the atmosphere in contact with the gold. Molten gold absorbs oxygen and then some hydrogen, or *vice versa*, with the result that their union causes "boiling," and globules of metal are projected from the mass and carried away by the effluent gases. The globules are generally small, some of those recovered in experimental work being much less than 0·001 mm. in diameter. Such globules show no signs of settling even in a still atmosphere, though they could doubtless be recovered by the Cottrell process (*q.v.*).

A similar loss occurs during pouring.

Pouring is usually done by hand, the crucible being held in a sling carried by two men one of whom controls the rate of pour of the metal. The stream should be steady and regular and fed at such a rate when pouring into moulds with comparatively small openings that molten metal is supplied to each succeeding layer as it starts to solidify. In the case of ingots the pouring may be faster. It is often desirable to pour the metal either through

---

[1] *44th Ann. Report of the Royal Mint*, 1913, p. 37.

[2] See p. 7.

[3] T. K. Rose, *Trans. Inst. Min. and Met.*, 1919, **28**, 135.

the outer fully-combusted zone of a luminous gas flame, or through a ring of luminous flames from a burner, in order that air may not be absorbed into the stream and oxidation may be prevented. A bright finish is obtained on fine gold ingots by playing an acetylene flame on their top surfaces. In order to reduce the viscosity and to allow dissolved and entrapped gases sufficient time to be liberated, the pouring temperature should be about 150° C. above the melting point of the alloy.

The moulds should be given a very light coating of a high flash-point oil.

Defects.—The chief defects in cast ingots, which are of greater importance if the metal is to be worked, are (1) surface blemishes, (2) porosity, (3) slag inclusions, (4) dissolved gas. The first may often be detected by visual examination and may be prevented in great measure by the use of a boric acid or boric acid-bone ash slag, by more oil on the moulds, or by raising the casting temperature. Porosity is usually due to the entrapping of gases within the solidifying metal or the contraction of the latter during cooling. Air and oxygen are the commonest entrapped gases, and the use of deoxidisers may be beneficial. Such deoxidisers will not remove absorbed furnace gases, however. Zinc improves the surface of gold-silver-copper ingots and it does not form dangerous brittle compounds with any of the three elements. Silicon is unsuitable as a deoxidiser as it produces extreme shortness in the alloy, attributed by some to the formation of a very low melting gold-silicon eutectic (359° C.[1]). Phosphorus in controlled small quantities is a good deoxidiser, and has also the effect of permitting increase in grain size due to the removal of the Cu_2O, giving freedom for the particles to grow. Care must be taken not to allow the crystal growth to pass beyond control, or it may result in the " orange peel " effect often noticeable in over-heated metal. Phosphorus may increase the tendency to tarnish. Calcium boride is also said to be a good deoxidiser.[2] Metallic impurities may also segregate in the crystal boundaries and become a source of weakness.

Sweep.—Crucibles that have been used for gold melting, furnace linings, and all materials that are likely to contain gold, are crushed, washed carefully to remove coarse particles, dried and then sampled and assayed. The dust is then usually sold to smelters.

---

[1] de Capua, *Atti Accad. Lincei*, 1920, **29**, 111.
[2] Capillon, *Amer. Inst. Min. Met. Eng.*, 1930, T.P. 282.

CHAPTER III.

CHEMISTRY OF THE COMPOUNDS OF GOLD.

Compounds of Gold.—Gold is characterised chemically by an extreme indifference to the action of all bodies usually met with in nature. Its simpler compounds are formed with difficulty, and decompose readily, especially when heated. The result is that gold is found in nature chiefly in the metallic form, and the mineralogist has, therefore, few compounds to consider. Gold also forms complex compounds, especially with cyanogen and sulphur. These do not exhibit the ordinary reactions of gold, and are in particular not so readily reduced. The gold in them is not present as an elementary ion, but forms part of complex ions.

Gold forms two series of simple compounds, having the general formulæ AuR and AuR_3, while doubtful compounds corresponding to AuR_2, AuR_4, and AuR_5, have been stated to exist. The two undoubted series are denominated *aurous* and *auric* respectively.

Gold Monochloride or Aurous Chloride, AuCl.—This salt is prepared by heating auric chloride at 190° C. in air for ten hours.[1] It can also be prepared in solution by running a solution of $HAuCl_4$ into an excess of a solution of sulphurous acid. The gold chloride is decolorised, showing that the auric chloride has been reduced to aurous chloride. After some time the solution becomes turbid and metallic gold is precipitated. Lenher[2] found that colourless AuCl is formed in solution by the reducing action of sulphurous acid on the double chlorides of gold and any one of the following metals—sodium, potassium, copper, magnesium, zinc, cadmium. The reaction is quantitative, according to the equation :—

$$AuCl_3.MCl_2 + SO_2 + 2H_2O = AuCl.MCl_2 + H_2SO_4 + 2HCl$$

In this way sulphurous acid can be used as an agent for the volumetric determination of gold. An arsenite can be used instead of sulphurous acid, thus :—

$$AuCl_3.MCl_2 + Na_2HAsO_3 + H_2O = AuCl.MCl_2 + 2NaCl + H_3AsO_4$$

Diemer states that the double chlorides require a considerable excess of sulphurous acid for complete reduction with the precipitation of metallic gold. One molecule of NaCl, KCl, or NH_4Cl to one molecule of AuCl is sufficient to ensure the formation of the double salts, but AuCl.NaCl is rather unstable. The double salt is more stable if an excess of NaCl is present. About 40 molecules of $CaCl_2$ can take the place of 1 molecule of NaCl in preventing the separation of metallic gold by the action of sulphurous acid in slight excess.

[1] T. K. Rose, *J. Chem. Soc.*, 1895, 67, 902.
[2] *J. Amer. Chem. Soc.*, 1913, 35, 546 ; see also Diemer, *op. cit.*, 552.

Aurous chloride is non-volatile and unaltered at ordinary temperatures and pressure by dry air, even when exposed to light, but begins to decompose at temperatures above 160°C., and the decomposition is complete if it is heated at 175° to 180° C. for six days, or at 250° C. for one hour. It combines with chlorine at the ordinary temperature, forming auric chloride. Its density is 7·4. Water converts aurous chloride into a mixture of gold and auric chloride, thus $3AuCl = 2Au + AuCl_3$. The same reaction takes place slowly at the ordinary temperature in moist air. Alcohol also causes decomposition, but more slowly than water. Aurous chloride is a citron-yellow amorphous powder. It is soluble in ammonia, and on the addition of HCl the white crystalline unstable substance $AuNH_3Cl$ (aurous-ammino chloride) is precipitated (Diemer). It is soluble completely in solutions of potassium cyanide and sodium chloride, and with separation of some gold when acted on by a solution of potassium bromide.[1] Various other complex ammino-compounds are produced by reactions with ammonia at higher temperatures. Small percentages of auric chloride may be removed from aurous chloride by solution in ether.

Auric Chloride or Gold Trichloride, $AuCl_3$.—Gold combines directly with chlorine to form auric chloride. Rose [2] found the action to be reversible in an atmosphere of chlorine at all temperatures between 100° and 278° C. The pressure of chlorine in a closed vessel reaches 760 mm. at about 250° C. Rose observed an attack by chlorine on gold at all temperatures from 15° to 1,100° C., the rapidity of absorption reaching its maximum at 225°. At higher temperatures the effective attack becomes slight, though still perceptible at 1,100°.

Auric chloride is decomposed in air by heat. The action is slow below 180° C. but rapid at higher temperatures. Aurous chloride results first and then gold. In an atmosphere of chlorine auric chloride begins to volatilise at 180° C. Volatilisation is rapid at 300° C., but becomes slow again at higher temperatures, especially at 600° to 800° C. By passing a current of chlorine over gold at 1,000° C., continuous though slow distillation of gold chloride takes place.[3]

These properties explain the loss of gold and its appearance in the covering slag in Miller's chlorine process of refining (*q.v.*).

Petit [4] obtained a 79 per cent. yield of auric chloride by heating gold leaf to 245°-250° C. in a current of chlorine at a pressure somewhat above that of the atmosphere. He also found the dissociation of the trichloride practically irreversible. The measurements of total pressures are satisfactory between 100° and 250° C. if taken at progressively rising temperatures. Petit [5] found that the heat of formation of the trichloride from aurous chloride and chlorine rises from 16·8 to 20·6 kilogram-calories between 181° and 202° C. and then falls to 15·6 between 229° and 250° C. At 249° C. the dissociation pressure is 760 mm.

If gold is treated with liquid chlorine in a sealed tube at ordinary temperatures, it is converted into a crystalline red mass, chiefly $AuCl_3$.

Gold placed in a beaker of chlorine water will be slowly attacked and pass into solution as gold chloride. As gold cannot displace hydrogen from

---

[1] Lengfeld, *J. Chem. Soc.*, 1902, 82, [ii.], 27 ; *J. Amer. Chem. Soc.*, 1901, 26, 324.
[2] *J. Chem. Soc.*, 1895, 67, 881.
[3] Krüss, *Ber.*, 1887, 20, 211.
[4] *Bull. Soc. chim.*, 1925, [iv.], 37, 1141.
[5] *Op. cit.*, p. 615.

water or hydrochloric acid it would appear that the reaction is molecular rather than ionic, and that the non-ionised chlorine molecule attacks the gold directly. Thus

$$2Au + 3Cl_2 \rightarrow 2AuCl_3.liq. (+ 54{\cdot}338 \text{ Cal.})$$

The usual method adopted for the preparation of auric chloride is to dissolve gold in aqua regia, and then to drive off the excess of acid by heat, adding HCl if necessary to maintain its excess over the nitric acid, keeping the temperature as low as possible. A brownish-red mass is thus formed, consisting of $AuCl_3$ mixed with more or less aurous chloride and hydrochloric acid. When heated at 95° to 100° C. until acid fumes no longer are evolved, the resulting chloride solidifies at 70°. It consists almost entirely of $HAuCl_4$, but contains from 0·5 to 1·0 per cent. of AuCl. On taking up with water, aurous chloride is decomposed into gold and trichloride, but the hydrochloric acid can only be eliminated by shaking with ether, which withdraws auric chloride from its solution in water. If an attempt is made to drive off the hydrochloric acid by heat, a partial decomposition of the trichloride results.

Auric chloride exists both in the anhydrous state and in combination with two equivalents of water, $AuCl_3.2H_2O$, when it occurs in orange-red crystals. The anhydrous salt has a brilliant-red colour, crystallising in needles belonging to the triclinic system, and melting at 288° C. in chlorine under a pressure of two atmospheres. It can be prepared by drying the hydrated salt at 150° C. The anhydrous and hydrated salts are both hygroscopic, and dissolve readily in water with elevation of temperature; they are also soluble in alcohol and ether, and in some acid chlorides, such as $AsCl_3$, $SbCl_5$, $SnCl_2$, $SiCl_4$, etc.

Solutions of gold chloride are also decomposed by heat.[1] The solutions are found to contain traces of hydrogen peroxide, the reaction being expressed by the equations—

$$AuCl_3 + 2H_2O = AuCl + 2HCl + H_2O_2$$
$$3AuCl = AuCl_3 + 2Au$$

Weak voltaic currents precipitate metallic gold on the cathode from a solution of the trichloride. Withrow has studied the rate of precipitation in the presence of potassium cyanide and of sodium sulphide, a rotating anode being employed.[2] The solution of auric chloride is also decomposed by many reducing agents, such as most organic substances, metals, and protosalts; heating the solution in every case hastens the decomposition. The reduction by organic matter is assisted by the action of light, which is especially efficacious in the presence of starchy and saccharine compounds, or of charcoal or ether. In some cases the presence of hydrochloric acid prevents or retards the action, but in other cases this effect is not observable. In direct sunlight ether deposits a bright mirror of metallic gold, but under ordinary conditions red colloidal gold (Faraday's gold) is set free. Alkalies also quicken the action of organic matter, and it may be said that all organic compounds reduce gold chloride on boiling in the presence of potash or soda, while Müller states that a mixture of glycerine and soda lye is one of the best precipitants for gold chloride, separating the metal completely

---

[1] Sonstadt, *Chem. News*, 1898, 77, 74.

[2] *J. Amer. Chem. Soc.*, 1906, 28, 1350; see also Neumann, *J. Chem. Soc.*, 1906, 290, [ii.], 764.

in highly dilute solutions. Priwoznik confirms this.[1] According to Krüss, if potash and soda are quite free from organic matter, they have no action on solutions of auric chloride, whether cold or hot. If a small quantity of organic matter is present, sub-oxide of gold is precipitated; if larger quantities are present, both metallic gold and sub-oxide are precipitated in the cold, but gold alone at boiling point. Measurements of acidity and conductivity during the titration of a solution of $HAuCl_4$ with NaOH have led Britton and Dodd [2] to believe that the former acts as a mixture of HCl and partially hydrolysed $AuCl_3$. The HCl is neutralised immediately, whilst the $AuCl_3$ reacts slowly with NaOH forming $NaAuO_2$. Alkali carbonates are without action on cold solutions, but if they are hot, then half the gold is precipitated as hydrate, while the other half remains in solution in the form of a double chloride of gold and the alkali. Gold is also precipitated from solution by acetylene.[3] Kindler [4] found that the separation was quantitative if the gold content did not exceed 1·5 per cent. About 12 per cent. of carbon dioxide and 86 per cent. of glyoxal are produced.

Charcoal precipitates gold. Avery [5] concluded that the reduction takes place in accordance with the equation:—

$$4AuCl_3 + 6H_2O + 3C = 4Au + 12HCl + 3CO_2$$

Occluded gases in the charcoal, such as hydrogen and carbon monoxide, will also reduce gold. The effect of the charcoal is diminished by use, but can be restored by heating to redness in the absence of air, or by the passage of a current of electricity, using the charcoal as the cathode.

Anhydrous auric chloride unites with carbon monoxide at 95° C. to give the product AuCl.CO, which is very sensitive to moisture.[6] If hydrogen is quite pure, it has no effect either on cold or hot solutions of auric chloride. Sulphur, selenium, phosphorus, and arsenic all precipitate gold on boiling with solution of the trichloride. Tellurium easily reduces gold chloride solution; the precipitation is complete at the ordinary temperature, but the tellurium must be finely powdered, or it becomes coated with gold, and further action is prevented. Silver telluride also reduces solutions of gold salts. In these reactions tellurium tetrachloride is formed, thus:—

$$4AuCl_3 + 3Te = 4Au + 3TeCl_4$$

The natural tellurides behave like tellurium in precipitating gold.[7] Many metals reduce chloride of gold, the action being, of course, most rapid in the case of the more highly electro-positive metals, such as zinc and iron. Lead sometimes gives fine dendritic plates of gold. Sulphuretted hydrogen completely precipitates sulphide of gold from both neutral and acid solutions, whilst phosphoretted, arsenuretted and antimoniuretted hydrogen, as well as H_2Te and H_2Se, all precipitate metallic gold. The lower oxides of nitrogen, nitrous acid, and many other "*ous*" acids and oxides effect the same decomposition. Arsenious oxide reduces gold chloride, but that produced

[1] *J. Chem. Soc.*, 1912, [ii.], **102**, 562.
[2] *J. Chem. Soc.*, 1932, [ii.], 2464, 2467.
[3] Mathews and Watters, *J. Amer. Chem. Soc.*, 1900, **22**, 108.
[4] *Ber.*, 1921, **54**, **B**, 647.
[5] *J. Soc. Chem. Ind.*, 1908, **27**, 255.
[6] Manchot and Gall, *Ber.*, 1925, **58**, **B**, 2175.
[7] Lenher, *J. Amer. Chem. Soc.*, 1902, **24**, 355, 918.

from the vitreous modification acts at four times the rate of the crystalline form. Precipitation by potassium nitrite in presence of sulphuric acid gives gold nodules easily collected.[1] Sulphur dioxide is a convenient reagent, and is often used in the laboratory, being almost equally efficacious in cold and hot solutions. The reaction is

$$2AuCl_3 + 3SO_2 + 6H_2O = 2Au + 6HCl + 3H_2SO_4$$

Various protosalts also reduce trichloride of gold. Ferrous sulphate is often used to detect the presence of gold in solution as chloride ; this reagent gives dilute solutions a pale blue colour by transmitted light, and brown by reflected light, owing to the formation of finely divided precipitated gold. The reaction is represented by the equation :—

$$2AuCl_3 + 6FeSO_4 = 2Au + 2Fe_2(SO_4)_3 + 2FeCl_3$$

A mixture of stannous and stannic chlorides gives a purple colour with chloride of gold even in very dilute solutions.

For the estimation of gold in dilute solutions of the chloride see Chapter XVIII.

Chlorauric Acid or Hydrogen Aurichloride, $HAuCl_4$.—Gold trichloride in the presence of free hydrochloric acid forms this compound, which crystallises out on evaporation in vacuo in long yellow needles, having the composition $HAuCl_4 + 4H_2O$,[2] and since gold chloride unites with many other soluble chlorides to form double chlorides, the hydrochloric acid compound is regarded as an acid. The acid may be used to precipitate cobalt and various alkaloids. It is more stable than gold trichloride. The chloraurates, having the general formula $M'AuCl_4$, or $AuCl_3.M'Cl$, are readily soluble bodies which can be crystallised, and which decompose with about the same readiness as chlorauric acid. They usually contain 8 to 12 molecules of water of crystallisation. The water content increases with diminishing atomic volume of the metal M'. Double chlorides are also formed with organic bases. The sodium salt (sodium aurichloride), $NaAuCl_4.2H_2O$, is used for "toning" in photography. It contains 49·45 per cent. of gold, as made in England, but only 23 to 30 per cent. as made in the United States.[3] The acetates and succinates are also used in photography.[4]

On adding silver nitrate to a solution of hydrogen aurichloride a brown precipitate is obtained,[5] according to the equation :—

$$HAuCl_4 + 4AgNO_3 + 3H_2O = Au(OH)_3.4AgCl + 4HNO_3$$

By the action of ammonia this is converted into fulminating gold (*q.v.*).

Aurous Bromide or Gold Monobromide, AuBr, is a yellowish-green powder obtained by heating the tribromide to about 115°. It is insoluble in water, but is decomposed by it, metallic gold and the tribromide being formed ; the change is especially rapid on boiling, and is hastened by the presence of hydrobromic acid. Aurous bromide is completely soluble in potassium cyanide ; and in ammonia, potassium bromide or HBr with partial separation of gold (Lengfeld).

---

[1] Jameson, *J. Amer. Chem. Soc.*, 1905, **27**, 1444.
[2] Schmidt, *J. Chem. Soc.*, 1906, **90**, [ii.], 862.
[3] Kebler, *J. Soc. Chem. Ind.*, 1900, **19**, 1038 ; Johnson & Sons, *ibid.*, 1901, **20**, 210.
[4] Mercier, *Brit. Journ. Phot.*, 1892, **39**, 354.
[5] Jacobsen, *J. Chem. Soc.*, 1908, **94**, [ii.], 601 ; *Compt. rend.*, 1908, **146**, 1213.

Auric Bromide or Gold Tribromide, $AuBr_3$, is produced by the action of a mixture of bromine and water on gold, particularly on the application of heat. The action of bromine on gold, however, causes the formation of a film of $AuBr_3$, which prevents further action. The film is removed by shaking the mixture.[1] The bromide may also be prepared by heating finely divided gold in sealed tubes with bromine and arsenic bromide, $AsBr_3$, to 126° C. Auric tribromide resembles the trichloride in most of its properties. It is volatile at 300° C. in an atmosphere of bromine (Meyer). It crystallises in blackish needles or scarlet plates, and forms intensely coloured brownish-red aqueous solutions, the presence of a mere trace of the salt in a solution being observable in this way. It is deliquescent, but is far less soluble in water than gold trichloride. Concentrated solutions, which may contain about 1 per cent. of the tribromide, are nearly black in colour. Auric bromide suffers decompositions similar to those noted in describing $AuCl_3$, its solutions being still less stable than those of the chloride. A solution of gold tribromide is gradually decolourised by sulphur dioxide, being completely reduced to the state of monobromide before any precipitate of metallic gold is formed. Double bromides analogous to the chloraurates exist.

Iodides of Gold are of little interest to the metallurgist. Aurous iodide, AuI, can be prepared by the action of iodine on gold above 50° C., the excess of iodine being removed by careful sublimation.[2] It is a white powder, turning green in air and decomposing at 190° C. The tri-iodide is supposed to be formed if gold is acted on by a solution of iodine in potassium iodide. Rapid dissolution ensues, and the solution is fairly stable. The complex iodides containing the radical (AuI_4) are more stable than the simple iodide AuI_3.

Aurous Cyanide, AuCN, is obtained by heating aurocyanide of potassium, $KAu(CN)_2$, with hydrochloric or nitric acid and washing with water, or by dissolving auric hydroxide in hydrocyanic acid. It is a lemon-yellow crystalline powder, insoluble in water, and unaltered by exposure to air. It is decomposed by heat, yielding metallic gold and cyanogen, and is soluble in ammonia, in yellow ammonium sulphide, in alkali cyanides, and in thiosulphate of an alkali. It is unattacked by the mineral acids, except by aqua regia, but is decomposed when boiled with potash, metallic gold being thrown down and aurocyanide of potassium formed.

Potassium Aurocyanide, $KAu(CN)_2$, is obtained by crystallisation from its solution, which is prepared by dissolving metallic gold, auric oxide, or aurous cyanide in a solution of potassium cyanide. It is also formed by adding potassium cyanide to an acid solution of gold trichloride. The solution for electro-plating purposes may be prepared by precipitating a solution of gold chloride with ammonia and dissolving the fulminating gold in potassium cyanide, by precipitating gold chloride with magnesia and dissolving the purified auric hydroxide in KCN, or by simply passing an electric current through a cyanide bath, with a gold anode. It forms a colourless solution in water, from which it can be crystallised as colourless, transparent rhombic octahedra. Cold water dissolves 15 per cent. and boiling water twice its weight of the salt. The aqueous solution, especially if hot, gilds copper or silver without the agency of a battery. Gilding is, however, generally effected by electro-deposition, using a gold anode. Gold is also

[1] Meyer, *J. Chem. Soc.*, 1909, **96**, [ii.], 321 ; see also Lengfeld, *J. Chem. Soc.*, 1902, **82**, [ii.], 27.

[2] Meyer, *Compt. rend.*, 1904, **139**, 733.

precipitated from the solution by zinc and many other metals. Precipitates are also formed on the addition of salts of zinc, copper, tin, iron, or silver, provided potassium cyanide is not present in excess. According to Lindbom, ferrous salts are without action on $KAu(CN)_2$, but oxalic acid, sulphurous acid, or mercurous chloride, Hg_2Cl_2, precipitate aurous cyanide from hot solutions.

Aurocyanides are decomposed by mineral acids, aurous cyanide being precipitated, and hydrocyanic acid evolved. Iodine, bromine, and chlorine are dissolved by $KAu(CN)_2$, and the iodine compound, $KAu(CN)_2I_2$, can be crystallised out. The aurocyanides of sodium, ammonium, barium, calcium, zinc, cadmium, and other metals have been prepared.

Potassium Auricyanide, $Au(CN)_3.KCN$, is formed by adding potassium cyanide to a neutral solution of trichloride of gold, the precipitate first formed being redissolved. The solution is completely decolourised, and on cooling deposits colourless crystals of $2[Au(CN)_3.KCN].3H_2O$. These effloresce in air, giving up two molecules of water ; and, on heating, the third molecule of water and some cyanogen are given off, aurocyanide of potassium being formed, and this in its turn is decomposed at a slightly higher temperature. Potassium auricyanide is somewhat soluble in cold water, and readily soluble in hot water. One or more of the cyanogen groups can be replaced by a halogen, *e.g.* iodine.

Aurous Oxide, Au_2O.—This oxide is said to be produced by decomposing aurous chloride, AuCl, or the corresponding bromide by potash in the cold (Berzelius) when a violet precipitate forms, which is blackish when moist, but greyish when dry. The fresh precipitate is soluble in alkalies and in cold water, forming an indigo blue solution, with brownish fluorescence, and on warming the solution slightly the corresponding hydrate is precipitated.

W. B. Pollard [1] has conducted many trials of earlier suggested methods for the preparation of aurous oxide, but comes to the conclusion that the substance hitherto believed to be aurous oxide may be more correctly described as a mixture of gold and auric hydroxide analogous to Purple of Cassius. He obtained no evidence that aurous oxide exists.

Buehrer, Wartmann and Nugent [2] have also concluded that the product of the reduction of dilute potassium bromaurate solution with SO_2 solution at 0° C. is not aurous oxide as was hitherto supposed.

Auric Oxide, Au_2O_3.—This is a black powder when anhydrous, and is precipitated from solutions of auric chloride in the form of a hydrate by the caustic alkalies, the carbonates of the alkalies, hydrates and carbonates of the alkaline earths or zinc, and also the carbonates of some of the heavier metals such as iron and manganese. The readiest method of preparing it is to add caustic potash, little by little, to a hot solution of gold chloride, until the yellow precipitate of auric hydroxide, $Au(OH)_3$, first formed, is dissolved to a brown liquid which contains potassium aurate, $KAuO_2$. Then a slight excess of sulphuric acid or some sodium sulphate is added, the precipitate filtered off, washed and purified from potash by being redissolved in concentrated nitric acid, and reprecipitated by dilution with water. On drying this precipitate in vacuo, the hydroxide, an ochreous powder, results.[3] The oxide can also be prepared by heating a solution of gold chloride with magnesia, which inhibits the formation of aurates and

[1] *J. Chem. Soc.*, 1926, [i.], 1347.
[2] *J. Amer. Chem. Soc.*, 1927, 49, 1271.
[3] See also Roseveare and Buehrer, *J. Amer. Chem. Soc.*, 1927, 49, 1221.

chloraurates, and washing the residue with nitric acid. It is a yellow, olive-green, or brown powder (according to the method of preparation), and becomes brownish or black on drying. It dissolves in potash solution, and the resultant unstable potassium aurate can be used for electro-gilding. If it is heated to 110° C., oxygen begins to be given off; at 160° AuO remains, and on heating for some time at 250°, metallic gold is left. Auric oxide dissolves in concentrated sulphuric and nitric acids, from which it is partly reprecipitated on boiling or on dilution. Double nitrates of gold and the alkalies have been obtained as crystals. Hydrochloric and hydrobromic acids dissolve the oxide forming the haloid salts, but hydriodic acid decomposes it on boiling, giving iodine and metallic gold. The oxide dissolves in boiling solutions of alkali chlorides, giving aurates and chloraurates, while it also combines with metallic oxides to form aurates.

It is easily reduced by hydrogen, carbon and carbon monoxide, with the aid of very gentle heat. Boiling alcohol or hot alcoholic potash reduces it, yielding minute spangles of gold which were formerly used in miniature painting. When heated with magnesium or sodium chloride under pressure decomposition occurs.[1]

Aurates.—The aurates of potash and soda have the general formula $Au_2O_3.R'_2O$ or $R'_2Au_2O_4$ assigned to them. They are readily soluble, crystallisable compounds, and are formed when alkalies are added in excess to solutions of gold chloride. The aurates of calcium, magnesium, and zinc are insoluble in water, but soluble in hydrochloric acid. With organic matter they yield explosive powders (Meyer).

Fulminating Gold is a compound of auric oxide with ammonia, $Au_2O_3(NH_3)_4$, which is formed by precipitating gold chloride with ammonia or its carbonate, or by the action of ammonia on the oxide. When prepared by the former method its composition is variable, but the fulminate is always a high explosive decomposing with violence at 145° C., or on being struck, and sometimes even spontaneously

Weitz [2] asserts that the reaction between ammonia and auric chloride results in the production of the compound $[Cl(NH_2)Au]_2NH$, which is hydrolysed on washing and thereby becomes more explosive. If strong ammonium chloride is used, the non-explosive diamino-aurichloride, $AuCl(NH_2)_2$, is formed. By heating the compound $2Au(OH)_3.3NH_3$—the product of hydrolysis mentioned above—either alone or with water, other complex explosive compounds are formed. The explosiveness is probably due to the tendency of the hydrogen and oxygen to combine being greater than the attraction of the nitrogen and the metal for these elements. Fulminating gold is decomposed without explosion by sulphuretted hydrogen, and by stannous chloride. It is a grey or buff-coloured powder, insoluble in water, but soluble in potassium cyanide, auricyanide of potassium being formed.

Sulphites of Gold.—Alkali sulphites, or sulphur dioxide, which reduce gold trichloride easily, do not produce the same effect on a solution of an alkali aurate. If sodium bisulphite is added to a boiling solution of sodium aurate ($NaAuO_2$) a yellowish precipitate is formed, soluble in excess of sodium bisulphite, and consisting of a double sulphite of gold and sodium, or *sodium aurosulphite*, having the composition $3Na_2SO_3.Au_2SO_3 + 3H_2O$. It is obtained pure by precipitating the corresponding barium salt with

[1] Lenher, *Econ. Geol.*, 1918, 13, 161.
[2] *Annalen*, 1915, 410, 117.

$BaCl_2$, and decomposing the precipitate with the minimum quantity of sodium carbonate. Double sulphites of potassium and ammonium with gold also exist. These salts are decomposed by acids, sulphite of gold being deposited, and also on boiling their aqueous solutions, but the addition of sulphuretted hydrogen or alkali sulphides has no effect on them.

Thiosulphates of Gold.—The extraction of gold from auriferous silver ores, when these were treated by "hyposulphites," depended on the formation of these compounds. The soluble double thiosulphates of gold with the alkalies and alkaline earths have the general formula $3R''S_2O_3.Au_2S_2O_3 + xH_2O$. The double compounds of gold with sodium, potassium, calcium, magnesium, and barium are all known. The sodium salt is prepared by adding a dilute solution of gold trichloride little by little to a concentrated solution of sodium thiosulphate, when the following reaction occurs:—

$$8Na_2S_2O_3 + 2AuCl_3 = Au_2S_2O_3.3Na_2S_2O_3 + 2Na_2S_4O_6 + 6NaCl$$

The double thiosulphate may be separated by precipitation with strong alcohol, with which it is also washed, or it may be purified by repeated solution in water and precipitation with alcohol. Thus prepared, it consists of colourless crystalline needles, highly soluble in water, but almost insoluble in alcohol. The solution, which possesses a sweetish taste, decomposes under the influence of heat, the action being much more rapid when nitric acid is present; metallic gold and sodium sulphate are formed. Gold, however, is not reduced from its solution as double thiosulphate by either stannous chloride, ferrous sulphate or oxalic acid, although sulphuretted hydrogen and alkali sulphides give a black precipitate of Au_2S_3. The addition of hydrochloric acid or of dilute sulphuric acid does not immediately cause an evolution of sulphur dioxide and a deposit of sulphur, as in the case of ordinary thiosulphates. Since, therefore, the double sulphite of soda and gold does not present the characteristics of either aurous salts or of thiosulphuric acid, it has been suggested that it contains a compound radical and has a composition expressed by either $Na_3S_4O_6Au$ or $Na_3S_4O_6Au + 2H_2O$. Precise details for the preparation of this compound are given by Gjaldback[1] and Brown.[2] The addition of any dilute acid soon effects the decomposition of this body in solution, gold sulphide being precipitated; the reaction is accelerated by heat. This double thiosulphate exists in combined fixing and toning photographic baths.

The double thiosulphates of gold with potassium, calcium, barium or magnesium present similar characteristics. If the barium salt is treated with the amount of sulphuric acid required by theory, a solution of the acid aurothiosulphate, $3H_2S_2O_3.Au_2S_2O_3$, is obtained, but it cannot be crystallised.

Finely divided gold is said to be soluble to a limited extent (*i.e.* 0·002 gram in 1,000 c.c. in 48 hours) in solutions of sodium thiosulphate of all degrees of concentration. The action depends on the oxidation of the gold by the air present in the solution, the soluble double thiosulphate, $Au_2S_2O_3.3Na_2S_2O_3 + xH_2O$, and caustic soda being formed.

The formation of this thiosulphate by the action of the sodium salt on gold sulphide is far more complete and rapid. In twenty-four hours at 15° C., 0·066 gram of gold, and in two hours at 65°, 0·117 gram of gold, was dissolved in dilute solutions.

---

[1] *Brit. Chem. Abs.*, 1927, **A**, 324.
[2] *J. Amer. Chem. Soc.*, 1927, **49**, 958.

The complex auro-thiosulphate compounds are of use in the treatment of tubercular diseases.

Gold Carbide, Au_2C_2, is formed by passing acetylene into aurous thiosulphate. It is explosive when dry, and is decomposed by hydrochloric acid forming AuCl and acetylene. When treated with water, it yields gold and carbon.[1]

Gold Chromate can be obtained as crystals by treating silver chromate with auric chloride. Sodium aurochromate containing excess of chromate is obtained by mixing solutions of sodium aurate and chromate. An excellent photographic toning bath is formed, giving purple to bluish tones.[2]

Gold Selenate is formed by dissolving gold in selenic acid. It forms small yellow crystals, insoluble in water but soluble in sulphuric, nitric, or hot selenic acid. By the action of hydrochloric acid on it chlorine is evolved and auric chloride and selenious acid are produced. It is decomposed by heat with the production of metallic gold. On exposure to light it becomes dark green and afterwards bronze-coloured.[3]

Other Compounds of Gold.—Arsenates, alkyl gold chlorides, mercaptides, and other complex organic compounds have been prepared.[4] Certain thio-organic compounds of gold are soluble, and can be employed in the ceramic, enamel, and glass industries for the deposition of the finest layers of bright metal on various substances.[5] Textile fabrics, printed with a gold salt and then treated with a reducing agent, assume a beautiful grey colour. When the grey fabrics are subjected to heat between rollers, red, purple, or pink colours are obtained.[6]

Silicates of Gold.—Gold has for centuries been used to impart colour to glasses, the method used being as follows:—A solution of chloride of gold is added to a mixture of sand with alkalies and alkaline earths or lead, and the whole is then fused, and colourless or yellow transparent silicates of gold thus formed. These are decomposed by being reheated gently to low redness, oxides of gold, or more probably metallic gold, being set free, and red or purple colorations thus obtained. The occurrence of silicates of gold in nature seems to be doubtful.

Experiments conducted by E. Cumenge[7] tend to show that the alkali auro-silicates, obtained in the wet way, may have played an important part in the formation of auriferous quartz.

Sulphide of Gold.—Gold sulphide is prepared as a brown or black precipitate by passing sulphuretted hydrogen through a solution of gold chloride. Gutbier and Dürrwächter[8] affirm that they found no evidence of the compounds Au_2S and AuS_2. By passing a rapid stream of H_2S into a 2 per cent. solution of chlorauric acid in hydrochloric acid a black precipitate consisting of Au_2S_3 and metallic gold was obtained. The relative amounts of these products depended on the temperature, concentration of the chlorauric acid and the rate of passing in of the H_2S. At low temperatures

---

[1] Mathews and Watters, *J. Amer. Chem. Soc.*, 1900, **22**, 108.

[2] *J. Soc. Chem. Ind.*, 1900, **19**, 1038; Thorpe, "*Dictionary of Applied Chemistry,*" 1922, iii, 459.

[3] Lenher, *J. Amer. Chem. Soc.*, 1902, **24**, 355.

[4] For references to the original papers, see Thorpe, "*Dictionary of Applied Chemistry,*" 1922, iii, 459.

[5] Pertsch, "*Friedländer's Fortsch. d. Teerfarbenfabrikation,*" 1894-97, 1324.

[6] Odenheimer, *J. Soc. Chem. Ind.*, 1892, **11**, 600.

[7] Frémy, *Ency. Chim.*, "L'or," vol. iii., p. 62.

[8] *Zeitsch. anorg. Chem.*, 1922, **121**, 266.

(−2° C.), a low concentration of chlorauric acid and with a very rapid stream of gas, the pure sulphide was obtained. Auric sulphide reacts more readily with auric chloride than with the chlor-acid.

Auric sulphide is soluble to some extent in a saturated solution of sulphuretted hydrogen, and is easily soluble in hot solutions of alkali sulphides or alkalies, or alkali sulphites, forming double salts, so that precipitation from alkaline solutions is never complete. It is readily decomposed into gold and sulphur by the action of heat, the decomposition being complete at about 200° C. Sulphide of gold is also dissolved at ordinary temperatures by potassium cyanide, and is slowly attacked by mercury with formation of mercury sulphide.

When finely divided gold is heated with sulphur and potassium carbonate, a double sulphide of potassium and gold is obtained, which resists a red heat and is soluble in water. It is used for the production of Burgos lustre in gilding china.[1]

Purple of Cassius.—This body was discovered by Cassius of Leyden in 1683. It contains gold and oxide of tin, and is used to colour artificial gems, porcelain, enamel, glass, and glazes, various shades of violet, red, and purple being thus obtainable. Several methods of preparation are used, of which the following is that employed at the factory at Sèvres [2] :—Half a gram of gold is dissolved in aqua regia composed of 16·8 grams of hydrochloric and 10·2 grams of nitric acid, and the solution is then diluted with 14 litres of water. To this solution is added, drop by drop, a solution of a mixture of dichloride and tetrachloride of tin, prepared as follows :—Three grams of finely divided tin are dropped, little by little, into 18 grams of aqua regia (constituted as above, with the addition of 5 c.c. water), the reaction is checked by cooling if it is too violent, and the solution of chloride of tin formed is allowed to cool. The precipitate of purple oxide thus obtained is finely coloured when it has been washed with boiling water. The purple precipitate obtained by Müller, by reducing chloride of gold with glucose in an alkaline solution containing tin oxide in suspension, and by various other methods not involving the use of tin,[3] differs from that prepared by the foregoing method in losing its colour at a red heat, while the true Purple of Cassius becomes brick red under such conditions. The true colour is seen when metallic tin acts on trichloride of gold, or when alloys of gold, silver, and tin are attacked with nitric acid. An alloy containing gold 2 parts, tin 3·5 parts, silver 15 parts is suitable.[4] Moissan obtained a finely divided mixture of stannic oxide, lime and gold, having the colour and properties of Purple of Cassius, by distilling gold-tin alloys in an electric furnace made of lime.[5]

Purple of Cassius when dry is insoluble in alkalies, but when moist it dissolves in water, very dilute acids and alkalies, and ammonia. The solution precipitates gold on exposure to light. When dry, no gold is removed from the purple by mercury. It loses its colour when heated, and becomes brick-red, but without evolution of oxygen.

The constitution of Purple of Cassius, of which the composition, by analysis, is variable, has been the subject of much discussion, but has not

---

[1] Thorpe, "*Dictionary of Applied Chemistry*," 1922, iii, 458.
[2] Frémy's *Ency. Chim.*, "L'or," vol. v., p. 63.
[3] Müller, *J. prakt. Chem.*, 1885, **30**, [ii.], 252.
[4] See also Schneider, *Zeitsch. anorg. Chem.*, 1893, **5**, 80.
[5] *Compt. rend.*, 1905, **141**, 977.

yet been finally determined. Some chemists have considered it to be a compound containing hydroxides of gold and tin. Debray regarded it as a lake of stannic acid coloured by finely divided gold. Müller confirms Debray's views, showing that fine purple compounds can be made with gold and magnesia, lime, baryta, sulphate of barium, etc., the colour depending on the presence of finely divided gold and not on the other constituent. Schneider [1] considered that Purple of Cassius, at any rate in its soluble form, is a mixture of the hydrosols of gold and stannic acid, and Zsigmondy [2] regarded it as a mixture of colloidal gold and colloidal stannic acid. This investigator prepared a red solution of metallic gold in water by mixing formaldehyde rapidly with a feebly alkaline boiling solution of gold chloride. A solution of about 0·005 gram of gold in 100 c.c. of water was thus obtained and concentrated by dialysis, until an intensely red solution containing 0·12 per cent. of colloidal gold was produced. Neutral salts, mineral acids, and alkalies precipitate the gold and an excess of alcohol changes the colour of the solution to dark violet, completely precipitating the metal, which retains the property of dissolving in water. If shaken up with mercury, these solutions are rapidly decolourised. The gold is also carried to the bottom by freshly precipitated lead sulphate and other precipitates. Gold purple of required composition and shade may be obtained by mixing solutions of colloidal gold and colloidal stannic acid and adding dilute acids or salt solutions. Precipitated Purple of Cassius prepared by Zsigmondy contained, after washing and ignition, 40·3 per cent. of gold and 59·7 per cent. of stannic acid.

Colloidal Gold.—Colloidal gold solutions (see "Faraday's Gold," p. 2) can be readily prepared in many ways, and may be red, blue, violet, or green in colour.[3] An interesting method is by the passage of carbon monoxide gas through a solution of auric chloride containing from 0·002 to 0·05 per cent. of gold.[4] According to Steubing [5] and Gans,[6] the various colours of colloidal gold solutions are due to the difference in form of the particles. The size of the red particles, according to Svedberg [7] and Zsigmondy, varies in different samples from 1 $\mu\mu$ to about 20 to 30 $\mu\mu$. The solutions are decolourised when shaken with animal charcoal, barium sulphate, powdered porcelain, fibres of filter paper, etc., but are protected from this by gum arabic.[8]

If an electric discharge is passed between two electrodes placed above but near to the surface of a solution of gold chloride, hydrosols of gold are produced by the decomposition of the chloride by the reducing gases formed in the vapour above the solution. If the discharge be passed between one electrode and the solution, the chloride is reduced by the hydrogen peroxide thus produced.[9]

Westgren [10] obtained fairly regular curves for a colloidal solution of gold by determining the distribution in horizontal layers under the influence

---

[1] *Zeitsch. anorg. Chem.*, 1893, **5**, 80.
[2] *Liebig's Annalen*, 1898, **301**, 29, 361 ; *J. Chem. Soc.*, 1898, **74**, [ii.], 522, 599.
[3] For details and references, see Thorpe's "*Dictionary of Applied Chemistry.*"
[4] Donau, *J. Chem. Soc.*, 1905, **88**, [ii.], 462.
[5] *Ann. Physik*, 1908, **26**, [iv.], 329.
[6] *J. Chem. Soc.*, 1912, **102**, [ii.], 508.
[7] *Ibid.*, 1909, **96**, [ii.], 645.
[8] Donau, *loc. cit.*
[9] W. Naumoff, *Koll. Zeit.*, 1923, **32**, 95.
[10] *Zeitsch. anorg. Chem.*, 1916, **94**, 193.

of gravity in a very thin rectangular glass cell, and then turning the latter through a right-angle and measuring the vertical distribution after a short interval.

Proteins precipitate colloidal gold according to the size of the charge and the magnitude of the colloid particles. Peptisation subsequent to precipitation is possible only when a certain critical size of particle is not exceeded. Proteins only slightly soluble in water yield strongly reversible precipitates with a gold sol, and also protect the subsequently peptised gel when a neutral salt solution is added.[1]

---

[1] *Chem. Zentr.*, 1926, **97**, ii., 9.

CHAPTER IV.

MODE OF OCCURRENCE AND DISTRIBUTION OF GOLD.

Dissemination of Gold.—Gold is distributed in minute quantities throughout the world. It is almost invariably found in association with silver in the same deposit, though their relative amounts vary considerably from place to place. In any one geological formation, however, there appears to be a general average relationship ; for example :—

	Au : Ag
Silver-lead lodes, Freiburg, Clausthal, . .	1 : 10,000 to 1 : 20,000
Silver lodes, Norway, ...	1 : 500
Comstock lode—gold-silver lodes, . . .	1 : 24
Tonopah, ...	1 : 100
Cripple Creek—gold-silver lodes, . . .	10 : 1

The reason probably lies in the fact that gold and silver have many chemical properties in common, which may be operative in deposition processes.

L. Wagoner [1] found minute quantities of gold and silver in a number of rocks taken from localities remote from veins or regions known to contain valuable minerals. Table XVII gives his results, which are assays made by cyanide, and do not pretend to give the exact values of the rocks, but only the amounts extracted. By calcining and grinding, one of the samples showed 20 per cent. more. The richest specimen contains over 17 grains of gold per ton.

J. E. Spurr [2] drew the deduction that igneous and metamorphic rocks contain more gold than sedimentary rocks, and that the order of richness of the sedimentary rocks in gold and silver may, on further investigation, prove to be (1) clays and shales, (2) sandstones, (3) limestones. The gold is no doubt derived in these cases from the sea, and is concentrated where much organic matter is present. It was found many years ago at the Royal Mint that coal from the North of England contained gold, and gold is also found in coal of Cambrian age in Wyoming. A very rich deposit occurs in a bed of lignite in Japan.[3]

F. C. Lincoln [4] gives many records of the occurrence of "primary" gold (as distinct from gold subsequently deposited by infiltration, etc.), in rocks and waters. The amounts for igneous rocks range from 0 to 5 grams per metric ton, the average gold content being 0·062 part per million, or 1 grain per ton. In sedimentary rocks, he estimates the average to be

[1] *Trans. Amer. Inst. Mng. Eng.*, 1901, **31**, 798.
[2] *Ibid.*, 1902, **33**, 288.
[3] Gowland, "*Non-Ferrous Metals*," 1930, p. 254.
[4] *Economic Geology*, 1911, **6**, 247 ; *Mining Mag.*, 1911, **5**, 71.

0·015 part per million, and in sea water 0·028 part per million, or nearly half a grain per ton.

TABLE XVII.

	Parts per Million.	
	Gold.	Silver.
1. Granite, California,	0·104	7·66
2. ,, ,,	0·137	1·22
3. ,, ,,	0·115	0·94
4. ,, Nevada,	1·130	5·59
5. Syenite, ,,	0·720	15·43
6. Sandstone, California,	0·039	0·54
7. ,, ,,	0·024	0·45
8. ,, ,,	0·021	0·32
9. Marble, ,,	0·005	0·212
10. ,, Carrara,	0·0086	0·201
11. Basalt, California,	0·026	0·547
12. Diabase, ,,	0·076	7·440

Gold in Sea Water.—The discovery of the occurrence of gold in solution in sea water was predicted by Percy,[1] and made by Sonstadt,[2] who states that it is far less than 1 grain per ton. Liversedge subsequently showed[3] that the amount of gold in sea water off the coast of New South Wales is from 0·5 to 1 grain per ton, or in round numbers from 130 to 260 tons of gold per cubic mile. Dr. Don[4] could not precipitate the gold from sea water either by charcoal, by insoluble sulphides, by metals or by a current of electricity. Wagoner[5] found 0·17 grain of gold per ton of water off San Francisco, and from 0·6 to 3·7 grains per ton of water from the depths of the Atlantic. In Chesapeake Bay, there was only 0·2 grain per ton. A statement has been made that about $\frac{1}{30}$ grain of gold per ton occurs in the water round the English coasts. P. de Wilde[6] found from 0 to 64 milligrams per ton of sea water, the water from the North Sea containing none.

Haber has made recent determinations as follows:—

South Atlantic, Lat. 48° S.,	0·008 mg. per m.³ (highest single value 0·044 mg. per m.³)
San Francisco Bay,	0·008 mg. per m.³
Newfoundland Bank,	2·25-8·46 mg. per m.³
Arctic regions,	0·04-0·047 mg. per m.³
,, ,, Surface ice,	4·84 mg. per m.³

[1] Percy and R. Smith, *Phil. Mag.*, 1854, (iv), **7**, 127.

[2] *Chem. News*, 1872, **26**, 159; *ibid.*, 1892, **65**, 131.

[3] *Roy. Soc. of New South Wales*, 1895, **29**, 335; see *Chem. News*, 1896, **74**, 146, 161, 166, 182, and 191.

[4] *Trans. Amer. Inst. Mng. Eng.*, 1897, **27**, 564.

[5] *Loc. cit.* and *Trans. Amer. Inst. Mng. Eng.*, 1907, **38** 704.

[6] "*Archives des sciences physiques et naturelles,*" Genth, 1905, **19**, 559-580; *J. Chem. Soc.*, 1905, **88**, [ii.], 532.

By allowing wood charcoal to adsorb gold chloride derived from sea water, Koch[1] obtained the figures 2·5 to 4 mg. gold per cub. metre.

Friedrich[2] examined many samples of the salt deposits of Germany, the product of crystallisation from ancient seas, and found generally unweighable traces of gold in them. In six samples gold was found in weighable quantities, varying from 3 to 13 milligrams per ton. Liversedge,[3] however, had previously found 2·03 grains of gold per ton of Stassfurt salt, 1·7 grains per ton in rock salt from Cheshire, and larger quantities in kelp and bittern. The form in which gold exists in sea water is unknown, and the problem of its profitable extraction is still unsolved in spite of much patient research.

Gold Ores.—Gold is obtained—(1) From quartz veins (also called lodes, reefs, or leads) in rock formations. In this division may be included replacement-deposits, disseminations in rocks, and also, for example, the marine deposits accumulated in shallow water, such as the conglomerates of the Transvaal. (2) From placers or the alluvial deposits of ancient and modern streams. Modern beach deposits and loose sands or gravels generally, may be included in this section. (3) From deposits worked chiefly for base metals and from which gold is obtained as a by-product. Gold is less easily transported than silver, and compared with copper and zinc is almost stationary.

One of the most striking differences between the ores of gold and those of all other metals lies in the extremely small proportion which the desired material bears to the worthless gangue with which it is accompanied. Occasionally hand specimens in vein stuff are found containing several per cent. of the precious metal, but these are of exceptional occurrence, and have no economic importance. The greater part of the vein gold now being produced is derived from ores containing only about one part of gold in 100,000. Placer deposits are usually much less rich than this; the average amount of gold contained in those now worked does not exceed one part in one million, and in California deposits of gravel with only one part of gold in fifteen millions have proved susceptible of successful treatment by hydraulic mining on a large scale.

(1) *Vein Gold.*—In this case the metal, whenever it is present in visible grains or masses, has sharp angular edges, although it is usually not distinctly crystalline. It frequently penetrates the rock irregularly in various directions, and is completely interwoven with, and attached to, the matrix, usually quartz, so that the metal cannot be separated from the rock without crushing the latter.

Quartz and gold may be deposited together over a considerable range of temperature, hence giving rise to several types of deposits:—(1) Deposition at high temperatures, distinguished by tourmaline, apatite, garnet, biotite and amphibole as typical gangue minerals. (2) Deposition near the surface at temperatures not exceeding 100° C. (3) Deposition at moderate temperatures, 200° to 300° C., associated with development at great depth. The first type is represented by Brazilian deposits, the second by those at Comstock and Tonopah, and the third by those in California and Eastern Australia. In the last variety the average fineness is 800, silver being the main adulterant.

---

[1] *Koll. Zeit.*, 1918, **22**, 1.
[2] *Metallurgie*, 1906, **3**, 627.
[3] *J. Chem. Soc.*, 1897, **71**, 298.

The gold in lodes is sometimes in the form of crystallisations, which are, however, exceedingly rare, although they occur more frequently in placer deposits. Arborescent branching and dendritic masses of crystalline gold are more common than single crystals in both quartz lodes and placer deposits. The crystalline forms met with have already been described, p. 7. In the Transylvanian lodes, gold occurs chiefly in thin sheets or plates, often as much as from half an inch to two or more inches in breadth. Such plates are rarely thicker than a visiting card, and are generally covered with crystalline lines and markings, revealing a distinct geometrical structure. Gold also occurs in wire-like forms, sometimes penetrating crystals of other minerals, such as calcite and dolomite.

It frequently happens that the gold in lodes, etc., is in a state of fine division and is not visible without magnification.

The country rock with which auriferous quartz reefs are associated generally consists of slates or schists, especially hydromica and chloritic slates. Gold also occurs sparingly in similar veins in granite and gneiss. Serpentine sometimes forms the walls of auriferous quartz lodes. Gold also occurs in the midst of rock formations without any obvious connection with quartz veins. Among such rocks are granite, aplite, gneiss, eurite, quartz-trachyte, syenite, andesite, basalt, diorite, gabbro, diabase, schists, porphyry and slate. These deposits seldom pay to work, and in that case can hardly be called gold ores.

Veins of auriferous quartz rarely occur except in association with eruptive rocks, such as dykes of diabase or diorite. So close is this association that we are led to believe that the eruptive rocks are the means by which the gold has been brought up towards the earth's surface, and from thence concentrated by slow aqueous action in the quartz-veins.

Some of these veins belong to a fairly deep-seated type of mineralisation, but on the other hand, other evidence from Comstock and New Zealand shows that gold-quartz deposits have a wide range of depth formation.[1]

J. E. Spurr writes [2]—" Native gold has been found in both basic and acid rocks. It has also been detected in the dark ferro-magnesian silicates of rocks of all degrees of acidity. The commercially-valuable concentrations of gold are generally connected with basalt, gabbro, diorite, phonolite, rhyolite or granite. They occur also in many different forms, as replacement-deposits in limestone, as disseminations in igneous and sedimentary rocks, as contact-deposits near intrusive masses, and in fissure-veins."

Gold-bearing replacement deposits have been found in limestone (Utah), quartzite (Nevada), and porphyry (Colorado, California).

(2) *Placer Gold and Nuggets.*—Placer gold differs from vein gold in that it usually occurs as small scales, pellets or rounded grains, the small pieces being called " gold dust," while large masses or nuggets are usually of a rounded mammillated form. The angularity is inversely proportional to the distance travelled. The crystal structure is often quite coarse, consisting of octahedral crystals with many penetration twins. The outside layers of water-worn nuggets frequently show signs of mechanical work and occasionally of incipient recrystallisation (see p. 9). The roundness of

---

[1] Rastall, "*Geology of Metalliferous Deposits*," 1923, p. 457.

[2] *Trans. Amer. Inst. Mng. Eng.*, 1902, **33**, 288, "Igneous Rocks as related to Occurrence of Ores."

the fragments is taken to prove that abrasion of the gold has been effected by attrition with water and grains of sand. "Quite commonly nuggets contain pieces of quartz, or they are sometimes embedded in quartz, which is always angular and just such as occurs in reefs, and not rounded and water-worn, as would be the case if the gold had accumulated while in the alluvial deposits. Limonite frequently occurs associated with the gold of nuggets, sometimes with, at other times without, quartz." [1]

The best conditions for concentration of gold in auriferous gravels are when the river gradient is moderate, say 30 feet per mile, under balanced conditions of erosion and deposition. Overloading causes irregularity in concentration. The distribution of gold in the gravel bed is usually irregular, except in small streams, the richer parts forming a narrow streak which follows a devious course, not necessarily in the deepest part.

The largest masses of gold yet discovered have been found in auriferous gravel. The "Blanch Barkley" nugget, found in South Australia, weighed 146 lbs., and only 6 ozs. of it were gangue; and one still larger, the "Welcome" nugget, from Ballarat, Victoria, weighed 2,195 ozs., or 183 lbs., being found at a depth of 180 feet. The largest nugget ever found was "The Welcome Stranger," of 2,520 ozs. gross weight, containing 2,284 ozs. of gold. It was discovered in 1869 at Moliagul, Victoria, in a rut cut by a dray wheel, only a few inches below the surface of the ground, and was associated with 68 lbs. of quartz. In all, thirteen nuggets of over 1,000 ozs., and 1,327 of over 20 ozs. in weight are recorded as having been found in Victoria,[2] but many others were not recorded. In Russia a mass of gold was found in 1842 near Miask, weighing 96 lbs. troy.

Minerals Associated with Gold.—The minerals most common in placer deposits are platinum, iridosmine, magnetite, iron pyrite, ilmenite (black sand), zircon, garnets, rutile and baryte; wolfram, scheelite, cassiterite, brookite and diamonds are less common. Diamonds are associated with gold in Brazil, and also occasionally in the Urals and in the United States. In auriferous quartz lodes the mineral most commonly associated with gold is pyrite. The pyrite is in rare cases replaced by pyrrhotite, and is often accompanied by complex sulphides, which may themselves be auriferous, the most important being chalcopyrite.[3] On the Far East Rand pyrrhotite, containing a small percentage of nickel, is more intimately associated with gold than the normal forms of pyrite. Young[4] gives a list of the minerals present in Banket ore. Mispickel is found in a number of auriferous deposits in many parts of the world, notably at Pestarena, in California, the Urals, Brazil, Honduras, and Matabeleland. Auriferous stibnite is also far from uncommon; the gold in it is, however, irregularly distributed. Among the localities in which this is found may be mentioned the Murchison Range, Mashonaland, Armida in New South Wales, and Kremnitz in Hungary. Galena and zinc blende are often found with gold; grey copper, tetradymite and similar sulphides are less commonly associated with it. Tellurides of gold are very widely distributed, occurring especially in Western Australia, at Cripple Creek in Colorado and in Transylvania.

---

[1] E. J. Dunn, *Memoirs of Geol. Survey of Vict.*, No. 12, 1912.
[2] E. J. Dunn, *loc. cit.*
[3] De Launay, "*The World's Gold*," London, 1908, p. 31.
[4] "*The Banket*," Gurney and Jackson, London, 1917.

Gold is associated with selenium at Redjong Lebong in Sumatra. In gossans limonite occurs with gold as commonly as pyrite in unoxidised ores, and other metallic oxides, especially ilmenite and magnetite, although not containing gold, are often found with it. Blanchand[1] records that at Edie Creek, New Guinea, there is enrichment in the drift near exposures of lode where the latter is highly manganiferous and where little or no limonite is present. There must also be mentioned the frequent association of gold with calcite, and to a less extent with dolomite. Among other minerals occurring with gold are tourmaline, molybdenite, pyrolusite, tetradymite, uranium ochre, roscoelite, vanadinite, crocoite, wollastonite, gypsum, zircon (white sand), monazite (yellow sand), and alunite.[2] The recognition of minerals as indicators of the presence of gold has been much discussed, and for particular localities valuable information may be obtained in this way. Speaking generally, however, there is no mineral, except gold itself, which infallibly indicates the presence of gold, and the absence of which denotes the absence of gold.

Gold in Pyrite.—As stated above, the sulphides present in auriferous quartz usually contain gold; the gold in such an ore is usually in part quite free, disseminated through the quartz, in which visible grains of the metal often occur, and in part locked up in the pyrite, whence but little can in general be extracted by mercury. Dr. Don found that in many lodes in Australia, traces of gold at least were present wherever pyrite could be found and absent when no pyrite could be detected.

In seeking to explain the behaviour of gold in pyrite during different treatments, various theories as to the form in which it exists have been propounded. The balance of evidence seems to be in favour of the theory that gold exists in pyrite in the metallic state. Although the metal is generally invisible in undecomposed crystals of pyrite, it becomes visible when such crystals are oxidised either by air and water in nature, by means of nitric acid, or by being roasted or subjected to deflagration with nitre. As a result of such decomposition, particles of bright, lustrous gold, angular and ragged in shape, but of considerable size, often become apparent. These particles may be separated from the oxides of iron by washing; and the use of nitric acid, followed by panning, is frequently resorted to in order to detect gold in pyrite. Moreover, although usually invisible, gold can sometimes be seen in unroasted pyrite. As long ago as the year 1874, Richard Daintree and J. Latta found specimens of cubical pyrite,[3] in which gold could be seen under a microscope gilding the cleavage planes of the crystals. Again, G. Melville-Attwood, on examining crystals of auriferous pyrite from California in 1881,[4] found that the faces of the crystals were gilded in some places, and that here and there little specks or drops of gold occurred, partially imbedded in the pyrite. These films were too thin to be detected by an ordinary lens, so that it did not seem surprising that such impalpable material could not be taken up by mercury. Louis Janin, Jr.,[5] found crystals of pyrite in a porphyritic gangue from the Republic of Colombia, which had

---

[1] *Eng. Min. Journ.*, 1933, **134**, 365.

[2] T. A. Rickard, *Trans. Inst. Mng. and Met.*, 1898, **6**, 194; Louis, "*Handbook of Gold Milling*," p. 7, gives an almost complete list of minerals occurring with gold.

[3] *Proc. Royal Soc. of New South Wales*, 1874, **8**, 25; "Iron Pyrites."

[4] "*Precious Metals in the United States*," 1881, p. 604.

[5] *Mineral Industry*, 1892, **1**, 249.

gold in small globules on their surfaces. It has long been known too that crystals of pyrite are often found adhering to an amalgamated plate, the particles of gold on their surfaces having been amalgamated. It seems likely, in view of all these facts, that most if not all of the gold is in the metallic state. It has been contended that it is disseminated mechanically through the mass of pyrite, but it seems more probable that the interior of the crystal is not auriferous, the deposition of the gold being superficial, so that the enrichment of the pyrite is confined to its crystalline faces and cleavage planes. Strong evidence of the richness of the outside of the crystals of pyrite in the Banket ore of the Transvaal is given by Caldecott.[1]

According to Head [2] gold may occur in pyrite in relatively large fragments, or down to specks 5 microns in diameter (≡ 2,000 mesh) on crystallographic faces. The liberation of the finest particles would involve too costly grinding. In cyanide residues Thurlow's method [3] of roasting to convert the pyrite to oxide and then dissolving the oxide in HCl containing stannous chloride left the gold undissolved; and there were indications of flat lamellar particles, which might have occurred originally on crystallographic planes in the pyrite.

Pyrite is a compact, non-porous mineral, and therefore does not permit solutions to penetrate to any contained gold that is not liberated by a crushing process.

H. L. Smyth, W. Lindgren, and W. J. Sharwood [4] examined a number of specimens, and found that under the microscope the gold could be seen at the surface of the pyrite or deposited in minute cracks in it, but never entirely enclosed within it.

Tellurides.[5]—The tellurides are the only definitely known natural compounds of gold with other elements. They form under widely varying conditions and are not generally present in deposits formed under conditions of high temperature and pressure, although Rastall [6] disputes this. They decompose readily above the water level, the tellurium being partly carried away as soluble compounds and partly fixed as tellurite (TeO_2) and tellurate of iron. Selenium may be present in deposits obtained from ascending waters.

Calaverite is a bronze-yellow gold telluride having a yellowish-grey streak and usually containing 2 or 3 per cent. of silver. It occurs in Western Australia and in certain mines in California and Colorado. The mineral has no cleavage. It is named after Calaveras County, California, where it was first found (Rickard). E. S. Dana gives the analysis by Genth as gold 40·7 to 40·9 per cent., tellurium 55·9 to 56 per cent., silver 3 to 3·5 per cent., corresponding to the formula $(AuAg)Te_2$.[7] The compound $AuTe_2$ can be formed by fusing the two elements together, but not

---

[1] *J. Chem. Met. and Mng. Soc. of S. Africa*, 1903, **4**, 196; *Trans. Inst. Mng. and Met.*, 1904, **14**, 51.

[2] *U.S. Bur. Mines Rep. Invest.*, 3226.

[3] *J. Chem. Met. and Mng. Soc. of S. Africa*, 1932, **32**, 311.

[4] *Mng. and Sci. Press*, 1906, **93**, 58, 226; 1907, **94**, 117.

[5] See T. A. Rickard, *Trans. Inst. Mng. and Met.*, 1899, **8**, 74-78; A. G. Charleton, "*Gold Mining and Milling in W. Australia*," pp. 104-107; Carnot, *Compt. rend.*, 1901, **132**, 1298.

[6] "*Geology of Metalliferous Deposits*," 1923, p. 468.

[7] T. A. Rickard, *Trans. Amer. Inst. Mng. Eng.*, 1900, **30**, 712.

readily in the wet way. The following analyses of calaverite are given by Charleton :—[1]

TABLE XVIII.

	Gold.	Tellurium.	Silver.	Total.
1. From Kalgoorlie (E. Simpson).	41·37	57·27	0·58	99·22
2. From Kalgoorlie (J. C. H. Mingaye).	41·76	56·64	0·80	99·20
3. From Kalgoorlie (P. Krusch).	58·63	37·54	2·06	98·23
4. From Cripple Creek (W. F. Hildebrand).	38·95	57·27	3·21	99·43
5. From Cripple Creek (F. C. Knight).	40·14	56·22	3·63	99·99
6. From Cripple Creek (Genth).	40·70	55·89	3·52	100·11

The densities were as follows :—

No. 1, ... 9·311
No. 2, ... 9·377
No. 6, ... 9·043

No. 3 contained also 0·29 per cent. of copper, 0·09 per cent. of iron, 0·07 per cent. of nickel, 1·13 per cent. of selenium and 0·10 per cent. of sulphur. The names given are those of the analysts.

According to Rickard, calaverite can be most readily distinguished from pyrite by its being easily cut by a knife and by rarely occurring in any other than a massive form.

Krennerite is similar in composition and appearance to calaverite but is orthorhombic. Its specific gravity is higher than that of sylvanite (8·35 against 7·9 to 8·3). It occurs in Transylvania and Cripple Creek.[2]

Sylvanite, called also *graphic tellurium*, is a telluride of gold and silver, supposed to correspond to $(AuAg)Te_2$. It sometimes contains antimony and lead in addition. It is usually steel-grey to silver-white, but is sometimes nearly brass-yellow in colour. The cleavage is perfect in two directions. The arrangement of the crystals sometimes bears a resemblance to writing characters, whence the name graphic. The following analyses are of sylvanite from the Black Hills, South Dakota (I.) and Cripple Creek (II.) [3] respectively :

	Au	Ag	Te	Total
I., ...	7·64	32·39	59·96	99·99
II., ...	5·61	34·23	60·16	100·00

Sylvanite occurs also at Kalgoorlie, in Transylvania, and in Calaveras County, California. Analyses of Kalgoorlie sylvanite prove it to contain gold 25 to 28 per cent., silver 13 to 15 per cent., and tellurium 56 to 60 per cent.[4] T. A. Rickard states that sylvanite is the characteristic telluride of

[1] "*Gold Mining and Milling in W. Australia,*" 1903.
[2] Chester, *Amer. J. of Sci.*, 1898, 5, [iv.], 375.
[3] F. C. Smith, *Trans. Amer. Inst. Mng. Eng.*, 1897, 26, 485.
[4] Charleton, *op. cit.*

Cripple Creek, and of Boulder County, Colorado, just as calaverite is of Kalgoorlie.[1]

Petzite is a telluride of silver, Ag_2Te, in which the silver is partly replaced by gold. A specimen from the Golden Rule Mine, according to Genth, contained tellurium 32·68 per cent., silver 41·86 per cent., and gold 25·60 per cent. It occurs in West Australia, Transylvania, Chili, California, Colorado, and Utah. It is black or steel-grey in colour, and is slightly harder and more brittle than sylvanite. It has no cleavage and its fracture is conchoidal (Rickard). Charleton quotes[2] the following analysis, made by F. W. Grace, of Kalgoorlie petzite:—Gold 24·64, silver 40·47, tellurium 34·60, mercury 0·29 per cent. The specimen had a specific gravity of 9 and a hardness of 2·5 to 3.

Nagyagite or Foliated Tellurium, a sulpho-telluride of gold and lead, with a few per cent. of antimony, is remarkable for being foliated like graphite, which it also resembles in its colour, a blackish lead-grey, and in having a hardness of from 1 to 1·5 only. Its density, however, is above 7. It occurs in Transylvania, and contains, according to one analysis, tellurium 32·2, lead 54·0, gold 9·0 to 13·0 per cent.

Kalgoorlite is an iron-black mineral with sub-conchoidal fracture and occurs at Kalgoorlie, Western Australia, and in Colorado and Transylvania. Its analysis corresponds to the formula $HgAu_2Ag_6Te_6$.[3] The analysis by J. C. H. Mingaye gave gold 20·72, silver 30·98, mercury 10·86, tellurium 37·26, copper 0·05, sulphur 0·13 per cent. The proportion of mercury is very variable. The mineral is massive and brittle, with a brilliant metallic lustre (Charleton).

Coolgardite is a sesquitelluride $(AuAgHg)_2Te_3$, from Kalgoorlie, described by Carnot.[4] According to Spencer[5] and Liveing,[6] kalgoorlite and coolgardite are not true mineral species, but mixtures of coloradoite, HgTe, with the other tellurides. At Kalgoorlie, at the Cripple Creek district, Colorado, in Transylvania, in Boulder County, Colorado, and in many other localities, the value of the ore depends on the tellurides of gold contained in it.

At Cripple Creek, where the deposits are of late Tertiary age, it is the rule to find the tellurides in crystallised form, whereas at Kalgoorlie they are almost without exception in massive form. The auriferous deposits at Kalgoorlie belong to two types: (1) Found in altered igneous rocks, especially quartz diabase, which are gold-bearing wherever they are sufficiently sheared. (2) Found in the calcareous schists, and of lower grade than the first type. The age of the Kalgoorlie deposits is pre-Cambrian.[7] As far as the Cripple Creek area is concerned, Lindgren and Ransome[8] conclude that the vein-forming epoch was brief and that the telluride ores were formed by alkaline solutions originating from deeper igneous masses, and accompanied by exhalations of carbon dioxide and nitrogen. Free gold resulting from the decomposition of primary tellurides in the oxidation zone occurs in four characteristic forms: (1) Lustreless, mustard-coloured, blotches or coatings

---

[1] *Trans. Amer. Inst. Mng. Eng.*, 1900, 30, 712.
[2] *Op. cit.*, p. 107.
[3] Pittman, *Records Geol. Survey, N.S.W.*, 1898, 5, 203.
[4] *Loc. cit.*
[5] *J. Chem. Soc.*, 1903, 84, [ii.], 378; *Mng. Mag. and Pacific Coast Miner*, 1903, 13, 268.
[6] *Eng. and Mng. J.*, 1903, 75, 814.
[7] See also Lindgren, *Mng. and Sci. Press*, 1907, 114, 472.
[8] *U.S. Geol. Surv.*, Prof. Paper, 54, 1906.

(*mustard gold*). (2) Filmy coatings of minute crystals in cracks and crevices (*flake gold*). (3) Larger aggregates of matted crystals (*sponge gold*). (4) Spots, stars and irregular splashes; thin coatings in fissures and cracks.[1] In Western Australia the oxidation zone is usually poorer than the primary zone. The oxidised portions of telluride ores also contain lemon-yellow tellurium-ochre or dioxide of tellurium (Transylvania and Boulder County) which is usually, however, converted into tellurite of iron, discovered by Knight.[2] *Hair* or *moss gold* may be produced artificially by heating gold telluride with silver foil at 500° C.[3] The dull-looking gold from the oxidised ore of Mount Morgan, Queensland, has been recognised as having been derived from tellurides. Mustard gold occurs at Kalgoorlie in yellowish splashes like yellowish clay, and can be distinguished by burnishing (Rickard). The telluride ores of Kalgoorlie and Cripple Creek contain pyrite, zinc blende and tetrahedrite disseminated through them. Among the minerals associated with tellurides of gold, fluorite and roscoelite are especially notable, according to Rickard.[4] Fluorite characterises the telluride ores of Cripple Creek and Boulder County and roscoelite, a brownish-green hydrated mica containing vanadium, is found at Kalgoorlie, and especially in the tellurides of Boulder County. Calcite is characteristic of the Kalgoorlie telluride ores. Sylvanite, petzite, calaverite, and nagyagite occur in Canada.[5]

Composition of Native Gold.—Native gold generally contains more than 99 per cent. of gold and silver taken together, and from 0 to 1 per cent. of copper, iron, etc. The colour becomes paler with increase of silver. The minimum gold content is usually about 600 parts per 1,000. Native gold of high value is usually coloured yellow, but nearly white electrum occurs naturally in a number of localities, and the proportion of silver may exceed half the weight of the mixture. Native silver is usually free from gold.

According to Gowland,[6] placer gold in Japan ranges from 620 to 904 in fineness, vein gold in Japan from 566 to 926, and placer gold in Korea from 764 to 873.

Placer gold is usually finer than that derived from lodes, containing a smaller percentage of silver, and its fineness usually increases with the distance of transportation and with decreasing size of the grains. Fisher[7] suggests that these effects are due to an electrolytic corrosion process which results in silver being removed from the gold-silver alloy, the associated gold then being re-deposited on the surface of the nuggets and flakes as a thin film of fine gold. The average composition of the placer gold formerly obtained in California was gold about 885, silver 112, base metals 3. Australian placer gold averages about 950 fine. In the Wiluna area the fineness is 400 to 580 parts per thousand, the remainder being silver. M'Connell[8] has shown that Klondyke nuggets were 60 to 70 parts per thousand finer on the outside than in the inside. Yukon gold dust varies from 600 to 965 fine, but in any particular locality the coarser the particles the higher the assay value.[9]

[1] Beyschlag, Vogt and Krusch, "*Ore Deposits*," 1916, vol. ii., p. 594.
[2] T. A. Rickard, *Trans. Amer. Inst. Mng. Eng.*, 1900, **30**, 708.
[3] Beutell, *Zentralblatt für Mineralogie*, 1919, 14.
[4] Rickard, *loc. cit.*
[5] Cairnes, *Can. Mng. J.*, 1911, **32**, 215.
[6] "*Metallurgy of the Non-ferrous Metals*," 1930, p. 255.
[7] *Trans. Inst. Min. Met.*, 1935, **44**, 337.
[8] *Geol. Survey of Canada*, 1907, p. 979.
[9] H. I. Ellis, *Eng. and Mng. J.*, 1914, **98**, 817.

Gold is occasionally found alloyed with copper, and sometimes also with iron, bismuth, lead, mercury, tin, antimony, palladium, or rhodium. Rhodium gold from Mexico was found to be of the specific gravity 15·5 to 16·8 and contained 34 to 43 per cent. of rhodium. The native alloy of palladium, gold and silver from Porpez contains 85·98 per cent. of gold, 9·85 palladium and 4·17 silver. A native amalgam of gold is found in small yellowish crystals of specific gravity 15·47 in the native mercury of Mariposa in California, containing gold 39·02 to 41·63 per cent., and the rest mercury. An amalgam of gold and silver, found in small soft white grains at Choco, New Granada, contains 38·39 per cent. of gold, 5·00 of silver and 57·40 of mercury.[1] *Maldonite*, from Maldon, Victoria, contains gold 64·5, bismuth 35·5 (Louis).

Geographical Distribution of Gold.—All the districts from which the ancients derived their gold have long since ceased to yield any appreciable amount of the metal, and the days of activity have been almost forgotten.

Gold is found in some of the streams of Cornwall, in lodes and river gravels near Dolgelly and in other parts of Wales, in Sutherlandshire, near Leadhills in Scotland, and in the County of Wicklow. On the Continent of Europe, gold is most abundant in Hungary and Transylvania, where the gold occurs in quartz lodes contained in eruptive rocks of Tertiary age, chiefly propylite, porphyry, diorite and granite. The minerals occurring with the gold are galena, blende and pyrite. Gold mines have long been worked in Tyrol. In Germany the gold obtained is chiefly derived from the smelting of argentiferous galena, in which small quantities of the more precious metal are contained. In Italy the only important mines are those of Pestarena and Val Toppa in North Piedmont, near Monte Rosa. Gold is also found in the sands of the Rhine, the Reuss, the Aar, and other rivers, and in small quantities in Sweden and Finland. A little gold comes from Spain (Corunna, etc.), and Serbia.

The gold-bearing districts of Russia are (1) the Urals, (2) Eastern and Western Siberia, whilst an insignificant amount is also derived from Finland and from the Caucasus.

In India gold is produced in the Kolar goldfield, Mysore, in Southern India. A little gold also comes from the Presidencies of Madras and Bombay. In China, Korea, and Japan considerable quantities of gold are produced. Some gold is obtained from the auriferous sand of Bokhara. Among other gold-producing districts of Asia and the adjoining islands may be mentioned Annam and other countries in French China, the Malay Peninsula, the islands of Borneo, Celebes, Sumatra, Mindanao in the Philippines and New Guinea.

In the U.S A. the chief producing States are California, Colorado, Dakota, Montana, Alaska, Arizona, and Utah, but smaller amounts come from many other States. Canada (Yukon Territory, British Columbia, Nova Scotia and Ontario) produces large quantities of gold, and gold ores are also found in various parts of Mexico, Colombia, Bolivia, Chili, Venezuela, Brazil, Peru, and the small States of Central America. The production of several of these countries was formerly much larger than it is at the present day, the reduction being especially marked in the cases of Brazil and Venezuela.

Gold is somewhat widely distributed in Africa, the chief sources of production in former times being the placer deposits of the Gold Coast and Abyssinia. The discoveries of auriferous conglomerates in the Transvaal

[1] Rammelsberg, "*Mineralchemie*," p. 10.

since 1884 have converted that region into the most important gold-producing country. Rhodesia and West Africa are large producers, and some gold comes from Egypt, Madagascar and the Soudan.

Gold is found in Australia, Tasmania and New Zealand. In Western Australia, Kalgoorlie or East Coolgardie is by far the richest goldfield.

Geological Age of Gold Deposits.—Deposits of gold occur in the formations of every geological age and in most countries of the world, although it may be said generally that the most important zones of gold production in the world to-day are in pre-Cambrian rocks.[1] At each centre of gold mining the conditions are different from those existing elsewhere, and even within the narrow limits of one particular district the strata bearing the precious metal are often of very divergent geological and mineralogical constitution.

Rickard [2] gives the following table, showing the varied distribution of gold, as illustrated by the principal mining districts of the world :—

TABLE XIX.

Period.	Rock.	District.	Region.
Quaternary.	Andesite.	Monte Christo.	Washington.
Tertiary.	Eruptive.	Cripple Creek.	Colorado.
Cretaceous.	Sandstone.	Verespatak.	Transylvania.
Jurassic.	Amphibolite Schist.	Mauposa.	California.
Triassic.	Limestone.	Raibl.	Carinthia.
Permian.	Conglomerate.	Stupna.	Bohemia.
Carboniferous.	Shale.	Gympie.	Queensland.
Devonian.	Conglomerate.	Witwatersrand.	Transvaal.
Silurian.	Slate and Sandstone.	Bendigo.	Victoria.
Cambrian.	Slate and Quartzite.	Waverly.	Nova Scotia.
Algonkian.	Schist.	Homestake.	South Dakota.
Archæan.	Granite and Schist.	Lake of the Woods.	Ontario.

The prominent goldfields are associated mainly with the intermediate igneous rocks and with granites and tonalites. The auriferous Archæan schists are principally derived from molten magmas, and the ancient veins have shared the metamorphism of these schists. Gold is often found in association with andesitic rocks of Tertiary age, but the precise connection between them is a question to which no direct answer has been given. No important goldfield now being worked occurs in sedimentary rocks of age more recent than Cretaceous, although there are many igneous formations belonging to the Tertiary period. Generally speaking, the sequence of the gold-bearing strata throughout the ages is complete.

Nevertheless, some generalisations may be made by a comparison of the petrological characters of formations on various continents, and sometimes, though not in every case, by observations upon the contained fossils. This latter mode of correlation is not applicable in the case of Archæan rocks, which are unfossiliferous.

T. Lindsley [3] classifies gold deposits in two broad divisions—(1) Tertiary deposits; (2) deep-seated deposits. He discusses both classes from the

[1] Mackintosh Bell, *Can. Min. J.*, 1931, 52, 246.
[2] *J. Chem. Met. and Mng. Soc. of S. Africa*, 1907, 7, 311.
[3] *Eng. and Mng. J.*, 1914, 97, 1043.

point of view of persistence in depth. The Tertiary formations exhibit a depreciation of primary precious metal contents with depth, while in the deep-seated type there is a much less marked decrease.

Malcolm Maclaren[1] divided gold-bearing areas into Primary and Secondary. By a Primary auriferous province he signifies " one which shows no prior state of combination and has had no former locus in space." Such are the auriferous sulphides and many free gold deposits. A Secondary province is " one which has been derived from some earlier deposits, either from sulphide or telluride ores or from gold quartz veins." The following table shows Maclaren's[2] general classification :—

CLASSIFICATION OF AURIFEROUS DEPOSITS.

Primary.

Connected with extrusion of intermediate or basic igneous rocks (andesites or diabases)	Archæan	Schistose rocks: West Australian (Kalgoorlie, etc.), India (Kolar) (Hutti), Rhodesia.
	Pre-Cambrian	Arising from intrusion of diabase and diorite dykes through Archæan schists: West Australia, India (Dharwar), South Africa (Pilgrim's Rest, Witwatersrand, and Barberton), Guianas, Appalachian Fields, and Eastern Canada.
	Tertiary	Andesitic goldfields: Northern Chili, Peru, Colombia, Mexico, California (Bodie), Nevada, Utah, Colorado, Unalaska, Japan, Sumatra, Celebes, New Zealand, Transylvania.
Connected with extrusion of acid rocks of granodioritic type	(Palæozoic)	—Urals.
	Permo-Carboniferous	—Eastern Australia and Tasmania.
	Jurassic	—Western North America: Alaska, Oregon, and California.

Secondary.

Deposits produced or modified by chemical agencies	Free gold in original sulphide and telluride veins	*a.* Arising from decomposition of auriferous sulphides and tellurides by acid waters or by other tellurides below the zone of oxidation.
		β. Arising from decomposition of sulphides and tellurides in zone of oxidation.
	Placer gold	—In part.
Deposits produced by mechanical agencies	Placer gold in part.	

A. **Primary Deposits connected with Extrusion of Basic Igneous Rocks** :—

(1) *Archæan.*—All the members in the Archæan group are schistose in nature, although it is often difficult to determine whether they have originated under sedimentary or igneous conditions. It is in the igneous amphibolites of this period that the oldest known forms of auriferous deposits occur. The rocks are unfossiliferous, and can only be correlated from observations

[1] "*Gold*," p. 42.
[2] "*Gold*," 1908, p. 44. The table is reproduced by permission of the *Mining Journal.*

on their petrological characteristics. The best defined group of Archæan rocks is found in countries bordering on the Indian Ocean. In the Lake Superior district of Canada similar formations are associated with large copper deposits.

In India a complex series of Archæan schists and highly metamorphosed rocks called Dharwars [1] occurs. It consists of boulder beds or conglomerates, pebbly grits, quartzites, limestones, argillites and chloritic schists. They stretch from Bombay through Mysore, and are apparently sedimentary. In some places their structure suggests formation in running water. The most characteristic rock in the whole series, and one which is observed in all the regions associated with Archæan gold deposits, is a well-banded contorted hæmatite and magnetite quartz rock. In the Dharwar rocks themselves there appear to have been two periods of vein formation and gold deposition, the older of which is said to have been one of dynamic metamorphism. The rocks have been subjected to great strain. Gold is found in the interior of the lumps of ore, and this fact has been advanced as evidence of the simultaneous deposition of gold and quartz. In some cases the hornblende schist has been penetrated by intrusive granites, and here the walls of the dyke have become changed into diorite. This would seem to have an important bearing on the occurrence of gold in seemingly metamorphosed igneous rocks.

In West Australia there are Archæan rocks existing under similar conditions to those found in India. Gneissoid granites overlaid by greenstone schists (amphibolites and hornblende) are the primary rocks of the country. The characteristic magnetite is present and runs parallel with the foliation and direction of the main schistose belts. Two distinct types of formation are noticeable in this district—(1) Lode formation, as at the main foci of gold ores in the State—viz. Kalgoorlie, Kanowna and Peak Hill: the auriferous rock runs imperceptibly into the barren country rock, and the limits can only be determined by assay values; (2) Quartz veins of the blue and white varieties.

Gold-bearing Archæan rocks are not so common in South Africa as in other parts of the world. The economically important gold deposits occur in rocks of more recent date. The Barberton series in Swaziland may be correlated with the Archæan schists in Australia and India, which have been described already. They consist of chlorite and talc schists, argillites and, finally, the characteristic hard, banded hæmatite. The latter withstands denudation more than the other beds, and thus stands out in relief across the country.

In America, the Appalachian, South Dakota and Brazilian fields belong to the Archæan period. The sedimentary rocks in the first area consist of the Talladega series of slates, quartzites, conglomerates and dolomites, while the igneous rocks are complex green schists, diorites and gneisses. The Homestake Mine in Dakota lies in a field of highly metamorphosed Archæan schists impregnated with auriferous pyrite, and containing many lenticular masses of gold-bearing quartz. Some doubt has arisen as to the true age of the rocks in Brazil, but they are more usually referred to as Archæan. There is no mention of igneous intrusions in the Brazilian formations. The gold is in lenticular masses or in auriferous lines of iron ore in itaberite.

(2) *Pre-Cambrian.*—The Pre-Cambrian gold-bearing strata are usually

[1] J. M. Maclaren, *Rec. Geol. Survey, India*, 1906, 34, 96.

found either in, or together with, Archæan rocks. Two important exceptions to this generalisation have been noted at Nullagine in West Australia, and in the Witwatersrand in the Transvaal, where conglomerates occur which are porous and offer but slight resistance to the passage of auriferous solutions.

In India the younger Pre-Cambrian doleritic dykes are found to penetrate the Archæan Dharwar series, and may be correlated with the Cheyair group of the lower Cuddapah system. Where subsequent movements have burst the Archæan fissures, it is possible to find older blue and younger white quartz side by side. The latter often contains graphitic matter, which has probably been taken up from a carboniferous band during its intrusion.

In West Australia there is again the penetration of vertical doleritic dykes into older schists. Acidic dykes occurring in the same district are barren. In the North it is found that where doleritic dykes are cut by other dykes and faults, they become auriferous, and the gold is found for a few feet on either side. In the intersection of white quartz veins with doleritic intrusions there is said to be a close parallel with the Banket reefs of the Rand, where the auriferous conglomerate occurs as lenticular masses and contains gold in white quartz veins, and also interspersed in the matrix. The veins are much richer than the conglomerate.

The theory of the origin of gold on the Rand is discussed later (see p. 99). Doleritic dykes break through the Barberton series, forming banded shoots similar to those in West Australia. The geological age of the Witwatersrand series is not certainly known, though they are reputed to be older than the Devonian rocks of the Cape system, and younger than the Swaziland system.

The American fields in which the gold is present in rocks of Pre-Cambrian age are almost identical with those previously mentioned. In South Carolina the country rock consists of an altered Archæan muscovite schist, which is often metamorphosed for some distance from the intruding Pre-Cambrian dyke. It seems probable that the controlling factor in the formation of the gold deposits in this dyke has been the heat of the igneous magma. In Colorado the hornblende schists have replaced mica schists in many instances. The Klondike region, according to Tyrrell,[1] is underlain by Pre-Cambrian schistose rocks which have been very much folded. Outliers of sedimentary rocks of Eocene age occur round the margin of the district, indicating a depression of the surface at that period. After Eocene times the land was raised and has suffered no subsidence since.

(3) *Tertiary.*—The Tertiary goldfields are andesitic in character, ranging from the Eocene to the Pliocene eras. They are distinctly modern, following the lines of modern volcanic activity, the "Pacific Circle of Fire."

It is characteristic of the Tertiary andesitic goldfields that the andesite tends to become propylite, in which chlorite and epidote have replaced the ferro-magnesian silicates, and quartz, chlorite and epidote have replaced the felspars. Tellurides are also characteristic, and are usually set in a matrix of quartz.

Maclaren considers that Tertiary andesitic goldfields have probably derived their gold from Pre-Cambrian deposits, either by leaching or by fusion. Lindgren[2] is of the opinion that the gold in Tertiary lodes is derived from the leaching of Pre-Cambrian placers by magmatic or meteoric waters, which have been caused to circulate after the intrusion of the andesites.

[1] *Econ. Geol.*, 1907, 2, [iv.], 343

[2] *Econ. Geol.*, 1911, 6, 247.

In America the Tertiary fields pass from Mexico through Western North America, and belong to the Miocene period. At Pachuca the precious metal occurs only in andesitic tuffs, overlying Cretaceous sediments which probably were broken up in the middle Tertiary period. The Comstock and Cripple Creek regions belong to this period. At Cripple Creek, Pre-Cambrian schists are occasionally found in the granite in association with diabasic dykes. Oligocene and Miocene volcanic rocks prevail in the auriferous area, and the earliest of these consist mainly of andesitic breccias, and in a few cases of lavas and tuffs. It is probable that hydrothermal agencies have been active in the region. Phonolitic breccias and dykes succeed the andesitic rocks, and nepheline basalts have been formed by the intrusion of basic magmas into the former. The veins contain fluorite, which is unusual in andesitic rocks. Tellurides occur in the porous breccias.

According to Rickard,[1] the rocks in Colorado are Post-Cretaceous and Pre-Pliocene. Following the Cretaceous period in this district there occurred an interval of volcanic upheaval and a large outpouring of lava forming brecciated rocks in surrounding water, dykes traversing older strata, and volcanoes as at Cripple Creek and Silver Cliff. Volcanic eruption was very vigorous in Eocene times, and continued into the Miocene. Finally, it was succeeded by a period of intense underground thermal activity. So far as is known, no igneous rock associated with auriferous deposits in this district is older than Tertiary.

The belt of vulcanism is continued in Japan and Formosa, the gold areas in which are now not very productive, though they were greatly exploited by the Dutch and Portuguese merchants in the fifteenth and sixteenth centuries.

In New Zealand there is abundant evidence of the junction of two great Pacific axes of folding and faulting in the Hauraki Peninsula. The gold-bearing veins occur in breccias of andesite and dacites, in which propylitisation has been extensive, and which probably form the Upper Eocene beds, the oldest eruptions of Tertiary age in this district. Obscure rocks of Mesozoic and Palæozoic age underlie these formations.

B. **Primary Deposits connected with Extrusion of Acid Rocks.**—These occur in three principal regions—(1) Urals, (2) Western North America, and (3) Eastern Australia, which contain the chief placer gold deposits of the world. They all present a characteristic mingling of igneous rocks. It is probable that in Tertiary times the auriferous areas were elevated, since the gold occurs free and in coarse grains, due to the exposure to washing and sorting agencies.

(1) *Palæozoic.*—The deposits in the Urals are supposed to be of late Palæozoic age.

(2) *Jurassic.*—The Californian belt in Western North America includes Rossland, Oregon and Alaska. Lindgren[2] has shown that there is a close relationship between the veins in this area and the metamorphic series of early Palæozoic-Jurassic ages. They consist of altered slates, sandstones, limestones and quartz porphyrites. The gold quartz veins are closely connected with the grano-dioritic rocks. The Great Mother Lode is one of the richest deposits in this belt.

(3) *Permo-Carboniferous.*—The Eastern Australian fields exhibit intrusions

---

[1] *J. Chem. Met. and Mng. Soc. of S. Africa*, 1907, 7, 311.
[2] *Bull. Geol. Soc. Amer.*, 1895, 6, 225.

which are older than Triassic, and it is probable that the maximum enrichment of the auriferous rocks occurred in the Carboniferous period. In the Snowy River and Mitta Mitta valley, Lower Devonian porphyrites are found, giving evidence of early activity. The gold deposition in the Mount Morgan, Gympie, Lucknow, Wood's Point and Walhalla districts is referred to the Permo-Carboniferous period. The character of the rocks at Mount Morgan is sedimentary, which is, however, an exception to the general rule that usually sedimentary rocks of Ordovician and Silurian age are found in the South, while igneous rocks occur in the North. The latter contain much graphite which was probably assimilated by the magmas from an adjacent Carboniferous formation during intrusion. At Ballarat the origin of the gold is connected with granitic dykes intruded in late Palæozoic times into the vertical Ordovician slates and quartzites. Tertiary basic dykes penetrate through the Silurian slates. It is stated, however, that these are not associated with the deposition of the precious metal, but that the original lode formation began at the time when the neighbouring granite was extruded at the end of the Silurian period, and before the Devonian sediments were formed.

Auriferous quartz veins are visible at the summits of parallel anticlines —saddle reefs—at Bendigo. The country rock consists of Ordovician black clay slates.

C. **Secondary Deposits.**—These are correlated in point of time with the primary deposits from which they have been derived. Subsequent action, however, may not have occurred immediately on the formation of the primary rocks, and processes such as filtration and the chemical action of gases may have proceeded long after the original deposits were laid down. Thus it becomes increasingly difficult to assign secondary rocks to their respective geological periods, and correlation must be based on a study of the primary country rock, which may be found in the vicinity.

Placer Deposits.—According to Maclaren, all the important placers were formed in Tertiary times, and are thus comparatively recent. No placers, the gold of which has been derived from the primary contents of igneous rocks, have proved of much value. A few belong to the Cretaceous and Eocene periods, but the greater proportion has been referred to the Recent, Pliocene and Miocene. The conglomerates of Tallawang, New South Wales, are most probably Permo-Carboniferous. In some instances considerable doubt has arisen as to the true geological horizon of placer deposits. Thus Lindgren [1] describes the conglomerates at Mine Hill, California, as Jurassic, and Dunn [2] places those at Klamath, Oregon, in the Cretaceous period. Fairbanks [3] says that in both cases the gold has been deposited by infiltration of solutions, and that the previous deductions as to geological age are in error. Usually it is found that placers have been formed subsequent to any period of great gold deposition, but it has been pointed out by Lincoln [4] that Cambrian placers are rare, although the Pre-Cambrian period is known to have been one of active lode formation. A well-known example of an ancient placer is the Cambrian basal conglomerate of Black Hills, South Dakota, which covers, unconformably, Pre-Cambrian schists and gold-bearing quartz veins, and is overlain by

[1] *Amer. J. Sci.*, 1894, 48, 275.
[2] *12th Ann. Rep. Cal. State Mineralogist,* 1894, p. 459.
[3] *Eng. and Mng. J.*, 1895, 59, 389.
[4] *Econ. Geol.*, 1911, 6, 247.

quartzite. The conglomerate is 2 to 30 feet thick, and carries rounded gold grains of undoubtedly detrital origin.[1]

Origin of Gold Ores.—The old theory that the quartz of veins was originally in a molten condition and was ejected from below into fissures is no longer maintained. A theory now strongly advocated is that the materials forming the veins have been transported in aqueous solution and precipitated where they occur. In certain cases, superheated vapours may have played a part. The theory of magmatic segregation has also to be considered. One view is that the solutions found their way downwards from above; while the *ascensional* theory and the *lateral secretion* theory have both been advocated. The last-named theory found its principal supporter during many years in Prof. F. von Sandberger, who pointed out that the gangue of many lodes varies in composition if the nature of the rocks through which the lodes pass is changed, and claimed to have proved by analysis that the materials forming vein-stone are derived from the adjacent country rocks. He stated, moreover, that such minerals as augite, hornblende, mica, and olivine, which are essential constituents of crystalline rocks, contain small quantities of the heavy metals occurring in veins.[2] Although Sandberger did not try to detect gold in the silicates, this metal is not likely to be an exception. Prof. A. Stelzner objected to these conclusions, urging that small quantities of the sulphides of the heavy metals were probably mechanically mixed with the crystals of minerals which Sandberger had analysed in the belief that they were pure. Stelzner advocated the retention of the ascensional theory, which alone affords a satisfactory explanation of the difference in composition observable in neighbouring lodes passing through the same rocks, and apparently formed at different periods. The two theories are, however, not contradictory, and perhaps neither need be entirely rejected, the solutions being supposed to pass more or less freely in the plane of the lode after they have been impregnated.[3]

Don has stated [4] that the gold in Australia and New Zealand if present in country rock is invariably contained in pyrite. He found that the amount of pyrite and its richness in gold diminished with increase of distance from an auriferous lode, and the gold soon disappeared. In the various minerals of igneous rocks, augite, hornblende, mica, etc., no gold could be detected. He concluded, therefore, that the gold in these lodes was not derived from the country rock adjacent, but that the latter was impregnated with gold from the lode by solutions rising from a depth greater than that of any of the rocks exposed at the surface. L. Wagoner [5] doubts the accuracy of these statements, as he has found gold in rocks in Western America apparently free from sulphides.

Lenher states [6] that the alkali sulphides play an important part in the transportation of gold, but only in non-oxidising zones, and from these sulphide solutions the metal is precipitated by oxidation or acidification. Chloride solutions deposit their gold on contact with reducing agents such

[1] Lindgren, "*Mineral Deposits*," p. 212.

[2] "*Untersuchungen über Erzgänge*," Wiesbaden, 1882 and 1885. Useful abstracts are given in Phillips' "*Ore Deposits*" and in Le Neve Foster's "*Ore and Stone Mining*."

[3] See also Posepny, "On the Genesis of Ore Deposits," *Trans. Amer. Inst. Mng. Eng.*, 1893, **23**, 197.

[4] "The Genesis of Certain Auriferous Lodes," *Trans. Amer. Inst. Mng. Eng.*, 1897, **27**, 564.

[5] *Trans. Amer. Inst. Mng. Eng.*, 1901, **31**, 798.

[6] *Econ. Geol.*, 1918, **13**, 161.

as pyrite, ferrous minerals or organic material. Low-grade quartz ores consist usually of a bed of quartz containing very finely divided gold, and are generally of remarkably uniform composition. The theories usually expounded do not account for the presence of silver with gold, as the sulphide of silver is insoluble in alkali sulphide solutions. Hence there would appear to be the incongruous necessity of two theories for the simultaneous deposition of the two metals. The presence of silver may be due to the fact that it will precipitate gold owing to the electrochemical relationship, or to the fact that silver oxide is decomposed at lower temperatures than oxide of gold. Hence breakdown of the latter implies that of any associated silver oxide.

From laboratory experiments Lenher deduces that compounds of gold are more resistant to high temperatures under the pressure of expanding steam than at atmospheric pressure. Calcium and magnesium carbonates—both plentiful in rocks and solutions—precipitate the hydroxide of gold from chloride-bearing liquors. Then, if the temperature is raised above 310° C. under the pressure which steam would exert at those temperatures (approximately 100 atmospheres), crystals of metallic gold appear. Magnesium and sodium chlorides tend to depress the decomposition point, while calcium chloride inhibits it. The existence of silver in native gold may be accounted for by the solubility of silver chloride in alkali chloride solutions.

J. E. Spurr [1] traces certain quartz veins directly to igneous magmas as the result of a process of magmatic segregation, the veins being the outward extensions of pegmatite or other igneous dykes. The magmatic concentration is succeeded by aqueous concentration acting underground by dissolution and precipitation, and on the surface chiefly by mechanical means, in both cases producing ore bodies.[2]

The conglomerate gold ores of the Rand have been considered by some authorities to be placer deposits.[3] Gregory contended [4] that the banket was a marine placer in which gold and black sand (magnetite with some titaniferous iron) were laid down in a series of shore deposits. The gold was in minute particles, and it was concentrated by the wash to and fro of the tide, which swept away the light sand and silt, while the gold collected in the sheltered places between the larger pebbles. The black sand deposited with the gold has been converted into pyrite by infiltrating waters, and at the same time the gold was dissolved and redeposited *in situ*. Mellor [5] favours a theory involving the deposition of gold in alluvial gravels as a delta deposit rather than a shore gravel, but recognises dissolution and subsequent deposition. Schwarz [6] considers the placer theory the least hypothetical.

L. C. Graton [7] takes exception to the placer theory of the origin of Rand deposits and contends that the gold was deposited in the conglomerates by hydrothermal solutions emanating at depth from a magmatic reservoir.

---

[1] *Trans. Amer. Inst. Mng. Eng.*, 1902, **33**, 288.

[2] A full discussion of the subject is given in Thomas and Macalister's "*Genesis of Ore Deposits*," and by F. H. Hatch, Presidential Address, *Trans. Inst. Mng. and Met.*, 1914, **23**, xli.

[3] See G. F. Becker, "The Witwatersrand Banket," *18th Ann. Rep. U.S. Geol. Survey*, part v., 1897. E. T. Mellor, *Mng. and Sci. Press*, 1914, **108**, 781.

[4] *Trans. Inst. Mng. and Met.*, 1907, **17**, 2-41 ; also Becker, *loc. cit.*

[5] *Trans. Inst. Mng. and Met.*, 1916, **25**, 226.

[6] *South African Min. Journ.*, 1915, **24**, 469, 491, 536.

[7] *Eng. Min. Journ.*, 1929, **127**, 733.

In support he points out (1) the highly silicified character of the conglomerates, (2) the presence in them of quartz, pyrite, sericite and chlorite, minerals commonly found in gold deposits of magmatic origin, (3) the intimate association of gold with small seed-like bodies of carbon at the bottom of the conglomerates which, owing to the relative specific gravities of gold and carbon, does not fit in with the placer theory.

The infiltration theory of the origin of the gold in the Rand conglomerates has received more general acceptance.[1] According to this view, after the deposition and cementation of the pebble beds, the gold and pyrite have been introduced, dissolved in solutions flowing freely through the banket, which offers less resistance to their passage than the adjacent country rock, the process being thus similar to the mineralisation of ordinary veins. The deposition of pyrite from solution appears to have taken place by replacement of other material, such as silica, which has been removed in solution. The mechanism of the deposition of the gold appears to be less certain.

Lindgren[2] adduces the following arguments against the infiltration theory—(1) the absence of channels formed by the solutions, (2) regular distribution of gold in the conglomerate with sometimes a concentration in the lower layers, (3) practical confinement of gold to the conglomerates, though the quartzites are equally permeable, (4) the conglomerates were deposited in an alluvial plain skirting a schistose area carrying lenticular gold quartz veins and would thus certainly contain some placer gold.

Young[3] abandons the idea of contemporaneous precipitation and infiltration and accepts the placer theory.

A third theory—that of sub-aqueous deposition—postulates that the gravels which were eventually to become the bankets were laid down off the shore of a continent and the gold was dissolved in the waters adjacent. The arguments supporting this theory are mainly hypothetical.

An important suggestion as to the deposition of gold in quartz reefs generally has been made by Hatschek and Simon, based on their work on siliceous gels.[4] They placed in test tubes silicic acid solution mixed with gold chloride, and after the solutions (sols) had become gelatinous (forming jellies or gels), solutions of reducing agents were poured on the top, or the test tube placed in a large vessel containing a reducing gas. Among the reducing agents tried were oxalic acid, ferrous sulphate, sulphur dioxide, carbon monoxide, illuminating gas, graphite, charcoal and crude petroleum. In some cases the reducing agent was diffused in the gel and a solution of gold chloride allowed to come in contact with it. It was found that in a gel consisting of silicic acid in which a gold salt is uniformly distributed a reducing or precipitating agent has the following effects :—

(*a*) If the reducing agent is hypotonic in relation to the salts in the gel (*i.e.* if the concentration reckoned in molecules is less in the reducing agent than in the gel), the gold will leave the gel and will deposit in aggregates,

---

[1] See Hatch, "*Types of Ore Deposits*," San Francisco ; *Mng. and Sci. Press*, 1913, **107**, 1019 ; C. B. Horwood, "The Rand Banket," *Mng. and Sci. Press*, 1913, **107**, 563, 604, 647, etc. Further references are given in the articles mentioned. L. C. Graton, *Econ. Geol.* (*Supp. to No.* 3), 1930, **25**. Beyschlag, Vogt and Krusch, "*Ore Deposits*," 1916, vol. ii., 106.

[2] "*Mineral Deposits*," p. 241.

[3] "*The Banket*," 1917.

[4] *Trans. Inst. Mng. and Met.*, 1912, **21**, 451.

usually of crystalline form, at the area of contact of the reducing agent with the gel.

(*b*) If the reducing agent is hypertonic in relation to the salts in the gel, the gold will be precipitated within the gel, in crystals or reddish-brown amorphous particles, not always uniformly throughout the gel, but often in distinct layers parallel to the surface of contact, the distance between one layer and the next one increasing with the distance from this surface. Carbon also appears to be precipitated by hydrocarbons, recalling the fact that graphite is often present in gold and silver deposits.

The frequent banded structure of auriferous quartz is brought to mind by these results, and the occurrence of gelatinous silica in reefs has been occasionally noted. Chloride solutions frequently occur near the surface, and might penetrate downwards in lodes, carrying gold in solution. Gold can undoubtedly be transported in the form of chloride, which may readily arise from the oxidation of pyrite in the presence of chloride-bearing solutions, especially if manganiferous or other highly oxidised ores are present. These solutions may then penetrate the earth's crust until $FeSO_4$ is again formed, due to the formation of hydrates and reduction in acidity. As pointed out by Sulman, however, [1] solutions rising from below are more likely to contain alkali sulphides in which gold might be dissolved, and the enrichment of siliceous gels in veins might be carried out by the action of such solutions. There is no reason to suppose that the effects observed by Hatschek and Simon are limited to chloride solutions, although these may have played an important part in the secondary enrichment in gold of ore deposits.

Von Veimarn [2] found that gold chloride was slowly reduced in silicic acid without the presence of a special reducing agent. He mixed together dilute solutions of $NaAuCl_4$ and Na_2SiO_3, and observed a gradual change in colour from yellow to colourless, then rose, lilac and blue. He supposed that an unstable silicate of gold was formed and afterwards decomposed with spontaneous reduction of the gold. A precipitate of silicic acid formed after about a year contained disseminated gold in particles too fine to be seen under the microscope, even as ultramicroscopic appearances.

Boydell [3] affirms that laboratory methods for the preparation of gold sols have no counterpart under geological conditions. The concentration of gold sols rarely exceeds 0·008 per cent. and this is only obtained by dialysis. They are extremely sensitive to the presence of electrolytes. Bastin [4] has shown, however, that they may be protected by gelatin, glue and colloidal silica. Dispersion of gold during erosion and transportation will be assisted by the action of any peptising agent present, *e.g.* silica, and in this state transportation of the metal may occur until conditions suitable for its deposition are met. As the concentration of the gold in the colloidal state is small, a very small number of nuclei available for growth will be present.

The deposition of gold by naturally occurring sulphides has been the subject of much research. Among these studies may be mentioned those of A. D. Brokaw [5] and of Chase Palmer and E. S. Bastin,[6] who proved that nearly all the sulphides and arsenides which commonly occur in ore deposits

---

[1] *Op. cit.*, p. 466.
[2] *Zeitsch. Chem. Ind. Colloide*, 1913, **11**, 287 ; *Mng. and Sci. Press*, 1913, **107**, 309.
[3] *Trans. Inst. Min. Met.*, 1924-25, **34**, 188.
[4] *Wash. Acad. Sci.*, 1915.
[5] *Mng. and Sci. Press*, 1913, **107**, 309.
[6] *Loc. cit.* ; *Trans. Amer. Inst. Mng. Eng.*, 1913, **45**, 224.

are capable of reducing gold from a solution of its chlorides, although at different velocities. The apparent preference of gold for chalcopyrite and tetrahedrite rather than for pyrite in deposits carrying these three minerals may be due to differences in their reducing power.

Morris [1] subjected gold compounds to the action of water at high temperatures and pressures in a steel bomb. He found that gold hydroxide in water gives free gold at 322° C. Sodium and magnesium chlorides lower this temperature, whilst calcium chloride raises it. Gold chloride in water does not decompose below 370° C.; whilst in the presence of sodium, calcium or magnesium chloride decomposition is delayed until a temperature of 450°-460° C. is reached owing, probably, to the formation of double chlorides. Calcite and magnesite become plated with gold when heated to 310° C. with gold chloride solutions, but in the presence of other chlorides a temperature of at least 500° C. is required. It is possible that reactions similar to these may have occurred in the earth's crust, resulting in the deposition of gold.

The origin of the gold in *placers* was long a vexed question. One theory advanced at least as early as 1864 was that nuggets were formed by accretion. Gold dissolved in water containing acids and salts was supposed to be redeposited around certain centres in drift gravels previously formed. According to this view nuggets and gold dust might still be growing when found. Some facts appear to support this view. There is the occurrence in placer deposits of drift wood containing gold which has replaced, or been deposited on, the woody fibres. [2] Placer gold is of higher fineness than the lode gold found near it. Some nuggets found in Australia were much larger, it was asserted, than any masses of gold encountered in veins. The conclusion was reached that placer gold could not have been derived from the erosion of lodes. In any case it did not seem to the accretionists to be necessary to assume that the processes of dissolving gold in underground waters and reprecipitating it in lodes must have ended there. [3]

The view is now generally accepted, however, that placer gold has resulted from the erosion of older auriferous deposits, the nuggets being transported entire.[4] The structure of nuggets (see p. 9) gives no support to the theory of accretion. The fineness of the placer gold is explained by the supposition that the silver and base metals formerly contained in the gold have been in part dissolved in natural waters, in which gold is less soluble. This action should make the surface layers of nuggets richer in gold than the interior and nuggets from New Guinea, examined by R. Law,[5] and elsewhere have this characteristic. Law observes that the grains and small nuggets from New Guinea were at least 95·0 per cent. fine on the surface. They were similar in appearance to alluvial gold from other sources. The grains were porous, and in some instances could be torn apart in the fingers. A cut or abraded surface showed a much lower fineness than the outside. Microscopical examination revealed that the gold was crystalline, and in some instances good dendritic and fern-leaf forms were observed. The grains were usually agglomerations of smaller grains, each consisting of a white nucleus and a fine gold exterior. Law suggests that the conditions involved in the production

---

[1] *J. Amer. Chem. Soc.*, 1918, **40**, 917.

[2] Daintree, *Trans. Amer. Inst. Min. Eng.*, 1893, **22**, 312; *Trans. Inst. Min. Met.*, 1912, **21**, 456.

[3] See also Gross, *Jahrbuch d. Radioaktivitat und Elektronik*, 1918, **15**, 280.

[4] Fisher, *Trans. Inst. Min. Met.*, 1935, **44**, 337; "*Mineral Deposits*," M'Graw-Hill, 1933.

[5] R. Law, *57th Ann. Report Royal Mint*, 1926, p. 103.

of these small nuggets were as follows: (*a*) the original gold in the quartz was of low fineness, probably about 56 per cent. fine, (*b*) the gold was detached from the quartz and acted upon superficially by solvents, and (*c*) coalescence of the fine gold surfaces to form larger nuggets was caused by the physical action of running wa er which brought the surfaces into contact.

Fisher [1] suggests an electrolytic corrosion process in which silver is removed and gold is subsequently re-deposited. In support of this view he urges that his micro-examination of auriferous pebbles from the Bulolo area of New Guinea indicates that the gold has been derived mechanically from gold-quartz veins or lodes which have been subjected to denudation. The purity of the alluvial metal increases as the rivers are descended. This is due in part to an increase in the proportion of redeposited fine gold caused by a decrease in the size of the grains, and the more prolonged action of the water on the gold-silver alloy; part is attributed to the entrance at various points of new supplies of gold that are purer than the up-river products.

Alluvial gold from the Yukon similarly has the appearance of being from 900 to 980 fine, but on being melted it is found to be only from 750 to 850 fine.

It is denied by Lindgren [2] that heavier masses of gold occur in placers than in quartz veins and he states that almost all large nuggets have been found in superficial deposits near to the outcroppings of rich veins. Nuggets sometimes have quartz embedded in them.

Emmons [3] holds the view that gold is soluble in superficial water only when the latter contains free chlorine—an unlikely contingency. He states [4] that those deposits in which enrichment in gold has taken place are manganiferous, and as enrichment usually occurs in a downward direction, instead of by superficial removal and accumulation, both gold placers and outcrops rich in gold are generally associated with non-manganiferous minerals. Where chloride solutions, iron sulphides and manganese compounds exist, and where there is no very effective precipitant present, gold placers are rarely formed and outcrops of gold ores are likely to be poorer than deposits that lie deeper. In cases where lodes are fractured gold will migrate downward in solution and the metal may be precipitated by one or other of the minerals it encounters.

Zemczuzny [5] has attempted to produce synthetical nuggets. To him the existence of nuggets of practically constant composition (30-40 per cent. Ag (electrum), and 72-80 per cent. Ag (custellite)) with no intermediate composition up to the pure metals, made their crystallisation from a molten state appear improbable. By using various artificial solutions he found that silver is reduced about seven times more slowly than gold, so that natural solutions, it would seem, should first precipitate crystals high in gold, with practically a non-variant content of silver. He finally concluded that nuggets are formed by the simultaneous reduction of silver-gold solutions by substances acting either directly or catalytically in the presence of primary magmatic crystallites. They are then subjected to deformation by powerful earth movement and eventually disintegrated, to be carried down into the placers.

---

[1] *Trans. Inst. Min. Met.*, 1935, 44, 337. [2] "*Mineral Deposits*," p. 26.
[3] *U.S. Geol. Surv., Bull.* 65, 1917, p. 305.
[4] "*Principles of Economic Geology*," 1918, p. 408.
[5] *J. Russ. Phys. Chem. Soc.*, 1923, 54, 5; *Chem. Abs.*, 1923, 17, 3468.

The gold in the gravels of the Klondike region differs from ordinary placer gold. It is often in large nuggets, including much quartz, and is usually rough and but little water-worn. It is of low standard, and has resulted from the erosion of auriferous rocks by glaciers which have carried the broken materials into the valleys and left them as moraines. In this region the grains of gold are flat, roughly elliptical plates, more or less smooth on both surfaces. This shape would not result from travel of the gold along with the gravel down the creek beds, but rather from the pressing and polishing action of the gravel, flattening out the grains and elongating them in the direction of its passage over them.[1]

Lindgren [2] observes that in all the great placer regions the concentration has generally been preceded by deep secular decay of the surface, and the final result of weathering is a loose, ferruginous detritus, easily washed and containing gold that is recovered simply. This gold occurs as rough irregular grains, and has a fineness but slightly greater than that of the primary vein. To such concentrations Stelzner has applied the term *eluvial*.

Rastall classifies placers as follows: (1) Residual placers (eluvial), resulting from weathering of rock *in situ*, with removal of soluble material, but with little or no mechanical transport. (2) Dry or "Eolian" placers of residual deposits formed by weathering in arid regions, the lighter particles being removed by wind. (3) Stream placers, normal type of alluvial deposit. (4) Marine placers, gravels and sands laid down on the sea shore by waves and currents.

---

[1] "*Report of Supt. of Mines, Ottawa,*" 1902, [vi.], p. 17.
[2] *Loc. cit.*

CHAPTER V.

TREATMENT OF PLACER DEPOSITS.

Introduction.—The deposits grouped together under the name of "placers" comprise sands, gravels, or any loosely coherent or non-coherent detrital beds containing gold. They have accumulated owing to the action of running water, in the beds of rivers, or on the adjoining inundation plains, or on sea beaches. They fall naturally into two groups, between which no strict line of demarcation exists. These are—

(1) Shallow or modern placers, which are usually in or near existing rivers, and have not yet been covered by other deposits. In certain Arctic districts, the gravel beds have been formed by the action of glaciers.

(2) Deep level or ancient placers, which now lie buried beneath an accumulation of debris or coherent rock, the rivers by which they were formed having often been deflected into other channels by more or less extensive changes in the physical geography of the district in which they existed. Beach deposits occur in each subdivision.

Auriferous sands are found in the beds of most rivers which flow during any part of their course through a region composed of crystalline rocks. If the rivers have rocky beds, gold may be found in the crevices, caught in natural riffles, and the whole may subsequently be covered by beds of sand.

The gravels may contain boulders of any size, or may shade off into fine sand, while sandy clays, especially if on the bed-rock, are frequently very rich. In the Urals, the placer deposits often consist of heavy clays, while others are formed of water-worn fragments of auriferous quartz, talcose and chloritic schists, serpentine, greenstone, etc. Gold occurs under very various conditions in these deposits. It may occur in the grass roots on inundation plains, or near the surface of the gravels in river beds, or dispersed through the whole thickness of a stratum. More commonly, however, the lowest part of the superficial beds, just above "bed-rock" (the country rock of the district), is richest. In hollows, cracks, and crevices of the bed-rock, or, if it is soft and decomposed, in the substance of the upper part of the rock itself, gold occurs in the greatest quantities. In pipeclays just above bed-rock, in Victoria, it was not uncommon to find 12 ozs. of gold or more in a single tubful of "dirt," and similar rich bed-rock deposits were found in early days in California. The depth at which bed-rock is found varies greatly; it may crop out at the surface, or it may be buried beneath hundreds of feet of gravel, and great variations occur even in a single district.

Methods of Washing the Gravel.—*The Pan* (see Fig. 34) is usually made of stiff sheet-iron, is flat-bottomed and circular, with a base of 8 to 10 inches in diameter. The sides slope outwards at an angle of about 45° to the bottom, and the depth of the pan is from $2\frac{3}{8}$ to $3\frac{1}{2}$ inches. A riffle is a useful addition, formed by the thickening or bulging inwards of the side, situated about half-way up the latter and running about half-way round the pan. Grooves to act as riffles are also sometimes added in the angle of the pan and on the sides. The inside of the pan is kept smooth and free from rust. The method of using this pan embraces several operations. First, it is filled to about two-thirds of its capacity with pay-dirt, of which it then contains about one-tenth of a cubic foot. It is then placed at the bottom of a water-hole or convenient stream, and the dirt is thoroughly broken up with both hands, care being taken not to leave any lumps of clay. As soon as the contents of the pan are reduced to the consistency of soft mud, the pan is grasped with both hands a little behind its greater diameter, inclined away from the operator, raised until the dirt is only just covered with water, and shaken sideways, while a slight oscillatory circular motion is also imparted to it. The mud and fine sand are soon obtained in suspension in the water, and gradually pass over the far edge, which is lowered more and more, until little but the stones, coarse particles of sand, black sand, and gold is left. The larger stones lie on the top and are removed by hand. The final stage consists in

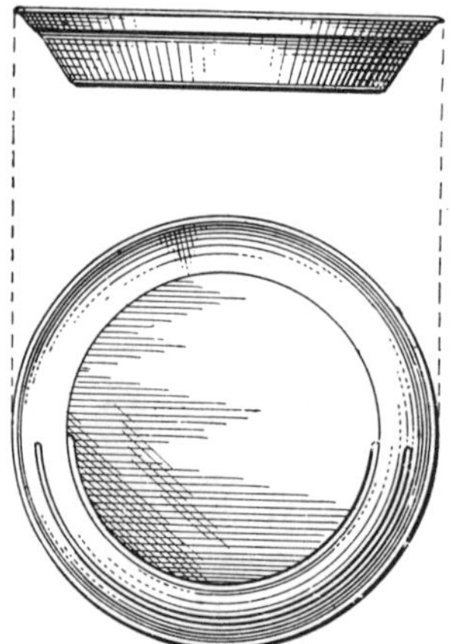

Fig. 34.—Miner's Pan.

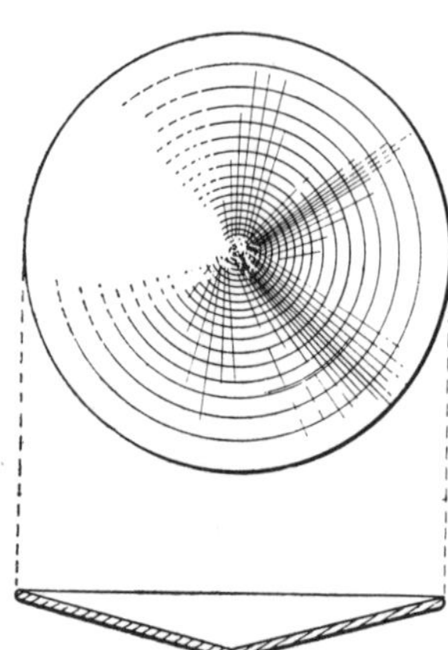

Fig. 35.—Batea.

lifting the pan with a little water in it, and by a movement of the wrist, something like that used in vanning, causing the material to be spread out by the water in a comet shape, in the angle of the pan. The separation of the gold is also sometimes effected by merely running the water round the angle of the pan. The "colours"—*i.e.* yellow specks of gold—are seen at the extreme head of the comet, and also occur in the succeeding inch or two, mixed with the black sand, while the quartz-sand forms the remainder of the tail and is scraped or washed off. The gold is separated from the black sand by (*a*) amalgamation with mercury, or (*b*) drying and blowing away the black sand, a wasteful process. Liquid amalgam is readily separated from sand, and the mercury is then driven off by heat.

The Batea (see Fig. 35) differs from the miner's pan in not having a flat bottom. It is of wood turned in a lathe, about 12 inches in diameter, conical, or more rarely basin-shaped, and about $1\frac{1}{2}$ inches deep in the centre, so that the angle at the apex is about 150°. The gold collects at the lowest point and clings to the wooden surface under conditions when it would slide over iron. The best material for the batea is mahogany cut with the direction of the grain vertical to the surface of the implement. It had its origin in South

America, and is especially favoured by the negro race. It is interesting to note that in some modern gold mills a mechanical batea has been incorporated in the flowsheet to trap residual concentrates.

The Cradle or *Rocker* consists of a rectangular wooden box, about 3 feet long and 18 inches wide, resting on two rockers (D, Fig. 36) similar to those used for infants' cradles. The shape of the walls is shown in Fig. 36, which is a section of the apparatus. The method of using it is as follows :—

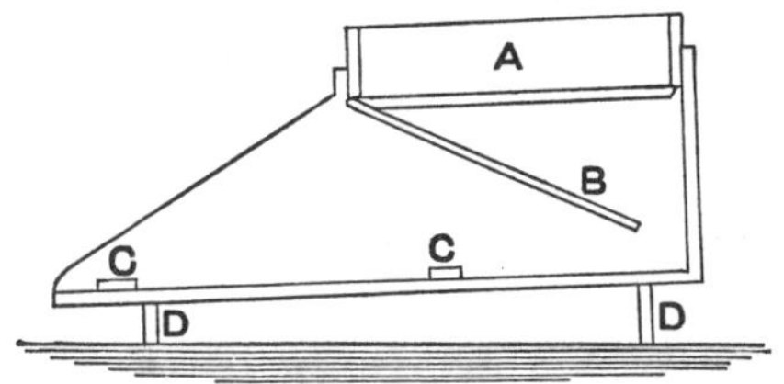

Fig. 36.—Cradle or Rocker.

The gravel is shovelled into the riddle-box, A, the bottom of which consists of $\frac{1}{2}$-inch mesh screen ; the workman rocks it and pours on water. The dirt is disintegrated and carried through the riddle, and falls on the apron, B, which consists of blanketing, canvas or wood. Here some fine gold is caught, and the dirt then passes out from back to front over the bottom, which is slightly inclined towards the front, and the coarse gold, black sand, etc., is caught in two or three riffles, C, consisting of transverse strips of wood each of about 1 inch in height, behind which mercury is sometimes placed to assist in retaining the gold. When a clean-up of the cradle is desirable, the riddle is removed, the apron is taken out and washed in a bucket, and the accumulations behind the riffles are scraped out and panned. Most of the fine gold in the gravel is lost by the cradle. According to Richards[1] the cradle is used in cleaning-up sluices and quartz mills.

The Long-Tom consists of a sluice-box or trough (A, Fig. 37) about

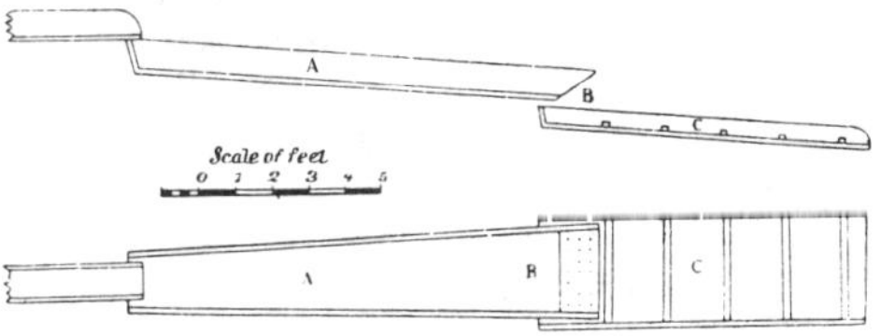

Fig. 37.—Long-Tom.

12 feet long, 20 inches wide at the upper end, and 30 inches at the lower end, and 9 inches deep, with an inclination of about 1 inch to the foot. The lower end of the trough is cut off at an angle of about 45° and closed by a screen of sheet-iron, B, in which a number of half-inch holes are punched, so that the fine dirt is allowed to pass through while the stones are retained. Below the screen is the upper end of the riffle-box, C, which is usually about 9 feet long, 3 feet wide, and at about the same inclination as the upper trough. It is fitted with several riffles, which are sometimes supplied with mercury. In working, a stream of water enters at the upper end of the

[1] "*Ore Dressing*," 1903, p. 723.

sluice-box, into which gravel is continually shovelled. The coarse gold only is caught, and the machine is not suitable for washing other than small quantities of rich dirt where there is a plentiful supply of water. The material caught by the riffles is scraped out occasionally and panned, but the riffle-box is too short for close saving of the gold.

The Puddling-tub.—When water is scarce, the long-tom is inadmissible, and the puddling-tub is resorted to. This is particularly well adapted for washing clays, and is still used to disintegrate lumps of clay encountered in sluicing operations. It consists of one-half of a barrel into which dirt is dumped and stirred up with water by means of a rake, until all the clay is held in suspension, when a plug a few inches from the bottom is removed, and the slime run off. The operation is repeated until the tub is filled with gravel and sand to the level of the plug-hole and this residue is then shovelled out and washed by the pan, the cradle, or by sluicing.

The cradle, long-tom and puddling-tub are now little used except in new districts for a short time.

Appliances used in Siberia for washing auriferous gravel are described by Perret.[1]

The Siberian Trough.[2]—In Siberia, in the Urals and in the valleys of

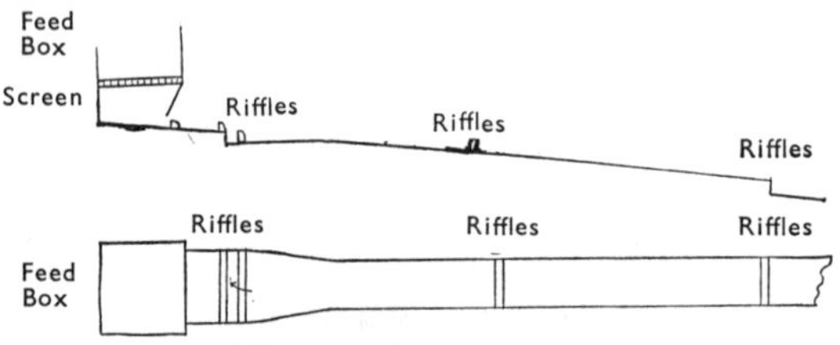

Fig. 38.—Siberian Trough.

the Obi, the Yenisei and the Lena, individual workers still exclusively use a trough, which differs from the long-tom mainly in requiring more constant attention on the part of the operator, and which resembles the old German buddle. The trough consists of a rectangular box open above and at one end.

When sandy gravels are being treated, the bottom of the disintegration box (Fig. 38), which is about 40 inches square, is made of a perforated screen of wood or sheet-iron, having holes of from $\frac{1}{2}$ inch to 1 inch in diameter. The dirt is shovelled into this box, and, contrary to cradle-practice (see p. 107), if the gold is present in fine flakes mercury is added in the feed-box, the amount depending on the richness of the auriferous material as determined by assay, the proportion used, however, being never more than 10 of mercury to 1 of gold. Water is directed upon the charge in the box, either by pipes from a reservoir or more often by pumping, and the fine material is carried through the screen and falls on to the table, while the pebbles are collected by hand and thrown away. If clay is being treated, no screen is used, the lumps are puddled in the box, and the mud carried over by an overflow of water. The table is slightly inclined, about 20 feet long, and, for the greater

---

[1] *Trans. Inst. Min. and Met.*, 1912, **21**, 647.

[2] For further particulars see the account given by Cumenge and Fuchs in Frémy's *Ency. Chim.*, "L'or," part iii., 1st section, 12; and also Levat, "*L'or en Siberie Orientale,*" Paris, 1897.

part of its length, is about 20 inches wide. It is furnished with five riffles, of which two near the top are about 2 inches high, while the others are of less height. The disintegration of the sand is completed on the table with the aid of a small rake continually used by the workman. When disintegration is complete, the stream of water is diminished in amount, and the workman continues to rabble the sands which have accumulated above the riffles, pushing the contents of the lower riffles up the table again, until the water runs clear, and little except pyrite is left behind the riffles where the so-called "grey concentrate" accumulates. This is either concentrated further on the same table, or removed and worked on a smaller table. In either case the stream of water is still further reduced, being graduated so as to carry away the last particles of quartz, together with all materials of moderate weight, such as garnets, rutile, tourmaline, etc., and even all the fine pyrite. If mercury has not been added previously, it is sprinkled on before this last operation, unless the gold is very coarse, when no mercury is added at any stage of the proceedings.

The "black concentrate" obtained behind the riffles consists entirely of amalgam, magnetite, and the large grains of pyrite. The final operation, by which the amalgam is separated, is the most difficult, and requires the greatest amount of skill on the part of the operator. This material is worked on the same table with very little water, with the aid of a small rake, or more often with the hand of the workman, who kneels down by the trough for the purpose. Finally, all the pyrite having been washed away, the magnetite is removed with a magnet, and the amalgam collected. The tailing from the black concentrate is treated over again, together with the grey concentrate.

The apparatus just described treats about 500 lbs. of sand at one time, and can be worked by one man, but usually gives employment to four people (frequently three of whom are women), who can treat about 5 tons of sand per day. The degree of success attained depends largely on the skill of the workman; in Siberia and Russia the art is handed down from father to son, certain families devoting their whole lives to the work during many generations. These workmen attain such a degree of dexterity in the use of the trough, that practically the whole of the valuable contents of the gravels treated is extracted by them, but the work is only suited to those who are content with small earnings.

The washing of the samples obtained in prospecting work is effected in Siberia usually by means of the *Washherd* (Fig. 39), and in the Urals by means of the *Stanok* (Fig. 40).[1] The illustrations are self-explanatory.

A description of the *Butara* washing trough, which resembles a long-tom, and is used in the Urals, with illustrations and figures as to working costs, etc., is given by J. P. Hutchins.[2] The grade of this trough is 4 feet in 12 feet, an unusually steep grade, for which there is no good reason.

The Sluice.—Sluices are constructed of "boxes," each of which resembles the upper part of the long-tom. The bottom of each box is made of rough boards, about 12 feet long, cut 4 inches wider at one end than at the other; the total width is usually from 16 to 30 inches, while the sides are 8 to 20 inches high, a portion of this being freeboard. In certain instances where the bulk of material was extraordinarily large, the width was increased

[1] Leon Perret, *Trans. Inst. Mng. and Met.*, 1912, **21**, 660. Figs. 39 and 40 are reproduced with the permission of the Institution of Mining and Metallurgy.

[2] *Mining Mag.*, 1914, **10**, 52.

to as much as 12 feet and the depth to 4 feet. The depth and width of a sluice depend on the volume and velocity of the stream, and these again are dependent on the amount and size of material to be treated. The narrow end of the box fits into the wide end of that next below it, and so a sluice, made up of hundreds of boxes, can be readily put up or taken down and moved to another locality. In general, boxes are designed to allow the larger pieces to roll, while the finer material is carried in suspension.

In all but the simplest and cheapest sluice boxes, extra strips of wood are affixed to the sides so as to protect them from the wear caused by the grinding action of the stones and gravel carried through by the current.

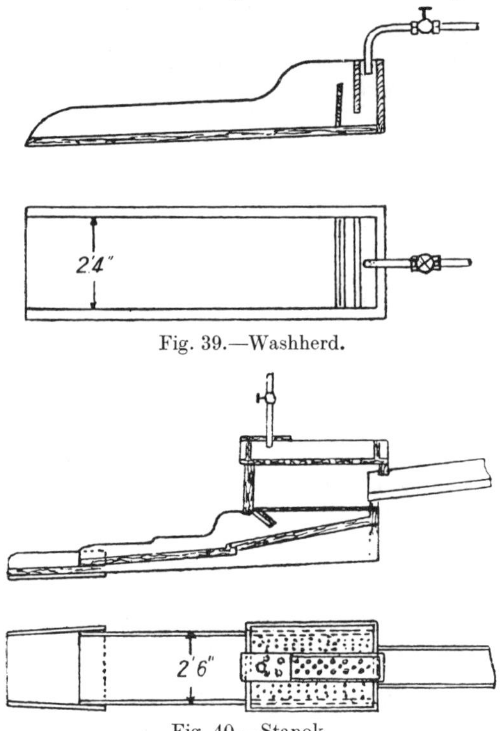

Fig. 39.—Washherd.

Fig. 40.—Stanok.

The bottom is provided with riffle bars to catch the gold. Many different kinds of riffle bars are used.[1] The simplest are strips of cut wood or fir poles 3 inches in diameter and about 6 feet long. They are usually placed longitudinally, and are wedged in the boxes at a distance of 1 or 2 inches apart by means of transverse bars, so that two sets of riffles are placed in each box in the manner shown in Figs. 41 and 42, which represent the whole of one box and parts of two others. Other forms consist of angle-iron or spaced square timbers. The depressions or riffles proper thus formed between the bars intercept heavy particles passing down the sluice, such as gold, mercury, amalgam, pyrite, etc., which gravitate to the bottom of the stream. Sometimes the riffles are placed transversely.

[1] See *Mining Mag. and Pacific Coast Miner*, 1905, II, 123, article on "Washing Plants and Riffles," by C. W. Purington.

The length of the sluice varies with the consistency of the gravel, the fineness of the gold and the fall of the ground. It must be sufficient to complete the disintegration and then to catch the gold. The grade of the sluice is measured by inches per box, so that a grade of "12 inches" means one of 12 inches in 12 feet. The usual grade is about 6 inches per box, but it varies according to requirements.

Tough, tenacious, clayey, or cemented gravels require higher grades to effect their disintegration than loose material. The velocity of the water current for clayey material may reach 12 feet per second. There must be sufficient grade to enable the water to carry away the gravel, but on the other hand fine gold is lost if the current is too rapid.

A reduction of the grade lessens the duty of the water, so that if the supply of the latter is short the grade is made as steep as possible, consistent with saving a fair proportion of the gold. A steep grade reduces the necessary length of the sluice, as disintegration takes place sooner. The upper part of a sluice is sometimes made of higher grades, or with narrower boxes than the lower part, which is occupied solely in catching the gold. When this is done additional supplies of water are introduced at the point where the change is made.[1]

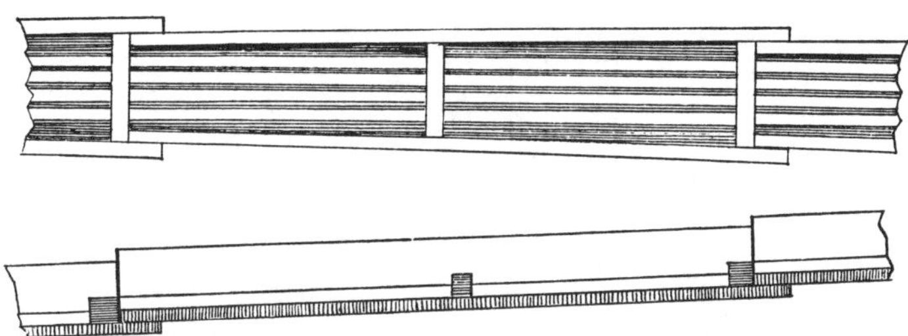

Figs. 41 and 42.—Sluice Boxes.

TABLE XX.—CAPACITY OF SLUICES.[2]

Width (ins.).	Depth of flow (ins.).	Grade per cent.	Water flow, Cub. ft. per min.	Cub. yds gravel per 24 hrs.
10-12	6-7	4·16	45	67·5 to 135
12-14	10	6·2	100	150 to 300

Width (feet).	Flow (miner's inches).	Flow (cub. ft. per min.).
3	200- 600	300- 900
4	400-1200	600-1800
5	1000-2500	1500-3750
6	2000-4000	3000-6000
8	3000-5000	4500-7500

[1] *Eng. Min. Journ.*, 1908, **86**, 1257.
[2] "*Elements of Mining*," G. J. Young, 1923, p. 408.

For a *drop*, the sluice terminates in a "grizzly," or inclined grating made of parallel iron bars placed longitudinally to the stream. The water and fine stuff pass through the grizzly and fall into a sluice below. The larger stones or boulders roll down the inclined bars, and are shot outside the sluice. The higher the fall, the more effectively it acts in causing disintegration. Sometimes boulders are retained to help in breaking up the gravel.

A *mud-box* is merely a wide part of the sluice, at which a man is stationed to break up and puddle the lumps. A better appliance is a trommel or a pan with a mechanical stirrer.[1]

The *undercurrent table* (Fig. 43). A grizzly with bars placed close together allows most of the water and fine material to pass through, while the coarse stuff is carried over and falls into the main sluice below. The fine material is discharged on to a wide table. A number of check-boards

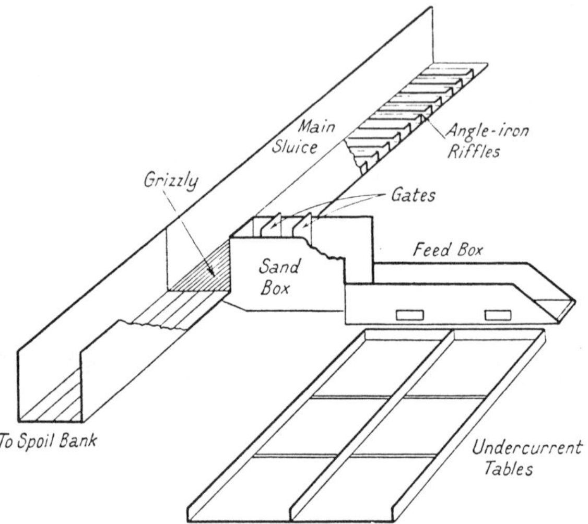

Fig. 43.—Undercurrent Table (*Truscott*).

helps to distribute the stream evenly over the width of the table which is usually of higher gradient than the sluice, so that a broad shallow stream flows over its surface. It is supplied with riffles and mercury, and is intended to catch fine gold and amalgam. The tailing from the undercurrent is discharged into the main sluice below the drop.

The Use of Mercury in Sluicing.—Mercury is added at the head of the sluice after washing has been in progress for a sufficiently long time for all leakages to have been stopped, and for the lowest depressions to have been filled in with sand. The mercury is sometimes sprinkled into the sluice, and sometimes poured behind the riffles. The amount added varies with the richness of the gravel, enough being added to dissolve the amalgam formed. It is carried down the sluice and lodges in the riffles, the greater part being retained in the first few boxes. Fresh supplies are introduced every few hours at the head of the sluice, and sometimes at various points

[1] Purington, *Mining Mag. and Pacific Coast Miner*, 1905, II, 16.

lower down the sluice also ; in particular, mercury is added to the under-currents, as it is especially valuable in catching the finer particles of gold which would otherwise be lost, whilst coarse gold can in great part be saved without mercury. Amalgamated copper plates are sometimes used.

The Clean-up.—The length of each "run," at the end of which the boxes are cleaned-up, varies, according to the richness of the gravel, from a day to a whole season, but is usually a week. The upper part of the sluice, which retains most of the gold, is usually cleaned-up more frequently than the remainder. A clean-up is begun by discontinuing the supply of gravel, and letting the water continue to flow until it passes through the sluice quite clear. The first six or eight sets of riffle bars are then taken up, and the sand, mercury and amalgam washed down, all the latter being caught by the first riffle left in. It is scooped out thence by a wooden ladle or iron spoon into a bucket, and the rich sand is collected and panned. The next few riffle bars are now taken up, and so on. Alternatively the work may be begun on several sections at the same time. Lastly, the whole sluice is carefully searched over, and particles of amalgam or mercury picked out with spoons, penknives, etc., from every crevice where they have lodged.

The amalgam thus collected is stirred with fresh quicksilver in "amalgam kettles" or buckets, and the black sand and other foreign matter skimmed off. It is then strained through chamois leather or drilling, liquid mercury passing through and pasty amalgam being retained by the skin. The amalgam is well squeezed and then retorted. The retorts used in large well-conducted enterprises are similar to those in use in stamp mills, described in Chapter VII. Amalgam obtained as the result of operations on a small scale, however, is often merely heated on a shovel over an ordinary fire, the mercury being driven off and lost.

Fly Catchers were invented in Australia for the purpose of catching the fine particles of gold, which, successfully evading the riffles of all sluices, float down on the surface of the rivers. These devices consist of weirs constructed on piles driven into the river bed, and stretching across from bank to bank of the river. Boards covered with blanketing or coarse gunny-sacking are attached to the weirs and collect all particles floating on the surface of the water. At intervals the blankets are taken up and washed in a tank.

Dry-Blowing.—If no water can be obtained, it is sometimes profitable to concentrate pay-dirt by winnowing, tossing it in a pan until the lighter particles have been blown away, and finishing with mouth-blowing. It is, of course, a wasteful method of concentration.

Among machines used for the purpose, the simplest consist of flat screens supported on a frame and shaken by hand, the material falling through being winnowed by the wind. In other contrivances, a bellows is added and worked by the same hand-mechanism by which the screens are shaken.

In some of the machines a blast of air is used to keep the sand partly in suspension, while it is moved by gravity down an inclined table which is furnished with riffles. The auriferous material must be quite dry or perfect disintegration cannot be accomplished. In a typical dry washer, the gravel is first made to pass through shaking screens, the mesh of which is adapted to the character of the material. The object of the screen is to eliminate the larger fragments, which are usually barren. The shaking screen delivers the material on to an inclined table formed of a wire-screen, covered with light canvas or some similar material through which are forced pulsating blasts of air. These sudden puffs throw up the sand and let it settle again alternately, and as a result the light material works down the table, while the gold is retained by the riffles, being too heavy to be tossed over them by the air.

Dredging.[1]—In dredging, gravel is raised from the bottom of the river and delivered into a barge (also called scow, pontoon or hull), and the material is there washed, the gold extracted, and the tailings sluiced back into the river or stacked on the bank. Gravel beds not in rivers are also dealt with. This method of recovering gold has made remarkable progress, and its extension to the working of all flat placers, including those which are at some distance from the nearest stream, has completely changed the aspect of shallow alluvial mining. Large quantities of material can be worked at a low cost, and without filling the rivers with débris and causing damage to agricultural lands down stream. Dredges of the largest type excavate from 10,000 to 12,000 cubic yards per day. The consumption of water is small. Ideal conditions for dredging should comprise a comparatively level surface, ample water, not too large boulders, little clay, few buried timbers, ample volume, depth not too great, and fair value.

Dredges may be divided into three classes, according as (1) suction pumps, (2) continuous chain-bucket elevators, or (3) a crane and bucket or shovel are used to raise the gravel,

(1) *Suction Pumps.*—In this system, a centrifugal pump draws material through a large suction hose reaching to the bottom of the river and delivers the gravel to sluice boxes. The pumps [2] are usually of the single-stage type and differ from similar water pumps only in details. The impeller is wider and set to calculated curves, and the space between the impeller and the casing is greater. The suction pipe is 6 to 12 inches in diameter. Manganese steel is used extensively for the casings and liners, the latter being renewable when worn. The pump may be erected on the bed-rock of a deposit, or on the lowest level of ground being worked, on a pontoon in swampy ground or in river channel deposits. Any ground can be successfully handled by a gravel pump unless there are boulders larger than the bore of the pump. Boulders may be removed by a grizzly placed immediately in front of the pump sump. The ratios of water to gravel should range between 2 : 1 and 5 : 1.

It is difficult to regulate the relative amounts of gravel and water raised by suction pumps, the latter tending to be in great excess. The suction pipe is soon worn out, especially by coarse gravel, and the power required per ton of gravel is considerably greater than in the case of ladder-bucket dredges. Nevertheless, dredges of this type have been in successful operation on river banks. A common method is first to sink a pit. The sides are then sluiced down by jets of water. The mud is sucked up through a hose pipe until a large hole, say 200 feet square, has been made, when the dredging apparatus is moved and a second pit excavated, the tailing being discharged into the first pit. The material handled ranges from fine sand to boulders 8 inches in diameter.

These machines are useful in comparatively small areas and where alluvial deposits are patchy. They are comparatively easily transported. In general, however, they are not used when bucket dredging is possible.

---

[1] A good account of the treatment of all placer deposits is given in *U.S. Bureau of Mines Information Circular* 6786, 1934. See also *U.S. Bur. Mines Bulletin* 127.

[2] *Eng. and Min. J.*, 1933, 134, 184.

Proposals for the improvement of suction dredges include the strengthening of the parts (pump-shell, runners and liners) which are most liable to wear and breakage, the provision of a cutter for loosening compact gravel at the intake, and, most important of all, an increase in the diameter of the suction pipe from the ordinary 10 or 12 inches to 24 inches, in order to allow larger stones to pass through.

(2) *Chain-Bucket Dredges.*—These are in general use. They are made in great numbers both in New Zealand and in Western America.

The simplest and cheapest form of the chain-bucket type of dredge is the sluice-box dredge,[1] on which the gold is caught in a long sluice run, into which the buckets tip their burden. The other type is the screen dredge, which is fitted with a revolving trommel, through which the material raised is washed on to tables. The diagram, Fig. 44, shows a Risdon screen dredge at work. In whatever way the gravel is washed the excavating is done by means of an endless chain of buckets passing round two tumblers, which are placed at the ends of a long ladder or frame. The ladder and bucket line extend through a well hole running from near the centre to the bow. The ladder is of such a length that, when the buckets are working at the bottom of the water, it hangs at an angle of not more than 45°, as in this position the buckets give the best working results. The depth to which a dredge can dig is proportional to the length of the digging ladder and to the height of the bank carried. In a dredge digging 60 feet below water level and working against a 20 foot bank, the total digging depth is 80 feet. A recent dredge has a 200 foot ladder and digs to a depth of 110 feet.[2] The lower tumbler is, in modern dredges, an idler or sheave, not power-driven. It consists of ordinary cast steel or, better, of manganese steel. The pitch of the tumbler is usually a little larger than six times that of the buckets, and this ensures even wear, each succeeding bucket sitting on a different part of the surface. Cast-iron bushings are mostly used in the lower tumbler bearing, but some operators still prefer babbitt metal, which does not tend to fracture, although its life is shorter. Sleeves may be shrunk on to the shaft, also to take up wear. The ladder rollers, on which the chain runs, consist of chilled cast-iron or steel. The upper tumbler is usually hexagonal, and acts as a positive driving sprocket for the bucket chain. It is usually provided with a series of wearing plates which are renewed from time to time, but they are not altogether satisfactory in compensating for the increase in the pitch of the buckets. In California a practice is springing up of casting the shaft with the tumbler in order to obtain increased strength and eliminate trouble from loose keyways, etc. Both six- and eight-sided tumblers are used for large dredges. Manganese, nickel, chrome, and other special steels are much used in dredge construction for the tumblers, buckets, rollers, spindles, etc., which are subject to heavy wear.

" The ladder is suspended at its lower end from a heavy gantry on the bow of the dredge hull, the raising and lowering of the ladder controlling the digging depth. The elevation of this ladder is controlled by a winch, called the *ladder hoist winch*, situated on the port side of the dredge. The lateral motion of the buckets is secured by swinging the entire dredge on one of the 'spuds' as a pivot. This is done by means of lines run from two drums, the 'swing winch,' on the starboard side, connected through

[1] E. S. and G. N. Marks, *Trans. Inst. Mng. and Met.*, 1906, 15, 453; see also Gardner and Shepard, *General Electric Review*, May, 1914, 17, 436.

[2] Lewis, *Can. Min. Journ.*, 1935, 56, 106.

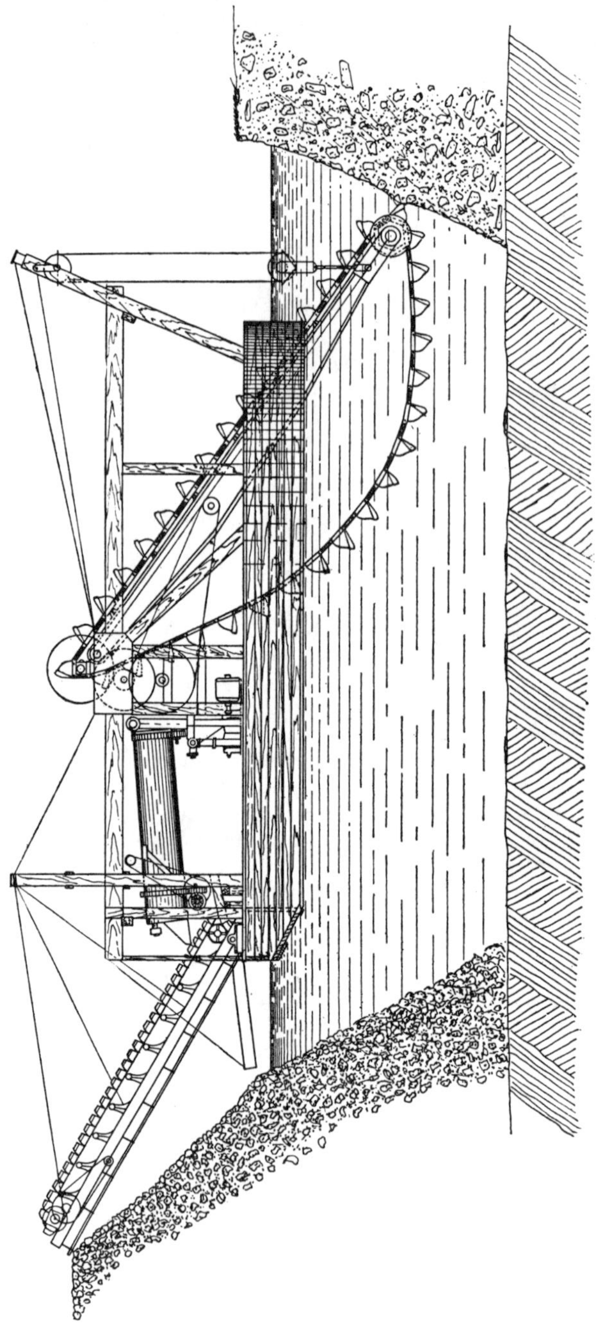

Fig. 44.—Risdon Gold Dredge.

sheaves on the port and starboard bow to 'dead men' on the banks. The continuously revolving bucket line is thus swung across the face of the bank for a width of from 190 to 300 feet, and at the extreme lateral limit of the swing the ladder is dropped and the side swing repeated. The revolving buckets thus terrace away the bank until bed-rock is reached, when the ladder is raised, the dredge stepped forward by swinging on alternate spuds, and the whole operation recommenced. These spuds, which are single sticks of lumber on the smaller dredges and of structural steel on the larger and more powerful dredges, are at the stern of the dredge and suspended from the stern gantry. By alternately raising and dropping these two spuds as the dredge swings, it is stepped ahead much as a stiff-legged man might walk." [1]

The spuds take the thrust caused by digging and also the torsional stress when the dredge swings from side to side. Two types of steel spuds are in common use, one with a rectangular hollow section and the other very similar but with a centre web. The latter is found not quite so satisfactory as the former and is more difficult to repair. The bearing pressure, which used to be concentrated on three or four rivets only, is distributed more uniformly by making a swivel casting which bears against the spud at all positions for a distance of 3 or 4 feet. The spud is made to bear on to a casting several feet wide, covering the whole depth of the hull. Bolts instead of rivets are used on spuds and all parts which are subjected to heavy and varying stress. They can be more readily tightened-up and remain tight for a longer time. Wherever possible it is the commoner practice to install two digging spuds. If there is only one, it comes at the end of the screen and interferes with the sequence of operations. It is also in the way when repairs to the screen and driving gear have to be carried out, and when it needs attention the whole dredge must stop. The length of the spud is determined principally by the depth of the digging. Anchorage by spuds is more rigid than by mooring lines and prevents the rocking of the barge, which interferes with the gold-saving tables.

The buckets are of steel plate, with a wide mouth and a shallow back to facilitate discharge. Their capacity varies from 3 to 16 cubic feet, but the tendency is towards larger buckets. In the Californian type of dredge the buckets are placed close together, following one another with no connecting link between (close-connected bucket chain, see Fig. 45), and in the New Zealand type of dredge the buckets are spaced to form an intermittent chain (open-link bucket chain, see Fig. 46). The former type is said to be better in loose free-running ground, and the latter is said to be better in hard ground or when dealing with timber and boulders. The bucket for a Californian dredge consisted originally of three parts, bottom, hood and lip, but it is now customary to make the hood and bottom in one piece as a manganese or nickel-chrome steel casting. If the latter steel is used for the main part, manganese steel inserts may be employed for those portions taking the most wear. A rivetless lip has recently been tried.[2] The arch of the lip should not be too flat as that induces crushing. The chief cause of bucket-line trouble is the elongation of the pitch of the buckets and decrease of the effective pitch of the tumblers, due to wear on the pins. These should therefore be of sufficiently large diameter and correctly heat-

[1] W. H. Gardner and W. M. Shepard, *General Electric Review* (New York), 1914, 17, 438.

[2] *Eng. and Min. Journ.*, 1934, 135, 489.

treated. The back eye of the bucket and sometimes the front eyes are provided with bushes to take up wear. A "save-all" sluice box catches the drippings from the buckets. A winch and drums are used for anchorage lines (when spuds are not employed) and for working the elevators. The power is electrical where possible. A heavy water pump is required.

"The ascending buckets, full of gold-bearing material, are dumped over the upper tumbler into a hopper, lined with heavy wearing bars, where the ore is subjected to high pressure sprays of water. The gravel is then delivered to the screen, a revolving cylinder sloping toward the stern and lined with perforated steel plates, in which it is continuously played upon by jets of water and is completely disintegrated. The sand, fine gravel and gold particles are washed through the perforations into the distributor under the screen which serves to distribute the mass properly on the gold-saving

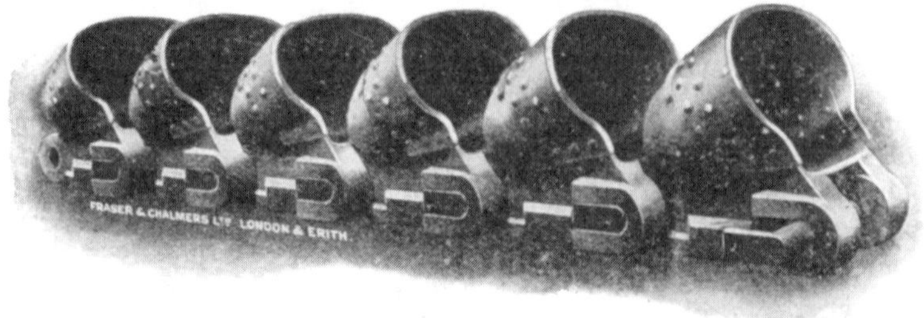

Fig. 45.—Close-connected Bucket Chain.

Fig. 46.—Open-link Chain with Solid Links.

tables. These are in two banks, an upper and a lower, and are composed of fore and aft and thwart-ship sluices, lined with steel-shod sugar-pine riffles, where the gold is caught and amalgamated with mercury."[1] A careful adjustment of feed to water is necessary. In cold weather particles of gold are among the first to accumulate ice in the sluice, thus causing losses among the flaky pieces. "The waste sand is delivered from the tail sluices at the stern of the dredge, some of it being deposited around the spud points to enable them to obtain a firmer hold, and the remainder dumped at some distance behind the dredge. The boulders and heavy gravel, which do not pass through the perforations, fall from the rear end of the screen into a chute and are thence delivered on to the stacker belt. This endless belt carries the material up a long stacker, which is hung from the stern gantry of the dredge, and the waste is finally dumped far to the stern, forming the extensive 'rock piles' characteristic of a dredging field."[1]

The proper selection of a dredge is of vital importance. In general if

[1] Gardner and Shepard, *loc. cit*

the ground is deep and large, with uncemented gravel, the larger the dredge the lower the operating cost. If the ground be shallow and narrow, a large dredge obviously cannot be used. Under certain conditions, however, smaller boats may be preferred to larger ones. For instance, lower first cost, working expenses and easier maintenance are all points in favour of dredges of small capacity.

The bucket-line and tumblers probably present the greatest difficulty and must be built in relation to the ground to be worked. For example, different types of buckets would be required for working in gravel and clay, or where there are large boulders. Other factors mentioned by Janin [1] are as follows :—(1) Buckets handling boulders should have low backs so that as they come round the lower tumbler to the horizontal position the boulders will not crush the back of the preceding bucket. (2) Buckets should not be filled so full as to cause excessive spilling, and the lip should not be much higher than is determined by the natural angle of repose of the material. (3) The width of the bottom of the bucket should vary directly with the length of bucket-line to be used and with the hardness of the material being dug. Square bottoms are becoming more common in order to increase the bearing surface at the upper and lower tumblers and thus prevent wear. (4) Buckets for clayey ground should be wide and short to facilitate dumping.

While no great changes have occurred in the general lines of dredges for many years, considerable improvements in detail have been made in order to reduce costs of recovery to a minimum. The following are particulars of a dredge built for use in Colombia.[2] The hull, which is 100 feet long, 50 feet wide and 9 feet deep, is built of steel plates and the equipment, except for pumps, is on or above deck. The superstructure is designed to afford support to the buckets and also to stiffen the hull along the centre line where the greatest stress occurs. The bucket-line consists of 59 close-connected cast manganese steel buckets, each of 9 cubic feet capacity, passing over manganese steel tumblers. The lower tumbler (see Fig. 47) is cylindrical, 57 inches in diameter, and has a 30-inch face. The upper or driving tumbler (Fig. 48) is hexagonal, 26 inches in diameter, with a 29-inch face. The driving tumbler has recesses in the centre of each face to accommodate corner pieces in manganese steel, so that on the occurrence of any wear, these can be pushed outwards and held so as to fit the worn chain —in effect increasing the pitch of the tumbler. Renewable plates covering the faces of the hexagon were previously used.

The top or driving tumbler revolves at 3·5 to 4 r.p.m. The bucket chain is thrown in and out of gear by a powerful internal clutch on the driving pulley which is operated from the control room. The buckets travel at the rate of 22 to 24 per minute, giving an output of about 500 yards per hour. The ladder carrying the chain of buckets is 65 feet long and 64 inches wide and consists of two webs built up of plates, angles and channels. The buckets travel on a line of rollers along its upper edge. When the ladder is at an angle of 45° excavation can be carried to a depth of 22 feet.

The buckets are discharged into a chute leading to an inclined rotary screen, 90 inches in diameter and 42 feet long, turning at 7 r.p.m. The screen is driven by a friction drive at the upper end. The holes in the screen vary from $\frac{5}{16}$ to $\frac{5}{8}$ inch in diameter and the undersize passes to a catchment

[1] *Trans. Amer. Inst. Mng. Eng.*, 1911, **42**, 855.
[2] *Engineering*, 1925, **119**, 697.

box, below which the first gold saving is effected. A 12 inch centrifugal pump delivers 3,000 gallons of water per minute to the screen. From the catchment box the wash passes to transverse launders on each side of the screen, and from these it is distributed over three pairs of concentrating tables (total area, 2,000 square feet) running fore and aft. Dressing water

[By courtesy of Messrs. Hadfield's Ltd., Sheffield.

Fig. 47.—Bottom Tumbler for Gold Dredger.

[By courtesy of Messrs. Hadfield's Ltd., Sheffield.

Fig. 48.—Top Tumbler for Gold Dredger.

up to 3,000 gallons per minute is supplied from another pump. Riffles, matting and mercury traps are used for saving the gold. The tables are pivoted at their upper ends and supported at points about two-thirds down their length by wire ropes whose resilience imparts a slight but rapid vibratory motion to the tables. They are housed in heavy expanded metal, and brilliantly illuminated in order to prevent tampering with the concentrates by unauthorised persons. The oversize from the screen is regarded as worthless except as washed ballast. A "save all" or miniature concentrator traps all material that fails to get into the screen.

Semi-Diesel engines are employed to drive the whole of the plant.

Fig. 49 shows one of the largest British dredges recently constructed. The hull is 276 feet 5 inches long, 65 feet wide and 11½ feet deep, and carries 132 manganese steel buckets each of 12 cubic feet capacity. The dredging depth is 130 feet below the water line. The ladder is 195 feet long between centres, and works in a well extending through the forward part of the hull. Raised mud is delivered through sluices into trommels and then on to riffles, launders or concentrating tables. The dredge is moved across the face of the deposit, as required, by head and side lines operated by a mooring winch. To minimise the inconveniences of excessive slack, side swing and spillage on the chain of buckets when worked at great depths, a caterpillar track has been introduced which carries the bucket chain for a considerable part of its return journey. A special gantry head has been erected at the forward end of the dredge to take the weight of the bucket line when the latter is moved across the face, thus avoiding strain on the hull. The tailings can be disposed of to the side or to the rear at a considerable distance from the working face.[1]

Dredges now in use in California fall into three general classes:—[2] (1) The smaller unit ranging in bucket size from 3 to 7½ cubic feet. It is sectionalised and is useful for working small areas. (2) Units of 9 to 18 cubic feet bucket capacity for digging to depths of 30 to 80 feet below water. They are suitable for larger areas. (3) An 18 cubic feet bucket dredge of all-steel construction. It digs to 100 feet or more below water level. This type of dredge, as lately devised, has an interesting feature in a bucket idler at the mid-point of the ladder, which permits digging at the greater depths (see below).

Dredges with pontoon hulls are proving useful on account of the ease with which they are assembled and dismantled.

The latest Yuba Dredge, No. 17, built for work on the south side of the Yuba River, California, is of all-steel construction.[3] The hull is 233 feet 9 inches long, 68 feet wide and 11 feet 6 inches deep. Two heavy trusses run the entire length of the hull, one on each side of the well hole. The total weight is 1,300 tons. The ladder is 200 feet long between tumbler centres, 13 feet 3 inches deep and 5 feet 6 inches wide. In order to decrease the catenary pull caused by the great weight of the buckets on the return line, a bucket idler, designed by O. B. Perry, is employed on the under side of the ladder. This idler is 10 feet in diameter and weighs 40 tons. It is placed near the mid-point of the ladder length, and divides the catenary into two parts. By so doing it takes care of its part of the weight and the live load of the lower half of the bucket line. There are 126 buckets in operation, each of 18 cubic feet capacity. They are made of manganese steel, and are equipped with a cast stop near the back eye to regulate the amount of backlash when passing over the idler. Each bucket weighs 2½ tons. The lower tumbler is of nickel chrome steel, 5 feet 8 inches diameter, and has manganese steel side liners. The upper tumbler is of high-carbon

---

[1] *Engineering*, 1932, 133, 154.
[2] C. M. Romanowitz and G. J. Young, *Eng. and Min. Journ.*, 1934, 135, 486.
[3] *Eng. and Min. Journ.*, 1934, 135, 338.

Fig. 49.—Large Modern British Dredge.

[F. W. Payne.

chrome steel cast integrally with the shaft and is provided with forged and heat-treated nickel chrome steel heel plates and manganese steel ear or corner plates. Discharge from the buckets passes to the main hopper, which is lined with abrasion-resisting steel $\frac{3}{4}$ inch thick, and is provided with a bottom of 4 × 6 inch nickel chrome steel bars. From this screen the material passes to a 50 × 9 feet revolving screen of which a length of only 37 feet is perforated. The screen plates are of abrasion-resisting steel with perforations $\frac{3}{8}$ to $\frac{1}{2}$ inch and $\frac{1}{2}$ to $\frac{5}{8}$ inch. The stacker belt is made of seven-ply rubber, 44 inches wide, and is carried on a ladder 250 feet long. The latter is supported on a double three-point suspension, and is swung from a gantry at the stern. The undersize discharge from the screen is divided between two superimposed double banks of table sluices. Each sluice is 2 feet 8 inches wide and is equipped with standard riffles and mercury trap riffles at suitable points. The total sluice area is 9,000 square feet, approximately 350 to 500 square feet per cubic foot bucket capacity. The upper bank of sluices discharges over the stern; the lower discharges into the pond. The two steel spuds are 38 × 60 inches in section, 70 feet long and weigh 40 tons each. They are worked by a gantry at the stern and spud lines to a winch in the control house.

A new all-steel dredge has recently been completed by the Natomas Company.[1] Several features new in dredge construction include:—(1) Direct-driven bucket line and individual bow line winches and ladder hoist winch. (2) A power plant in the interior of the hull to generate direct current for the motors driving the bucket line and bow line winches. (3) Air-controlled brakes on the stern line winch and ladder hoist winch. (4) Electrical travelling jib and trolley cranes. (5) Speed reducers for driving all mechanical units.

(3) *Crane and Bucket Dredges.*—In these dredges a single large bucket or shovel, with a capacity of 1 or 2 cubic yards, is filled with gravel and hoisted on board a barge by some form of crane. The bucket is sometimes of the Priestman grab type, opening at the bottom as it is dropped on to the gravel bed, and closing automatically when the chains begin to raise it. A large shovel is also sometimes used by which large boulders up to 5 or 6 tons in weight can be handled with ease. The Priestman grab bucket does not close completely if a stone lodges between its jaws, and comes up partly empty, but in cases where large boulders, tree trunks, etc., are dispersed through the gravel, it might be used with advantage, instead of the ordinary ladder-bucket, which cannot deal with such obstructions. The wear and tear is said to be less with a Priestman dredge than with the bucket dredge.

It is considered by some that "dipper-dredges" are useful under the following conditions:—(1) Where the ground is somewhat shallow; (2) in small areas where a cheap dredge is required; (3) where the material is rough, with boulders and stumps: (4) where the ground is mixed with more or less clay, as the dipper will relieve itself notwithstanding the adhesiveness of the material.

In places unsuitable for dredges, *bucket-scrapers* are sometimes used.[2] This is a single bucket of about 1·5 cubic yards capacity on a crane, loaded by a drag line. The gravel is dumped by the bucket into a hopper, which

---

[1] *Eng. and Min. Journ.*, 1935, **136**, 270.

[2] Janin, *Trans. Amer. Inst. Mng. Eng.*, 1911, **42**, 855.

feeds a trommel screen. The machine rests on rollers, by which it is moved on a plank track. It has been tried by Purington in Siberia.

Gold-Saving Apparatus.—The available space on a dredge is so small that long sluices are out of the question, and it is, therefore, necessary to disintegrate the gravel thoroughly in revolving trommels or on shaking or travelling screens, and to make the gold-saving tables as wide as possible, so as to reduce the speed and depth of the stream of gravel running over them, especially where fine or scaly gold is abundant. The best size for the holes in the trommels is usually considered to be within a range from $\frac{5}{16}$ to $\frac{5}{8}$ inch. All material not passing through is regarded as worthless, and is not treated on the tables. The inclined tables may be covered with coir matting or plush,[1] or with calico, covered with cocoa matting, on which is laid a sheet of expanded metal, the raised edges of which form a series of very effective riffles. Hungarian riffles, or bars somewhat resembling T-shaped rails, placed transversely, are also used.[2] Mercury is placed in the riffles.

A trommel described by E. S. Marks and G. N. Marks[3] is 17 feet long and 54 inches diameter inside. It is constructed of steel plates with perforations ranging in four successive sections or rings from $\frac{5}{16}$ inch to $\frac{1}{2}$ inch in diameter. The pitch aft of the screen is 1 inch to the foot. Water is supplied by a 9-inch pipe perforated for 12 feet opposite the perforated rings of the screen with four rows of $\frac{1}{2}$-inch holes at 4-inch centres. The finer material undergoing treatment is washed through the screen perforations on to the gold-saving tables running athwartships. The tables in this case were 14 feet long and 12 feet wide, made in four widths of 3 feet each, stepped one above the other with division plates between them. The pitch of the tables is usually about 1 in 10, and they discharge into a chute which runs out over the stern of the dredge, either into the water or on to the bank. The oversize from the trommel is either passed direct to a tailings elevator set at an angle of about 30°, or is first treated in a sluice box fitted with calico, cocoanut matting and riffles to save gold, as in the case of the tables.

Rotating inclined trommels up to 9 feet in diameter and 50 to 60 feet long are now used. They have $\frac{3}{4}$-inch thick plates with perforations $\frac{3}{8}$ to $\frac{1}{2}$ inch diameter. Owing to punching, the diameter on the outside is $\frac{1}{16}$ inch larger than that on the inside.

Marks[4] also describes a sluice box which may be used instead of a trommel and tables. The sluice may be 40 to 50 feet long and 3 to 6 feet wide, with a pitch varying from 1 in 8 to 1 in 10. The bottom of the sluice is laid with perforated plates and riffles overlying the cocoanut matting and calico throughout the entire length. In order to give variation in treatment as many different classes of riffles as possible, such as angle irons, venetians, perforated plates and crimped or diamond-shaped expanded metal, are used in the sluice. The riffles are prevented from becoming packed or choked. Undercurrents are sometimes used.

The sluice boxes or tables are cleaned-up weekly, the riffles or expanded metal being lifted out and the matting and calico washed in tubs. Most of the gold is found in the first few feet, which are sometimes cleaned-up every few hours. The washings or concentrate may be run through a small

---

[1] J. P. Smith in *Rept. of Dept. of Mines, New Zealand*, 1899-1900, p. 51.

[2] For illustrations of these riffles, see Purington, *Mining Mag. and Pacific Coast Miner*, 1905, 11, 21, 127.

[3] *Trans. Inst. Mng. and Met.*, 1906, 15, 471.

[4] *Loc. cit.*, p. 479.

sluice box, 15 inches wide and 12 feet long, called a *streaming-down* box (Marks). It is laid with silk plush or green baize. The final product is amalgamated, sometimes in a clean-up pan or small tube mill. The original concentrate may also be treated by amalgamation, either by running it over amalgamated plates or by treatment with mercury in a clean-up pan or a rotating barrel.

The latest dredge at Bulolo, New Guinea, has fine jigs, in place of the usual long riffled launders, following the primary trommel. The concentrate from the jigs is treated in a tube mill, while the tailing is discharged through a tailing launder. From the tube mill the concentrate passes to amalgamators and the tailing from these to a Kraut [1] flotation cell. The concentrate from this is smelted. Tailing from the flotation cell goes to waste.

Disposal of Tailing.—The tailing was formerly always discharged into the water again at some point beyond the stern of the dredge. It is now often piled on the bank to prevent any risk of mixing it with untreated gravel. Coarse and fine tailings are usually piled up separately, and when flat inland placers are being dealt with, as described below, the separated tailing occupies 50 or 60 per cent. more space than the undisturbed ground. The bucket elevators for raising the tailing are shown in Figs. 44 and 49.

Treatment of Flat Inland Placers—"Paddock Dredging."—The advantages of dredges in handling gravel caused their use to be extended from the beds of rivers to all flat placer ground wherever situated. The dredge is placed in a reservoir which is filled with water. The machinery is then erected and dredging begun, the tailing being discharged at the stern, so that the reservoir and dredge move gradually forward together. By piling gravel round the reservoir and letting in more water, the dredge is raised and can be made to work its way up sloping ground. Clear water may be kept flowing into the reservoir to prevent it from becoming too muddy for close gold saving.

Difficulties in Dredging.—The difficulties of raising the gravel due to the occurrence of tree trunks or large boulders are obvious. Nevertheless, the chief difficulty is to save the fine gold in the gravel. The losses of gold are often large. In one New Zealand dredge, it was found as the result of treating measured quantities of tailings that only 60 per cent. of the gold was saved. Lumps of clay from the overburden and water carrying clayey matter in suspension are responsible for some loss of gold. The best method of avoiding such loss is to strip off the overburden of clay before attempting to raise and wash the auriferous gravel.

The bottom on which the gold-bearing material rests is also of importance. A hard bed-rock prevents the gold lodged in its crevices from being removed by the dredge. Soft bed-rock partly decomposed *in situ* is, on the other hand, an excellent bottom for dredging. A soft tenacious clay underlying the gravel causes loss of gold in washing. A. F. J. Bordeaux [2] mentions the great difficulty in dredging in French Guiana due to a tenacious clay, which must be puddled by hand in the sluices to free the gold contained in, or picked up by, it. In a dredge the clay sticks tightly to the buckets and goes round and round in spite of all water jets. It is necessary to dislodge it with shovels.

One of the greatest sources of expense in gold-dredging is the repair of

[1] *Chem. Eng. Min. Rev.*, 1933, 26, 130.
[2] *Trans. Amer. Inst. Mng. Eng.*, 1910, 41, 587.

the bucket-line, including the tumblers. Stones carried between the tumblers and the buckets break or bend the latter.

Untreated wooden hulls last only 8 to 10 years. If the timber is impregnated with creosote, however, the life may be extended to 15 years. A good circulation of air through the hull is an aid to long life.

In general tie rods and not timbers should be used to withstand tensile forces. Timbers should not support the screens, the drippings from which will cause rotting. All heavy timbers should be heavily capped at each end in order to stop decay and also ensure a better bearing surface on neighbouring timbers without crushing.

A great many modern dredges are built of steel. In this case the fire hazard is minimised. Sectionalised construction is employed where transportation of large heavy masses is impossible or expensive.

The horse-power installed on gold dredges varies, of course, according to the capacity of the bucket, dredging depth, etc. An average distribution of power is as follows :—[1]

Bucket chain	33 per cent. of total power.
Water pump	30 ,, ,, ,,
Winch	15 ,, ,, ,,
Screen	10 ,, ,, ,,
Elevator	12 ,, ,, ,,

Eckart [2] describes the practical methods which he has found suitable in the erection of a dredge.

Chellson gives tables showing the probable costs, based on experience at a gold placer area of 76 acres, of working medium-sized placers, and the average profit that may reasonably be expected.[3]

Prospecting Dredging Ground.—Janin [4] draws attention to the necessity for the thorough prospecting of any proposed dredging ground, and summarises the following conditions which should be investigated :—

(1) Value, character and distribution of gold content.
(2) Depth, character and quantity of ground to be worked.
(3) Character and contour of bed-rock.
(4) Water level and available water supply.
(5) Costs of fuel, power obtainable, possibilities of obtaining hydroelectric power.
(6) Labour, transportation, supplies.
(7) Surface contour and timber growth.
(8) Operating costs.
(9) Cost of land, royalties, titles, etc.
(10) Climatic conditions.
(11) Taxes, duties, stability of foreign Governments, etc.

The value of the ground is tested by drill holes or shafts, and attention paid to any help that may be obtained from gulches, old prospects, pits and exposed faces. Shafts usually yield more reliable information than drills, permitting larger samples and a greater area of examination. They are, however, subject to the existence of favourable conditions, whereas drills

[1] Kempe, "*Engineer's Year Book*," 1932, p. 2647.
[2] *Min. Mag.*, 1935, **52**, 265.
[3] *Eng. and Min. Journ.*, 1934, **135**, 441.
[4] *U.S. Bur. Mines*, 1918, *Bull.* 127.

may be sunk in any class of ground.[1] The material issuing from the holes is sampled and panned. The residues, showing "colours," are usually preserved in a small dish and afterwards amalgamated, the mercury subsequently being removed by nitric acid or by distillation. Fouling of the mercury may be due to the presence of grease, iron oxide, arsenic or other impurities. A little caustic potash or common salt will often assist amalgamation. In all the sampling operations careful note should be made of any unusual occurrence or appearance.

In prospecting a large area a few preliminary holes are first sunk, to indicate the general run of the deposit. Their position will often be determined by surface features or old workings. Further holes are then made, usually placed in square divisions over the area and staggered so as to give indications over as wide an area as possible. Where finds are made the number of holes per unit area will naturally increase in order to determine the extent and direction of the valuable portion. In arriving at an estimate of the gold content of any area it is customary to take a percentage, usually 75 to 80 per cent., of the value indicated by the drillings as the amount recoverable by dredging.

Hand drilling is usually undertaken by some type of Empire drill, in which an auger bit makes the first hole into the ground and is followed by the drilling tool proper. As the core accumulates it is washed to the surface where it is collected and sampled. Power drills of the Keystone type are often used. In these, percussion on the drill rods is obtained by means of a small petrol engine. An unusual type of drill for use in frozen ground is described by Janin.[2]

Hydraulic Mining.[3]—This method of working consists in breaking down banks of gravel by the impact of powerful jets of water, and passing the disintegrated material through a line of sluices, without the agency of hand labour. The chief requisites for the successful application of hydraulic mining are:—

(1) Large quantities of auriferous gravel, not less than 30 feet in thickness, and not overlaid by any appreciable thickness of barren material, which would necessarily be passed through the sluices with the pay-dirt. The gravel treated need not be rich, a mean yield of less than 1 grain of gold per cubic yard being often enough to furnish profits if the operations are on a sufficiently large scale.

(2) A plentiful and uninterrupted supply of water throughout considerable portions of the year.

(3) Sufficient fall in the ground so that (*a*) the water may be delivered under the pressure of a head of from 100 to 300 feet, (*b*) the tailings can be easily carried away to a large dumping ground, which is, most conveniently, either the sea or a large and rapid river.

The Supply of Water.—The amount of water required is very large. The water is delivered at a convenient height above the workings into a penstock or "pressure-box," consisting of a small wooden or ferro-concrete reservoir from which the pipes take their origin. In many cases the reservoirs and ditches are owned by separate companies, or, as in New Zealand, by

---

[1] For the methods adopted in sinking shafts or drill holes the reader is referred to standard works on mining and ore dressing.

[2] *Loc. cit.*

[3] The invention of hydraulic mining is ascribed to Edward Mattison, who used the method in 1852 on a small scale.

Government, and the water is sold to the miners by measure. The unit in New Zealand is a "government head," and in the United States a "miner's inch," following the system in vogue in Spain and Italy. The amount of water that will flow through an orifice 1 inch square, cut in a board 1 inch thick, under a head of water that varies with the custom of the locality, but is usually from 4 to 8 inches, is called a miner's inch. The amount of flow in twenty-four hours is called a "twenty-four-hour inch," and similarly there are ten-hour and twelve-hour inches. The quantity of water delivered varies with the head of water used and the form and size of the orifice for delivery. Thus the amount delivered from an orifice 25 inches long and 2 inches wide is reckoned as 50 miner's inches, although it will be more than fifty times as much as the delivery from an orifice 1 inch square. The twenty-four-hour inch under a head of 7 inches amounts to about 2,230 cubic feet.

The water is conveyed from the pressure-box by pipes of sheet iron or steel. Sharp bends in them are avoided, as the flow of water is checked thereby. They are liable to collapse if the level of the water in them is

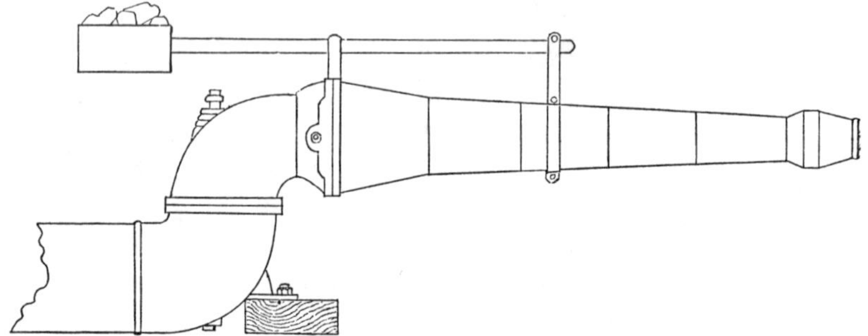

Fig. 50.—Hydraulic Giant or Monitor.

reduced or the pipe-line rises above the hydraulic gradient, and a partial vacuum formed inside; hence they are fitted with valves, which are constructed so as to admit air freely from without. An average rate of flow of water is 3·5 feet per second.[1] The water is discharged through a nozzle called a "giant" or "monitor" (Fig. 50). The nozzle may reach a diameter of 11 inches, and work under a head of 200 feet. For high pressures nozzles should have a long snout; the shorter or bull nose type is used for lower pressures. The interior of the monitor barrel should be provided with long water vanes to put the water in train before it reaches the nozzle.

Breaking Down the Bank.—The nozzle is placed as near to the bank as possible, consistent with the safety of the workers, so as not to waste too much of the initial velocity of the stream of water. Consequently, lofty banks are not advantageous, and if they exceed 200 feet they are usually worked in terraces of 100 feet or so in height.

The jet is, if possible, delivered unbroken against the face of the bank, as its disintegrating power is thus kept at its maximum. However, in some cases, where it is cemented, the gravel is too hard to be economically broken down by the water alone, and blasting is then resorted to, a drift

[1] For details of pipe-line design, see "*Mining of Alluvial Deposits,*" W. E. Thorn and A. W. Hook, 1929, p. 36.

being run into the bank, and cross-cuts made at the end in which the powder is placed.

It is good practice to move the whole face of the bank through the sluices at one operation, thus maintaining the fine and coarse materials in their proper proportion and permitting an easy re-start after a stoppage.

Washing the Gravel in the Sluices.—The sluices in which the gold is caught are constructed on exactly the same principles as those already described, but are larger and of more massive construction, in accordance with the great quantities of gravel to be handled and the continuous nature of the work. They are made of wood or steel or a combination of both. The sluices are commonly called "flumes," but it is better to restrict the use of this word to a conduit for carrying water only. The sluice boxes used in hydraulic mining, though, as usual, only 12 feet long, may be up to 20 feet wide and from 2 to 3 feet deep, with a gradient varying from 1 in 30 to 1 in 20; they are lined with heavy planks on the sides, and the pavements are made of more durable material than is usual in shallow placer sluicing, wooden blocks, rocks, or T railroad iron being most usually employed. The wooden blocks are from 12 to 30 inches square, and from 8 to 13 inches deep. They are usually made of one of the softer varieties of pine which "broom up" under friction, and thus present a better catching surface. The blocks are cut across the grain of the wood, and are set side by side across the sluice, each row separated from the next by strips of wood to which they are nailed, while they are also kept in position by the side lining which is placed upon them. The interstices in the block pavement act as gold catchers, and are filled with small stones, or, with less advantage, allowed to fill up with gravel when washing begins. On account of the rapid wearing away of the wood, much of the gold and amalgam caught is scooped out and carried off again. Wooden block riffles last only a few weeks when in heavy work, but are easy to take up and put down again in cleaning-up; they are discarded when worn so as to be only 4 to 6 inches thick.

Rock pavements are made of those boulders which are most easily obtained in the particular district. Basalt is generally used, oval stones 15 or 18 inches long and from 9 to 12 inches thick being selected and placed on end, with a slight slant in the direction in which the current flows. They are held in place by wooden planks, which divide the sluice into compartments, so that if one stone works loose the pavement as a whole is not affected. The interstices as before are filled with gravel. Rock pavements are very durable, lasting from three to six months, but require more grade to the sluice, and occasion loss of time in cleaning-up and repaving the sluices. Consequently, they are never used near the head of a sluice, where cleaning-up is a frequent operation, but are often used for the lower parts of sluices, where they sometimes alternate with block riffles, and are especially suited for tail-sluices which are only cleaned-up once a year. Rock pavements cost less than other forms of riffles.

Iron riffles, which usually consist of T-iron rails, are placed longitudinally in the sluice, closely packed side by side. They present a large amount of space available for catching the gold and amalgam, last well, present little resistance to the current (so that the grade may be low while the duty of the water remains high), and are easily taken up and put down. They are, therefore, generally used at the head of the sluices. Though their first cost is higher than that of wooden blocks, they are more economical in the end, owing to the saving of time in cleaning-up and to their longer life.

In other instances where the sluice is necessarily very short, most of the stones are first separated from the gravel, and the finer material is then passed over a sluice paved with transverse angle-iron riffles, placed with the hollow side facing down stream (Fig. 51, in which the arrow shows the direction of the stream); these iron riffles are placed 2 or 3 inches apart. To catch fine gold there may be a false bottom to the sluice, formed by a perforated iron plate through which some of the water and the finest particles of the gravel fall on to cocoa-nut fibré matting, laid on the true bottom of the sluice.

Mechanical sluice boxes, in which the material is kept agitated so that the gold may settle, have also been described.[1]

The sluice is often divided into two by a median longitudinal partition, so that one side may be at work while the other is being cleaned-up or repaired. Both sides may be worked when water is plentiful. There are usually unpaved rock-cuts above the sluice, leading to it from the places undergoing the process of piping. These rock-cuts are rarely supplied with mercury, and very little gold is usually caught there.

The length of the sluice depends on the cost of construction and of maintenance, as compared with the value of the gold saved owing to

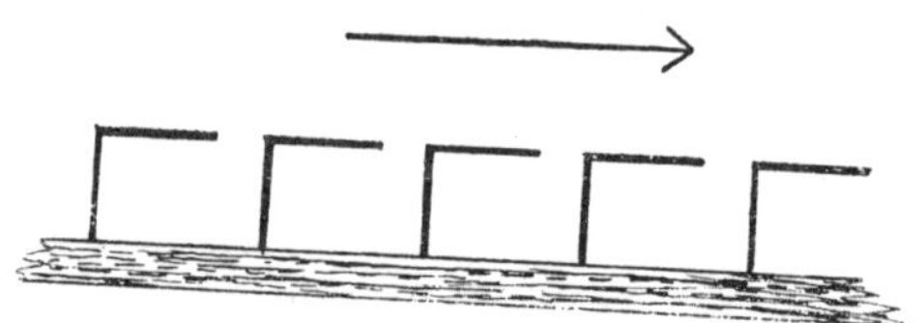

Fig. 51.—Angle-iron Riffles in Sluice.

increased length of the system. The length may be diminished by a plentiful use of drops, grizzlies and undercurrents. Coarse gold is, of course, soon caught, but fine gold may successfully evade all the riffles of a long sluice. The average length of the sluice is about 1,000 feet. The maximum size of material admitted to the sluice box varies according to the nature of the material being treated. The most difficult stuff to treat is fine heavy alluvium not containing much stone, as it requires more water and a steeper gradient.

Steep grades effect disintegration rapidly, thus shortening the length of the sluice, and enable all but the largest rocks to be sluiced, but less gold is then caught and a more plentiful use of undercurrents is necessary. It is considered necessary to have a sufficient depth of water to cover the largest boulders to be sluiced, but the deeper the water, the more difficult it is to save gold.

The "duty" of the miner's inch varies from about 1 to 5 or more cubic yards of gravel.

The Use of Mercury.—Mercury is added daily during the run in gradually lessening quantities, the object being to keep the mercury uncovered and clean at the top of the riffles. The feeding is regulated by the appearance of the amalgam in the sluice, the additions being made in the riffles near the head-box and in the undercurrents. The loss of mercury is usually from 5 to 10 per cent. of the amount used per run. When cemented gravels

---

[1] *Min. Mag.*, 1928, **38**, 144.

are being treated, owing to the extra amount of trituration required, the loss may be as high as 30 per cent. These losses are the more serious, for the reason that amalgam is more easily lost than pure mercury, so that a heavy loss of mercury denotes a heavy loss of gold.

Cleaning-up does not differ from that already described when dealing with smaller sluicing work. Usually from 50 to 95 per cent. of the total yield of amalgam is caught in the first twenty or thirty boxes, which are cleaned-up frequently.

The bullion obtained by retorting the amalgam from the sluices is finer than that from quartz mills, although California placer gold is often as low as 850 fine. The remainder is mainly silver, but copper, lead, iron, and some of the minerals existing in the gravel also occur. The amalgam from the head of the sluices yields finer gold than that caught lower down and in the undercurrents.

The tail-sluices usually terminate on the side of a cañon, in a river, or in the sea, care being taken not to encroach on agricultural land in settled areas, and not to cover up other valuable mineral land.

Treatment of Mine Dump.—At Golden Horseshoe (W. Australia) a dump assaying 1·65 dwt. per ton has been treated by sluicing with hydraulic monitors, agitation with spent cyanides, filtration in Oliver filters, and ultimate recovery by the Merrill Crowe zinc dust precipitation system[1] (see Chap. XIV). The cost is 24·85 pence per ton.

Treatment of Cement Gravel.—In many cases the gravel from deep leads won by drift or shaft mining is cemented by iron oxides or clay into a conglomerate which is too tough to be easily disintegrated in the sluices. It is then passed through "cement mills," which closely resemble the stamp battery to be described in the next chapter. The gold may be caught on amalgamated copper plates, and then passed to a short line of sluice boxes.

It is in some cases an advantage to keep compacted gravels exposed to the air during a few months before washing them, as they "slack" and disintegrate under the influence of the weather, and subsequently are more easily treated, while for a similar reason, tailing is sometimes impounded, and re-washed after some time has elapsed. The disintegration of cemented material, which has been "slacked" by exposure to the weather, is usually completed in a *cement-pan*. This is a cast-iron pan with perforated bottom, and with a gate in the side for the removal of boulders, which are mostly barren and are separated from the auriferous material by this system, instead of being crushed and mixed with it, as is the case when stamp-mills are used. In the pan, four revolving arms, furnished with plough-shares, break up the gravel, which is carried through the apertures in the bottom by a stream of water, and falls into the sluice. A pan 5 feet in diameter and 2 feet in depth will treat from 40 to 120 tons per day, according to the nature of the gravel.

The Hydraulic Elevator.—In this machine a jet of water under high pressure forces water, gravel, and boulders up an inclined plane, and delivers them all at the head of the sluice, which may be as much as 100 feet above bed-rock. The differences in construction between the machines made in Australia, New Zealand, and the United States are only matters of detail. They consist essentially of an upraise steel pipe, having a diameter of

---

[1] *Eng. Min. Journ.*, 1932, 133, 10; *Aust. Inst. Min. Met.*, 1933, No. 91, p. 113; *Inst. Mng. Met.*, 1936, *Bull.* 379.

from 12 to 24 inches, which terminates below in an open funnel; a hydraulic nozzle, delivering water under the pressure given by a head of from 100 to 500 feet, projects into this funnel, and sand and gravel can also enter round the sides or through a special orifice. The inclination of the upraise pipe is usually from 45° to 65°. The top of the upraise pipe is turned over and terminates above a sluice, into which the gravel falls and is washed in the ordinary way. The Evans elevator made by the Risdon Iron Works, San Francisco, is shown in Fig. 52. A is the orifice through which the water for the jet is forced, B the main suction opening, and C and D two auxiliary suction openings, which can be connected with pipes of any length and serve for raising water or fine material. The pipe is greatly contracted at the throat or lower end of the elevator, into which the nozzle discharges. The nozzle and the throat are sunk in a sump excavated in the bed-rock, and the gravel is washed down by any means (usually by a jet from an ordinary hydraulic nozzle) into this sump. The entrance to the upraise pipe is protected by a coarse grating, which prevents large stones, pieces of wood, etc., from entering it. The force of water is enough to complete the disintegration of the gravel during its passage through the upraise pipe, so that a short sluice is enough to effect the washing proper. If the excavation is carefully arranged, it may be kept funnel-shaped, so that the elevator, once placed in a sump, may be worked there permanently without being moved. When the pit is large enough, the washing may be done inside it, only the tailing being raised to the surface by the hydraulic elevator. The head of water required varies according to the vertical height through which the gravel must be raised; a head of about 60 feet is required for every 10 feet of vertical upraise.

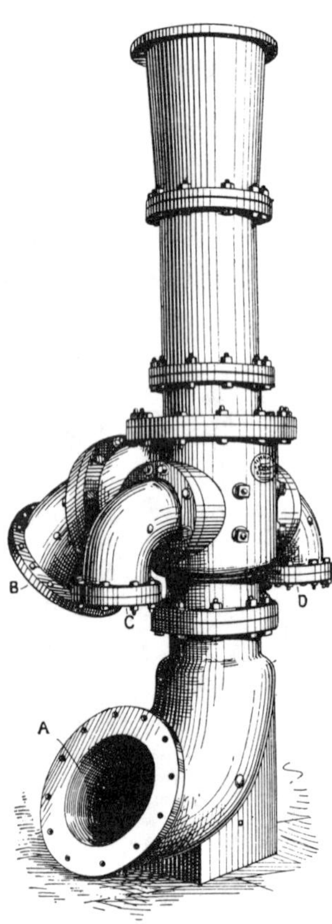

Fig. 52.—Hydraulic Elevator.

Wherever the necessary head of water is available, the hydraulic elevator is recognised as a good method of working flat placers, river-bars, etc., or any deposits which are either below the water level of the district, or which have not sufficient fall for the disposal of the tailings by gravity.

CHAPTER VI.

PRIMARY ORE CRUSHING.

Primitive Methods of Crushing and Amalgamation.—In all countries, when the richest alluvial deposits have been worked out, efforts have been made to extract the gold from the various hard, auriferous materials met with in veins. The earliest machines used for the purpose in Egypt were stone slabs, on which the gold quartz was broken by stone-hammers,[1] hard stone rubbing mills with mullers for coarse crushing, grinding mills or querns from 18 to 20 inches in diameter for fine grinding, and stone inclined tables, on which the particles of rock were washed away from the gold. The washing was completed in broad, flat dishes.[2] Hollowed-out stone mortars for grinding gold ore have also been found in Wales, Central America, the Pyrenees, and Transylvania. In order to separate the pulp from the uncrushed lumps, the Egyptians, in common with other races in ancient times, employed sieves, but in extracting the gold from auriferous sands they used raw hides, on which the flakes of gold were entangled.[3] These devices closely resemble those recently in use in many parts of the world.[4]

The date of the first use of mercury for amalgamation is unknown, but it has no doubt been used for this purpose for the last 2,000 years. The earliest mention of quicksilver itself appears to occur in the works of Aristotle,[5] who speaks of it as fluid silver, and in those of Theophrastus, about B.C. 300 ; but Diodorus, in the account just mentioned,[2] does not refer to its use. Only a few years later, however, Vitruvius,[6] about B.C. 13, described the manner in which, by the help of mercury, gold was recovered from cloth in which it had been interwoven, and in Pliny's time the separation of gold from its impurities generally by the same means was well known.[7] It is probable that this knowledge was never afterwards entirely lost, although the references to it in the Middle Ages are very scanty. For example, Geber,[8] in the thirteenth century, was aware that mercury would dissolve considerable quantities of gold and silver, but not earthy materials, and Theophilus, the monk,[9] in the eleventh century, carefully described the method of washing the sands of the Rhine on wooden tables, the final

---

[1] Gowland, " Huxley Mem. Lecture," *J. Roy. Anth. Inst.*, 1912, 42, 256.

[2] *Diodorus Siculus*, 3, 13. Booth's translation (London, 1700, p. 89) is given in Hoover's "*Agricola*" (London, 1912), p. 279.

[3] Beckmann, "*History of Inventions*" (London, 1846), vol. ii., p. 334.

[4] An interesting account of some primitive methods of treating gold quartz employed by the Chinese is given by Henry Louis in *Eng. and Mng. J.*, 1892, 54, 629, from which it appears that these methods bear points of resemblance to those of the Egyptians.

[5] *Meteorologica*, iv., 8, 11. Hoover, "*Translation of Agricola*," p. 432.

[6] *Vitruvius*, lib. vii., cap. viii. For translation, see Hoover, "*Agricola*," p. 297.

[7] Pliny, *Naturalis Historiæ*, lib. xxx., cap. vi., sect. 32. Quoted in full in Percy, "*Metallurgy, Silver and Gold*," p. 559.

[8] Salmon's "*Geber*," cap. xlvii.

[9] *Theophili*, lib. iii., chap. xlix.

operation consisting in treating the concentrates with quicksilver for the removal of the gold. The extraction of silver by mercury on lines precisely similar to those used in the patio process is described by Biringuccio,[1] and had apparently been a secret art for some time previously.

Mercury was introduced into Mexico as a means of extracting the precious metals by Bartolomé Medina in the year 1557,[2] and its use doubtless soon spread to Peru and the neighbouring countries. When Barba wrote in 1639 there were three amalgamation machines in use in Peru [3]—viz., the *mortar* (used in the Tintin process), the *trapiche* or Chilian mill, and the *maray*. The *mortar* was hollowed out in a hard stone, the cavity being about 9 inches both in diameter and in depth. The ore was triturated with water and mercury in this mortar with the aid of an iron pestle, while a stream of water, flowing through, carried away the crushed particles. The slimes were afterwards roughly concentrated, but most of the gold remained in the mortar with the mercury.

The *maray*, although equally primitive, was probably of greater capacity ; it made use of the principle employed in the bucking hammer. About the year 1825 Miers [4] saw it in Chili. It consisted of a flat or slightly concave stone, 3 feet in diameter, on which lay a spherical boulder of granite about 2 feet in diameter. This was rolled to and fro by two men seated on the ends of a long wooden pole firmly fixed to the boulder. Ore, water, and mercury were ground together in this machine, and then washed down.

The Peruvian *trapiche* had a circular bed of hard stone, but only one stone runner, which was driven by mules. The Chilian mill was used to prepare ores for treatment in the *arrastra*, which was not mentioned by Barba, and may perhaps be regarded as an outcome of the *trapiche*.

Among other ancient methods of amalgamating ores, the *tina* and *cazo* systems may be mentioned. The *tina* system, largely used in Chili, at least as late as the year 1853,[5] is a modification of that much used in the Tyrol. In Norway and Chili charges of crushed ore were introduced into wooden tubs (*tinas*) furnished with cast-iron bottoms and stirred up with water and mercury by means of revolving iron agitators, comparable to the mullers of the modern amalgamating pan. After the agitators had been at work for a few hours, the mercury was drawn off and squeezed through calf-skin or canvas bags. The *cazo* process, introduced in Peru in 1609,[6] was similar, but the bottom of the vessel was of copper and the pulp was boiled. These processes were used in treating ores rich in silver.

The *Kröncke* process [7] introduced in 1860 was a modification of the cazo process. In 1908 a plant was completed to operate on the gold ores of Hog Mountain, Alabama, by amalgamation with mercury in the presence of copper sulphate, salt, and iron, in a large 5- by 22-foot tube mill. The method was similar to those used in the Kröncke and patio processes. The grinding, chemical treatment and amalgamation went on simultaneously. The pulp was subsequently passed over plates. This method was devised

---

[1] "*De la Pirotechnia*" (Venice, 1540), lib. ix., cap. xi., fol. 142. For translation, see Percy, "*Metallurgy, Silver and Gold,*" p. 560.

[2] Georg Agricola, "*Bermannus . . . übersetzt von Friedrich August Schmid,*" 1806, p. 49.

[3] Barba, "*Arte de los Metales,*" lib. iii., cap. xxi.

[4] "*Travels in Chile and La Plata,*" vol. ii., p. 398.

[5] Percy, "*Metallurgy, Silver and Gold,*" p. 567.

[6] Barba, "*Arte de los Metales.*" *Loc. cit. ant.*

[7] See Collins, "*Metallurgy of Silver*" (1900), p. 69.

after treatment of the ore (red oxidised quartz) by cyanide had been tried and abandoned.[1]

The *Arrastra* was in use in America about 1557. It was a circular, shallow, flat-bottomed pit, 10 to 20 feet in diameter, surrounded by a circular wall of stones, and paved with hard, uncut stones. In the centre revolved a vertical shaft carrying two or four horizontal arms, to each of which was attached a heavy stone by thongs of hide, or by chains. The ore was ground and amalgamated with mercury and water. The stones were moved by mules walking round outside the arrastra. Doubtless the arrastra will continue to be used, occasionally, but only in prospecting.

The Stamp Battery.—The stamp battery, evolved, no doubt, from the pestle and mortar, was not introduced until a comparatively recent date. Beckmann[2] states that mortars, mills, and sieves were used exclusively in Germany throughout the whole of the fifteenth century, and in France stamps were unknown as late as the year 1579. They were first applied to the gold industry at the beginning of the sixteenth century. In 1512 Sigismund Maltitz, "rejecting the dry stamps, the large sieves and the stone mills of Dippoldswald and Altenberg, invented a machine which crushed the ore wet under iron-shod stamps."[3] Detailed descriptions and drawings of the apparatus were given by Agricola[4] in 1556, from which it appears that the dry-stamp batteries in use at that time consisted of three or four square wooden stamps with square iron heads, working on a flat surface without sides (*i.e.* no mortar-box) and raised by levers or cams set in the axle of a water wheel. The wet stamps had a wooden mortar-box with an iron bottom, and an iron screen set in an opening at the end of the mortar through which the water flowed carrying the crushed ore.

In 1767 M. Jars saw stamp batteries in use in the Hartz[5] which resembled those described by Agricola. Even then only a single screen of brass wire 12 inches square delivered the product of three stamps, and in several other districts in Germany screens had not been adopted. The screen was completely protected from the splash of the stamps, so that the sieving of the ore was very slowly effected by a current of water flowing through.

Later, mercury was charged with the ore into the battery, and amalgamated copper plates placed in the mortar, so as to catch gold. No mention of these practices appears to have been on record before stamp batteries began to work in California in 1850, although they had possibly been adopted in Georgia before that time.

The use of the copper plate was probably suggested by the experience in Mexico and South America of the working of the *cazo* process, in which amalgam tended to adhere to the copper sides of the vessel. Nevertheless, the exact date and locality of the introduction of the copper plate remains a matter for conjecture. It was soon found that the plates worked better from the start if they were coated with mercury before they were placed in position.

The scouring action of the pulp on the plates is, however, always great, and becomes more violent in proportion as the stamps are larger. The plate on the feed side was discarded, and that below the screens

---

[1] T. H. Aldrich, jun., *Trans. Amer. Inst. Mng. Eng.*, 1908, 39, 578.
[2] "*History of Inventions*," vol. ii., p. 334.
[3] Agricola, "*De re Metallica*," p. 246; Hoover's translation, p. 312.
[4] "*De re Metallica*," lib. viii., p. 220; Hoover's translation, pp. 284, 287.
[5] "*Voyages Métallurgiques*" (Paris, 1780), vol. ii., p. 309.

curved away so as not to be struck directly by the splash from the stamp.

Later, the tendency was towards the discontinuance of inside amalgamation, and the abolition of gold-catching plates in the mortar. On the Rand the introduction of coarser crushing in the battery, followed by regrinding, assisted this movement.

Before 1850 it was the practice, in districts other than Transylvania, to treat the pulp after it had left the mortar mainly by passing it over inclined tables covered with blankets,[1] much in the same way as Jason may have seen the golden sands worked in Asia Minor. The sands accumulating on the blankets were washed off at intervals and ground in mills with mercury. In 1923 blankets were introduced on the Rand.

Amalgamated copper plates over which the pulp flowed were tried, after those placed in the mortars had been proved to be beneficial, but in Western America were at first almost everywhere rejected,[2] probably owing to heavy erosion and to the great depth of the stream of ore and water made to flow over them. When long afterwards, about the year 1870, this mistake began to be rectified, the value of the plates was soon recognised in California.

The pulp is led over the surface of the inclined plates in a thin stream, not more than a quarter of an inch deep. The pulp does not run down in a regular stream, but in a series of little wavelets which tumble over and over, and are supposed to bring every part of the pulp successively in contact with the amalgamated surface. The catching powers of the plates are thus practically independent of the tendency of the particles of gold or amalgam to sink to the bottom of the stream. Gold however is caught more easily in proportion as it contains less silver (and is of higher density), so that when the particles of metal contain a large proportion of silver, and are, therefore, of comparatively low density, the yield on the plates is generally poor. The plates are wiped down with rubber or brushes as often as required, and the gold recovered in the usual way from the amalgam thus collected.

The ordinary method of reduction and amalgamation of gold ore in a stamp battery consists of the following operations:—

1. The ore is broken down to a moderate size in a rock-breaker.

2. It is then fed with water into the mortar-box of a stamp mill, where it is pulverised to the required degree of fineness. The blows of the stamps splash the water and pulp against screens set in the side of the mortar, the finely divided ore being ejected in this way.

3. On issuing from the battery, the pulp is allowed to run over a series of inclined, amalgamated copper plates, by which some of the gold is amalgamated and retained.

4. The residue or tailing is subjected to further treatment, usually either by cyaniding or concentration.

5. At intervals the gold amalgam is wiped or scraped off the copper plates, the excess of mercury separated by squeezing through filter bags of chamois leather or canvas, and the pasty amalgam thus obtained retorted in order to distil off the mercury.

6. The retorted bullion is melted in crucibles with the necessary fluxes, and cast into bars, which are then sent to a refinery.

---

[1] See p. 233.

[2] See G. Küstel, "*Nevada and California Processes of Silver and Gold Extraction,*" p. 61. (San Francisco, 1863.)

The following is a general description of the machinery employed in primary reduction of ore :—

Ore Bins.—For mixed ore as received from the mine, it is an advantage to install large bins of sufficient capacity to hold several days' supply for the stamps, in case of breakdown or other delays in the mine. Bin floors were formerly made of planking, laid with the lengths in the direction of the slope. The slope is at least 45° in order to enable the ore to move downwards by gravity. The wear on the inclined bin floor is, however, very great, and the floor is generally protected by steel plates or bars, or the whole bin may be made of steel plates. The ore is discharged through a sliding door, which is also well protected against wear. Ore bins with flat bottoms have greater capacity, but necessitate an additional handling of part of the ore. In this case the bottom of the bin can be made of wood throughout.

A shoot from the mixed ore bin door leads to a grizzly, through which the fine ore drops into the main battery ore bin. The larger pieces of rock are discharged either (1) into a coarse ore bin, or (2) upon the platform by the side of the rock-breaker and on a level with its mouth, into which they are shovelled by hand, or (3) on to a travelling belt which feeds the rock-breaker continuously.

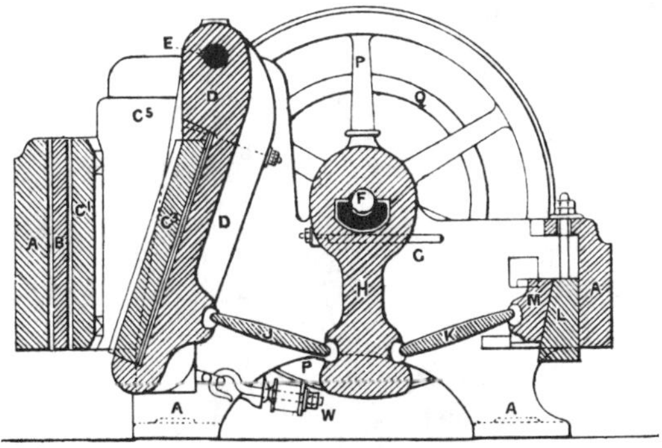

Fig. 53.—Blake Crusher.

Rock-breakers.—There are two classes of these machines in use, viz. :—(*a*) Jaw crushers with reciprocating motion, and (*b*) Gyratory crushers. The Blake and the Dodge crushers are representative of the former class, and the Gates and Traylor crushers of the latter.

The *Blake Crusher* is shown in section in Fig. 53. The rock is crushed between the stationary jaw, B C^1, and the swinging jaw, D C^3, which is pivoted at E, and moved by the eccentric, F, through the toggles, J K. The jaw plates, C^1, C^3, are usually of manganese steel or some other special cast steel, suitable for resisting hard wear ; they have longitudinal corrugations. The machine works at about 250 or 300 revolutions per minute. At each revolution the moving jaw is advanced towards the other, and the lumps of rock which have dropped down between the jaws are broken ; as the moving jaw recedes, the fragments slip lower down and are further crushed at the next advance, and this process is repeated until the ore is small enough to pass out at the opening at the bottom. The distance between the jaws at the bottom limits the size of the fragments, and this distance may be regulated at will by moving the wedge, L, or by changing the length of the toggles, J K. The jaws wear most at the centre and lower end, which is partly compensated for by reversing. The "Osborne" composite jaw, in use on the Rand, is designed to meet this difficulty.[1] The capacity of the machine is great, being about 300 tons of ordinary rock per day of twenty-four hours in the case of the machine whose dimensions at the mouth are 20 inches by 10 inches, when the lower edges of the jaws are set to approach within 1½ inches of each other. The power required for this is stated to be 14 H.P. In large "coarse breakers" on the Rand 20 to 30 H.P. is required for a breaker dealing with from 40 to 60 tons of ore per hour. Jaw breakers of this type are used on the Rand for coarse crushing to 4 or 5 inches maximum, as they make little "fines."

In the *Dodge Crusher*, shown in Fig. 54, the moving jaw is pivoted below instead of above. The effect of this arrangement is to make the product more uniform in size, and as there is little or no motion of the movable jaw at the delivery aperture, this may be made as narrow as desired, so that a

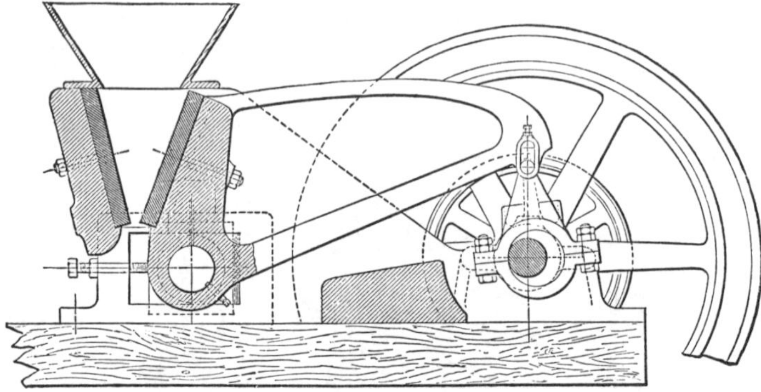

Fig. 54.—Dodge Crusher.

finer product can be obtained, although it is at the expense of capacity. The Dodge crusher is more particularly recommended for fine crushing in concentration works, or where the product is to be subsequently passed through rolls.

Authorities differ as to the relative advantages of the two positions of the pivot. R. H. Richards decides[2] in favour of placing the pivot above, assigning as a reason that the work of crushing is in that case more evenly distributed, whereas if the pivot is below, most of the power is used near the throat or discharge aperture in recrushing particles which have already been reduced sufficiently (*i.e.* "choke" crushing).

In later Allis Chalmers jaw crushers the jaw plates are smooth, allowing either closer setting for a fine product or 50 to 60 per cent. greater capacity with the same setting. The movable jaw plate is straight for about two-thirds the distance down from the top and is then curved with a contour nearly parallel to the face of the fixed jaw. Thus the plates are almost parallel in the lower third of the crushing zone and the crushing capacity is increased at the fine end.

[1] C. O. Schmitt, "*Rand Metallurgical Practice*," vol. ii., p. 36 (Griffin & Co., 1919).
[2] "*Ore Dressing*," vol. i., p. 35.

The *Gates Crusher* is shown in Fig. 55, and consists of a vertical spindle of forged steel, revolved by a bevel wheel and a bevel pinion. The spindle is

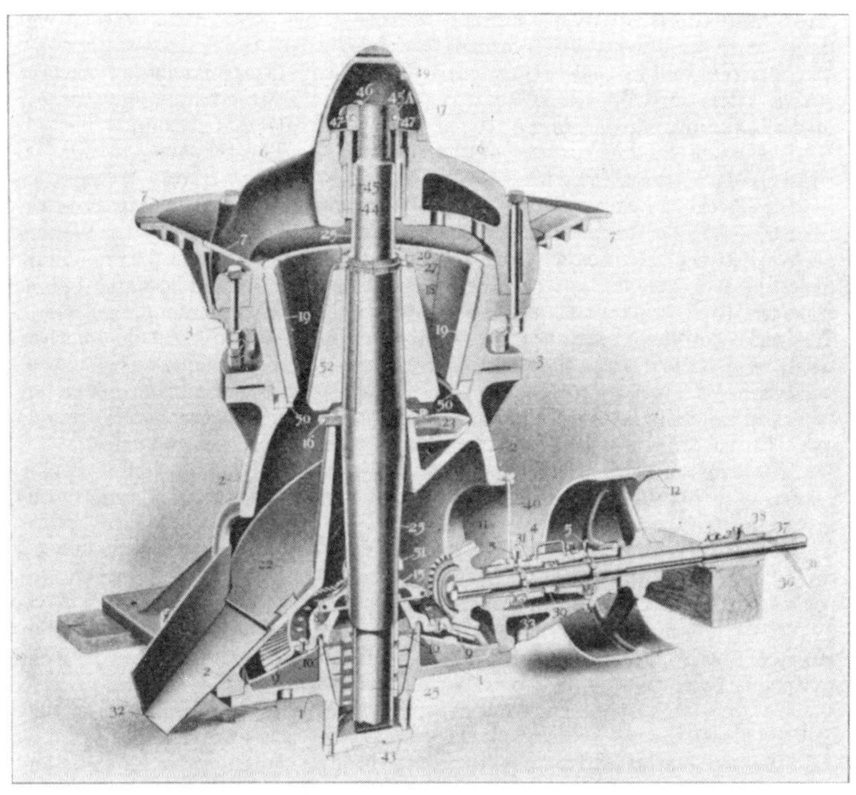

Fig. 55.—Section of Style K "Gates" Rock and Ore Crusher.

LIST OF PARTS.

1. Bottom Plate.
2. Bottom Shell.
3. Top Shell.
4. Bearing Cap.
5. Oil Nipple and Cap.
6. Spider.
7. Hopper.
8. Eccentric.
9. Bevel Wheel.
10. Brass Wearing Ring.
11. Bevel Pinion.
12. Band Wheel.
15. Oil Bonnet.
16. Dust Ring.
17. Dust Cap.
18. Head.
19. Concaves.
22. Wearing Plates.
23. Upper Wearing Plate.
25. Main Shaft.
26. Upper Ring Nut.
27. Lower Ring Nut.
31. Countershaft.
32. Spout.
33. Oiling Chain.
36. Outboard Bearing Base.
37. Outboard Bearing Cap.
38. Outboard Bearing Oil Nipple and Cap.
39. Countershaft Collar.
40. Dust Door.
43. Bottom Plate Cover.
44. Spider Bushing.
45. Shaft Bushing.
45A. Shaft Split Nut.
46. Gib Key for Split Nut.
47. Clamp Bolts for Split Nut.
49. Oil Plug in Dust Cap.
50. Concave Supporting Ring.
51. Main Shaft Collar.
52. Zinc Keys—Poured in Place by Owner.

Key Concave.
Narrow Key Concave.
Babbitting Sleeves.
Babbitting Mandrel for Countershaft Bearing.
Side Dust Door.
Side Dust Door Handle.
Split Nut for Clamp Bolt.
Oil Feed and Drain Pipe.
Oil Pump.
Telltale Pipe for Eccentric.
Filling Pipe.
Telltale Pipe for Countershaft.
Headcenter or Core.
Mantle.

set eccentrically in the bevel wheel, and has a gyratory motion when rock is being crushed. At the top of the shaft is a breaking head, and the shell surrounding this is lined with twelve concave pieces which form the crushing faces, and consist of manganese steel or other special hard cast steel. The faces wear rapidly and are frequently renewed.

The *Newhouse Crusher* (Fig. 56).—This is a gyratory machine for heavy work. It is normally suspended by three cables E from the main structure of the building, thus absorbing the shock and vibration of crushing.

The main casting consists of three parts: (1) The two-armed spider A, which is high enough to permit the feeding of large pieces of ore. It carries a ball and socket bearing F for the crusher main shaft and within it is the fulcrum. (2) The top shell, which is attached by bolted flanges to the spider above and the shell below. The top shell is lined with chilled iron or manganese steel segments G. The liners are symmetrical about a horizontal plane through their centre and may be reversed when the lower ends are worn. The curved crushing surfaces reduce choking to a minimum. (3) The bottom shell, which carries an eccentric bearing, and has heavy lugs to which the suspension cables are attached. Ore discharges on to white iron liners I in this shell and then passes through a discharge spout. The eccentric is carried on a thrust bearing bolted just above the bottom of the bottom shell.

The breaking head H is conical and has a renewable chilled iron or manganese steel mantle. The eccentric to which the head is attached is rotated by a vertical electric motor above the spider.

A special seal between the main shaft and the bottom shell prevents leakage of oil and the admission of dust to the eccentric bearing. Lubrication is provided by a circulating oil system.

The crushing is done between the continually advancing and retreating faces of the head and the main liners of the casting, being controlled by the extent of the gyratory motion imparted to the head.

The receiving opening may be up to 14 inches and the discharge opening on the close side up to 3 inches, according to the size of the machine.

The high-carbon driving shaft passes through the hollow main shaft and is connected at one end to the driving motor and at the other to the eccentric drive plate.

TABLE XXI.—CAPACITY OF NEWHOUSE CRUSHERS (SHORT TONS PER HOUR).

Width of Feed Opening (inches).	Width of Discharge Opening on Close Side (inches).						
	¼	½	1	1½	2	2½	3
5	15	27	48	72	...	...	...
10	...	45	80	109	140	...	...
14	..	110	210	300	375	450	530

Gyratory crushers are of far greater capacity than machines with reciprocating motion, and have accordingly come into use where a large output is required.

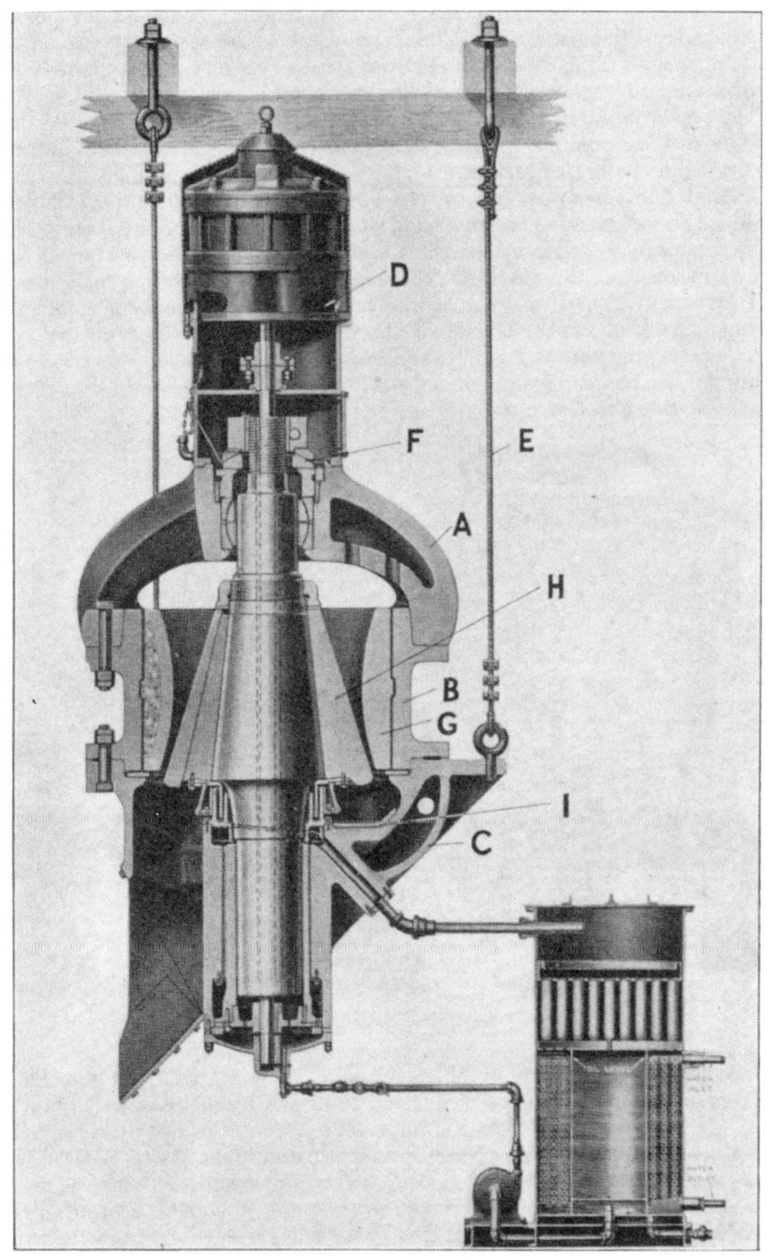

[*Allis-Chalmers.*

Fig. 56.—Newhouse Crusher.

On the Rand, gyratory breakers are used as "fine breakers," taking the product of the coarse breaker after it has been passed through a trommel and sorted, and breaking it to a maximum of about 1¾ inches. The cost of fine breaking in gyratories is about 1½d. per ton, as against approximately the same amount for cone crushers and 2½d. per ton for jaw and disc crushers.

The power required is ¾ to 1 H.P. per ton.

It is now customary to place the rock-breaker in a separate building distinct from the battery house.

A grizzly or screen of steel or iron bars, set 1 or 2 inches apart, is often employed to separate the fine material, which is passed straight to the stamps. Washing trommels are in use on the Rand. When the stamp battery is used only to crush the ore, which is subsequently treated in other machines, it is of great advantage to separate the fine product of the rock-breaker by sieving, instead of passing the whole through the stamps. This arrangement increases the output and prevents unnecessary sliming of the ore. The product of the rock-breaker is mixed with the original fines, and led by means of a shoot direct to the automatic stamp feeders.

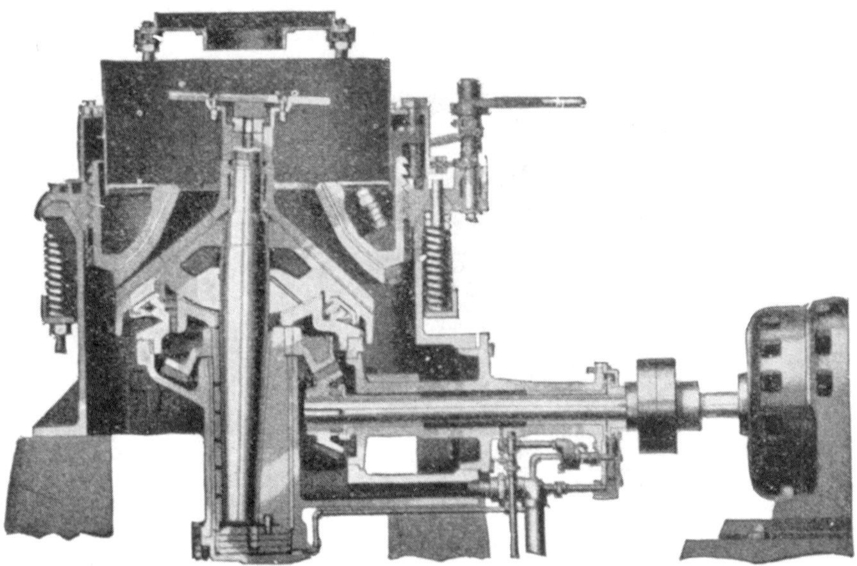

Fig. 57.—Section of Symons Cone Crusher.

The *Symons Cone Crusher*, Figs. 57 and 58, is stated to be capable of effecting a larger single-stage reduction than the ordinary jaw or gyratory type. It is in use for feeding stamps, but during the last few years has also come into favour for the intermediate and final stages of crushing. It has some resemblance to a gyratory crusher, but rotates at a greater speed. The main shaft is supported on a large single spherical bearing placed immediately under the crushing head, instead of being suspended from a spider, thus permitting free discharge from, and easy entry into, the machine. The bearing takes the whole of the load, and has a sealed cover protecting it from grit and dust. A gyrating distributor plate attached to the head ensures uniform distribution of material in the crushing cavity.

A long gear-driven eccentric carries the main shaft. The rotation of the former causes the head to gyrate, the centre of gyration being in the crushing head. Lumps of ore entering the crushing zone at the top of the bowl are cracked by short powerful strokes. Lower down the stroke is longer and less powerful, thus causing the ore to be crushed more rapidly in the finishing stages and quickly discharged without choking. Over-crushing is therefore a minimum. A large reduction ratio is made possible and this is enhanced by the curved shape of the bowl.

In the lower part of the crushing zone the surfaces of the head and bowl are parallel, and the intervening space is deep enough to ensure that with a suitable speed of gyration no piece of ore can pass through it without suffering two or three impacts from the head. The size of the issuing material is therefore fixed by the distance between the bottom edges of the head and bowl when they are closest together.

Fig. 58.—Symons Cone Crusher showing Crushing Cavity.

An adjusting ring on the outside is held down to the main frame by springs which are flexible enough to permit the moving parts to rise if pieces of iron or similar foreign material enter with the lump ore. Coarse threads on the inner side of the adjustment ring connect with corresponding threads on the outside of the bowl. The manganese steel liner within the bowl can be adjusted for wear, or change in size of product, whilst the machine is running, by means of a windlass and chain which cause the bowl to rotate in its threads. Each size of machine has two ranges of capacity due to the fact that it can be fitted with a coarse or fine crushing bowl according to the work required.

The rock entering the feed opening drops first on to the distributing plate, and then on to the head, being crushed against the bowl. Due to its eccentric motion the head recedes from the bowl during part of its circuit, thus permitting the crushed particle to drop and come into contact next with head and bowl when they are still closer together. Successive pinches of this nature as the particles fall lower between the jaws, and afterwards in the parallel zone, reduce the pieces to a dimension equivalent to that of the smallest interval at the closed side of the crushing zone. They are then ready to be liberated as the cone recedes (see Fig. 58).

A Symons short-head cone crusher (fig. 59) has been designed on similar principles to the older type, but having a larger intake opening. It will crush the product from the larger machine down to $\frac{1}{10}$ inch size. More crushing is done at the top and the more steeply angled jaws give faster crushing and larger capacity. The parallel zones at the bottom are longer and give better sizing. This machine aims to replace rolls for preparing the feed to ball mills. As a rule the short-head machine operates in closed circuit.

Fig. 59—Symons Short-Head Crusher.

Choke crushing is not ordinarily possible with a Symons crusher. Where this crushing is needed a second gyratory crusher with a suspended shaft is recommended. Such is the Traylor machine, in which there is a curved bowl similar to that in the Symons, but otherwise the mechanism is that of the usual gyratory. The suspension of the shaft limits the head movement to a smaller circle of gyration than is the case in the cone crusher. The ratio of reduction is still large enough however to enable crushing of the ore from the primary breaker to $\frac{1}{2}$-inch size.

Crushers are usually driven by means of a belt and pulley, but direct connection to slow-speed motors has also been tried. The tex-rope, consisting of a number of V-shaped endless rubber belts running on grooved pulleys, is better than the flat belt. The grip of the former is great enough to allow a reduction to 30 per cent. of the distance usually arranged between pulley centres for a flat belt. Thus greater safety, economy of space and easier adjustment of stretch by moving the motor on its rails are possible.

The following figures show the power required for Symons cone crushers.[1]

TABLE XXII.

Size of Crusher (ft.).	Max. Feed. (Ins.)	Max. Product. (Ins.)	Avge. Power. Kw.	No Load Power. Kw.
3	$2\frac{1}{2}$	$\frac{1}{2}$	23·5	6·5
7	6-8	$\frac{3}{4}$	128	27
7	6	$\frac{3}{4}$	99	24
7	12	1	80	19

Table XXIII gives a comparison, as regards the grading of the product and the capacity, between jaw crushers and slow and rapid gyratory crushers. The increased capacity of the last two, obtained with economical power consumption, will be noted.

TABLE XXIII.—Capacities of Jaw and Gyratory Crushers.

Type of Crusher.	Grade of		Capacity. Tons per hour.
	Feed.	Product. Per cent. $\frac{3}{4}$ mesh.	
30 in. × 15 in. Jaw.	−8 in. + 2 in.	30	40
Slow gyratory (200 r.p.m.)	,,	40	60
Rapid gyratory (730 r.p.m.)	,,	70	70
Ditto	− 10 in.	40	140

The Stamp Battery.—A stamp is a heavy iron, or iron and steel, pestle, raised by a cam keyed on to a horizontal revolving shaft, and let fall by its own weight. Stamps are ranged in line in groups of five stamps each, which have a mortar box in common. Fig. 60 represents the side view, and Fig. 61 the front view, of a ten-stamp battery, with the amalgamating tables removed to show the foundation timbers or mortar blocks, A.[2] Fig. 62[3] (see folder) gives the details of a Transvaal ten-stamp battery (Simmer Deep).

The foundations are of the highest importance. If they are badly made through carelessness or false economy, the efficiency of the battery is greatly decreased, and it soon shakes itself to pieces. The blow of a stamp is partly spent in crushing the ore, and partly in producing a concussion or jar acting

[1] *Min. Ind.*, 1930, 39, 673.

[2] For full details of construction of stamp-mills, see Richards, "*Ore Dressing*," vol. i., pp. 144-230; also, "*Rand Metallurgical Practice*," vols. i. and ii. (Griffin), from which many of the details in the preceding and following sections are taken.

[3] Schmitt, "*Rand Metallurgical Practice*," vol. ii., p. 140.

on the framework and foundations. The amount of energy used up in the latter way depends largely on construction. In preparing the ground for the foundations, the earth is removed until bed-rock is reached if possible, and the latter is then carefully smoothed and covered with a layer of concrete. The wooden mortar blocks of from 6 to 18 feet length are placed upright in

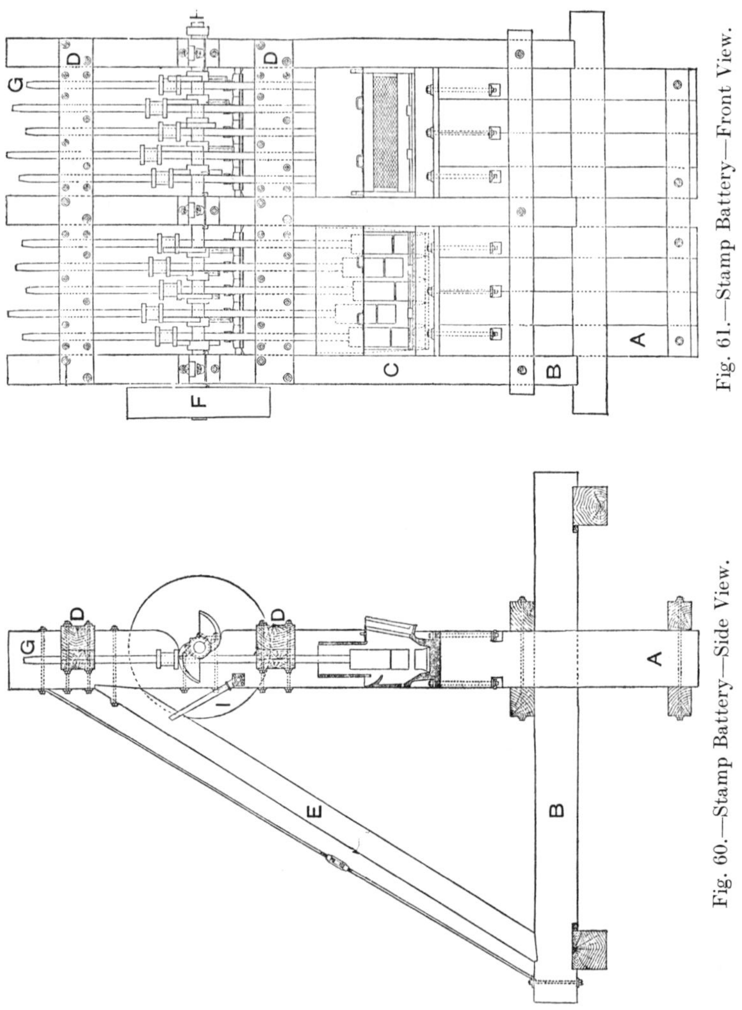

Fig. 61.—Stamp Battery—Front View.

Fig. 60.—Stamp Battery—Side View.

this trench, and the space round filled up with sand, or, as in the Transvaal, solid masonry is built round the blocks. Massive concrete foundations are now almost universal (see Fig. 62). Heavy iron anvils or mortar blocks are also used on concrete foundations. The *framework* is made of wood, iron, or steel. It consists of the massive cross sills, B (Fig. 60), on which rest the battery- or king-posts, C, and the braces, E. The cross sills rest on horizontal mud-sills, placed parallel to the line of stamps. The posts are held together by the guide- or tie-timbers, D. The mud-sills are shown below the cross sills. Frames of a number of different designs are in use. In some mills on the Rand the king-posts have been made of reinforced concrete.

The Mortar.—The mortars are made of cast-iron or steel and differ in shape according to the nature of the ore and the corresponding modifications made in the course of treatment. An ordinary mortar is about 4 feet 7 inches long, 50 inches high, and 12 inches wide on the inside at the level at which the dies are set. The bottom is from 3 to 11 inches in thickness, and has a heavy flange cast on it by which it is bolted to the mortar blocks. For heavy stamps, mortars are larger and heavier,[1] the bottom being 15 inches thick. The mortar blocks are tarred over, all cracks in them having been filled with sulphur, and are then covered with three thicknesses of blanket, carefully coated with tar on both sides. The mortar is placed on these blankets and securely bolted down. This arrangement lessens the chance of the mortar working loose, the jar being diminished. A sheet of rubber, $\frac{1}{4}$ or $\frac{3}{8}$ inch thick, is used instead of the blankets in many mills. Cast-iron or steel lining plates and false bottoms take up the wear inside the mortar and are renewed when worn out.

The width of the mortar varies from 10 to 14 inches at the level of the bottom of the screens. Narrow mortars are best fitted for rapid discharge, but there is more frequent breakage of the screens.

Mortars are cast in one piece, including the housings. The roof of the mortar is made of 2-inch planking, through which holes are cut to admit the stems of the stamps and the water pipes.

Low mortar boxes with practically open fronts were later adopted in consequence of the increase in weight of stamps. They give greater accessibility.

The *depth or height of the discharge* is the distance, measured vertically, from the top of the die to the lower edge of the screen opening. It varies from $1\frac{1}{2}$ to 10 inches, according to the nature of the ore and the general scheme of working. Adjustable battery-screens keep this height constant, in spite of the wearing of the dies. The screen-frame is supported on a removable wooden chuck-block, which is replaced by one of less height when the dies wear down. The height of the die may also be regulated by periodically introducing liners below the die as it wears down. The greater the height of discharge the longer the ore is retained in the mortar and the finer the product. Where coarse crushing is used as a preliminary to fine grinding in other machines, the height of discharge is about 2 inches. Where inside amalgamation was practised a greater height of discharge was maintained.

The lip of the mortar projects some inches, and to it is bolted a cast-iron *apron* about 14 inches long, which may carry an amalgamated copper plate, the *apron-plate.* The pulp is dashed through the screen against a splash-board and falls thence on the apron-plate.

When the mortar is in place, the *dies* are put into it, a layer of sand being often introduced first. The dies consist of two parts, the *footplate* or *base* and the *die* proper or *body.* Fig. 63 shows, in plan and elevation, one of the many forms of dies in use; here the base is octagonal; it is 1 or 2 inches thick, and $9\frac{1}{2}$ to 10 inches across at E. The body is cylindrical,

[1] See "*Rand Metallurgical Practice*"—Smart, vol. i., p. 50; Schmitt, vol. ii., p. 81.

7 or 8 inches high, and of the same diameter as the shoe. On the Rand the standardised die has the following dimensions in inches :—

D	E	F	H
9	9½	9½	7
9½	10	10	8
10	10¼	10¼	8

Shoes and dies are made of steel, and last longer than the iron shoes and dies formerly in use. Chrome steel and manganese steel have been used for

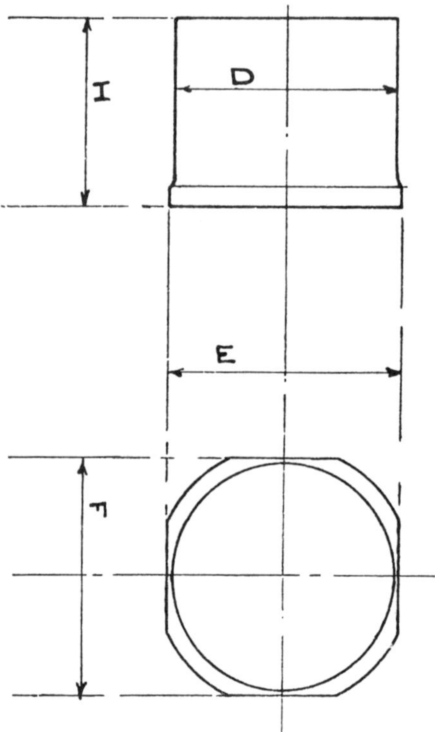

Fig. 63.—Standard Die, Plan and Elevation.

shoes and dies. Shoes and dies are now cast locally from battery scrap. The carbon content of the shoes should not exceed 0·90 per cent. and that of the dies 1·00 per cent., to give the best wear compatible with absence of breakages. Wraight [1] has commented on the following materials for stamp battery shoes and dies :—(*a*) *Cast iron* is cheapest, especially if a foundry

[1] *Trans. Inst. Min. Met.*, 1921, 30, 201.

is attached to the mill. Both shoe and die require to be heavily chilled on the wearing surface. A soft shank is desirable on the shoe.

(*b*) *Forged or cast steel*—an electric furnace engaged on remelting and recasting scrapped shoes and dies gave a product of the following average percentage analysis :—

C	0·64-1·18	P	0·02-0·07
Si	0·12-0·35	Mn	0·35-0·67
S	0·02-0·06	Cr	0·02-0·20

(*c*) *Alloy steels.*—Chromium up to 0·5 per cent. confers beneficial properties on forged steel shoes. Manganese steel has also been used.

It is suggested that the best combination for resistance to abrasive impact would probably be a manganese steel shoe with a chrome, chrome-vanadium or nickel-chrome die. When the body is worn down to within from ½ inch to 1 inch of the base, the die is replaced. Dies wear more slowly than shoes, since they are protected by a layer of pulp over an inch thick. The dies are all renewed together, as it is important that those in the same battery should be of equal height, otherwise one or more will become almost bare of ore, and a disastrous pounding result. If a die breaks, it is not replaced by a new one, but by one worn to the same extent as the others in the battery.[1]

Dies require to be hardened and tempered very evenly, to prevent uneven wear. A die worn down at one side causes a diminution of output, and may result in a broken stem. According to E. E. Aulsebrook,[2] the three centre dies in a mortar wear down at the back faster than at the front, on account of the size of the material which is fed at the back being larger than that which is crushed at the front. For this reason he recommends dies to be turned round once a month. He also states that dies often wear unevenly because the stamps are out of centre.

The *cam-shaft* is constructed of wrought iron or forged steel. It is usual to have a separate cam-shaft driven by a 25 H.P. or 50 H.P. motor for each five or ten stamps, which have thus a separate driving pulley. The cam-shaft is placed at a distance of from 5 to 10 inches from the stem-centre, and is 9 to 10 feet above the mortar bed. The bearings rest on supports often attached to the battery posts, generally on the discharge side. The cast-iron cam-shaft bearings and supports at the City Deep Mill are shown in Fig. 64.[3]

The *cams* are made of cast steel. The double cam, illustrated in Fig. 65, is now in general use. Sometimes cams are cast in two pieces which are bolted together, so that when one is worn out, it can be replaced without first removing the other cams on the shaft, but these sectional cams work loose, and are not much used. Cams are either right- or left-handed. Both are used in the same battery to equalise the cam thrust, the stamps being rotated in opposite directions. The shape of the cam face is the involute of a circle slightly modified at the end so as to stop the upward motion gradually. The radius of this circle is equal to the distance between the centres of the cam-shaft and the stem, which depends on the height to which

[1] For the method of changing shoes and dies, see G. O. Smart, "*Rand Metallurgical Practice*," vol. i., p. 65.

[2] *J. Chem. Met. Mng. Soc. S.A.*, 1904, 5, 12.

[3] Schmitt, "*Rand Metallurgical Practice*," vol. ii., p. 124.

the stamp is to be lifted, so that the curve of the cam varies with the drop.[1] A cam should last several years.

Fig. 64.—Cam-shaft Bearing (City Deep).

[1] For a full discussion of the cam curve, see H. Louis, "*Handbook of Gold Milling*," 1894, pp. 479-491.

The old-fashioned keyed cam has now been replaced in many mills by the *Blanton cam*, which will take any position on the cam-shaft. To secure it two holes are bored in the cam-shaft, and two pins dropping into these hold fast a semi-circular tapering wedge or "bushing." The cam slips over the bushing and tightens itself in working. When it is necessary to take off a cam, a slight blow on the back edge with a hammer loosens it instantly. The rifled Blanton cam (see Fig. 66)[1] is still more convenient than the older form. The bushings are abolished, and the cam-shaft made with ten taper faces, so that its cross-section is like a ratchet-wheel with ten

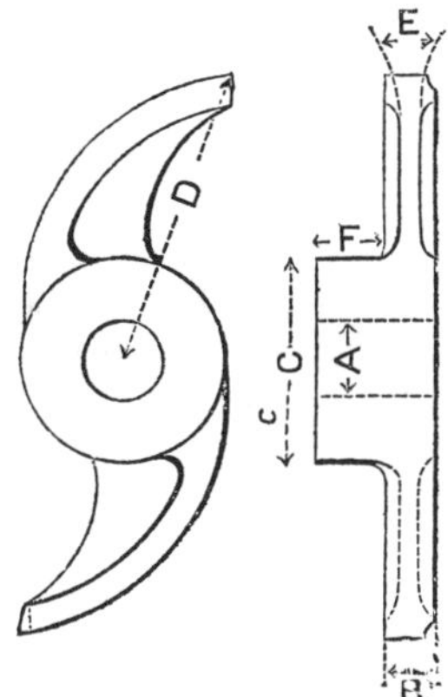

Fig. 65.—Cam, Front and Side Elevation.

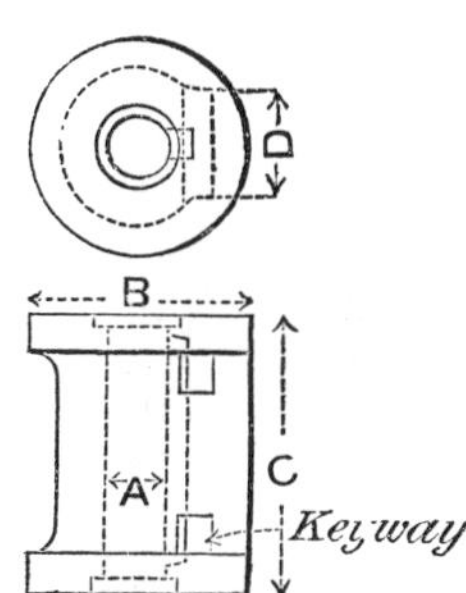

Fig. 67.—Tappet, Plan and Elevation.

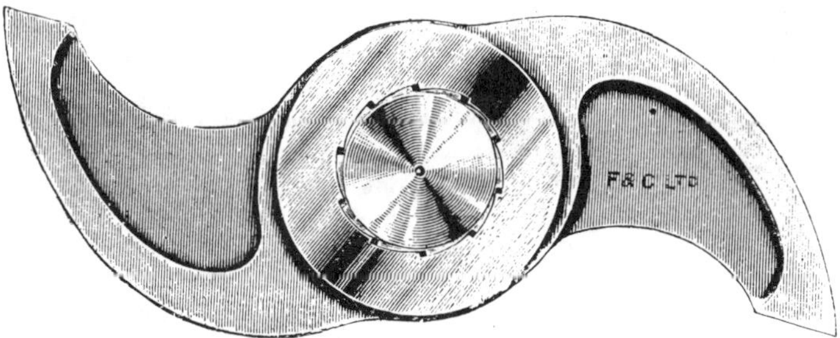

Fig. 66.—Cam with "Blanton" Rifling.

teeth instead of being circular. The bore of the cam has ten corresponding faces, and the cam slips on in any one of ten positions. It is then tightened sufficiently to hold it in position until put into operation, when it tightens itself further on the shaft in proportion to the work it has to do. It was superseded by the Blanton gib.[2]

The cam-face works against the collar or *tappet*, shown in plan and section in Fig. 67, which is bored out to fit the stem of the stamp. The tappet is usually made of hard fine-grained cast steel, and is fitted with a wrought-iron gib, which is pressed against the stem by two or three keys

[1] Schmitt, "*Rand Metallurgical Practice*," vol. ii., p. 113.
[2] White, "*Rand Metallurgical Practice*," vol. i., p. 479.

behind it, thus binding the tappet firmly on the stem while, at the same time, admitting of rapid adjustment to another position. With heavy stamps two or more gibs are used. A tappet much in use has three keys and three gibs. The material of which the tappet is made must be tough, to stand the outward pressure of the key, and hard, to take the wear on the face, without being brittle so that it chips at the point. Daly has invented a modified tappet with two keys and a safety arrangement for driving them home.[1] The cam always strikes the tappet a heavy blow, and this could be diminished by a change in its design, such as that proposed in Behr's cam (see p. 157). The weight of the tappet should not exceed 12 to 15 per cent. of the total weight of the stamp.[2] Adjustment of the tappet on the stamp-stem is required every few days to allow for the wear of the shoes and dies.

The entire end surface of the tappet comes in contact with the cam-face, by which the stamp is raised and, at the same time, rotated. The effect of this is, that the shoe does not strike the ore in the mortar in exactly the same place twice in succession, and the wear of its face is made more uniform. The greater part of the revolution takes place during the raising of the stamp, but the latter does not quite cease to rotate as it falls, and a slight grinding action on the ore has been noticed by many observers. The amount of rotation varies with the fall, the extent to which the cam and tappet are greased, and the state of wear of their surfaces. Tappets can be reversed when one end is worn out. Some millmen assert that tappets may be broken by the cam if keyed too tightly to the stem.

As the cam-thrust is not applied at the centre of the stamp, there is always a considerable side pressure, which greatly increases the friction in the guides and wears out the latter, besides causing a loss of power. Another result of this is that the stamp tends to be inclined when it is released, and so the blow on the die is given slightly to one side—*i.e.* the side of the die on which the cam works. Consequently, there is a tendency for this side to wear down more quickly than the other. To obviate these disadvantages, cams have been introduced with a wide hub, and the two blades set one at each end of the hub, so that they work on opposite sides of the stamp and cause it to revolve in different directions at each successive uplift.

The *pulley* on the cam-shaft (F, Fig. 61) is made of wood on cast-iron or cast-steel centres, which are keyed to the cam-shaft; if iron alone were used, the rapid succession of jars, caused by the dropping of the stamps, would soon cause the material to break. A tightener pulley on the belt driving the cam-shaft is often used, by which the stamps can be put in motion or stopped without interfering with the driving power.

The stamp itself consists of three parts, the stem, the head or boss, and the shoe. The *stem* (G, Fig. 61) is made of wrought iron or steel, or steel in the middle portion with wrought-iron ends, and has both ends tapered for a length of 6 or 8 inches to fit the heads, so that, if one end is broken off, the stem can be inverted and the other end used. The *head* and *shoe* are made of equal diameter—viz., about 8 or 9 inches. The head is of cast iron, or in more recent practice of hard fine-grained cast steel, about 15 to 24 inches long (or in later practice on the Rand 48 inches long, when the stem is reduced to 12 feet), and has a tapered socket at each end, the upper

[1] White, "*Rand Metallurgical Practice*," vol. i, p. 476.
[2] Schmitt, "*Rand Metallurgical Practice*," vol. ii., p. 107.

one for the stem and the lower for the tapered shank of the shoe. When these are driven into their respective sockets, into which a few strips of wood are inserted to keep the two metal surfaces from touching each other, a few blows by the stamp bind them securely together, no other fastening being necessary. Slots are provided at the base of the two sockets, through which wedges may be driven to force out the shoe or stem when necessary. G. O. Smart has designed a head (Fig. 68) [1] in which a hole drilled or cast through the entire length facilitates removal. The head lasts several years, being rarely ruptured.

Standard dimensions for stamp heads, tappets and cams are given by White.[2]

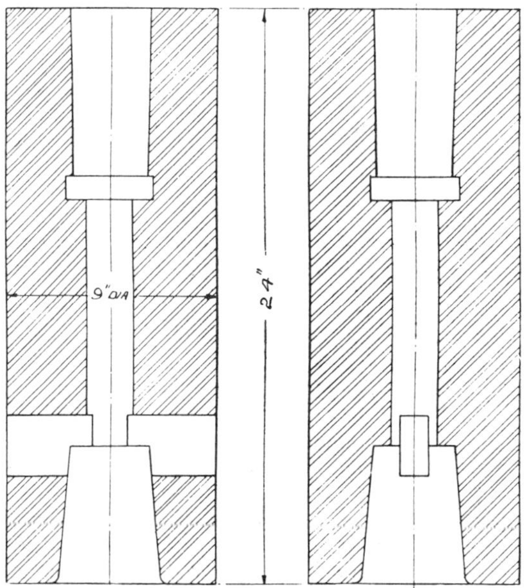

Fig. 68.—Stamp Head with Axial Opening for removal of broken stems.

The *shoe* (Fig. 69) [3] consists of two parts, the shank, which fits into the head, and the shoe proper or butt. It is made of hard forged steel, or latterly on the Rand of scrap steel obtained locally, with the shank made softer than the butt in the tempering. The diameter of the shank is about half that of the butt, but a somewhat thicker shank is now used to prevent breakages at the neck. The shoe is replaced when the butt, which is from 9 to 18 inches in length when new, has been worn down to about 1 inch in length. To keep the total weight of the stamp constant, several sizes of heads are sometimes used in one mill, the heavier heads taking partly-worn shoes. "Chuck-shoes" are inserted between heads and shoes with

---

[1] "*Rand Metallurgical Practice*," vol. i., p. 56.
[2] "*Rand Metallurgical Practice*," vol. i., pp. 476-7.
[3] Schmitt, "*Rand Metallurgical Practice*," vol. ii., p. 99.

the same object. Compensating weights are affixed to the stem as the shoe wears.[1]

The relative weights of tappet, stem, head and shoe, which together make up the stamp, vary considerably. There is an advantage in increasing the diameter of the stem, as one of small diameter tends to spring and bend from the blow of the cam, or when the stamp falls, and to wear the guides rapidly. There is also an advantage in diminishing the length of the stem, as, owing to its elasticity, the effect of the blow when the stamp falls is partly expended in compressing the stem momentarily. For the same reason it has been found advantageous to make the head longer and heavier. Heavy stamps are 2,000 lbs. or more in weight. The introduction of the long head on the Rand raised its weight to 900 lbs., or 45 per cent. of the total weight, while that of the stem fell from 46 to 27 per cent.[2]

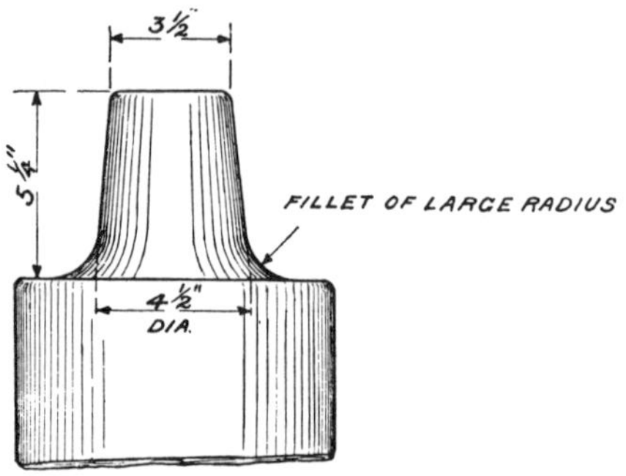

Fig. 69.—Shoe for Heavy Stamp.

The stamp stems are guided in boxes bolted to the wooden guide-timbers, which also serve to hold the battery posts together. There are two of these *guide-timbers* (D, Fig. 61), one within 2 or 3 feet of the top of the battery posts, and the other about 6 or 7 feet lower. The depth of each guide block is about 15 inches. The guides may be made of wood or metal. Wooden stem-guides wear the stems more rapidly than metal ones, in spite of a higher expenditure on lubricants. The guide-beams are sometimes pierced with large square holes in which bushes of wood, with the grain parallel to the length of the stamp, are placed fitting the stem exactly. In this way, the guide-beams themselves are preserved from wearing out. Sectional guides, consisting of a series of iron keys enclosing wooden bushings, are also used. In this case each stem has a guide to itself and the bushings can be renewed by hanging-up the one stamp without stopping the other stamps in the battery. The wooden guide block has practically become obsolete, however, and various forms of individual iron guides are in common use.[3]

Each stamp is provided with a *finger-bar* or *jack* (I, Fig. 60) made of wrought iron, or wood protected with iron, and carried on a separate *jack-shaft*, which is supported on cast-iron brackets attached to the king-post. The jack is for the purpose of raising the stamp and hanging it up out of reach of the cam. When this is to be done, a strip of wood or wrought iron (the *cam-stick*), an inch or more thick, is laid on the cam as it rises, and the

[1] Smart, "*Rand Metallurgical Practice*," vol. i., p. 65.
[2] Schmitt, *ibid.*, vol. ii., pp. 103, 104.
[3] "*Rand Metallurgical Practice*," vol. ii., p. 474.

stamp is thus raised an inch higher than usual, so that the jack can be slipped in under the tappet with the free hand. The stamp is thus suspended above the cam and can be repaired without stopping the others, while it can only be released in a manner similar to that in which it was hung up. As a precautionary measure, jack-shafts should not be put too close to the stamps, otherwise the fingers stand nearly upright and tend to slip. The finger caps should also occasionally be examined to see that the angle is forward instead of backward, this latter tending to insecurity of the fingers. Above the stamps there is a double rail, on which is a tackle block carriage (*crawl, crab*); by this the stamp, etc., can be lifted up for repairs. When the stamps are to be set up, the head is put on the die, strips of wood or other packing put into the head socket and the stem dropped into the latter. The stamp is then raised and dropped on to the shoe, the shank of which is surrounded by strips of wood for packing. The parts are soon wedged firmly together by raising and letting fall the stamp a few times.

The height of "drop" of stamps varies from 4 to 18 inches, and the number of drops per minute varies from 30 to over 100. These depend on one another to a great extent, an increase in the height of the drop being necessarily accompanied by a diminution in the number of drops per minute. With a drop of $8\frac{1}{2}$ inches, about 95 blows per minute can be obtained, the tappet then just having time to fall after leaving one face of the cam, before the other begins to raise it. The actual height of drop, as pointed out by Smart, is the "set" height less the thickness of the ore on the dies after the stamp has fallen. As, within certain limits and under certain conditions, an increase of speed results in an increase of yield of pulverised ore, efforts have been made to raise the number of blows per minute. This entails, however, the closest control of the conditions for successful amalgamation.

D. B. Morison discusses the question of height of drop from the point of view of speed of crushing.[1] The time occupied in seconds in the fall of the stamp, if all friction, including the resistance of the air, is left out of account, is $\sqrt{\frac{2H}{G}}$, where H = height in feet, and G = 32·2. For a fall of 8 inches this amounts to about 0·2 second, but owing to the friction between the stamp and the guides and the resistance due to the pulp, the actual time taken is about 0·225 second (see Fig. 70). The velocity at the time of impact and the force of the blow are correspondingly diminished. The time required in raising the stamp is somewhat greater, owing to the imperfect method necessarily employed. The shape of the cam causes the stamp to start upwards at a certain velocity immediately the cam meets the tappet, and to maintain the same velocity until near the end of the stroke. With a properly constructed cam, designed to give a drop of 8 inches, the vertical component of the velocity of the cam is, for 100 drops per minute, about 2 feet per second. If the cam were suddenly removed from the tappet when moving at this velocity, the stamp would continue to rise against gravity for about $\frac{3}{4}$ inch, neglecting friction.

The inherent defect of the cam is that it strikes the tappet a tremendous blow in the effort to lift the stamp at the full velocity from the start, and it is this blow which is the cause of much of the intense noise and vibration felt in every stamp mill.

---

[1] "Gravitation Stamp Mills for Quartz Crushing," *Trans. North-East Coast Inst. of Engineers and Shipbuilders*, 1896-7, 13, 221.

Fig. 70 shows a curve traced out by a pencil attached to a 900-lb. Sandycroft stamp, the ordinates representing vertical movement of the stamp and the abscissæ being marked in tenths of a second.

The dotted curve shows that at first, in spite of the power of the

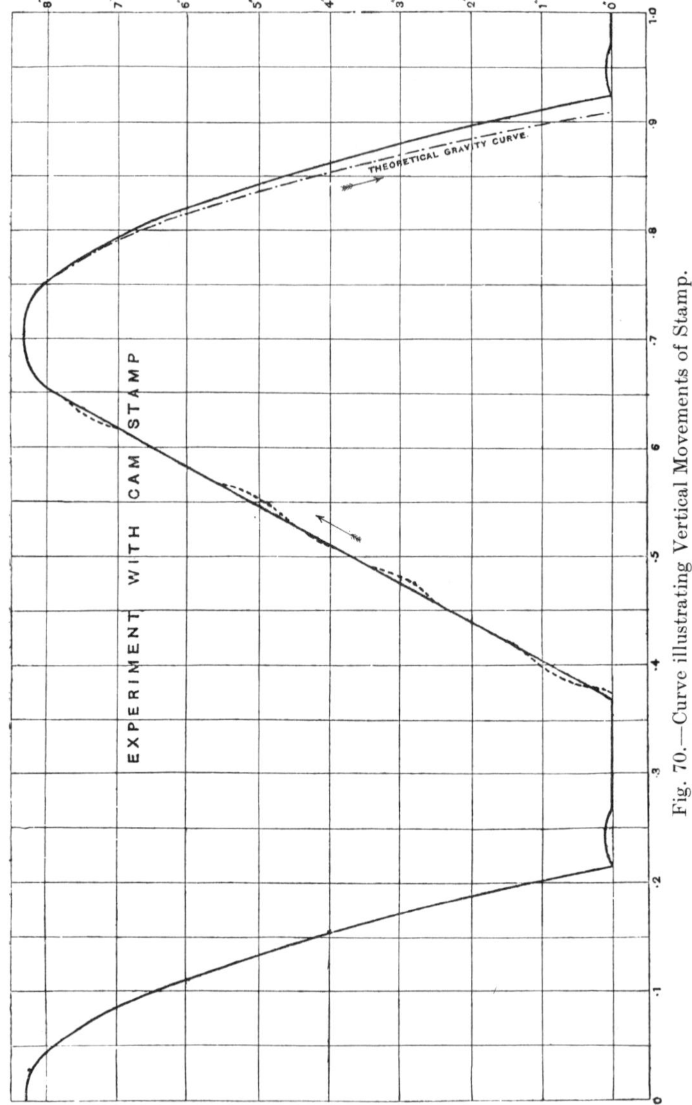

Fig. 70.—Curve illustrating Vertical Movements of Stamp.

blow, which tries the cam-shaft severely, the movement of the cam is checked a little, probably in part by distortion of the cam. It rapidly recovers itself, and by elastic distortion in the opposite direction lifts the stamp faster than the normal rate. These alternations succeed each other during the whole lift, after which the cam presumably recovers itself in the time that elapses before it strikes the next blow.

Space-time diagrams have also been taken in actual practice on the Rand,[1] and show that at 100 drops per minute the total time of each cycle of 0·6 second is divided into rise 0·25 second, fall 0·22 second, rest 0·13 second. The Rand engineers continue, " the stamp, owing to the shock on the cam striking the tappet, bounds off at a greater velocity than that due to the speed of the cam, but the energy so imparted is not sufficient for the total rise ; hence the cam overhauls the stamp and imparts a fresh blow, this time less violent, but still resulting in a bound on the part of the stamp. This is repeated several times, until the full height is attained." The height of the rebound of the stamp on striking the ore varies with the thickness of the layer of ore on the die. Behr's cam, which is modified in shape so as to begin to lift the stamp gradually and afterwards to increase the speed of lift, reduces the shock and wear, but also reduces the number of blows per minute.[2]

Screens.—The screens are set in wooden or iron frames, which slide in grooves cast in the mortar, and are keyed to it. The joints are made tight by blanketing. Screens are made either of steel or brass wire-cloth, sheet-iron or steel, or tin-plate, in which holes are punched, either round or consisting of slots (from $\frac{1}{4}$ to $\frac{1}{2}$ inch long) ranged parallel or inclined to each other. The width or diameter of the holes, formerly as little as $\frac{1}{80}$ to $\frac{1}{12}$ inch, is now generally much greater. On the Rand in 1926 the screen apertures were rarely less than $\frac{1}{3}$ inch, and in some cases as wide as $\frac{3}{4}$ inch or 1 inch. Using $\frac{1}{2}$ to 1 inch screens the grading of the pulp was as little as 20 per cent. of — 90 mesh.[3] The stamp duty was correspondingly high, averaging 11 tons per day. At Modder Deep it was 24 tons, and at Modder B, with Nissen stamps, 29 tons. Slots discharge well, but there is a great loss of discharge area in the use of punched iron, and wire screens are generally used.

The " diameter " of the square apertures, the thickness of wire, the percentage of discharge area, and the number of holes per linear or square inch are all factors required to be known by the millman using any particular screen.[4] Taylor meshes are taken now as standard.[5]

The area of the screen is usually from 3 to 4 square feet per battery, the height from the bottom to the top of the screen being from 8 to 10 inches. On the Rand, the screen openings are usually 52 inches long and 10 inches deep. Smart [6] states that the portion of the screen area principally utilised is the lowest 3 inches, the upper part of the screen receiving only a small amount of the splashed pulp.

The screen frame is often made double, with two sizes of screens mounted, to facilitate a rapid change of screen size if a tube mill goes out of action.

The screens are placed at an angle which varies somewhat but is never far from 10°. Smart, however, observes [7] that, with ore containing much

---

[1] "*Rand Metallurgical Practice,*" vol. ii., p. 111.

[2] *Loc. cit.*, p. 115.

[3] White, *op. cit.*, vol. i., p. 480.

[4] For details of these, see G. T. Holloway, *Trans. Inst. Min. and Met.*, 1905, **14**, 95; also Report of Committee, *J. Chem. Met. Mng. Soc. S.A.*, 1906, **6**, 393.

[5] See *Rep. Inst. Min. Met.*, 1930, and also *British Standards Institution Specification for Test Sieves*, No. 410—1931.

[6] "*Rand Metallurgical Practice,*" vol. i., p. 74.

[7] "*Rand Metallurgical Practice,*" vol. i., p. 74.

wood or other débris, a vertical screen is advantageous in keeping the apertures from choking. In wet stamping, screens are usually placed on one side of the mortar only—viz., that opposite the feeding side. In cases where the discharge is required to be as rapid as possible, the screen area is sometimes increased, and double discharge (front and back) mortars have been made, but have not been used much, except for dry crushing. Caldecott, in fact, has shown [1] that double discharge mortars do not give a higher stamp duty, and result only in an increase in the amount of water required.

Although iron is the material usually employed for screens, it is sometimes preferable to use copper if the water is acid.

The rate of wear of the screens depends greatly on their position, being more rapid with a shallow than a deep discharge, and more rapid in a narrow than a wide mortar. The battery is hung up now and then, so that a thorough inspection of all the screens may be made, and those that are broken replaced.

Wraight [2] suggests the following percentage compositions for stems, screens and cam-shafts.

TABLE XXIV.

	Cam-shaft.	Battery Screens.	Battery Stems.	
			(a)	(b)
C, .	0·28-0·32	0·8-1·0	0·15-0·25	0·20-0·3
Si,	Up to 0·3	Up to 0·25	Up to 0·15	Up to 0·25
S, .	≯ 0·04	0·04	≯ 0·06	≯ 0·04
P, .	≯ 0·04	0·04	≯ 0·06	≯ 0·04
Mn,	0·35-0·45	0·5	0·6	0·35-0·55
Cr,	0·55-0·65	...	...	...
Ni,	3·25-3·75	...	...	2·5-3·5

The question of screens in gravity stamps has been exhaustively treated by F. W. Kendree.[3] The factors of choice, operation, wear and construction are discussed.

Order of Fall of Stamps.—The aim is to keep the ore equally distributed through the mortar, so that each stamp shall do the same amount of crushing, although it is inevitable that the middle stamps should be more efficient than the end ones in discharging the ore. The order generally used is 1, 3, 5, 2, 4.

Feeding.—Self-feeders are universally employed in modern mills. The art of feeding consists in keeping the depth of pulp on the dies constant throughout the battery, for, if the dies are insufficiently covered with ore, less crushing is done, while a greater concussion must be taken up by the stamp and by the die, mortar, etc. If the die is quite bare this concussion is so great that the stem may be bent or broken, and the shoe and die battered. On the other hand, if the ore is too deep in the mortar, there is so thick a cushion that much of the force is taken up in compression without crushing it; whilst, besides the reduction of output, the head, in these circum-

[1] *Trans. Inst. Mng. and Met.*, 1910, **19**, 57.
[2] *Loc. cit.*
[3] *Min. Mag.*, 1931, **44**, 370.

stances, sometimes becomes detached from the stem, which is broken or battered by the next blow. The maximum capacity is obtained with "low feeding," the depth of pulp on the dies being about 2 inches or less.

One advantage of even feeding is that a larger proportion of gold is caught, owing to the more regular and even flow of the pulp from the battery.

There are many automatic feeders of different designs. *Hendy's Challenge Feeder*, shown in Fig. 71, is a typical machine. It is constructed so that the tray, A, below the sheet-iron hopper, B, is revolved in an almost horizontal plane by a gear-wheel placed below it. The ore is fed into the hopper and the amount passing to the tray, A, is regulated by a sliding door.

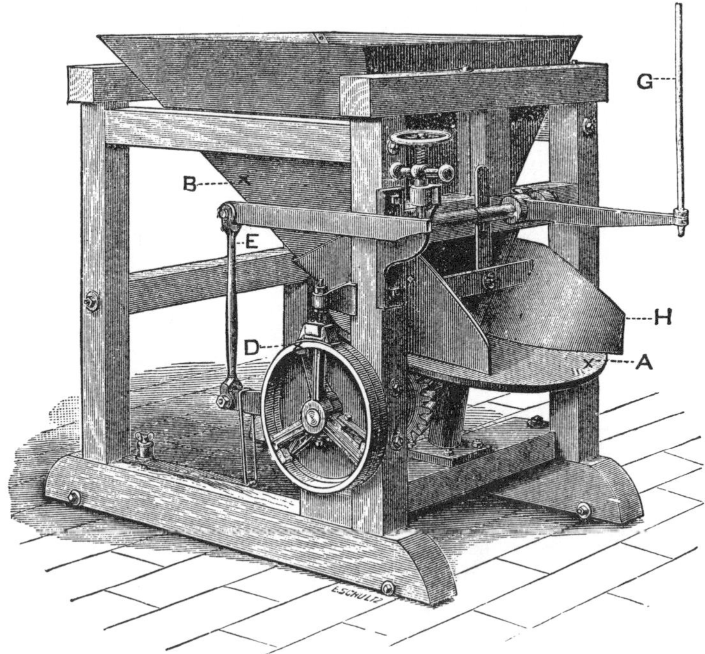

Fig. 71.—Challenge Feeder.

The gear-wheel is set in motion by a friction grip, D, placed on the outside of the frame, and actuated through the lever, E, by the bumper-rod, G, against which the feed tappet strikes. This is a collar clamped to the centre stamp-stem.[1] At each partial rotation a given quantity of ore is scraped off by the stationary wings or side plates, H, resting on the tray, A. This amount of ore is regulated by the condition of the mortar. The machine is especially adapted for very wet or sticky ores. In general one feeder is sufficient for a battery of five stamps (delivering up to 75 tons per day), more ore being fed to the middle stamps, where most work is done, than to the end ones. The feeders are usually suspended from above instead of being supported from below (Fig. 71A). On the Rand, owing to the large amount

---

[1] For G. O. Smart's improvement on the ordinary method of actuating the feeder, see *J. Chem. Met. Mng. Soc. S.A.*, 1906-7, 7, 133.

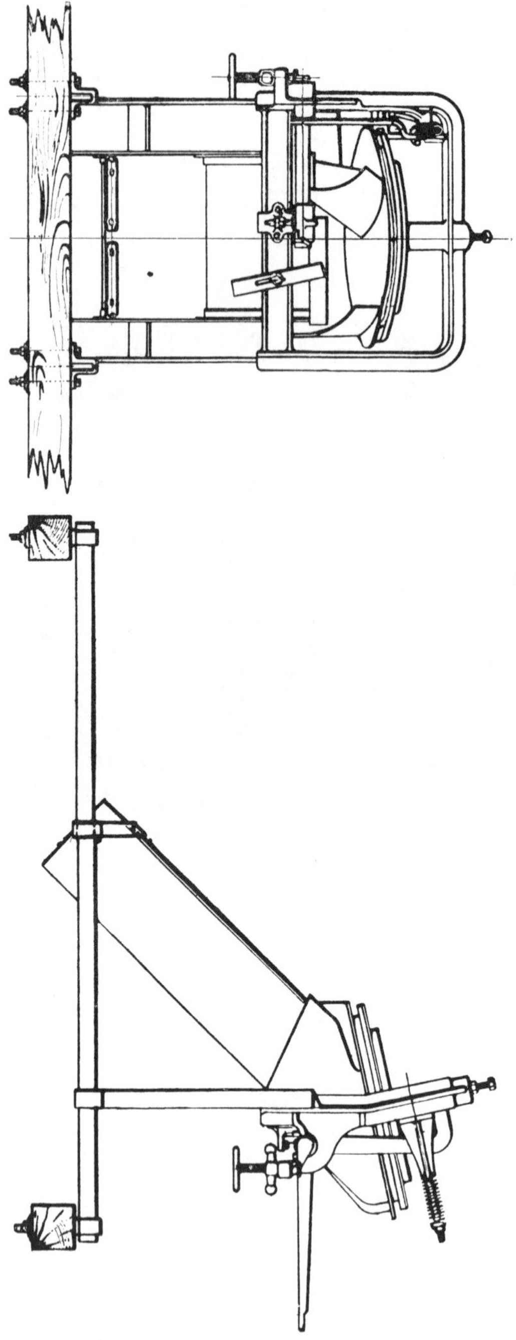

Fig. 71A.—Fraser & Chalmers' Suspended Nelson Feeder.

of rock fed into the stamps, endless balata belt feeders, mounted on two or three rollers, which deliver material to all five stamps, are employed. These belts are actuated by a lever operated by the centre stamp.[1]

The *Nelson Feeder* (Fig. 71A) is similar in many respects to the Challenge feeder. It is suspended from a cross member behind the battery. The movement of the table is caused by successive blows communicated to the operating lever from one of the stamps.

On some plants *apron feeders* are preferred. These consist (Fig. 72) of two side frames made of 10-inch channel irons, with cross members 8 inches wide. The apron is of $\frac{1}{4}$-inch steel plate mounted on two strands

[*Fraser & Chalmers.*

Fig. 72.—Apron Feeder with Bin Gate and Worm Reduction Gear.

of combination chain and acts in one direction only. It passes over two sprocket wheels, the hindermost being driven through gearing. A clutch allows the feeder to be stopped without interfering with neighbouring plant. A standard unit is 30 inches long and 14 inches wide.

Other feeders are the roller and the shaking chute.

Water Supply.—The water is usually supplied to the stamps by horizontal pipes passing just above the top of the housing of the mortar-box, or through the front of the box above the head, with one or two feed pipes or, better, five to each battery. The Rand engineers have added[2] five 1-inch holes

[1] "*Rand Metallurgical Practice,*" vol. i., p. 474.

[2] F. T. Pitt, *J. Chem. Met. Mng. Soc. S.A.*, 1907, 8, 373; Caldecott, *Trans. Inst. Mng. and Met.*, 1910, 19, 57; "*Rand Metallurgical Practice,*" vol. ii., p. 87.

11

along the back of the mortar-box level with the feed chute, with the nozzles of the water pipes pointing so as to discharge the water against the faces of the dies from the back, in order to clean the dies of fine material and increase the output. In front of the battery there is sometimes another

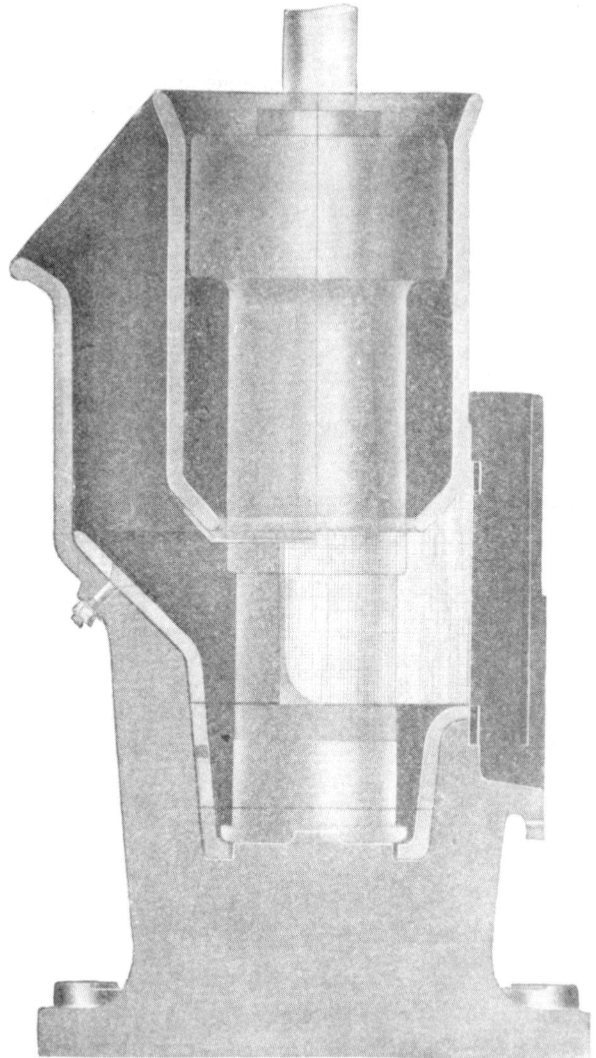

Fig. 73.—Nissen Mortar-Box.

pipe of about half the size, to supply water to the tables to help carry off the pulp. The feed pipes are often pierced with pin-holes, so that the water is supplied as a number of fine jets, in order to keep the stems, etc., clean. The water pipes require to be cleaned out frequently (Smart). The water should be supplied under a constant head, but even then requires continual adjustment in amount to follow the changes in the screen used.[1]

The amount of water used per ton of rock is from 4 to 10 tons, on the Rand the average being 7 tons or 1,400 gallons (Caldecott) for the comparatively fine crushing formerly in use, and about 4½ to 5 tons in modern practice with 1-inch screens.

Richards gives [2] the average amount of water used in 21 mills as 2·77 gallons per stamp per minute, or 6·68 tons of water per ton of ore crushed.

Besides varying with the method of crushing adopted, the amount of water varies with the nature of the gangue, clayey ores requiring more. As a rule, the more rapid the output, the less water per ton of ore is required in the battery. Coarse crushing requires less water in the battery, but, on the other hand, more has to be added on the plates. The amount of battery water per ton is increased by over 20 per cent. by the employment of double-discharge mortars. The amount of water to be added on the plates varies with their grade, as well as with the density and size of the particles of crushed ore. It should be only just enough to prevent the pulp from accumulating on the plates, as any excess over this tends to check amalgamation and to scour the plates.

The Nissen Stamp.—The Nissen stamp [3] (see Figs. 73 and 74) [4] is a gravity stamp mill, each stamp having its own cylindrical mortar-box. The latter is made of special steel with manganese steel liners. The screen has the large area of 3¾ to 4 square feet, and passes half-way round the mortar. The cams, shoes and dies are all made of chrome steel. The dies have round bases fitting into recesses in the mortar-box, and are thus kept central. The stamps usually weigh from 1,750 to 2,000 lbs. each. Excessive vibration in the cam-shaft is prevented by providing caps for the bearings, as shown in Fig. 74. It is obvious that with a single stamp in each mortar, any stamp may be stopped without interfering with the others.

Some Nissen stamps were installed on the Rand and in Rhodesia about 1911, and the last one was added in 1918. It was claimed that they require less power than ordinary stamps for the same output and are cheaper to install and keep running. Prentice [5] quotes an instance where 45 tons of ore per day were crushed through a screen of 1 inch mesh with a single Nissen stamp of 1,910 lbs. weight. He adds that the cost of crushing to ½ inch size was 3½d. per ton.

The Tremain Steam Stamp [6] works in sets of two in a mortar which has front and side screens of 417 square inches area (see Fig. 75). The upper ends of the stems are pistons moving in steam cylinders. The stem, acting as a piston rod, is about 1½ inches less in diameter than the piston, so that the stamp is lifted by steam acting on an annular ring ¾ inch wide, but is forced down by pressure on the whole area of the piston. At 100 lb. pressure the stamps work at 200 drops per minute, but at lower pressures the speed is less. The same crushing effect is obtained with a 450 lb. stamp as would be obtained with a gravity stamp of 800 to 1,000 lb. dropping 8 inches. The steam power required for the mill is 10 to 15 h.p. The feed should be

---

[1] Schmitt, "*Rand Metallurgical Practice*," vol. ii., p. 134.
[2] "*Ore Dressing*," vol. i., p. 222.
[3] Nissen, *J. Chem. Met. Mng. Soc. S.A.*, 1911, **12**, 111.
[4] Schmitt, "*Rand Metallurgical Practice*," vol. ii., pp. 145, 149.
[5] *Trans. Inst. Min. Met.*, 1935, **44**, 479.
[6] C. E. Parsons, *Mng. and Sci. Press*, 1908, **97**, 386.

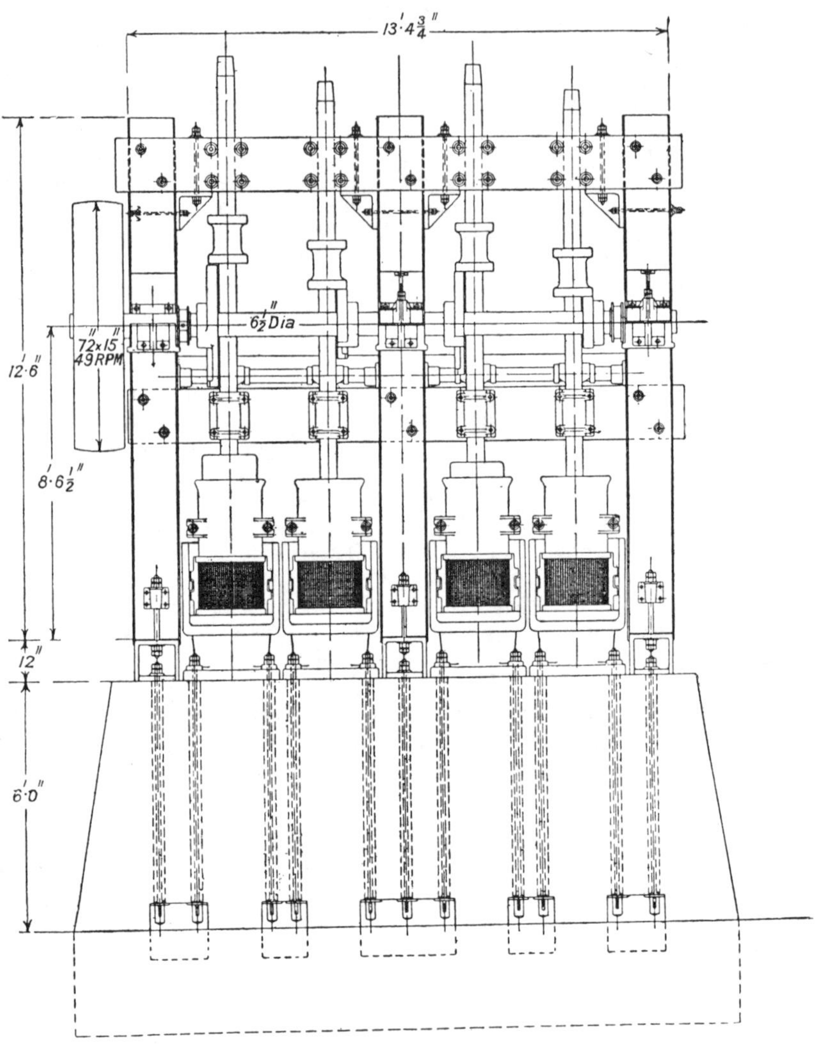

Fig. 74.—Nissen Stamp Battery (Elevation).

about 1 inch cube. The whole plant is light and can be sectionalised for mule-back transport, so that it is suitable for work during development and on small mines. Over 40 such mills were in use in Southern Rhodesia

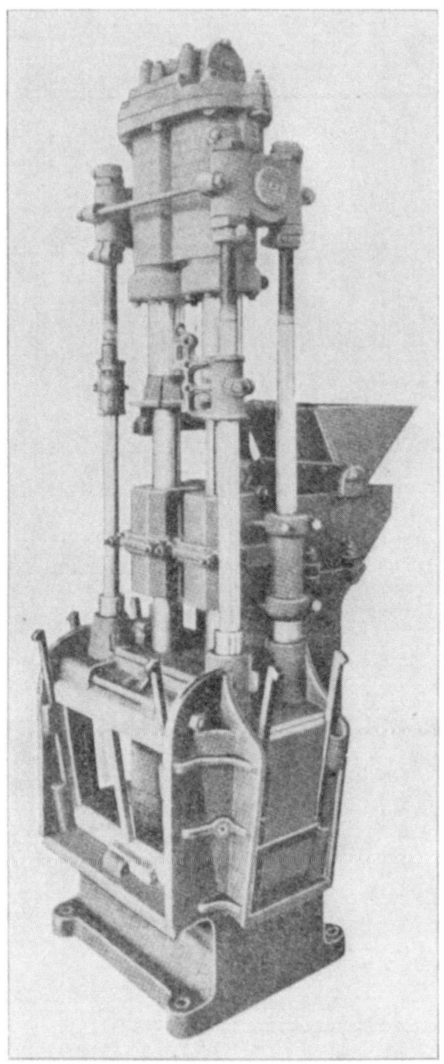

Fig. 75.—Tremain Steam Stamp Mill (*Allis Chalmers*).

in 1908. The capacity of the two stamps is usually about 10 or 12 tons per day with 20-mesh screens.

Arrangement of the Mill.—Fig. 76 represents a mill in which the ore is delivered from the ore-cars through a grizzly on to the rock-breaker floor, and thence by a shoot to the automatic feeders of the stamp battery; the pulp, after passing over the plates, is conveyed by sluices to the double row of concentrating tables which are shown standing back to back on the lowest floor.

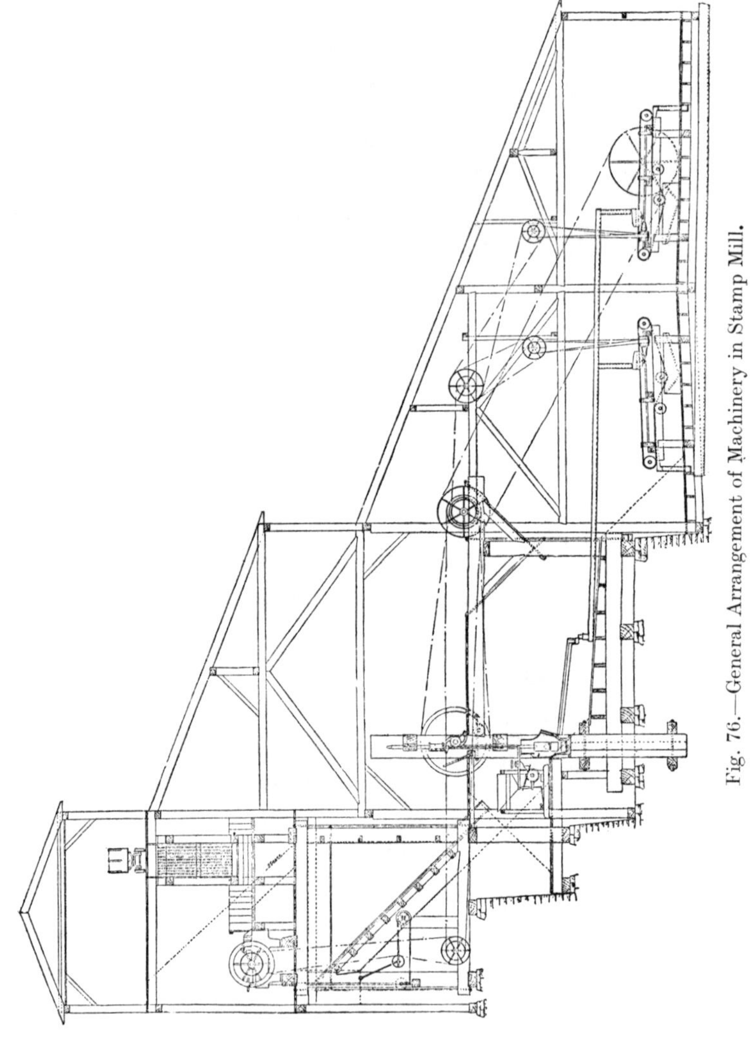

Fig. 76.—General Arrangement of Machinery in Stamp Mill.

In Fig. 76 the amalgamating plates are placed close to the stamps in the old-fashioned way. Many years ago Pitt [1] suggested that the plates should be separated from the battery by a space of several feet, to facilitate sampling and repairs, delivering the pulp at the plates by launders. Later,

---

[1] *J. Chem. Met. Mng. Soc. S.A.*, 1904, 5, 409.

on the Rand, the plates were placed in a separate building in some mills to avoid vibration.

Fig. 77 is a photograph taken along the mortar box platform of a modern

[*Fraser & Chalmers.*

Fig. 77.—Battery of High Duty Stamps (2,200 lb. falling weight).

mill. In this instance no amalgamation plates are used and the pulp proceeds in launders direct to secondary crushers.

CHAPTER VII.

AMALGAMATION.

Amalgamated Plates.—These plates are usually made of copper $\frac{3}{16}$ to $\frac{3}{8}$ inch thick. The copper should be of the best quality, and, if it is hard, it must be annealed before applying the mercury, so as to make it absorbent. The surface is cleaned until quite bright and free from all traces of tarnish and grease. When clean, it is rubbed with a mixture of fine sand, cyanide and mercury by means of a brush, the cyanide preventing the recommencement of oxidation. The sand is sometimes omitted. More mercury is sprinkled on and wiped over with the brush or a piece of rubber until the surface is pasty. In order to make the plates efficient from the start, they are usually coated with silver amalgam before being laid down in the mill. The amalgam is rubbed on with a piece of india-rubber, the plate being wetted with a solution of sal-ammoniac or potassium cyanide to keep it bright. The whole operation is called *setting* the plates. About 16 ozs. of mercury and 6 ozs. of amalgam are usually enough for setting a plate 16 feet by 4 feet 9 inches.

The simplest method of making silver amalgam is to reduce clean precipitated silver chloride with hydrochloric acid and iron nails, and after washing and picking out the nails, to rub the finely-divided silver with mercury in a porcelain mortar.[1] The pure amalgam, of buttery consistency, is applied to the copper plates after the usual scouring, and some two or three weeks before the plate is put into commission. The plate is frequently dressed with more amalgam during this period. It absorbs the amalgam, and tarnishing on exposure to the air is thus prevented. After milling has begun, the coating of silver amalgam is gradually removed from the plates as gold amalgam is scraped off, but by the time the former is all gone its place will have been taken by a permanent and equally efficient coating of gold amalgam.[2]

A more usual method of preparing the plates in California is to coat them with *electro-deposited silver*. After being silvered, the plates have the mercury applied to them. They absorb a large amount of mercury, catch gold well, and are little trouble to keep clean. The plates need not be re-silvered, except after scraping and sweating in the " clean-up," as they become coated with gold amalgam in the course of time. About 2 ozs. of electro-deposited silver per square foot of copper plate are required.

The *plates* are fastened to a wooden table with copper nails, or wooden clamps, or by wedges driven into the raised edges of the table. The battery tables are heavy and are unconnected with the battery frame, in order to avoid excessive vibration due to the stamps. Possibly some vibration is advantageous. The table is as wide as the battery (4 feet 9 inches to 5 feet), and from 6 to 20 feet long. A length of 2 or 3 feet of plates of

[1] E. H. Croghan, *J. Chem. Met. Mng. Soc. S.A.*, 1909, **10**, 43.
[2] W. A. Caldecott, *ibid.*, 1908, **9**, 142.

the same width, the *apron plates*, if used, is interposed between the battery and the tables proper, on to which there is a drop of 2 or 3 inches. An observation platform or a launder now takes the place of the apron plates. Launders are best lined with rubber or white cast-iron in sections and 1 inch thick. Fig. 78 shows amalgamating tables and copper plates as formerly used on the Rand.

Drops of 2 or 3 inches are also in use in some mills on the tables proper, but they may cause scouring, and, according to MacFarren, should not exceed $\frac{1}{2}$ to $\frac{3}{4}$ inch in height.

The use of outside battery plates is advantageous with fine crushing, but when the screens are of 8-mesh (64 holes per square inch) or coarser,

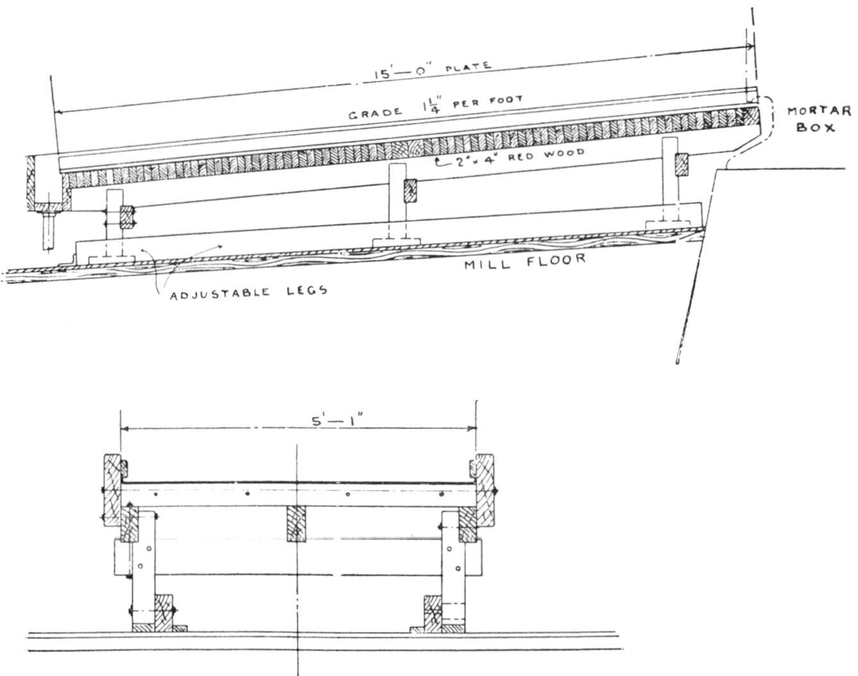

Fig. 78.—Amalgamating Plate for Stamp Batteries.

the scouring action of the pulp prevents the gold from being caught. Where coarse crushing is practised, battery plates have been discarded. On the Rand, the coarse battery pulp was for a time passed through tube mills and then over three plates, each $4\frac{1}{2}$ feet by 12 feet. Two plates were considered to be enough, but the third plate was added as a measure of precaution and to enable two plates to be in operation whilst the third was being cleaned up or dressed. This transference of the plates from stamp mill to tube mill had the advantage that the former was not stopped while the plates were dressed. The size of the plates was also considerably reduced. It was a step in the direction of getting rid of the plates altogether.

Where amalgamation is employed with ball or tube mills the plates follow the last mill, or in exceptional cases, where the gold is coarse, they may be placed between the grinding stages. Usually a series of plates is employed and recovery begins as soon as practicable.[1]

The practice of feeding mercury into the battery, although still frequently pursued, has been more generally discontinued. At the Guysborough mill, Nova Scotia, amalgamation inside the mortar-box, but without the use of plates, is employed. Mercury is fed in by means of a wooden spoon.

Treatment of the Plates.—In order to keep the plates in proper condition so that successful amalgamation may be maintained, the closest watch must be kept over them. It is not considered desirable to put on the plates as much mercury as they will hold, since, if the amalgam is too fluid, losses are sustained by scouring, but, on the other hand, if the amalgam becomes too hard and dry from absorption of gold and silver, further amalgamation is checked and fresh mercury must be added.

The amalgamated plates are *dressed* as frequently as is necessary, the length of time allowed to elapse between two such operations depending partly on the richness of the ore. To dress the plates, the flow of pulp is stopped and water used to wash off the loose sand. The "black sand," consisting of grains of gold adhering to particles of pyrite, is swept off by brushing from the bottom to the top of the plate, and kept separately for grinding with mercury. The plates are then rubbed with a hard brush to soften and distribute the amalgam, and fresh mercury is added where there are hard spots. The amount of mercury put on the plates should be enough to keep their surfaces in a pasty condition, but not enough to gather into liquid drops or to run off. Sometimes cyanide is added in dressing the plates, to assist in the removal of stains. As cyanide causes a loss of gold by dissolution, its use is not recommended, and if any chemical is considered necessary, a 10 per cent. solution of hydrochloric acid suffices and is innocuous (Smart). Usually in dressing the plates no amalgam is removed, but sometimes a partial clean-up is effected at the same time, the surplus amalgam being wiped off with rubber squeegees, working from the bottom to the top.

Discoloration of the Copper Plates.—The plates often become stained owing to the corrosive action of the water and pulp, a yellow film being formed on the surface of the metal. Those which have been silver-plated are less liable to become dirty than the others; whilst when a plate has become covered with a thick layer of amalgam, it is not readily discoloured. The presence of carbonic acid in the water is also harmful, but the addition of lime to the water neutralises the acid substances and diminishes the tarnishing. Where lime is not added, the yellow, brownish, or greenish discoloration, the so-called "verdigris," appears in spots and spreads quickly, especially on new plates. These stains must be at once cleaned, as the stained part catches little or no gold. The chemicals used for the purpose have been sal-ammoniac and potassium cyanide. The pulp flow is stopped, the plates rinsed with clean water, and a solution of sal-ammoniac applied to the stained parts with a scrubbing brush, and left covering them for a few minutes in order to dissolve the oxides. It is then washed off. The use of cyanide is now unusual, as it has been found to cause serious losses by dissolving gold. Soda is used to remove grease, and hydrochloric acid for spots caused by oxide, etc. These reagents are effective and harmless.

The merest trace of any kind of grease or oil is very prejudicial to successful amalgamation, forming a film over the plates and over the little

---

[1] *U.S. Bur. Mines*, 1932, *Bull.* 363.

globules of mercury, and thus preventing contact between them and the particles of gold.

Effect of Temperature on Amalgamation.—By an increase in temperature the wetting of the gold by the mercury and the "catching" of it by the plates is facilitated, as is the coalescing of the globules of mercury. This may be due mainly to the reduction in the surface tension of the mercury. Increase of temperature also causes an increase in the rate of absorption of mercury by the gold and by the plates, owing to an increase in the rate of diffusion. On the other hand, undesirable effects in raising the temperature are the increased solubility of harmful salts and a corresponding increase in the precipitation of base metals into the mercury. This effect both hinders the proper action of the mercury and leads to its loss. An increase of temperature softens the amalgam on the plates, and may cause loss by scouring. Hence Smart states[1] that the temperature of the feed water should not exceed 80° F. When the water is too cold, however, the amalgam becomes hard and powdery and loss may again occur, whilst its catching powers may be seriously reduced. For this reason, the feed water in many mills is warmed by steam in winter. It is generally admitted that sudden variations in the temperature of the water should be avoided.

Inclination of Plates.—The grade or slope of the plates varies with the nature of the ore to be treated, heavy pyritic ores requiring a higher grade than light quartz, while the coarser the crushing the steeper must be the grade. The copper tables have usually an inclination of $1\frac{1}{4}$ to 2 inches per foot. The narrower plates have a lower slope in order to avoid increased scouring action. With heavily sulphuretted ores a grade of from 2 to $2\frac{1}{2}$ inches per foot is used. The steepness of the grade is of great importance, as on it, and on the amount of water supplied, the attainment of the necessary contact between the ore and the plate depends. When the pulp is flowing properly, it travels down in a series of little waves and ripples, and, in consequence of the friction between the plate and the film of water in contact with it, the upper portions of these little waves travel faster than the lower parts, so that the motion becomes one of tumbling over and over. As a result of this, if the plate is long enough, every particle of pulp comes in contact with the amalgamated surface.

Muntz Metal Plates.—Muntz metal (which consists of copper 60 per cent., zinc 40 per cent.) differs from copper in not requiring to be covered with gold- or silver-amalgam before it begins to do good work. Moreover, the amalgamated surface is very superficial, since the mercury does not sink in so far as it does into a plate composed of pure copper, so that only a small quantity of mercury is required to cover it. Cleaning-up is therefore easy and rapid, no iron instrument being necessary, but rubber being always sufficient. These properties make it particularly valuable for custom mills, where it is desirable to catch as much as possible without mixing the amalgam obtained from two parcels of ore crushed in succession. On the other hand, as it holds little mercury, it cannot absorb much gold, and must be cleaned-up at frequent intervals.

The mercury on Muntz metal plates does not suffer so easily from "sickening" as that on copper plates; it has been suggested that this is due to the electrolytic action of the copper-zinc couple, which sets free nascent hydrogen, and so reduces the compounds of mercury and other

---

[1] "*Rand Metallurgical Practice*," vol. i., p. 72.

metals which have been formed. It follows that Muntz metal plates are preferable for ores containing large amounts of heavy sulphides or arsenides. The verdigris does not appear when Muntz metal is used, and such discolorations as occur on these plates can be removed by dilute sulphuric acid. It is stated, however, that, in the treatment of highly acid ores, which have been weathered for some time so that they contain large quantities of soluble sulphates, or in cases where the battery water contains acids, copper plates are less affected than Muntz metal, over which a scum is rapidly formed. In the Thames Valley, N.Z., Muntz metal is preferred in spite of the extremely acid nature of the water and ore.

In dressing new Muntz metal plates the following method is adopted in New Zealand :—The surface of the plate is scoured with fine, clean sand, then rinsed with water, and washed with a dilute (1 to 6) solution of sulphuric acid. Mercury is applied and rubbed in with a flannel mop until it " wets " the surface of the plate in one or more places, after which the mop is given a circular movement, passing through these spots, until the amalgamation of the surface spreads from them over the whole plate.

Shaking Copper Plates.—A shaking copper plate has been recommended to be used either below or in place of the ordinary amalgamating tables, especially in cases where these do not appear to give good results. A short rapid shake keeps the sand from packing, and amalgamation may be well performed with a grade of only $\frac{1}{4}$ to $\frac{1}{2}$ inch per foot; alternatively the amount of water needed with the pulp may be greatly reduced and better contact thus obtained. A plate is fixed to a light wooden frame which is moved by a crank-shaft, revolving 180 to 200 times per minute, placed on one side, with a throw of 1 inch at right angles to the direction of the flow of the pulp. In some mills, a longitudinal shake is given to the plate instead of this side shake. The frame may be hung on rods from above, or on four short iron springs, forming rocking legs. The width of the tables is made as great as possible, while the length is of less importance, as, the thinner the current of pulp flowing over them, the better the chance of the gold particles coming in contact with the plates and being retained. It is advantageous to add an amalgam and mercury-saver. A simple device for this purpose is to nail a strip of wood, half an inch thick, across the copper plate near the top, thus forming a shallow riffle, the angle of which is soon filled with sulphides and coarse sand, which are kept in agitation by the movement of the table. Shaking plates are effective, but tend to produce excessive flouring, with a high loss of mercury.[1] Pearce and Thomas state that they are no longer in use on the Rand. Fig. 79 shows a complete shaking table.[2]

Mercury Traps.—Another method of saving mercury and amalgam, which would otherwise be lost in the tailing, consists in the application of mercury wells or riffles. A mercury well consists of a shallow gutter filled with mercury, over the surface of which the pulp flows or through which it is forced to pass by suitable machinery. Such wells or traps are usually placed between the successive plates, the pulp dropping from the end of a plate on to the surface of the mercury in the well, and then passing on to the next plate. Large particles of amalgam are caught in mercury traps, but finely-divided or floured mercury or amalgam is carried over and escapes (Smart).

---

[1] *U.S. Bur. of Mines*, 1932, *Bull.* 363.
[2] Schmitt, "*Rand Metallurgical Practice*," vol. ii., p. 185.

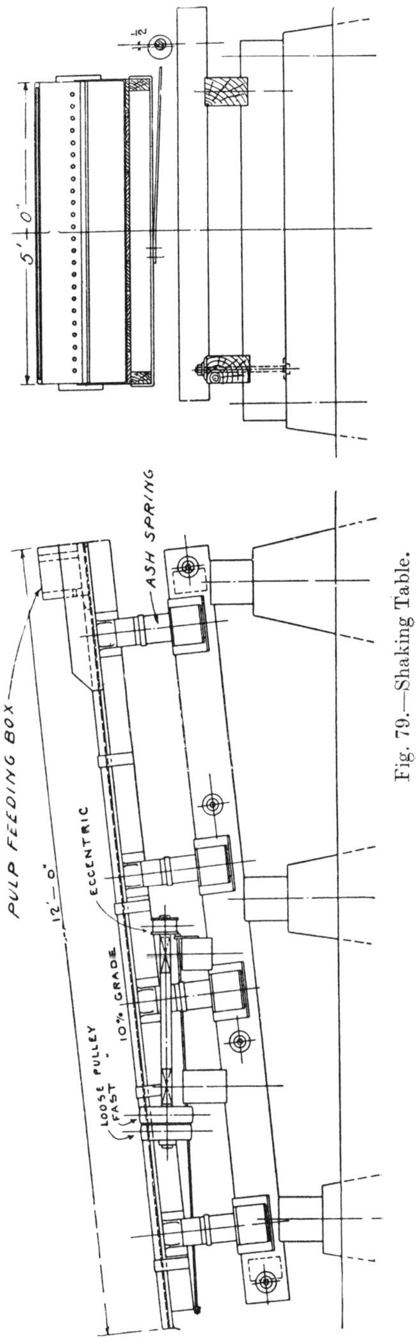

Fig. 79.—Shaking Table.

On the Rand,[1] "the pulp overflowing the lower edges of the amalgamated plate falls into a short transverse launder secured to the table frame, and sloping slightly from either side to the centre. At this point the tailing pulp escapes through a short vertical pipe, the upper end of which projects about an inch above the bottom of this transverse launder, which constitutes the first amalgam and mercury trap. The vertical pipe usually delivers into another (circular) mercury trap (Fig. 80),[2] after which the overflow enters the main tailing launder."

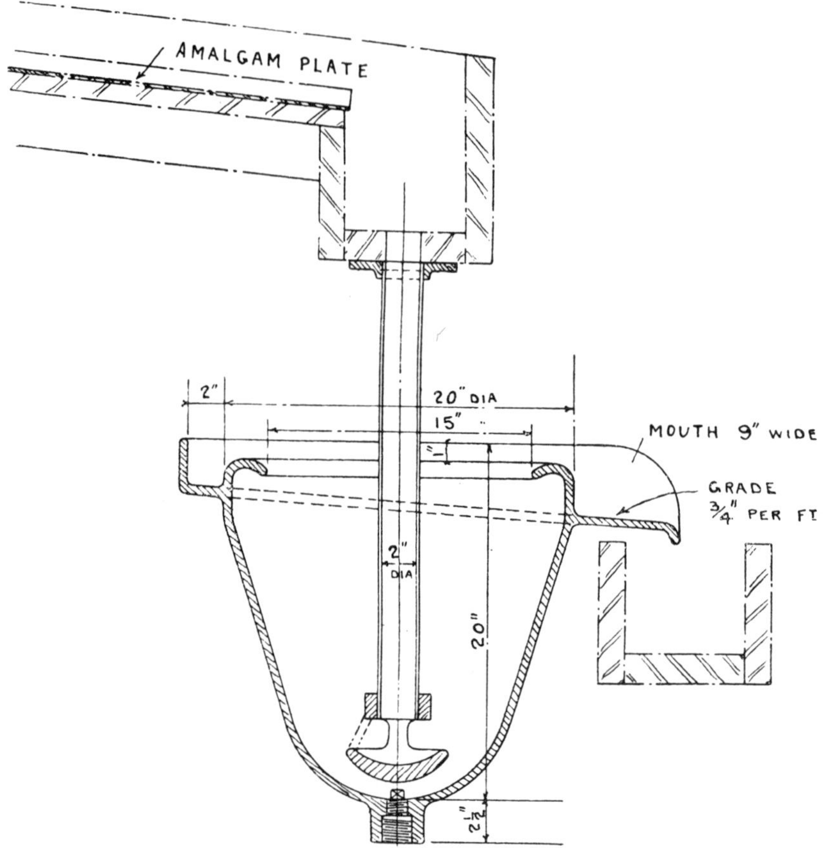

Fig. 80.—Circular Mercury Trap.

Galvanic Action in Amalgamation.—In amalgamation not only are free metals absorbed, but the dissolved salts, and, to a less extent, the insoluble compounds of the heavy metals are reduced and amalgamated, chiefly by galvanic action. The copper of the plates, or the iron of the pan, constitutes the positive element, and all metals less oxidisable than this reacting metal are reduced by it, and are then amalgamated by the mercury.

[1] G. O. Smart, "*Rand Metallurgical Practice*," vol. i., p. 76.
[2] *Ibid.*, p. 76.

In this way iron reduces both lead and copper, although, if these are present in the form of undecomposed sulphides, this action will be very slight. If lead is introduced into the amalgam, the latter becomes very pasty, and is subjected to considerable losses. Copper has an equally harmful effect. In some mills, this galvanic action has been increased by the passage of a weak electric current through the charge. The amalgamated plates or the walls of the pan are the negative pole, while the positive pole is formed of a plate of graphite, lead or iron dipping into the pulp. Such methods are attended with the best results when dealing with ores containing little or no copper, lead, etc., since in these cases the strength of current can be increased without any ill effects, and the mercury kept clean.

Experiments in California [1] showed that with a small current good results were obtained by adding a solution of mercuric chloride to the pulp, when mercury is electro-deposited on the amalgamation plate. The amalgam formed is tenacious and bright, resisting scouring. The method may be used in the sluices of placer workings.

The Clean-up.—In cleaning-up, the stamps are hung up, two batteries at a time; the screens and dies are all taken out and washed in tubs; and the "heading" or contents of the mortar, consisting of the pulp, sulphides,

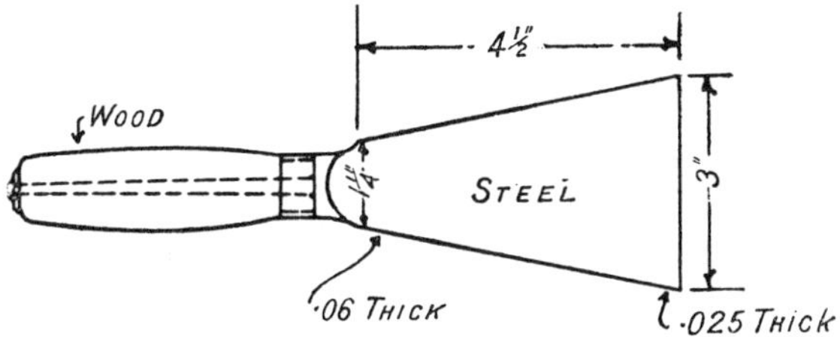

Fig. 81.—Scraper for Removing Amalgam from Plates.

and pieces of iron and steel, amounting in all to a quantity sufficient to fill two or three buckets, is carefully scraped out and panned or fed into the mortar of one of the other batteries, which has not yet been cleaned-up. In California, the heading from the last batteries is panned, the iron removed with a magnet, and the remainder ground with mercury in the clean-up pan. The amalgam does not accumulate evenly on the plates, but in ridges and knots which serve as nuclei for the collection of more. It is not advisable to allow the coating of amalgam to become very thick, since, although the plates catch better as the amalgam accumulates, losses may be experienced by scouring. The removal of amalgam takes place as often as necessary, depending on the richness of the ore.

The operation usually consists in the removal of the black sand, followed by a vigorous brushing over all parts of the plate with a stiff brush, mercury being sprinkled on at the same time and the amalgam thoroughly softened throughout. The amalgam is then removed either by a sharp-edged piece of hard rubber or by scraping with broad flexible steel scrapers, such as that shown in Fig. 81, care being taken not to scratch the plates.[2] In some mills scraping is not thought advisable on the ground that it may lead to too close removal of the amalgam, leaving the plate liable to discoloration.

[1] E. E. Carey, *Mining J.*, 1909, **85**, 617.
[2] "*Rand Metallurgical Practice*," vol. i., p. 83.

The scraping of the plates usually takes from ten to fifteen minutes for each battery.

The amalgam obtained from the plates, mercury wells, or sluices, is rarely clean enough for immediate retorting; it is usually found to contain mixed with it grains of sand, pyrite, magnetite, and other minerals, together with fragments of iron and other foreign substances. The skimmings from mercury wells are still more impure. These materials must be purified by grinding with fresh mercury in a clean-up pan, and washing, before they can be passed to the retort.

In small mills the dirty amalgam is ground in a mortar by hand with fresh mercury and hot water, until it is reduced to an even thin consistency, when the dirty water is poured off, and the mercury poured backwards and forwards from one clean porcelain basin to another until the pyrite, dirt, etc., have risen to the surface, when they are skimmed off. The skimmings obtained are put back into the mortar, and re-ground by themselves with fresh mercury. The clean mercury is then squeezed through canvas or wash-leather, when the greater part of the gold and silver contained in it, together with about one and a-half times its weight in mercury, remains in the bag, the rest of the mercury, with a small quantity of the precious metals dissolved in it, passing through. The amalgam is now often squeezed in powerful hydraulic presses.[1] A small hand amalgam press is shown in Fig. 82. The separated mercury is found to contain, however, a little fine amalgam, and should therefore be kept for clean-up room work.

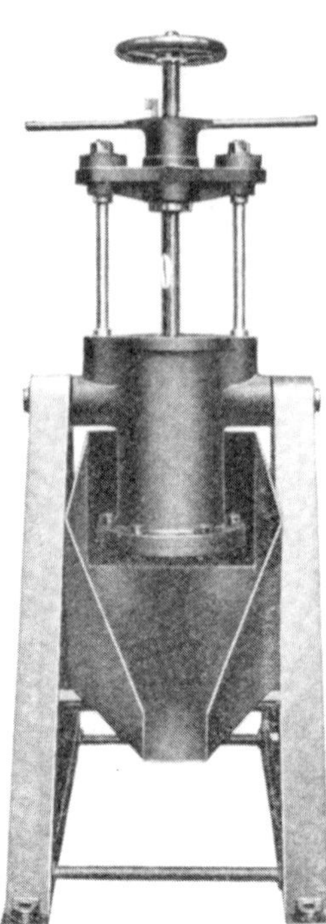

Fig. 82.—Hand Amalgam Press.

In large mills a revolving clean-up **barrel** (Fig. 83) is often employed to mix the amalgam. The barrel may be 3 feet in diameter and 4 feet long, and revolves twenty times a minute; the charge is 700 lbs. of amalgam and 20 lbs. of mercury, or more if the amalgam is very rich. The black sand is also treated in the barrel. The contents are kept alkaline with lime. A dozen or more iron balls or pieces of iron, such as worn-out battery shoe shanks, are put into the barrel, together with sufficient mercury to yield a fluid amalgam, when grinding is complete, and enough water to make a thin pulp. The use of the iron is to help to mix the amalgam and mercury, but it causes some loss by flouring. After being revolved for from two to twelve hours, or until the amalgam has run together and the sand ground to slime, the barrel is opened and washed out with water, the tailing being run over rough concentrating devices, the lighter portion rejoining the pulp in the mill circuit while the heavier is re-treated. The washings from the barrel are also run over amalgamated plates and through a mercury well or some other form of amalgam-saver (a large power-driven batea

[1] White, "*Rand Metallurgical Practice*," vol. i., p. 481.

being used on the Rand). The amalgam is scooped out of the barrel and squeezed.

The clean-up **pan** is also extensively used. It consists (Fig. 84) of an

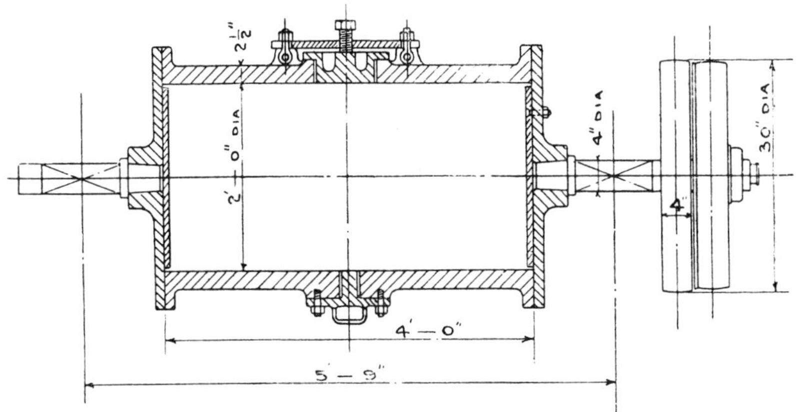

Fig. 83.—Amalgam Clean-up Barrel.

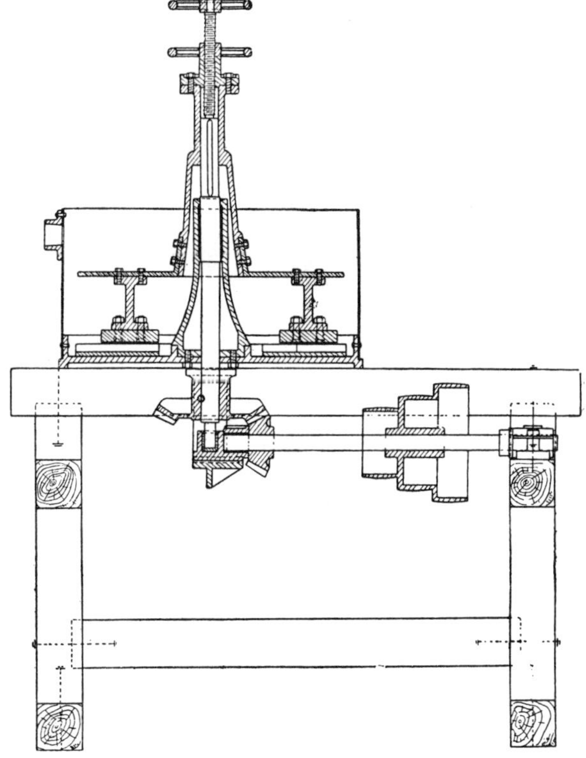

Fig. 84.—Clean-up Pan.

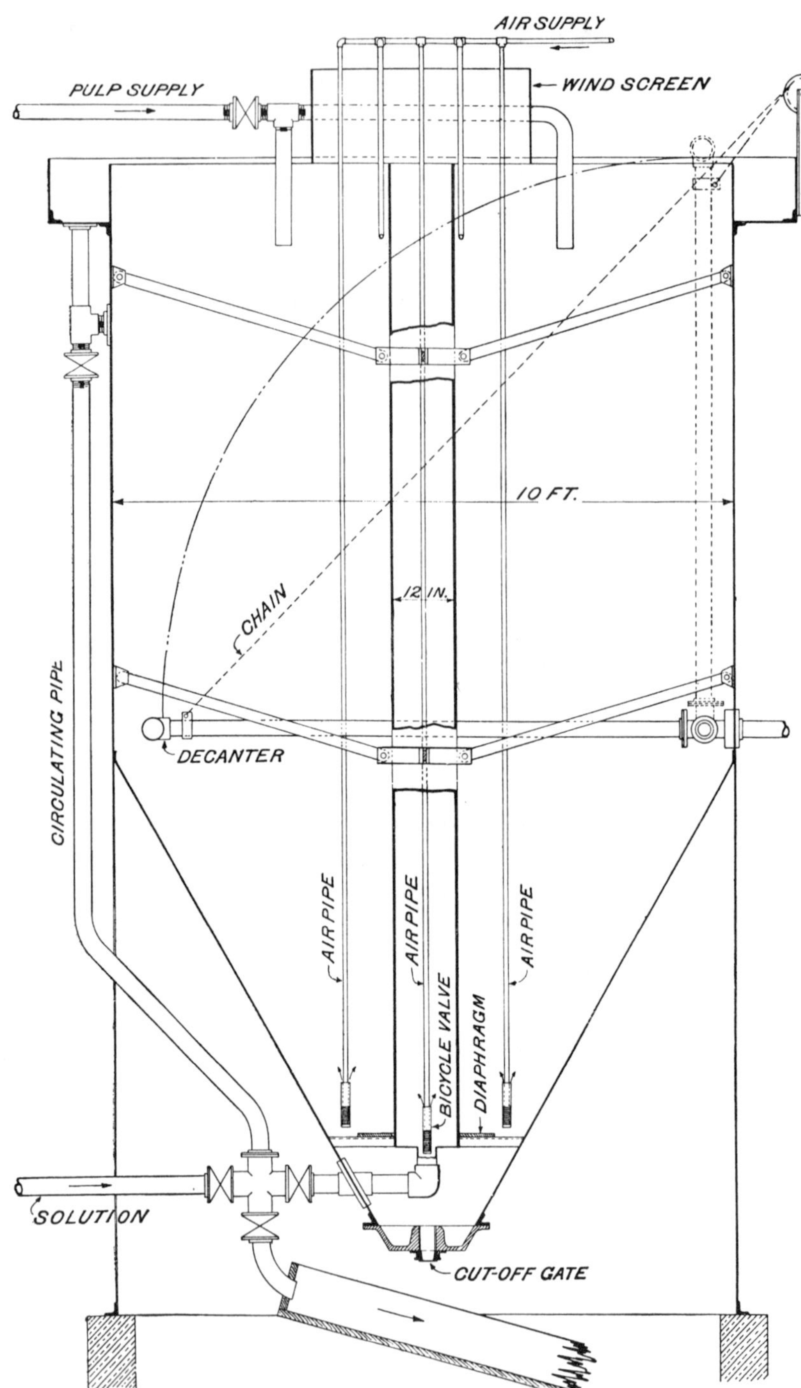

Fig. 85.—Air Agitation Vat for Treatment of Slime resulting from the Crushing of Battery Sand.

iron pan 5 feet in diameter and 14 inches deep. Wooden or iron shoes are attached to the arms, which make from twelve to fourteen revolutions per minute. Iron shoes are considered better for brightening or polishing the particles of gold contained in the pyrite, and so rendering them fit for amalgamation. The charge for this pan is about 300 lbs. of impure amalgam, mercury, concentrate, skimmings, etc. The charge is made into a pulp with water and ground for three or four hours, after which more mercury is added, and mixing carried on for a few hours longer, before the pulp is diluted, settled, and discharged. The tailing suspended in the water is usually passed over amalgamated plates, and is then often caught in settling pits, and either sold or subjected to further treatment in the mill, as it is frequently of high value. The mercury is squeezed through canvas, and the amalgam retorted.

On the Rand, "the tailing from pans, barrel and batea, and the poorer sand, is fed to the clean-up tube mill, which is the final clean-up grinding machine, and through which all tailing obtained during the clean-up should be finally passed."[1] The clean-up tube mill is a miniature tube mill,[2] about 6 feet long and 4 feet in diameter, fed by hand. The outflow passes over a small shaking amalgamated plate, and thence to a settling box. The overflow from the box is slime, which is treated with cyanide in a small air agitation vat (Fig. 85),[3] and thence, after decantation, passes to the ordinary cyanide slime plant.

The position from which the greater part of the gold is obtained in a clean-up varies according to the ore and the method of treatment. Amalgam from near the top of the tables is richer than that from the bottom. Coarse gold forms a richer and stiffer amalgam than fine gold, for the reason already given on p. 39.

Steaming and Scaling Amalgamated Plates.—In spite of ordinary cleaning-up, amalgam gradually accumulates as a hard scale on the plates, and is removed periodically by steaming, immersing in boiling water, or otherwise heating them so as to soften the amalgam before they are scraped. On the Rand the plates are generally steamed and scraped every three or four months. According to Smart, "steaming is carried out by placing a wooden cover over the whole surface of the plate and introducing steam through a $\frac{3}{4}$-inch pipe in the middle for from 10 to 15 minutes. The cover is then removed and the amalgam scraped from the whole of the surface before the plate cools."[4] Mercury is then applied to the plate while it is still warm. The plates are sometimes "sweated" in California by heating them over a wood fire before they are scraped. After scaling and sweating, the plates may require replating, especially if they are scoured with sand to help in the removal of the amalgam. In course of time they wear out, the copper becoming brittle and worn into holes, but they usually contain enough gold when discarded to pay for a new set.

Several methods are in use for recovering the gold from old plates. For example, they may be dissolved in nitric acid, when the gold is left nearly pure. A more economical method of detaching the gold, much used in Australia, is described by W. M'Cutcheon,[5] as follows:—The plate is placed

---

[1] Smart, "*Rand Metallurgical Practice*," vol. i., p. 88.
[2] See Chap. VIII.
[3] Smart, *loc. cit.*
[4] "*Rand Metallurgical Practice*," vol. i., p. 84.
[5] Private communication.

on the hearth of a reverberatory furnace, or on a fire made with logs in the open air, and the mercury expelled at a gentle heat. The plate then appears to be more or less coated with gold on one side. This surface is treated with hydrochloric acid for eight or ten hours, and the plate is then replaced on the hearth and exposed to a dull red heat until well blackened. On plunging it into cold water, the gold now scales off, and is collected and freed from copper by boiling in nitric acid. If the temperature in the furnace is too high, the gold sinks into the copper at once, and the copper must then be dissolved.

A method in use on the Rand for scaling plates before putting them aside is as follows:—A mixture of ½ lb. sal-ammoniac, ½ lb. nitre, ½ lb. hydrochloric acid and a pint of water is applied to the plate with a soft brush, and allowed to stand for about fifteen minutes. Afterwards the plate is heated over a good fire. When it has become quite black, which will be in about half an hour, it is dipped into a bath of water, when the scaling can be washed off. Some millmen drive off the mercury before

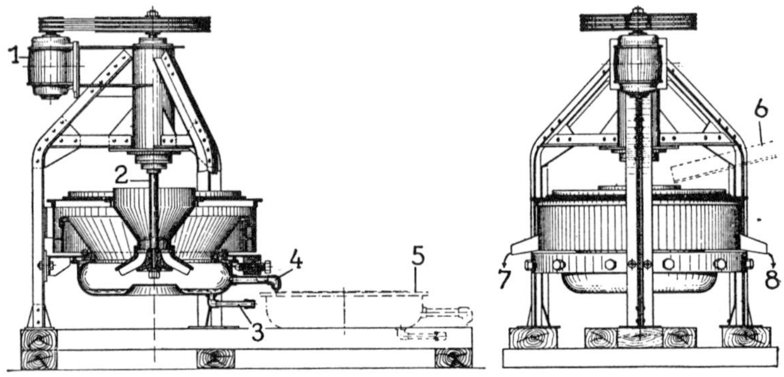

Fig. 86.—Centrifugal Amalgamator.

1. Motor.
2. Spindle.
3. Mercury run-off.
4. Clean-up run-off.
5. Pan removed for cleaning.
6. Feed.
7 & 8. Pulp overflow.

applying the above mixture; others do so afterwards. Should all the scaling not come off, the parts needing it are treated over again. The scalings are afterwards collected and mixed as follows:—1 part scalings, 1 part sulphur, 1 part borax, and 1 part sand. This is melted in a crucible and the gold extracted. Smart[1] gives the charge for the crucible as, scales 100 parts, borax 50, sand 25, and manganese dioxide 8 parts, yielding gold bullion 800 to 900 fine.

The gold amalgam which has "accumulated" on the copper plates is in part responsible for the poor returns from new mills. The usual absorption is about ¼ oz. of gold per square foot of plate, most of which is taken up in the first fortnight. After a long run some plates in a Rand mill yielded by scaling 36·8 ozs. of fine gold, the total area of the plates being 191 square feet. Halse found in a mill in Columbia that from 3 to 5 ozs. of gold per square foot of electro-silvered plate accumulated as scale in about two years.

[1] *Loc. cit.*

Old copper plates are usually melted down and sold to refineries, where they are useful for mixing with gold and silver bullion in making up the alloy for parting.

Amalgamation in Revolving Barrels.—This method is very old. Its practice in Norway was described by Schlüter in 1738.[1] It has never been entirely abandoned and its use was revived for the treatment of "black sand" from stamp batteries. More recently it has been used to treat blanket concentrate and also flotation concentrate. The barrel is similar to that shown in Fig. 83, p. 177, the dimensions being according to requirements. The charge of concentrates in a thin pulp, mercury and steel balls is rotated at about 4 r.p.m. for some hours and discharged through the manhole. The amalgam is recovered on amalgamated plates or some form of concentrating table and the tailings cyanided.

The Tyrolean Mill.—In Hungary, Transylvania and the Tyrol, amalgamation mills are employed having some resemblance to the amalgamation pans described in Chapter VIII. In a recent form, the centrifugal amalgamator (Fig. 86), the pulp is fed into the centre of the containing chamber through the launder 6 and passes through into the lower compartment. A definite quantity of mercury is added. The spindle 2 carries the central cage, which rotates at high speed causing agitation of the mixture during which amalgamation of the gold particles occurs. The mercury and amalgam are finally run off through the pipe 3 and sent to be squeezed and retorted, while the pulp issues through pipe 4 into the settlement vessel 5.

Retorting.—The solid amalgam, which is retained in the canvas or wash-leather filters, usually contains from 30 to 45 per cent. of gold and silver, according to the state of division of the gold present in the ore, and also to the degree of care exercised in squeezing out the excess of mercury. For separating the gold from the mercury there are two kinds of retorts in general use—the *pot-shaped* retort, which is sometimes cast with trunnions to swing on supports, in small mills; and the *cylindrical* retort in larger mills. Figs. 87 and 88 show respectively *cylindrical* and *pot-shaped* amalgam retorts. The pot shaped retort can, if necessary, be heated over an assay furnace or forge fire, or in a fire built on the ground, when it is placed on a tripod stand. In the latter case the fire is lit at the top and burns slowly downwards. The pot-shaped retort is not filled to more than two-thirds its capacity, and must be heated very gradually at first. In Figs. 89, 90[2] the retort furnace in use on the Rand is shown; the cylinder is 5 feet 3 inches long and 12 inches in diameter and is set with a slight fall towards the outlet. It carries a number of trays in which the amalgam is packed.

The pasty amalgam is rolled up into balls or kneaded into cakes, and squeezed into the pot-shaped retort. It is often rammed down with a bolt-head, although this course is deprecated by some metallurgists, who prefer to leave the amalgam as spongy and open in texture as possible, believing that a uniform product is thus obtained more rapidly at a lower temperature and without so much loss. In the horizontal retort the amalgam is placed in iron trays divided into compartments by partitions. In either case, the retorted metal is prevented from adhering to the iron, either by laying it on three or four thicknesses of paper, the ashes of which remain beneath the amalgam, or by covering the iron trays with a coating of lime white

[1] "*Gründlicher Unterricht von Hütte-Werken,*" p. 211.

[2] Smart, "*Rand Metallurgical Practice,*" vol. i., p. 92.

or fireclay. The mercury is condensed in cooling tubes passed through water; the loss is usually very small, and may be taken as being about one grain of gold per pound of mercury.

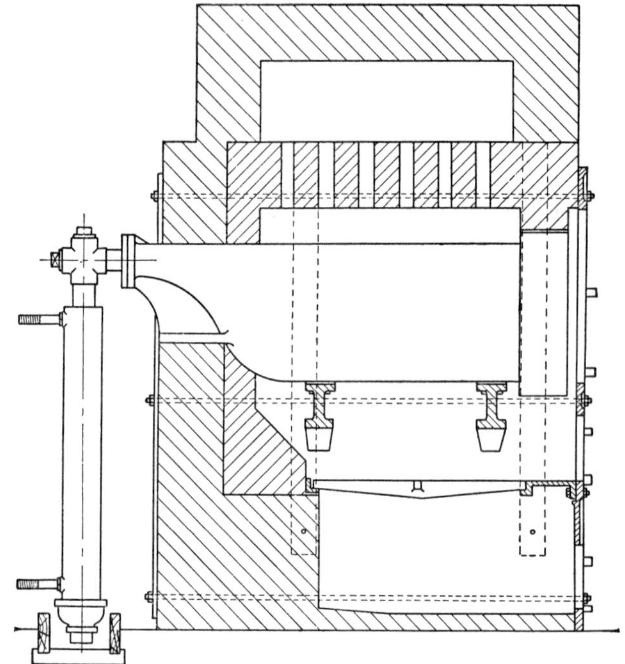

[*By courtesy of Messrs. Fraser & Chalmers.*

Fig. 87.—Horizontal Amalgam Retort.

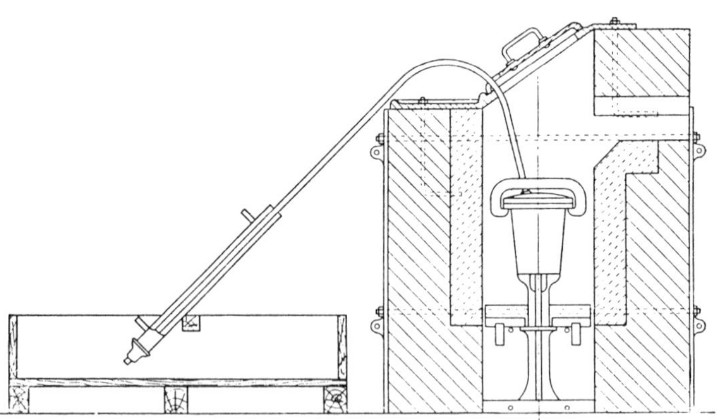

[*By courtesy of Messrs. Fraser & Chalmers.*

Fig. 88.—Vertical Amalgam Retort.

The charge is heated slowly until the boiling point of mercury is reached, when the fire is checked, and the retort kept at an even temperature for

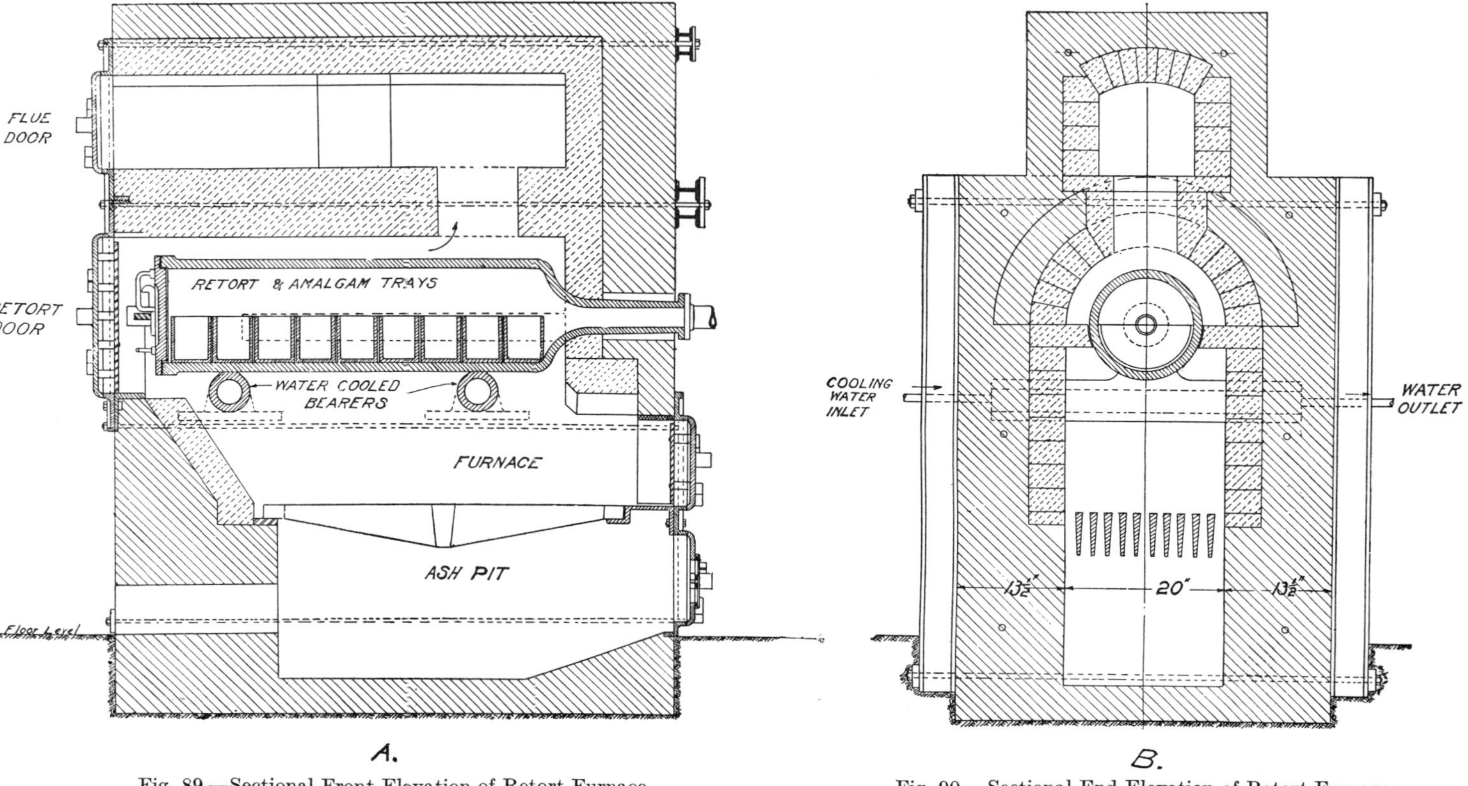

Fig. 89.—Sectional Front Elevation of Retort Furnace.

Fig. 90.—Sectional End Elevation of Retort Furnace.

three or four hours, or until the bulk of the mercury has been driven off. The retort is then raised gradually to a bright red heat to expel the remainder. After cooling, it is opened, the trays withdrawn, the retorted metal loosened by a chisel, if necessary, and turned out on a table.

In retorting amalgam containing considerable quantities of base materials, there is a danger of the vent being choked up by condensation of solid material. The retort should be so arranged that a rod can be passed through the condensing pipe so as to clear it of obstructions, if necessary. The front of the retort is luted on carefully with chalk or wood-ashes and salt, and firmly clamped with an asbestos joint, so as to be quite tight, otherwise a loss of mercury is incurred. In all retorts the lid is turned and ground so as to fit on perfectly. The condensing pipe should not have an open end dipping freely into water, as in that case a sudden cooling of the retort would cause the water to be sucked in, and an explosion would occur. The open end of the pipe may be enclosed in a bag of sacking or rubber immersed in water.

Pyles [1] describes a method whereby the fumes are blown out of the retort by compressed air when distillation is completed. The exit pipe is carried outside the building. By attaching an air-ejector suction to the condenser and delivering the vapour into a water-sprayed receiver, Walker [2] recovers an extra ounce of mercury per 1,000 ozs. amalgam.

The retorted metal is porous and from 500 to 950 fine in gold, the remainder being, in general, chiefly silver, with base metals and sulphides in smaller quantities. The gold is melted in crucibles with sand, carbonate of soda and borax, and suffers a further loss in weight, due to the slagging off of oxides, earthy impurities, etc., and to the volatilisation of a small quantity of mercury, which is obstinately retained until the melting takes place. Richards states [3] that if the amount of mercury is to be reduced below 1 or 1½ per cent. in retorting, a white heat must be used, by which the retorts are damaged and soon worn-out. Additions of nitre, corrosive sublimate, etc., are not to be recommended. The melting of bullion is dealt with fully in Chapter XVII. The loss of weight in melting is given by Richards [4] as 1·5 per cent. at the Homestake Mill and 7 per cent. in the Caledonia Mill. It is given by Smart as 0·5 to 1 per cent. on the Rand.[5] Part of this loss consists of gold and silver retained by the slag which is either remelted or passed to the stamp mill or clean-up barrel.

Loss of Mercury.—The loss of mercury in milling is due to (1) "flouring," or minute mechanical subdivision, and (2) "sickening," or extreme subdivision caused by chemical means.[6] In the latter case, a coating of some impurity is formed over the minute globules of mercury, which are thereby prevented from coalescing, from taking up gold and silver, or from being caught by the plates and wells, as the coating prevents all contact between the mercury and other bodies. The impurity may be an oxide, sulphate, sulphide or arsenide of some base metal, either originally present in the

[1] *J. Chem. Met. Mng. Soc. S.A.*, 1920, **20**, 22.

[2] "*Rand Metallurgical Practice*," vol. i., p. 484.

[3] "*Ore Dressing*," 1903, vol. ii., p. 782.

[4] *Op. cit.*, p. 784.

[5] "*Rand Metallurgical Practice*," vol. i., p. 93.

[6] Richards points out ("*Ore Dressing*," p. 751) that there is no advantage in trying to draw a distinction between these two terms, which was apparently first done by T. A. Rickard ("*Stamp Milling of Gold Ores*," 1897, p. 216). Richards holds the view that the globules are in all cases prevented from re-uniting by a film of some foreign substance. The term "deadening" of mercury is also used.

mercury, or taken up from the ore by it. The base metals usually present in mercury are rapidly oxidised in the air, especially in contact with acid water. The metallic oxides thus formed are not soluble in mercury, and float on its surface. Impure mercury, when used to amalgamate the plates, causes their discoloration by oxidation of the dissolved metals. One of the impurities in mercury most to be feared is lead, as the amalgam of this metal tends to separate out of the bath of mercury in which it is dissolved and rise to the surface.

Sickening of the mercury is also promoted by base minerals present in the ore. Many sulphides are harmful, their action being, however, much less energetic than the compounds of arsenic and antimony. Arsenical pyrite produces a large amount of black sickened mercury, the action being especially energetic if the pyrite is partly decomposed. Sodium amalgam acts beneficially when arsenic is causing loss of mercury.

Sulphide of antimony also breaks the mercury into black powder, some sulphide of mercury being formed if there is any trituration, whilst the antimony forms an amalgam. The action of sodium amalgam on this mixture is of no avail, as sodium sulphide is formed, more antimony amalgam produced, and sulphuretted hydrogen set free, the results on the amalgamation of the gold being very disastrous. Bismuth sulphide acts similarly, but with less rapidity. Graphite and finely divided carbonaceous matter also appear to have a deleterious effect.

Floured mercury is perfectly white in appearance, like flour. If it is examined with a lens, it is seen to consist of a number of minute particles—many of them microscopic—each of which is perfectly bright and pure, shining like a mirror. They are prevented from coming into contact and coalescing by being surrounded by films either of air or of some transparent foreign substance. Floured mercury is readily carried away and lost in the tailings, but if passed through and agitated with a large body of clean mercury much of it is at once absorbed in the mass. The loss through flouring is experienced in the milling both of refractory and free-milling ores. The effects of grease and also of talc, serpentine, clay, hydrous silicates, and carbonaceous material in subdividing mercury are doubtless due to mechanical action only.

Properties and Purification of Mercury.—Mercury solidifies at — 39° and boils at 357° C., but is slightly volatile even at ordinary temperatures. Its specific gravity is 13·59, its electrical conductivity is 1·6, and its thermal conductivity 1·8, if the values for silver be taken as 100. Pure mercury is unaffected by the air at ordinary temperatures, but is slowly oxidised if heated to about 350° C. When mercury is impure from the presence of other (oxidisable) metals, these are rapidly oxidised in air, forming powdery scales. It is not acted on by hydrochloric acid, and is almost unaffected by dilute sulphuric acid, but with hot concentrated sulphuric acid it forms $HgSO_4$. Mercury is dissolved even by cold dilute nitric acid, and is rapidly dissolved by hot nitric acid. It is dissolved by aqua regia with the formation of mercuric chloride, $HgCl_2$. Pure mercury will roll down an inclined surface without forming a pronounced "tail" and without leaving any streak behind it. If a blackish film is left behind, the mercury requires purification.

When agitated with oil, fats, turpentine, many organic substances, sulphur, etc., mercury is split up into minute globules ("flour") not easily re-united. Vegetable or animal oils cause more flouring than mineral oils. Coalescence of floured mercury is effected by the action of certain reducing agents, such as water and sodium, the passage of an electric current, or with some loss by the action of nitric acid.

The vapour of mercury has a poisonous effect (salivation) on the animal system. Among the remedies are cleanliness, fresh air, acid foods, abstention from alcohol, and potassium iodide as medicine.

Purification of Mercury.—It is very important to use pure mercury in ordinary amalgamation processes, so as to reduce the losses as far as possible. The purification may be effected by distillation with lime and iron filings. The iron filings decompose sulphides and prevent bumping. An addition of charcoal powder is mentioned by Richards, its use being to prevent the formation of volatile oxides. Lead, cadmium and zinc pass over in part with the mercury, but are oxidised in a current of air. Zinc, tin, copper, and iron may be removed by shaking with dilute hydrochloric acid ; shaking with mercurous nitrate, ferric chloride or potassium dichromate and strong sulphuric acid has also been recommended. If mercury is covered with dilute nitric acid (one part of acid to three parts of water) it is gradually purified, especially if stirred occasionally ; mercurous nitrate is formed and acts on the base metals. A more rapid way of removing base metals is to pass a stream of air through mercury covered with dilute nitric or sulphuric acid. The base metals are rapidly oxidised by the air and dissolved by the nitric acid. L. Meyer [1] lets the mercury fall through a long column of mercurous nitrate, for which nitric acid may be substituted. J. H. Hildebrand [2] recommends that the falling mercury should be broken up by passing it through muslin.

Sodium amalgam is used to revivify sickened mercury, or to keep it in good condition. It is prepared by heating a basin or iron flask containing mercury to about 150° C., and dropping in pieces of sodium not larger than a pea, one by one. Each addition causes a slight explosion and a bright flash of flame. The sodium may be added with less loss and less danger to the operator if the mercury is kept at a somewhat lower temperature and the sodium stirred into the mercury with an iron pestle or pressed below its surface with a spatula. When about 3 per cent. of sodium has been added to the mercury, the reaction becomes less active and the amalgam is poured out upon a slab or shallow dish, allowed to cool and solidify, and then broken up and kept in stoppered bottles under naphtha. When it is necessary to revivify a quantity of mercury, a few small pieces of the amalgam are added to it and stirred in, or are previously dissolved in clean mercury before being added to the impure stuff. The strong affinity of sodium for oxygen, chlorine, etc., enables it to reduce the oxides and other compounds of the base metals which are coating the mercury globules. The base metals are redissolved by the mercury which is then in good condition to take up the precious metals or to be caught on amalgamated surfaces or in riffles ; but the mercury is not really purified and the base metals in it are soon oxidised again. The sodium hydroxide passes into solution, neutralising part of the acidity of the water in the pulp. Sodium amalgam is not much used except in amalgamating-pans or in mercury wells or riffles, or in cleaning retorted mercury—*i.e.* wherever large bodies of mercury can be directly acted on by it.

Examples of Amalgamation Practice.—At the *Mysore* Gold Mine gyratory crushers following the jaw breakers have been discarded owing

[1] *Zeitsch. anal. Chem.*, 1863, **2**, 241.

[2] *J. Amer. Chem. Soc.*, 1909, **31**, 934 ; *J. Chem. Met. and Mng. Soc. of S. Africa*, 1909, **10**, 224.

to their high cost of upkeep.[1] Stamps, each weighing 1,156 lbs., are used and are provided with outside amalgamation tables. Wooden piles set in concrete are preferred for battery foundations. Sixteen-mesh screens are employed. Mercury is fed into the mortar-box at intervals according to the richness of the ore and the state of the plates. The pulp leaving the plates falls into a 6 inch deep riffle containing 30 lbs. of mercury and having, on its discharge edge, a blanket-covered board to catch stray amalgam. Amalgam is squeezed in drill instead of chamois leather, as it gives a harder ball. The average percentages of recovery of amalgam are as follows :—On the plates 82, in die sand 12, in riffles 4, in boxes 2. The average percentage of gold in the amalgam is 43, and the fineness of retorted bullion 932/1,000 gold and 62/1,000 silver. The pulp passes on to the cyanide plant.

Homestake.—A recent installation of stamps is in the South Mill of the Homestake Company. Each stamp weighs 1,550 lbs., and the feed is from a Symons cone crusher set at $1\frac{1}{2}$ inches. The capacity of the mill is 23 tons per stamp per 24 hours through a screen $\frac{3}{4}$ to $\frac{7}{8}$ inch aperture. The mill product is dewatered in cones and then ground in a 6×12 foot rod mill to which mercury is added. The pulp next passes to a simple launder amalgamator which has replaced the plates. Finally it is re-ground in 5×14 foot ball mills, treated in amalgamators and sent for cyanidation.[2]

Goldenville, N.S.—Stamp milling and plate amalgamation are still carried on at Goldenville, N.S.,[3] on a siliceous gold ore carrying 85 per cent. coarse free metal. Gold also occurs in the accompanying sulphides. The ore is crushed to $1\frac{1}{2}$ inches in jaw breakers, and then in stamp mills to pass a 24-mesh screen. The stamps weigh 840 lbs. each and have a 6 inch drop. Water is supplied under pressure in jets of $\frac{1}{8}$ inch diameter directly on to the dies but below their surface, in order to keep them clear of sand and to improve the crushing efficiency. The rising water also has a classifying action, and the two effects combined increase the rate of crushing and help the gold recovery, the metal sinking between the dies. Inside amalgamation without inside plates is practised. Mercury is fed to the mortar every hour in $\frac{1}{4}$ inch droplets, the number depending on the condition of the outside plates. The outside plates are of silver-plated copper, and there are 17·33 square feet for 5 stamps. From these plates the pulp goes through an amalgam trap. The plates are cleaned periodically with borax, soap and water.

---

[1] *Kolar Goldfield Min. Met. Soc., Bull. 21*, 1928.
[2] *Min. Ind.*, 1934, 42, 640.
[3] E. H. Henderson, *Trans. Can. Inst. Min. Met.*, 1935, 38, 222.

CHAPTER VIII.

INTERMEDIATE AND FINE GRINDING.

Discussion of Coarse and Fine Crushing.—In general some of the gold is distributed through an ore in an extremely fine state of division, and the finer the crushing the more gold is laid open to the attack either of mercury or of cyanide. This consideration points to the desirability of "sliming" all the ore, if it can be done cheaply enough. On the other hand, the gold is generally in part contained in sulphides, tellurides, etc. These form the richest part of most ores, and retain their gold with greater obstinacy than the other constituents of the ore. They are also the most brittle, so that if the whole ore is reduced to a fine state of division the sulphides, etc., are converted into an impalpable slime. In this condition they cannot readily be saved by gravity concentration.

Perhaps the most useful definition of *slime* is "that portion of the crushed ore which, owing to its physical condition, cannot be leached by percolation under the action of gravity." The "physical condition" is defined by H. A. White [1] as "minutely subdivided condition and the presence of colloidal substances." Such material does not subside readily in water.

Allen [2] is of the opinion that ultra fine grinding invariably leads to the "activation" of the ore particles by reason of the vast increase in surface area exposed by comminution and to the bringing into play of physico-chemical forces that exercise a powerfully disturbing effect on the stability of a solution containing cyanide and precious metal.

Before the introduction of the cyanide process, when the general methods of treatment were crushing and amalgamation, followed by concentration and the treatment of the concentrate by roasting and chlorination, the aim was to avoid the formation of slime. At that time the value of crushers giving a uniform product seemed especially great, and rolls were strongly advocated as against stamps, which produced a greater proportion of very finely divided material. Later, when the Kalgoorlie ores began to be treated by dry crushing and roasting, the uniform product of ball mills was found to be better than the partially slimed product of Griffin mills, because of the greater difficulty of roasting the latter.[3] When ores are subsequently to be roasted there is little advantage in fine grinding, because in the course of roasting the particles usually become porous, so that chlorine or cyanide can penetrate into their interior and dissolve gold which has not been exposed on the surface of the grains.

In the case of ore which is treated without roasting, the finer the state of division the higher the percentage of gold that can be extracted. The method

---

[1] "*Rand Metallurgical Practice,*" vol. i., p. 189.
[2] *Trans. Amer. Electrochem. Soc.*, 1931, 60, 67.
[3] W. Evan Simpson, *Trans. Inst. Mng. and Met.*, 1904, 13, 22.

of regrinding the product of the stamp battery or other machines was therefore developed, resulting in a higher extraction in many cases. This was followed by attempts to increase output by using coarser battery screens, and also to use a method of stage reduction in successive machines, rather than one of great reduction in size in a single machine.

For example, it was established that for such hard, brittle ores as those of the Rand it is advantageous to reduce the ore to 1¾ inches in rock-breakers before feeding it to the stamps, or their successors, the gyratory crushers, where it is crushed to a screen size of ½ to 1 inch, and to grind to about 100 mesh in tube mills.

Improvement in the efficiency of mills dealing with low-grade ore would appear to lie as much in better machinery as in changes in fundamental metallurgical processes. At the Howey and Malartic mines it was found that grinding costs could be reduced, without interfering with the high extraction by cyaniding, by crushing only to 40 mesh. Greater attention is being paid to the critical grinding mesh, resulting in the lowest grinding costs consistent with recovery.

In grinding pyritic gold ore in a ball mill the softer pyrite is rapidly removed from the circuit by the classifier. The heavy sand containing gold is reground and the latter becomes highly comminuted owing to the dismembering action of quartz particles embedded on its surface. Re-classification of the overflow of the primary classifier results in a low-grade sand, with the larger portion of the values in the slimes. Hence sometimes the overflow is agitated and the sands of this secondary classification leached, giving decreased grinding costs.

Selective treatment of the pyritic portion of the banket ore has often been advocated and has been adopted beneficially on some new reduction plants, either through the agency of Johnson concentrators or of multiple-stage classification and closed circuit stage grinding. The coarse quartz is confined to the primary tube mills and is not present in the secondary mills to hinder the effective grinding of the finer quartz and pyrite.[1] There appears also to be a selective concentration of pyrite in the closed circuit systems.

White does not accept Thurlow's conclusions and suggests that the gold exists on the interfaces between the pyrite and quartz and that up to 60 per cent. of the gold is definitely associated with the quartz.

It is interesting to note that as early as 1906 G. A. and H. S. Denny,[2] when dealing with the proposals for a plant to produce a single-product pulp in which the whole of the values were to be recovered by cyanidation, suggested "the design of the new plant would be entirely different . . . from a plant erected to-day." The stamp battery is left out in this design, the principles being stage crushing and stage grinding, the cost of such plant being cheaper than that of one including a stamp battery.

Energy Consumed in Crushing.[3]—Attempts have been made to make use of the "laws" of Rittinger and Kick, and adapt them to ore crushing, but neither of them provides a perfectly satisfactory basis for calculation. Rittinger's law is to the effect that the energy absorbed in crushing is proportional to the new surface produced or to the reduction in diameter; and

---

[1] Thurlow, *J. Chem. Met. Mng. Soc. S.A.*, 1932, **32**, 316.

[2] *Proc. S.A. Assoc. of Eng.*, 1905-6, **11**, 239.

[3] Stadler, *Trans. Inst. Mng. and Met.*, 1910, **19**, 478; Gates, *Eng. and Mng. J.*, 1913, **95**, 1039; 1914, **97**, 795; Taggart, "The Work of Crushing," *Trans. Amer. Inst. Mng. Eng.*, 1914, **48**, 153; Speak, *Trans. Inst. Mng. and Met.*, 1914, **23**, 482.

Kick's law, as stated by Stadler, is that "the energy required for producing analogous changes of configuration of geometrically similar bodies of equal technological state varies as the volumes or weights of these bodies." It is generally agreed, largely because of work done in America,[1] that Rittinger's law is the more correct criterion to apply in crushing calculations.

When the energy required for crushing an ore has been estimated, data can be obtained for approaching from the side of theory the problem of determining the relative efficiency of various crushing machines, *e.g.*, stamps, rolls, pans, and tube mills; some help may thus be given in the selection of the machine to be adopted for a definite purpose in a particular case. At present this selection is difficult, and mistakes are sometimes made, theory giving no certain guidance.

Gross and Zummerley[2] describe a device which they have used for determining the work input to a laboratory ball mill.

Grinding studies on actual mill machines demonstrate the high efficiency in the use of energy for the production of new surfaces in the initial stages, and the large waste of energy in useless grinding that produces very fine particles.[3]

Tube Mills.—Fine grinding of coarsely crushed ore is effected in pans, tube mills, or conical pebble mills, such as the Hardinge mill.

Tube mills are good machines for fine grinding, and on the Rand are still the only machines used for grinding the ore to the degree of fineness required for subsequent cyaniding. They consist of revolving cylinders rather more than half-filled with pebbles, by the impact of which, in falling, coarse particles of sand are crushed fine.

More recently large pieces of the hard ore have been used in place of pebbles. Where these have not been available, steel balls have been employed. In this case slower revolution speeds are usually required. E. H. Rose[4] has advanced arguments for the use of cubes instead of balls in tube mills, on the supposition that they offer more contact surface for trapping and grinding the particles of ores. Various other shapes have been tried, but the sphere has proved most effective.

The *cylinders* consist of steel plate with cast-iron or steel heads, which are strengthened by radial ribs. Fig. 91[5] shows a tube mill with cast-iron ends. Dished ends have in some cases been superseded by plane ends.

The variation in size of tube mills used in different parts of the world is considerable.[6] At Kalgoorlie some of the earlier tube mills were 3¼ feet by 13 feet. In North America 5 feet by 18 feet mills were popular. On the Rand the standard size is 5½ feet by 22 feet. In 1914, short tube mills only 6 feet long and 7 or 8 feet in diameter were introduced. 20 feet by 6½ feet mills, driven by 250 h.p. motors, have been erected recently on the Rand, and at East Geduld seven mills measure 16 feet by 8 feet, being driven by 350 h.p. motors.

---

[1] *Amer. Inst. Min. Met. Eng. Tech. Pub. Nos.* 46, 126, 127; *Min. and Met.*, 1929, **10**, 447; *U.S. Bur. of Mines, Report on Investigations No.* 3223; G. Martin, *Trans. Ceram. Soc.*, 1925, **25**, 51.

[2] *U.S. Bur. of Mines*, 1931, *Report on Invest. No.* 3056.

[3] *U.S. Bur. Mines, Report on Investigations Nos.* 2948, 2951, 2952.

[4] *Eng. Min. J.*, 1926, **122**, 95.

[5] Dowling, "*Rand Metallurgical Practice*," vol. i., p. 106.

[6] See Gieser, *Eng. and Mng. J.*, 1914, **97**, 463, for complete data.

The cylinders of tube mills can be made in sections for transportation on mule back.

The cylinder is carried on two hollow *trunnions*, one at each end. The pulp enters through one of these (left-hand end in Fig. 91), and is discharged through the other.

The *lining* consists usually of iron or steel, and its shape and constitution should be such that the wear is uniform and the minimum of scrap is rejected when the liner wears out. The lining used originally was of silex or flint blocks about 6 inches thick, cemented in. On the Rand, where local chert was used for lining, the life was about 80 days. The use of steel or iron liners has now become general practice. Their life varies from 140 to 330 days and their cost from 9s. to 25s. per day for a standard tube mill. For a 6 foot 6 inch × 20 foot mill the cost is roughly 25s. per day and the life 140 days. Manganese steel linings lasted fifteen months at Kalgoorlie, and chilled iron liners from seven to ten months, but the great cost of manganese steel made it less economical than hard iron.[1]

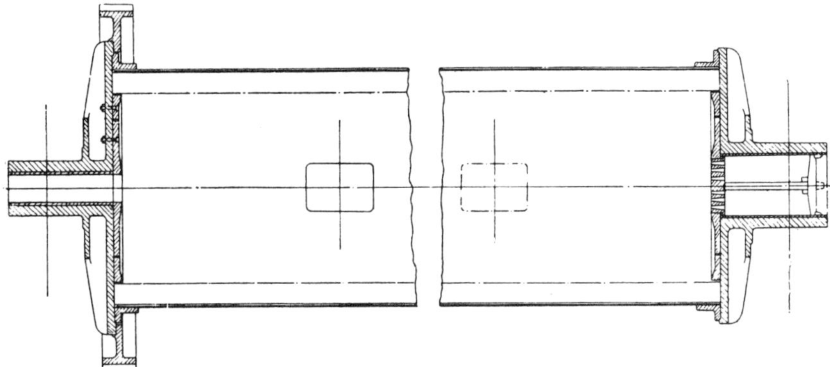

Fig. 91.—Section of Tube Mill.

Iron-ribbed liners were introduced first at El Oro (Gieser), and are in wide use. The ribbed form is designed to reduce the slip of the pebbles, which was very great with smooth iron liners, and not inconsiderable with new silex linings. When the pebbles slip, they are not carried high enough for effective impact in falling (see p. 197), and the output of the mill falls off. Nevertheless, smooth iron liners were preferred at Tonopah, Nevada,[2] for very fine grinding. The plates may be made thicker at one edge than the other, giving a wave-formation. The particles then fall from higher to lower plates.

Ribbed, corrugated or honey-combed steel liners absorb more power, but increase the efficiency of the mill. The speed of rotation may be reduced and power saved without loss of efficiency in the case of some of these liners.

The Osborn liner used on the Rand (Fig. 92) [3] consists of a series of slightly wedge-shaped iron bars, 4 × $1\frac{1}{4}$ to $\frac{3}{4}$ inches in section, spaced at from $2\frac{1}{2}$ to 5 inches apart at the base round the periphery of the tube mill. They are held in place by iron wedges $2\frac{1}{2}$ to 3 inches wide by $\frac{1}{2}$ to $\frac{3}{4}$ inch thick.

[1] A. James, *Trans. Inst. Mng. and Met.*, 1904, 14, 98; Trewartha James, *ibid.*, p. 121.
[2] Megraw, *Eng. and Mng. J.*, 1913, 95, 414.
[3] Schmitt, "*Rand Metallurgical Practice*," vol. ii., p. 158.

The liner weighs about 10 tons for a standard mill. Its life is found to be much greater than that of silex blocks, and is said to be 300 days.[1] It can be replaced more rapidly than silex, and the unworn ribs near the outflow can be used again. The pebbles lodge between the ribs, take up the wear and fall out as the tube revolves. The Osborn liner has been improved by placing the radial bars upon the flats instead of between them. At one Rand mine it is the practice to use the Osborn lining in three sections, each 7 feet in length : (1) chrome steel, (2) ordinary steel, (3) mild steel, in order from the feed to the discharge end, so that they all wear out at the same time.

Block liners consisting of white iron or steel cast blocks, measuring approximately 21 × 8 inches, and varying in depth from 4 to 5 inches, are also used. They are either solid or honeycombed, the latter type serving to hold pebbles as in the case of the Osborn liner. The wearing surface of the solid block is arched to give a corrugated and lifting effect. The sides are ground so that only wedging is necessary to fix the blocks in position. Such a liner for a standard mill weighs roughly 13 tons.

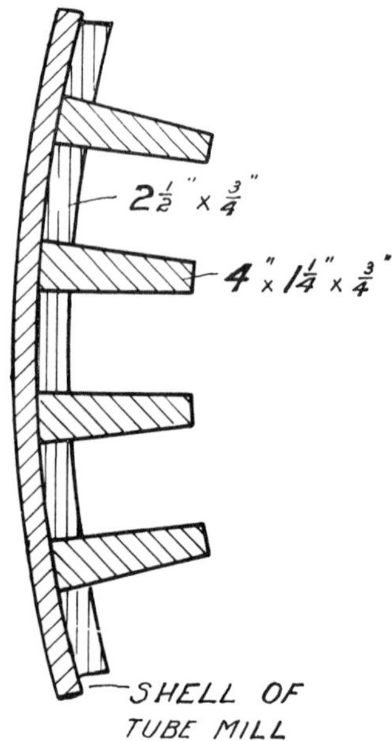

Fig. 92.—Osborn Liner.

A corrugated or wave type of tube mill liner has recently been introduced in South Africa. The separate segments are ground in the foundry to straighten the edges and splay them so that the radial lines are correct to gauge. When the ring is complete keys are driven in the last joint to tighten the ring and hold it permanently in place. The liner is made ¾ inch thicker at the inlet end of the mill where most wear occurs.[2]

The El Oro liner (cast-iron segments), the honey-comb liner and the Gibson liner (short pegs of steel set in cement) have also been used on the Rand.

The *pebbles* usually consist of flint, or steel balls. The size varies up to 3 or 4 inches in diameter. The larger the pebbles the coarser may be the ore fed in. For dry grinding, small pebbles of 1 to 2 inches in diameter are preferable. On the Rand pieces of banket ore about 4 inches in diameter are fed in instead of pebbles. The amount required may be as much as 2½ per cent. of the total tonnage milled.[3]

At West Springs large pebbles are used in the primary mill and −1½ inch pebbles (rejects from the primary mills) plus scrap steel in the secondary. The secondary mills yielded the best results when run at the same speed as the primary mills (25·6 r.p.m.).[4]

The mill is kept about half filled with pebbles, the upper level of which

---

[1] Gieser, *Eng. and Mng. J.*, 1914, 97, 465, 466.
[2] *Eng. and Mng. J.*, 1933, 134, 25 ; *Min. Mag.*, 1933, 48, 119.
[3] W. R. Dowling, "*Rand Metallurgical Practice*," vol. i., p. 116.
[4] A. James, *Eng. Min. World*, 1930, 1, 15.

varies from 3 inches above the centre line of the mill (as in S. Africa) to as much as 7 inches below.

It has been calculated that 40,000 pebbles in a mill deliver 1,200,000 blows per minute. The weight of pebble load for a 22 foot mill at all heights and diameters and the method of calculation are given by Caldecott.[1] The quantity of pebbles to be added daily may be taken as 13 tons for a 5 foot 6 inch mill, 20 tons for a 6 foot mill, 33 tons for a 6 foot 6 inch mill and 55 tons for an 8 foot mill.

Tube mill pebbles are obtained from the run of ore by placing a trommel or grizzly at the discharge end of the sorting belt.[2] The trommel is better as an intermediate size of pebble can also be obtained. In modern plants it is usual to adjust the rock breaker so that the pebbles will not be too large. Movable grizzlies can be withdrawn when a sufficient amount of pebbles has been obtained.

A mixture of balls and pebbles has been suggested as being better than pebbles alone.[3] Thus in one Rand mill where the load consisted of 75 per

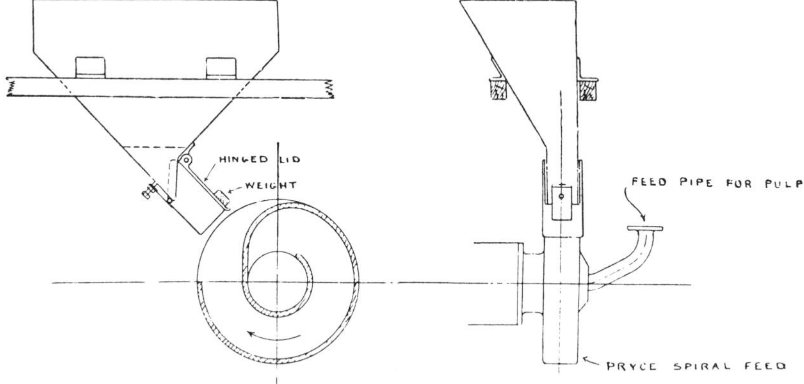

Figs. 93, 94.—Pryce Pulp and Pebble Feeder, with Thomas's Hopper.

cent. pebbles and 25 per cent. of $2\frac{1}{2}$ inch steel balls, there was an increase of 28 per cent. in —200 mesh material for an increase of 17·5 per cent. in power consumption and 18 per cent. in that of balls. 3-inch balls are generally used for primary and $2\frac{1}{2}$-inch balls for secondary grinding. Their cost works out at about 2d. per ton of original ore. On some mines scrap steel, *e.g.* discarded drill steel in 4-inch lengths, is used in place of balls.

There are two *manholes* in the shell for use when the mill has to be relined or repaired.

The *feeding* of ore and of pebbles into a tube mill is effected through one of the trunnions, which is up to 10 inches in internal diameter. Several different methods of feed have been proposed. One of these, designed by L. Pryce, is shown in Figs. 93, 94.[4] It consists of two parallel discs 22 inches diameter and 7 inches apart. A spiral between the discs connects an opening

---

[1] *J. Chem. Met. Mng. Soc. S.A.*, 1913, **13**, 363.

[2] Prentice, *Trans. Inst. Mng. Met.*, 1935, **44**, 479.

[3] King, *J. Chem. Met. Mng. Soc. S.A.*, 1931, **32**, 38. See also Wartenweiler, *ibid.*, 1932, **32**, 262.

[4] Dowling, "*Rand Metallurgical Practice*," vol. i., p. 161

at the periphery with the centre. The pebbles, fed from a hopper, designed by J. E. Thomas, into the peripheral opening, are carried to the centre and pushed into the hollow trunnion by an Archimedean screw. The pulp coming from a classifier or dewaterer enters as shown. The most common type on the Rand now is a simple hopper which passes pebbles up to 10 inches in size. The inlet end is protected by cast plates or small forged steel blocks, the plates lasting about 120 days and the blocks about 300 days.

The *discharge* takes place through the other trunnion, which is also hollow and is of greater diameter, so that the outflow is about 5 inches lower than the inflow. The outflow is through a perforated plate (right hand of Fig. 91), flush with the end liner, with holes expanding towards the outflow to prevent choking.

Neal's discharge, introduced at El Oro in Mexico, is an alternative to the perforated plate. It consists of a baffle placed inside the tube mill, and a reverse screw which returns to the tube mill any pebbles which might escape the baffle, although it permits of the free escape of the water-borne pulverised pulp. Peripheral discharge has been tried, but though suitable in dry crushing, is not much used in wet crushing, owing to excessive wear of the liners and large consumption of pebbles.

Fragments of pebbles and worn-out pebbles too small to assist in the crushing, pass out with the pulp, and are separated from it by a circular screen or pebble-catcher of ¼ inch mesh, through which the pulp passes into a launder, while the pebbles are delivered at the end of the screen.

Prentice states that on the Rand the discharge is now aided by scoop elevators. A grid plate, 2 to 3 inches thick and having apertures which allow the free exit of the pulp but retain the pebbles, is fitted to the discharge end and the pulp passes through this to lifters by which it is elevated to be discharged through the trunnion. The greater the radius of the grid and lifters the lower is the level of the pulp in the tube mill at the discharge end. Hence the pebbles do not fall into a mud bath but on to other pebbles covered by sand, thus increasing the grinding efficiency by some 15 per cent. There is also less likelihood of a back-flow of pulp. Liner and pebble consumption are appreciably increased, and additional power is required. Where stamp mills are not used, the feed to the tube mills is coarse and the apertures in the grid have been enlarged to permit worn pebbles up to 1½ inches diameter to be discharged. These are separated from the pulp by a pebble eliminator, usually a ¼-inch screen attached to the discharge end of the mill. From 2 to 40 tons of pebbles are thus removed per mill per 24 hours, according to the system of crushing in vogue. The rejected pebbles if small are returned to the tube mill feed, or may be screened. The larger pieces are used for grinding in the secondary or tertiary circuit and the intermediate size is fed to a crusher or to the stamp mill. The outlet end plates and the discharge screen plates last about 120 days.

Thurlow introduces a small number of cast-iron balls at the discharge end to break up pieces of pebble that are constantly being discharged, and thus does away with a pebble eliminator.

A scoop discharge (Fig. 95) is described by Dowling.[1] Four curved arms, having their concave portions facing the direction of revolution of the mill, are bolted to a flat plate which is perforated with ½ to ¾ inch holes, except

---

[1] *J. Chem. Met. Mng. Soc. S.A.*, 1915, **15**, 214. See also White, "*Rand Metallurgical Practice*," vol. i., p. 488.

near the arms, where it is solid, to give pockets which carry the pulp to discharge into the outlet trunnion. The curved arms are placed between the screen and the mill. The diameter of the screen regulates the depth of the pulp in the mill. The tube ends are protected from wear by a cast-iron ring between the screen and the tube walls. The screen diameter ranges from 24 to 36 inches, and depends on the loading of the mill and the speed of running. The mechanical efficiency of a mill fitted with scoop discharge is about 10 per cent. The increased capacity is roughly 20 per cent. with a 24 inch scoop to 25 per cent. with a 36 inch scoop. The pebble consumption varies with the scoop diameter. A 30 inch scoop doubles the tonnage of pebbles that must be fed, reduces the lining life by 30 per cent. and needs a 125 to 150 h.p. motor to drive the mill.

The *speed of revolution* of tube mills varies with the diameter. Krupp uses the formula $\frac{32}{\sqrt{D}}$ for the correct number of revolutions per minute, where D = internal diameter in metres. The Davidsen formula is $\frac{200}{\sqrt{D}}$, where D is the diameter in inches. These two formulæ are almost identical and would give about 26 revolutions per minute for a mill of 5 feet diameter. White's original formula was $\frac{34 \cdot 22}{\sqrt{D}}$, where D is in metres.

From a long series of later experiments on tube mill conditions, White draws the following conclusions, which apply to Rand ore :—

(1) For a standard tube mill, loaded to the centre, and dependent on the observed angle of running slope, the optimum speed of rotation for capacity would be 20 or 21 r.p.m. Generally for any tube mill r.p.m. = $\frac{150}{\sqrt{D}}$ to $\frac{168}{\sqrt{D}}$ or 56 to 62 per cent. of the critical speed. (D = diameter in inches.)

(2) The weight of balls in a tube is $0 \cdot 0582\, L\, D^2$ tons for steel and $0 \cdot 0202\, L\, D^2$ tons for banket, where L and D are the length and diameter in feet, respectively.

Davis [1] confirms White's results on the best running speed for tube mills.

White gives figures based on theoretical calculations for the speed of tube mills under varying conditions [2]:—

TABLE XXV.

Diameter, ins.	R.P.M. for Max. Efficiency, Load 35 per cent.	R.P.M. for Max. Capacity for Coarsest Feed, Load 50 per cent.	R.P.M. for Max. Capacity for Fine Feed, Load 55 per cent.
42	33·1	34·4	37·5
45	32·0	33·2	36·2
48	31·0	32·2	35·1
51	30·0	31·2	34·0
54	29·2	30·3	33·1
57	28·4	29·5	32·2 (Stand.)
60	27·7	28·8	31·4
63	27·1	28·1	30·6
66	26·4	27·4	29·9
69	25·8	26·8	29·3
72	25·3	26·3	28·7
84	23·4	24·3	26·5
96	21·9	22·7	24·8

[1] *Trans. Amer. Inst. Min. Eng.*, 1920, **61**, 250, 295.
[2] "*Rand Metallurgical Practice*," vol. i., p. 491.

The peripheral speed inside the lining at 32 revolutions, with new linings 6 inches thick, would be 452 feet per minute, and with worn linings only 1 inch thick it would be 536 feet per minute, the diameter inside the shell being $5\frac{1}{2}$ feet. As the lining wears, the number of revolutions per minute is reduced to keep the peripheral speed approximately unchanged. There is a certain peripheral speed which is most advantageous with a particular ore.

The *power* required varies with the rate of feed (load) and speed. A Rand tube mill, 22 feet long by $5\frac{1}{2}$ feet diameter, requires from 90 to 120 h.p., increasing as the liners wear, owing to increased size and capacity of the mill. Prentice states that the average power consumed by tube mills amounts to 18 kwh. per ton of -100 mesh product obtained. According to Davidsen the formula P (in h.p.) $= 0{\cdot}15$ N, where N $=$ capacity of mill in cubic feet, gives the power required, but this is too low for Rand practice. Whereas, formerly, the tube mills were belt-driven, direct-drive through a reduction gear, with the driving mechanism placed at the more convenient end, has become standard practice.

The *capacity* (output) of tube mills depends on the hardness and size of the material supplied to them, and on the fineness of the product. The amount of grinding done and the fineness of the product depend on the rate of feed and the length of the tube, as well as on the hardness of the ore. On the Rand, a product only fine enough to pass through a 100 mesh screen is aimed at. A standard tube under average conditions will make about 150 tons of $-$ 100 mesh material per day; a 6 foot 6 inch $\times$ 20 foot mill about 230 tons. The average feed of solids to give this result is:—standard mill, 450 tons; 6 foot 6 inch mill, 900 tons; 8 foot mill, 1,300 tons.

The calculation of the tonnage fed to the mills is ascertained [1] from the grading samples taken by each shift from every tube mill. The following formula is used:—$T = \frac{100\ D}{c - b}$, where T $=$ tons of dry solids fed per tube per 24 hours, D $=$ tons of -90 product per tube per 24 hours, b $=$ percentage of $-$ 90 mesh product in tube mill feed, c $=$ percentage of $-$ 90 mesh product in tube discharge.

In 1919, at Brakpan Mines,[2] a duty of approximately 135 tons of $-$ 90 mesh material per 24 hours was obtained in a standard tube mill (22 feet $\times$ 5 feet 6 inches). The feed was the undersize of a $1\frac{1}{2}$-inch grizzly. The grinding charge was 12 to 15 tons of reef pebble and the reject pebble discharge was about 25 tons. As 40 per cent. of the run of mine ore passed through a $1\frac{1}{2}$ inch grizzly, it was shown that a large proportion of mine ore could be sent direct to the tube mills.

Davis, Willey and Ewing [3] showed that, using a standard tube mill on East Rand ores, and adopting a feed subjected to single-stage crushing, an output of 145 to 190 tons of -90 mesh per 24 hours could be obtained. Of this tonnage 72 per cent. was -200 mesh size. The product is suitable for direct cyanidation in an ordinary filter plant, yielding a poorer tailing at a reduced cost.

A recent investigation of ball and pebble milling is based on the power consumed. Capacities are expressed in terms of surface-tons per hour. Various formulæ suitable for practical application were obtained.[4]

---

[1] "*Rand Metallurgical Practice*," vol. i., p. 493.
[2] Prentice, *loc. cit.* [3] *Min. and Met.*, 1925, 6, 245.
[4] A. M. Gow, M. Guggenheim, A. B. Campbell and W. H. Coghill, *Amer. Inst. Mng. Met. Eng.*, 1934, *Tech. Pub.* 517. Also *Trans. Amer. Inst. Mng. Met. Eng.*, 1934, **112**, 24.

The correct proportion of *water* to ore depends on the specific gravity of the ore, on its coarseness, on its hardness and on the amount of the feed. There should be no "free" water for the balls to fall through, but there should be enough water to make the particles of ore adhere to the pebbles, so that when impact occurs there are some particles between the pebbles ready to be crushed. The right proportion of water is 24 to 32 per cent. with Rand ore using the usual amount of feed. The amount of water giving the best results is determined correctly to 0·1 per cent. and adhered to rigidly. In dry tube milling a small percentage of moisture is very injurious, 1 per cent. of moisture reducing the capacity in certain cases by one-half.[1]

The *cost* of tube milling in the Central Mining Group mines on the Rand is given by Prentice as varying from 3·7d. to 13·8d. per ton milled. The average cost on an all-sliming plant is about 15·5d. per ton of —100 material produced. On sand and slime plants it is 11·3d.

Absorption of Gold in Tube Mills.[2]—Even in the best practice on the Rand where blanketing is used in the tube mill circuit and where the ore is crushed in cyanide, considerable amounts of gold are held up in the mills, especially between the linings and the tube shell. Experience shows that mere stoppage and restarting of tube mills causes a considerable discharge of very fine gold. Gold absorption is also found to increase as the amount of moisture in the tube delivery decreases. At Ooregum the liners are laid in cement and the "lock-up" is said thereby to be considerably reduced. In the Brunswick Mill, California, a raised backing around the bolt holes of the liners of the ball mills allows the pulp to circulate behind the liners and prevents the gold from collecting there.

In cleaning a tube mill the bars are loosened,[3] the mill then being rotated with water for about 10 minutes. The bars are taken out, scrubbed with steel brushes, and the mill swept out with similar brushes. The concentrates recovered are valuable and are amalgamated. It has been shown that the oftener a tube mill is dismantled the more the lock-up gold that is recovered.

Theory of Tube Mills.—According to H. Fischer,[4] the tube mill does its work mainly by impact, not by grinding. Glass drums and drums with gratings at the ends were constructed, and the action in the interior observed. One such drum 1 metre (39·3 inches) in diameter was filled with flint balls to a height of 450 mm. (17·7 inches) and rotated. Prof. Fischer found that, at the slow speed of rotation of 21 to 23 revolutions per minute, the balls rolled slowly down the slope. At 32 revolutions per minute (the correct speed for a tube of this diameter, according to the formulæ already given) the charge of balls had become looser, and their bulk was considerably more than half the capacity of the drum. At 34 revolutions (Fig. 96) the balls (A A_1, etc.) next the drum were carried up without sliding or rolling on the drum, until at a certain height they separated from it and were projected outwards in a curve, falling near the other side of the charge of balls. At this speed each ball fell separately, crushing and spattering the cushion of ore between it and balls that had previously fallen. The

[1] Durant, *Eng. and Mng. J.*, 1912, 94, 111. "*Mineral Industry*," 1912, 21, 939.
[2] H. A. White, *J. Chem. Met. Mng. Soc. S.A.*, 1930, 31, 161.
[3] *Kolar Gold Fields Min. and Met. Soc.*, 1933, 7, 10.
[4] *Zeitsch. Ver. deut. Ing.*; 1904, 48, 437.

other layers of balls described similar but shorter courses, and a hollow space, D, continually varying in size and shape, was always recognisable. The blow struck by each ball of the outer layer was a sliding one, the relative velocity of two balls at the moment of impact being 3·5 metres per second in the line joining their centres, and 1·2 metres per second in the direction at right angles to this. These velocities give some measure of the respective effects of the impact and grinding actions.

H. A. White agrees[1] with Fischer that the pulverising action is due almost entirely to actual impact of the falling balls in dry crushing, but points out that when the mill is half full of water the effect of impact is of less importance. A ball falling into 2 or 3 feet of water will not strike the bottom with enough force to do much crushing, and in this case it is probable

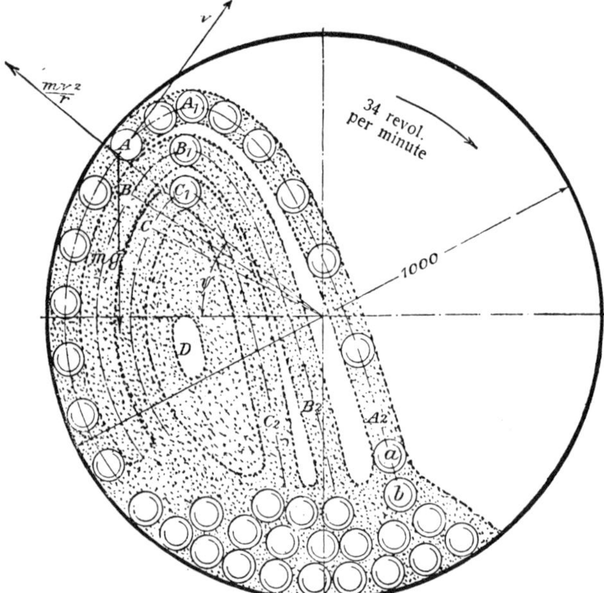

Fig. 96.—Diagram showing Action of Tube Mill.

that the grinding action between the balls will do a greater proportion of the work.

White finds that the average fall of the balls is at a maximum, provided they half fill the mill, when the number of revolutions per minute, N, is equal to $\frac{34 \cdot 22}{\sqrt{D}}$, where D is the diameter of the mill expressed in metres. White's result, obtained by calculation, gives the same speed as that at which Fischer found by experiment each ball fell separately. A somewhat faster rate is now used on the Rand, and, consequently, it appears to be probable that the theoretical investigations were not made without error.

H. A. White[2] adds :—" The following phenomena may be observed when

[1] *J. Chem. Met. Mng. Soc. S.A.*, 1905, 5, 290.
[2] *J. Chem. Met. Mng. Soc. S.A.*, 1930 31 4.

a tube mill loaded to the centre line with balls or pebbles is revolved at gradually increasing speed :—

" (1) The ascending balls are lifted up and those descending are depressed until a certain angle, depending on the kinetic energy of repose, is reached by the ball-pulp surface.

" (2) Thereafter balls roll down from the top to the bottom intermittently, the lack of continuity being due to the difference between the static and kinetic angles of friction, which may vary over a wide range.

" (3) When the speed reaches 10 or $15 \times \frac{1}{\sqrt{D}}$ r.p.m. (D = diameter in inches of tube inside the lining less diameter of one ball) pebbles follow each other continuously, with only two or three layers moving, and the stage known as 'cascading' is reached, where the balls merely roll over one another. At this point the upper boundary of the balls is roughly a straight line.

" (4) As the speed increases to 60 to $90 \times \frac{1}{\sqrt{D}}$ r.p.m., the angle of slope increases gradually, more layers of balls are falling, the top curve is pronounced and of a parabolic shape, and the bottom heap of bouncing balls shows a curved upper outline, flattening out to a nearly horizontal line at higher speeds.

" (5) With still greater speeds, up to 100 to $150 \times \frac{1}{\sqrt{D}}$ r.p.m., the balls are still packed closely together. The angle of slope increases from 50° to 75°, the top curved line becomes a parabola reaching to the centre of the tube, the width of falling layers of balls may reach nearly half way from the centre to the circumference, the curve separating the stationary balls on the rim from those sliding down approximates to a parabola and the bottom heap shows a flatly curved outline with a tangent 5°–10° to the horizontal.

" (6) At speeds of 150 to $200 \times \frac{1}{\sqrt{D}}$ r.p.m., or 60-80 per cent. of the critical speed, the outer parabola reaches the lower limit of the falling balls and covers completely the heap of bouncing balls. 'Cataracting' is now in full action, the balls being thrown into the air and falling freely."

In ball milling there is a definite trend towards higher speeds and greater density of pulp. Higher speed increases ball and liner consumption, but this is offset by an increase in capacity. Experiments have been tried with mills running above the critical speed—*i.e.* when the balls are just carried round by centrifugal force and the first layer of balls adheres to the tube mill rim. For a standard tube, 5 feet 6 inches diameter, with 3 inch linings and 2 inch balls, the critical speed is 35·11 r.p.m. In present practice the speed is usually 53 to 86 per cent. of the critical.

In any mill operating at more than critical speed ball slippage is essential for grinding, requiring smooth linings, a small ball load, a small feed load (necessitating also a means of discharging the ground material from near the circumference) and regulated pulp dilution.[1]

[1] *Amer. Inst. Mng. Eng., Tech. Pub.* 375, 1931.

Hardinge Conical Mill.—It has been urged that, in the case of the ordinary cylindrical form of tube mill, particles which are crushed to the required fineness near the feed end of the mill go gradually forward to the discharge end, using up needlessly the energy of the falling pebbles, which could otherwise be employed in crushing the larger fragments still remaining. Moreover, both ends of the tube are loaded with the same quantity and size of pebbles. When these become worn a considerable amount of energy is expended by the larger pieces in acting upon the smaller ones. Hardinge suggests that in the perfect machine the crushing should theoretically be done in stages,[1] the larger pebbles crushing the larger particles of ore, leaving the smaller pebbles to finish the pulverising to the mesh for which the mill is designed. Also, as it is reduced, the fine material should be removed, leaving the space available for larger particles to come under the influence of the pebbles.

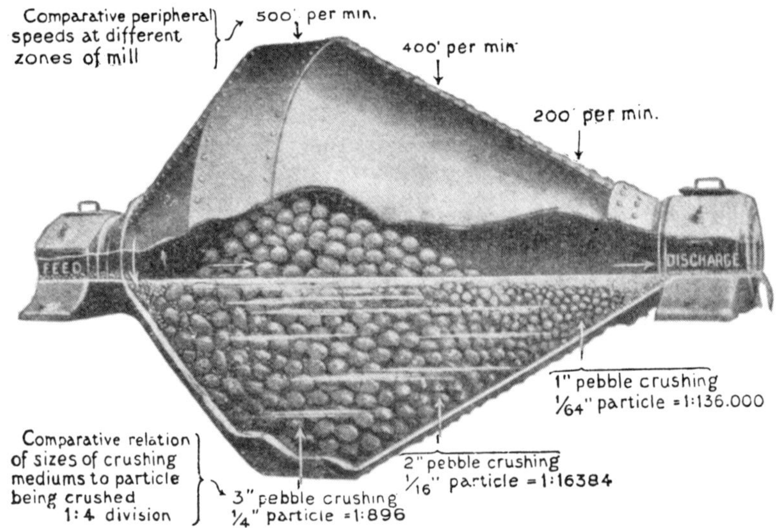

Fig. 97.—Hardinge Conical Pebble Mill.

The Hardinge conical mill [2] using pebbles or balls, was designed to avoid these possible defects in the tube mill, and is said to be adjustable for producing an all-slime product with a minimum of coarse grains, or, on the other hand, for granulating without the formation of an excess of slime. In practice it has been found to be better suited to the latter purpose than to the former.

The conical mill consists essentially of two hollow cones (see Fig. 97) the rims of whose bases are attached by a short cylindrical section, and whose apices are formed into hollow trunnions. The shell is made of heavy steel plate.

---

[1] *Trans. Amer. Inst. Mng. Eng.*, 1913, 45, 194.

[2] *Trans. Amer. Inst. Mng. Eng.*, 1908, 39, 336. "*Electrochem. and Metallurgical Industry*," 1909, 7, 47. *Eng. and Mng. J.*, 1907, 84, 925

Rabone [1] gives the following figures for dimensions and capacities of Hardinge ball mills :—

TABLE XXVI.

Cylindrical Section.		Ball Load, Lbs.	H.P.	R.P.M.	Capacity, tons per hour.	
Diameter, Feet.	Length, Ins.				½″ to 100 mesh.	¼″ to 200 mesh.
4½	16	4,500	18	30	1	¾
5	36	9,000	45	28	2½	2
8	48	34,000	135	21	8	6
10	72	80,000	400	18	30	24

The ore is fed in by a scoop or spiral feed, attached to the trunnion at the apex of the cone with greater slope. No screens are used at the discharge end. The axis is inclined slightly to the horizontal and according as the obliquity is large or small the discharge is quicker or slower and the crushing coarser or finer. The foundations on which the supports rest are usually made of concrete with foundation bolts as in ordinary stamp mill practice. The lining of the mill and the character of the pebbles used are much the same as those ordinarily employed in tube mills. Steel balls are replacing pebbles in most instances, and in this case the lining is of chrome steel or cast-iron plate.

Crushed ore of ½ inch size is fed to the mill. The size of the mill feed is regulated by means of ½ or ⅜ inch screens, mechanically or electrically operated. Gyrex, Tyler and Niagara screens are typical of the mechanically-operated and the Hum-mer of the electrically-driven.[2] Opinion is tending more and more to the use of rolls to give finer feed to ball mills. This cuts down the circulating load in the latter, which, with coarse feed, sometimes up to 2 inch, is found to consist mainly of material coarser than 10 mesh. Automatic feeders ensure regular feeding.

The work of crushing progresses as follows :—As the ore enters it comes into contact with the balls, and, when crushed to a certain degree, passes on between slightly smaller balls, which by a sizing action have found positions on the incline of the outlet end. The charge maintains a gradual progression towards the exit and over-crushing is avoided. The ore is crushed partly by the impact and partly by the grinding action of the falling balls. The peculiar feature of the machine is the gradual sizing of both the balls which are crushing and the material which is being crushed. The difference in the size of the particles is equalised by the diminished fall and the reduced speed at the periphery. Gradation in size is dependent on the amount of feed, the inclination of the mill, and the rapidity of the discharge.

Hardinge mills may be run singly, in tandem or in series. Ore may be passed straight from the rock breakers to a ball mill, after which it is classified and the oversize further reduced in a pebble mill whose dimensions are determined by the degree of fineness required. The oversize from the pebble mill is classified and returned to the latter for re-grinding.

[1] "*Flotation Plant Practice,*" 1936, p. 35.
[2] J. C. Farrant, Private Communication.

To overcome the hold-up of gold which may occur in the ball mill classifier circuit, *hydraulic gold traps* are often placed after the ball mill exit to intercept the gold. These are used at Montezuma Apex and at Siscoe, recovering respectively 75 and 80 per cent. of the free gold in the original ore [1] (see Fig. 98). Jigs are also used for the same purpose.

Balls.—Balls should resist sliding and abrasive attrition, and therefore the harder their surface the longer the service. Manganese steel, however, is said to be comparatively soft and yet gives good results in tube and Hardinge mills.[2]

Wraight quotes percentage analyses made on three balls from different sources :—

C, ...	0·84	1·17	1·21
Si, ...	0·42	0·40	0·80
S, ...	0·031	0·034	0·031
P, ...	0·031	0·067	0·098
Cr, ...	0·06	0·03	0·06
Mn, ...	0·56	0·89	10·97
Hardness { Brinell (diam. of ball impression),	3·5 mm.	3·3 mm.	4·1 mm.
Hardness { Scleroscope,	45	55	27

The heat-treatment is of vital importance in the case of manganese steel.

A high grade of forged ball reduces the amount of pure iron found at later stages in the circuit, where it is a harmful impurity.

The centre of a cast ball is spongy and crumples when worn. The use of flattened balls is said to assist grinding, and probably this is due to the density of the iron being rendered more uniform in their manufacture.

Rabone [3] states that ball consumption is as follows :—

TABLE XXVII.

	Consumption—lb. per ton of ore ground.		
	Coarse grinding, 65 mesh.	Medium grinding, 100 mesh.	Fine grinding, 200 mesh.
Chrome steel, .	1·0	1·5	2·0
High-carbon steel,	1·5	2·0	2·5
Cast-iron, .	2·0	2·5	3·0

The average factory cost of the balls is roughly in the ratio chrome : high-carbon : cast-iron = 100 : 95 : 75.

The wear of liners is about 25 per cent. of the ball consumption when both are of the same material.

At a Colorado mill, flint pebbles have replaced steel balls in conical

[1] *Trans. Amer. Inst. Mng. Met. Eng.*, 1934, **112**, 561.
[2] *Trans. Amer. Inst. Mng. Eng.*, 1915, **53**, 437.
[3] "*Flotation Plant Practice*," 1936, p. 38.

mills, as the abraded iron appeared to inhibit the flotation of —200 mesh material. A mixture of iron balls and flint pebbles is also being tried.[1]

The **Marcy Ball Mill** is of the grate discharge type. It consists of a cylindrical shell connected to feed and discharge heads all made of cast semi-steel. The trunnions are hollow, and are lined with manganese steel to protect them from wear ; they run in self-aligning bearings. Within the feed end trunnion is a deep spiral screw to assist the entrance of the ore ; the discharge trunnion has a smooth interior. The feed is by means of a scoop fitted with a renewable manganese steel lip. The lining of the body consists of overlapping manganese steel ribs, 3 to 5 inches in thickness at the deepest part. Within the discharge head and trunnion is an alloy steel grate having a number of sections secured by clamp bars fitting over the junctions between the sections. Spacing and lifting ribs placed radially are cast in one piece with the head and are parallel with the clamping bars. They serve to maintain a clearance of several inches from the end of the mill. This space is divided by the ribs into a similar number of compartments to that of grates. Each compartment is open at its apex to the discharge end trunnion. The bars forming the grates are made of chromium steel and are welded into sections, tapering slightly towards the discharge side to prevent choking. The distance between the bars varies from $\frac{3}{16}$ to $\frac{3}{8}$ inch, and the slots are directed so that they present an almost continuous opening to the pulp as the mill revolves. The discharge is into lifting compartments behind. Radial clamp bars keep the grates clear of balls and large lumps of ore.

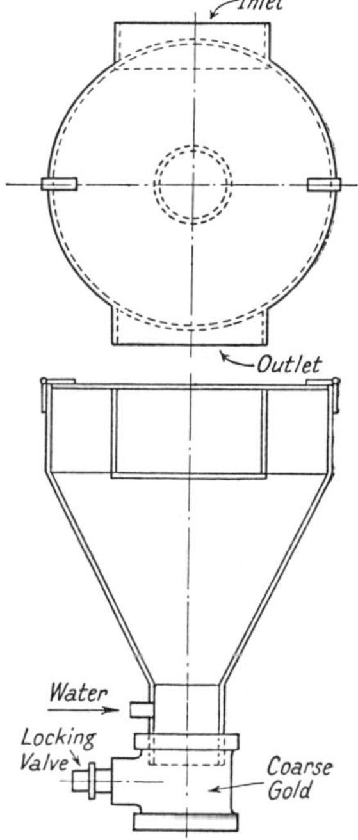

Fig. 98.—Hydraulic Gold Trap.

The initial level of the balls in the mill is some inches below the level of the axis and the ball load, speed, etc., are such that the ore is small enough to pass through the grates as it reaches the discharge end. During working, the pulp level falls from the trunnion height at the feed end to a point only a few inches above the periphery at the discharge end. Hence grinding is most intense at the feed end and diminishes towards the discharge owing to the increasing rate of travel of the ore. About a third to a quarter of the capacity is filled with thick pulp having 70 to 75 per cent. of solids. This low charge of pulp tends to increase the rate of travel through the mill.

The cylindrical grate mill, owing to its low pulp discharge, which reduces overgrinding at the discharge end, usually requires less power per ton of ore than a conical mill when used in coarse grinding. In fine grinding, however, the conical mill is usually the more economical. Whereas a Hardinge mill can be relined without lifting the shell from the trunnion bearings, a grate mill has to be almost dismantled.

[1] Locke, *Min. and Met.*, 1935, 16, 28.

The Marcy mill is suitable for reducing coarse ores to 10 mesh in one pass, or will regrind the product from primary crushers up to 50 mesh, in which case a classifier or screen should be used in closed circuit. In conjunction with a modern highly efficient mechanical classifier the issuing mesh may be increased to 100. The alloy steel liners are consumed at the rate of about 0·5 lb. per ton and the forged steel balls at 0·3 to 0·7 lb. per ton.

A 30 × 36 inch Marcy mill in Idaho is running at 70 r.p.m., or 135 per cent. over the critical speed.

Rabone [1] gives the following capacities and sizes of Marcy mills :—

TABLE XXVIII.

Diam., feet.	Length, feet.	H.P.	R.P.M.	Capacity, tons/24 hrs.		Ball Load, lbs.
				½″ to 48 mesh.	¼″ to 100 mesh.	
3	2	10	35	11	5	1,000
5	4	40	29	69	27	5,000
8	6	225	22	550	220	28,000

Capacity of Ball Mills and Grade of Products.—The amount and fineness of ore passing through a ball mill depend on :—

(1) *The Hardness and Size of the Ore.*—The latter is usually about 1 inch cubes.

(2) *Rate of Feed.*—If this be slow, the ore particles receive a greater number of blows and are therefore crushed finer. The general practice, however, is to feed fast and return the oversize, after classifying, for regrinding.

(3) *Moisture Content.*—This should be just sufficient for the pulp to coat the balls, leading to less wear on the balls and liners.

(4) *Sizes of Balls.*—The usual sizes are 1 to 7 inches diameter. The best results are obtained when the balls occupy from 0·3 to 0·4 of the space in the mill.

(5) *Speed of Mills.*—This is less than the critical speed and is a function of the internal diameter. Modern mills operate at 0·6-0·8 of the critical speed. The power required is directly proportional to the speed. If the size of the feed is decreased, the size of the balls may be correspondingly decreased and their number in the mill increased, thus securing a greater grinding surface. The speed may then be reduced to give more rubbing action. The efficiency of grinding is enhanced in this way.

(6) *Open or Closed Circuit.*—A closed circuit tends to keep the mill at full capacity and more efficient working is obtained.

(7) *Type of Mill.*—There is fairly evenly divided opinion as to whether the quick discharge grate mill or the centrally discharging mill is the better.

(8) *Size of Mill.*—The tendency is towards large diameter—up to 9 foot mills.

Fahrenwald and Lee [2] state that the power required to operate a ball

[1] *Min. Ind.*, 1932, 41, 581 ; "*Flotation Plant Practice,*" 1936, p. 30.
[2] *Amer. Inst. Min. Met. Eng.*, 1931, Tech. Pub. 375.

mill partly loaded with balls varies with the angular speed of rotation of the mill, the ball load and the ball size. Each mill has corresponding speed, ball size, feed size, and ball and feed loads to produce the best results. Wet grinding, in these writers' opinion, is more efficient than dry grinding, allowing larger charges owing to a low coefficient of sliding friction. The maximum efficiency and output of the mill are obtained at speeds above the critical. It is estimated that the average diameter of the feed grains should be such that they are just nipped at an angle of about 17° between the balls. If the maximum number of grains is reduced below the best size, the efficiency of the mill is reduced. It is therefore better not to attempt too great a reduction in a single mill. The permissible pulp dilution for the greatest mill output depends upon feed size, ball load and mill speed, among other factors.

Stamps v. Ball Mills.—It has been claimed on behalf of ball mills, that for any tonnage that warrants the use of a 6 foot or larger ball mill, the latter followed by a tube mill is preferable to stamps. The ball mill appears to do less crushing per unit of power than the stamp, but may have a smaller percentage of lost time, requires less labour, is quieter, takes up less room and is simpler. The stamp, on the other hand, has an advantage with small tonnages that it can crush to slime from a very coarse feed.

Robertson,[1] however, gives the following comparison of costs of the working of a ball mill and the equivalent crushing efficiency in stamps, showing the greater economy of the latter in some respects:—

One Ball Mill, 8 × 6 feet.		28·14 California Stamps (1,000 lbs.)	
	d.		d.
Power, 121 h.p.,	3·70	Power, 84·7 h.p.,	2·59
Steel balls, 2·5 lb. at 2·33d. per lb.,	5·82	Shoes, dies, screens,	1·83
Liner wear, 0·20 lb.,	1·55	Maintenance, repairs,	2·11
Maintenance, repairs, spares,	0·64		
Per ton,	11·71	Per ton,	6·53

He found that the ball mill with 10·64 tons load of balls would grind 215 tons of Mysore ore per 24 hours to 8 mesh, which is equivalent to 28·14 stamps at a duty of 7·64 tons per stamp per 24 hours using 8 mesh screen. The results indicated that with another crushing machine prior to the ball mill better results would be obtained.

In districts like Rhodesia, which is a typical small-scale gold mining territory, small Hardinge mills have in many cases been installed, on account of cheapness, in place of the usual 5-stamp battery.

Marcy Rod Mill (Fig. 99).—This mill is similar in many respects to the Marcy ball mill, but without grates. It is suitable for soft friable ores. Its length is about twice the diameter. It has a trunnion at the feed end, and a tyre running on two friction supporting rollers at the other end. This enables a low pulp discharge to be maintained, reducing overgrinding. Instead of balls or pebbles, high-carbon steel rods, 2 to 3 inches in diameter and a few inches shorter than the mill, are used. Rods give more rolling

[1] *Kolar Gold Fields Min. and Met. Soc.*, 1928, *Bull.* 23, 144.

and less cascading action than balls, and hence the shattering effect is not as great. The pieces in the feed to the rod mill should be less than 1 inch mesh size and should not consist of excessively hard material. They are not ground so fine as in a ball mill. The rod mill works best with a thick pulp, and gives a distinctly granular product which is fairly uniform and rarely overground. The most efficient reduction in size in the machine is usually from $\frac{1}{2}$ inch to 20 mesh.

The lining consists of waved alloy steel sections. Special liners are fixed to the feed and the discharge ends. The rods, which would ordinarily bind or cross in a flat-ended mill, are prevented from so doing by truncated heads which slope away from the rods towards the mill openings. At the same time internal pockets are produced into which feed material and crushed product may settle without interfering with the normal rolling of

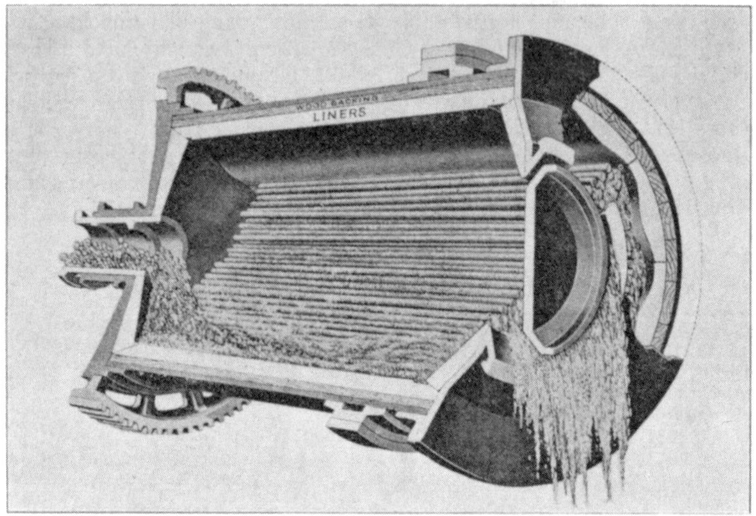

Fig. 99.—Section of Marcy Rod Mill in operation.

the rods. The truncated heads also cause the rods to move away from the ends of the mill, and the wear on the end liners is reduced. The lateral motion imparted to each rod assists the grinding action of its roll. The discharge end has a large opening, allowing the finished pulp to flow freely away, and also facilitating the renewal of rods, liners, etc. A hinged splash door covers the housing at the discharge into which the pulp flows. The door may be opened for interior inspection while the mill is running. The feed is by an ordinary type of scoop. The rod mi.l has occasionally been used for dry grinding. A 5 × 10 foot mill is rated to wet grind 330 tons of medium hard rock to 35 mesh per 24 hours, and 200 tons to 65 mesh. The circulating load is usually low.

General Considerations of Grinding—Classification Combinations.—Two systems are employed in fine grinding :—

(1) *Open Circuit.*—The feed is at such a rate as to produce the amount

of finished product required. If concentration follows such grinding, the mineral is quickly separated from the gangue. This system is now much less used than the following owing to the tendency to overgrind.

(2) *Closed Circuit.*—The grinding in the mill is not completed in one passage of the ore. The latter is classified,[1] the finer material passing to later processes and the coarser returning to be reground. This is practically standard practice in every mill, and the advantages of the concentration in the open circuit are derived from the use of a flotation cell, jig or corduroy blanket between the grinder and the classifier to remove coarse gold early. The mill is supplied with ore at a greater rate than if on open circuit, the time of passage is much reduced, and a smaller proportion of the pulp entering the classifier is in the finished state. The aim is the separation of sand from fines as quickly as possible. The system results also in diminished power costs. At Lake Shore (Ontario) closed circuiting increased the capacity by 45 per cent. and reduced the power consumption by 10 per cent. The steel consumption in the mill decreased from 6·5 lbs. to 3·2 lbs. per ton.[2]

The grinding efficiency of a ball or tube mill increases with the rate of feeding. Hence the increase in the latter due to the introduction of a classifier leads to greater economies and greater capacity. For instance, at the Tough Oakes Gold Mine (Ontario) the use of two classifiers in closed circuit with a given mill increased the capacity by 28 per cent. above that obtained with one classifier.

The classifier in effect controls the whole system. It loads the mill with circulated sands so that the utmost effective work results, with the least destruction of the balls or pebbles and liners.

With the recent introduction of heavily constructed classifiers having large raking capacities up to 1,000 tons of sand per day per foot of rake width, higher circulating loads have become possible. This in turn has resulted in the use of large trunnions and feed scoops in the mills and improvement in the pulp circulating systems. Ball mills having extra large feed and discharge trunnions have handled circulating loads of 1,800 per cent., *i.e.* a returned classifier tonnage of sand 18 times as great as the new ore feed.

Multi-Stage Grinding Operations.—These represent an endeavour to grind by economical steps with reasonable power requirements, instead of doing the whole grind in one operation. The number of steps is governed by the proportion the final size bears to that of the original feed. If a greater reduction in diameter than 5 to 1 is attempted at any stage there is a probability of efficiency loss.

Primary, secondary and tertiary grinding are now practised in some mills, each stage having its own classifier—usually of the bowl type. Latterly the practice has arisen of adding a preliminary desliming classifier between the last coarse crushing stage and the primary closed circuit grinding stage. Initial slime is therefore removed very early and suffers no further comminution, nor does it interfere with other crushing; but this is not always to be recommended.

The Hadsel[3] Mill (Figs. 100, 101) combines crushing and grinding in one process and is operated normally in closed circuit with a counter-current

---

[1] See Chap. IX.

[2] Dorr and Anable, *Trans. Amer. Inst. Min. Met. Eng.*, 1934, 112, 161.

[3] See also Hall, *Trans. Amer. Inst. Min. Met. Eng.*, 1934, 112, 15.

classifier. It consists of a steel drum or bucket wheel type of elevator, 10 to 36 feet in diameter, revolving in a shallow tank through which water is continuously flowing. Around this are two steel tyres running on friction rollers. Power for rotation is applied by chain and sprocket, the former encircling the drum. Inside the drum, near the bottom but above the buckets and water level, there is a series of stationary anvils or breaker plates set at a slight angle to prevent rock from resting on them. Both ends of the drum were open in the original model, but are now closed, and the supports for the breaker plates are outside the framework. As the wheel rotates, the ore, which is fed in at any convenient point above the buckets, is carried up by them and eventually falls on to the plates. The cascading causes the rock lumps to crush one another. The mass then slides into the buckets and the process starts again. When the particles are crushed sufficiently they remain more or less in suspension and flow out of the side of the wheel as a pulp, thence over an adjustable baffle on to amalgamation plates. A mill 16 feet

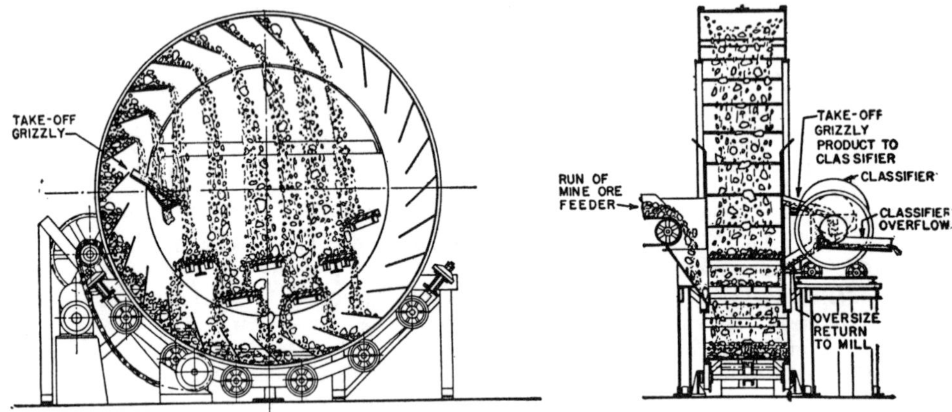

Figs. 100, 101.—Principle of Operation of the Hadsel Mill.

in diameter and 5 feet wide absorbs 40-50 h.p. at full load and is said to crush 50 tons per 24 hours from 8 inch size to a product approaching 85 per cent. of – 200 mesh material. It rotates at about 4·25 r.p.m. Wear and tear on the buckets amounts to about 0·1 to 0·2 lb. per ton milled.

The Chilian Mill.—This machine is also known as the "edge runner." It consists of a circular cast-iron pan, to which are bolted wrought-iron sides. Runners or crushing rollers travel round a cone in the centre of the pan. The track or die-ring on which they run consists of sections of cast steel, and the tyres of the crushing wheels are also of cast steel.

Chilian mills are of two kinds [1] :—

1. High-speed—with die-rings of small diameter (4 feet to 6 feet) and crushing wheels also small (2 or 3 feet, with 4 to 6 inches face). The speed is 30 to 40 revolutions per minute.

2. Low-speed—with die-rings of large diameter (6 to 10 feet) and large, heavy crushing wheels (7 to 8 feet, with 20 inches face). They run at 8 to 12 revolutions per minute.

[1] H. A. Megraw, *Eng. and Mng. J.*, 1910, 90, 967; 1913, 96, 18, 821.

High-speed mills are commonly used for regrinding, following rolls, breakers and stamps. Low-speed mills are better adapted to amalgamation,[1] and are fed with coarse material, using a large volume of water in the process of crushing.

Fig. 102 shows a typical Chilian mill with the casing cut away so that the rollers may be seen.

The *Akron mill* is described by Eaton as a typical instance of a Chilian mill of *high speed* [2] (see Fig. 103).

The rollers, three in number, are heavy solid castings, weighing 3,000 lbs. each, and are carried on trunnions which fit into boxes on the drive head over the central spindle. The rollers can assume any angle when passing over uneven ore. The centre of the roller shaft is placed a little higher than

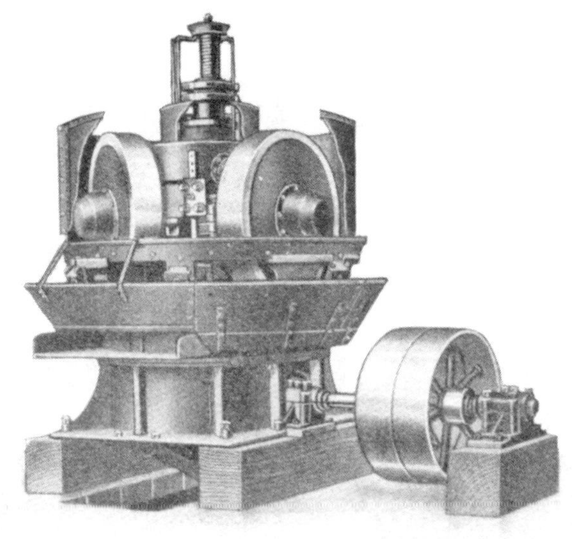

Fig. 102.—Chilian Mill.

the centre of the trunnions, thus allowing the centrifugal force to increase the crushing capacity. The correct position of the rollers is to have them inclined slightly inwards at the top.

The total screen area is 1,800 square inches, and is divided into five sections round the mill. A semi-circular launder, placed round the mortar and inside the splash plates, serves to carry the pulp towards the apron.

When new, the die-ring, which fits into the mortar, is 1 inch below the screen. It is 5 feet in diameter, and has a cross-section of 7 × 4 inches.

Ore is fed in from an annular casting fixed to the cover, through three 3½-inch pipes, one directly in front of each roller and after the corresponding "plough."

---

[1] A. Maclaren, *Eng. and Mng. J.*, 1911, 92, 305.
[2] *Trans. Inst. Mng. and Met.*, 1911, 20, 161.

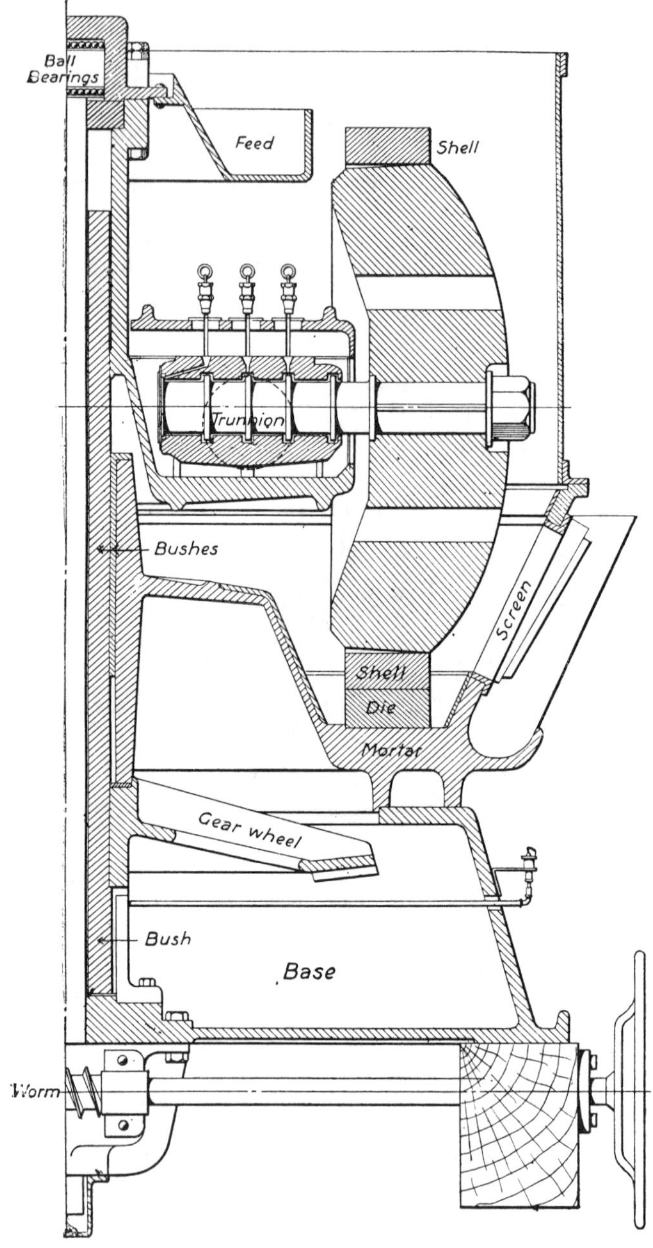

Fig. 103.—Section of Akron Chilian Mill.

Argall, in discussing [1] the work of Akron mills, says that the reduction of the ore results from two causes :—

1. The direct crushing effect due to the weight and speed of the rollers.

2. The grinding or abrading effect due to the rollers being constrained to travel in a circular path.

The smaller the diameter of the die the greater is the twisting effect on the ore particles ; with a die of greater diameter, the action is similar to that of small thin cylinders rolling between the faces of the rollers and dies. The abrasion thus becomes less, and the action of low-speed mills accordingly approximates to that of rolls, and the ore is not finely ground.

The Huntington Mill.—The Huntington Roller Mill, here described as a typical roller mill, is best suited for the fine crushing of ores which are not

Fig. 104.—Huntington Mill.

too hard, or in re-grinding coarse tailing. It consists of an iron pan, at the top of which a ring, B (Fig. 104), is set, and attached to this are three stems, D, each of which has a steel shoe, E, fastened to it. The stems are suspended from the ring and are free to swing in a radial direction, as well as to rotate round their own axes, whilst the whole ring, B, with the stems and shoes, revolves round the central shaft, G. The shoes or rollers are thus driven outwards by centrifugal force and press against the replaceable ring die, C. In front of each roller is a scraper, F, which keeps the ore from packing. The rollers are suspended with their bases at the distance of 1 inch from the bottom of the pan, which can also be replaced when worn. The lowest part of the screen is situated a little above the top of the rollers, and outside it there is a deep gutter into which the ore is discharged, and from which a passage leads to the amalgamated plates. The ore and water being fed into

[1] *Min. Mag.*, 1911, 5, 365.

the mill through the hopper, A, generally by an automatic feeder, the rotating rollers and the scrapers throw the ore against the sides, where it is crushed to any required degree of fineness by the centrifugal force of the rollers acting against the ring die. If inside amalgamation is to be attempted, from 17 to 25 lbs. of mercury are placed in the bottom of the pan, the clearance below the rollers permitting them to pass freely over the mercury without coming in contact with it, so that it is not stirred up and "floured," but the motion is such as to bring the pulp in contact with the quicksilver. The speed of the mill is from 45 to 75 revolutions per minute. The ore should be broken in rock-breakers to a maximum size equal to that of a cobnut, before being fed in. The free gold is in great part amalgamated and retained by the mercury at the bottom of the pan, the remainder being caught on the plates outside the mill. The chief advantages supposed to be gained by the use of Huntington mills instead of stamps are reduced first cost, saving of power and less cost of renewals. The mill requires to be set to work in an intelligent manner by experienced and skilful hands, and watched carefully.

In an account of Huntington mill practice at Kalgoorlie,[1] von Bernewitz gives a list of ten 5-foot mills in use there in 1912. In these mills, the percentage of gold recovered by amalgamation, in the cases where it was stated, varied from 20 to 60 per cent. The ores consisted of mixtures of oxide of iron, ferruginous clay and quartz. Punched screens equal to about 30-mesh wire screens were used in all cases. The speed of the mills was from 68 to 74 revolutions per minute, and the daily capacity was from 30 to 80 tons of ore per mill, according to the hardness of the ore. The pulp was collected and the sand cyanided.

The high cost of repairs in the Huntington mill limits its use. The objection to the mill is not so much high wear, but that certain parts of the mill wear out in one spot, so that a large casting, little worn in other places, has to be scrapped (Semple). Suggestions are made by C. C. Semple to reduce the excessive wear and to make replacement of parts cheaper.[2] The absence of sliming effect is in favour of the use of the mill for re-grinding tailing for gravity concentration.

Amalgamation Pans.—An amalgamation pan consists of a circular cast-iron pan about 5 feet in diameter, provided on the inside with a renewable false bottom of cast iron (*a*, Fig. 105)—constituting the lower grinding surface—and a "muller," or upper grinding surface, *d*, attached to a vertical revolving spindle, *g*, which is set in motion by bevel wheels, *t*, placed below the pan. The mullers are shown resting on the cast-iron dies, *c*, which protect the bottom from wear, whilst replaceable shoes attached to the lower surface of the mullers are also shown. The muller grinds to impalpable pulp ore which has been already reduced to a coarse powder and also mixes the ore with mercury, introduced into the bottom of the pan, and so amalgamates the gold and silver. The muller can be raised by rotating the hand-wheel and centre screw, *j*, on the top of the spindle, so that only circulation and mixing of the charge take place.

In work on gold ores the use of amalgamating pans was formerly mainly limited to re-grinding skimmings, blanket sand, and concentrate obtained in working a stamp mill.

Pans were more recently used at Kalgoorlie in grinding sulpho-telluride

[1] *Mng. and Sci. Press*, 1912, **105**, 618; "*Cyanide Practice*, 1910 *to* 1913," p. 177.
[2] *Trans. Amer. Inst. Mng. Eng.*, 1911, **42**, 602.

ores. These pans were of the Wheeler type.[1] They were usually 5 feet in diameter, revolving at 45 to 60 turns per minute and requiring about 5 horse-power. They ranged up to 8 feet in diameter.[2] They had heavy mullers, which could be raised or lowered as usual. " A set of shoes and dies lasts from four to six months on roasted ore, and when half-worn down a compensating weight of some 600 lbs. is put on the top of the plate. They grind about 10 tons per day. The side feed is satisfactory and is almost universal "[3] (the alternative being central feed).

M. G. F. Söhnlein [4] found a 5-foot pan efficient for fine grinding small quantities of sand concentrate in Bolivia. The pulp was de-watered to contain 48 per cent. of solid material and fed centrally through the cylinder

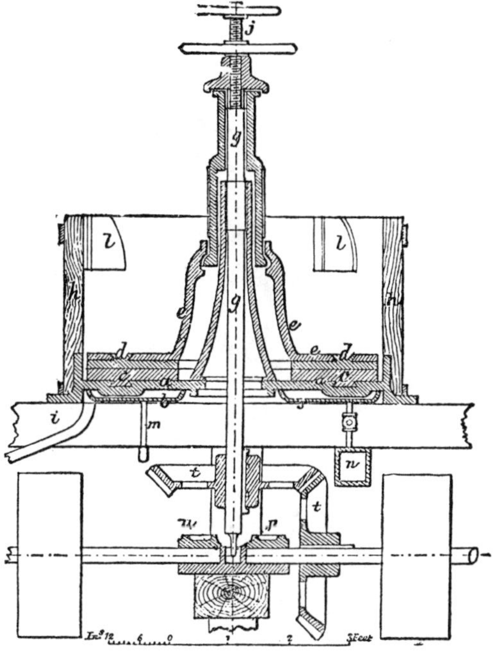

Fig. 105.—Amalgamation Pan.

around the central column, so that it all passed under the mullers, and was discharged after the single passage. At one passage about 30 per cent. of the feed was sufficiently ground. The oversize was returned from a Dorr classifier. The speed of the mullers was 60 revolutions per minute. The feed was 38 tons of sand from Overstrom tables, the quantity of sand actually ground to pass 200-mesh being 30 tons per day. The material was not hard (consisting of 50 per cent. of quartz, felspar and slate and 50 per cent. of iron oxide obtained by roasting pyrite).

---

[1] For description of the original Wheeler pan, introduced in 1862, see J. A. Phillips, " *Gold and Silver*," 1867, p. 397 ; Schnabel and Louis, " *Metallurgy*," 1905, vol. i., p. 793.

[2] Gowland, " *Non-Ferrous Metals* " (Griffin), 1930, p. 383.

[3] Von Bernewitz, *J. Chem. Met. Mng. Soc. S.A.*, 1909, 10, 222.

[4] *Eng. and Mng. J.*, 1913, 96, 581 ; see also E. E. Wann, *ibid.*, p. 1183, and Söhnlein, *ibid.*, 1914, 97, 822.

G. A. and H. S. Denny[1] give some of the advantages of using pans for fine grinding instead of tube mills, as follows :—

(1) Their accessibility.
(2) The possibility of internal amalgamation.
(3) The smaller capacity of the unit.
(4) The saving of time in the renewals of working parts.

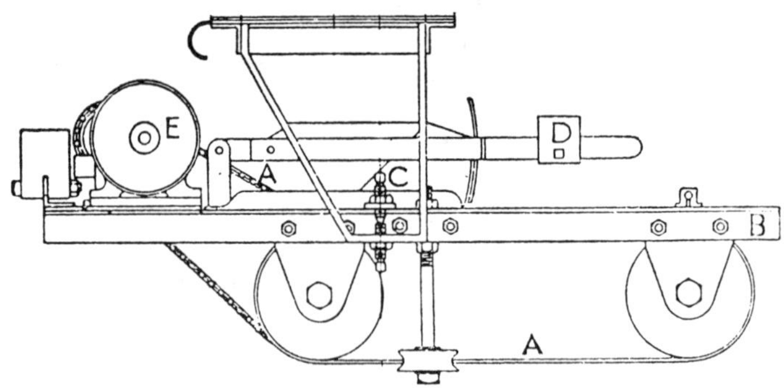

Figs. 106, 107.—Hardinge Constant Weight Feeder.

Nevertheless, tube mills are preferred to pans for fine grinding almost everywhere.

Amalgamation of Concentrate in Pans.—Among special forms of pans designed to treat concentrate is the *Berdan Pan*.[2] This is a shallow annular

[1] *Min. Mag.* (New York), Sept. 1905, p. 180.
[2] For full description and illustrations, see Louis, "*Gold Milling*," 1894, pp. 343-6.

basin about 4 feet in diameter, surrounding a cone which is attached to a spindle set at an angle of about 15° to the vertical. The spindle is rotated by bevel gearing at 20-30 r.p.m., and carries the basin round with it. In the annular basin are one or two loose iron balls which remain at the lower side of the cone when the pan revolves. The pulp is fed in with mercury at the higher side, and is ground by the balls and discharged over the lower edge of the pan. The capacity of the Berdan pan is from 1-2½ tons per day. It is used chiefly in Australia but is described in operation in California by G. J. Young.[1]

The Hardinge Constant Weight Feeder (Figs. 106, 107).—This machine is used to ensure the regular feeding of a ball, tube or rod mill. It consists of a travelling belt A (Fig. 107) on a framework B which is suspended on two pivoted points C. Any variation in the weight on the belt moves the frame, which in turn actuates the feed gate. The required weight is set by means of a balancing weight D. If the weight on the belt increases, the belt and frame are depressed, which shuts the gate slightly until the correct rate of feed is restored, when the gate is automatically returned to its normal setting. A revolution counter registers the total weight of feed. The motor E driving the belt is also carried on the frame.

---

[1] *Eng. and Min. J.*, 1931, **132**, 195.

CHAPTER IX.

GRAVITY CONCENTRATION IN GOLD MILLS.

Concentration.[1]—The object of concentration is the separation of the heavy valuable mineral from the light worthless gangue. Complications are often introduced by the fact that various base minerals must be separated from one another, an ore being subdivided into several products. Many gold ores, however, only require separation into two parts—the " concentrate " in which the precious metal is contained, and the " tailing," which is thrown away.

All the concentrating machines depend for their action on the effect of a difference of densities on the fall of bodies in a fluid. The fluid employed in almost every instance is water, although several machines have been devised in which air is used as the concentrating medium. The fall of solid materials in still water takes place according to two laws, one applicable to very shallow water, through which the particles fall with increasing velocity, while the other is true when the depth of the water is considerable, so that the particles for the greater part of their course proceed at their maximum velocity. In shallow water the fall is almost entirely according to density, so that those machines which utilise only the first instants of the fall will have great efficacy in concentrating.[2] It is this fact which has necessitated the use of shallow currents in concentrating tables, sluices, etc. In almost all these machines the fine sand and slime are brought into suspension in water, and the liquid is then run over an inclined surface. The deposit of sand, which is thus formed on the table, tends to become enriched in heavy minerals, because the stream moves faster at the surface of the water, where the lighter particles remain, than it does next the bed, where the heavy particles have settled. The deposit is continually worked up and brought again into suspension by a rake or broom, or by a series of shakes or blows imparted to the apparatus, so that the effect mentioned above is repeated frequently. If the stirring up is violently performed all the slime and very fine particles are kept in suspension in the water and carried away and lost, a slow stream of water and very slight agitation being favourable to their retention in the deposit which is formed. When the stream of water is rapid and voluminous, fine material, whether heavy or light, is swept away and lost in the tailing, whilst if too small a stream of water is used, much worthless sand is deposited with the concentrate. It follows that the amount of water used must be

---

[1] For a full discussion of ore dressing, see the standard text-books on the subject, such as those by Richards, Truscott, Taggart, etc. In this chapter brief descriptions of processes and plants that are of present interest in gold milling are given.

[2] The matter is complicated by the quantity of solid matter present, as free settling differs from hindered settling. See, among other papers, G. G. Bring, *Jern-Kontorets Annalen*, 1906, p. 321 ; Richards and Locke, "*Mineral Industry,*" 1907, **16**, 970 ; Taggart "*Handbook of Ore Dressing,*" 1927, p. 552 ; Schiller and Naumann, *Z. ver. deut. Ing.*, 1933, **77**, 318.

regulated according to the work which it is proposed to do. Frequently it happens that clear water must be added to the pulp to dilute it sufficiently. On the other hand, it often happens that the pulp is too thin, and water is then removed by means of pyramidal boxes, cones or other machines (*de-waterers*), described in the following.

A. A. Hirst [1] and Finkey [2] have examined thoroughly the principles involved in the separation of particles by virtue of density difference.

Other operations, which it is often of the utmost importance to perform before concentration, are *classifying* and *sizing*. The necessity of sizing is obvious when it is remembered that a shallow stream of water swift enough to carry down fine sulphides, might be powerless to move a pebble of quartz. The usual method of classifying is based on the varying rates of fall of particles through a deep column of water, or on the different movements of particles when in an upward-moving column of water, which are dependent on the same properties. In this way *equal-falling* particles are obtained together, and since a sphere of galena is equal-falling with a sphere of quartz of from 1·5 to nearly 4 times the diameter, according to the absolute sizes,[3] it follows that classifying has very different results from sizing. Screens produce classes of particles of nearly equal size regardless of their respective specific gravities. Classifiers produce classes of equal-falling particles, the sizes of which depend on their respective specific gravities. Nevertheless, classifying, when efficiently performed, is of great assistance as a preparation for treatment by shallow-stream concentrators, and especially to separate sand from slime before treatment with cyanide, and to prepare material for feeding into tube mills. The placing of a classifier in closed circuit with a tube mill or similar crushing machine is now common practice. It ensures uniformity of product delivery and it relieves the tube mill from receiving material already crushed finely enough.

Sizing by Screens.—Screening to equal sized particles is a desirable preparation for concentration on most machines, but in the past has not usually been employed on material finer than about 8 mesh, owing to difficulty in working and to wear and tear of fine screens if used wet in the usual way. The two ordinary types of screens are the cylindrical revolving and flat shaking forms. For the finer sizes it has been usual to substitute hydraulic classifiers—*i.e.* boxes with ascending currents of water, described below—for screens.

Several devices have been brought forward for using much finer screens than were formerly considered practicable. Steeply inclined and rapidly jarred flat screens, with sprays of water, are reported to give good results ; and very fine slightly inclined screens submerged in water (with jarring motion, and a special construction for delivering the coarser oversize into clear water) have been recommended. The advantages of this last type are stated to be that no extra wash water is used, screening is more perfect under

---

[1] *Trans. Inst. Min. Eng.*, 1929-30, 79, 463. *Ibid.*, 1933, 85, 236.

[2] *Bull. Sch. Mines and Metallurgy, Missouri*, 1930.

[3] R. H. Richards, "*Ore Dressing*," 1st edition, 1903, p. 471. Richards states ("*Mineral Industry*," 1896, 5, 705) that in practice, with particles between 10-mesh and 60-mesh the diameter of the quartz is from 2¼ to 3¾ times that of the galena, varying with the absolute size. See also *Trans. Amer. Inst. Mng. Eng.*, 1907, 38, 210 ; "*Mineral Industry*," 1897, 6, 694 ; 1907, 16, 969.

Note.—For a discussion of Fine Grinding and Classification, see Dorr and Anable, *Trans. Amer. Inst. Min. Met. Eng.*, 1934, 112, 161.

water than with washing sprays, and the wear of screens is reduced to a minimum. Such screens have been tried as fine as 60 mesh—that is, with 3,600 holes to the square inch.

The Hum-mer Screen.—In recently designed mills, screening of the crushed ore is an established feature. The old type of stationary or shaking screen, with comparatively few vibrations per minute, and the trommel, have been replaced by the rapidly vibrating type, of which the Hum-mer is an example. Its usual place is prior to the secondary crusher, and its function to by-pass material already fine enough, so that the work of the crusher is reduced.

In this machine (Fig. 108) a woven wire screen, 4 × 5 to 10 feet, composed of stout wires, or rods made of a hard-wearing alloy, is stretched across a

Fig. 108.—Hum-mer Electric Screen, Type 400.

fixed sloping rubber-covered frame. It is gripped along the ends by hinged side plates or strips having hooked edges that grip a hook strip on the screen. The tension on the screen can be adjusted by means of screwed studs at the upper end.

Vibration is imparted in the older types by a mechanism attached to two opposite sides of the framework and to the middle of the screen. More than one such mechanism may be used if the length of screen warrants it. An alternating field is generated in a solenoid coil by an external supply of 15-cycle single-phase A.C. current. This field causes a to-and-fro motion of the armature which, being fixed to the screen, imparts this motion at the rate of 30 vibrations per second. A jerk is given to the screen at every back stroke when the frame hits a stop, the larger pieces thereby being thrown upwards, leaving the finer material to pass through the screen apertures. The amplitude and number of vibrations can be closely controlled. A screen delivering a $\frac{1}{2}$ inch product and sloping at roughly 30° to the horizontal will deal with a feed of 5 tons per square foot per hour.

In a later type of machine (Fig. 108) an electric vibrator is placed at each corner of the screen, the pairs at the top and bottom being yoked together.

Hydraulic Classifiers were introduced by Rittinger in the middle of the last century for use in the Hartz,[1] and from their shape were known as *spitzkasten* or *pointed boxes*. These boxes have the shape of inverted pyramids, the stream of unsized pulp entering at one side and flowing out at the other, whilst there is also a small discharge at the apex. The current slackening on entering the box, the heavier and larger particles in suspension at once begin to settle, and, escaping the influence of the current, fall quietly to the bottom of the box, where they are discharged.

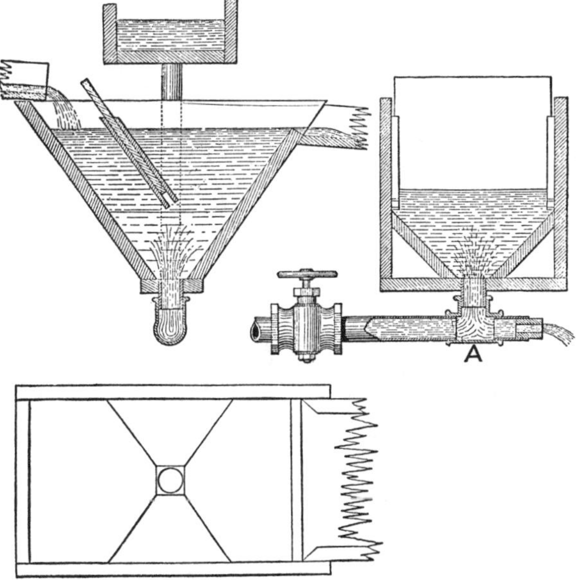

Fig. 109.—Pointed Box with Ascending Current.

Equal-falling particles are carried away through the aperture in the apex by muddy water containing material of all sizes down to the finest slime, and it was to eliminate this material that the *ascending current* was introduced. This is a current of clear water, which enters at the apex of the box in greater quantity than can be discharged by the outflow near the same spot, so that there is an upward current of water into the box. The result is that no muddy water is discharged below, but only the particles of ore which have weight enough to drop through the ascending current, so that by regulating the strength of this, any desired class of ore can be obtained. In Fig. 109, a form is shown in which the sliding partition assists the settling, by causing the pulp to pass downwards, rapid surface currents across the box to the overflow being thus prevented. The discharge of the heavy particles is effected through A, the clear water pipe itself, by the arrangement shown. The launder

[1] "*Aufbereitungskunde*" (1867); Richards, "*Ore Dressing*," vol. i., p. 439.

above the box supplies the clear water current, and shows the head of water used, which must be kept constant to ensure uniformity of results.

Cone Classifiers.[1]—These were introduced on the Rand. They are

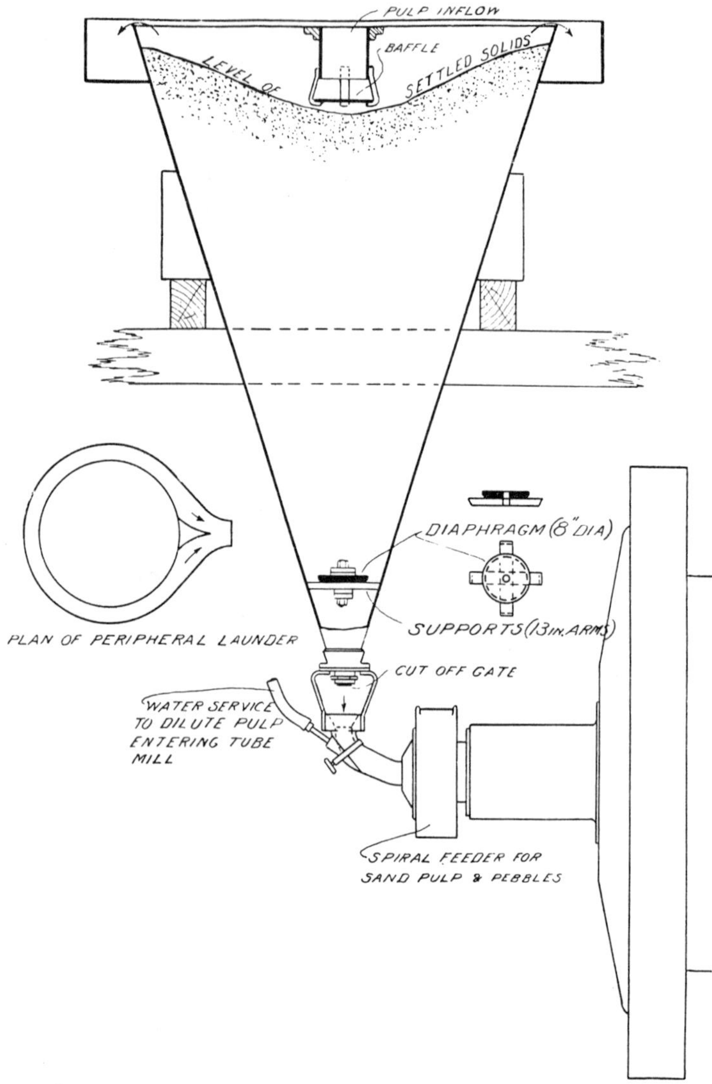

Fig. 110.—Caldecott's Hydraulic Cone Classifier, with Diaphragm.

large cones 5 to 8 feet in diameter at the top and from 7 to 10 feet deep (see Fig. 110).[2] The cones are built of $\frac{3}{16}$-inch sheet steel with the lower 18

[1] W. A. Caldecott, *J. Chem. Met. Mng. Soc. S.A.*, 1909, 9, 312; Dowling, "*Rand Metallurgical Practice*," vol. i., p. 99.

[2] Dowling, "*Rand Metallurgical Practice*," vol. i., p. 100.

inches of the cone of cast iron, to resist the hard wear of that part and to facilitate renewals. The pulp enters from a launder through a large central pipe, which delivers about 12 inches below the level of the pulp, and is supplied with a circular baffle plate placed horizontally about 3 inches below the open end of the pipe. The fine pulp overflows all round the cone into an annular launder. The underflow passes through a nozzle at the apex of the cone, and passes vertically downwards into the inlet of the tube mill, or into launders, as horizontal or inclined pipes tend to choke. The nozzle is regulated with a cut-off gate (see Fig. 111).[1]

The cone classifier is kept nearly full of sand by means of a circular diaphragm, or baffle, due to Caldecott, about 8 to 10 inches in diameter,

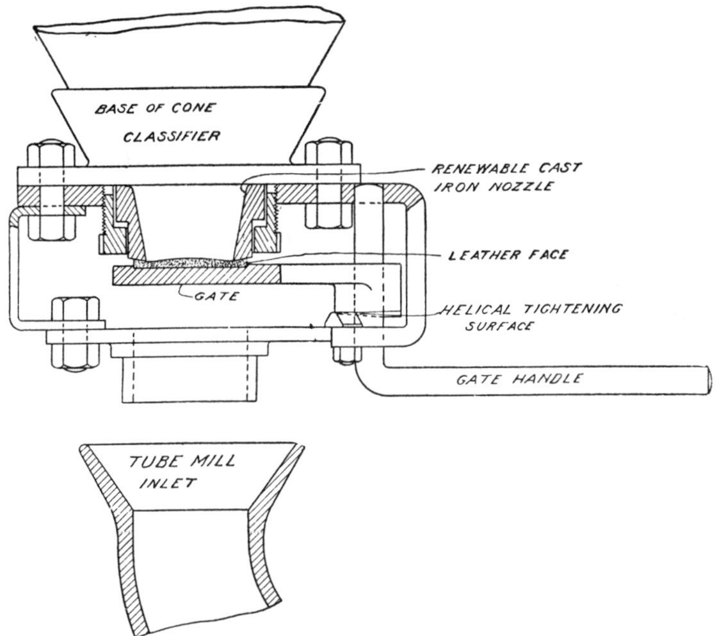

Fig. 111. Adams' Cut-off Gate.

placed in the axis of the cone near the apex. The diaphragm prevents the sand from settling and forming a channel in the middle, but allows it free passage in the annular space round the diaphragm, and ensures a steady flow of coarse material through the nozzle. Later it was often omitted as it tends to choke during a temporary stoppage. The level of the sand is kept constant under varying inflow by opening or closing a sliding shutter or gate at the nozzle. Coming from below so great a depth of sand, the underflow contains little water, say 25 to 28 per cent., and issues very slowly even with a free vertical discharge. A large outlet is required, the diameter of the nozzle being from $1\frac{3}{4}$ to $2\frac{1}{4}$ inches. Separate de-watering appliances are not required. In working, coarse grains settle near the centre and fine grains near the periphery.

[1] *Ibid.*, p. 101.

The capacity of these cone classifiers is considerable. A cone 8 feet in diameter at the top and 10 feet deep is capable of delivering from 400 to 600 tons of sand per twenty-four hours at its underflow, from pulp composed of 44 per cent. slime and 56 per cent. sand,[1] the product thus obtained carrying rather less than 30 per cent. moisture, against about twenty times that amount in the original pulp.

The objection has been made to the Caldecott cone classifier that bubbles of air are carried down by the inflow and are still rising through the liquid at the outflow, thus interfering with efficiency. There is also the difficulty that the thick spigot product becomes coarser if the amount of feed is increased.[2] This entails close watching of the underflow. A jet of water below the diaphragm has been tried and considered an improvement. The fine sand, too, in a finely ground pulp does not settle quickly enough to prevent the liquid from breaking through to the apex. Any attempt to cut down the underflow to obviate this reduces the output. The Caldecott cone has generally passed out of use.

An automatic discharge control for cones used on the Kolar Field has been designed by Hocking.[3] It is said to produce a uniformly dense pulp.

According to White [4] two forms of conical classifiers were in general use on the Rand in 1926 :—

(1) The *Brazier separator.*—This is a cone having an adjustable nozzle for discharging a thin circular stream of clear water horizontally near the apex of the cone. It delivers the sand free from slime.

(2) A similar classifier in which extra clear water is added outside the cone. A second cone is used if it is desired to remove the slime even more completely. In this method a much larger discharge aperture can be used for the slow stream of thick sand pulp, and the added water is easily controlled. An automatic regulator closes off the cone when the bed of settled sand sinks below a stated level, thus obviating the necessity of delivery by a gate. The proportion of sand to slime is about 40 : 60. It is usually necessary to put the final slime through return sand cones to complete the separation of leachable sand. These cones have about two-thirds the area of the original ones, and are operated with a thick underflow. They are run with a much smaller depth of settled solids.

More recently, Wartenweiler [5] has introduced the "P. & P." (pipe and plate) device for washing the sand with water inside the cone. A stream of water is played on a steel plate near the outlet, thus forcing the sand to pass through a sheet of water and so yield up most of the adherent slime to the overflow (see Figs. 112, 113). The slime separated in this manner contains from 4 to 10 per cent. of + 200 mesh material. About 40 to 50 per cent. is true colloidal slime.

It is important that some safety arrangement—possibly in the shape of cones—should be interposed between the overflow of the main tube mill cones and the final stream of pulp to the cyanide plant, so that no coarse particles may escape during a stoppage.[6]

---

[1] J. E. Thomas, "*Rand Metallurgical Practice,*" vol. i., p. 152.
[2] Robertson, "*Mineral Industry,*" 1912, **21**, 944.
[3] *Trans. Inst. Min. Met.*, 1925, **34**, [ii.], 73.
[4] "*Rand Metallurgical Practice,*" vol. i., p. 496.
[5] White, *loc. cit.*
[6] "*Rand Metallurgical Practice,*" vol. i., p. 487.

At Siscoe, Quebec,[1] where the gold is in a free state except for a small proportion locked up in the sulphides, a special box type of hydraulic concentrator is placed at the discharge end of each ball mill. Mill solution added through pipes keeps the sludge well stirred up. The lighter materials overflow to the classifier, while the heavier, including fine gold and small nuggets, remain at the bottom and are cleaned out daily. Amalgamation increases the recovery to 78·5 per cent., blankets immediately after the

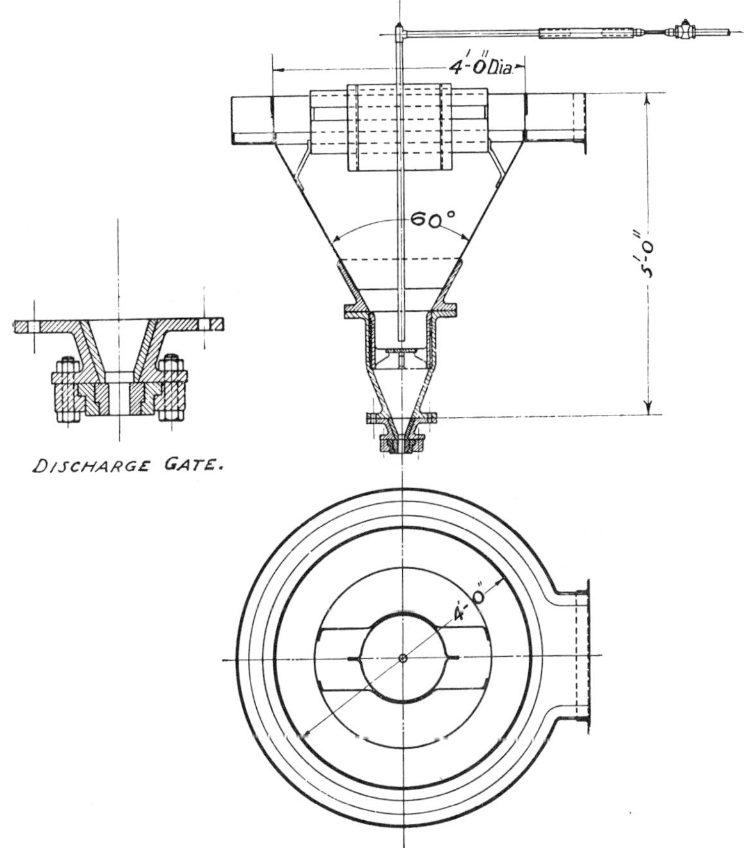

Figs. 112, 113.—Hydraulic Return Classifying Cone (*Rand Mines, Ltd.*).

classifiers recover 7·0 per cent., and cyanidation by grinding in cyanide solution and precipitation by standard methods produces 13·0 per cent.

Fahrenwald,[2] in a study of hydraulic classification, has arrived at the following conclusions :

" For a given combination of mineral particles confined to form a bed and under a condition of agitation to produce a dynamic state in the bed,

---

[1] *Eng. Min. Journ.*, 1932, 133, 319.
[2] *Min. and Met.*, 1926, 7, 437.

the nature of the segregation of the particles is dependent chiefly, if not wholly, on the velocity of water caused to flow up through the agitated particle bed.

"The best separation of an ore product in gravity concentration results from a preparatory treatment in which the 'just suspension' volume of water is used, followed by a final treatment in which zero or a very small volume of rising water is used."

The Dorr Classifier, with the lower end removed, is shown in Fig. 114. It consists of a settling box in the form of an inclined trough open at the upper end, in which mechanically-operated rakes are placed to remove the heavy material as fast as it settles, the liquid and slime overflowing at the closed end. It has a large raking capacity and is an excellent classifier as well as de-waterer.

Fig. 114.—Dorr Duplex Classifier.

The rakes in single, double or quadruple rows, are suspended from bell cranks connected with levers which terminate in rollers. The latter press against cams attached to the crank shaft. The rakes are lifted and lowered at opposite ends of the stroke by the action of the cams transmitted through the levers to the bell cranks. More recently the cams and rollers have been replaced by eccentrics, cranks and linkage. The horizontal motion is produced by cranks. The pulp is fed across the centre of the trough. The sand settles to the bottom, and is pushed up the inclined bottom of the trough by the scrapers. After emerging from the liquid the sand is washed and discharged from the open end of the machine with about 26 per cent. moisture.

The slime is prevented from settling by the flow of the liquid and by the agitation caused by the reciprocating scrapers. It overflows at the lower end of the machine. The agitation, while ample to prevent the slime from settling, is not sufficient to cause the sand to overflow with the slime. The more vigorous the stirring, however, the larger are the particles that are prevented from settling.

The height of fall in this machine is small. The capacity is $3\frac{1}{2}$ to 8 tons of solids per hour in four to six parts of water, and the power consumption varies from 1 h.p. to 8 h.p.

The general character of the pulp and the amount of colloidal or clayey matter in it affect the efficiency of classification. Satisfactory classification is impossible if the pulp is too thick.

Dorr Bowl Classifier.—Where an extremely clean oversize, a fine overflow (say 90 per cent. of − 200-mesh) or a large capacity is essential, a bowl classifier may be installed. It consists (Fig. 115) in the addition of a circular

Fig. 115.—Dorr Bowl Classifier.

shallow bowl, having revolving rakes, to the ordinary rake classifier. The plough arms rotate at a speed of $\frac{1}{2}$ to 8 r.p.m. according to the load carried and the degree of classification required. The feed enters at the centre of the bowl (see Fig. 115). The fines overflow the periphery into a collecting launder, while the coarse material settles to the bottom and is raked to a central discharge opening, whence it passes into the reciprocating rake compartment for re-classification in the ordinary way. Wash water is introduced across the centre of the main tank, and flows in an opposite direction to the solids up through the centre of the bowl. The rising pulp in the bowl centre is not so much affected by the rake surge as in the standard type.

More recently, a Pohlé air lift has been used to lift back into the classifier feed those grains which accumulate between the opening from the bowl and the rake compartment. These particles interrupt the smooth working of the classifier if they are not removed.

Hubler and Martin[1] give the following comparison between classification in an 8-foot 60° cone and a 7-foot Dorr classifier with 8-foot bowl and 3-foot rake :—

TABLE XXIX.

Mesh.	Weight, Per cent.	
	Cone.	Bowl.
+ 200	34·0	54·5
− 200 + 325	21·0	16·5
− 325	45·0 (Cone sand 60 per cent. solids.)	29·0 (Bowl sand 81 per cent. Solids.)

These figures show the much better removal of the slime from the sand in the Bowl classifier.

A Dorr classifier unit of unusual size has been installed in the primary grinding circuit of the Wright Hargreaves gold mill. It is 12 feet wide, 25 feet long and in closed circuit with a 9 × 7 foot ball mill handling 600 tons per day of ore which ranges up to ⅜ inch in size. The raking capacity is 10,000 to 12,000 tons per day, and the overflow − 48 to − 65 mesh. The secondary circuit, grinding to 2 per cent. of + 200 mesh, uses bowl classifiers 16 feet wide, with bowls of 18 feet diameter.[2]

At the McIntyre mine Denny[3] used the bowl classifier more as a concentrator in order to collect the pyrite carrying the gold. The pyrite required finer grinding than the quartz. It accumulated in the rake return product and was then returned to a tube mill to be ground to 200 mesh. The quartz was eliminated from the circuit at a much coarser size.

Another classifier, the **Dorr Hydroseparator,** has been used at Lake Shore. The underflow from a previous thickener passes to this machine, from which material overflows fine enough to go direct to flotation. The coarser residue returns to be ground in tube mills in closed circuit with bowl classifiers. The overflow from the latter joins the hydroseparator overflow and proceeds for flotation.[4]

The **Maclean Spiral Classifier** used on the Rand is a cylindrical pan the bottom of which is an upright truncated cone. Pulp is fed in tangentially at the side and, after traversing two-thirds of the outer channel, it meets a weir leading to the remaining third which is open. The fines pass over the weir, while the sands are scraped up the sides of the cone by a spiral rake, revolving in the direction of the pulp flow, to the aperture at the top.[5] (See Fig. 116).

---

[1] *Can. Mng. Journ.*, 1934, 55, 311, 351.
[2] *Eng. Min. Journ.*, 1933, 134, 440.
[3] *Can. Mng. Journ.*, 1931, 52, 622.
[4] *Min. Ind.*, 1931, 40, 624.
[5] Wartenweiler, *Trans. Amer. Inst. Min. Met. Eng.*, 1934, 112, 772.

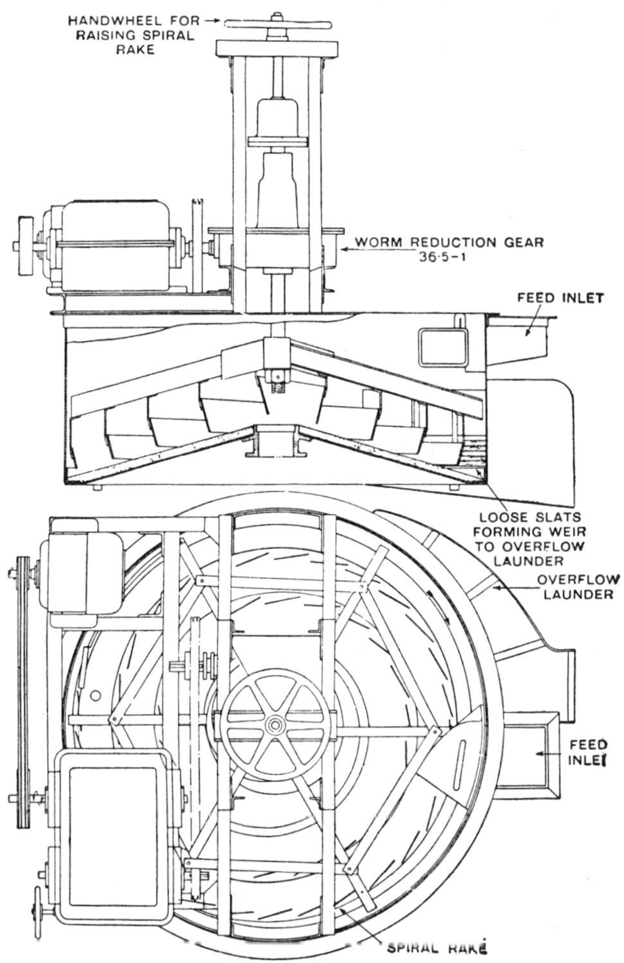

Fig. 116.—Maclean Classifier.

The **Akins Classifier** (Fig. 117) is a trough with curved sides mounted on a steel frame. Helical screws, 16 to 78 inches in diameter, revolve in the trough. The trough is inclined downwards towards the slime discharge end, where there is an overflow launder. Each screw consists of a hollow shaft having heavy welded steel arms that carry the blades forming the screw. Two complete screws are on each shaft. The feed enters below the surface of the pulp at a point on the side between mid-way and the overflow end. The top of the weir at the overflow end normally is above the shaft but below the upper edge of the spiral blade. In a more recent development the top of the weir is considerably above the top of the spiral at the overflow end. Side overflow weirs are provided where extremely fine, high density or high capacity overflows are required.

Fig. 117.—Akins Classifier.

Efficiency of Classification.—Newton's method for measuring the performance of the closed circuit classifier is based on sieve analyses,[1] using the formula :—

$$E = \frac{10{,}000\,(b - a)\,(a - c)}{a\,(100 - a)\,(b - c)}$$

Where E = overall efficiency of the classifier,
a = percentage finished product in classifier feed,
b = percentage finished product in classifier overflow,
c = percentage finished product in classifier dragover.

The efficiency of the classifier as a remover of finished product, assuming that everything in the overflow is finished, is given by

$$E_1 = \frac{10{,}000\,(a - c)}{a\,(100 - c)}$$

White gives the following formula for calculating the efficiency of classification :—

$$E = \frac{(I - O)\,(U - I)}{(I - I^2)\,(U - O)}$$

Where I = fraction of oversize in inflow,
O = do. in overflow,
U = do. in underflow.

Fahrenwald[2] asserts that sieve analyses may lead to many fallacies in efficiency calculations. Closed circuit classifier efficiency may be better determined by testing samples in a vertical elutriator.

Weinig[3] gives the following formula for classifier efficiency :—

$$E = 1 - S$$

where S = decimal part of undersize in total sand return to ball mill.

---

[1] *Amer. Inst. Min. Met. Eng.*, 1930, Tech. Pub. 275.
[2] *Trans. Amer. Inst. Min. Met. Eng.*, 1930, 87, 82.
[3] *Min. and Met.*, 1934, 15, 291.

In practice a correction has to be made depending on what is acceptable as a finished overflow. If b is percentage oversize on a certain mesh in an acceptable overflow, then we have to consider a proportionate amount of oversize in the undersize of the sand return. Hence $E = 1 - \frac{100}{100 - b} S^1$, where S^1 is the decimal part that is finer than that certain mesh in the sand return.

Factors Controlling Classification.—These are [1] :—

(*a*) *Slope.*—This should be $2\frac{1}{2}$ to $3\frac{1}{2}$ inches per foot for primary and $1\frac{1}{2}$ to $2\frac{1}{2}$ inches per foot for secondary grinding. Too great a slope will cause slipping of the sand load.

(*b*) *Feed Rate.*—This is governed by the overflow capacity of the necessary fineness, depending on the nature of the ore.

(*c*) *Dilution.*—This operates over a relatively narrow range, otherwise the increased flow would operate against efficient classification. Good classification cannot take place in too thick a pulp. A check may be obtained by specific gravity or actual dilution tests. Uniform dilution is most desirable. A steady-head water tank giving uniform water pressure in the supply lines is a useful adjunct.

(*d*) *Speed.*—This is governed mostly by the capacity required. The faster the speed the greater the agitation and the coarser the overflow. If there is no agitation, slimes tend to settle in the sand and be returned to the mill. An ideal classifier gives minimum agitation at the overflow weir, and maximum agitation and washing of the sands as they travel upwards.

(*e*) *Reagents.*—These may easily interfere with classification by bringing into play properties that are needed in a subsequent operation, *e.g.* flotation. Dispersing and flocculating agents have opposite effects on classification, and some reagents, *e.g.* sodium sulphide, render it completely ineffective. A dispersant may be an aid with a siliceous ore carrying low amounts of sulphide, but a distinct detriment with heavy sulphide ores.

(*f*) *Primary Slimes.*—Lack of these causes surging, rendering constant mesh overflow difficult. Too much slime tends to take sand with it in the overflow and enforces dilution.

(*g*) *Specific Gravity of the Ore.*—The higher the specific gravity of the ore, the faster must the rakes travel to hinder the settling of particles already sufficiently fine.

The control of a classifier is mostly effected by watching the density of the overflow. This may be determined in any simple apparatus as

$$\frac{\text{weight of 1 volume of overflow}}{\text{weight of equal volume of water}}$$

According to Donoghue [2] the following formula serves to determine the

[1] Hubler and Martin, *Can. Min. Journ.*, 1934, 55, 311. Irwin, *ibid.*, p. 400.
[2] *Can. Min. Journ.*, 1934, 55, 531.

percentage weight of solids in the classifier overflow if there is no simultaneous concentration :—

$$S = \frac{(P - 1)\,k}{P}$$

Where S = percentage by weight of solids,
P = specific gravity of pulp,
k = constant depending on specific gravity of ore treated.

The degree of separation between fine and sandy material in a mechanical classifier is determined by three factors[1] :—(*a*) Rake speed, (*b*) outflow dilution, (*c*) slope of tank bottom. The greater the rake speed, the lower the dilution, and the steeper the slope, the less fine material is separated from the sand. The closer the separation the finer the coarsest particle in the overflow. To separate fine sand the speed of the rakes must be slower and the slope of the tank less. Capacity is thereby reduced. With identical speed, dilution, slope, separation and ore, the overflow and raking capacities are proportional to the width of the classifier tank.

Lay-outs for Grinding and Classification.—Hubler and Martin[2] and Playford[3] give a number of typical grinding-classifier layouts and discuss the application and advantages of each. At Beattie gold mines three-stage selective grinding is employed. An alkalinity of 0·004 per cent. is best for classification and also for subsequent metallurgical treatment.

Generally the flow sheet depends on the size of the plant and the characteristics of the ore. For a small plant single-stage grinding with the classifier in closed circuit with the grinding mill is usually sufficiently economical. On larger plants two- and three-stage grinding, with a mechanical classifier in closed circuit at each stage, will probably be desirable and cheapest. In primary grinding the classifier should normally have a large sand raking capacity, with vigorous agitation; in secondary and tertiary circuits high overflow capacity and low agitation are desirable.

Large circulating loads decrease the percentage of finished material returned to the mill in the sands. Consequently, mill capacity is increased and overground material is diminished. Dilution and alkalinity must be carefully watched under these conditions. There is rarely any concentration of heavy metallic minerals of value in the classifier, because the majority of such are reduced in size in the grinding mills more readily than the gangue.

Classifiers are mostly used in closed circuit with ball, tube or rod mills, taking their partially ground material and returning the oversize. A higher rate of feed is possible when fine material is continuously removed.

Figs. 118, 119 and 120 show three types of circuit in which classification is employed in conjunction with ball and tube or rod mills. The first is of a simple nature, while the other two indicate arrangements when two-stage grinding is employed.

While the principle of removing finished material from grinding circuits as early as possible in order to obtain maximum grinding efficiency is sound, the use of classifiers for preliminary de-sliming, and also as intermediate classifiers in 2-stage and 3-stage ball mill grinding, has shown some disadvantages in certain instances. The finished material, owing to its

[1] Dorr and Anable, *Trans. Amer. Inst. Min. Met. Eng.*, 1934, 112, 161.
[2] *Can. Min. Journ.*, 1934, 55, 351.
[3] *Proc. Aust. Inst. Mng. Met.*, 1933, 91, 439.

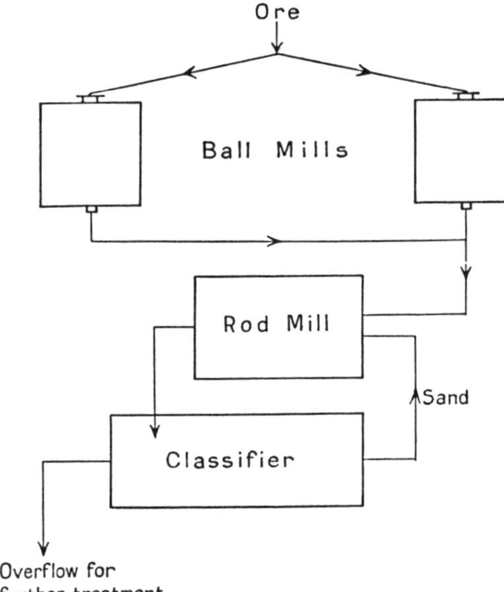

Fig. 118.—Type I Classification Circuit (Full mill feed together with circulating load is ground in the secondary mill).

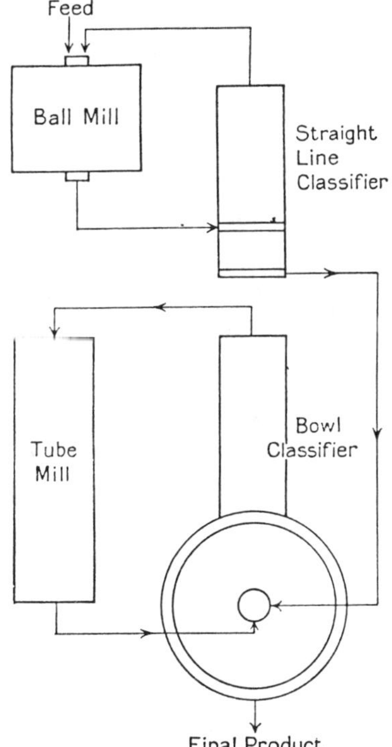

Fig. 119.—Type II Closed Tube Mill Classifier Circuit.

make-up from various sources—preliminary and intermediate classifiers and the final grinding stage—shows irregularities in quantity and dilution,

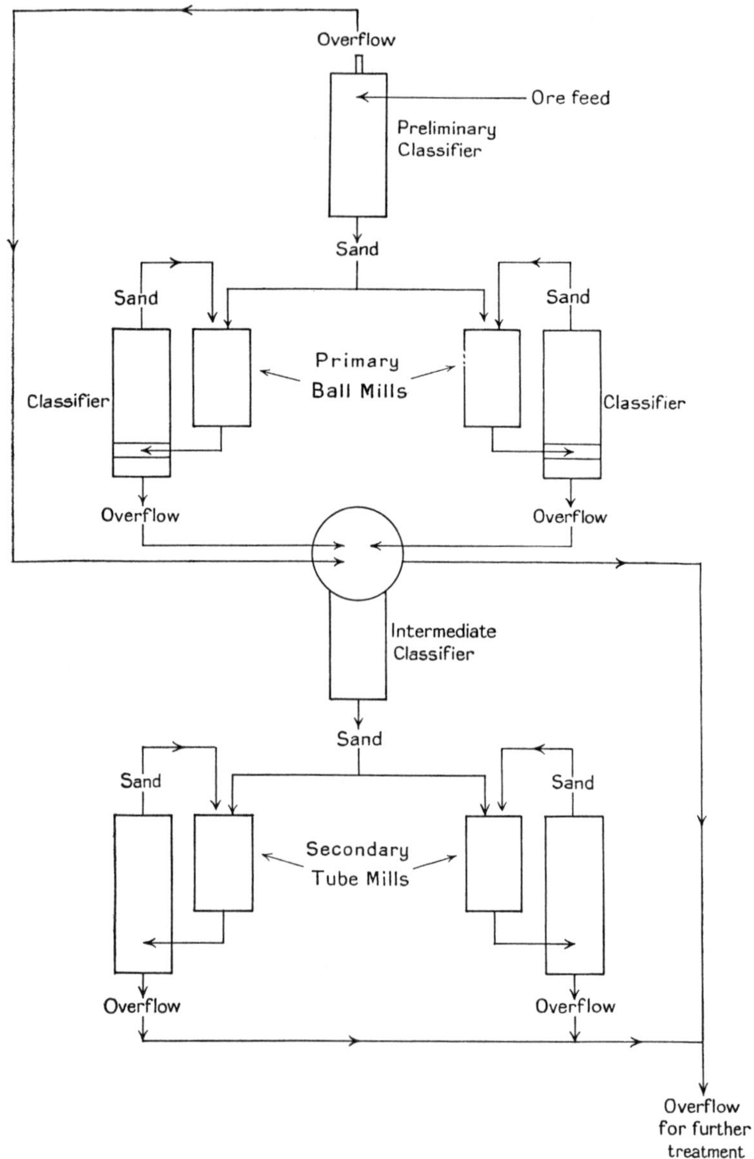

Fig. 120.—Type III Classification Circuit.

resulting in violent surges, especially in the second and third stages, leading to uneven grinding. This was attributed to the early removal of primary slime, which had certain lubricating properties that kept the pulp uniform in the succeeding stages.[1] The remedy so far found effective is a return to two-stage grinding without preliminary de-sliming, and a finished product coming only from the second stages.

An automatic control of the amount of material fed to a closed circuit has been devised by Dorr. It is based on the principle that the power required by a classifier varies in proportion to the tonnage raked. A demand meter in the classifier motor circuit provides the electrical contact between the sand load limits within which the control works. If the power drops below the lower limit, denoting a decrease in the sand tonnage, the circuit is completed and less dilution water is added to the classifier. Conversely, the water supply is increased when the sand tonnage increases. Thus the total feed to the circuit, including both new feed and returned sand, is kept automatically at a definite amount.

Early Concentrating Machinery.—One of the oldest and most primitive machines employed in the concentration of fine sand by means of a shallow stream of water was the German *buddle*, which has a distinct but imperfect resemblance to the Long Tom described on p. 107. Canvas tables were placed below these buddles in Germany, and probably suggested the use of *blanket-strakes* or *tables*, which were still in general use in the early days of the goldfields of the United States and Australia. The rough surface of the blanketing seems to be particularly efficacious in catching and holding thin plates and spangles of free gold or sulphides, which are readily washed off smooth surfaces by a current of water. The blanketing was usually in strips 16 or 18 inches wide, and several feet long, and was nailed or stretched on wooden frames.

At Morro Velho, Brazil,[2] the framework supporting the blankets was hung on pivots above a shallow tank. When it was necessary to clean them, the framework was tilted so that the upper surface of the blanket was inclined downwards, and the mineral washed off its surface by a hose. Bullocks' skins as well as blankets were used. The concentrates were amalgamated in revolving wooden barrels.

After the introduction of plate amalgamation, blankets were employed mainly to scavenge and catch any gold not retained on the plates or shaking tables. Later, they again became competitors of plate amalgamation. They do not catch as much free gold as plates, but they cost considerably less to operate.

Blanket Strakes at Morro Velho (see Fig. 121).—The present mill at Morro Velho mine[3] has been in operation since 1895, and concentration on blanket strakes has been continuously in use there, as well as at all other mills in Brazil. The ore is complex and contains much arsenopyrite (with which the gold and silver are associated) and pyrrhotite, besides other heavy minerals. The gold and silver occur in a natural alloy of gold 77 per cent., silver 23 per cent., which is in great part not amenable to amalgamation owing to its "rusty" nature. The ore is crushed by Californian stamps of 850 to 1,000 lb. weight, through battery screens of 50 and 60 mesh. The height of discharge is $7\frac{1}{2}$ inches. The pulp is fed on to strakes (Fig. 122) consisting of wooden tables or "decks" covered with coarse cotton duck nailed down. The canvas surface of each deck is 12 by 3 feet with an inclination of $1\frac{3}{8}$ inches per foot run. The strakes are washed every half-

[1] *Min. Ind.*, 1932, **41**, 581.
[2] J. A. Phillips, "*Mining and Metallurgy of Gold and Silver,*" 1867, p. 210.
[3] J. H. French and H. Jones, *Trans. Inst. Mng. Met.*, 1933, **42**, 189.

hour by a " washer car " running on rails above them (see Fig. 123). There is a second line of strakes, partly for the tailings from the first line and partly

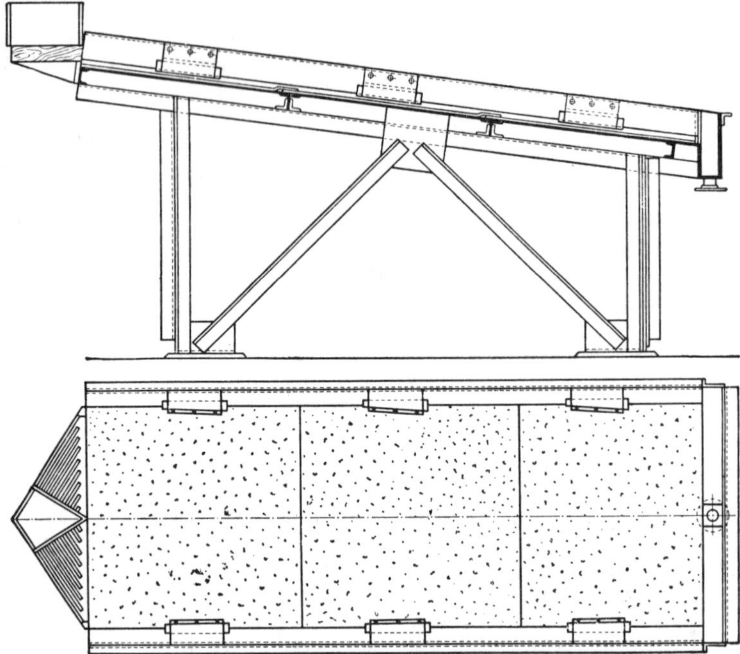

Fig. 121.—Blanket Strake.

Fig. 122.—Line of strakes ; stamp batteries on left. Morro Velho, Brazil (*French and Jones*).

for the product of tube mills. The concentrates are collected in settling tanks and passed over more canvas strakes for re-concentration. The second product is next fed to travelling-belt canvas "vanners," which are similar to Frue vanners (*q.v.*), but without a side shake and having canvas instead of rubber (Fig 124). The final concentrate consists of 50 per cent. bullion

Fig. 123.—Washer car on strakes. Strake partly tilted in position for washing. Morro Velho, Brazil (*French and Jones*).

with 50 per cent. heavy minerals, a remarkably successful result. It is melted with manganese dioxide, Chili saltpetre, sand and borax. The yield is approximately 60 per cent. of the total content of the ore at a cost of 2½d. per ton. The tailings go to the cyanide plant and the tail sands from that are again concentrated on canvas strakes. The substitution of James tables (p. 243) for canvas strakes was completed in 1935, with improved results.

Fig. 124.—Continuous Canvas Table.

Blanket Concentration on the Rand.[1]—About 1920 a trial was made of blanket tables in place of the amalgamated plates which at that time received the product of the tube mills. As a result the plates were soon discarded at all the mills and replaced by blankets which were fastened on the same or similar tables placed as before to receive the pulp from the tube mills. The pulp contained 30 to 40 per cent. of + 90 mesh material. The actual material used as blanketing was corduroy or " pulp sifting " cloth, consisting mainly of cotton, with the ribs spaced more widely than in corduroy used for clothing. This material is strong and closely but not coarsely woven, with a short stiff nap akin to the pile on some floor rugs, and is placed on the tables in strips, each length overlapping the succeeding one

[1] J. C. Wartenweiler, *J. Chem. Met. Mng. Soc. S.A.*, 1923, **23**, 150.

by a few inches. The strips adhere when wet without being fastened down, but air bubbles must be removed from beneath them. The ribs are placed transversely to the flow of the pulp and their high side faces the stream. The tables are about 5 feet wide and 10 or 12 feet long. The slope is determined by actual trial, but 1½ to 2 inches per foot is the usual fall. About 1 square foot of corduroy is allowed for each 1½ to 3 tons of ore passing over the table per 24 hours.[1]

The pulp flows in a thin stream, uniform in depth, across the width of the corduroy. The projecting ridges of the cloth cause rippling of the pulp as it flows downward, and the ridges, acting as riffles, retain a concentrate consisting largely of pyrite and containing free gold. The corduroy is removed and washed when necessary, every 4 to 6 hours having been mentioned as suitable. One of the difficulties is that the hollows in the corduroy are soon filled up, and a smooth surface formed, so that frequent washing is necessary. The tailings from the corduroy tables pass to the cyanide circuit.

The concentrate thus obtained is fed to a rather flat Wilfley table (*q.v.*) where it is collected in a filter box, the tailing being returned to the mill circuit. The former is then ground and amalgamated in cast-iron or steel barrels. The charge of concentrates, steel balls or heavy scrap steel, water, lime and mercury is revolved slowly for 2 to 12 hours. Any type of grinding and amalgamating pan is also suitable, *e.g.* the Berdan pan. On the Rand the barrel discharge is sent to a mechanical batea where the amalgam is retained. The overflow from the batea passes to Wilfley or similar tables, or to a small corduroy table, where osmiridium is separated, and the final residue goes to the tube mill circuit. At some plants osmiridium is recovered in considerable quantities.

The drop in recovery in the grinding section under this treatment amounted at some mines to as much as 10 per cent., but the total recovery was not affected.[2] Blanket recoveries are 25 to 60 per cent. of the gold content of the ore contained in a concentrate, which, after dressing, is 0·05 per cent. of the ore tonnage.

The advantages obtained by the use of corduroy instead of amalgamated plates are as follows:—(1) The risk of mercurial poisoning is reduced. (2) The personal skill involved in dressing the plates is not required. (3) Amalgam no longer gets into the tube mill circuit. (4) The daily recoveries agree better with the differences between the values of the heading and tailing. (5) There is a saving of labour. (6) The gold formerly locked-up in the amalgamated plates and in the amalgam going free in the circuit is more rapidly recovered and brought to account, so that there is saving of interest. (7) There is less risk of loss by theft. Amalgam must remain unprotected, but wire netting can be fixed round the corduroy tables without interfering with the supervision of the flow of pulp. It may be added that corduroy combines the advantages of blanketing with the additional concentrating effect of the ribbing.

According to H. A. White, neither amalgamated plates nor corduroy tables are necessary with Rand ores if they are ground fine enough. Corduroy, however, brings some of the gold to account more quickly, and thus obviates its being locked-up in tube mill circuits.

The position of the blanketing at the Modder East Mine in 1923 is given

---

[1] Von Bernewitz, *Eng. and Mng. J.*, 1935, 136, 63.
[2] T. K. Prentice, *Trans. Inst. Mng. Met.*, 1935, 44, 479..

in Fig. 125. The flow sheet of the treatment of Rand corduroy concentrate is given in Fig. 126.

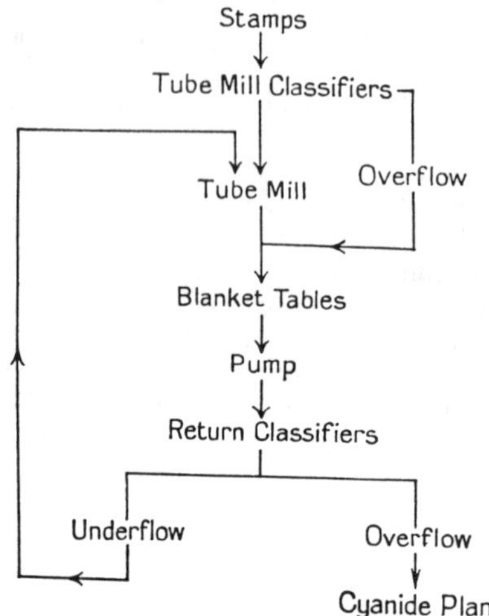

Fig. 125.—Position of Blanketing, 1923 (Modder East Mine).

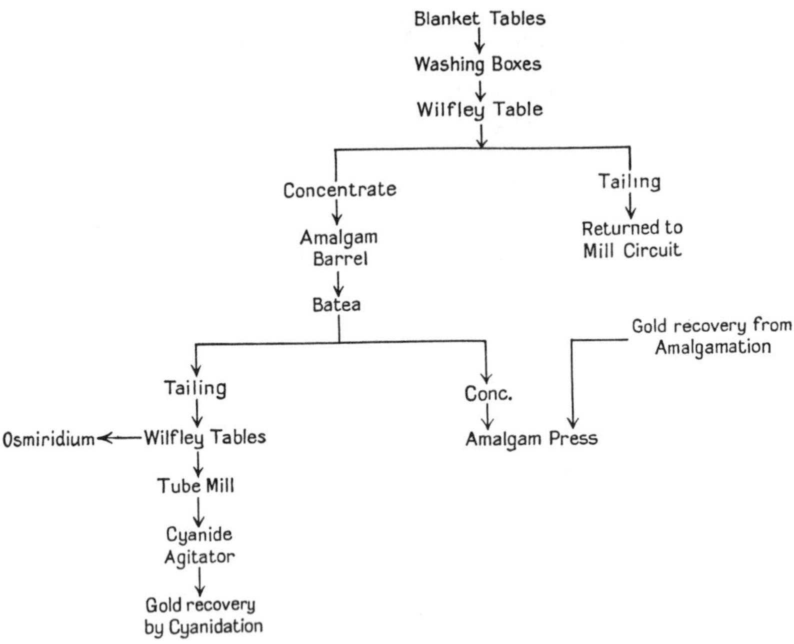

Fig. 126.—Treatment of Blanket Concentrate.

Blanket Concentration on Other Fields.—At Ooregum,[1] on the Kolar gold-field, the inclined wooden tables formerly used to hold the amalgamated plates

[1] A. F. Hosking, *Kolar Gold Fields Mining and Metallurgical Soc.*, 1930, 5, No. 27, p. 71.

have been formed into two strakes by a wooden partition down the middle. These have a 16 per cent. fall. The pulp from the stamps passes through a 9-mesh screen at the rate of 13 tons per day per stamp. The strakes are stepped at 3-foot intervals. Indian black blanket is used. The top, middle and bottom sections are changed every 1, 2 and 3 hours respectively. The ratio of pulp to water is about 1 : 8 and about 65 square feet of blanket are used on each table. The consumption over the whole mill is about 853 square feet per month.

It was found that black blanketing loses its nap completely after 16 days, whereas corduroy retains its corrugations to the last. On comparative tests Hosking found that corduroy gave 83·5 per cent. greater extraction than the blanket. In Mysore, Bangalore carpets or mats have been substituted for blanketing with beneficial results.

At Mysore 65 per cent. of the total gold value in the stamp mill product is caught on blanket strakes. The pulp then passes to the tube mill circuit with further blanket concentration. The blanket concentrates are treated in a Wheeler pan with a spitzkasten overflow.

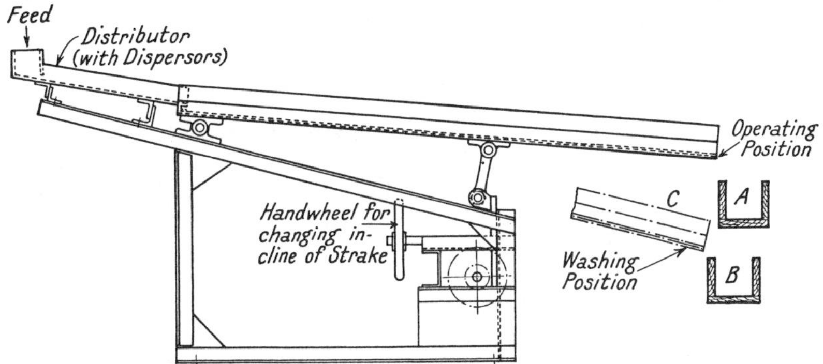

Fig. 127.—Adjustable Rubber Blanket Strake (*Knapp and Bates*).

At the Montezuma Apex mine [1] the concentrate from the blankets and the hydraulic traps is again concentrated on a Wilfley table, the tailings from which are returned to the mill circuit, and the valuable product amalgamated in an amalgamation barrel, excess mercury being added and grinding continued for 2 to 3 hours. The contents are discharged on to a mechanical batea. The overflow from this passes along a launder fitted with amalgamation plates and then into a mercury trap whose overflow in turn goes to a Wilfley table. Care is taken that no mercury returns to the mill circuit. The combined concentrates are sent to the refinery.

Copley says of strake recovery of gold from antimonial ore in S. Rhodesia :—" Width of strakes is more important than their length, and if for any reason their width must be curtailed, it is then better to install tables with 6 inch steps. The fall should be $1\frac{1}{2}$ inches per foot, and a convenient width is 30 inches. The blankets should not be more than 36 inches long. Canvas is superior to blanket for catching fine gold." [2]

In the Kalgoorlie area up to 30 per cent. of the gold is extracted by

---

[1] Gayford, *Trans. Amer. Inst. Min. Met.*, 1934, **112**, 565.
[2] *J. Chem. Met. Mng. Soc. S.A.*, 1921, **21**, 118.

blanket concentration either before flotation, before treatment with bromocyanide or before direct cyanidation.

Messrs. Knapp and Bates [1] recommend the use of a para rubber strip on

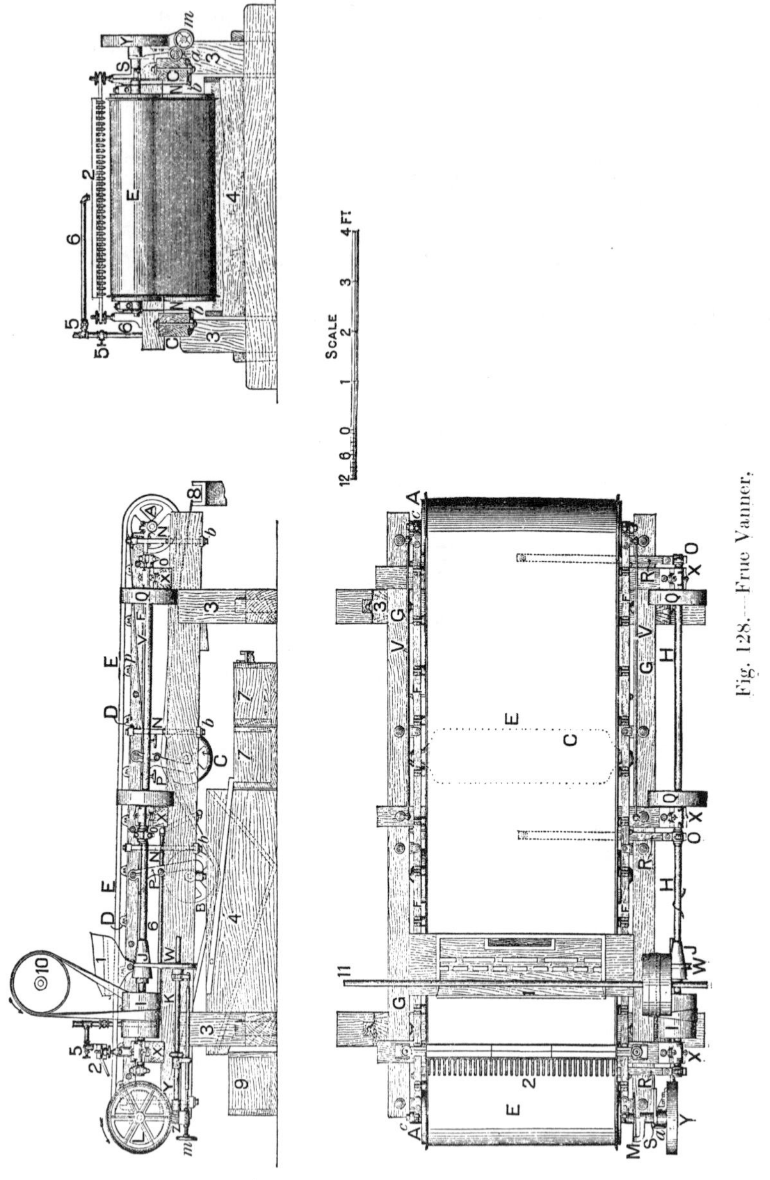

Fig. 128.—Frue Vanner.

the tables instead of the ordinary blanket or canvas (see Fig. 127). The roughened surface of the rubber serves to catch the heavy particles. The

[1] Private communication.

table is constructed so that when working it has a normal slope of about 1½ inches per foot, but when it is to be cleaned, the pulp current can be by-passed to another circuit and the whole table tilted to a steeper angle (C). The concentrates may then be washed off into a separate launder (B) situated below the ordinary pulp launder (A) without removing the rubber strip. This saves the time usually expended in rolling and washing ordinary blankets.

The **Johnson Concentrator** is used in some mills on the Rand instead of corduroy tables. It consists of a rotating drum with a rubber lining in which riffles and grooves are arranged as in the rifling of a gun. The sulphides collect in the riffles and as they are carried upward they are removed by sprays. This machine, according to von Bernewitz, recovers 44 per cent. of the pyrite and 55 per cent. of the gold in a concentrate weighing 8 per cent. of the total ore treated. The free gold is removed on blanketing and the pyrite ground and treated with cyanide together with the mill pulp.

The Frue Vanner.—This machine (Fig. 128) consists of an endless rubber belt with side flanges mounted on a frame, with its upper surface slightly inclined to the horizontal, and subjected to two movements, a slow constant longitudinal movement and a slight and rapid side shake. The belt is slowly revolved in the direction of its length and *up* the incline ; thus its surface is constantly travelling. The pulp is delivered near the upper end of the belt. A stream of clean water is applied at 2, and the gentle side shake keeds the sand in a state of agitation. The water distributor, 2, is placed about a foot above the pulp distributor, and delivers small jets of water, 3 inches apart, over the entire width of the belt. The heavy particles of mineral settle through the sand and cling to the belt, when they are carried up by it past the small jets of water and deposited in the collecting tank, 4, while the lighter gangue is carried down by the stream and delivered into the tailings launder, 8. R, R, R are three flat-steel spring connections bolted underneath the cross-pieces of the frame, F, and attached to the cranks of the shaft, H. These springs give the quick lateral motion—about 200 per minute. The speed of the uphill travel of the belt varies from 2 to 12 feet per minute. The capacity per day for a 4-foot vanner is about 6 tons of material, fine enough to pass a 40 mesh screen. Sizing of the pulp is usually omitted.

The Wilfley Table is shown in Fig. 129. It consists of a smooth shaking surface, of which Fig. 130 is a diagrammatic plan. A sheet of linoleum is tacked down over the whole table and is fully exposed at D. A number (usually 46) of wooden riffles shown at E are placed on the top of the linoleum and are nailed down. The longest riffle runs the whole length of the table, and the shortest riffle is about as long as the feed box. The riffles are ¼ inch high at the "head end," A F, and taper down to nothing at the "concentrate end," C G. They are about ½ inch wide, the shorter ones being narrower, and are separated by ⅞-inch spaces, so that the distance between centres is about 1⅜ inches. The fall from the back, A B C, to the front, F G, can be adjusted by wedges. The rise from the "head end," A F, to the concentrate end, C G, is usually set at about ½ inch. The shaking mechanism is situated at the end A F, and consists of a movement towards C G, beginning slowly and uniformly accelerated by means of a toggle joint, and a return movement towards A F effected by a spring, beginning suddenly and uniformly retarded. The effect of this vanning motion is to a certain extent similar to that of a bump delivered at the end, C G, the concentrate moving uphill towards G. The length of the stroke is about ¾ inch and the number about 240 per minute. The pulp is supplied from the wooden feed box, A B, through a number of small holes. Clear water is supplied from the trough, B C, also through small holes. The light, worthless material is washed down sideways towards F G by a flowing sheet of water, while the heavy concentrate is thrown towards the upper end, C G, against gravity, by a longitudinal jerking action, and eventually falls into a launder. A middle product passes over near G in the last 16 inches of the front, falling into a special launder, whence it is returned to the feed box or to other machines.

Fig. 129.—Wilfley Table.

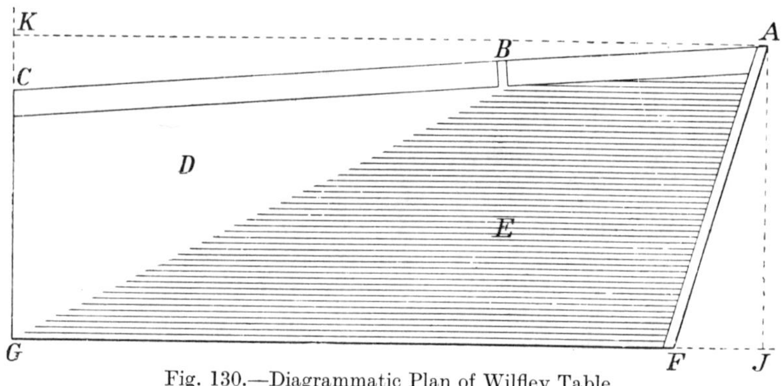

Fig. 130.—Diagrammatic Plan of Wilfley Table.

The size of the pulp treated on this table is usually from 16 to 30 mesh. The capacity is given as varying from 12½ to 30 tons per day. The power required is about 1 h.p. per machine.

Flury [1] records an improvement in extraction by cyaniding following concentration of the sand on Wilfley tables after a preliminary classification of the battery pulp. He attributes this to the removal of a clay film on the sands by the washing action of the water, thus making the concentrate cleaner, to the removal of graphite on the tables, and to the elimination of soluble salts which consumed zinc, cyanide and lime.

[1] *Min. Sci. Press*, 1921, 123, 771.

The James Table.—The *James Sand Table* (Fig. 131) is used for the classification of sands (40 to 200 mesh) received immediately from the stamp mills (thus replacing blanket tables as at Morro Velho). It consists of an almost flat deck (15 × 4 feet) which has a slight well-shape at the feed end and slopes away gradually where the concentrate discharges. The deck is covered with linoleum and is mounted on 42 flexible carriers which secure it to a steel framing. It is given a vibratory motion (stroke $\frac{5}{8}$ to $\frac{3}{4}$ inch) by a special vibrator, in the direction of a line passing across the corners of the deck. The riffling is almost parallel to the same direction, and is fastened with copper nails. From the feed distributor the pulp is spread in a thin even layer and the vibration assists in separating the heavier particles and conveying them to the discharge end, while the lighter particles or gangue are washed away to the front. The water-solid ratio in the feed should be about 4 to 1. Between one and four gallons of clear wash water per minute are required. The capacity of the table depends on the type of ore and may vary from 10 to 50 tons per day.

Fig. 131.—James Sand Table.

The James Slime Table (Fig. 132) differs from the sand table mainly in the construction of the deck covering. It is 15 feet long × 6 feet wide, and consists of several planes inclined at small angles to one another and intersecting the stroke line of the table diagonally. In this way it is intended that the material shall have a double treatment on the respective planes before being discharged. In Fig. 132 A is the major plane. The pulp is distributed along 10 feet of its upper edge and flows down about 4 feet of the plane. The fine particles settle and stratify according to their gravities. At the lower edge of A the minerals are arrested by the plane B, which is at a slightly less angle, while the water and gangue pass over. Again the gangue is checked by the plane C and, being held up, acts as a buffer, compelling the concentrates to follow the lower edge of A. The end of C is 4 feet from the discharge end of the table, and from this point B increases in width gradually until the table edge is reached, thus affording a larger area on which the concentrates are again spread in a thin layer for further washing. The impurities from the second section of B flow down the inclined plane D together with middlings and gangue from C, are again arrested by the planes E and F, and then re-treated as on the original planes A, B and C. The capacity is 10 to 15 tons per day. The water-solid ratio and the amount of clean water necessary are similar to those for the sand table.

Fig. 132.—James Slime Table.

Some other tables of the riffle type are in use, *e.g.* Deister concentrators. Since 1914 there has been a tendency towards multi-deck riffle tables.

Harz Jigs.—These machines differ in principle from all those previously described, inasmuch as the particles are separated by their fall through a somewhat deeper column of water than is the case on inclined tables, while a series of blows from below, causing waves moving upwards, continually brings the particles into suspension, and allows them to drop again. The initial period of the fall in water, during which the motion depends chiefly on density, is thus continually reproduced, and the result is a perfect separation of heavy from light particles of ore when working on any materials except the finest pulp. These machines are rarely used on gold ores. Recently, however, a new jig has been introduced to follow the tube mills at West Rand Consolidated.[1] A screen frame is suspended at the top of a large cone, 8 feet diameter, and fixed to it at the circumference by a flexible rubber joint. An overhead movement gives the jigging vibration. The feed is at the centre, the tailings overflowing round the circumference. A spigot discharge is obtained from the cone apex. One jig is allotted to each tube mill. The concentrate is dressed on cleaning jigs and James tables, ground in ball mills and cyanided separately. Free gold is recovered on blankets in the cleaning and grinding circuits. The residue is used for sulphuric acid manufacture.

[1] Wartenweiler, *Trans. Amer. Inst. Min. Met. Eng.*, 1934, 112, 775.

CHAPTER X.

FLOTATION.

Introduction.—The Flotation Process had its origin in the discovery by J. S. Elmore [1] that residuum oil, if mixed with ore pulp, would froth and rise to the surface, carrying with it sulphides and metallic particles. H. F. Collins [2] suggested the reason was that smooth lustrous particles were wetted by the oil, but the oil would not adhere to rough porous particles which absorbed water. The method was applied in 1900 to the concentration of an auriferous copper ore at Glasdir, near Dolgelley, in Wales, and 69 per cent. of the gold was floated.

The progress of the method was marked by the development of froth-formation on two lines. In *vacuum flotation* the air bubbles are formed in a vacuum by being drawn out of the porous ore particles and from solution in the water. The froth disappears on the re-establishment of atmospheric pressure. In *froth flotation*, a stable froth is formed under atmospheric pressure with the aid of an oil or other frothing reagent. It is in this latter direction that the greatest advances have been made, and this method is the only one described below.

In froth flotation air is drawn or blown into the mixture of ore, oil and water which is agitated to disseminate the oil and the air bubbles. The air bubbles adhere to the film of oil, which is formed on the surface of the valuable particles, but do not adhere to the particles of gangue.[3] The amount of oil has been gradually reduced in modern practice until an addition of only two drops of pine oil per ton of ore has been found to achieve the results. The two drops may be contrasted with the weight of oil equal to that of the ore which was used in Elmore's Glasdir practice. The film is in fact ultra-microscopical.

Many different oils have been tried and also a large number of synthetic organic compounds, some of which produce similar or even better results than the original natural oils. Other chemicals are used for special purposes, such as to produce an increase in the stability of the froth, and to aid the dispersion of the agglomerations of ore particles.

Flotation made great progress with other minerals before gold ores were treated, but it has been widely applied to the latter in the last dozen years. At the present time (1936) flotation is included in the flow sheets of a large number of gold mills, and machines are being installed in dredges to save the fine gold which has hitherto been lost. Flotation is capable of recovering the finest mineral particles, so that the production of slime in crushing, although bad for gravity concentration, is much less detrimental in flotation. It has also displaced gravity concentration in recovering gold from certain complex ores which are not amenable to cyaniding. Finally, it has made

---

[1] *Trans. Inst. Mng. Met.*, 1900, 8, 379. [2] *Ibid.*, p. 394.

[3] This simple statement may be taken as a working hypothesis. Various theories of the cause of the selective adherence of the air are given later.

amalgamation and cyaniding more profitable in many instances when worked in conjunction with them.[1]

The factors to be considered in the flotation of gold ores are as follows :—(1) The size of the particles and the character and the rate of feed of the ore. (2) The type of machine to be used. (3) The type of circuit. (4) The reagents to be used. (5) The ratio of water to solids. (6) The amount of agitation and aeration required. (7) The temperature. (8) The *p*H value of the circuit. (9) The nature of the water supply.

(1) **Size, Character and Rate of Feed.**—The feed to flotation cells should be maintained constant in density, in quantity, in value, and in the size of the particles. Particle size is best controlled in the grinding circuit ; density is taken care of in the classifiers and thickeners ; quantity may be regulated by thickener output and by the provision of surge tanks which come into operation during irregular running of the mill. Conditioning tanks, placed before the flotation unit, provide storage and an opportunity for the agents to react.

Grinding should free all the valuable mineral from the gangue. In general, froth flotation will recover minerals from pulp if the ore particles are not more than 0·15 to 0·18 mm. in diameter, whilst they may sometimes be as much as 0·25 mm. in diameter. Recovery approaching the maximum is to be expected with grains from 0·03 to 0·1 mm. in diameter.

Flotation fails with coarse particles owing to their greater weight and the inability of the air bubbles to carry them to the surface. In this connection —and for gold ores particularly—the mechanical types of frothing machines are superior to the pneumatic ones. The froth in the former is denser and more easily holds the particles once these are incorporated in it. Occasionally fine material is also lost ; this may be due either to its adherence to the falling gangue, thus causing it to become coated with slime, or to inferior flocculation of the mineral.

Gaudin[2] illustrates by the following figures the necessity for adequate grinding in the all-flotation treatment of pyritic gold ores :—

TABLE XXX.

	Fineness of Grinding, per cent of − 200 mesh.					
	38	46	57	79	89	98
Tailing, Au ozs./ton, .	0·066	0·050	0·037	0·032	0·022	0·015
,, , Ag ozs./ton, .	2·26	2·07	1·80	1·55	1·47	1·43
Concentrate, Au ozs./ton,	5·76	6·11	5·73	4·67	4·18	4·63
,, , Ag ozs./ton,	109·9	117·8	109·7	101·7	89·7	95·5
Gold recovery, per cent.,	85·0	88·5	91·5	92·1	94·4	96·6
Silver recovery, per cent.,	75·6	77·7	81·0	83·6	84·4	86·2

[1] For complete studies of the flotation process, see the following :—
"*Flotation,*" A. M. Gaudin, McGraw-Hill Publishing Co., 1932.
"*Flotation,*" Mayer and Schrantz, S. Herzel, Leipzig, 1931.
H. L. Sulman, *Trans. Inst. Min. Met.*, 1919-20, 29, 44 ; 1929-30, 39, 548.
"*Flotation Plant Practice,*" P. Rabone, London, 1936.
Wark and Cox, *Trans. Amer. Inst. Min. Met. Eng.*, 1934, 112, 189-302.
Taggart, Guidice and Ziehl, *Trans. Amer. Inst. Min. Met. Eng.*, 1934, 112, 348-381.

[2] "*Flotation,*" 1932, p. 330.

If the feed is comparatively rich, the operation of the machines is more straightforward ; the concentrate is richer, but at the same time the tailing also tends to become richer, necessitating vigorous scavenging. A heavy mineralised froth gives less opportunity for gangue to be retained. On the other hand, the nature of the gangue influences the course of flotation materially. For instance, clayey material, which may pass into colloidal suspension, is more difficult to free than sandy or granular material. Although attempts have been made to separate out this slimy material at an early stage, the conclusion has been reached that it is beneficial if not above a critical quantity. Hence adequate control of the grading of the feed is desirable.

The rate of feeding should be determined for each ore. It will usually be found that there is a critical range within which recovery is at or near the maximum. Underloading, particularly at the scavenging end of a series of cells, is common.

(2) **Flotation Machines.**—The mechanical types are probably best suited to gold ores, as they give a better conditioning effect in the cells themselves owing to the violent agitation which can be obtained. They can handle coarse and heavy sulphides, whilst giving every gold particle a chance to float. Better results appear to be obtained, too, in the flotation of oxidised or free gold ores with this type of machine than with pneumatic machines. The machines are also better suited, though to a less extent, to the treatment of auriferous sulphide ores if a high pulp level, *i.e.* a shallow froth, be maintained, and if the froth can be rapidly removed.

The best flotation machines are of large capacity and have low power consumption. When treating heavy or coarse ores they should allow the ore a free course. Room and light are essential around these machines in order to facilitate adjustments and a due observance of the froth by the operator.

In the selection of a flotation machine the following factors should be borne in mind—power consumption, maintenance, ease of operation and repairing, capacity per unit of space, positive circulation, and flexibility. In the design of the plant the following points should be taken into account :—

(1) The possibility of making a final tailing by flotation alone, and the recovery and ratio of concentration so effected. If the ratio be high, then shipment of the concentrate instead of bullion may be considered.

(2) Whether amalgamation should precede flotation or whether the flotation concentrate should be amalgamated in a grinding pan.

(3) Whether gravity concentration can be employed in preventing too rich a tailing escaping from the mill.

(4) The economic grinding limit at which values are liberated sufficiently for treatment.

(5) The possibility of using flotation before cyanidation.

The machines most commonly used for gold flotation will now be described.

The Subaeration Machine.—This machine has been used extensively for the flotation of gold ores, principally because it is well adapted to the earlier robust methods involving vigorous agitation with heavy oils, resulting in the production of a sturdy froth.

A section of the Minerals Separation (or M.S.) Subaeration Machine is shown in Fig. 133. It consists of a series of square cells in each of which rotates an impeller A attached to the lower end of a vertical shaft B. Any number of cells, but usually ten to twelve, can be employed. Rabone[1] quotes the following range of dimensions :—

TABLE XXXI.

Diameter of Impeller, ins.	Speed of Impeller, r.p.m.	Section of Cell, in.^{2}	Volume of Cell, ft.^{3}	H.P. per Cell.	Capacity, tons per 24 hrs. (W/S = 3/1)
12	450	24	8	0·7	140
15	360	30	16	1·3	250
18	300	36	27	1·8	360
24	220	48	64	3·0	820

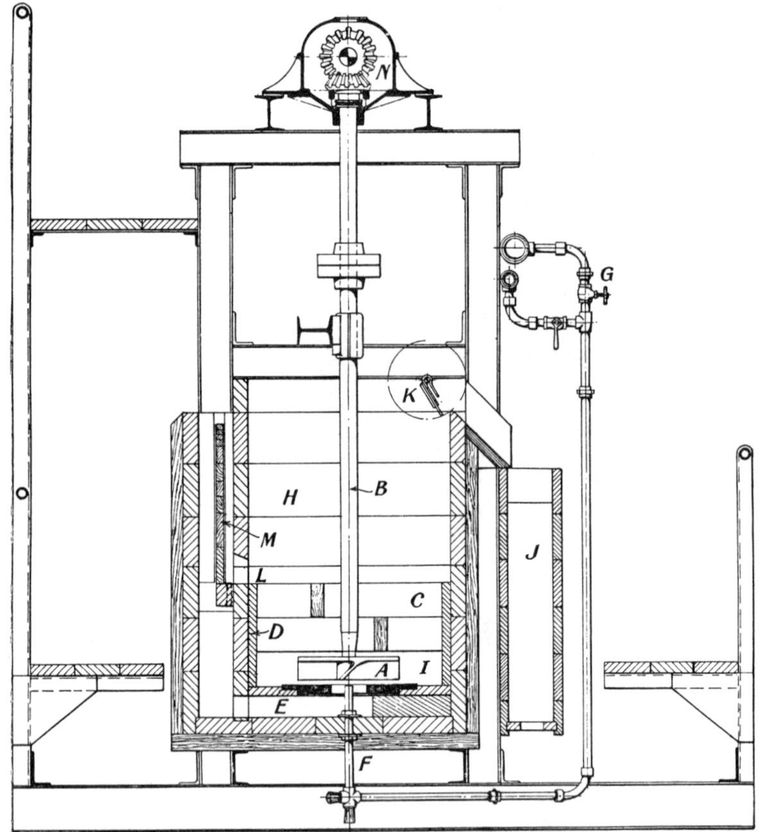

Fig. 133.—End Sectional Elevation of Minerals Separation Subaeration Flotation Machine.

Two varieties of impeller (*i.e.* agitator) are in use, constructed as follows :—(*a*) Four blades at right-angles to one another are inclined at 45°

[1] "*Flotation Plant Practice*," Mining Publications Ltd., 1936, p. 86.

to the horizontal. Over the whole is a circular plate, which is fixed to the shaft and blades. (*b*) An enclosed pump with curved blades inclined at 45° has an aperture on the underside which serves as an intake. The inclination of the blades is such that the pulp is projected upwards.

Diagonal baffles C placed immediately above the impeller divide the cell into a lower turbulent zone I and an upper quiet zone H. The baffles may be covered with a species of wooden grid. Removable hardwood, rubber or cast-iron liners D protect the interior face of the agitation zone from excessive wear. Pulp is admitted through a passage E under the bottom liner and is drawn up by the rotation of the impeller through a circular aperture in a plate immediately below the impeller. Air from a low-pressure blower enters through this same aperture from a pipe F passing through the bottom of the cell.

The pulp on entering the chamber I is thrown into great agitation by the centrifugal action of the impeller, assisted by the square shape of the cell. Simultaneously the entering air is broken up into small bubbles which mix thoroughly with the pulp. The amount of air admitted can be regulated by a valve G on the inlet; it is essential that every particle of ore may have the opportunity of contact with an air bubble.

From the lower compartment the mixture is forced by the pumping action of the impeller past the baffles into the upper quiet zone H. The bubbles, attached to mineral particles, rise to the surface, while gangue particles fall. A froth collects in the upper portion of the cell and is scraped over the side into a launder J by a revolving paddle K. The gangue eventually passes through a slot L at the rear of the cell above the baffles over the discharge weir M, the height of which can be adjusted. The level of the contents of the cell is thus controlled. The discharge falls by gravity into the feed passage leading to the next cell, whence it is again drawn up for further treatment by the impeller of that cell. The cycle of operations is then repeated. The number of repetitions is governed by the number of cells in the machine, the discharge from the final cell being tailing.

The impeller shafts may be driven by "quarter-twist" belts from a line shaft at the rear of the machine, driven in turn by a belt from a ground motor. More recently a drive through bevel gearing N from a line shaft immediately above the cells has been adopted, while probably the most satisfactory method is to fit each vertical shaft with a grooved pulley and drive this by means of one or two tex-ropes from a vertical motor.

Read[1] has shown in tests on mechanical machines that, (1) with a constant peripheral impeller speed the power increases with an increasing diameter of the impeller; (2) the depth of the pulp in the cell has a marked effect on the power and the volume of air drawn into the cell; (3) the 18 inch impeller is the most efficient; (4) the power consumption varies directly with the specific gravity of the pulp; (5) the coverplate should not be less in diameter than the impeller which it covers.

The Denver "Sub-A" Machine.—In this machine, Fig. 134, the impeller (*i.e.* agitator) is attached to a central driving shaft A in a square cell which contains a lower zone of agitation, E, and an upper quiet zone in the shape of two spitzkasten, F. The froth flows over two opposite sides of the cell. The feed pipe enters through the third side and passes down on to

---

[1] *Proc. Aust. Inst. Min. and Met.*, 1933, No. 91, 377.

the hood D; the fourth side carries an adjustable weir for the tailings overflow. The impeller, B, is a circular disc attached to the lower end of the shaft. It carries a number of upright blades, C, and is covered by a stationary hood D through which pass a stand-pipe G, the feed pipe and two middling-return pipes I. The hood is also perforated to permit the pulp to recirculate through the zone of agitation. The pulp enters through the feed pipe (and also possibly through one or both middling pipes if middlings have to be re-treated) and falls by gravity on to the impeller, whose action throws it outward into the bottom agitated zone. At the same time air is drawn down the stand-pipe by suction created by the impeller, and is broken up by pulp

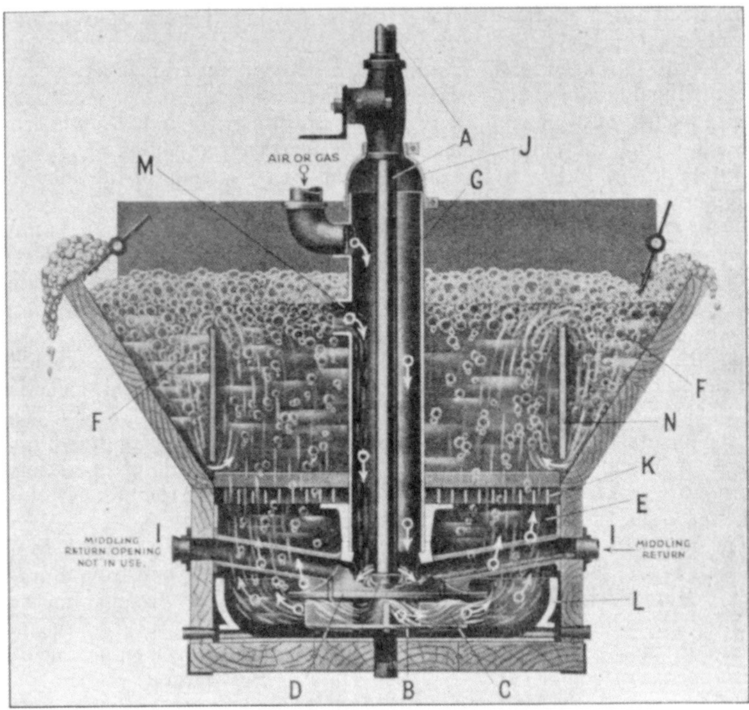

Fig. 134.—Section of Denver Subaeration Flotation Machine.

falling over the tops of the impeller blades. If additional air is required it is introduced from an external blower through holes near the top of the stand-pipe. In this case the top of the stand-pipe must be sealed from the atmosphere by a rubber cap J. Four vertical baffles N break the swirl of the pulp and a grid K below the baffles confines the agitation to the lower chamber. A cast-iron liner L, curved at the corners, serves to protect the lower part of the cell from wear and also assists in projecting the pulp upwards.

In the upper, less agitated zone, the bubbles coated with mineral rise and form a froth, while gangue particles drop out. The former is scraped away by revolving paddles into a concentrate launder (not shown), while the latter pass over an adjustable weir (also not shown) to the next cell. Middling products that require further treatment may be returned for agitation in the same cell through a regulated opening M in the stand-pipe just below the level of the froth, or may be directed to flow by gravity to any other cell.

The impeller shafts are driven by "quarter twist" belts or tex-ropes as mentioned in the description of the M.S. machine.

A machine with a low head has recently been introduced. This has steel supporting framework, gives better light and visibility for operation, and has reduced by one-third the head-room required. It also differs from the older model in having one spitzkasten instead of two, so that the froth overflows at one side of the cell only.

The following data refer to different sizes of cell (Rabone) :—

TABLE XXXII.

Designation.	Section of Cell in.^{2}	Volume of Cell without Spitz, ft.^{3}	Capacity, tons/24 hrs.	H.P. per cell.	Speed of Impeller, r.p.m.
A	20	5	15-40	0·6	550
B	27½	15	75-150	1·8	370
C	37½	35	250-400	3·2	290
D	43	45	300-750	3·8	252

The capacity depends on the type of ore, time of treatment, and water-solid ratio of feed. The following table shows the capacity in dry tons per 24 hours per cell :—

TABLE XXXIII.

Time of Treatment.	8 mins.			12 mins.			16 mins.			20 mins.		
W/S ratio,	2/1	3/1	4/1	2/1	3/1	4/1	2/1	3/1	4/1	2/1	3/1	4/1
A	10	7	5	6½	4½	3½	5	3½	2½	4	3	2
B	29	20	16	19½	13½	10½	14½	10	8	11½	8	6½
C	73	51	39	48	34	26	36	25	20	29	20	16
D	102	71	55	67	47	36	50	35	27	40	28	22

The low-head type of Denver "Sub-A" machine is made in the form of a unit cell for use in the grinding circuit.

Fig. 135 shows one of these machines operating in a tube mill and classifier circuit, and Fig. 136 shows the hydraulic cone attachment for the discharge of coarse free mineral, including gold, which is too large to be floated. The introduction of these cones at the base of the Denver subaeration unit cells following the primary tube mills at McIntyre Porcupine has increased the recovery of coarse gold at this point from 60 to 75 per cent. of the total values.

It is becoming common practice to build up complete multi-cell flotation machines from unit cells. Any size of machine can be made by mounting the required number of cells on a common foundation and connecting the tailing discharge of one cell to the feed inlet of the next. The advantages of this form of construction are (1) erection is simple and rapid, (2) extra

Fig. 135.—Combination Flotation Cell and Hydraulic Cone, operating in a Tube Mill-Classifier Circuit.

cells can be added at any time if extra capacity is needed, (3) the circuit can be altered with the minimum of expense and of loss of time to meet changing ore conditions.

The *Fagergren Flotation Machine* (see Figs. 137, 138 and 139) is a high-speed mechanical agitation machine and consists of a "stator" or stationary

Fig. 136.—Denver Unit Flotation Cell, showing hydraulic cone attachment for discharge of coarse free mineral.

Fig. 137.—Fagergren Flotation Machine.

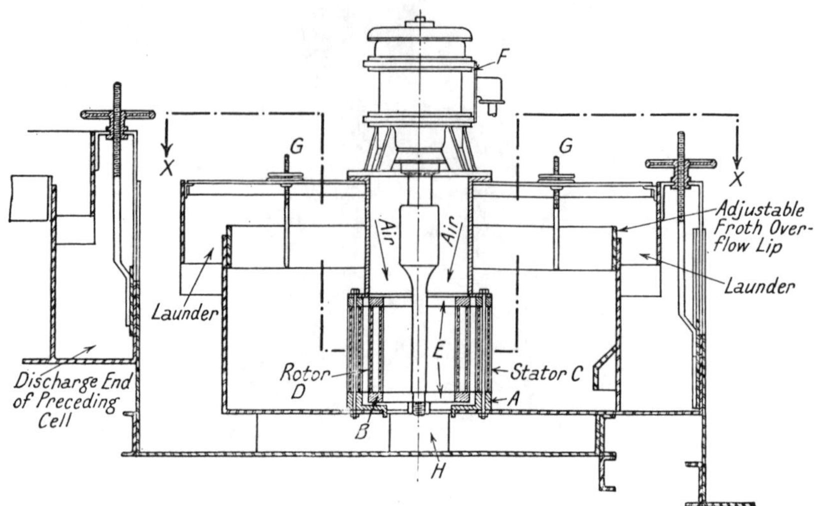

Fig. 138.—Section of Fagergren Flotation Machine.

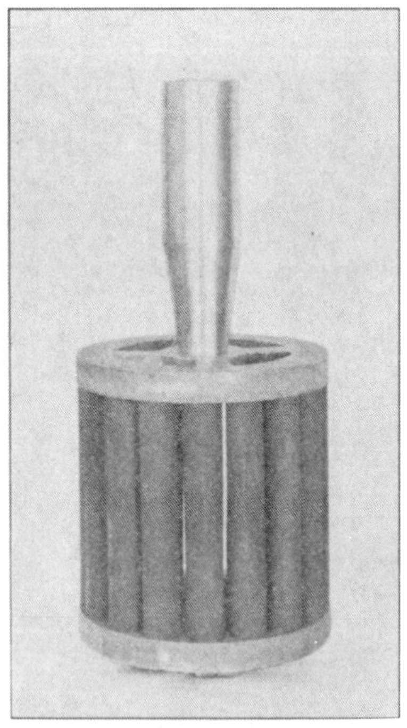

Fig. 139.—Rotor of Fagergren Flotation Machine.

cage C (Fig. 138), comprising rubber-covered cylindrical iron rods or spacers fixed between top and bottom annular rings A, and a " rotor " or rotating cage D (see Fig. 139), also having rubber-covered cylindrical spacers, but attached to upper and lower bladed impellers E which form the top and bottom of the rotor, and which by the set of the blades elevate the pulp from the feed opening H and draw in air from the top. A direct-coupled motor F supplies power to the rotor. The pulp enters each machine through ports in the bottom, whence it is elevated by suction of the rotor, mixed with the air drawn in from above, and dispersed horizontally by centrifugal force into the circular tank. The spacers on stator and rotor cause intense agitation and aeration. The froth collects in the quiet zone on the top of the liquor in the tank and overflows into the launder round the entire periphery. The depth of pulp in the machine is controlled by an adjustable overflow lip operated by the handwheels G. Several machines may be operated in series.

The Kraut Flotation Machine originally consisted of a long open red wood or welded steel tank, subdivided into a number of cells and mounted on a suitable steel frame. In each cell was suspended a conical stationary vessel within which a hollow rotor or agitator revolved. The outer surface of the rotor had a series of projecting helices and its wall was provided with slots for the admission of air. The interior communicated with the atmosphere through a hollow shaft by means of an air passage controlled by a valve. A bell-shaped vessel, open at the bottom and having a series of internal baffles, overhung the upper parts of the cone and rotor.

The pulp entering the cell passed into the space between the lower cone and the rotor, and was lifted rapidly by the helical riffles to the top. Discharge was into the space below the bell. In the initial upward movement the particles became increasingly dispersed, allowing space for the air to enter *via* the control valve through holes in the rotor wall. The baffles within the bell prevented the air-pulp mixture from rotating, and its downward movement was arrested, permitting the bubbles carrying the minerals to rise to the surface as a froth. This eventually overflowed over the concentrate discharge lip. An adjustable gate served to regulate the depth of froth. The partly de-mineralised pulp circulated back to the bottom intake of the same cell, or passed on to the next cell for re-treatment. At the discharge end of the machine a movable weir was employed to control the pulp level, an auxiliary sand discharge being provided to prevent the accumulation of sands in the cell.

In a more recent construction (see Fig. 140) the rotor element D consists of four rubber-covered steel staves assembled on steel spiders attached to a hollow steel shaft E. Segments of helices having an increasing lead from the bottom up are arranged on the outside surfaces of these staves and constitute the pumping element of the machine. Narrow vertical slots are left between the staves and ports in the hollow vertical shaft and connect its interior with that of the rotor. The rotor itself is mounted within a rubber-covered steel cylinder B which has holes at the lower end to admit pulp. The cylinder does not extend to the top of the rotor and thus allows a free discharge of pulp. A stationary conical hood C lined with rubber and extending slightly below the pulp level in the tank covers the top end of the rotor and the cylinder. A series of radial baffles G inside the hood stops the whirl of the pulp as it is discharged from the rotor. A single motor drives two spindles.

All cells have a common pulp level, but each has its own adjustable froth lip. The pulp enters at A, near the bottom of the first cell, and advances by displacement from cell to cell. Most of it is discharged over the weir I, but

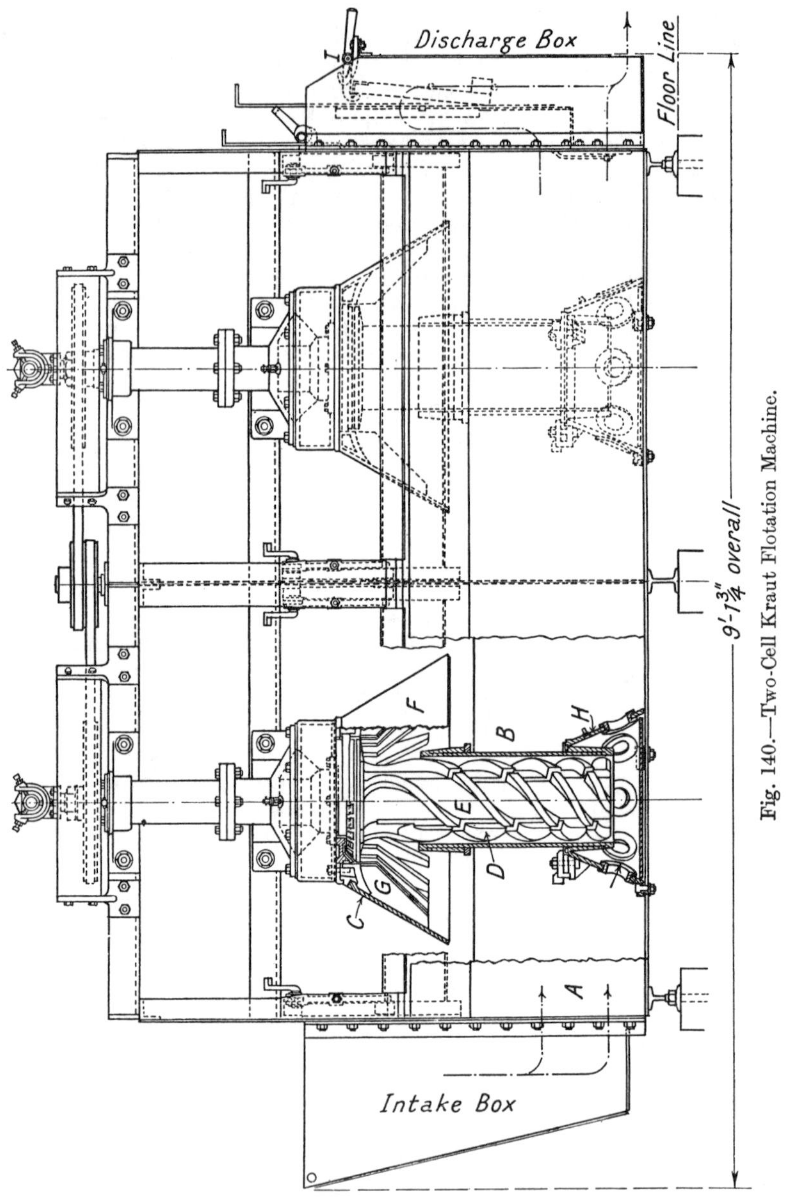

Fig. 140.—Two-Cell Kraut Flotation Machine.

there is an extra sand discharge at the bottom of the last cell through which a portion of the pulp may be discharged if necessary. The amount of pulp admitted between the rotor and the surrounding stationary part at each revolution is determined by the lead of the helix at the bottom of the rotor. The increasing lead upward causes the pulp to rise at an increasing velocity. Spaces are thus left in the flow into which air, entering initially through the control valve in the hollow spindle, is induced. An intimate mixture of air and pulp results, and is discharged at the top of the rotor among the baffles inside the hood, which steady the agitation. It then passes quietly into the pulp mass where the froth is formed.

Sabin [1] quotes an instance in which, if the Kraut machines were worked sufficiently hard, the froth concentrated in the overflow launder. By placing weirs from the first launder into a second launder, to catch the upper concentrated portion of the froth, a better separation was effected, and the tailings were found to be lower in value than was the case in any other method.

(3) **Flotation Circuits.**—Usually the flotation section in a flow sheet is comparatively simple and its aim is not so much the obtaining of an absolutely clean concentrate as the recovery of all the minerals with which the gold is associated in as clean a concentrate as possible. A straightforward unit incorporating roughing and cleaning is, in most instances, all that is necessary, though in some cases the concentrate from each cell is sent to the cyanide

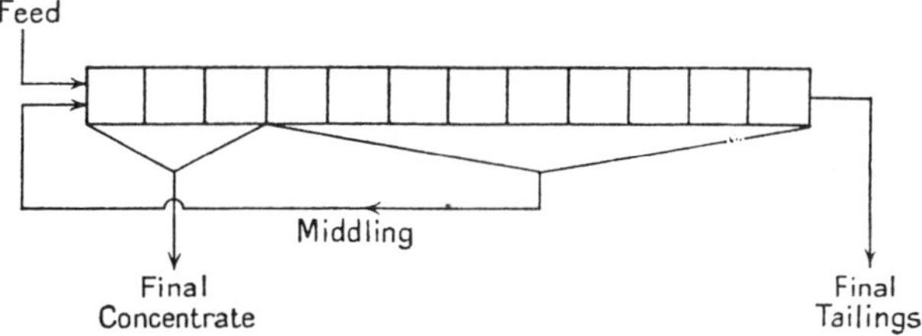

Fig. 141.—Simple Flotation Circuit No. 1.

section without cleaning, and in others a middling product is returned to the head of the machine.

Two simple circuits commonly employed are described by Rabone [2] :—

(1) A finished concentrate is obtained from the first few cells and the remaining cells are employed as scavengers to clean up the tailing from the former. The middling froth from the scavengers rejoins the feed to the first cell. This method may be employed when the gangue is relatively unfloatable. It requires more careful control than when a separate cleaning circuit is used. (See Fig. 141).

By careful regulation, however, mainly by restricting the air in the earlier cells, it is possible to obtain a concentrate from these which may be clean enough to send direct to the filters. The froth layer in the primary cells is deep, so that entangled gangue particles may be induced to drop out. At the discharge end the weirs are raised and the amount of aeration is increased, thus giving a scavenging action, ensuring that the minimum of valuable material escapes.

(2) This circuit is as a rule preferable to No. 1 and easier to regulate. A very clean concentrate is obtained. The primary or roughing froth is diluted to a W/S ratio of 6 : 1 to 10 : 1 and transferred to the cleaner cells, which, in number, are about one-quarter to one-third of those in the rougher circuit. The tailing from the last cleaner cell is mixed with the primary feed.

[1] *Eng. Min. Journ.*, 1935, 136, 68. [2] "*Flotation Plant Practice*," 1936, p. 116.

The final concentrate is thus produced from a pulp containing much more mineral than gangue, and there is less likelihood of the latter entering the froth. Moreover, changes in feed, as regards both character and rate are, in practice, more or less automatically controlled in the cleaner circuit (see Fig. 142).

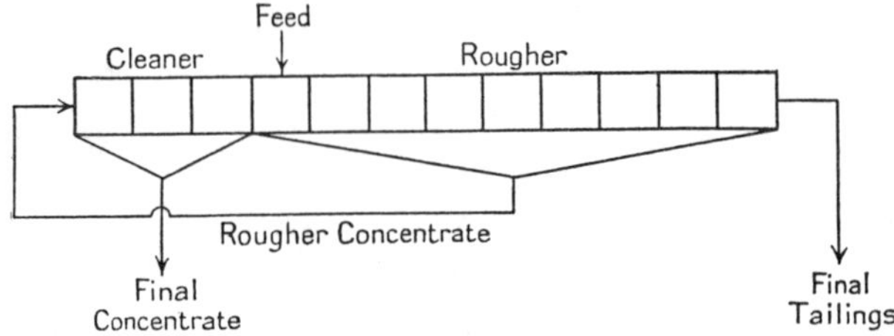

Fig. 142.—Simple Flotation Circuit No. 2.

(4) **Reagents used in Flotation.**—The addition of small quantities of oil, certain chemicals and a mass of air bubbles is made in flotation of mineral particles to produce the following effects [1] :—(1) To form a froth. (2) To stabilise the froth. (3) To coat the mineral particles with an ultra-microscopical layer of oil or equivalent chemical. (4) To cause attachment of the air bubbles to the oiled particles, thus assisting them to rise. (5) To assist in intensifying the froth formation. (6) To assist in intensifying the gungue depression.

For efficient separation it is necessary that all foreign material be more or less completely wetted by water, while the particles of valuable mineral be completely "wetted" by oil or organic material which is as nearly immiscible with water as possible. The mineral particles are then raised in the froth, the amount and character of which determine its carrying and holding power and therefore the grade of the concentrate.

Oil emulsions in water in which no oil is taken into solution do not froth. Where there is partial solution, frothing is less than if the portion soluble in water existed alone.

Taggart [2] gives the following solubilities of some flotation agents in water when in the proportion 1/4000 :—

	Percentage Dissolved.
Kerosene,	0·5
Olive oil,	3·0
Fuel oil,	4·0
Linseed oil,	5·0
Crude oleic acid,	5·5
Coal tar,	6·5-29·0
Coal tar oil,	12·5-35·0
Eucalyptus oil,	68·0
Pine oil,	79·0
Oil of wintergreen,	89·0
Cresylic acid,	92·0
X-cake,	96·0
Phenol (pure),	100·0
Alkaline xanthates,	100·0

[1] For a full discussion of flotation reagents, see B. W. Holman, *Trans. Inst. Min. Met.*, 1929-30, 39, 588 ; Rabone, "*Flotation Plant Practice*," 1936, p. 61.

[2] "*Handbook of Ore Dressing*," 1934, p. 840.

A *frothing agent* may be more or less selective. It disseminates through the pulp a mass of small air bubbles. The froth should be stable, and of bulk sufficient to carry the mineral and to overflow the machine regularly. It should be capable of being broken down readily to recover the mineral. There should, however, be no excessive frothing during the transportation of the pulp in the mill circuit.

In causing a froth most reagents reduce the surface tension of water after greater or less solution in it. The most effective reagents have the greatest influence on the surface tension, and in general have high molecular weight. In a homologous series of chemical compounds the higher members are best. If, however, the surface tension be reduced too much, the water layers around the air bubbles will not be sufficiently strong to hold the mineral particles. Soap, for instance, acts in this manner.

As a rule frothing agents are sparingly soluble in water. Pine oil and cresol are in common use. The amounts required may vary from 0·05 to 0·15 lb. per ton for pine oil and 0·1 to 0·5 lb. per ton for cresylic acid.

Dean and Hersberger [1] give the following properties of an ideal frothing reagent :—

(1) It must form in low concentrations a copious but not too persistent froth.

(2) It must be insensitive to hydrogen-ion concentration of the pulp, *i.e.* it must froth equally well in an acid or alkaline medium.

(3) It must be insensitive to salts, even in high concentration.

(4) It must be absolutely non-collecting to both sulphides and non-sulphides.

(5) Its frothing properties must not be affected by collecting reagents, including soap.

(6) It must readily emulsify and disperse any insoluble collecting agent likely to be used.

The investigators mention several organic materials as being among those that are nearest to ideal frothing reagents, and give examples of what they term " emolsol reagent X-1." It permits the use of collectors hitherto not possible, *e.g.* insoluble paraffin oils.

The froth should be sufficiently stable to permit its removal from the flotation machine. The more closely the finely ground particles are packed together, so that there is little interspace or liquid medium, the stronger is the cohesive force holding the froth together. Coarse ore demands a stiffening agent in the froth, such as tar, wood creosote or petroleum oil.

Typical examples of frothing agents used for gold ores are pine oils, cresylic acid, " Aerofloat," hardwood and coal tar creosotes. The first three are the most used at the present time.

In practice it is important to add the correct amount of frothing agent. " Over-oiling " causes numerous relatively small bubbles, which are persistent; a large amount of solids passes over the lip due to lack of opportunity for the gangue to settle, and recovery is lowered. " Under-oiling " causes a watery froth, leading to greater use of compressed air and again a larger proportion of gangue is forced over the lip.

Collecting agents are organic materials that attach themselves to certain minerals in preference to other constituents when present together in a water

---

[1] *Amer. Inst. Min. Met. Eng., Tech. Pub.* 605, 1935.

pulp. The power of water to wet these minerals is thereby diminished and floatability is increased.

These agents may be solid or liquid. Oils,* which are rarely selective and need much agitation for emulsification, were once extensively used. They have now been replaced almost wholly by the purely chemical reagents, though the presence of a very small quantity of one of the oils may in certain instances tend to give a firmer froth. Insoluble agents are brought into contact with, and ultimate adherence to, the mineral particles by mechanical stirring. Where the reagent is soluble, the phenomenon is mostly brought about by chemical adsorption. Typical examples of the former class are oleic and other fatty acids, fish oil, castor oil, linseed oil, petroleum oils, less soluble essential oils, creosotes, tars, etc. Those of the latter group are α-naphthylamine, thiocarbanilide, metal xanthates, pyridine, quinoline, etc. The amount of reagents of the second class necessary to give satisfactory recovery is of the order of 0·1 to 0·3 lb. per ton of ore.

The completely soluble chemicals of this class are sometimes called *promoters*, and may be placed in three groups :—

(*a*) Xanthates, the most powerful of the promoting agents. They have the general formula CS.SM.ORad., where M signifies sodium or potassium and Rad. one of the ethyl, propyl, butyl or amyl groups. These materials attack the most highly flocculated surface in addition to assisting in the flotation of the common sulphides. Selectivity can be effected by careful control. In general the xanthates with ethyl and propyl groups are less powerful but more selective than those with groups of greater molecular weight. In certain instances, in order to have greater recovery and greater selectivity at the same time, xanthates from each group may be used simultaneously. Xanthates usually give superior results in slightly alkaline circuits. They are added to the pulp in the form of 10 to 20 per cent. aqueous solutions.

(*b*) Thiocarbanilide (diphenylthiourea, $CS(NH.C_6H_5)_2$), which is nearly insoluble in water but soluble in *o*-toluidine (" T.T. mixture "), can be added as a 15 per cent. solution. The usual practice, however, is to add it as a solid to the ball or tube mill before the xanthates. It has comparatively little action on pyrite. Consumption normally ranges from 0·05 to 0·2 lb. per ton of ore.

(*c*) Phosphocresylic acid or dicresyldithiophosphoric acid, $HS.SP(OC_6H_4CH_3)_2$ ("Aerofloat"), usually contains some free cresylic acid, which is also a frothing agent. Like thiocarbanilide, it is less sensitive, and has less effect on pyrite, than xanthate, but is often used in conjunction with xanthate for the flotation of ore containing pyrite. It works best in a neutral circuit. Consumption may range from 0·1 to 0·4 lb. per ton of ore. It is corrosive and must be handled carefully. The sodium or potassium salt is a soluble solid.

Activators belong to the promoter class, and make more floatable those minerals which do not float well without them or have been thrown down by a

---

* *Note.*—For flotation purposes the term "oil" is of wider application than that normally accepted, being applied to any organic liquid which may have an oily feel or appearance. It includes such materials as fatty acids, *e.g.* oleic, kerosene, amyl acetate, and other substances of similar nature.

depressor. Copper sulphate is a good example of an activator. Lime or soda-ash has no material effect, in spite of the precipitation of copper hydrate. Indeed the two are normally added together. Copper sulphate and xanthate should not be added together as they result in the formation of an insoluble copper xanthate which is inactive. About 0·25 lb. of activator per ton in the form of a 20 per cent. solution is usually effective. Sodium sulphide is also used as an activator for gold ores associated with oxidised minerals, such as copper and lead carbonates or the oxides of iron and manganese, or where the gold is mixed with pyrite. Sodium sulphide is a better sulphidising agent than the calcium salt, which depresses the gold.

Modifiers assist in the wetting of the gangue particles and cause them to sink. Lime and soda-ash are the commonest, and they have the further effect of neutralising acidity. This latter function may give rise to the formation of salts, such as calcium sulphate, which may affect flotation and which therefore have to be controlled carefully or removed before the promoting reagent will work. Accurate pH determinations are necessary in order to ascertain alkalinity. The usual figure is 7·0 to 8·5. Soda-ash is a better deflocculator than lime, and has a direct action in depressing sulphides, so that it is really employed to a greater extent for preventing the flotation of unwanted minerals than in precipitating gangue. The consumption of either agent, depending mainly on the acidity, may rise to 6 or 7 lb. per ton of ore.

Sodium silicate is also used, but it is decomposed by some inorganic salts and has the disadvantage that it disperses the slime in the gangue so thoroughly, if much of it is used, that the fine part of the tailing cannot subsequently be settled in thickeners. A maximum amount of 1 lb. per ton of ore is usual.

Dispersion agents cause the gangue particles to sink and to separate from the mineral particles. Common agents in this category are sulphuric acid, lime, copper sulphate, soda-ash, sodium silicate,[1] sodium hydrate. A simultaneous effect of these reagents sometimes observed is the flocculation of the mineral.

Depressors reduce the floatability of one or more of the minerals in a mixture that ordinarily floats as a whole. On their proper use depends to a great extent the success of differential methods of flotation. Most of these materials also act as dispersors. Others, *e.g.* alkaline dichromates, react chemically with some sulphides and not with others, while the greater number of depressors seem to cause the deposition of a compound on the mineral surface.

The commonest depressors are lime, sodium cyanide and zinc sulphate. Lime, in particular, throws down pyrite and, if added in sufficient quantity, subsequent addition of xanthate has no effect.

Leaver and Woolf[2] have investigated the flotation of gold in minor quantities in large-scale copper concentrators. They conclude that lime is an active depressant for free gold during flotation if the alkalinity in the solution exceeds $p\text{H} = 10$. Hence the avoidance of excess lime in the flotation of base metal ores will increase gold recovery and, provided the pyrite is crushed finely enough to liberate its gold, the latter will separate.

---

[1] Shorey, Patek and Roland, *Amer. Inst. Min. Met. Eng.*, 1934, *Tech. Pub.* 559, discuss the use of sodium silicate as a deflocculator.

[2] *Amer. Inst. Min. Met. Eng.*, 1931, *Tech. Paper* 410. *Min. Mag.*, 1931, 44, 251.

Cyanide is a depressor for pyrite. In the treatment of the latter after cyanidation and prior to flotation, sulphuric acid may be added to neutralise the lime. Collecting and frothing agents are then added ; or soda-ash may be used to remove calcium hydroxide, followed by copper sulphate, ferrous sulphate and butyl xanthate.

When cyanide is used as a depressant in the flotation of heavy minerals gold tends to be dissolved. Anderson [1] recommends filtration of the flotation tailing, cyaniding of the cake and precipitating the solution with charcoal (lack of free cyanide precludes zinc precipitation).

Foul air, smoke, etc., do not help flotation, as the carbon dioxide they contain precipitates the lime in the pulp.

Conserving agents protect the flotation reagents from attack by substances present in the ore pulp.

In actual practice a combination of " Aerofloat " (phosphocresylic acid or dicresyldithiophosphoric acid) in conjunction with a dithiophosphate and potassium or sodium ethyl xanthate is generally used. Pine oil may be added to increase the amount of froth. In order to regulate the alkalinity of the pulp to a pH value of 7·0 to 7·5 [2] (*i.e.* neutral to faintly alkaline), lime or soda-ash may be used. Soda-ash is the better, as lime tends to depress free gold and pyrite. The addition of too much sodium silicate for depressing the gangue may, in certain instances, also result in carrying down gold. Starch, to the extent of 0·1 lb. per ton of ore, is preferable to sodium silicate as a deflocculator. Copper sulphate up to 0·25 lb. per ton is useful in speeding-up flotation.

Rabone [3] gives the following amounts of reagents for use with ores containing free gold and gold-bearing pyrite :—

"Aerofloat 25," [4] . . .	0·20-0·5 lb. per ton.
Pine oil, ...	0·01-0·03 ,,
Higher xanthates, . . .	0·02-0·05 ,,
Dithiophosphates, . . .	0·02-0·05 ,,
Copper sulphate, . . .	0·10-0·25 ,,
Soda-ash, ...	to maintain an alkalinity equivalent to pH = 7·0-7·5.

Free gold [5] floats alone in the absence of sulphides. Very small amounts of reagents are needed. It has been shown that on a rich Californian gold ore, when all oil and grease have been removed from both ore and flotation machine, a froth assaying some £2,700 per ton with 60 per cent. extraction could be obtained by using only soda-ash, silicate and " pentasol xanthate " The addition of two drops of pine oil per ton increased the assay values of the concentrates to £5,000 per ton.

0·5 per cent. of coke is said to be a desirable addition for entraining the gold in the froth.

---

[1] *Can. Min. Journ.*, 1934, 55, 269.

[2] For a discussion on pH values, see "*Hydrogen Ions*", H. T. S. Britton (Chapman and Hall).

[3] *Loc. cit.*

[4] The composition of this reagent is not published. It is said to be cresylic acid treated with 25 per cent. of P_2S_5 to make phosphocresylic acid plus free cresylic acid.

[5] *Min. Ind.*, 1932, 41, 590.

Leaver and Woolf [1] have investigated the factors relating to the flotation of free gold in milling ore and arrived at the following conclusions :—

(1) Clean metallic gold of — 60 mesh can be floated successfully from a siliceous gangue containing pyrite with the usual reagents used for sulphide ores. Gold coarser than 40 mesh is not usually floatable. Organic collectors, *e.g.* xanthates, are better than coal tar oils. With coarse feed a high pulp density should be used.

(2) Gold does not float so readily or so rapidly as most of the mineral sulphides.

(3) Copper sulphate does not increase the recovery, but increases the rate of flotation of — 60 mesh gold particles.

(4) Lime is a depressor in gold flotation.

(5) Subaeration machines are more effective than straight mechanical types.

(6) Sodium sulphide in general retards the flotation of free gold.

(7) Talcose or carbonaceous slime has strong flotative properties. Ordinary dispersing agents, *e.g.* sodium silicate, are ineffective.

(8) Clayey slime does not float, but remains in suspension and coats mineral particles.

(9) Ferrous and manganous slimes consume much reagent, and cause low recovery. Hæmatite, however, appears to give no trouble.

(10) Starch is the most effective depressant for all the foregoing varieties of slime. Crude potato, rice or flour starch made by boiling potatoes, rice or flour, with water or 0·1 per cent. KOH solution for 30 minutes is as effective as fine starch, but should be added as a solution to the ore pulp. The solubilities of these varieties are as follows :—

	Starch in Solution, per cent.
C.P. soluble starch,	100·0
Flour,	71·0
Rice,	40·0
Potatoes,	11·0

(11) Starch has a selective action. It acts first on the slime, then, if sufficient is present, on the sulphides and metallic gold, either by complete wetting or by making a very brittle froth. Each ore is an individual problem in this respect.

0·1 to 0·2 lb. starch—or its equivalent in the cheaper reagents—per ton of ore is an average quantity required.

Vendensky and Duschak [2] have also found that under properly controlled conditions soluble starch exerts a powerful depressing effect on certain readily flotable constituents of the gangue, such as talc, serpentine and carbon. At the Idaho-Maryland mill a 4 per cent. starch solution with 1 per cent. sodium hydroxide is used.

Lange [3] offers the further conclusions :—

(1) Soda-ash produces a depressing effect.

(2) A sulphuric acid circuit, when employed at a flotation density of 4 to 1 on an ore containing a small amount of sulphides, assists flotation. In pulps of higher density it has a marked depressing effect.

(3) Sodium sulphide is a depressor of gold.

[1] *U.S. Bur. Mines, Rep. Invest.*, 3226, 1934.

[2] *Eng. Min. Journ.*, 1932, 133, 334.

[3] *Eng. Min. Journ.*, 1935, 136, 116.

(4) Sodium silicate has a slight depressing effect.

(5) "Pentasol xanthate," amyl xanthate, and a combination of "pentasol xanthate" and thiocarbanilide, were the collectors which gave the best results on the free gold used in the investigations.

(6) Flotation of free gold is best done in circuits of high pulp density.

A new reagent used in the second series of rougher cells in the Beattie gold mill, Quebec, is rhodamine or diethylmetamine-phenolphthalein.[1]

General Considerations on Flotation Reagents.—The amount of flotation reagent required depends on the following factors :—

(1) The percentage mineral content of the ore.
(2) The amount of solids in the pulp.
(3) The fineness of the grinding.
(4) The treatment of the pulp subsequent to the addition of oil.
(5) The nature of the minerals in the ore.
(6) The kind of reagent.

Generally speaking, chemicals are preferable to oils, though in some instances oils are the better. Chemicals can be bought according to specification. They are more readily dispersed in the pulp than oils, and produce a higher-grade froth concentrate, which is more readily broken down later and is more amenable to filtration.

The point at which each reagent is added is highly important. Small portions of particular reagents are usually added at various selected points, instead of all at one place, thus affording better control and more selective action. Extremely stable reagents may be added in the grinding mill to ensure thorough dissemination, whereas those that soon decompose or whose maximum effect is transient are best added in the flotation machine. The amount of reagent in use should be in proportion to the amount of ore, and should also bear some relationship to the water in the pulp. When the amount of water is varied, the action of fixed quantities of reagents is unreliable. Hence the W/S ratio should always be carefully regulated.

Reagent feeders should be accurate, permit of easy adjustment to meet varying conditions, and in general be designed for a definite class or type of reagent. Thin oils and solutions may be fed through valve-controlled feeders ; more viscous materials through pumps with a regular and controlled stroke, or a small elevator. Solid materials are most satisfactorily added in the form of solution, or as a fine powder which may be brought to the cells on a small belt or a shaking plate. Lime can be added solid or as milk of lime.

(5) **Water/Solid Ratio.**—The ratio of water to solid should be about 3/1 or 4/1. Lower ratios generally cause low recovery and poor concentrate. Higher ratios give a watery froth and similarly poor returns. With the latter, aeration has to be more intense in order to have good frothing, and there is thus the danger of gangue also passing over. In the cleaner cells the W/S ratio is usually less than in the roughers, but the grade of feed, and therefore of concentrate, is higher, so that there is little danger of gangue becoming entangled, especially if the aeration is modified accordingly. For every ore there is a critical ratio which yields good flotation results, and this should be so controlled by hydrometer or other means as to be as nearly constant as possible.

---

[1] Locke, *Min. and Met.*, 1935, 16, 29.

(6) **Amount of Agitation and Aeration.**—Both the time of contact with reagents and the time of agitation are most important factors in flotation. They are intimately connected with the efficiency and capacity of the plant.

Usually oils require greater agitation than chemical agents, entailing, of course, greater power consumption. Recovery can sometimes be improved by extra agitation even if the nature of the reagents remains constant.

The amount of air used, and its pressure and method of introduction, vary greatly in the several types of machines. The air may serve merely for bubble formation, or agitation, or for both. Most concern is usually felt in providing sufficient air to keep the machines working actively and the limit of the amount necessary has received but little attention. It has been suggested that modification of the air by the introduction of other gases or of atomised materials might, in some instances, be beneficial.

(7) **Temperature.**—Usually a rise in temperature increases the efficiency of the process.

(8) ***p*H Value or Hydrogen-ion Concentration of the Pulp.**—The *p*H value denotes the alkalinity or acidity of the pulp, and its regulation is an important factor, especially when such reagents as α-naphthylamine, thiocarbanilide and xanthates are used. The neutral figure is 7·0. For normal working in slightly alkaline circuits the range is 7·0 to 7·5.

Colorimetric testing outfits by which *p*H values may be measured against definite standards are now obtainable.[1]

(9) **Water Supply.**—The chemical analysis of the water should be known, especially the hardness (alkalinity) and the amount of organic material present. Where water is scarce and has to be used over again, it may in time take up large quantities of salts gathered from the process, which may have a direct effect on recovery. Corresponding corrections are then necessary.

Cost of Flotation.—Amalgamation is cheap, because it requires a minimum of grinding and supplies. Cyaniding is dear because of the fine grinding necessary and the cyanide consumption, which is large when compared with the reagent consumption in flotation. Initial capital cost is also less in flotation than in cyaniding.

Gaudin[2] gives the following example, which is illustrative of these factors :—

Gold Ore.

Au,	0·48 oz. per ton.
Ag,	2·0 ,, ,,
Pb,	0·05 per cent.
Fe,	2·5 ,,

Ore reserves are good for 8 years. Working days per year, 350. Haulage $3 per ton of dry concentrate.

Pb	worth	6 cts.	per pound.
Cu	,,	13 ,,	,,
Ag	,,	50 ,,	ounce.

Cyanidation.

Yields 96 per cent. extraction of gold and 30 per cent. of silver.
First cost of cyanide plant $1,000 per ton of ore treated daily.
Operating cost $1·75 per ton of ore.

---

[1] For a full discussion of the interpretation of *p*H values, see Sulman, *Trans. Inst. Min. Met.*, 1932, 41, 354.

[2] "*Flotation*," McGraw-Hill Publishing Company, 1932, p. 479.

Flotation.—Gold recovery 94 per cent. in a concentrate consisting of the following constituents :—

Au,	. 7·20 oz. per ton.	Cu,	. 0·8 per cent.
Ag,	. 25·8 ,,	S,	. 40·2 ,,
Pb,	. 7·4 per cent.	Zn,	. 1·2 ,,
Fe,	. 36·0 ,,	Insol.,	. 6·0 ,,

First cost of flotation plant, $600 per ton of ore treated daily.
Operating cost, $1·25 per ton of ore.

The particulars required in order to assess the relative advantages of the two processes in this example are shown in the following tabulation :—

TABLE XXXIV.

	Cyanidation $.	Flotation $.
(1) Annual amortisation cost per ton daily capacity,	155·00	93·00
(2) Amortisation cost per ton treated, . . .	0·44	0·26
(3) Total cost per ton treated, ...	2·19	1·51
(4) Value of bullion per ton of ore, ...	9·82	...
(5) Value of concentrate at concentrator, per ton of concentrate, ...	...	142·82
(6) Ratio of concentration, ...	...	15·96
(7) Value of concentrate at concentrator, per ton of ore,	...	9·14
(8) Net returns per ton of ore, ...	7·63	7·63

The two processes show a similar return, but it may be anticipated that flotation would be preferred as improvements in its technique are likely to be greater than those in cyaniding ; moreover the capital cost and consequent amortisation charges are much less when flotation is employed.

Examples of Practice.

Gold Flotation at the Lake View and Star Mine.[1]—The gold is here present as free gold, as tellurides, and in association with pyrite. The ore is ground in three stages in ball and tube mills. Corduroy tables are employed after each stage and catch about 25 per cent. of the total gold in the ore. " Sodium Aerofloat " (0·035 lb. per ton) is added to the ball mills and thiocarbanilide (6·05 lb. per ton) to the tube mills. The discharge from the third stage of crushing contains 95 per cent. of − 200 mesh material. It is pumped, together with a small quantity of pine oil (0·044 lb. per ton) to a surge tank, where more " Aerofloat " is added. Flotation is then applied in three 10-cell Denver machines, forming the roughing circuit (see Fig. 143).

A finished concentrate is taken from the first one or two cells. The froth from the remaining cells passes to a 10-cell Minerals Separation unit for cleaning, a small quantity of sodium silicate being added to depress the gangue. The cleaner tailing returns to the tube mill circuit, but that from the roughers is discarded, while the combined concentrate, representing 5 to 6 per cent. of the original ore, is dried, roasted and cyanided. A total extraction

[1] T. B. Stevens, *Min. Mag.*, 1933, 49, 201.

of 90 per cent. of the precious metals is obtained at a cost of about 16 shillings per ton of ore. The tailing assays about 2 shillings per ton.

A conical cowling hood has since been fitted to each flotation cell.[1] It has a rectangular base with a top opening 2 feet in diameter. Thereby a deeper froth bed can be made. The breaking of the froth due to the swirl of the liquor is also prevented. This swirl was attributed to a small quantity of pyrite in the feed. Starting from the feed end the hoods are stepped down ½ inch from cell to cell. The machines so equipped are said to have 100 tons per day greater capacity, yield a cleaner tailing, and deal with a pulp whose density would be too great for the earlier type of machine.

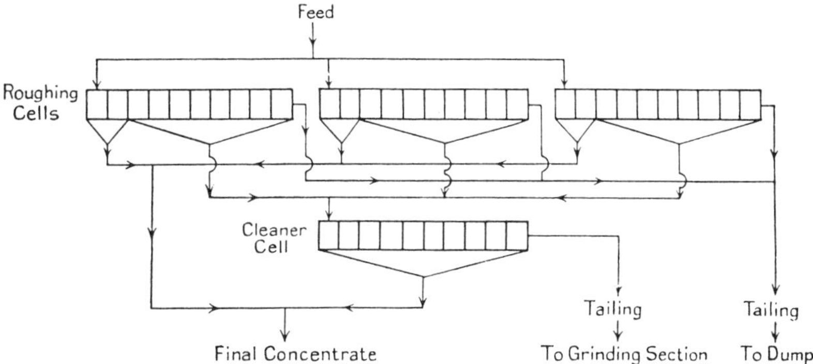

Fig. 143.—Flotation Circuit at Lake View and Star Mine.

Treatment at Wiluna.[2]—The ore is a complex sulphide containing pyrite, arsenopyrite, some telluride and copper, silver and gold. The treatment is practically the same as that at the Lake View and Star and consists of (1) crushing in gyratory and Symons crushers, (2) fine grinding in Hardinge ball and Ruwolt tube mills, (3) flotation in subaeration cells, (4) filtration of flotation concentrate in Oliver filters, (5) drying concentrate in Louden driers, (6) roasting in Edwards furnace, (7) cyanidation and gold recovery.

The consumption of reagents is approximately as follows: Copper sulphate 0·3 to 0·5 lb., potassium ethyl xanthate 0·3 to 0·5 lb., per ton of ore. Half of the reagents is added in the primary Hardinge mill and the remainder in the secondary tube mills. Pine oil (0·6 lb. per ton) is added to the pump suction taking the thickener underflow.

Fresh water is used to break down the concentrates and to remove dirty salt water. The concentrates are then subjected to triple filtration, the first and second operations being for washing and the third for drying. The concentrates are dried until they contain 5 per cent. moisture in Louden driers heated by the flue gases from the roasters. The roasting is done in oil-fired Edwards duplex furnaces. The roasting temperature is 600°-700° C. and must be closely controlled.

Dust from the flues under the driers is mixed with the drier discharge. That from points ahead of the driers is charged into a roaster. Any dust settling in the main flue is sluiced with water and then thickened. The installation of a cyclone separator is expected to increase dust recovery to

[1] *Proc. Aust. Inst. Min. and Met.*, 1933, No. 91, 104.
[2] *Ibid.*, 131.

over 90 per cent., which then can be fed to the roasters. The cyanided residues are weathered in shallow dams for about two years and then re-leached with strong cyanide solution for one day, with weak solution for 2½ to 3½ days, and with water for one day. During the last 24 hours a 10 lb./in.^{2} vacuum is applied to help draw out the solutions. 6 lb. CaO and 1·1 lb. NaCN per ton of residue are consumed. The gold solution is returned to the main mill circuit.

92 per cent. of the gold is recovered by flotation.

McIntyre Porcupine.—Fine grinding and total cyanidation have given place to flotation, regrinding of the concentrate to 325 mesh, and then cyaniding. Briefly, the ore is ground to 65 mesh (55 per cent. of — 200) in five tube mills, each in closed circuit with a flotation cell and a Dorr classifier. The pulp discharge from the tube mills is screened, the undersize going to the cells and the oversize to the classifiers. Xanthate and " Aerofloat " are added to the mills. The sand from the classifier returns to the tube mill. To the pulp overflowing the classifier additional " Aerofloat " is added and the mixture is then treated in the primary flotation unit of eight 6-cell Denver subaeration machines in parallel. A final clean-up of the tailings from these machines is carried out in a secondary system of six similar units. The concentrate from both series of machines is thickened, filtered, reground in cyanide solution and then treated in the cyanide plant, while the tailings from the secondary unit go to waste.

Tippett [1] describes the flotation of telluride ores in cyanide solution at the *Independence Mill* (which is typical of Colorado practice). The ore was crushed to 30 mesh in ball mills carrying solution — 0·25 lb. cyanide and 1·00 lb. protective alkalinity per ton—and then sent to the flotation machines. A mixture of 90 per cent. fuel oil and 10 per cent. pine oil was added in the ball mills at the rate of 0·2 to 0·4 lb. per ton of ore. The fuel oil broke down the natural froth tending to form, and the pine oil was effective in producing a new froth carrying the tellurides. The lime depressed the pyrite. The concentrate amounted to 1 per cent. of the feed and was thickened. It was then reduced by grinding with twice its weight of alkaline cyanide solution in a tube mill carrying ½ to 2 inch balls. The charge was sealed to prevent the admission of air and rotated for about 2 hours. It was then air-agitated and treated with 1 lb. per ton of highly alkaline cyanide solution during 10 days. An extraction of 96 per cent. was thereby obtained. The flotation tailings were cyanided in the usual way.

At the *Golden Cycle Mill, Colorado,*[2] tellurides and sulphides are floated in water, and 70 to 80 per cent. of the gold is recovered. The concentrate is roasted in Edwards furnaces with crude ore and subsequently cyanided.

Practice in Roumania.—The recovery of gold at Dealulkruzii in Roumania was formerly effected by amalgamation and hearth roasting. The introduction of flotation has raised the recovery from 55 to 90 per cent.[3] The flotation concentrates represent only about one-thirtieth of the weight of the raw ore ; hence transport costs are small, the concentrates usually being sold to a local smelting works.

At Brad, 92 per cent. of the values are recovered by flotation and the concentrate is treated with cyanide solutions without previous roasting. Gold bullion and marketable iron pyrites are obtained.

---

[1] *Eng. Min. Journ.*, 1927, **124**, 181.
[2] *Eng. Min. Journ.*, 1932, **133**, 448.
[3] Quittkat, *Metall u. Erz*, 1929, **26**, 400, 509.

The flotation of Roumanian ores is also discussed by Badescu.[1]

Hoge Mill (1933).—The ore consists of quartz embedded in granodiorite. The gold is native and is associated in a very fine form with about 5 per cent. of base metal sulphides, in which pyrite predominates. Recently tellurides have

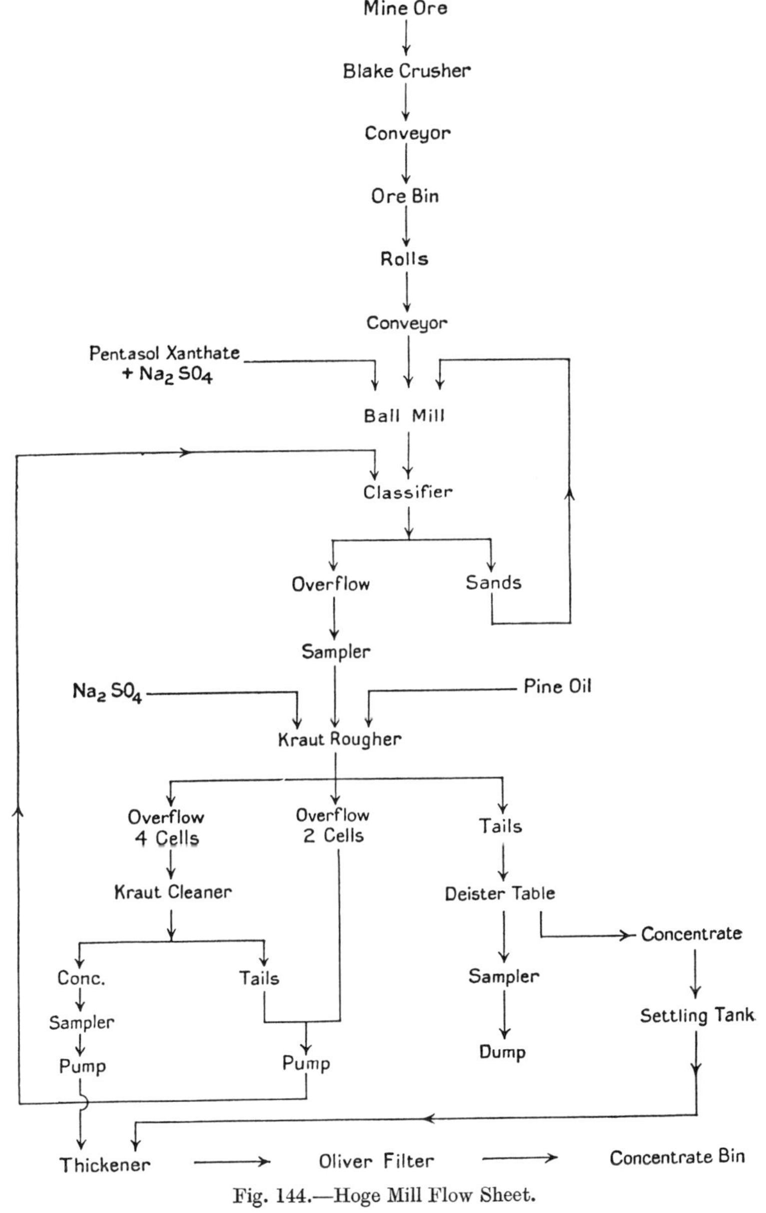

Fig. 144.—Hoge Mill Flow Sheet.

[1] *Analele Minelov Romania*, 1931, 14, 179, 239.

been found. The ore is fed through a 1½ inch grizzly to a 16 × 24 inch Blake crusher, and then *via* a storage bin to 10 × 30 inch rolls. The product is ground in a 6 × 3½ foot Marcy ball mill with ¼ inch grate discharge, working in closed circuit with a 56 inch Dorr duplex classifier. A product is obtained containing 70 per cent. of —200 mesh material. The overflow from the classifier goes to the flotation department which contains a 6-cell Kraut rougher and a 2-cell Kraut cleaner machine. Tailing from the former goes to waste after passing over a Deister concentrating table. The overflow from the first four roughing cells flows by gravity to the cleaning unit, the concentrate from which is pumped to a 16 foot thickener and thence to a 3 × 2 foot Oliver filter. The cake is stored and afterwards shipped to a smelter. The overflow of the last two rougher cells and the tailing of the cleaner are returned to the classifier for re-treatment.[1] "Pentasol xanthate" is the principal promoter and is added in the ball mill at the rate of 0·02 lb. per ton. The "pentasol xanthate" tends to decompose, however, and therefore a second slightly more stable and insoluble agent has been used. The pine oil is added at the head of the rougher circuit. A solution containing 1 per cent. of sodium sulphate is fed to the ball mill and the rougher cells to prevent oxidation of the iron sulphide. The solution is believed to act also as a buffer in maintaining a constant pH value in the liquid. The flow sheet is shown in Fig. 144.

The modern flotation mills on the *Mother Lode, California*, employ a pair of gyratory crushers for primary crushing, stamps, rolls or Symons cone crushers for secondary crushing, and a ball mill operating in closed circuit with a mechanical classifier for fine grinding. It was here that the fact emerged that normal amounts of reagents may result in high tailing losses and that there is a critical point with a minute quantity of reagents where high extraction and high-grade concentrates are obtained.[2]

At the *Brunswick Mill* in Idaho, 60 to 70 per cent. of the gold was formerly recovered by amalgamation and the remainder by flotation. The ore was crushed by stamps to 40 mesh and the pulp passed over amalgamation plates. The plate tailing was further ground to 80 mesh in a 7 × 5 foot ball mill in closed circuit with a Dorr classifier. The overflow from the latter was pumped to a 6-cell flotation machine and the rougher concentrate was re-treated in two cleaning operations. The tailing from the cleaner cells rejoined the flotation feed. Finished concentrate passed to a cone-thickener and thence to an Oliver filter. A starch solution was added in part ahead of the rougher cells and the remainder to the cleaner cells. Other reagents included xanthate 0·1 lb., cresylic acid 0·15 lb., and copper sulphate 0·1 lb. per ton. Quite recently, amalgamation has been eliminated and the coarse gold recovered on a table of the Wilfley type. The table tailing goes to the flotation circuit.[3]

Ore Treatment at the Premier Mine, British Columbia.[4]—The approximate composition of the ore is as follows :—

Au, .	0·02 to 1·33 ozs. per ton.	Cu, .	0·005 to 0·28 per cent.
Ag, .	5·15 to 42·9 ,, ,,	Fe, .	3·6 to 15·2 ,,
Pb, .	0·20 to 2·45 per cent.	S, .	1·65 to 18·05 ,,
Zn, .	1·2 per cent.	Insol., .	57·6 to 91·1 ,,

[1] *Eng. and Min. Journ.*, 1933, 134, 422.
[2] *Min. Ind.*, 1930, 39, 683.
[3] *Min. and Met.*, 1932, 13, 175.
[4] *Min. Mag.*, 1933, 48, 178.

The gangue is highly siliceous and contains pyrite.

The present method of operation is straight flotation. The mine ore is divided on a 12 inch grizzly into high-grade ore and mill ore. The former is crushed to 3 inch size in a Blake crusher and then sent direct to smelters. The mill ore is also crushed to 3 inch size, freed from iron by electromagnets, and then passed to a screen with 1 inch openings. The oversize is again crushed to 1 inch to unite with the undersize, two-thirds of which then pass to a Hardinge ball mill and the remainder to a Marcy mill, each running in closed circuit with a Dorr classifier. The overflow from the classifier in each instance passes to a 4 foot 6 inch cone with 75° slope, equipped with a Premier flotation cell and sand discharge box working in closed circuit with a 5 × 8 foot tube mill as a regrinding machine. The Premier cell takes out a coarse concentrate as early as possible, which is sent to a 12 foot K. and K. (Kraut and Kollberg) flotation machine,[1] where it is cleaned and sent to the thickener for collecting concentrate. The overflow from the cones proceeds to another set of three Premier cells in series. The tailing from the final cell is sent to an 8 foot cone which discharges into a 4 foot 6 inch Dorr classifier in closed circuit with a 6 foot × 16 inch Hardinge mill. The concentrate from all the Premier cells is pumped to the K. and K. machine.

The overflow of the 8 foot cones is pumped to a surge box and then divided between two 12-cell 24 inch subaeration M.S. machines in series and two double spitz K. and K. machines in series. The concentrate from the first four cells of No. 1 M.S. machine is pumped to the concentrate collecting thickener. The froth from the other eight cells joins the froth from No. 2 machine and is returned as middlings to the 4 foot 6 inch cones above the regrinding mills ; there it is re-classified and re-ground. The concentrate from the first K. and K. machine is pumped to the 12 foot K. and K. cleaner cell. The tailing from the first K. and K. machine is fed to the second similar one, whose concentrate is returned with the M.S. middlings to the 4 foot 6 inch cone. The tailing from the second K. and K. machine joins the tailing from the second M.S. machine and is sent to waste.

The concentrate collecting tank is a 30 foot Dorr thickener, whence the discharge is pumped to an 18 foot thickener. The thickened material then passes to two Oliver filters. The cake drops into a concentrate bin.

Cyanamide and soda-ash are mixed dry and fed to the ore at the gyratory crusher. " Aerofloat " is fed to all grinding units. A pH value of 7·2 to 7·8 is maintained.

Flotation of Rand Ores.—An attempt in 1924 to apply flotation prior to cyaniding at the Modder Deep Mine was not successful.[2] A. King [3] has conducted experiments on Simmer and Jack ores and has demonstrated that flotation cannot give a tailing low enough in values to discard without further treatment. A large amount of the gold remaining in the siliceous tailing is, however, easily soluble in cyanide, and this treatment, it is estimated, would be profitable. Potassium ethyl xanthate and pine oil were used as reagents, and for ore assaying 2·9 dwt. per ton the pH value had to be maintained above 7·2 in order to get both a good froth and a good recovery. Regrinding of the pyritic concentrate to 325 mesh and continuous air agitation to reduce the ill-effects of iron from the balls of the ball mill are indicated as necessary.

---

[1] *Trans. Can. Inst. Mng. Met.*, 1933, 36, 28.

[2] H. Smith, *Eng. Min. Journ.*, 1926, 122, 175, 215.

[3] *J. Chem. Met. Mng. Soc. S.A.*, 1934, 35, 136.

When treating current sand residues it was found that the addition of copper sulphate, ferrous sulphate or sulphur, added as activators in a slightly acid pulp (pH = 6 to 6·6), overcame the inhibiting effect of lime and cyanide on the pyrite.

White [1] issues a warning that flotation reagents may interfere with the precipitation in zinc boxes, and asserts that "to make certain that flotation could be economically applied to our sand and slime residues (*i.e.* on the Rand), even when finer grinding has been pursued close to the economic limit, a market for the pyrite . . . is necessary."

Wartenweiler [2] states that at the Government Areas Mine a 1,000 ton plant is being erected for treatment of sand, the concentrates from which will be cyanided separately. The results of tests were as follows :—

(1) Original sand value, 5·00 dwt. per ton.

(2) Flotation concentrate, 4 per cent. by weight, containing 80 per cent. of the gold and 95 per cent. of the pyrite, leaving a sand tailing of 1·04 dwt. gold per ton and 0·15 per cent. pyrite.

(3) Regrinding the concentrate to − 325 mesh and cyaniding for 24 hours gave a gold extraction of 98 per cent., with a 2 dwt. per ton residue.

(4) Cyaniding of the sand tailing gave a residue of 0·20 dwt. per ton.

(5) The total residue thus becomes 0·27 dwt., which compares with a present residue of 0·47 dwt. per ton.

Wartenweiler also quotes test results from the Central Rand and Bird Reef as follows :—

TABLE XXXV.

Central Rand.—Value 19·94 dwt. per ton. Ground to + 100 = 13·1 per cent. ; − 100, + 200 = 36·3 per cent. ; − 200 = 50·6 per cent. Flotation agents—xanthate and pine oil.

	Percentage Weight of Product.	Gold Content, dwts./per ton.	Percentage of Original Gold Content.	Dwts./per ton, Original.	FeS_2, per cent.
Concentrate,	5·56	252·4	70·35	14·03	57·47
Middlings, .	3·44	61·8	10·65	2·13	16·40
Tailings, .	91·00	4·16	19·00	3·78	0·45

Bird Reef.—Value 7·26 dwt. per ton.

	Percentage Weight of Product.	Gold Content, dwts./per ton.	Percentage of Original Gold Content.	Dwts./per ton, Original.	FeS_2, per cent.
Concentrate,	6·4	80·1	70·6	5·12	43·2
Middlings, .	3·2	31·4	13·8	1·01	17·0
Tailings, .	90·4	1·25	15·6	1·13	0·23

At *Beattie, Quebec*, the ore is very hard and the gold is mostly associated with pyrite in a finely divided condition. Arsenopyrite and other sulphides, together with magnetite, ilmenite and chalcopyrite, are present in much

[1] *J. Chem. Met. Mng. Soc. S.A.*, 1934, 35, 144, 148.
[2] *Ibid.*, p. 148.

smaller quantities. In order to liberate the fine gold the following steps are taken [1] :—

(1) Fine crushing to ¼ inch. (2) Relatively coarse primary grinding. Flotation of low-grade concentrates in primary and rougher circuits. Discard of the bulk of the material as a coarse tailing. (3) Regrinding and refloating of the low-grade rougher concentrate. (4) Combining the concentrate obtained from the floating of the reground rougher concentrate with primary concentrate and regrinding both products so that 95 per cent. is of — 325 mesh, then refloating this material to get a final concentrate.

The ore is crushed to — 10 inches in a jaw-breaker underground and then passed over a roll grizzly feeder to a second jaw-crusher set at 4 inches. The discharge from the latter, together with the grizzly undersize, is screened through a ¼ inch screen. The undersize goes to the mill ore bins, the oversize being returned to the crushing plant to be crushed and screened again, whence the oversize goes to a 4/1 standard cone crusher set at ¾ inch and the undersize to a 4 foot Symons short-head crusher set at ¼ inch. Both cone crushers are in circuit with screens. Only 5 or 6 per cent. of the material is — 100 mesh.

The crushed ore is sent to Hardinge ball mills (see Fig. 145) in closed circuit with Dorr classifiers. The overflow of the latter contains 65 per cent. of — 200 mesh pulp and gravitates to two 6-cell subaeration flotation machines in parallel. The froth from cells (1) and (2) is thickened, reground in a tube mill in closed circuit with a bowl classifier, whose overflow contains 95 per cent. of — 325 mesh material. The reground primary concentrate is floated to produce a final concentrate and a cleaner tailing. The latter is then retreated in a 2-cell subaeration machine, the concentrate from which is reground in the tube mill and the tailing sent to the primary ball mills.

The primary tailing from cell 6 in the 6-cell machines is re-treated in two parallel 10-cell subaeration machines, producing a final tailing and a low-grade concentrate. This latter is thickened and reground in a ball mill which is in closed circuit with a Dorr classifier. The overflow from the classifier contains 85 per cent. of — 200 mesh material. It gravitates to a 4-cell flotation machine. The concentrate from this goes to the primary ball mills and the tailing is sent to an air cell. This makes a low-grade concentrate, which joins the 10-cell concentrate, and a low tailing which goes to waste.

The final concentrate contains 20 per cent. solids, and is thickened and filtered in a Dorrco filter down to a moisture content of 12 per cent. The cake is then passed through a rotary drier to reduce the moisture to 5 per cent. before going to the smelter.

The primary circuit has an alkalinity of 0·004 per cent. due to free sodium carbonate; above this alkalinity classification is retarded and its efficiency falls. The primary concentrate regrinding circuit is, however, kept neutral.[2]

The following flotation reagents are used :—

To Grinding Circuit, lb. per ton of ore :

Barrett No. 4 flotation oil, ...	0·05
Soda-ash, ...	1·00
Copper sulphate, ...	0·10

[1] *Trans. Amer. Inst. Min. Met. Eng.*, 1934, 112, 691.
[2] See *Can. Min. Journ.*, 1934, 55, 351.

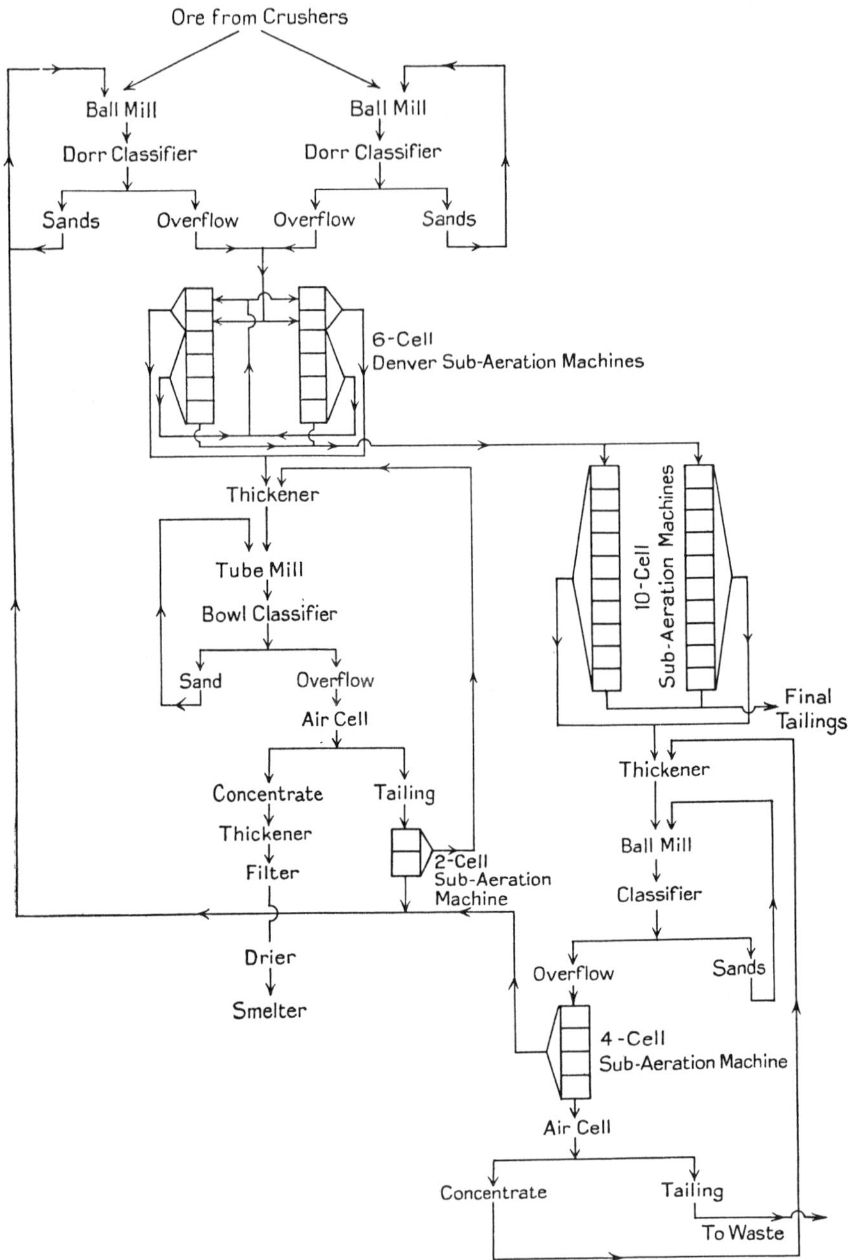

Fig. 145.—Flow Sheet of Concentrator Plant. Beattie Gold Mines, 1934.

To Primary Flotation, lb. per ton of ore :

Sodium ethyl xanthate, ...	0·07
Pine oil, ...	0·05

To Rougher Flotation, lb. per ton of ore :

Rhodamine B soln., ...	0·02
Pine oil, ...	0·02

A new 200-ton plant for the treatment of concentrates is planned as follows (see Fig. 146). It is intended to save the cost of transporting the concentrate.

Instead of a $2\frac{1}{2}$ oz. per ton concentrate a 1·2 oz. concentrate is to be produced in the primary cells of the present flotation mill and sent to the cyanide plant. The concentrate from the rougher circuit is reground in a tube mill and then returned to the primary ball mills. The 1·2 oz. concentrate is reground in a tube mill in closed circuit with a bowl classifier having an overflow all passing a 325 mesh screen. This is then agitated in a Wallace machine, and afterwards thickened, the thickener overflow going to waste. The underflow is filtered and washed on a rotary filter, the cake from which is repulped with cyanide solution, and agitated in a Dorr machine having a Denver agitator. The agitator discharge passes to another thickener, the overflow going to clarification prior to precipitation and the underflow being triply filtered in Oliver filters. In each case the cake is repulped, and then washed with barren solution in cases Nos. 1 and 2 and with water in case No. 3. The cyanide residue is returned to the concentrator for flotation treatment, giving a low-grade concentrate with 0·70 oz. gold per ton, 38 per cent. iron and 40 per cent. sulphur. A method is being devised to treat this concentrate locally without recourse to smelter treatment at a distance.

Mountain Copper Company, Big Canyon Mine.[1]—The ore is a hard silicified amphibolite schist containing 3 dwt. of gold per ton. 20 per cent. of the gold is free, the rest being associated with pyrite and pyrrhotite. The ratio of concentration is roughly 14 : 1.

The ore passes to a primary jaw-crusher *via* a grizzly. The undersize of the latter and the product of the jaw-breaker go to the secondary Symons cone crusher, which reduces the size to $\frac{3}{8}$ to $\frac{5}{8}$ inch. The material then passes to ball mills working in closed circuit with Dorr classifiers, and trials are being made with the introduction of unit flotation and of a hydraulic trap between the ball mill and the classifier in order to catch coarse gold as soon as possible. The 60 mesh overflow from the classifier is pumped to a storage tank, whence it falls by gravity to a conditioning tank employing a Dorr agitator working at 1 r.p.m. From here the pulp passes to a five-cell 56 inch Fagergren unit. Concentrates from any cell may go direct to the pump sump, and the middling to a 36 inch cleaner machine. The tailing from the latter goes to the two-compartment sump into which the classifier overflow goes. The cleaner concentrates join the rougher concentrates or are retreated in the head cell. The rougher tailings are led over a concentrating table or corduroy tables. Concentrates gravitate into a Dorr thickener and then to an Oliver filter. The cake from the latter is sent to the smelter.

---

[1] Huttl, *Eng. Min. Journ.*, 1935, **136**, 216. Averill, *Chem. Eng. Min. Rev.*, 1931, **24**, 87.

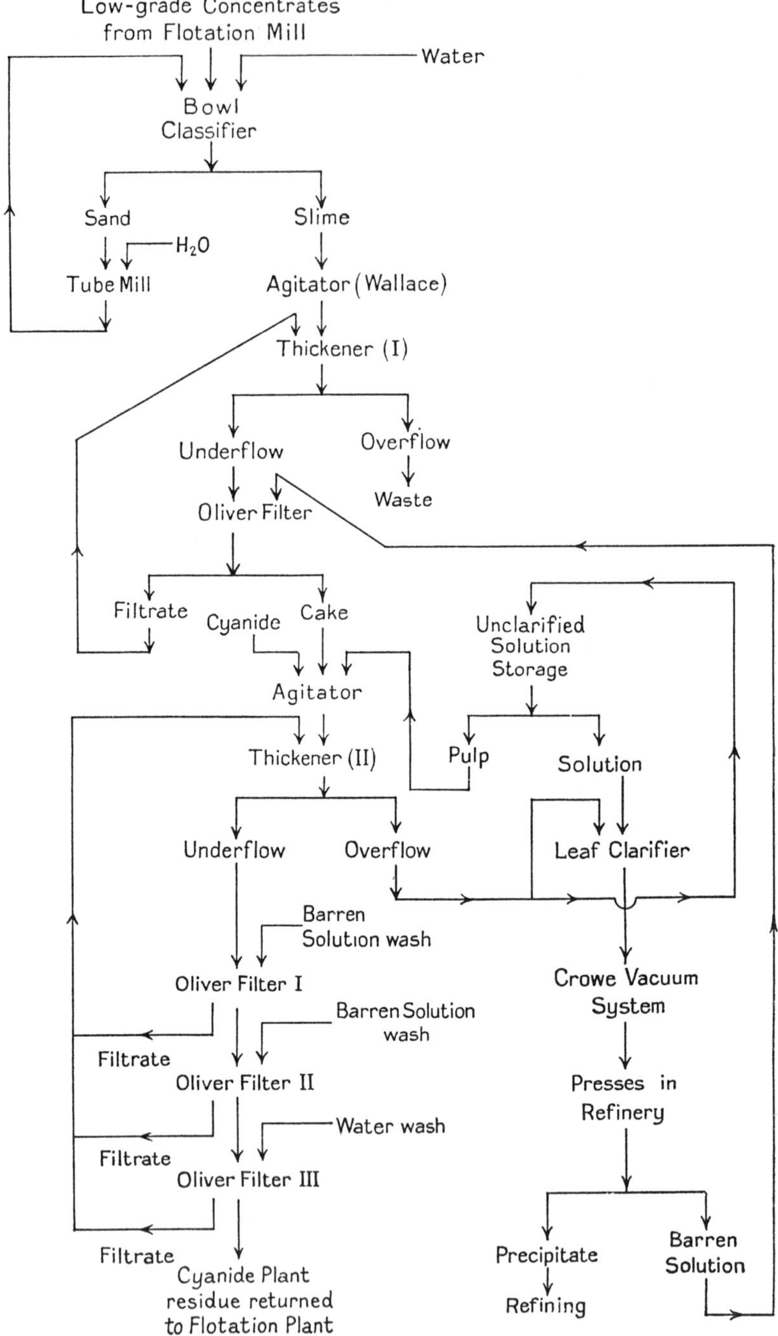

Fig. 146.—Flow Sheet of Cyanide Plant, Beattie Gold Mines.

The Flotation of Cyanide Tailings.—The cheapness and efficiency of flotation have made it attractive for the treatment of dumps of tailings containing small values hitherto irrecoverable by the older processes. This has become more and more important during the period of high prices of gold in terms of sterling. The residual cyanide and lime left in cyanide tailings have, however, an inhibiting effect on flotation of the pyrite and hence of the gold. The difficulty may be overcome either by washing the tailings, or by the addition of some "cyanicide" such as copper sulphate, or an acid.

Special Problems of Flotation.—Among studies of special problems of flotation, the work on carbonaceous gold ores and on gold-copper-iron ores may be mentioned.

Leaver and Woolf [1] observed that certain *carbonaceous gold ores* are not amenable to cyaniding owing to the reprecipitation of the gold. This results in poor recovery and high cyanide consumption. They tried as flotation reagents sodium amyl xanthate, 0·25 lb., cresylic acid, 0·05 to 0·125 lb., and pine oil, 0·05 to 0·10 lb., per ton. The results obtained were as follows :—

TABLE XXXVI.

Slime.	Product.	Gold Assay, per ton.	Gold Recovery, per cent.
1.	Feed,	0·0665	...
	Concentrate,	1·62	69·0
	Cleaner Tailing,	0·40	18·2
	Rougher Tailing,	0·009	...
2.	Feed,	0·099	...
	Concentrate,	2·05	85·3
	Cleaner Tailing,	0·28	7·2
	Rougher Tailing,	0·008	...
3.	Feed,	0·1375	...
	Concentrate,	1·345	82·5
	Cleaner Tailing,	0·055	4·0
	Rougher Tailing,	0·023	...
4.	Feed,	0·0433	...
	Concentrate,	0·555	72·7
	Cleaner Tailing,	0·040	11·1
	Rougher Tailing,	0·008	...

Gaudin [2] discusses the treatment of *gold-copper-iron ores* as follows :—

(1) If the gold is entirely associated with the copper sulphides, or is free and finely divided, selective flotation of the copper mineral at low *p*H values collects the gold in the copper concentrate.

(2) If the gold is associated with sulphides, partly with copper minerals and partly with worthless iron sulphides, several alternative flotation schemes can be followed, *e.g.*

[1] *U.S. Bur. Mines Rep. Invest. Serial* 2998 (1930).
[2] "*Flotation*," 1932, p. 326.

(*a*) The production of a low-grade combined concentrate, especially if the iron content of the ore is not too high.

(*b*) The production of a copper concentrate and also a copper-free concentrate of gold and sulphides, with subsequent cyanidation of the latter.

(3) If part of the gold is associated with copper minerals and the rest is free and coarse, amalgamation should be used before flotation.

(4) If part of the gold is associated with copper minerals and the rest is finely dispersed in the non-sulphide gangue, flotation followed by cyanidation of the flotation tailing is suitable.

In treating copper-gold ores total recovery is more important than the grade of the concentrate. Consequently frothing and collecting agents may be used more freely, and iron-inhibiting agents more sparingly, than in the treatment of purely copper ores for base metals. In these latter the precious metals make only a small contribution to the total value of the ore, and are of secondary consideration.

The Value of Microscopic Study in Flotation.—The use of the microscope and a general mineralogical study of the crude ore, and also of the products at every stage, are invaluable in preparing a logical flow sheet. Questions as to whether ores having different characteristics should be treated individually, or mixed to form a uniform product, are thus answered. Such studies will often show whether a middling product should or should not be re-crushed before treatment, where this product should be fed back into the mill circuit, when various middling products may be combined for treatment, and whether concentrates may be enriched and tailings further impoverished.[1]

An example of the value of microscopic study is afforded by the work of Oldright and Head[2] on the various products from a flotation mill treating a complex lead-zinc ore containing gold. Their conclusions are as follows:—

(1) Gold in place is usually bright and clean. Little deformation of the particles takes place in the earliest stage of grinding.

(2) The larger particles are flattened and deformed into irregular shapes as grinding continues. Little shearing or puncturing is in evidence.

(3) Continued grinding and attrition due to pounding on hard material cause shearing of the flattened shapes into smaller pieces.

(4) Constant grinding causes non-metallic material to be embedded in the roughened surfaces of the gold.

(5) Some of the gold becomes entirely coated with slime, and acquires the characteristics of the coating in not becoming attached to bubbles. Hence it does not float. Apparently the gold selects a non-metallic slime, which continues to build up.

(6) Very fine particles of sulphide have been found in the gold in the lead concentrate. The gold may, therefore, have floated because of the sulphide on its surface.

(7) Grinding in closed circuit causes the heavy minerals to be ground preferentially. Particles of high density, such as metallic gold, cannot overflow a classifier unless they are associated with a lighter mineral, or are exceedingly fine, or both.

---

[1] See Quittkat, *Metall und Erz*, 1929, **26**, 509.
[2] *Eng. Min. Journ.*, 1933, **134**, 228.

(8) Means should be adopted for recovering the gold from the classifier circuit while it is relatively coarse. The residue could be treated by gravity concentration or flotation.

The influence of *finer grinding* on losses in flotation has not hitherto been properly emphasised,[1] but may be traced by microscopical analysis. In the case of gold ores where the metal is associated with pyrite, the percentage of gold extracted appears to increase steadily with increase of fineness of grinding. For example, at the Teck Hughes plant any difference in grinding that diminishes the amount of — 400 material increases the loss of gold in the tailing. Overgrinding has, however, been proved to be detrimental in some instances.

General Discussion of Flotation.—The choice between amalgamation, cyaniding or flotation, or some combination of these, will be determined for any particular ore principally by the character of the ore, the distance from a smelter or refinery where concentrates can be treated, and the total cost of treatment. Generally speaking, amalgamation is most suitable for the recovery of coarse, free gold, cyaniding for finely divided gold, which can be freed by fine crushing, and flotation for gold contained in sulphide, selenide or telluride ores which are somewhat difficult to treat by ordinary methods. Exceptions occur in which cost becomes the determining factor. Amalgamation alone rarely suffices. If it is used, it is commonly followed by cyaniding or flotation. Where amalgamation and flotation are carried on together, they are used in that order. Otherwise the reagents used for flotation may coat the gold particles and prevent amalgamation.

There has been a rapid increase in the application of flotation to gold ores since the change from acid to alkaline circuits and from oils to synthetic organic chemical reagents. Flotation supplements rather than supplants the older processes, and there is considerable advantage in withdrawing as much gold as possible from the pulp at the earliest possible stage of treatment. Finer grinding, using cone crushers or rolls for a greater proportion of the reduction than formerly, is becoming more common, and primary grinders are less used.

The processes which embrace flotation and are available for gold ores include the following:—

Simple Flotation may be used for ores in which gold is associated with base metals that interfere with the usual methods of extraction, and also for small concerns that merit no large capital outlay. The concentrate is sold to a smelter.

Amalgamation and Flotation are employed where losses occur in amalgamation. These can usually be considerably reduced by flotation, sometimes without the cost of additional grinding.

Flotation and Cyanidation of the Concentrate are mainly used for ores containing gold in the free state and in pyrite, entailing low milling costs, less capital expenditure and the specialised treatment of a high-grade concentrate.

Flotation and Cyaniding of the Tailings are specially employed for ores containing "cyanicides," which are removed in the concentrate and can be dealt with separately.

Cyaniding and Flotation are applicable to ores containing finely

[1] *Eng. Min. Journ.*, 1933, 134, 228.

disseminated gold, especially in the gangue, and where the mineralised portion is not very amenable to flotation.

Gold ores which are suitable for concentration by flotation divide themselves naturally into three classes :—

(*a*) Those in which the gold is present as a secondary constituent in base metal complexes, such as those containing copper or lead-zinc in various combinations. The gold is often in a fine state of division, down to colloidal dimensions, and therefore invisible to the eye. Usually the gold is associated with pyrite and chalcopyrite The amalgamation method applied to these ores entails losses, and cyaniding is not without difficulty, especially if copper, arsenic or antimony is present. Concentration by gravity does not always ensure a high recovery. The gold is recovered with the base metal complexes by standard methods of flotation, modified in order to ensure the economic separation of the gold. Rabone [1] points out that if, for instance, the recovery of the gold involves lowering the grade of the copper with pyrite, the fall in value of the concentrate due to the pyrite must be counterbalanced by a rise in value due to the gold content if profit is to be made.

(*b*) Those in which the gold is the primary mineral and base metals are not present in sufficient quantity to be valuable. The gold in these instances is associated with such minerals as pyrite or other sulphides of iron, arsenopyrite, stibnite, or cupreous sulphides. Two methods of attack are available, viz. :—(1) Cyaniding, which may necessitate grinding to slime in order to release the gold ; or (2) flotation, with the recovery of a concentrate containing most of the sulphide minerals, which is re-ground and then treated with cyanide, and a tailing which is cyanided separately.

Where the gold occurs finely dispersed within the pyrite, cyaniding produces good results only after extremely fine grinding, and if the gold cannot thus be freed from the pyrite, a roasting operation was once considered essential. This is now avoided in many cases by flotation of the sulphides, at much lower cost. In cases where roasting is necessary, *e.g.* with ores containing tellurides, stibnite or copper sulphides, flotation usually reduces the amount of material to be so treated. Where impurities harmful in cyaniding—cyanicides—are present, flotation may be used for their removal. Direct treatment of gold telluride ores for instance is difficult ; cyanide consumption is excessive and recovery low. The readily floatable tellurides, however, can be concentrated into a small rich bulk, and the remaining pulp treated by standard cyaniding methods. The small amount of concentrate is roasted in the usual way.

(*c*) Free-milling ores, or those containing very little of the sulphide minerals, with the gold occurring free in a simple gangue, are usually treated by cyaniding, after amalgamation or blanket concentration. The application of flotation to these ores has not been proved as in the case of those previously mentioned. More attention is now being paid to the extraction of free gold by gravity methods before attempting to make a flotation concentrate and so avoid the possibility of losing gold in the flotation tailing. Any dilution of the pulp thereby occasioned may be remedied by thickening before flotation.

Kraut [2] has, however, stated that gold particles passing 30 mesh but retained on 40 to 60 mesh screens can be recovered by flotation. Flotation may

---

[1] "*Flotation Plant Practice*," 1936, p. 135. *Min. Mag.*, 1934, **51**, 137.
[2] *Min. and Met.*, 1935, **13**, 195.

accordingly be more profitable than cyaniding for such gold in small plants on account of its lower cost. In this case it is necessary to crush the ore only to a size sufficient to ensure the separation of the gold from the gangue, thus avoiding classification into sand and slime or grinding to an extremely fine state of division. The only additional cost is the thickening of the pulp after amalgamation and before flotation, though this may be avoided if the operations are reversed, a scheme which is not favoured.

In larger plants dealing with free-milling ores cyaniding and flotation are carried on together, the latter forming an additional stage between grinding or amalgamation and the cyanide extraction plants. It is an ideal process for conditioning a pulp for cyanidation, since it recovers the gold which is difficult to extract by cyaniding without very fine grinding and prolonged agitation in cyanide solution. In this way recovery is improved and costs usually diminished owing to the coarser grinding that can be employed. Gold too coarse for quick cyanidation must be recovered in the earlier stages by amalgamation, on blanket strakes, or by flotation. At the McIntyre Porcupine plant [1] a single flotation cell has been introduced into the grinding circuit in order to recover coarse gold at the earliest opportunity. The tube mill discharge, to which flotation reagents have already been added, passes a 4 mesh screen. The oversize goes to the classifier, and the undersize, diluted to a water-sand ratio of 1 : 1, is treated in a Denver subaeration cell which catches about 60 per cent. of the gold and silver in the concentrate. Most of the coarse gold is collected in a small hydraulic cone placed immediately below the impeller (see p. 253, Fig. 136). It is assumed that the swirling action of the impeller has a classifying effect, preventing the settlement of all but the heaviest particles. Each day the tube mill discharge is temporarily diverted to the classifier, while water is run through the flotation cell to remove gangue. The cone residue is removed through a plug valve, and the recovery here has increased the total recovery of the grinding circuit to approximately 75 per cent.

It has been suggested more recently [2] that in place of the unit flotation cell intended to catch coarse gold at the earliest opportunity after the crushing to coarse mesh, a special type of jig should be interposed between the ball mill and classifier.

Krebs [3] states that in treating highly oxidised ores, or even ores showing little or no oxidation but containing a considerable amount of colloids, the adsorption of gangue slime appears to be the chief cause of difficulty in the flotation of free gold.

Theories of Flotation.—While flotation has progressed mainly on empirical lines, theoretical considerations as to the reasons for the observed phenomena are becoming better known and further advance has thus been made more probable. No theory has yet been advanced which is universally acceptable or applicable to the known facts. The following theories, however, have been brought forward during the last thirty years :—

(1) *The Electrical or Cataphoresis Theory.*—According to this, the mineral particles carry an electrical charge of an opposite sign to that on the air bubbles, to which they accordingly become attached. The gangue particles bear a charge similar to that on the bubbles and consequently are repelled.

---

[1] *Eng. and Min. Journ.*, 1933, 134, 477.
[2] E. A. Knapp, Private communication.
[3] *Min. and Met.*, 1935, 16, 225. See also Pallanch, *Min. and Met.*, 1935, 16, 177.

(2) *The Gas Theory.*—This postulates an adsorbed gas layer on the mineral particle, to which the air bubble readily adheres.

(3) *The Chemical Theory.*—This assumes that flotation reagents react chemically on the surface molecules of the minerals.

(4) *The Adsorption Theory.*—According to this the collector molecules are adsorbed on the surfaces of the minerals which the collector can cause to be floated. The forces involved are too weak to act on gangue particles. Grouping of the atoms of the collector molecules is assumed.

(5) *The Contact Angle Theory.*—This postulates that flotation is related to the hysteresis of the contact angle.

CHAPTER XI.

DRY CRUSHING.

Dry Crushing is employed only with ores which are to be roasted before treatment with cyanide or by amalgamation. It is now carried on in but few plants, the most notable instances being in Western Australia. Its most serious disadvantage is the amount of dust produced. The introduction of flotation is tending to render dry crushing rarer still.

Drying.—If an ore is to be finely crushed by means of rolls or other dry crushing plant it should be dried previously. The screens soon become clogged by damp ore, especially if it is argillaceous. The moisture is reduced to 1 or 2 per cent., and the removal of water of hydration by heating to about 250° C. makes it much easier to screen hydrated ores.

The old method of drying was to spread the ore, after the large lumps had been removed by a grizzly and crushed to 1½ inch size, on large flat areas heated from below by flues from the roasting furnace. The floor was usually covered with iron plates. After being dried, the ore was shovelled up and passed to the crushing mill. This has been superseded by the adoption of inclined, continuous-discharge, revolving iron cylinders. The ore is passed through these, and is dried by the products of combustion of a fire, which are also passed through them, but in the reverse direction. One such cylinder, 3 feet in diameter and 18 feet long, will dry from 30 to 40 tons of ore per day, at a small cost for fuel and power.

Rolls.—Rolls are specially adapted for intermediate crushing, taking a product with a maximum diameter of about 1 inch, and reducing it to a size passing through a screen of 12 or 16 mesh to the square inch. Like other dry crushing machines, they are better suited for soft friable material than for either hard or caking (clayey) ores.

Rolls are cylinders, between two of which pieces of ore are drawn and crushed by compression. The main driving power is usually applied by a belt to the shaft of one roll, the other roll being driven only with sufficient force to ensure that the rollers will always take hold of the ore, and also to keep them in motion when no ore is passing between them. Geared rolls, however, are sometimes used for coarse crushing. The rolls have crushing tyres made of alloy steel or of chilled iron. These can be replaced when they are worn out. In some rolls the crushing strain is taken up by springs, which press the rolls towards one another; when particularly hard fragments are passing through the rolls, the latter are forced apart against the action of the springs. The hopper is designed to spread the ore evenly across the crushing face, and the rolls, screens, elevators, etc., are all securely boxed in with a wooden housing in order to prevent loss by floating dust. Rolls for moderately fine crushing are usually from 12 to 16 inches across the face, and from 22 to 36 inches in diameter, but for coarse crushing larger rolls, 36 to 54 inches in diameter, with faces 15 to 28 inches wide, are used.

Richards points out [1] that ore may be crushed by rolls in two ways, according to the rate of feed, speed of rolls, etc. If the speed is high and the feed light, each particle of ore is crushed separately between the rolls. In this case, which he calls "free" crushing, there will be a maximum of coarse and a minimum of fine material produced. In "choke" crushing, the ore is fed in a thick stream between the rolls, so that the particles crush each other and a maximum of fine material is produced. Part of the power will be used in compressing the loosely-packed stream. According to Richards free crushing is the more advantageous course, provided that a very fine product is not required.

Roll crushing of gold ores is being revived to some extent after a period of disuse. With the trend towards finer feeds to ball mills, $\frac{1}{4}$ inch or less, rolls form the last step in crushing, usually following the Symons cone crusher. The rolls are in closed circuit with vibrating screens. Where, however, large tonnage is to be handled, or where a still finer feed to the ball mills is desired, large rod mills are usually preferred to rolls.[2]

The rolls installed recently at the McIntyre Porcupine mill are 78 inch diameter by 18 inch face. They are choke fed, and set at $\frac{1}{8}$ inch, reducing the material to $\frac{3}{16}$ inch. Each roll is driven by a 150 h.p. motor at 125 r.p.m., giving a peripheral speed of 2,550 feet per minute with new shells. Spring pressure on the roll face is 40,000 lbs. per linear inch. The chrome shells on the rolls are $8\frac{11}{16}$ inches thick. They last 90 to 100 days and wear to $2\frac{1}{2}$ inches thickness before they are renewed. As they wear in the centre, emery stones are used to grind down the edges.[3] A dust collecting system collects 5 tons of dust containing 82 per cent. of $-$ 325 mesh material every 16 hours. The dust assays roughly 8s. per ton.

The rolls in the Hoge Mill reduce the 2-inch crusher product to $\frac{1}{2}$ inch size for the ball mill feed. While stamps were found ideal for ores carrying their principal values as free gold fit for amalgamation, it was felt that where only a minor part of the values was in such a form, rolls would be better. The initial cost of rolls is lower than that of stamps.

After a critical study of existing data, Finkey derives the following formula for peripheral speed of rolls—

$$V = \frac{1{\cdot}27\sqrt{D}}{\sqrt[4]{\left(\frac{D+d}{D+ud}\right)^2 - 1}}$$

where V = peripheral speed in metres per second, D = diameter of rolls in metres, d = diameter of particle to be crushed in millimetres, u = reduction ratio, *i.e.* ratio of space between rolls to diameter of feed.

Ball Mills for Dry Crushing.—The *Krupp* or *Krupp Grüsonwerk* Ball Mill [4] consists of a cylindrical drum rotated by belting, but geared down by toothed wheels (see Fig. 147). The interior of the drum consists of from 5 to 10 overlapping grinding steel plates, *a*, *b*, so arranged that the balls drop a few inches in passing from each plate to the next. The half, *a*, of the grinding plates is perforated with $\frac{1}{2}$-inch tapered holes, and the other or main grinding

[1] "*Ore Dressing*," p. 98.
[2] *Min. Ind.*, 1931, 40, 615.
[3] *Trans. Amer. Inst. Min. Met. Eng.*, 1934, 112, 632.
[4] See "Ball Mill Practice at Kalgoorlie," von Bernewitz, *Mng. Mag.*, 1911, 5, 139.

half, b, is bent inwards towards the axis to raise the balls above the next succeeding plate. The crushed ore passes through the perforations in a, and falls upon a coarse sieve, J, constructed of finely perforated steel plates,

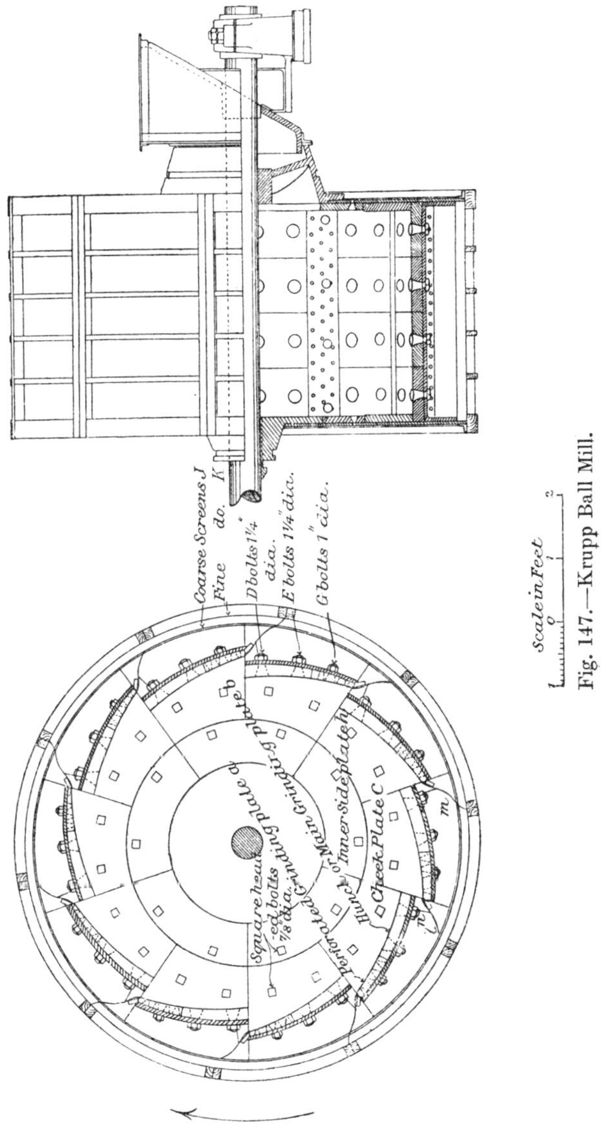

Fig. 147.—Krupp Ball Mill.

which protect the outer fine sieves, K, of phosphor bronze or steel wire gauze. The coarse particles retained on both inner and outer sieves, which enclose the crushing drum and rotate with it, are returned by elevator plates, l, m, to the interior of the drum through perforated plates, n, guarding apertures between the grinding plates. The material, which need not be

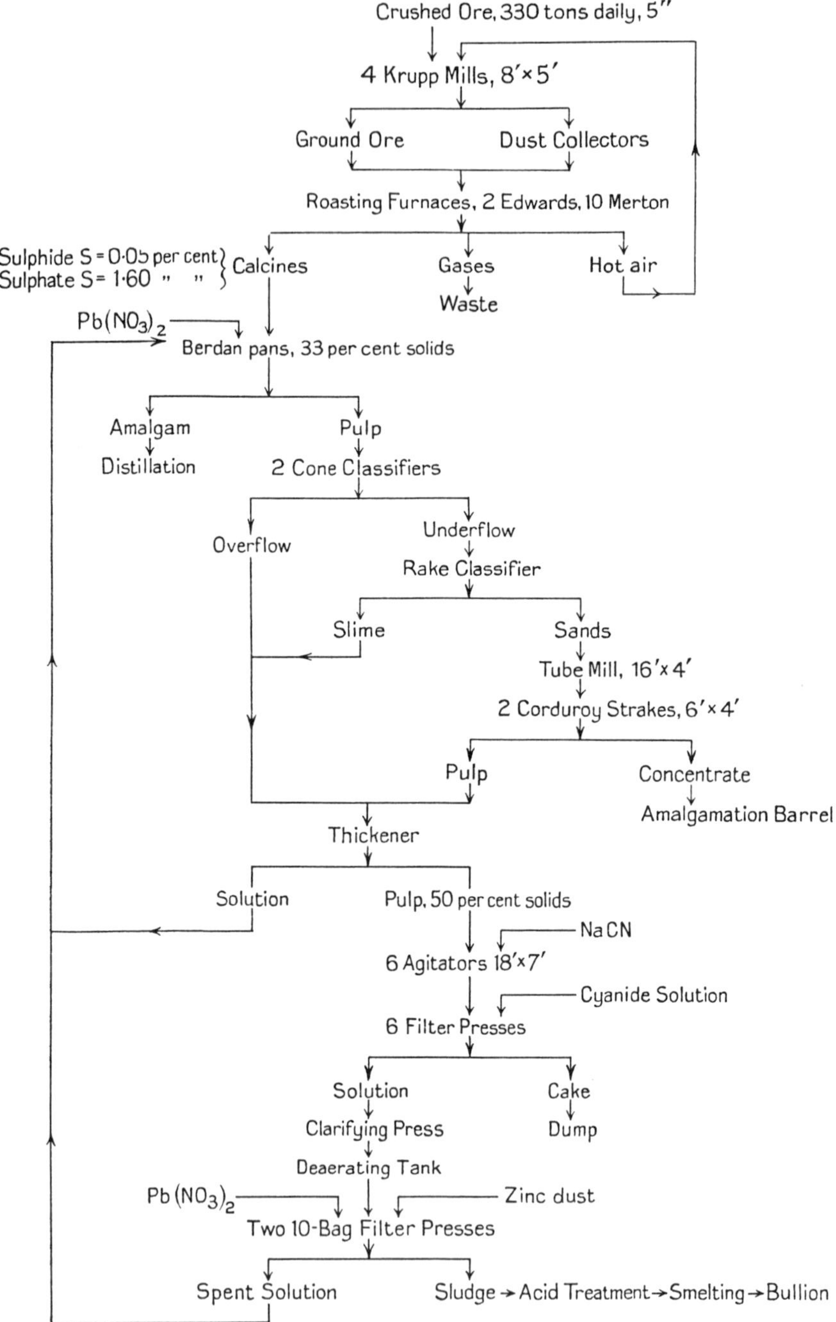

Fig. 148.—Flow Sheet, South Kalgoorlie Mill.

smaller than 2 or 3 inches in diameter and is sometimes larger, is fed into the drum from a hopper at the side. The balls are of chilled cast iron or cast steel and of various sizes, the largest being about 5 inches in diameter. As they wear down, others are added. The material which passes through the fine sieves falls into a hopper which is continuous with the dust-proof casing of sheet iron. The interior of the sides of the drum is lined with chilled iron or steel plates. A mill 7 feet 6 inches in diameter, containing 20 to 22 cwts. of steel balls, had an output of about 1½ tons of Kalgoorlie ore per hour, using a 20-mesh screen.

Example of Practice with Dry Ball Mills.—According to O'Malley,[1] the South Kalgoorlie Mill (see flow sheet, Fig. 148) gives the best results obtained in milling the Kalgoorlie ores, containing pyrite and tellurides. The process is essentially as follows:—Jaw-breakers crush the ore to 5 inch size, and this is fed to 8 by 5 foot Krupp ball mills, the inner screen of which has ¼ inch apertures and the outer 800 apertures per square inch. The ball load consists of 2¼ tons of 5 inch steel balls, and the output per mill is 90 to 100 tons per day. The moisture content is restricted to 2 per cent., and hot air from the roaster flues is passed into the mills to help the screening. Each mill requires 60 h.p. The wear of the balls and liners is 1·25 lbs. of steel per ton. There is much dusting.

The ground ore is roasted in two Edwards duplex roasting furnaces (*q.v.*), which treat 75 to 80 tons per day, and in 10 Merton furnaces each taking 20 tons daily. The calcined ore contains 1·65 per cent. total sulphur. Water is added in the proportion of 2 : 1 and the pulp is amalgamated in Berdan pans, where 25 per cent. of the gold is recovered. Lead nitrate, amounting to 0·15 lb. per ton of ore, is added to the spent solution to precipitate the soluble sulphides. The pulp is next classified in cones, the underflow reground in tube mills, and then passed over corduroy strakes sloped at 1/6. The pulp is next agitated for eight hours in 0·05 per cent. cyanide solution with vigorous aeration, and filtered. The filtrate is clarified in a second filter press, the cyanide strength brought up to 0·06 per cent. and 0·015 lb. of lead nitrate added per ton of ore before passing to the Merrill precipitation plant. Recovery of gold amounts to 93 to 95 per cent.

---

[1] *Chem. Eng. Min. Rev.*, 1934 26, 163.

CHAPTER XII.

ROASTING.

Roasting.—With the discontinuance of the chlorination process and the progress of other methods of treating telluride ores by cyanide, it has become unusual to roast such ores, except at Kalgoorlie. Where flotation is employed to recover sulphide concentrates containing gold, these are usually roasted prior to cyanidation, however.

The object of roasting is the expulsion of the sulphur, arsenic, antimony and other volatile substances existing in the ore, and the oxidation of the metals left behind. For this purpose the ore is heated in a furnace, through which a current of air is passed.

In cyanide treatment both sulphates and sulphides are deleterious. Soluble sulphides are active "cyanicides," attack zinc freely, and, under certain conditions, can eliminate all oxygen from the solutions. Acid sulphates are "cyanicides" if not quickly neutralised. Calcium sulphate precipitates in cooling solutions and forms incrustations in pipes, launders, thickeners and filters. It is formed by interaction between the limestone gangue and the sulphides during roasting. It impedes filtration and special provision has to be made for periodical washing and acid treatment of filter cloths.

Roasting is used as a preliminary to cyanidation to remove the sulphur or tellurium and to prevent the waste of cyanide by "cyanicides". The ore is roasted "dead" because the insoluble ferric oxide, Fe_2O_3, finally formed, is not attacked by cyanide. Telluride ores are roasted with the object of rendering the gold soluble, telluride of gold being only very slowly attacked by cyanide. As sulphur is always present in telluride ores, the roasting is continued until practically the whole of the sulphur is expelled.

Although roasting is carried on until the ore is said to be "dead," it is not practicable to eliminate the whole of the sulphur, and a small percentage both of insoluble and of soluble sulphur can be found in roasted ores. It is not usual to leave more than about 0·1 to 0·15 per cent. of insoluble sulphur in the ore, but much larger percentages of soluble sulphates are often left undecomposed in roasted ore.

The size of material to be roasted is governed by the method employed and the subsequent processes to which the ore will be subjected.

The following remarks on the decomposition of the various minerals present in complex ores may be of use in assisting the student to understand the reactions which proceed in the roasting furnace :—

1. *Pyrite*, FeS_2.—On heating this compound sulphur is volatilised, the reactions being probably expressed thus :—

$$FeS_2 = FeS + S$$
$$3FeS_2 = Fe_3S_4 + 2S$$
$$7FeS_2 = Fe_7S_8 + 6S$$

The sulphur burns to SO_2 (an exothermic reaction), which is partly converted by the heated quartz and iron oxide, by catalytic action, into SO_3, uniting with the free oxygen present. The ferrous sulphate formed by this sulphur trioxide is split up by the heat, and the ferrous oxide (FeO) is converted into ferric oxide (Fe_2O_3), which gives the ore a red colour when cold. Some basic sulphates always remain undecomposed. If the temperature has been too high, or if the charge is kept too long in the furnace, especially when not freely exposed to the air, some magnetic oxide is formed, thus :—

$$3Fe_2O_3 = 2Fe_3O_4 + O$$

The presence of magnetic oxide makes the ore darker in colour.

W. E. Greenawalt[1] states that the dark magnetic oxide can be reconverted to the red sesquioxide by subjection to a lower temperature and an abundant supply of air.

Other common sulphide minerals act similarly to pyrite, but with the exception of chalcopyrite, they do not yield sublimated sulphur. The ignition and incandescence temperatures are higher than for pyrite.

2. *Chalcopyrite.*—Copper sulphate is formed and then decomposed at a higher temperature, giving off SO_3 and leaving a mixture of cuprous and cupric oxides.

3. *Galena*, PbS.—The presence of this mineral in any but small quantities is very detrimental, as both lead sulphate and lead silicate (formed by its decomposition in the presence of silica) are very fusible, and, at the temperature required to split up copper sulphate, cause the ore to become pasty and form lumps. Roasting must be performed very slowly and cautiously to avoid this effect.

4. *Arsenopyrite*, FeAsS.—Some sulphide of arsenic is volatilised and then oxidised when it reaches the air. If roasting proceeds at a low temperature and with a limited supply of free oxygen, arsenious oxide, As_2O_3, is formed and volatilised. At higher temperatures, with a good supply of air, the conditions are favourable for the formation of arsenic oxide, As_2O_5. This unites with the bases present forming arsenates, which are not volatile and not decomposed by heat. Some arsenates are always formed in roasting. They may be reduced and the arsenic volatilised by stirring in coal dust, and the presence of ferrous oxide, cuprous oxide and sulphur dioxide assists the action. Sulphur trioxide slowly decomposes arsenates at a red heat. A large proportion of pyrite in the charge helps to eliminate arsenic, and charges are sometimes mixed with this end in view.

At Morro Velho the sand after cyanidation of the concentration tailings is weathered for about six months and roasted in Edwards furnaces. Volatilised arsenious oxide is collected in a series of flues.

At Black Hills, South Dakota, a refractory ore containing 0·5 per cent. of arsenic is roasted at a low temperature, moistened with 10 per cent. of water, 30 to 40 lbs. of lime per ton added, then allowing to stand for 3 days before cyaniding.

5. *Antimonial Sulphides* are still more difficult to deal with, the antimonates formed being less easily decomposed than arsenates. Their formation is avoided in the same manner. Charges tend to become pasty. Steam is said to be useful, giving rise to the reaction

$$Sb_2S_3 + 3H_2 = 2Sb + 3H_2S$$

[1] *Eng. and Mng. J.*, 1905, 78, 145.

Leaver and Woolf,[1] in investigating cyanide extraction of gold from arsenical and antimonial ores, arrive at the following conclusions :—

(1) Maximum cyanide recovery may be expected following a 30 minute roast at 450° C.

(2) High-temperature roasting (600° C. and above) results in the formation of arsenates and antimonates which lock up the gold so that it is not soluble in cyanide solution. Ferrites may also be formed, and they inhibit cyanide solution.

(3) At 450° C. the arsenic and antimony minerals are dissociated, releasing the gold and rendering it available for cyanidation.

(4) Direct cyanide treatment of the calcine from the low-temperature roast results in excessive loss of cyanide and fouling of the solution. Lime is therefore added either as a preliminary wash or to the calcine, and the latter allowed to stand for three days before cyanide treatment.

6. *Blende*, ZnS, forms oxide and sulphate of zinc, of which the latter can only be split up at a very high temperature. At a bright red heat a basic sulphate is formed which is converted to oxide at a white heat. If blende is roasted at a high temperature and with a plentiful supply of air, sulphate of zinc is not formed to a large extent.

7. *Calcium Carbonate*, $CaCO_3$, is decomposed at a red heat, CO_2 being given off and caustic lime, CaO, left in the charge. The change is slow at 600° (low red heat) and rapid at 800° C. (full red heat). Magnesium carbonate is similarly decomposed.

In roasting calcareous ores containing sulphides, the lime is converted to $CaSO_4$ if sufficient sulphur be present. Between 700° and 900° C. the $CaSO_4$ is liable to be reduced by carbon monoxide and carbon :—

$$CaSO_4 + 4CO = CaS + 4CO_2$$
$$CaSO_4 + 4C = CaS + 4CO$$
$$CaSO_4 + 2C = CaS + 2CO_2$$

Hydrogen, methane and ethane may also exert a reducing action. Calcium sulphide when roasted loses approximately 32 per cent. sulphur and forms a product containing both $CaSO_4$ and CaO. In the presence of oxygen $CaSO_4$ begins to decompose at 1,250° to 1,300° C. and in the presence of silica gives off SO_2 at 1,000° to 1,300° C. The reaction

$$CaS + FeSO_4 = CaSO_4 + FeS$$

takes place at a low temperature.

8. *Tellurides* containing gold are fusible far below a red heat, and heavy losses of gold may occur through absorption by the furnace bottom. The melted telluride of gold may remain in the ore, and in that case at a red heat the tellurium is partly volatilised and partly oxidised to the oxide TeO_2, which also sublimes. Spherical beads of gold remain behind and are difficult to dissolve. *Selenides* behave similarly.

9. Metallic *Gold*, *Silver*, etc., are fused at high temperatures, forming spherical beads which are difficult to dissolve. The temperature of the ore should never be allowed to exceed 1,000° C. for this reason.

Clarke and Moore [2] conclude that in the treatment of flotation con-

---

[1] *U.S. Bur. Mines*, 1928, *Tech. Paper No.* 423.
[2] *Sch. of Mines, West Aust.*, 1931, *Bull.* 6, 19.

centrates from the Wiluna Mill, the roasting occurs in three stages: (1) Sublimation of arsenious oxide and partial oxidation of pyrite (340°-400° C.). (2) Further oxidation of pyrite and ignition of sulphur (450° C.), causing a temperature-rise to 500° C. without external heat. (3) Swelling of the charge (500°-600° C.), probably due to liberation of carbon dioxide. Temperatures should be carefully maintained at each stage to avoid sintering due to the formation of arsenates, or the volatilisation of gold. Up to 6 per cent. of gold is lost in the second stage, due possibly to the presence of chlorine. 0·08 per cent. cyanide solution gives good extraction.

Consumption of chemicals decreases as the finishing temperature of the roast increases. When the latter is 800° C. the consumption is 0·5 lb. cyanide and 5 lb. lime per ton.

Baking of Gold Ores.—Gardiner[1] describes an unusual method of baking auriferous banded ironstones in Rhodesia, whereby the gold in hydrated oxides was exposed. By ordinary methods cyanide treatment of the oxidised ores gave very poor recoveries, entailing high lime consumption. The temperature of baking was between 200° and 320° C.—much lower than for ordinary roasting—and an inclined cylindrical drier was used. The period of travel was 30 minutes and the issuing ore was black. Afterwards the ore was subjected to cyanidation. By this treatment recovery was increased from 60 to over 90 per cent.

Allen[2] advocates a reconsideration of the advantages of heating and drying an ore preparatory to treatment. The removal of water tends to destroy colloidal conditions, inhibits adsorbent action, increases the friability of crystalline particles, simplifies settling, thickening and filtering, removes some interfering elements, and modifies the action of certain refractory constituents.

The Use of Salt in Roasting.—Roasting with salt to chloridise material which would otherwise absorb chlorine when the ore came to be "gassed," was carried on when the chlorination process of extraction was popular, but is now practically obsolete.

The chemical action of the salt is due to a double decomposition between it and the sulphates of the heavy metals, by which sulphate of soda and the chlorides of the heavy metals are produced. The following general equation approximately represents the reaction:—

$$2NaCl + RSO_4 = RCl_2 + Na_2SO_4$$

Chlorine is also set free by the action of sulphuric anhydride on salt, and the presence of water vapour induces the formation of much hydrochloric acid gas. These gases act directly on the several constituents in the ore, forming chlorides and oxychlorides.

Silver, copper and other metals in an ore roasted with salt form chlorides. Ferrous sulphate, acted on by salt at a red heat in the presence of moist air, yields hydrochloric acid and chlorine, which act on the gold, while ferric sesquioxide and sodium sulphate are produced. Ferric chloride, Fe_2Cl_6, is also produced at the same time and is volatilised. Cupric chloride, $CuCl_2$, is easily decomposed into cuprous chloride, Cu_2Cl_2, and free chlorine, or into the oxychloride, Cu_2OCl_2, and free chlorine. The vapours of $CuCl_2$ thus give rise to further supplies of nascent chlorine available for chlorination.

[1] *Min. Mag.*, 1920, **20**, 109. *J. Chem. Met. Mng. Soc. S.A.*, 1920, **20**, 109.
[2] *Trans. Amer. Inst. Min. and Met. Eng.*, 1934, **112**, 543.

Arsenic and antimony form volatile chlorides which are decomposed by means of oxygen and water vapour, yielding arsenious and antimonious acids and nascent chlorine or hydrochloric acid. In passing off, however, the volatile compounds carry away with them varying proportions of gold and silver, which, as a rule, are not recoverable in the dust chambers. Cupric chloride is especially active in causing these losses.

The presence of basic minerals in the charge is advantageous as they may cause nascent chlorine to be set free in all parts of the furnace. On the other hand, the loss of gold is increased by any increase in the quantities either of silver or of the base metals, in the former case the time of the roasting being prolonged. The best chloridising effect is obtained in a highly oxidising atmosphere, so that very little sulphur is required in the ore, and, if much is present, the practice of eliminating the greater part before adding the salt is not likely to be attended with any diminution in the percentage of silver chloride formed. Moreover, any water vapour in the furnace gases promotes the formation of hydrochloric acid. When salt is used in roasting, the ore is often allowed to cool slowly in heaps after being withdrawn from the furnace. If treated in this way, a higher percentage of the silver, etc., is found to be chloridised than if the ore is wetted down at once or even spread out to cool in a thin layer. On the other hand the losses of gold by volatilisation are increased by slow cooling. Chlorine continues to be evolved for a long time after the withdrawal of the charge has taken place, the heaps smelling strongly of the gas.

A new application of salt roasting has been described by Richards.[1] The ore is about 80 per cent. oxidised, contains both gold and silver, about 15 per cent. iron and 1·2 per cent. sulphur, besides small quantities of lead, copper, zinc, arsenic and antimony. It was not amenable to ordinary methods of direct cyanidation, amalgamation, concentration or flotation. A 300 ton plant has been built, after preliminary pilot testing, on the following broad lines. Before crushing, the ore is dried to 7 per cent. moisture. After crushing in various stages in jaw-crushers, a gyratory machine and rolls, with intermediate screening in Hum-mer screens, the oxidised ore is sent to the mixing plant mostly as 8 to 10 mesh material. A sulphide ore, which comes from a different part of the mine, is crushed separately in rolls to $\frac{3}{8}$ inch and then to 100 mesh in a ball mill. Weighed charges of oxide, sulphide and salt (about 6 per cent.) are mixed in concrete mixers and then roasted in a series of Holt-Dern blast roasters, each 5 feet 8 inches $\times$ 10 feet 6 inches in area and 30 inches deep, supplied with blast at a pressure of 12·5 inches of water. The upper 8 or 10 inches of the charge are caked and left in when the charge is dropped in order to form the basis for a new charge and to start its ignition. The sulphide ore in burning provides sufficient heat by the exothermic oxidation of the sulphur to continue the calcination. Finely divided charcoal gave equally good results as sulphide ore. Moisture is closely controlled, as a dry charge will not burn. The roasting temperature is 675° C. and is maintained within a narrow range. The calcines are sent to the leaching plant while still hot and dumped into water at the bottom of wooden tanks fitted with the usual grids and discharge pipes and each holding 150 tons. Water is added just to cover the calcines. More water is then turned on and the discharge cock opened. The first 70 tons of effluent contain much salt and in them are dissolved some gold

---

[1] *Eng. Min. Journ.*, 1928, 126, 325, 370.

and silver chlorides. This solution is sent to Pachuca tanks for recovery of gold by agitation with cement copper in a slightly acid solution. Copper and silver are recovered, after removing the gold, by precipitation with scrap iron. The remainder of the solution from the wooden tanks contains only traces of gold and silver and is sent to waste until it gives no precipitate with ammonia. About 4 tons of water per ton of ore are used. A sulphurous acid wash is then given. This acid is produced by scrubbing the furnace gases with water and contains 0·1 to 0·15 per cent. SO_2 and some HCl. The wash, which amounts to 4-5 tons per ton of ore, is kept on for 24 hours and is a vital step. Without it cyaniding cannot proceed. The end of the acid wash is indicated by the effluent being acid to methyl orange and showing no copper and little iron with ammonia. A final short water wash is given to remove acid. All acid effluent goes to waste. Cyanide leaching then proceeds. About $1\frac{1}{2}$ tons of quicklime are spread over the charge, a water wash for six hours is given, the charge drained and cyanide solution, 0·08 to 0·1 per cent. KCN, turned on. The leaching rate is heavy for 24 hours as the gold and silver dissolve readily. The effluent is pregnant solution, which is agitated by means of compressed air with milk of lime in the storage tank to precipitate calcium sulphate, and then filtered in a Butter's filter. The solution shows no alkalinity to phenolphthalein. The residue is drained, washed again for 12 hours, drained for 48 hours, and after 72 hours washed with barren solution, followed by water. 8 tons of solution, of which 4 are subsequently precipitated, are used per ton of ore. If the calcium sulphate be not precipitated it causes trouble in filtration and goes also into the press with the precipitate. Precipitation is effected by the Merrill Crowe process, giving a precipitate averaging 70 per cent. in gold. The consumption of reagents is approximately as follows—salt 6 per cent., cyanide $1\frac{1}{2}$ to $1\frac{3}{4}$ lb., lime 50 to 60 lb. per ton of ore.

Losses of Gold in Roasting.—In the oxidising roasting of ordinary auriferous pyrite, a loss of gold takes place when the operation is carried on so rapidly that fine particles are carried off mechanically by the draught. But it is extremely difficult to prevent all mechanical loss by dusting, which is caused by even a moderate draught.

The losses of gold which are sustained when salt is added to the furnace charge may be very great, amounting possibly to 8 or 10 per cent. or even more, depending on the nature of the ore and the conditions of roasting. The loss is greatly reduced by diminishing the quantity of salt, and by reserving it until the dead roasting is nearly complete.

In the chloridising roasting of a Mexican ore, consisting mainly of magnetite and pyrite with 3·5 to 7 per cent. of chalcopyrite, C. A. Stetefeldt found the losses of gold to be from 42·8 to 93 per cent. of the total gold content. The volatilisation of the gold was due to the copper chlorides. The loss increased with the quantity of these chlorides formed and volatilised. The presence of copper chloride is not the only possible cause of loss, however, since an ore consisting of hard white quartz, intimately mixed with about 7 per cent. of calcite and a little pyrite, lost 70 to 80 per cent. of its silver, and 68 to 85 per cent. of its gold, when roasted with 5 per cent. of salt. When subjected to an oxidising roast, no loss of gold took place.

The temperature used in chloridising roasting must be very carefully regulated, the loss of gold being increased far more by high temperature than by a lengthening of the time in the furnace. Moreover, the salt must be reduced to the least possible quantity.

Fig. 149.—Edwards Furnace.

The variation of the loss in different ores which are treated precisely alike is doubtless due partly to the presence or absence of metals forming volatile chlorides which carry off the gold, and partly to the physical condition of the latter, the volatilisation being greater if it is in a state of minute subdivision.

The Edwards Furnace (see Figs. 149 and 150).—The Edwards furnace is made in two patterns, the stationary and the tilting. The stationary "Duplex" furnace is built of brick with a flue under the hearth. It has one or two side fire-boxes and one at the end. The length varies, but is usually from 100 to 120 feet, and the width from 9 to 12 feet. There is a number of rotating rabbles, which are hollow and water-cooled. The spindles of the rabbles pass through the arch of the furnace, and above the arch are fitted with crown cog-wheels which gear with bevel pinion cog-wheels, carried on two lines of shafting running longitudinally along the top of the furnace. The rabbles are geared right and left alternately so that adjacent rabbles rotate in opposite directions. The circular paths of the feet of the rabbles overlap, and there is little or no dead space on the hearth. The ore moves

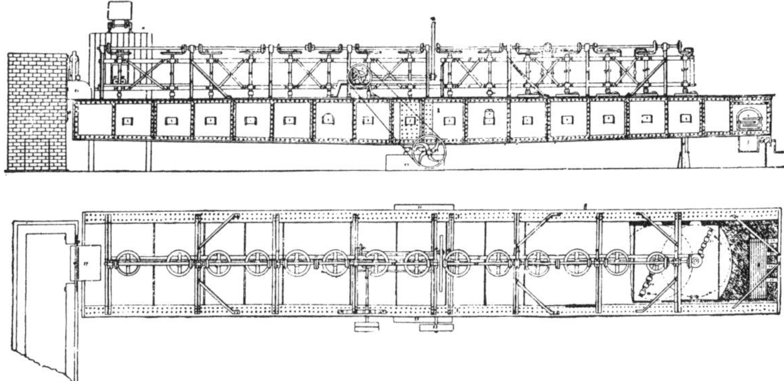

Fig. 150.—Edwards Tilting Furnace.

down the inclined hearth in a zig-zag fashion under the influence of the rabbles. The shoes of the rabbles are cast iron, and when worn out can be easily replaced on the feet without cooling the furnace. In the older furnaces there was only one line of rabbles, instead of two. The width was 6 feet, and the amount of ore roasted at Kalgoorlie (containing about 6 per cent. of sulphur) was 12 to 18 tons per day. The duplex furnaces roast from 60 to 100 tons of Kalgoorlie ore per day in preparing it for treatment by the cyanide process.

The hearths at the Associated Mine furnaces [1] were $11\frac{3}{4} \times 107$ feet, the fall 0·31 inch per foot, and the 52 rabbles moved at 2·62 revolutions per minute. There were three fire-boxes, and fuel consumption averaged 11 per cent. of the ore roasted, which amounted to 95 tons per day in each furnace. The sulphur in the ore amounted to 5·5 per cent. If bad work was detected the furnace was stopped for a time to heat up.

Bruhl [2] states that at Prestea the concentrates were charged into an Edwards furnace by means of a belt conveyor leading into a hopper feeding

[1] Von Bernewitz, *Mng. and Sci. Press*, 1911, 102, 661.
[2] *Eng. Min. Journ.*, 1918, 105, 368.

directly into the furnace. At about the fourth rabble from the feed end the charge gave off white fumes and began to spark. It became hotter as it progressed and finally was at a bright red heat. It was not allowed to clot, as this gave imperfect roasting. Later a new furnace on the regenerative principle was installed. The gases passed over the charge and beneath the hearth before going to the stack.

For customs treatment of Cripple Creek ores the Edwards furnaces are of the "Duplex" type with a live roasting area of 1,500 square feet, and equipped with coolers 570 square feet in hearth area. There are three fire-boxes for each roaster. The temperature of the ore at the discharge of the roasters is 485° C. and of the coolers 280° C. The most efficient rabble speed is 6 r.p.m.[1] The consumption of teeth is 0·16 lb., of arms 0·07 lb., and of spindles 0·03 lb., per ton of ore. All rabble arms are water-cooled, the incoming water being at 25° to 34° C. and the outgoing at 45° to 64° C. There are 27 pairs of rabbles in the furnace and 11 pairs in the cooler. The average time of passage of the ore is about 6 hours. Coal is used as fuel. The maximum roasting temperature is 850° to 900° C. The draught is kept at 0·2 inch of water to allow a free escape of gases and therefore a more oxidising atmosphere. The flames from the fire-boxes are given an excess of air to keep them from being smoky.

The tilting furnace (Fig. 150) is made of iron externally and lined with brick. It consists of two iron girders 63 feet long, resting on pivots at the centre, and the movement of ore can be regulated by tilting the furnace from the horizontal to any desired angle to suit the ore, thereby increasing or decreasing the rate at which the ore passes through the furnace. The tilting furnace is also made in duplex form. The output for the simplex pattern is from 8 to 20 tons per day, and for the duplex about double.

The Merton Furnace (Fig. 151) consists of three superimposed hearths and a fourth hearth in front of them at a lower level. The main hearths are 23 feet 9 inches long and 8 feet wide over all. The height of each arch is 16½ inches in the middle and 9 inches at the side. The ore is charged in on the upper hearth, stirred and moved forward by rotating rabbles attached to vertical shafts, and delivered in succession to the middle and lower hearths and finally to the finishing hearth. There are four shafts, with a rabble attached to each shaft on each of the three floors. The hearths are horizontal, having no fall. The capacity of the furnace when treating Kalgoorlie sulpho-telluride ore is said to be from 18 to 25 tons of ore per day.

In a more recent type of Merton furnace, the shafts on the primary hearths are hollow and water-cooled, like that on the finishing hearth, and the finishing hearth is made of greater area, being 7 feet long. The furnace is said to treat from 20 to 33 tons of telluride gold ore per day at Kalgoorlie, bringing the sulphur from 3 to 5 per cent. down to 0·01 per cent. with a fuel consumption of 14 per cent.

At the Associated Northern Blocks, Kalgoorlie,[2] six Merton furnaces were treating daily 120 tons of ore containing 3 per cent. sulphur. The hearths were 30 feet by 6½ feet, and most of the elimination of the sulphur was on the third or lowest hearth. The consumption of fuel was 14 per cent. by weight of the ore.

A more recent development in Australia is the **Leggo furnace,** which

---

[1] Blomfield and Trott, *Trans. Amer. Inst. Min. and Met. Eng.*, 1918, 60, 118.
[2] Von Bernewitz, *loc. cit.*

employs the same principle of rabbling sweeps revolving in overlapping circles. It is made in multiple-hearth straight-line type and also in turret form with four or six rabbles on each hearth.

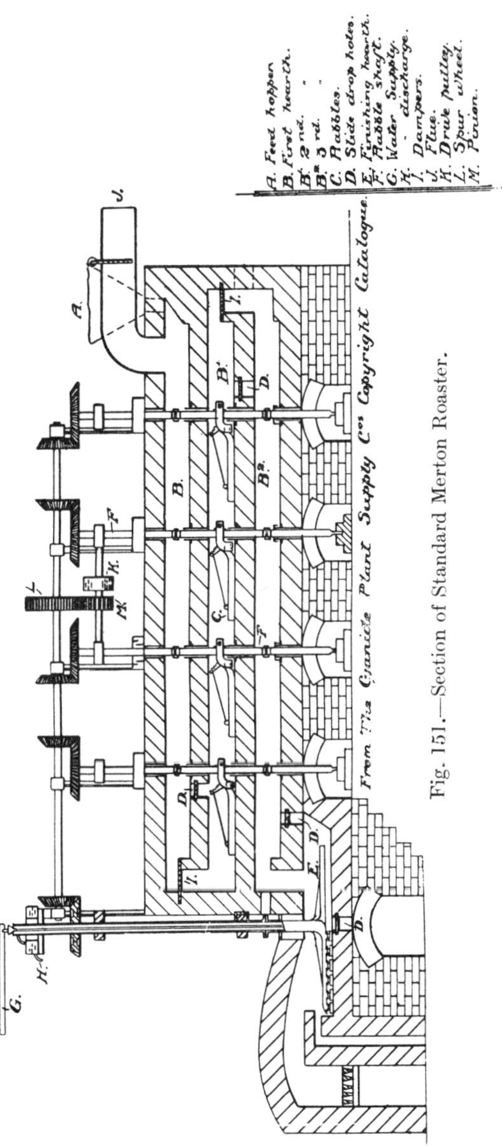

Fig. 151.—Section of Standard Merton Roaster.

In the Edwards, Merton and Leggo furnaces the moving parts are light in weight, rabbles can be replaced easily, and the low arches tend to give superior roasting.[1]

[1] Liddell, "*Handbook of Non-Ferrous Metallurgy*," 1926, i, p. 310.

CHAPTER XIII.

THE CYANIDE PROCESS. CHEMICAL REACTIONS.

Introduction.—The cyanide process will always be indissolubly connected with the names of its discoverers, J. S. MacArthur and R. W. and Wm. Forrest. The solvent action of potassium cyanide on metallic gold and silver has long been known, but it was believed that the use of an electric current in conjunction was needed to quicken the action, which was otherwise too slow to be of any practical value. Until the surprising results of the process were made public, metallurgists were disposed to regard the use of cyanide as a solvent for gold in ores as chimerical, on account of the instability of potassium cyanide and the slowness of its action on gold, combined with its high cost and poisonous character.

The process consists essentially in attacking gold ores with dilute solutions containing less than 1 per cent. of alkaline cyanide, a slight excess of caustic soda or lime being added to ores rendered acid by the oxidation of pyrite, and then precipitating the precious metals with zinc. This process was introduced at Karangahake, New Zealand, in 1889, and on the Rand in 1890.

The process was originally applied to the residue of gold ore after amalgamation, but was later used in many instances for the direct treatment of ore.

It is necessary to use dilute solutions, which have little action on pyrite and many other base minerals, but readily dissolve gold and silver. Certain compounds, such as those of copper, destroy cyanide and interfere with extraction. These compounds are called "cyanicides."

The cyanide used was at first impure potassium cyanide, then mixtures of potassium and sodium cyanide, and later sodium cyanide. Two brands of sodium cyanide in use for gold extraction are the British Cassel, containing 97·5 per cent. NaCN, and the Czecho-Slovakian with 91 per cent. NaCN.

Calcium cyanide of American or Canadian origin, containing an equivalent of 49 per cent. NaCN, is also in use.[1] When gold is precipitated from its solution in calcium cyanide, some calcium hydroxide is formed and is useful in replacing lime which would otherwise be added to correct acidity. From every lb. of calcium cyanide there is produced about 0·4 or 0·5 lb. of calcium oxide. One drawback to its use is the extra cost of transport owing to its low grade. It appears, however, to give satisfactory extraction in spite of the presence of some soluble sulphide. Buchanan[2] gives the

---

[1] Landis, *Trans. Am. Electrochem. Soc.*, 1920, **31**, 653. *Chem. and Met. Eng.*, 1920, **22**, 265. Freeman, *Can. Chem. Met.*, 1922, **6**, 129.

[2] *Trans. Amer. Electrochem. Soc.*, 1931, **60**, 93.

following average percentage analysis of calcium cyanide of the "Aero" brand :—

CN,	25·44	≡ 45·03 $Ca(CN)_2$ or 47·94 NaCN
Cl^-,	19·79	
Ca^{++},	32·13	
Na^+,	12·11	
CN^-,	0·88	
SiO_2,	0·87	
R_2O_3,	1·95	
Free Carbon,	2·81	
CaC_2,	1·90	

Although sodium and calcium cyanides have replaced potassium cyanide, owing to their cheapness, it is still common practice to refer to their standard in terms of the potassium salt. Thus 97 to 98 per cent. NaCN is designated as 129 to 130 per cent. cyanide or potassium cyanide.

Hamilton [1] states that the dissolving effect of a simple soluble alkaline cyanide is proportional to the CN^- content, and this holds true also for low-grade cyanide, manufactured from calcium cyanamide and containing a large proportion of $CaCl_2$ or NaCl.

Action of Potassium Cyanide [2] on Gold and Silver.—Dr. Wright, of Birmingham, discovered in 1840 that metallic gold is soluble in potassium cyanide when a current of electricity is passing through. Bagration, in 1843,[3] studied the action of cyanides on plates of gold in the absence of a current of electricity, and noted that they were slowly dissolved. Faraday, in 1857,[4] pointed out that gold-leaf is dissolved by a dilute solution of the salt, and also showed that if the gold floats on the surface of the liquid, so that one side of the leaf is in contact with the air, while the other is bathed by the solvent, the action is much more rapid than if the metal is completely submerged. White,[5] confirming this, found that gold-leaf floating in 6-10 per cent. cyanide solution is dissolved much more rapidly than sunk pieces. With 0·01 per cent. solution there is less difference. Elsner had previously, in 1846, furnished some evidence [6] that the presence of oxygen is required for the solution of the gold. W. A. Dixon found, in 1877,[7] that although cyanide by itself was slow in dissolving gold, its action was hastened by the addition of alkaline oxidising agents. He also mentions calcium hypochlorite, potassium ferrocyanide and manganese dioxide as hastening the action.

On evaporating the solution, colourless octahedral crystals of auro-potassium cyanide, $KAu(CN)_2$, are formed, which may be viewed as being a double cyanide, produced as follows :—

$$4Au + 8KCN + O_2 + 2H_2O = 4KAu(CN)_2 + 4KOH$$

[1] M. Hamilton, *Min. Ind.*, 1931, **39**, 465.

[2] In general, when potassium cyanide is named, it may be taken to include sodium cyanide and mixtures of the two salts. Also in the equations given in this chapter, K and Na are interchangeable, except from the point of view of ionic dissociation.

[3] *Bull. Acad. Sci. St. Petersbourg*, 1843, **2**, 136.

[4] *Proc. Roy. Inst.*, 1857, **2**, 308.

[5] *J. Chem. Met. Mng. Soc. S.A.*, 1919, **20**, 2 ; 1934, **35**, 1.

[6] *J. prakt. Chem.*, 1846, **37**, 441-446.

[7] *Proc. Roy. Soc. of N.S.W.*, Aug. 1, 1877 ; *Chem. News*, Dec. 20, 1878, p. 293.

This equation is often referred to as "Elsner's equation," but Elsner himself does not give it.

According to the equation given above, 130 parts by weight of KCN in the presence of 8 parts of oxygen suffice for the solution of 197 parts of gold. This has been proved by J. S. Maclaurin [1] to be the case in all carefully conducted experiments.

According to G. Bodländer,[2] the chemical action in the dissolution of the gold is as follows :—

$$24Au + KCN + 2H_2O + O_2 = 2KAu(CN)_2 + 2KOH + H_2O_2$$

and subsequently

$$4KCN + H_2O_2 + 2Au = 2KAu(CN)_2 + 2KOH$$

the ultimate result being as in the "Elsner" equation given above.

Evidence has been adduced that a substance is formed reacting like H_2O_2, but the action represented by the equation given by Bodländer may be limited to an insignificant part of the whole mass.

Janin [3] suggested that hydrogen is liberated, thus :—

$$2Au + 4KCN + 2H_2O = 2KAu(CN)_2 + 2KOH + H_2$$

Ramsay [4] also contends, from the change in potential of gold in cyanide solutions, that hydrogen is evolved when gold is dissolved in cyanide. The weight of evidence, however, is against this view.

Barsky, Swainson and Hedley [5] have compared the free energies of the equations said to represent the solution of gold in cyanide, viz., Elsner's, Bodländer's and Janin's, with the respective equilibrium constants. They have found that the Elsner equation best fits the case and that the reaction will proceed to completion in the presence of oxygen—*i.e.* until the metal or the cyanide has been removed. Elsner's equation is considered not strictly correct, however, as it does not show the intermediate formation of hydrogen peroxide. Bodländer's equation, from the same reasoning, is found to be possible and, it is suggested, may represent the partial state of affairs.

It has been suggested by Christy,[6] and Julian and Smart,[7] that the dissolution of gold and other metals in cyanide depends primarily on the electromotive force exerted. H. F. Julian [8] attached great weight to the formation of local voltaic circuits. Christy contends [9] that the rate of dissolution of gold in cyanide depends on two factors, the electromotive force of gold in cyanide solutions, and the number of hydroxyl ions in the solution. Michailenko,[10] however, denies the latter contention, and states that OH ions

---

[1] *J. Chem. Soc.*, 1893, 63, 724.

[2] *Zeitsch. angew. Chem.*, 1896, p. 583 ; see also *J. Chem. Met. Mng. Soc. S.A.*, 1903, 4, 273.

[3] *Min. Ind.*, 1892, 1, 239.

[4] *Chem. Eng. Min. Rev.*, 1932, 24, 236.

[5] *Trans. Amer. Inst. Mng. Met. Eng.*, 1934, 112, 660.

[6] "The E.M.F. of Metals in Cyanide Solutions," *Trans. Amer. Inst. Mng. Eng.*, 1900, 30, 864.

[7] "*Cyaniding Gold and Silver Ores*," (Griffin, 1921).

[8] Report Brit. Assoc., 1905, p. 369. See also Allen, *Trans. Amer. Inst. Mng. Met. Eng.*, 1934, 112, 546.

[9] *Loc. cit.*

[10] Michailenko and Meshtscherjakoff, *J. Russ. Phys. Chem. Soc.*, 1912, 44, 567 ; *Mng. and Sci. Press*, 1912, 112, 500.

do not accelerate the velocity of solution, but that an excess of them has a retarding influence.

Allen[1] observes :—" Pure gold is doubtless insoluble in a pure solution of cyanide in oxygen-free water, but perfect purity and absolute uniformity are unattainable under commercial conditions. A minute degree of contamination is sufficient to cause galvanic action, stimulating dissolution. Gold in cyanide solution in a glass vessel is only slightly attacked, but dissolution is active if the metal is in contact with iron. The action diminishes, however, unless the iron surface remains exposed, or polarisation is prevented, and oxygen is effective for this purpose. Elsner's equation ignores galvanic action and subsidiary chemical changes that may have an important influence on commercial extraction (for example, pyrite is electronegative to gold, as Julian demonstrated)."

A. F. Crosse has shown that gold is also dissolved in HCN (or acidulated KCN solution) in the presence of oxygen. In this case auricyanides are formed instead of the usual aurocyanide.

Influence of Oxygen.—It is clear that oxygen or some other substance acting similarly is necessary to assist cyanide of potassium in dissolving gold. The oxygen unites with the hydrogen and reduces the loss of cyanide as hydrocyanic acid gas. It was formerly supposed that any oxygen coming in contact with the solution would be at once taken up by the cyanide, forming cyanates, but it is now well known that oxygen and cyanide remain side by side without rapid union between them taking place. Cyanates are without action on metallic gold. The most noticeable effect of extra aeration of solution is a substantial reduction in cyanide consumption.[2] In cyanide solution oxygen is dissolved to the extent of about 8 mg. per litre, or say $4\frac{2}{3}$ dwt. per ton. The solubility is practically unaffected by increasing the cyanide up to as much as 2 per cent. NaCN.[3] Wallace[4] draws attention to the drop in extraction due apparently to loss of dissolved oxygen in mill solutions. In a vacuum no gold dissolves in cyanide solution, and on this Crowe[5] based his well-known precipitation process.

The rate of dissolution of gold in cyanide solutions is proportional to the amount of oxygen dissolved in the solution—or, in other words, to its partial pressure. Hence to attain the greatest rate of dissolution the solutions must be fully aerated. "The decomposition of the potassium cyanide is facilitated by the affinity of the potassium for the free oxygen present, so that nascent cyanogen is liberated to combine with the metallic gold" (Caldecott). It has, however, been frequently pointed out that in the interior of a mass of ore undergoing treatment the conditions are not favourable for the maintenance of a sufficient quantity of oxygen in a free state. Certain constituents of the ore tend to unite with it, and further absorption of free oxygen from the air is extremely slow. Hence the time required for the treatment of a charge is many hours, or even days, although under favourable conditions the gold could be dissolved in a few minutes, or at most in two or three hours. To supply the oxygen, various oxidising methods have been tried, such as the passage of a current of air

---

1 *Trans. Amer. Inst. Mng. Met. Eng.*, 1934, **112**, 546.
2 See *Eng. Min. Journ.*, 1934, **135**, 298.
3 Barsky, Swainson and Hedley, *Trans. Amer. Inst. Mng. Met. Eng.*, 1934, **112**, 671.
4 *Min. and Met.*, 1932, **13**, 522.
5 *Amer. Inst. Mng. Eng.*, 1919, Bull. 140, 1279.

through the solution, and the addition of various oxidisers. The charges are also sometimes drained, turned over, and transferred to other vats, or subjected to other treatment in order to aerate the damp ore.

Aeration of cyanide solutions has been attempted by the use of agitators, Pachuca tanks, turbo-mixers and "super-agitators." At Kirkland Lake the addition of sodium peroxide, 1·25 ozs. per ton of ore in the ball mill feed, has been beneficial. It is noted that despite the constant introduction of oxygen in the agitators the content of this element does not increase, but decreases by steps corresponding to the stages of gold extraction, until filtration, where the introduction of barren solution of high oxygen content restores it (see following table).

TABLE XXXVII.

Time.	Process.		Gold Extraction, per cent.
4 hrs.	Grinding circuit, . .	O_2 drops from 100 to 43·6 per cent.	80
5 ,,	Primary agitation, .	,, 43·6 to 40 ,,	4
2½ ,,	Thickening, ...	O_2 increases from 40 to 41·1 per cent.	
5¾ ,,	Secondary Agitation, .		4
2½ ,,	Secondary Thickening,		
22 ,,	Filtering and repulping,	O_2 increases from 41·1 to 65·7 per cent.	4
...	Final agitation of cake from primary filter, .	O_2 drops from 65·7 to 51 per cent.	...

De-oxidising Agents.—Organic matters, such as oil, grease and wood, are probable de-oxidisers and also cause frothing in the agitators. Small quantities of iron also consume oxygen. Wallace contends that hydrogen is produced by the interaction of certain metals (*e.g.* iron, copper, zinc, nickel) with cyanide. He finds that the iron particles from the grinding machinery are bright and recommends the removal of iron fragments by cleaning out all mills periodically, by passing all tube mill pulp through a central magnetic extractor, and by double filtration where the accession of barren solution introduces oxygen.

Paton [1] attributes oxygen consumption to the production of ferrous compounds from the same source.

Effect of Oxidising Agents other than Air.—Bettel [2] found that gold dissolves in the absence of oxygen if the crushed ore contains basic ferric sulphate, from which potassium ferricyanide is formed, the reactions being expressed thus—

$$(1)\ Fe_2(SO_4)_3 + 12KCN = 3K_2SO_4 + K_6Fe_2(CN)_{12}$$
$$(2)\ Fe_2(SO_4)_3 + 6KCN + 6H_2O = Fe_2(HO)_6 + 3K_2SO_4 + 6HCN$$

Bettel and Marais also found that gold leaf would not dissolve in a solution of potassium cyanide from which the air had been expelled by the passage of a current of hydrogen, and that the addition, in these circumstances, of either potassium dichromate, chromate, chlorate, perchlorate, nitrate or nitrite, or of ferric hydrate or bleaching powder, did not

[1] *Eng. Mng. World*, 1930, I, 200.
[2] *S. African Mng. J.*, 1897, May 8.

enable the gold to dissolve. The addition of pyrolusite gave a doubtful result, and lead dioxide caused very slow dissolution of the gold. On the other hand, gold dissolved slowly if chlorine, iodine dissolved in potassium iodide, or ferric chloride were added ; rapidly, if bromine were added ; and decidedly, if potassium ferricyanide or permanganate, sodium dioxide, hydrogen peroxide or barium dioxide were added. Michailenko and Meshtscherjakoff [1] confirmed most of these conclusions, and found that sodium dioxide exercised its maximum influence at a concentration of 0·02 of an equivalent (or about 0·2 per cent.). Morris Green [2] finds that potassium permanganate as such does not aid dissolution. Rusden and Henderson [3] experimented with the use of permanganate of potash, lead peroxide and barium peroxide as oxidisers in the treatment of black sand concentrates. These were added immediately after amalgamation, but before cyanidation. The first increased cyanide consumption slightly, the second a little more, whilst the third decreased it. None gave such results as to warrant wide use. White considered that such oxidisers might be used where high-valued concentrates were under treatment. In practice, when air is not used, potassium permanganate, sodium dioxide, bromine (Sulman-Teed process), and umber (Adair process)[4] have been mainly employed to assist in the dissolution of the gold. In general, artificial oxidation is not resorted to unless there is some reducing agent present in the ore or the water which absorbs the oxygen. The oxidising substances generally act as " cyanicides," destroying some of the solvent.

When bromine is used, it is in the form of cyanogen bromide (bromocyanogen, bromocyanide), as proposed by H. L. Sulman,[5] and used in the Diehl process. When this substance is added to cyanide solutions, the presence of oxygen is not required, and the rate of dissolution is markedly increased. The cyanogen bromide is set free as follows : [6]—

$$2KBr + KBrO_3 + 3KCN + 3H_2SO_4 = 3BrCN + 3K_2SO_4 + 3H_2O$$

A mixture of the bromide and bromate salts in the proper proportions is manufactured and sold. When required for use the bromocyanogen is made by agitation of the mixture with cyanide and sulphuric acid in a closed vessel. Its action in the treatment vat is supposed to be as follows :—

$$BrCN + 3KCN + 2Au = 2KAu(CN)_2 + KBr$$

Cyanogen chloride and iodide behave similarly, but are very unstable. For the effect on tellurides, see Chap. XV.

The addition of a small quantity of potassium mercuric cyanide, $Hg(CN)_2 \cdot 2KCN$ to ordinary cyanide solutions to accelerate their action has also been proposed by J. S. MacArthur and by N. S. Keith.[7] Mercury is electronegative to gold in cyanide solutions, and Skey [8] found that metallic

---

[1] *Loc. cit.*
[2] *J. Chem. Met. Mng. Soc. S.A.*, 1913, 13, 355.
[3] *Ibid.*, 1923, 23, 234.
[4] Adair, *ibid.*, 1908, 8, 331.
[5] *Trans. Inst. Mng. and Met.*, 1895, 3, 202.
[6] E. W. Nardin, *Trans. Australasian Inst. Mng. Eng.*, 1907, 12, 69; *Mng. and Sci. Press*, 1908, 97, 562.
[7] *Engineering*, 1895, 59, 379.
[8] *Trans. New Zealand Inst. Eng.*, 1876, 8, 334.

gold in contact with a solution of mercuric cyanide would rapidly dissolve and mercury be reduced. The mercury is precipitated on the surface of the parctiles of gold and forms amalgam, the equation being probably—

$$K_2Hg(CN)_4 + 2Au = 2KAu(CN)_2 + Hg$$

The accelerating action in practice is, however, inappreciable. Mercury dissolves in excess of cyanide, but less readily than gold ; thus, according to Julian and Smart,[1]—

$$Hg + 4KCN + 2H_2O = Hg(CN)_2 \cdot 2KCN + 2KOH + H_2$$

H. A. White [2] finds that gold when covered with mercury will dissolve in cyanide solutions of normal reduction-works strength at only about half the usual rate.

Rate of Dissolution.—By direct experiments Christy [3] found that solutions containing no more than 0·00065 per cent. KCN did not dissolve gold, and that for all practical purposes the cyanide of potassium solution ceases to act when its strength falls below 0·001 per cent. Above that strength, however, the solubility of gold increases rapidly when free oxygen is present.

J. S. Maclaurin found [4] that the rate of dissolution of pure gold, in the form of plates, in potassium cyanide solution passes through a maximum when proceeding from dilute to concentrated solutions. The maximum is reached when the solution contains 0·25 per cent. of KCN. The solubility of gold is very slight in solutions containing less than 0·005 per cent., but increases rapidly as the strength rises to 0·01 per cent., when the rate of dissolution is ten times as great as in the 0·005 per cent. solution, and about half as great as that in the 0·25 per cent. solution. The rate increases slowly as the strength rises to 0·25 per cent. Silver is also dissolved at a maximum rate in solutions containing 0·25 per cent. of cyanide, and the changes in the rate are similar to those noted above in the case of gold, the rates for silver being always about two-thirds of the corresponding rates for gold, or, roughly, in the same ratio as the atomic weights of the two metals. In both cases there is only a small change in the rate of solubility as the strength rises from 0·1 to 0·25 per cent.

From later studies, however, it would appear that, although Maclaurin's results are doubtless correct in the presence of unlimited oxygen, in practice the supply of oxygen is limited and strengths of cyanide much lower than 0·25 per cent. KCN are the most effective. Thus Julian and Smart state [5] that if the time factor is left out of account, as in cases where gold dissolves rapidly but can be washed out of the ore only slowly, the most efficient strength of solution is 0·07 to 0·09 per cent. KCN or 0·05 to 0·07 per cent. NaCN. The rate of dissolution is greatly reduced at high *p*H values.

White [6] has traced, mathematically and experimentally, the consequences of assuming that the rate of solution of gold in cyanide solutions under working conditions is mainly dependent on the rate of diffusion of oxygen.

---

[1] "*Cyaniding Gold and Silver Ores* " (Griffin) (1921), p. 88.
[2] *J. Chem. Met. Mng. Soc. S.A.*, 1923, 23, 170.
[3] *Trans. Amer. Inst. Mng. Eng.*, 1900, 30, 864.
[4] *J. Chem. Soc.*, 1893, 63, 724 ; 1895, 67, 199.
[5] "*Cyaniding Gold and Silver Ores*," 1921, p. 65.
[6] *J. Chem. Met. Mng. Soc. S.A.*, 1934, 35, 1.

The maximum solution rate is at about 0·027 per cent. KCN, equivalent to 0·020 per cent. NaCN, when saturated with oxygen.[1] When the cyanide concentration is lower than this, the diffusion rate will be the determining factor in a saturated oxygen solution. If there is greater concentration, the rate of diffusion and solubility of the oxygen will be hindered. The presence of excess oxygen, by means of increased pressure or of oxidising agents (which are also slow "cyanicides"), will raise the optimum cyanide concentration and the maximum rate of gold solution. In ordinary working conditions solutions are not saturated with oxygen; the rate of attack on the gold is thus diminished and the cyanide strength can be correspondingly reduced. From considerations based on experiment and on diffusion rates, it is deduced that in works practice the best cyanide strength for rapid action is between 0·02 and 0·03 per cent. KCN, but the exact strength can be determined only by actual trials for each case involved.[2]

The rate of diffusion depends partly on the *viscosity* of the mill solutions. F. C. Bond[3] advocates close attention to this factor. Viscosity is resistance to flow, or the reciprocal of fluidity. It will vary with the amount of dissolved salts and depends on the character of the ore, the grade of cyanide used, the lime and salts added, and the impurities originally present in the water. Bond gives a simple method of determining the viscosity of mill solutions, accurate to within ½ per cent.

Gold is dissolved by solutions of $Na_2Zn(CN)_4$, although the double cyanide remains unaltered so long as any simple cyanide is present. A solution of $Na_2Zn(CN)_4$ containing lime and with free access of oxygen is almost equivalent in dissolving power to one of equal strength of NaCN or KCN.

White found that when plates consisting of gold 870, silver 113 and base metal 17 (the average composition of Rand gold) were immersed in $Na_2Zn(CN)_4$ solution, the rate of dissolution of the gold increased steadily up to a strength corresponding to 0·027 per cent. KCN.

White found that contact of fine gold with carbon in the dark appears to increase the rate of dissolution in KCN, but attempts to make use of this fact have not been successful. When a solution of $Na_2Zn(CN)_4$ was used instead of KCN, the carbon contact was not advantageous below a strength equivalent to 0·032 per cent. KCN, and indeed disadvantageous up to that point.

Effect of Temperature.—Bagration[4] states that solution of metallic gold in potassium cyanide is facilitated by a rising temperature, and Loevy[5] mentioned that a Rand pyritic ore containing 50 to 60 per cent. of its gold in the free state yielded 10 to 14 per cent. higher extraction when heat was applied. The free gold was more affected by the heating of the solutions than the gold in the pyrite. An improvement in extraction of about $1 per ton on Tonopah ore, due to raising the temperature of the solution from 60° to 90° F., at a cost of 16 cents, is recorded by A. H. Jones,[6] and further details are given by von Bernewitz.[7] The increase in the consumption of cyanide, if any, is not given. Julian and Smart observed that up to 85° C. the solubility of

[1] *J. Chem. Met. Mng. Soc. S.A.*, 1919, 20, 1.
[2] *Op cit.*, 1934, 35, 9.
[3] *Min. and Met.*, 1926, 7, 15.
[4] *J. prakt. Chem.*, 1844, 46, 367.
[5] *J. Chem. Met. Mng. Soc. S.A.*, 1898, 2, 390.
[6] *Mng. and Sci. Press*, 1912, 104, 176.
[7] *Ibid.*, 1912, 105, 828.

gold in 0·25 per cent. KCN solution increases, but after that declines. H. Mayer[1] found the maximum reaction velocity in cyanide solutions to be at 80° C., but secondary reactions reduce the practical upper limit for solution of gold. Wraight[2] found that the use of hot air for the agitation of cyanide solutions resulted in better extraction, but attributed the effect to the increased activity of the oxygen.

In practice, the solutions are rarely heated. It is generally believed that an increase of temperature is accompanied by an increase in the consumption of cyanide, but exact data are lacking.

Dissolution of Base Metals.—The dissolution of mercury has already been mentioned on p. 304. Iron dissolves slowly in cyanide solutions, but is much less soluble than gold or silver.[3] The equation is probably as follows :—

$$Fe + 6KCN + H_2O + O = K_4Fe(CN)_6 + 2KOH$$

Zinc and copper dissolve readily in cyanide solution without the presence of free oxygen but with evolution of hydrogen, thus—

$$2Cu + 4KCN + 2H_2O = K_2Cu_2(CN)_4 + 2KOH + H_2$$
$$Zn + 4KCN + 2H_2O = K_2Zn(CN)_4 + 2KOH + H_2$$

Hence an excess of zinc dust is avoided in precipitation. Where zinc dust is used to precipitate copper from cyanide solution, a rapid and deleterious accumulation of zinc in solution soon occurs. The sodium-zinc cyanide formed is only a weak solvent for gold, but the addition of lime or soda liberates free cyanide and thereby raises the efficiency of the solution.

In the presence of free oxygen, however, gold dissolves in cyanide about six times as fast as zinc.[4] The relative rate of attack is in the ratio of the equivalents concerned :—

$$\frac{\text{Gold dissolved}}{\text{Zinc dissolved}} = \frac{2 \times 197{\cdot}2}{65{\cdot}38} = 6{\cdot}03$$

In the absence of oxygen, zinc, of course, easily displaces gold from aurocyanide solutions.

Lead and tellurium dissolve very slowly, and aluminium apparently not at all, but the latter is readily soluble in alkali ; arsenic (according to Julian), platinum and carbon do not dissolve in cyanide. Tellurium dissolves in alkali in the presence of oxygen.

Action of Cyanide on Tellurides of Gold.—Cyanide solutions act very slowly on tellurides of gold, but dissolve both elements, whether in a fused mixture or in native tellurides.

P. F. Thompson[5] is of the opinion that in Kalgoorlie ore the gold is reduced in the oxidation of the ion $Te^{=}$ to $TeO_3^{=}$. Thus :—$6Au^+ + Te^= + 3H_2O = 6Au + TeO_3^= + 6H^+$. Many of the Kalgoorlie specimens (according to Stillwell) contain laminations of metallic gold. In certain acid cyanide solutions metallic gold is thrown down and the tellurides are dissolved. Thompson advocates an electrochemical theory of the solubility of the tellurides and suggests that dissolved oxygen, or oxidising agents such as

[1] *Metall u. Erz*, 1931, **28**, 261.
[2] *Trans. Inst. Mng. Met.*, 1915, **25**, 59.
[3] Julian and Smart, "*Cyaniding Gold and Silver Ores,*" 1921, p. 84.
[4] H. A. White, "*Rand Metallurgical Practice,*" vol. i., p. 549.
[5] *Chem. Eng. and Min. Rev.*, 1933, **26**, 90.

bromocyanide, may assist by oxidising the telluride ion, thus leaving the gold ions free to form the complex $[Au(CN)_2]^-$. The rate of dissolution of the gold is greatly increased by the addition of bromocyanide.

"It is evident that there is a wide difference in the solubility of the gold in different telluride minerals, and even in the same minerals from different localities. Thus in Western Australia, where the gold is in a sulpho-telluride, the ore can be dealt with by fine grinding and long treatment, whereas the same treatment applied to certain Mexican and United States tellurides is less successful."[1]

In a cyanide solution from the Goldfield Consolidated Mill, von Schultz and Low found 0·00008 per cent. of tellurium and 0·0006 per cent. of gold.[2]

Haultain and Johnston [3] heated tellurides containing gold on a pyrex glass slip to red heat by means of a Bunsen burner. The metallic gold remained as bright particles in a lake of molten telluride. Johnston [4] considers that the tellurides in themselves are not responsible for any of the difficulties in the recovery of gold from the Kirkland Lake ores. He summarises his conclusions as follows :—

(1) Auriferous tellurides yield up their gold to cyanide if they are in a finely divided state and if excess lime is used.

(2) Sodium peroxide greatly reduces the time of treatment.

(3) The tellurides are very brittle, and of high density. They are therefore retained in the mill circuit for a long time, and become very finely divided (approximately — 1,600 mesh).

(4) Up to the present, gold-bearing tellurides have not been found in large quantities in mill tailings, or in concentrates recoverable from them.

Action of Cyanide on Compounds of Silver.—Silver chloride is readily soluble in cyanide, and the arsenate is also rapidly dissolved. Silver sulphide and antimonide are less easily acted on, but are not so refractory as metallic (cement) silver. The presence of copper salts appears to exercise a detrimental action on the solubility of silver sulphide.

Silver sulphide is more slowly dissolved than metallic gold, and strong solutions, containing up to 1 per cent. KCN or more are required. Caldecott has shown [5] that its dissolution by the primary reaction

$$Ag_2S + 4NaCN = 2NaAg(CN)_2 + Na_2S$$

is stopped by the sulphide of sodium, which reprecipitates silver if present in excess and soon establishes an equilibrium. The soluble sulphides may be removed by secondary reactions, the principal ones being

$$2Na_2S + 2O_2 = Na_2S_2O_3 + Na_2O$$

followed by

$$Na_2S_2O_3 + Na_2O + 2O_2 = 2Na_2SO_4$$

The other secondary reactions also involve the presence of oxygen, so that aeration is as important in the treatment of silver ores as it is in that of gold ores. In the treatment either of gold or silver ores soluble sulphides may be removed or rendered harmless by the addition of lead acetate, lead nitrate, litharge, or mercury salts.

[1] Julian and Smart, "*Cyaniding Gold and Silver Ores,*" 1921, p. 88.
[2] Von Bernewitz, "*Cyanide Practice, 1910 to 1913,*" p. 566.
[3] *Can. Inst. Min. Met.*, 1933, **36**, 217.
[4] *Loc. cit.*, p. 225.
[5] *J. Chem. Met. Mng. Soc. S.A.*, 1908, **8**, 266.

The arsenical and antimonial sulphides of silver (ruby silver, stephanite, etc.), are not readily dissolved by cyanide, and silver contained in native galena, tetrahedrite and zinc blende is not amenable to treatment.

Action of Potassium Cyanide on Metallic Salts and Gangue Minerals in Ores.—The ordinary gangue of most ores (viz., silica and silicates of the alkalies and alkaline earths) exercises no direct influence on the cyanide solution. The carbonates of the alkaline earths are also probably without influence. The decomposing effects of sulphides of the heavy metals vary with the physical state of the sulphides; hard undecomposed sulphides are very slowly attacked, but soft partly decomposed sulphides are more soluble, and in many cases act as "cyanicides." Arsenical ores behave similarly. Certain zinc minerals, including smithsonite, hydrozincite and zincite, dissolve in ordinary cyaniding in sufficient quantity to interfere with the efficiency of the solution. Soluble chromium ores stop precipitation.[1]

MacArthur and Forrest found that dilute cyanide solutions generally exercise a "selective action" in dissolving gold and silver in whatever form these may be present, in preference to sulphides or other salts of the base metals, and that gold dissolves in preference to silver. There are, however, exceptions to this rule. "For instance, cyanide of potassium solution has a strong tendency to dissolve precipitated sulphide of zinc, but its action on the natural sulphide of zinc, blende, is almost *nil*. The same holds for compounds of iron. The action of dissolved oxygen is all-important not only in dissolving gold but also other constituents of the ore."

J. S. MacArthur states [2] that "precipitated sulphide of copper is rapidly dissolved, and also a sooty form of the same substance occasionally met with as a mineral occurring in ores. On the other hand, fused copper matte is scarcely acted on at all, and in a great many cases the same may be said of the hard dense sulphides of copper usually found in nature. Sulphide of zinc exhibits the same differences of behaviour: the 'black-jack' concentrates of the Ravenswood Mine, Queensland, can be treated with good results, little zinc being dissolved. Again, oxide of copper, if freshly precipitated, is strongly acted on by the cyanide, but if it is heated to dull redness in a muffle it becomes insoluble, and a large excess of this material added to a gold ore makes no difference in the percentage of extraction, while the consumption of cyanide is not increased by its presence. The action of cyanide solutions on sulphide of silver is similarly dependent on its physical state."

Prof. Christy's experiments on the electromotive force of metals and minerals in cyanide solutions, compared with Ostwald's normal electrode, show [3] that many minerals tend to be less rapidly dissolved than pure metallic gold. For example, chalcopyrite has hardly more tendency to go into solution than pure pyrite, while bornite and copper glance have a strong tendency to set up electric currents and to dissolve. Pure chalcopyrite, galena, argentite, magnetic pyrites, fahlore, mispickel, blende, boulangerite, bournonite, ruby silver ore, stephanite and stibnite, when free from their oxidation products, are apparently little acted on by cyanide solutions, but no information is given as to the physical condition of the minerals tested.

Action of Potassium Cyanide on Oxidised Sulphides.—When pyrite, pyrrhotite or marcasite has been subjected to the action of air and water

[1] H. D. Bell, *J. Chem. Met. Mng. Soc. S.A.*, 1936, 36, 385.
[2] Private communication.
[3] *Trans. Amer. Inst. Mng. Eng.*, 1900, 30, 864.

(" weathering ") before treatment, compounds are formed which are more prejudicial to the solution than the sulphides. One change is the loss of an atom of sulphur, thus :—

$$FeS_2 = FeS + S$$

As much as 2 lb. of colloidal sulphur per ton has been found in oxidised ore from the Geduld mine on the Rand.[1]

Oxidation results in the formation of ferrous sulphate, $FeSO_4$, and free sulphuric acid. Further oxidation gives ferric sulphate, $Fe_2(SO_4)_3$, which eventually loses acid and becomes a basic sulphate, $2Fe_2O_3.SO_3$. Other basic salts of complex composition appear to be formed also, as well as ferric and ferrous hydroxides, $Fe(OH)_3$ and $Fe(OH)_2$. It is the ferrous salts and free acid which are deleterious to the cyanide solution.

It is generally agreed that marcasite is more to be feared than pyrite which, according to White,[2] may be oxidised directly to ferric hydroxide, giving no ferrous iron. He is also of the opinion that, in slightly alkaline water saturated with oxygen, the slow oxidation of pyrite yields primarily sulphite and thiosulphate, while pyrrhotite gives only thiosulphate. The thiosulphate does not appear to retard the dissolution of gold.

The oxidation of pyrite also occurs to some extent during treatment, on the surface layers of sand charges and during prolonged aeration and agitation of slime charges. However, pure pyritic concentrate, free from sand and metallic iron, oxidises but slowly. Pyrrhotite has been shown to have a detrimental effect during clarification by a sand filter (see p. 316).

Ferrous sulphide is produced by grinding pyrite between iron surfaces under acid conditions :—

$$FeS_2 + Fe = 2FeS$$

With alkaline solutions the effect is less marked. The reaction occurs in the stamp battery and steel-lined tube mills to some extent. The ferrous sulphide yields deleterious products, as it does if formed in weathered ore or tailing.

In the presence of oxidised copper and iron pyrites, the following reactions take place :—

(1) The free sulphuric acid liberates hydrocyanic acid,

$$H_2SO_4 + 2KCN = K_2SO_4 + 2HCN$$

The sulphuric acid also attacks silicate of alumina or magnesia present in the ore, producing gelatinous colloidal silicic acid which retards the settlement of slime, necessitating the heating of solutions.[3] The soluble sulphates can be precipitated by lime.

(2) Ferrous sulphate reacts with the cyanide, forming ferrous cyanide, which dissolves in the excess of potassium cyanide, so that it does not appear in the free state.

$$FeSO_4 + 2KCN = Fe(CN)_2 + K_2SO_4$$
$$Fe(CN)_2 + 4KCN = K_4Fe(CN)_6$$

[1] H. A. White, "*Rand Metallurgical Practice*," vol. i., p. 544.
[2] *Loc. cit.*
[3] W. A. Caldecott, *J. Chem. Met. Mng. Soc. S.A.*, 1907, 7, 217; "*Rand Metallurgical Practice*," vol. i., p. 383.

The potassium ferrocyanide, if sufficient acid be present, reacts with fresh ferrous sulphate forming a bluish-white precipitate.

$$FeSO_4 + K_4Fe(CN)_6 = K_2Fe_2(CN)_6 + K_2SO_4$$

This precipitate oxidises in the air to Prussian blue if free acid is present—

$$4K_2Fe_2(CN)_6 + O_2 + 2H_2SO_4 = 3Fe(CN)_2.2Fe_2(CN)_6 \text{ (Prussian blue)} + K_4Fe(CN)_6 + 2K_2SO_4 + 2H_2O$$

Both these precipitates are decomposed by potash or soda and, therefore, cannot be formed in their presence. The reactions may be represented as follows :—

$$K_2Fe_2(CN)_6 + 2KOH = K_4Fe(CN)_6 + Fe(OH)_2$$
$$3Fe(CN)_2.2Fe_2(CN)_6 + 12NaOH = 3Na_4Fe(CN)_6 + 4Fe(OH)_3$$

Consequently, if free acid is not present Prussian blue is hardly formed at all, as the solution becomes alkaline, and the precipitate is decomposed as fast as it is formed.

It follows from these reactions that if the blue colour of Prussian blue is visible in the vats or on the surface of the tailing heaps, waste of cyanide must have taken place.

(3) Ferric sulphates are decomposed by potassium cyanide, hydrocyanic acid being evolved and ferric hydroxide precipitated.

(4) A mixture of ferrous and ferric sulphates produces Prussian blue by reacting with potassium cyanide, ferrocyanide of potassium being formed at first as above ; the equation is—

$$3K_4Fe(CN)_6 + 2Fe_2(SO_4)_3 = 3Fe(CN)_2.2Fe_2(CN)_6 + 6K_2SO_4$$

Here again the waste of cyanide is reduced by keeping the solutions alkaline.

(5) Sulphate of copper, $CuSO_4$, acts differently from $FeSO_4$, cuprous cyanide, $Cu_2(CN)_2$, being formed, soluble in excess of KCN to give $K_2Cu_2(CN)_4$, a compound very prone to decomposition. Copper sulphate also gives a precipitate with potassium ferrocyanide, thus—

$$K_4Fe(CN)_6 + CuSO_4 = K_2CuFe(CN)_6 + K_2SO_4$$

Sulphate of copper acting on KCN or HCN appears also to produce nascent cyanogen, and so to promote dissolution of gold, thus :—

$$CuSO_4 + 2HCN = CuCN + H_2SO_4 + CN$$

Copper and zinc in the condition of hydroxides or carbonates are quickly dissolved in preference to the precious metals. If sulphates of these metals are formed in an ore containing limestone or clay, double decomposition occurs with the production of sulphate of lime or alumina, and oxides or carbonates of the heavy metals, which are dissolved by the cyanide, thus—

$$ZnSO_4 + CaCO_3 = ZnCO_3 + CaSO_4$$
$$ZnCO_3 + 2KCN = Zn(CN)_2 + K_2CO_3$$
$$Zn(CN)_2 + 2KCN = K_2Zn(CN)_4$$

Copper carbonate is a troublesome cyanicide. According to Clennell, the reaction which occurs is as follows :—

$$2CuCO_3 + 7KCN + 2KOH = K_4Cu_2(CN)_6 + KCNO + 2K_2CO_3 + H_2O$$

The cyanide consumption is greater, in general, where the copper minerals are oxidised than when they are sulphides.

(6) Ferrous hydroxide, when formed as above, is instantly dissolved in KCN, thus—

$$Fe(OH)_2 + 6KCN = K_4Fe(CN)_6 + 2KOH$$

The following reaction also occurs :—

$$2Fe(OH)_2 + O + H_2O = 2Fe(OH)_3$$

Ferric hydroxide, however formed, does not act on potassium cyanide.

(7) Ferrous sulphide is acted on as follows :—[1]

$$FeS + 6NaCN = Na_4Fe(CN)_6 + Na_2S$$
$$Na_2S + NaCN + O + H_2O = NaCNS + 2NaOH$$
$$2FeS + 9O + 3H_2O + 2CaO = 2Fe(OH)_3 + 2CaSO_4$$

Ferrous sulphide also reacts slowly to produce thiosulphate by oxidation, thus withdrawing oxygen from the circuit.

Ferrous sulphate produced by the oxidation of pyrite and adsorbed on the silica grains reacts very slowly with a saturated solution of lime. If free colloidal sulphur is present, it is dissolved readily in strongly alkaline cyanide solutions ; sodium thiosulphate is formed continuously until all the sulphur is removed. The sulphur is only slightly acted upon by the lead-zinc couple, but a coating of zinc sulphide is thereby formed on the zinc. If lime be sprinkled two days before the removal of an ore which has been allowed to stand and become oxidised by exposure in the stopes, then the gold lost in tailings appears to diminish.[2]

Pyrrhotite is regarded by Whitehead [3] and also by Goodchild and Wilder [4] as a particularly harmful "cyanicide," especially in alkaline solution. The latter investigators consider that it probably acts as a catalyst and thereby influences the effect of a quantity of material much larger than its own mass. Moreover there may, at the same time, be some decomposition of gangue minerals from the same causes. With oxygenated cyanide solutions, whether acid or alkaline, pyrrhotite enters into vigorous reaction and even the decomposition of a small quantity of the mineral leads to extensive effects. While pyrrhotite and iron separately react but slowly with dilute sulphuric acid, when they are together the reaction becomes vigorous. The action of pyrrhotite has been further discussed as follows.[5]

Pyrrhotite has the general formula Fe_xS_{x+1}. One atom of sulphur is very loosely held, while the remaining complex is easily oxidised to ferrous

---

[1] Caldecott, "*Rand Metallurgical Practice,*" vol. i., p. 387.

[2] H. A. White, *J. Chem. Met. Mng. Soc. S.A.*, 1920, **21**, 61.

[3] *Kolar Gold Fields Min. Met. Soc.*, 1929, Bull. 25.

[4] *Min. Mag.*, 1924, **31**, 208.

[5] See R. J. Lemmon, B. L. Gardiner, J. H. French and H. Jones, *Trans. Inst. Min. Met.*, 1933, **42**, 249, 263.

and ferric sulphates. The rate of decomposition of pyrrhotite by cyanide is greater than that of other common pyritic minerals.

$$Fe_7S_8 + NaCN = NaSCN + 7FeS$$
$$FeS + 4O = FeSO_4$$
$$FeSO_4 + 2NaCN = Fe(CN)_2 + Na_2SO_4$$
$$Fe(CN)_2 + 4NaCN = Na_4Fe(CN)_6$$

Where pyrrhotite is present litharge is sometimes added to the grinding mills :—

$$Fe_7S_8 + PbO = PbS + 7FeS + O$$

The ferrous sulphate resulting from the oxidation of the FeS will, if the alkalinity is strictly controlled, react as follows :—

$$FeSO_4 + PbO = PbSO_4 + FeO$$
$$2FeO + 3H_2O + O = 2Fe(OH)_3$$

i.e. without the production of ferrocyanides. Alkaline cyanide, as mentioned already, has little action on ferric hydroxide.

Arsenopyrite similarly causes a waste of cyanide when decomposing. Alkali sulphides are formed which absorb oxygen and necessitate the aeration of the solution. Stibnite is worse in these respects, and ores containing it are generally refractory.

Speaking generally, decomposing sulphides act as "cyanicides," destroying the cyanide, but copper ores are probably the most fatal in this respect. They also act as deoxidisers, necessitating much aeration, and also the removal of soluble sulphides by solutions of lead, mercury, etc. For the effect of lead sulphide, see p. 315.

In residues cyanide is quickly destroyed by oxidation to cyanate and by hydrolysis to ammonia and probably urea. Some sulphocyanide may be formed by the reactions :—

$$FeS_2 + 2O_2 = FeSO_4 + S$$
$$KCN + S = KCNS$$

Effect of Alkalies.—Since acidity of the ore causes decomposition of the cyanide, an obvious method of reducing the loss is to add alkali in some form. Lime is usually added either to the ore or to the mill service water; the amount required is carefully estimated, but an excess is added to give some amount of "protective alkali." With Rand ores an alkalinity equivalent to 0·03 per cent. NaOH gives the best aggregation, the most rapid settling and the most complete removal of colloidal sulphur (White). The "protective alkali" is kept as low as possible consistent with a reasonable consumption of cyanide.[1] One difficulty is that alkalies react with many ores forming deleterious compounds. On the other hand, alkaline sodium-zinc cyanide is a good solvent for gold. Wright [2] mentions that when dealing with a complex sulphide ore having a heavy cyanide consumption of 4½ lbs. per ton, any alkalinity in excess of 0·001 per cent. CaO was fatal to extraction. Treating 4 dwt. ore with 0·05 per cent. cyanide, an alkalinity of 0·005 to 0·01

---

[1] Robertson, *Trans. Inst. Min. Met.*, 1924, 34, [i.], 84.
[2] *Trans. Inst. Min. Met.*, 1933, 42, 239.

per cent. CaO caused gold extraction practically to cease. At Shamva compressed air is passed through lime water in a special tank. It is found that thereby a saving in lime and cyanide is effected.[1]

Lime water converts insoluble basic ferric sulphate into ferric hydroxide and calcium sulphate—

$$2Fe_2O_3.SO_3 + Ca(OH)_2 + 5H_2O = 4Fe(OH)_3 + CaSO_4$$

Also,

$$H_2SO_4 + Ca(OH)_2 = CaSO_4 + 2H_2O$$

$$FeSO_4 + Ca(OH)_2 = Fe(OH)_2 + CaSO_4 \text{ (minimum aeration)}$$

$$4FeSO_4 + 4Ca(OH)_2 + 2H_2O + O_2 = 4Fe(OH)_3 + 4CaSO_4 \text{ (full aeration)}$$

The calcium sulphate forms a saturated solution, and is deposited in the mill pipes and filter cloths, necessitating cleaning (Caldecott). The ferric hydroxide does no harm. Ferrous hydroxide dissolves in cyanide, forming ferrocyanides, and absorbs oxygen, as above, p. 311. Materials containing ferrous sulphate and ferrous sulphide are, therefore, not rendered innocuous by lime. During the ore treatment they are gradually converted into ferric salts by the action of air or cyanide or both. Alkali does not protect cyanide from copper salts.

In the Far Eastern Rand the natural water is alkaline due to magnesium carbonate and bicarbonate which have to be precipitated before the added lime becomes effective for normal alkalinity purposes.

Damp ore that has suffered long exposure underground shows the yellow colour of calcium polysulphide when treated with lime water.

" Caustic soda was at one time commonly used as a protective alkali, but besides being dearer than lime, it tends to introduce sulphides into solution from the ore and to retard flocculation and settlement or percolation, possibly by converting the colloids present into the 'sol' form of turbid suspension,[2] and likewise hinders precipitation, possibly by inducing more rapid formation of $Zn(OH)_2$ coating on the zinc shavings.[3] When much organic matter is present in tailings under treatment, the use of caustic soda assists the introduction into the solution of soluble organic compounds, which resemble a solution of soap in preventing the ready escape of hydrogen given off by the zinc sponge, so that the latter may rise with the froth above the solution level. Lime, in this respect as in others, displays a marked clarifying effect." [4]

Decomposition of Potassium Cyanide.—Hydrocyanic acid is one of the weakest acids known, and is expelled from its salts by all mineral acids and many organic acids. Wade and Panting [5] showed that by the action of sulphuric acid on potassium cyanide both hydrocyanic acid and carbon monoxide may be formed, and by varying the concentration of the acid a quantitative yield of either may be obtained. Harker [6] found, however, that

---

[1] *Trans. Inst. Min. Met.*, 1924, 33, 429.

[2] For a discussion of "gel" and "sol" forms of colloids and the influence of reagents on them as affecting slime settlement, see H. E. Ashley, *Mng. and Sci. Press*, 1909, 98, 831, and *Trans. Amer. Inst. Mng. Eng.*, 1910, 41, 380.

[3] *J. Chem. Met. Mng. Soc. S.A.*, 1898, 2, 319; Park, "*Cyanide Process*," p. 319.

[4] W. A. Caldecott, "*Rand Metallurgical Practice*," vol. i., p. 388.

[5] *J. Chem. Soc.*, 1898, 73, 255.

[6] *J. Soc. Chem. Ind.*, 1921, 40, 185T.

only a small proportion of hydrocyanic acid is evolved as gas when sulphuric acid reacts with potassium cyanide under atmospheric conditions. Greatly increased yields are obtained by working at lower temperatures and pressures. Carbonic acid gas decomposes potassium cyanide in the presence of water thus:—

$$2KCN + CO_2 + H_2O = 2HCN + K_2CO_3$$

The smell of hydrocyanic acid, noticeable whenever KCN or its solutions are exposed to air, is accounted for by this reaction. The cyanide is protected by the presence of a slight excess of alkali.

In the presence of air, potassium cyanide takes up oxygen, and is converted first to cyanate and then to carbonate—

$$KCN + O = KCNO$$
$$2KCNO + 3H_2O = K_2CO_3 + CO_2 + 2NH_3$$

These reactions are very slow, but become much more rapid if heat is applied. Strong solutions turn brown in the air. Clevenger and Morgan[1] found that the atmospheric decomposition of cyanide may be serious in solutions containing only a little "protective alkalinity," but if the latter is greater, the loss is not serious. Any cyanide lost by reaction with oxygen is in the form of cyanate, carbonate, urea or ammonia, and as such is irrecoverable.

In dilute solutions, potassium cyanide suffers hydrolytic dissociation, and is partly changed into HCN and KOH. It results from this that the passage of a stream of any neutral gas such as nitrogen through the solution causes an evolution of hydrocyanic acid, while the solution becomes alkaline. The equation is—

$$KCN + H_2O \rightleftarrows HCN + KOH$$

The extent of the hydrolytic dissociation is 1·12 per cent. of the salt in a solution containing 0·65 per cent. KCN and 2·34 per cent. of the salt in a solution containing 0·16 per cent. KCN.[2] In weaker solutions, the extent of the hydrolytic dissociation is greater. If the solution is boiled with acids or alkalies, hydrolysis of the cyanide occurs rapidly, ammonia and formates being formed, thus—

$$KCN + 2H_2O = NH_3 + HCO_2K$$

This equation does not represent the whole effect, as acetates and other organic substances are also formed. The reactions also proceed, although very slowly, if the solution is cold and neither acids nor alkalies are present. Hydrolysis is possibly strongest when crushing in cyanide in the tube mill, where the generated heat is greatest.

The destruction of cyanide by "cyanicides," as described in previous sections, is far more serious than this spontaneous decomposition in water.

Soluble Sulphides in Cyanide Solutions.—A small percentage of a soluble sulphide present in the cyanide solution greatly delays the dissolution of gold. Doubtless this is partly owing to the abstraction of oxygen from the solution by the sulphide, for gold sulphide is freely soluble in KCN, so that the surface of the metal should remain free from sulphide. Bettel, however, points out[3] that silver sulphide is far less soluble than gold sulphide, and

---

[1] *Mng. and Sci. Press*, 1916, **113**, 413.
[2] J. Shields, *Phil. Mag.*, 1893, **35**, 387.
[3] *South African Mining J.*, May 8, 1897.

that if native gold containing 20 per cent. of silver is treated, a film almost insoluble in (dilute) cyanide solutions may be formed. It is certain that some specimens of pure gold leaf dissolve with great difficulty if they have been previously dipped in sulphide solutions, or if traces of soluble sulphides or sulphocyanides are present in the solution. The difficulty disappears if the sulphides are removed, either by being precipitated with lead salts, or by the action of certain oxidisers.

Caldecott observes that the soluble sulphides are derived from commercial cyanide, and from the action of cyanide on sulphides of iron in the ore. They become converted into sulphocyanides at the expense of cyanide and oxygen, thus :—[1]

$$Na_2S + NaCN + O + H_2O = NaCNS + 2NaOH$$

The alkali sulphides are usually removed by dissolved salts of lead or mercury. Conversely an accumulation of base metals in the solution can be got rid of by the addition of soluble sulphides.

Dissolved lead salts precipitate soluble sulphides. Lead acetate reacts with cyanide solutions yielding hydrate or oxycyanide of lead, and this is dissolved as an alkali plumbite, $PbO.xNa_2O$. The reaction proceeds thus—[2]

$$PbO.xNa_2O + Na_2S = PbS + (x + 1)Na_2O$$
$$PbS + NaCN + O = PbO + NaCNS$$

so that the insoluble lead sulphide is also an absorbent of oxygen and cyanide. Any soluble lead remaining " passes into the zinc boxes, where it is deposited as metallic lead, thus constantly renewing the activity of the zinc-lead couple by forming fresh uncoated precipitating surfaces " (Caldecott). However, some may be wasted as lead carbonate, which is more insoluble than calcium carbonate (White).

G. H. Clevenger observes [3] that soluble sulphides are readily oxidised and seldom occur in mill solutions, but that lead salts may be advantageous :

(1) By removing sulphur ions and so accelerating the dissolution of silver occurring as sulphide.

(2) By retarding the formation of sulphocyanide and the separation of free sulphur, and so saving cyanide.

(3) By checking the fouling tendency of copper on the solution.

He also finds that—

(4) An excess of lead is disadvantageous, but this is less observable when litharge is used.

(5) Lead is more advantageous with silver or silver-gold ores than with gold ores.

Reprecipitation of Gold and Silver in the Leaching Vats.—If the solution is acid there is said to be some danger of a precipitation of gold previously dissolved, insoluble aurous cyanide being thrown down, according to the equation—

$$KAu(CN)_2 + HCl = KCl + HCN + AuCN$$

This, however, need not be feared as long as there is an excess of KCN, which must all be destroyed by the acid before the aurous cyanide can be

[1] "*Rand Metallurgical Practice*," vol. i., p. 388.
[2] *Op. cit.*, p. 387.
[3] *Mng. and Sci. Press*, 1914, 109, 635.

precipitated. Moreover, aurocyanhydric acid, $HAu(CN)_2$, is formed and remains in solution, AuCN being precipitated only as the HCN leaves the solution by evaporation.[1] There is danger in transferring a solution containing gold to a vat containing pyritic material. If the latter should contain any soluble salts of the heavy metals, insoluble salts are thrown down—*e.g.* :—

$$2KAu(CN)_2 + ZnSO_4 = K_2SO_4 + ZnAu_2(CN)_4$$

This trouble has occurred in the Kolar and Rand areas when mill sand containing small quantities of pyrrhotite was used for clarification purposes.

A. N. Mackay attributes reprecipitation of gold to the action of free alkali on fine sulphides,[2] by which soluble sulphides are produced. In one case in which the protective alkali amounted to 0·04 per cent. and the cyanide to 0·02 per cent., complete reprecipitation of dissolved gold took place in 18 hours. The slime in contact with the solution contained 0·5 per cent. of FeS_2. By reducing the alkalinity to 0·01 per cent., the reprecipitation was prevented.

"The presence of carbonaceous or decomposing organic matter in the sand or slime undergoing cyanide treatment is distinctly detrimental on account of its oxygen-absorbing properties and tendency to precipitate gold already in solution. Semi-burnt coal in commercial lime," dirt of various kinds, decomposing products of vegetation, or carbon in certain forms, may be "the cause in certain cases of reprecipitation of gold from cyanide solution when kept in contact with sand for a few days in a closed vessel; and further, its reducing capacity being stimulated by heat, serves to explain why in practice warming of sand solutions may result in less rapid dissolution of the gold content than with solutions at ordinary temperatures." [3]

Reactions in the Zinc Boxes.[4]—It was believed at one time that the precipitation of gold and silver by zinc was effected by simple displacement, according to the equation—

$$(1)\ 2KAu(CN)_2 + Zn = K_2Zn(CN)_4 + 2Au$$

This simple view is now generally discredited. The principal observed facts are as follows :—

(1) An excess of free cyanide favours the precipitation of gold.[5] Sometimes it is necessary to add cyanide direct to the solution in the zinc boxes. White has shown that with no free cyanide and no alkalinity zinc does not precipitate gold from aurocyanide, and only very little precipitation occurs with the zinc-lead couple in the absence of oxygen. Rapid precipitation occurred, however, when the solution contained 0·021 per cent. KCN and 0·0020 per cent. NaOH. Even then more zinc was dissolved than was equivalent to the gold brought down. The presence of an excess of

---

[1] Lindbom, *Bull. Soc. chim.*, 1878, 29, 422. See also "Cupriferous Ores—Treatment by Acid Solutions," Chap. XV.

[2] "Reprecipitation of Gold from Cyanide Solutions," *Trans. Inst. Min. Met.*, 1905, 14, 541.

[3] Caldecott, "*Rand Metallurgical Practice*," vol. i., p. 387.

[4] See W. A. Caldecott and E. H. Johnson, *J. Chem. Met. Mng. Soc. S.A.*, 1903, 4, 263.

[5] W. R. Feldtmann, *Eng. and Mng. J.*, 1894, 98, 126.

alkali and a deficiency of cyanide apparently favours the production of sodium zincate, thus—

$$Zn + 2NaOH = Zn(NaO)_2 + 2H$$

(2) An evolution of hydrogen takes place during precipitation, but free gaseous hydrogen does not precipitate gold. Very few hydrogen bubbles rise from the extractor boxes with the dilute solutions now used, however.

(3) Zinc dissolves in the solution, the amount depending on its strength. The waste of zinc is such that it may require 15 or 20 parts of zinc to precipitate 1 part of gold. Part of the zinc is disintegrated and remains with the precipitated gold.

(4) The precipitation is aided by the presence of metals, such as lead, which are electronegative to zinc in cyanide solutions. The lead increases the rate of dissolution of the zinc. It must, of course, be in contact with it, and generally is in the nature of a coating which forms a couple.

(5) It is necessary that the zinc should be of very great area for efficient precipitation, especially in solutions weak in cyanide. Zinc plates are unsuitable. Zinc shavings (or "thread") and zinc dust ("fume") are suitable.

The large area is required primarily to expose the largest possible surface to the gold solution in order to expedite precipitation. Further, "only a proportion of the zinc area usually does useful work, the remainder being rapidly coated with lime salts and zinc compounds (see p. 322), and thus put out of action. . . . The lead-zinc couple was first introduced to reduce the tendency of precipitated copper to form a smooth coherent metallic coating upon the zinc shavings."[1] Where organic matter is present in the solution, it may have the result of giving a continuous coating of lead on the zinc shavings, instead of scattered crystals, thus more or less isolating shavings from the cyanide solution. The removal of organic matter from mill solutions by sodium or calcium hypochlorite has been suggested.

"Hydrogen to some extent polarises the zinc by coating it with a thin film of gas, which prevents its actual contact with the gold in the solution; and no doubt the efficacy of a lead coating is in part due to the roughened surface, consequent on its use, assisting bubbles of hydrogen to form and escape."[2] Zinc shavings, by reason of their ragged edges, have a similar advantage.

(6) "In addition to gold, any silver, mercury, copper, cobalt and nickel present in solution as double cyanides are precipitated by zinc in the same way as gold, the copper being occasionally visible in the red coating produced on the shavings, whilst the zinc shavings are rendered brittle should much mercury be present."[3] Other elements, such as dissolved lead, tellurium and selenium, are also precipitated, the precipitate sometimes containing appreciable amounts of metals which previously existed in the ore and the solutions in such small proportions as to escape detection.

(7) "During the passage of the solution through the boxes some cyanide is consumed, and the free alkali in solution is increased, though in neither case is the amount very great in practice."[4]

---

[1] Caldecott, "*Rand Metallurgical Practice*," vol. i., p. 391.
[2] *Op. cit.*, p. 390.
[3] *Op. cit.*, p. 391.
[4] *Op. cit.*, p. 390.

(8) "Heat assists precipitation, but increases the waste of cyanide by decomposition." About as much gold is precipitated by zinc in twenty-four hours at 20° C. as in two hours at 80° C. In the latter case there is, according to MacArthur, "an enormous waste of cyanide by the formation of urea, which manifests itself by its strong unpleasant odour."[1]

(9) The proportion of dissolved gold which is precipitated in the zinc boxes diminishes as the solutions become weaker in gold. Hence some gold, say 0·02 dwt. per ton, is always left in the solution.

(10) The gold is redissolved if it becomes detached from contact with zinc, provided that oxygen is still present, and "air blown through gold-coated shavings immersed in cyanide solution raises the gold value of such solution. A similar effect can be produced by exposing such shavings to the air in a shallow vessel partly filled with solution."[2] The solutions leaving the boxes were practically free from oxygen even in early days.[3]

Precipitation is favoured by reducing conditions and prevented by oxidising conditions. It is thus the exact opposite of dissolution. Caldecott states the main reactions, thus :—[4]

$$(2)\quad Zn + 2H_2O = Zn(OH)_2 + 2H(+17{\cdot}4 \text{ cals.})^{5}$$
$$(3)\quad 2NaAu(CN)_2 + 2H = 2HCN + 2NaCN + 2Au$$
$$(4)\quad HCN + NaOH = NaCN + H_2O$$

He points out that the oxidation of zinc raises the temperature of the solution by amounts up to 1° F. in passing through the boxes. This oxidation can hardly take place except owing to contact with an element, such as lead, gold, mercury, etc., which is electronegative to zinc. It may be argued that any hydrogen is in all probability in the nascent state and mainly produced according to the reaction—

$$(5)\quad Zn + 4NaCN + 2H_2O = Na_2Zn(CN)_4 + 2NaOH + 2H$$

which must take place in the absence of metal to be precipitated, and, therefore, can hardly fail to be generally proceeding. This view of the method of production of hydrogen by zinc brings zinc precipitation into line with precipitation by aluminium, the essential fact in each case being the liberation of hydrogen during the d ssolution of the metal. In the case of aluminium, Moldenhauer found[6] that an excess of alkali was necessary, and this dissolves aluminium with the formation of an aluminate, hydrogen being given off. It follows that free cyanide should not be required in precipitation by aluminium. However, the equations for the precipitation of silver given by E. M. Hamilton[7] are the following :—

$$3NaAg(CN)_2 + 3NaOH + Al = 3Ag + 6NaCN + Al(OH)_3$$
$$2Al(OH)_3 + 2NaOH = Na_2Al_2O_4 + 4H_2O$$

---

[1] Private communication.

[2] Caldecott, "*Rand Metallurgical Practice*," vol. i., p. 393.

[3] A. F. Crosse, *J. Chem. Met. Mng. Soc. S.A.*, 1898, **2**, 402.

[4] "*Rand Metallurgical Practice*," vol. i., p. 389.

[5] There is also the reaction $Zn + 2NaOH = Na_2ZnO_2 + 2H$ to be considered, especially on the application of heat.

[6] Julian and Smart, "*Cyaniding Gold and Silver Ores*," p. 238.

[7] *Eng. and Mng. J.*, 1913, **95**, 936.

but for low-grade solutions, when caustic alkali and aluminium are both present in excess, he prefers—

$$2NaAg(CN)_2 + 4NaOH + 2Al = 4NaCN + 2Ag + Na_2Al_2O_4 + 4H$$

Aluminium does not react directly with cyanide.

Reaction No. (5) occurs the more readily the more cyanide is present, and if it is regarded as the essential reaction, the desirability of the presence of free cyanide is at once apparent. In this respect Dr. Caldecott's explanation is less obvious, and it even calls for a gradual increase in the free cyanide present during precipitation, which is the reverse of the observed facts. Moreover, it does not account for the well-known production of free alkali. If the zinc hydrate formed in equation (2) is dissolved in cyanide at the moment of formation, thus—

$$(6)\ Zn(OH)_2 + 4KCN = K_2Zn(CN)_4 + 2KOH$$

then the loss of cyanide and increase of alkali are accounted for and the necessity for the presence of free cyanide is explained as the need of keeping the metallic zinc clean. This explanation makes equations (2) and (6) identical with equation (5), which states the facts more simply, and gives the same evolution of heat.

MacFarren,[1] following J. E. Clennell, gives the precipitation of gold in the presence of cyanide as—

$$(7)\ KAu(CN)_2 + 2KCN + Zn + H_2O = K_2Zn(CN)_4 + Au + H + KOH$$

and in the absence of free cyanide as—

$$(8)\ KAu(CN)_2 + Zn + H_2O = Zn(CN)_2 + Au + H + KOH$$

making the precipitation an electrochemical displacement of K and Au by Zn, followed by the displacement of hydrogen in water by potassium. Equation (8) seems the more probable when it is remembered that the potassium ions in KCN itself are displaced by zinc in the presence of water. Christy also gives equation (8),[2] but instead of (7) he gives—

$$2KAu(CN)_2 + 3Zn + 4KCN + 2H_2O = 2Au + 2K_2Zn(CN)_4 + K_2ZnO_2 + 2H_2$$

The caustic alkali shown in equation (5), (7) or (8) is partly neutralised by carbonic acid from the air, and alkali carbonate is thus produced, which then reacts with lime in solution to precipitate calcium carbonate.

Further experimental data may decide the matter, but in its absence, always bearing in mind the nature of the help that theory can give to practice, Caldecott's main contention that nascent hydrogen is the agent of precipitation of gold may be accepted. The hydrogen which becomes molecular has escaped its true function and become inert. It would also tend to form an imperceptible layer on the surface of the zinc, which would thereby be protected from further action. Consequently, the sooner the gas is got rid of the better. The ragged edges of thread zinc, the rough surface of lead-

---

[1] "*Cyanide Practice,*" p. 159.

[2] " On the Solution and Precipitation of Gold," *Trans. Amer. Inst. Mng. Eng.*, 1896, 26, 735.

coated zinc and mixing with zinc dust are all beneficial in this respect. Circulation of the solution should enable a greater proportion of the nascent hydrogen to be used in precipitation. The evolution of hydrogen in zinc dust precipitation appears to be imperceptible, and it is probable that in this case the hydrogen is practically all used in precipitating gold, thus reducing the waste of zinc.

Crowe [1] believes that in precipitation two opposing forces, a precipitating force and a dissolving force, are continually working. The former is a reducing action, caused by solution of the zinc, thus creating an electromotive force and the evolution of hydrogen in the nascent state. The latter is an oxidising action and must be inferior to the former if precipitation is to occur. When it is removed—as by means of a vacuum—then weaker cyanide solutions and smaller amounts of zinc may be used.

Caldecott has observed that, if there is a rapid flow of solution in the boxes, particles of lead and gold and of mercury and gold may be detached from the zinc and float to the surface, buoyed up by bubbles of hydrogen gas. Such particles may be carried away and the gold redissolved.

With regard to the effect of electronegative metals, Caldecott states that the ordinary impurities of commercial zinc, such as lead, carbon, iron and arsenic, do not dissolve, but, like deposited gold and other metals, assist the zinc-lead couple to promote efficient precipitation. It has been suggested that as gold is electronegative to iron, gold might be deposited on the inside of steel extractor boxes, and on the trays of wire screening used to support the zinc shavings in the boxes. In practice, however, Caldecott finds that very little gold is deposited in this way, "probably owing to the fact that a protective coating of rust and calcium sulphate speedily forms on such iron surfaces and prevents actual contact of the metal with the gold-bearing solution."

Air is dissolved in water in amounts proportional to the pressure and inversely proportional to the temperature. At atmospheric pressure water dissolves 2 to 4 per cent. of air by volume, and this air, owing to the different absorption coefficients of oxygen and nitrogen, contains only 65 per cent. of nitrogen, with 35 per cent. of oxygen. At 6,000 feet above sea level, or a barometric pressure of 24·5 inches, water dissolves 6·74 mg. of oxygen per litre.

The oxygen dissolved in cyanide solutions reacts with the zinc of the extractor boxes or zinc dust presses, forming zinc hydroxide, which is partly or entirely dissolved by the free cyanide of the solution, thus causing waste of zinc and of cyanide. The oxygen also combines with the nascent hydrogen formed in precipitation and diminishes the quantity available.

The main reactions occurring during the solution of zinc in cyanide solutions in the presence of oxygen and with the concentrations of cyanide and alkali normally in use are, according to White :—[2]

$$Zn + 4KCN + O + H_2O = K_2Zn(CN)_4 + 2KOH$$
$$Zn + O + H_2O = Zn(OH)_2$$

Hence 6 mg. of oxygen per litre are equivalent to 0·0097 per cent. KCN and to 24·5 mg. of zinc per litre or 0·049 lb. of zinc per ton of solution. In the

[1] *Amer. Inst. Min. Eng.*, 1918, *Bull.* 140, p. 1279.
[2] H. A. White, *J. Chem. Met. Mng. Soc. S.A.*, 1919, **20**, 97.

absence of oxygen zinc may go into solution in accordance with the following equations :—

$$Zn + 4KCN + 2H_2O = K_2Zn(CN)_4 + 2KOH + 2H$$
$$Zn + 2KAu(CN)_2 = K_2Zn(CN)_4 + 2Au$$
$$Zn + Pb(CN)_2 + 2KCN = K_2Zn(CN)_4 + Pb$$
$$Zn + Na_2S_2O_3 = ZnS + Na_2SO_3$$
$$Zn + 2NaOH = Na_2ZnO_2 + 2H$$

In a solution containing 1·0 mg. of oxygen per litre and no excess alkalinity, the action is slow, but more than twice as much zinc is consumed as would correspond with the loss of oxygen, so that, unlike gold, zinc appears able to displace hydrogen in such solutions to a small extent. Where the solution had an alkalinity of 0·040 per cent. NaOH and an oxygen concentration of 7 mg. per litre the maximum effect occurred with 0·02 per cent. cyanide, showing a dissolution of 0·75 mg. of zinc per square centimetre per 24 hours. In contact with carbon or lead in solutions containing 0·0144 per cent. NaOH and 6 mg. of oxygen per litre, a white precipitate readily formed. With free cyanide up to 0·03 per cent. alkalinity variations have little effect on the solution of zinc, whether or not the zinc be in contact with carbon or lead. With a cyanide strength of 0·010 per cent., oxygen at 7·0 mg. per litre, varying alkalinities and the zinc in contact with lead, a marked fall in the solubility of the zinc occurred at 0·03 per cent. NaOH. There was precipitation of lead sulphide when the alkali was low, and the formation of white precipitate when it was high. There appears to be an inversion point in the electrolytic effect at a strength of 0·01 per cent. cyanide with more than 0·030 per cent. NaOH. Such high alkalinity might therefore endanger the work of the extractor boxes when using weak cyanide solutions. It was noted, however, that the alkalinity in stronger cyanide solutions can reach 0·03 per cent. NaOH (or the equivalent in lime), without danger in precipitation.

White goes on to say that the ideal of consuming only as much zinc as would correspond with the gold and silver precipitated calls for close regulation of the cyanide strength and alkalinity, as well as the total absence of oxygen. It is rarely achieved.

The average consumption of zinc in extractor boxes is at the rate of about 0·18 mg. per square centimetre per 24 hours. In the first three compartments, before the bulk of the oxygen is removed, the rate is about three times this figure.

White estimates that on the Rand, when zinc boxes were used, 0·1 lb. of zinc was consumed per ton of solution carrying 2 dwt. gold. The various causes of consumption were as follows :—

Gold and silver precipitation,	1·2 per cent.
Combination with dissolved O_2,	49·0 ,,
Evolution of hydrogen,	4·8 ,,
Taken for clean-up,	33·0 ,,
Precipitation of lead in "dipping," waste in cutting, etc.,	12·0 ,,

The "*white precipitate*" formed on the zinc shavings in weak and "medium" cyanide solutions, and tending to prevent good precipitation, has been the subject of several investigations. Williams noted in 1904 that no "white precipitate" formed when the free cyanide in the boxes exceeded 0·025 per cent., and White states that no "white precipitate" is deposited when the cyanide strength exceeds 0·0098 per cent. KCN, apart from that existing as $K_2Zn(CN)_4$. This agrees with the equation—

$$Zn + 4KCN + O + H_2O = K_2Zn(CN)_4 + 2KOH$$

and corresponds with 6 mg. of oxygen per litre, the agreed figure for South Africa.

The formation of "white precipitate" is mainly due to the reaction—

$$2Zn + O_2 = 2ZnO$$

White demonstrated that the amount of "white precipitate" in the zinc boxes is proportional to the amount of oxygen in solution. When no oxygen is present, no "white precipitate" is formed.[1]

Oxygen is removed from cyanide solutions either by a mechanical method (Crowe process) or chemically. The former is the more efficient and, according to Prentice, has been incorporated in all Rand plants built since 1923. A vacuum of 22 inches is employed, and the oxygen content is reduced from 6·0 mg. to 0·5 mg. per litre. The chemical method involves the use of finely divided iron and highly pyritic sand in the clarifiers, or else the addition of manganese sulphate or the tannin extract from wattle bark—both of which are rather expensive. The removal of oxygen in this manner is not so thorough as in the Crowe process. It can be improved by using two clarifiers in series. In any case care must be taken to prevent re-oxygenation before precipitation.

Among other substances mentioned by Caldecott as being deposited in the zinc boxes are zinc sulphide, silica and alumina dissolved from the ore and "presumably precipitated as zinc aluminate and silicate," calcium sulphate, especially when partly weathered ore is treated, calcium carbonate formed by the action of the carbonic acid of the air on dissolved calcium hydroxide, and minute particles of ore and especially slime mechanically deposited.

Zinc sulphide is probably formed by the reaction—

$$Zn + Na_2S_2O_3 = ZnS + Na_2SO_3$$

which takes place slowly in dilute solutions. The thiosulphate, to the extent of 0·005 per cent., is traceable to pyrrhotite in the ore. Adams [2] has shown that soluble silicates are precipitated by lead-coated zinc shavings, probably as colloidal zinc hydroxide and colloidal silica. They hinder contact with the gold solution and might adsorb some gold themselves.

One more point must not be forgotten, and that is that the soluble compounds are all more or less ionised, and consequently their ions will be present together in the solution in amounts depending on their respective degrees of ionisation, mass action having free play.

The Precipitation of Gold by Charcoal.—Carbon may enter into the cyanide process in two ways—(1) As a part of the original ore, when it may interfere in solution of the gold by cyanide by causing re-precipitation of the metal. Thus it may not be possible to employ the cyanide process,

---

[1] *J. Chem. Met. Mng. Soc. S.A.*, 1918, **18**, 292.
[2] *J. Chem. Met. Mng. Soc. S.A.*, 1920, **21**, 151.

or, at best, the consumption of the solvent will be high. (2) As an artificially prepared precipitant, instead of zinc, for the gold carried in the filtered cyanide solutions. The matter is considered here, although it is probable that such a method will not replace the one at present in use owing to its cost and the greater difficulty of removing the precipitated gold from the charcoal.

It has been shown that graphite from Natal and from Barberton does not precipitate gold from aurocyanide solutions,[1] but that graphite in West African schist may do so.[2] Dorfmann[3] considers that at Porcupine Crown Mine gold is removed from solution by the carbonaceous schist as its double cyanide salt which is adsorbed on the carbon surfaces.[4]

The power of charcoal to precipitate gold from cyanide was thought by Morris Green[5] to depend on carbon monoxide gas or hydrogen, or both, occluded in the charcoal.

Edwards[6] considers that the gold is held in the carbon in a combined state, probably as a carbonyl aurocyanide. The precipitating power of charcoal is increased in an acid solution, and is almost inhibited by small quantities of soluble sulphides, lime and thiosulphates. The charcoal should be finely ground in water after quenching from a red heat.

Allen[7] cites the following theories to account for charcoal precipitation :—

(1) *Occluded Oxygen* causes the formation of an insoluble gold cyanide and sodium cyanate.

(2) *Adsorption of* $Au(CN)_2$. This is supported by the following facts :—

(*a*) The loss of cyanide by titration agrees with theoretical considerations.

(*b*) Precipitation efficiency varies inversely with the concentration of the solution. A true adsorption curve may be plotted from the results.

(*c*) Precipitation is proportional to the surface exposed. Finer grinding gives better results.

(*d*) Increased activity is obtained by treating the charcoal so that occluded gas is driven off.

(*e*) Acid advances and alkali retards precipitation.

(*f*) The adsorbed material is insoluble in ordinary solvents, but is released by the addition of a solution of different surface tension, such as alcohol or sodium sulphide.

Chaney considers that two definite functions of the charcoal are involved : (1) due to " specific capacity " of an active carbon ; and (2) due to " capillary capacity " consequent on porosity. These may operate independently or together, and it is suggested that the second *may* be the cause of gold retention.

Gross and Scott[8] indicate that the precipitation on charcoal is not

---

[1] Green, *J. Chem. Met. Mng. Soc. S.A.*, 1912, 13, 84.
[2] Brühl, *Trans. Inst. Min. Met.*, 1914, 23, 82.
[3] *Trans. Inst. Min. Met.*, 1923, 33, 464.
[4] See also p. 316, "Re-precipitation of Gold and Silver in Leaching Vats," and "*Rand Metallurgical Practice*," vol. i., p. 387. Also Chap. XV., "Carbonaceous Ores."
[5] *J. Chem. Met. Mng. Soc. S.A.*, 1912, 13, 65.
[6] *Trans. Inst. Min. Met.*, 1918, 27, 277.
[7] *Met. and Chem. Eng.*, 1918, 18, 642.
[8] U.S. Bur. of Mines, *Tech. Paper* 378, 1927.

metallic because :—(1) The precipitate is invisible on the charcoal. (2) It does not possess the chemical properties of the metal. (3) The effect of acidity in the solution is not the same as with metallic precipitation. (4) The capacity of the charcoal is less than with metallic precipitation. Their other conclusions are :

(1) The mechanism of charcoal precipitation involves adsorption accompanied by a chemical change.

(2) Charcoal has less capacity for the precipitation of silver than of gold.

(3) The limit of charcoal precipitation appears to be about 2000 ozs. of gold per ton of charcoal.

(4) Little difference exists amongst charcoals from various woods.

(5) The heat-treatment of charcoal during manufacture is important, and quenching does not improve it.

(6) Pulverisation finer than 200 mesh does not add appreciably to the efficiency of the charcoal.

(7) Few impurities in the solution affect precipitation.

(8) The adsorbed gold is slightly soluble in boiling water and to a greater degree in hot cyanide solution.

(9) There is the possibility of adapting methods so that charcoal may be used for repeated precipitations. The charcoal used for precipitation from cyanide solution may be recovered by flotation.

Activation of Charcoal.—According to Gross and Scott [1] the modern theory of the activation of charcoal is that oxidation is necessary at a temperature high enough to decompose and drive off the inactive hydrocarbons that coat the charcoal surface. A charcoal thus activated is a better precipitant than non-activated charcoal. Hence it would appear that a reducing gas is not necessary for the chemical action.

Williams [2] states that heating out of contact with the air, in order to remove hydrocarbons, is essential to activation. He puts forward a theory of colloidal dispersion acting in unison with adsorption of $Au(CN)_2$ on the highly developed charcoal surface. It must be remembered, however, that solutes other than gold may be adsorbed on the charcoal surface and therefore the efficiency of precipitation will depend, to some extent, on the amount and nature of the other salts present, and selection may occur, *i.e.* triads, diads and monads will be adsorbed in that order.

King [3] says that charcoal is normally covered with a film of oxide that is difficult to remove. Its composition and properties vary with the temperature of activation of the charcoal. The oxide formed at 200°-500° C. has definite acidic properties. About 0·5 per cent. of the charcoal surface is occupied by oxalic acid. The formation of this acid is inhibited by intensive drying of the charcoal. Adsorptive activity appears to be greatest in charcoal heated to 450° C.

The Precipitation of Insoluble Cyanides of Gold.—In the absence of free cyanide, the sodium in sodium aurocyanide may be replaced by certain heavy metals, such as silver or copper, in which case the gold comes down as an insoluble double metallic cyanide. As the latter is soluble in alkaline cyanide, it is necessary to destroy the cyanide in the solution before the precipitation takes place, and consequently the method is not used in practice except in assaying.

---

[1] *Loc. cit.*
[2] *Min. Mag.*, 1923, 28, 139.
[3] *Journ. Soc. Chem. Ind.*, 1935, 54, 338. *Journ. Chem. Soc.*, 1933, p. 842.

Effect of Thiocyanates on Metallic Gold.—This has been investigated by H. A. White,[1] who found that potassium thiocyanate, formed by a reaction between cyanide and colloidal sulphur or thiosulphates, though inactive by itself, would dissolve gold if mixed with various oxidising agents, such as potassium permanganate, ferricyanide of potassium, dilute nitric acid, etc. Ferric thiocyanate also dissolved gold, but in this case it was supposed that the presence of dissolved oxygen was necessary.

The Stark process of treating dumps [2] was said to be explained by these results.

Testing of Cyanide Solutions.[3]

Free Cyanide.—The ordinary method of estimating the amount of free cyanide present in a liquid is by titration with a standard solution of silver nitrate. Silver cyan'de is formed, and redissolves in the excess of potassium cyanide until one-half of the latter has been decomposed. The equations are as follows :—

$$AgNO_3 + KCN = AgCN + KNO_3$$
$$AgCN + KCN = KAg(CN)_2$$

When one-half of the KCN present has been converted to AgCN, an additional drop of $AgNO_3$ solution causes the formation of a permanent white precipitate of AgCN. The amount of silver solution added is then read off, and the percentage of cyanide calculated. The equation of the end reaction is—

$$KAg(CN)_2 + AgNO_3 = 2AgCN + KNO_3$$

A few drops of a 10 per cent. solution of potassium iodide are often added, in accordance with a suggestion made by J. S. MacArthur, to make the end reaction sharper, and to prevent inaccuracy through the presence of ammonia or other substances in which silver cyanide is soluble. The results are, however, slightly too high, owing to the presence in mill solution of alkali and the double zinc cyanide. A. M'A. Johnston dissolves [4] 13·046 grammes of triple crystallised silver nitrate in distilled water and makes up to a litre. He titrates by running this into 100 c.c. of cyanide solution (previously filtered if turbid) containing one or two drops of potassium iodide solution. Then each c.c. of $AgNO_3$ solution used represents 0·01 per cent. KCN in the cyanide solution.

Total Cyanide.—This includes the free cyanide (NaCN or KCN), hydrocyanic acid and the cyanide in the double cyanide of zinc, $K_2Zn(CN)_4$. An excess of caustic soda (say 5 c.c. of a 10 per cent. solution) is added to 10 c.c. of mill solution, together with a few drops of the potassium iodide indicator and a drop or two of ammonia (Johnston). The solution is titrated with silver nitrate as before.

Protective Alkali.[5]—The solutions required are silver nitrate (as above);

---

[1] *J. Chem. Met. Mng. Soc. S.A.*, 1905, 6, 109.

[2] "*Rand Metallurgical Practice,*" vol. i., p. 394.

[3] For full details of analysis and the testing of ores, see J. E. Clennell, "*Chemistry of Cyanide Solutions*"; A. M'A. Johnston, "*Rand Metallurgical Practice,*" vol. i., pp. 322-379; MacFarren, "*Cyanide Practice,*" pp. 28-86.

[4] "*Rand Metallurgical Practice,*" vol. i., p. 323.

[5] For this account of the determination the authors are indebted to Mr. A. M'A. Johnston and to Dr. Caldecott. See also Green, *Trans. Inst. Min. Met.*, 1901-2, 10, 29; Robertson, *ibid.*, 1924, 34 [i.], 24.

phenolphthalein, prepared by dissolving 1 gramme of the powder in 100 c.c. of 60 per cent. alcohol or methylated spirits; decinormal oxalic acid, $N/10\ C_2H_2O_4$, prepared by dissolving 6·3 grammes of oxalic acid crystals, $C_2H_2O_4.2H_2O$, in distilled water and diluting to 1,000 c.c.; potassium ferrocyanide, $K_4Fe(CN)_6$, prepared by dissolving 10 grammes of $K_4Fe(CN)_6$ in 100 c.c. of distilled water.

The test is as follows:—Measure into a conical flask 100 c.c. of cyanide solution. Add just sufficient silver nitrate solution to give a permanent precipitate, and then 5 c.c. of the ferrocyanide solution and two drops of phenolphthalein. Titrate with decinormal oxalic acid till the phenolphthalein colour is discharged. Then each c.c. of oxalic acid solution used multiplied by 0·004 gives the percentage of alkalinity in terms of caustic soda. The portion of solution used to determine free cyanide may be afterwards used to estimate the protective alkali.

The use of potassium ferrocyanide in this determination is desirable, so as neither to include $K_2Zn(CN)_4$ as free alkali nor to neutralise free alkali present by excess of $AgNO_3$.

Oxygen in Cyanide Solutions.—*White's Method.*[1]—This depends on the darkening of pyrogallic acid (" pyro ") by oxygen. Water saturated with oxygen is placed in a 250 c.c. stoppered bottle and 0·1 grm. of solid " pyro " and 1 c.c. of 8 per cent. NaOH solution are added. The brown solution is matched with a solution containing the necessary amount of a brown dye, such as Diamond brown. For perfect matching it is necessary to add a little acid, methyl orange and potassium chromate. Suppose this colour corresponds to 8 mg. of oxygen per litre; then, by diluting the dye solution, colours corresponding to 7, 6, 5, etc., mg. per litre are obtainable.

The colour given to a mill solution by the same reagents is compared with the standards, and the oxygen is reported in milligrammes and half-milligrammes per litre. The colour in the mill solution fades slowly. A difficulty in the method is that it gives a violet or purple shade with foul solutions containing complex ferrous or ferric salts, which is difficult to match. Strengthening the solution with NaCN improves matters.

Tromp and Schilz[2] suggest that in White's method a minimum amount of calcium salt must be present in order to develop the full colour of the pyrogallol.

Hamilton[3] found that White's method gives too strong a colour at first, the colour then fading rapidly, and after six or seven minutes the solution begins to cloud and is soon too murky for comparison. He prefers caramel to Diamond brown for the standard colours. He saturates the mill solution with oxygen, adds " pyro " and soda, inserts the stopper, agitates to dissolve the crystals, and allows to stand for exactly six minutes. The colour now observed is taken as the standard colour for solution saturated with oxygen. It is matched with caramel solution and a series of colours prepared by diluting the caramel to represent ½, 1, 2, etc., milligrammes of oxygen per litre. A mill solution is then treated with " pyro " and soda in a similar bottle and left to stand for exactly six minutes before comparing it with the standards.

Method of Anderson and Dickson.[4]—Make the sample alkaline with NaOH

[1] *J. Chem. Met. Mng. Soc. S.A.*, 1918, **18**, 292, and 1919, **19**, 177.
[2] *J. Chem. Met. Mng. Soc. S.A.*, 1935, **35**, 313.
[3] *Eng. and Min. J.*, 1920, **110**, 116.
[4] *Chem. Eng. and Mng. Rev.*, 1927, **19**, 467.

and titrate with ferrous ammonium sulphate, keeping the ferrous hydroxide in solution with alkaline sodium tartrate. The indicator is one drop of Methylene blue, the colour being discharged. The sample is covered with a layer of paraffin oil.

Method of Weinig and Bowen.[1]—Titrate with a standard solution of sodium hydrosulphite. The indicator is Indigo blue, the colour being discharged by an excess of hydrosulphite.

When employing Weinig and Bowen's method Tromp and Schilz use an atmosphere of hydrogen instead of a cover of paraffin for protecting the solution. Stanley found the paraffin very permeable to oxygen.[2] Indigotin disulphonate has been used as an internal indicator in this method.

Tromp and Schilz [3] suggest a new evacuation method as a control for workshop methods. It depends on the removal of dissolved oxygen from the interfering substances in the solution.

White gives a table of the solubilities of oxygen in water under all conditions.[4] The amount of oxygen in milligrammes per litre of saturated

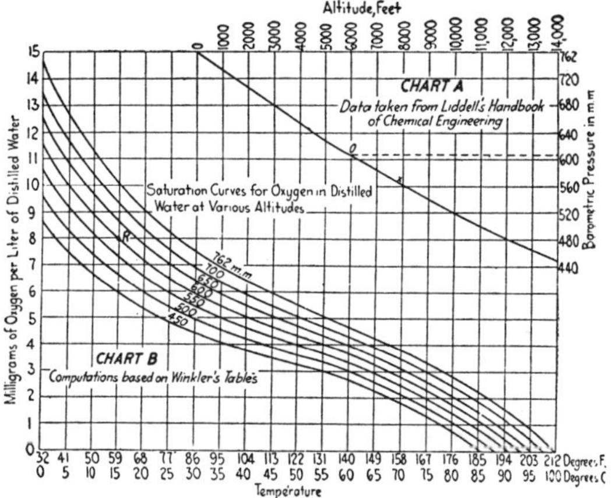

Fig. 152.

water is given in Roscoe and Lunt's table in Sutton's "*Volumetric Analysis*," 1911, p. 599. It varies with the temperature and barometric pressure. Weinig and Bowen's [5] chart is given here in Fig. 152.

Ferrocyanide.[6]—50 c.c. of the cyanide solution are treated with an excess of a solution of ferric sulphate made acid with H_2SO_4. Warm, filter and wash. The filtrate is used for the sulphocyanide test. The precipitate is treated with a slight excess of NaOH, again filtering. The filtrate is acidified with H_2SO_4 and titrated with potassium permanganate prepared by dissolving 0·316 grm. of the crystals in a litre of distilled water. Then

$$1 \text{ c.c. N/100 } KMnO_4 \text{ solution} \equiv 0{\cdot}00368 \text{ grm. } K_4Fe(CN)_6$$

[1] *Trans. Amer. Inst. Mng. Met. Eng.*, 1925, **71**, 1020.
[2] *J. Chem. Met. Mng. Soc. S.A.*, 1935, **35**, 313.
[3] *Loc. cit.*
[4] *J. Chem. Met. Mng. Soc. S.A.*, 1919, **20**, 15.
[5] *Loc. cit.*
[6] A. M'A. Johnston, "*Rand Metallurgical Practice*," vol. i., p. 327.

Sulphocyanide.[1]—The filtrate from the ferrocyanide determination is titrated with the same solution of $KMnO_4$. The red colour is destroyed and a permanent pink obtained. Then

$$1 \text{ c.c. N/100 } KMnO_4 \text{ solution} \equiv 0{\cdot}000162 \text{ grm. KCNS}$$

The method is inaccurate in the presence of thiosulphates.

Thiosulphates and Sulphites.[2]—Take 100 c.c. of solution for each and add one drop of methyl orange and a few drops of starch solution. Titrate both with normal sulphuric acid until red and add a further 1 c.c. of acid. To one solution add 10 c.c. of 10 per cent. solution of formaldehyde, then titrate both solutions with N/10 iodine to distinct permanent blue.

Formaldehyde destroys the sulphite and the smaller result gives the thiosulphate, while the difference between the two results gives the sulphite.

$$1 \text{ c.c. N/10 iodine} \equiv 0{\cdot}0158 \text{ grm. thiosulphate } (Na_2S_2O_3)$$
$$\equiv 0{\cdot}0063 \text{ grm. sulphite } (Na_2SO_3)$$

Zinc.[3]—Warm 100 c.c. of the solution and add sodium sulphide. Filter off the precipitate of ZnS and wash with hot water until free from soluble sulphide. The precipitate and filter paper are then transferred to a small stoppered bottle and shaken with a known volume of N/10 iodine solution. The excess iodine is estimated by titrating with sodium thiosulphate. Then

$$1 \text{ c.c. of N/10 iodine} \equiv 0{\cdot}00325 \text{ grm. zinc.}$$

Testing of Supplies.

Estimation of Zinc in Zinc Dust.[4]—The zinc is determined by the evolution of hydrogen. The fineness of the dust is examined by sieving; from 90 to 95 per cent. should pass through a 350 mesh sieve. The metallic impurities should be less than 0·3 per cent.

Place 1 grm. of zinc dust in a flask with a platinum sheet and 5 grms. of $FeSO_4$ crystals for catalysis. Add water and then 30 c.c. of 1 : 1 H_2SO_4, little by little. The gas passes over into a measuring flask. The temperature should be kept constant. Then

$$\text{Percentage of metallic zinc} = \frac{V \times (P - p) \times 0{\cdot}29196}{(1 + 0{\cdot}00367t)760}$$

Where V = volume of gas, P = barometric pressure, p = vapour tension of water at $t°$, t = room temperature (°C.)

Available CaO[5] in lime is found by titrating with H_2SO_4 using rosolic acid as indicator. The results are low unless any CO_2 present is neutralised with lime water, using phenolphthalein.

For the determination of gold and silver contained in cyanide solutions, see Chap. XVIII., and for the numerous other tests occasionally required in a mill laboratory, see the references cited in the footnote on p. 509.

[1] A. M'A. Johnston, "*Rand Metallurgical Practice*," vol. i., p. 327.
[2] *Op. cit.*, p. 534. [3] *Op. cit.*, p. 328.
[4] L. A. Wilson, *Eng. and Min. J.*, 1918, 106, 334. See also *Met. Ind.*, 1920, 16, 488.
[5] C. A. Meiklejohn, *J. Chem. Met. Mng. Soc. S.A.*, 1919, 19, 85, 201, 249.

Cyanide Poisoning.

Treatment in Case of Cyanide Poisoning.—A coating of oil or kerosene or the use of rubber gloves protects the hands from the action of the cyanide solution, by which a rash is sometimes developed. The hands should be washed thoroughly before meals, which should not be taken in the vicinity of any cyanide storage.

Early warning symptoms of cyanide poisoning are (1) Irritation and choking in the throat. (2) Increased difficulty of breathing. (3) Watering and bloodshot eyes. (4) Dizziness and heaviness of limbs. (5) Frothing at the mouth.

First Aid Treatment :—

(1) Must be given with the greatest promptitude.
(2) If the patient is conscious, give an antidote immediately and remove him to a pure atmosphere.
(3) If the patient is unconscious, remove him to a pure atmosphere at once.
(4) Place the patient in a recumbent position. Do not walk him about.
(5) If breathing shows signs of failing or has ceased, apply artificial respiration (Schafer's method) and continue without interruption.
(6) Administer oxygen containing 7 per cent. of CO_2 (or use "Novox" apparatus by Siebe, Gorman & Co.).
(7) Cut away clothing splashed with cyanide.
(8) Wrap patient warmly in blankets and apply hot-water bottles.
(9) If partial recovery is noticed, suggest a little deep breathing.

If cyanide solution is swallowed, anything that causes vomiting may be given, but the action of cyanide is extremely rapid and every second is of value. Freshly precipitated ferrous hydroxide, made by mixing magnesium oxide, caustic potash and ferrous sulphate, is a useful antidote. In the reaction which occurs ferrocyanide of potassium or sodium is formed, which is innocuous. The antidote recommended in S. Africa [1] is 30 c.c. of a 23 per cent. solution of ferrous sulphate and 30 c.c. of a 5 per cent. solution of caustic potash in separate phials. These are mixed in a mug and 2 grammes of powdered oxide of magnesium stirred in and the dose administered to the patient. A hypodermic injection of caffeine sodium benzoate (2 grns.) may be given if the symptoms are severe. The use of nitrate of cobalt and of sodium thiosulphate, both by mouth and as injections, has also been suggested. 20 minims (1·2 c.c.) or more of a 0·5 to 1 per cent. solution of cobalt nitrate, or of a 5 to 10 per cent. solution of sodium thiosulphate may be repeatedly injected under the skin. The cobalt salt is rapidly absorbed and a small quantity renders large quantities of cyanide inert.[2] Comparatively harmless cobalt cyanide or sodium sulphocyanide is formed.[3] As an alternative copious subcutaneous injections of a 3 per cent. solution of hydrogen peroxide are recommended. The patient may also be rubbed with camphor and alcohol, or cold water may be dashed upon the skin. Alcohol should not be taken. Kossa [4] suggests the administration of $\frac{1}{3}$ to $\frac{1}{2}$ litre of a 5 per cent. solution of potassium permanganate in cases of cyanide

[1] *J. Chem. Met. Mng. Soc. S.A.*, 1904, **4**, 679.
[2] Taylor, "*Principles and Practice of Medical Jurisprudence*," 1934, **2**, 683.
[3] Dixon Mann and Brend, "*Forensic Medicine and Toxicology*," 6th ed., p. 448 (Griffin).
[4] Taylor, *loc. cit.*

poisoning. Artificial respiration is only useful if it enables the victim to survive long enough for the cyanide to be broken down by oxidation. Administration of oxygen alone is said to have no special effect.[1]

If hydrocyanic acid has been inhaled, an inhalation of ammonia, or chlorine, or ether, should be administered, but respirators should be worn in cases, such as in the acid treatment of gold slime, where the evolution of poisonous gases is probable. Sieverts and Hermsdorf state that one part of hydrocyanic acid in 10,000 parts of air may be detected by exposing strips of filter-paper moistened with freshly prepared solution of copper benzidine acetate, the papers turning blue in a few seconds. Voille[2] has found that the ill-effects of an atmosphere containing hydrocyanic acid can be resisted for longer periods after doses of glucose taken by injection or the mouth.

Park records the warning that in the acid treatment of gold slime from arsenical ores, an evolution of arseniuretted hydrogen may occur, and the inhalation of this gas is very deadly, no antidote being known. Potassium sulphocyanide is also exceedingly poisonous, and ammonium thiocyanate in a somewhat less degree.[3] The fumes of molten cyanide appear to be more or less harmless.

---

[1] E. Leschke, "*Clinical Toxicology,*" 1934, p. 146.

[2] *Lancet,* July 10, 1926.

[3] For further particulars on cyanide poisoning, see "*Rand Metallurgical Practice,*" vol. i., p. 331 ; *Eng. and Mng. J.*, 1906, 82, 835.

CHAPTER XIV.

THE CYANIDE PROCESS. GENERAL METHODS.

THE method of treatment may be conveniently considered as being divided into seven distinct operations, viz. :—

(1) Preparation of the ore for treatment.
(2) The cyaniding of sand.
(3) The cyaniding of slime.
(4) Clarification.
(5) Precipitation of the dissolved gold.
(6) The treatment of barren solution.
(7) Conversion of the precipitated gold into bullion.

1. PREPARATION OF THE ORE.

In the treatment of gold ores by cyanide, it is necessary to crush finely, for otherwise too great a proportion of the gold remains locked up in the larger grains and escapes dissolution. Such crushed ore always contains a proportion of slime, and when this is allowed to remain intermixed with the rest, the mixture resists the passage of liquids through it. Many ores contain so much slime that a bed only 12 or 15 inches thick cannot be leached at a reasonable rate, even with a vacuum below the filter bed. The gold is usually more readily dissolved from slime than from less finely divided ore, and this constitutes a reason for separating sand from slime. Nevertheless equally good results may often be obtained by simple direct leaching as by more intricate operations, provided that : (1) there is an even distribution of the vat charge, to prevent a " break through " during leaching, and (2) if necessary the dry filling of vats is practised, to preserve the filtering aid of coarse material, and to prevent the formation of colloidal particles which increase the resistance to solution flow.[1]

The efficiency of displacement of dissolved gold from a pulp decreases as the ore is ground finer. This is due partly to the lessened permeability of the ore, and partly to the large increase of surface by comminution, whereby the stability of a solution containing cyanide and a precious metal is reduced. Colloidal slime absorbs gold solutions during milling in cyanide solution, and this is only slowly displaced during washing. The absorption may be reduced by roasting the ore. Adsorption of aurocyanide also occurs, and gold thus held is not easily removed by water.

Primary crushing is generally effected in the stamp battery or tube or ball mills, and secondary and tertiary reduction in tube mills or other machines. Sand is crushed more finely than formerly as a preliminary to treatment with cyanide.

Very fine grinding is best done in highly alkaline water to counteract the

---

[1] Allen, *Trans. Amer. Inst. Min. Met. Eng.*, 1934, **112**, 540.

action of "cyanicides," effect coagulation of slime particles, and permit the adsorption of valueless instead of valuable ions.[1]

Dry crushing is employed before direct *roasting*, which has been mainly applied to sulpho-telluride ores at Kalgoorlie and Cripple Creek. It is now rarely used. Roasting before cyaniding must be complete, as ores containing sulphides, when partially roasted (not quite "dead") contain ferrous salts and other cyanicides. It is noteworthy that unoxidised pyrite and sesquioxide of iron, Fe_2O_3, the final product of roasting, are alike without effect on cyanide solutions, but all intermediate compounds are cyanicides (see Chap. XIII). The efficiency of roasting is tested by adding cyanide to a clear solution obtained by shaking the roasted ore with water. If a discoloration appears the ore still contains soluble salts, which will destroy cyanide and make foul solutions.

In ordinary practice ores are not roasted. If they contain a fair proportion of amalgamable gold, especially if some of the gold is not very finely divided, they may be treated by amalgamation or concentration before being cyanided.

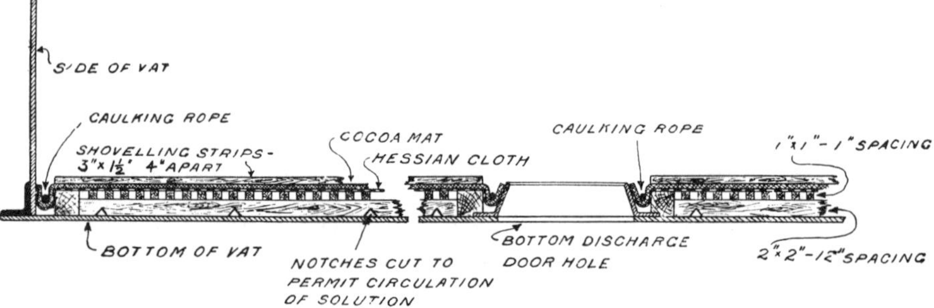

Fig. 153.—Filter Bottom in Sand Vats.

Two methods are used for treatment :—(1) Separation of the sand and slime with subsequent leaching of the sand and agitation of the slime with cyanide solution. (2) Grinding until all the ore is reduced to slime and the treatment of this single product by agitation with cyanide solution (all-sliming). An advantage of gravity leaching is that there is simultaneous enrichment of solution and impoverishment of ore.

The separation of sand from slime by means of classifiers has been dealt with in Chap. IX. Some other methods in use will now be described.

Collection of Sand.—In the Transvaal sand is collected for treatment in one of three ways :—

(*a*) By hose filling.
(*b*) By the "Butters and Mein" pulp distributor.
(*c*) By the "Caldecott" continuous collecting plant.

(*a*) *In hose-filling* the underflow from the classifiers passes through pipes and is fed into the collecting vats by one or more hose pipes, the position of which is changed from time to time, in order to distribute the pulp evenly. The vats are constructed of steel, and are from 25 to 50 feet in diameter and 6 to 9 feet deep. The sand settles to the bottom and the slime and water

[1] Allen, *Trans. Amer. Inst. Min. Met. Eng.*, 1934, **112**, 540.

overflow through discharge holes in the side of the vat. The level of discharge is raised from time to time as the level of the collected sand rises. When the vat is full, the residual water is drained off through the filter bed at the bottom of the vat. The construction of the filter bed is shown in Fig. 153.[1] The collected sand is usually transferred to other vats for treatment with cyanide.

(*b*) *The Butters and Mein Distributor* is shown in Fig. 154.[2] It is an

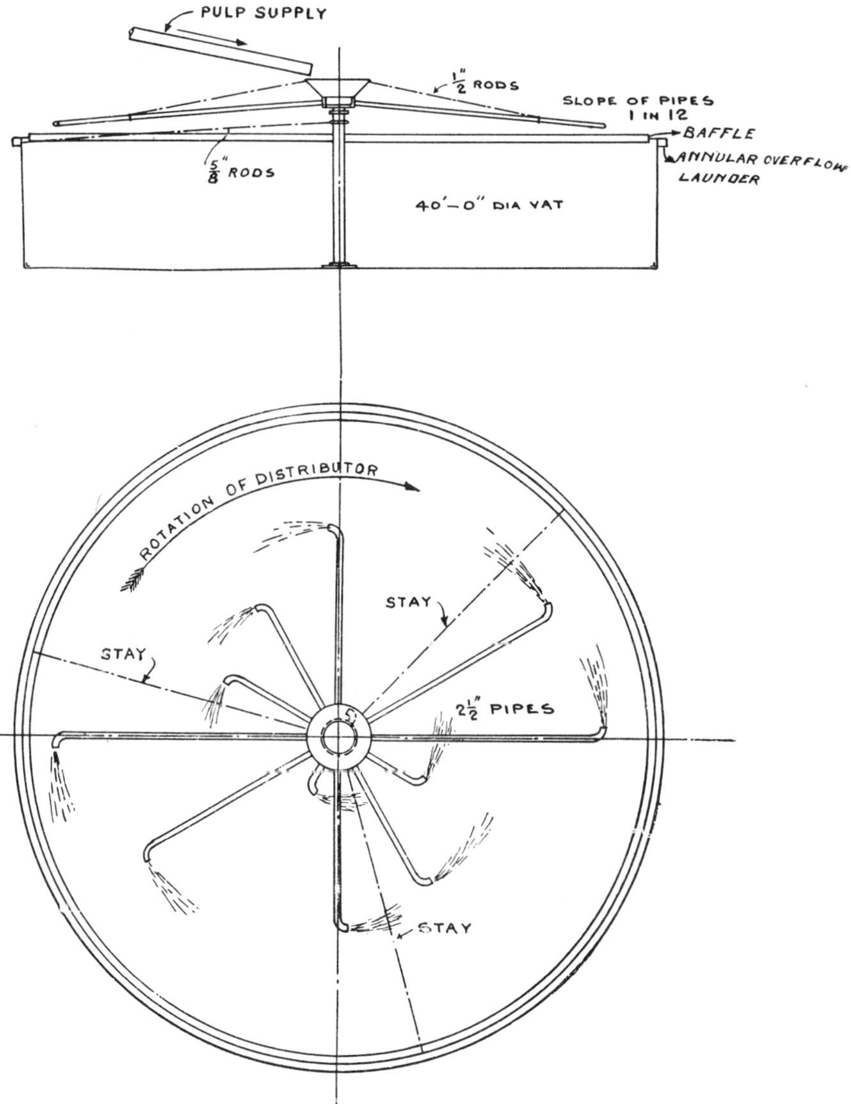

Fig. 154.—"Butters and Mein" Pulp Distributor for Filling Sand-collecting Vats.

[1] Schmitt, "*Rand Metallurgical Practice,*" vol. ii., p. 219. [2] *Op. cit.*, p. 221.

automatically revolving apparatus working on the principle of the garden sprinkler. Pipes of several different lengths conveying the pulp revolve round a central spindle, and spread the pulp evenly over the surface of the water with which the vat is originally filled. The sand settles to the bottom

Fig. 155.—Collecting Tanks, Champion Reef Cyanide Plant.

and accumulates there, while the slime overflows at the top. The collecting tanks of the Champion Reef cyanide plant are shown in Fig. 155.[1]

When the accumulation of sand in the vat reaches nearly to the top, the water is drained off and the ore discharged by shovelling, or by means of machinery, such as the Blaisdell vat excavator (see p. 339), in which steel discs attached to revolving arms push the sand towards the centre of the vat. In either case the sand falls through apertures in the bottom into

[1] Reproduced with the permission of the Cyanide Plant Supply Co.

ore-cars, or on to a travelling belt below. The sand is now in a suitable condition for leaching, containing only small quantities of slime. Sometimes the pulp collected in this way is treated directly without removal. In other mills, where the method of "double treatment" of the tailing is employed, the first solution of cyanide is directly applied to these vats, and, after draining, the pulp, wetted with cyanide solution, is transferred to the second treatment vats. This has the advantage of an additional aeration of the charge during transference. Double treatment is made cheaper by using two superimposed vats. As already stated, the usual method is to remove the collected sand to the cyanide vats for treatment.

At New Modderfontein, Bosqui has tried the method of direct leaching in the sand collectors without transference.[1]

At Mysore, during sand vat filling, 45 lbs. of lime (30 per cent. CaO) and 2½ lbs. of calcined fine manganiferous limestone are added every 2 hours, thus forming alkaline layers in the vat to improve the lime-washing of the sand before cyaniding.[2] The manganiferous material acts as an oxidiser and reduces cyanide consumption.

The slime overflowing from the collecting vats is treated with the rest of the slime which has previously been separated from the pulp by cone or other classifiers.

(*c*) *Caldecott's Continuous Collection.*—In this method collecting vats are not used and time is saved. One arrangement is shown in Fig. 156.[3] Here sand from which the surplus water has been drained on Caldecott's sand filter table is charged into the treatment vat. The sand filter table is merely a de-waterer, and consists of a rotary filter bed with a vacuum pump, and a stationary plough to remove the comparatively dry sand. The pulp is distributed at a point about 3 feet behind the plough, so that the sand makes almost a complete revolution on the table before being removed.

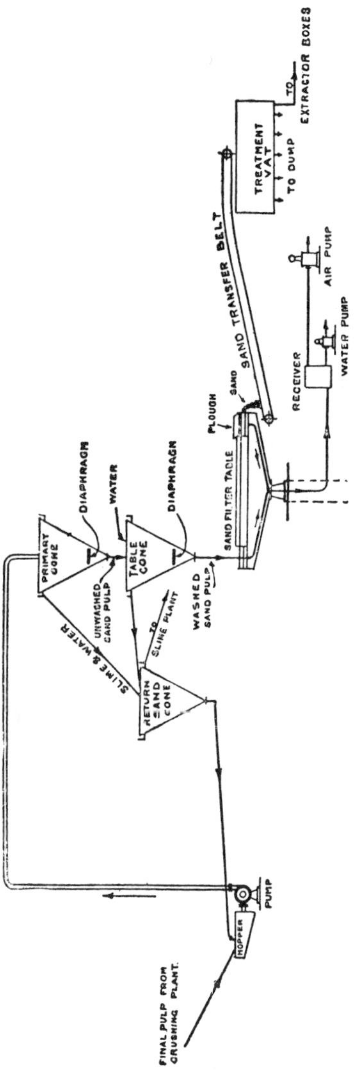

Fig. 156.—"Caldecott" Continuous Sand-collecting Plant.

[1] *Eng. Min. Journ.*, 1921, 111, 623.
[2] *Trans. Inst. Min. Met.*, 1925, 34 [ii.], 84.
[3] "*Rand Metallurgical Practice*," vol. ii., p. 224.

The sand filter tables at the Simmer and Jack Proprietary Mines in 1913 are shown in Fig. 157, which is from a photograph kindly sent by Dr. W. A. Caldecott. The tables are each 25 feet in diameter, with 3 feet breadth

Fig. 157.—Sand Filter Tables at Simmer and Jack Mine, 1913.

of filter, and a daily capacity of 750 tons of sand. They receive the thick sand pulp underflow from the diaphragm cone classifiers. The sand can be collected, aerated and leached in the same vat when these tables are employed and the dissolving of the gold by cyanide begins within half an hour of the ore being crushed in the battery. These vacuum tables have been abandoned in recent years.[1]

In some instances a thickened underflow is taken from the cones and conveyed with cyanide solution to the sand collectors, whence the overflow passes to cyanide solution tanks. In this case the whole treatment may be carried out in the collectors.[1] In other arrangements the sand is mixed with cyanide solution in a hopper and forced through pipes to the vat by a centrifugal pump.

2. The Cyaniding of Sand.

This is effected in round vats, constructed rarely of wood and mostly of concrete, or mild steel. Wooden vats are made of staves, 4 to 6 inches wide and 3 inches or more thick, held together by round iron hoops, with bottom planks 3 inches thick, fitting into slots in the staves. Sometimes in new mining districts square tanks are built, as being cheaper and more easily constructed, but they do not last long and are more difficult to keep tight.[2] The wood is usually covered with a coating of paraffin paint or a mixture of asphaltum and coal-tar. Steel vats are 40 to 55 feet in diameter, 7 to 15 feet deep inside, and, on the Rand, hold 400 to 850 tons each. The sides are of $\frac{1}{4}$-inch steel plate, and the bottoms are $\frac{5}{16}$ inch thick. The plates are riveted together and strengthened with angle iron at the top and bottom of the vat. They are painted inside and out to prevent rusting.

The dimensions of vats vary with the work to be done. In calculating the capacity of a leaching vat, the volume of a ton of collected sand on the Rand is taken as 21·5 cubic feet (C. O. Schmitt), and that of transferred sand as 26 cubic feet. When settled, clean Rand sand occupies about 23 cubic feet per ton (Caldecott).[3] If the material to be treated is such that percolation is difficult, the depth of the vat is kept small, and the diameter made as large as possible. A depth of 2 or 3 feet of dry-crushed ore, which always contains some slime, is usually as much as can be conveniently leached, and vacuum pumps are often added to expedite the work.

The false bottom is usually a wooden framework, constructed of boards pierced with numerous auger holes, or, in larger vats, of wooden slats crossing each other. The framework is covered by cotton twill, canvas or cocoa-nut matting. Thick canvas duck, resting on matting, forms a trustworthy filter-cloth. The filter-cloth is protected by wooden slats, which prevent it from being injured by workmen's shovels in clearing out the charge. The construction of a filter-bed is shown in Fig. 153.

An iron pipe communicates with the space below the false bottom, and conveys the liquid to the pumps or to the zinc boxes. The solution does not attack wood or iron ; brass and bronzes are attacked and corroded rapidly.

The vacuum may be obtained by a direct-acting pump, or by the use of a large boiler, in which a vacuum is created by a Westinghouse or other pump. As soon as the pressure falls to about half an atmosphere, the boiler is connected with the aperture of the vat below the filter-bed. The rate of leaching is often doubled by the diminution of the pressure, below the

---

[1] T. K. Prentice, *Trans. Inst. Min. Met.*, 1935, 44, 511.

[2] A. James, "*Cyanide Practice*," 1901, p. 19

[3] For the capacity of vats of various dimensions, see "*Rand Metallurgical Practice*," vol. ii., pp. 214 and 232.

filter-bed, to this extent, and in some cases it is increased from ½ or 1 inch to 7 or 8 inches of liquid (in the leaching vat) per hour.

The vats are filled to within a few inches of the top, and the charge is levelled by means of hoes. The amount of ore charged in is such that, after the solutions have been applied, the surface of the charge is about 10 to 12 inches below the rim. In levelling the ore, the labourer must not step into the vat or forcibly press down the ore, as irregular filtering is produced by this cause. The shrinkage of the charge on the addition of liquid is from 10 to 18 per cent. The ore is charged in as dry as posssible, but a few per cent. of moisture (up to say 15 per cent.) makes very little difference to the subsequent leaching.

The use of wash-water to remove soluble salts is not required with fresh material. Lime will possibly have been added in the battery or at any grinding stage, and is not often required at this point to neutralise acidity. A solution of lead acetate or nitrate (which is cheaper) is sprinkled on the charge before it is transferred to the treatment vat. "From 15 to 25 lbs. dissolved in water will be found sufficient to precipitate the soluble sulphides from a charge of 750 tons of current sand" on the Rand.[1]

The "strong" solution of cyanide is then run on. "From 25 to 30 per cent. by weight of the charge of strong solution of 0·05 to 0·1 per cent. KCN will be found sufficient to treat the majority of Rand sands and also to keep the solutions in circuit up to strength. The practice of closing the leaching cock and allowing the solution to saturate the charge thoroughly has the drawback of expelling the air. . . . A better method is to pump sufficient solution on to the charge to cover it to about a foot in depth, and as fast as this solution leaches down, to follow with more until the desired quantity has been added."[2] The charge is then drained and air drawn through by the vacuum pump. In some later plants a separate air supply is admitted to the solution on the suction side of the circulating pump.

The strong auriferous cyanide solution is usually conveyed at once to the zinc boxes. When the solution has drained away and aeration is complete, a medium-strength solution follows, and then a weak solution containing 0·02 or 0·03 per cent. KCN. The amount of the latter is four or five times that of the strong solution. The weak solution is drained off, and followed by a final wash with water or by an extremely weak solution such as waste liquors from the slime plant, in order to remove cyanide and soluble gold as completely as possible. In one large group of Rand mines 72 per cent. of the leach solution from the sand tank is precipitated and the remainder used as a first wash for a succeeding batch of ore. This method is applicable only where the sand is almost free from slime.

As frequently as possible during the process, the sand is permitted to drain to allow the entry of air. In some instances vacuum pumps are used to assist aeration. Other methods for obtaining sufficient aeration are to scrape the top of the sand in the tanks and to pump compressed air into the initial cyanide solution. At one plant on the Rand sand leaching has been modified[3] by the introduction of a small volume of air under pressure for 5 hours during the early periods, and drainage, assisted by vacuum, during the later periods. Thus a more thorough oxidation of the ferrous compounds has been obtained, together with lessening of cyanide consumption and

[1] J. E. Thomas, "*Rand Metallurgical Practice*," vol. i., p. 161.
[2] J. E. Thomas, *ibid.*, p. 165.
[3] Wartenweiler, *Trans. Amer. Inst. Min. Met.*, 1934, **112**, 776.

shortened treatment. At Homestake air is admitted after the sand has been drained.

The whole time of treatment in the sand vats on the Rand is four to seven days, depending on the capacity of the plant, and the ratio of solution to sand is about two to one by weight.

Although the strong solution on the Rand contains only 0·05 to 0·1 per cent. KCN, a strength of 0·5 per cent. or more may be necessary in the treatment of some ores, especially silver ores. Still stronger solutions were used in the early days of the process, but were found to be unnecessary.

For the first nine months of 1934 in the Central Mining Rand Mines the original sand value was 2·884 dwt. and the discarded residue 0·324 dwt. per ton, giving an extraction of 88·8 per cent. The grading of the sand was + 48 mesh, 5 per cent. ; − 48, + 100, 46·2 per cent. ; − 100, + 200, 35·4 per cent. ; − 200, 13·4 per cent.

At Randsburg, California,[1] a tailings dump was being treated in 1934 by leaching in rectangular steel tanks 100 × 14 × 6 feet, along which a drag scraper on a slack line was worked for loading and unloading. At one end the rails on which the scraper moved rose obliquely to the rim, and thus the pulp was dragged over the edge to a hopper. A burlap filter bottom was used. Lime was added at the rate of 3 lbs. per ton and cyanide (2·2 lbs. per ton of water) was pumped in through the bottom. The solution stood on the pulp for 24 hours, circulated for 2 hours, stood for 24 hours, and was then drawn off from below, barren solution being run in on top. Precipitation of the solution was by zinc shavings.

Sumps.—The solutions, after passing the zinc boxes, are stored in sumps, which are usually large steel tanks. These may be placed either at a lower or a higher level than the leaching vats. The former is the more convenient plan. A single storage tank to hold enough solution for one day's work is then placed above the leaching tanks, and is filled by pumping from the sumps once a day. As an alternative, a single sump of moderate size receives the liquid from the boxes, and as soon as it is nearly full its contents are pumped to a large vat above the leaching tubs.

Disposal of the Tailing.—This is sometimes sluiced out of the tanks by water under pressure. Where the supply of water is not large, or the fall of the ground is insufficient for sluicing out the tailing—*e.g.* on the Rand—it is removed by shovelling out or by the Blaisdell excavator. The bottoms of the vats are furnished with doors which are usually circular, constructed of steel, and about 16 inches in diameter. The sand falls into trucks, which are hauled to the dump, or on to a conveyor belt. On the Rand the sand residue is in some instances returned underground into old workings to prevent subsidences.[2] It may be transported by pumping through pipes or on trucks, as at the Robinson Deep (see Fig. 158, which is from a photograph kindly supplied by Dr. Caldecott). Here the sand is dumped from the trucks into an inclined (20°) tunnel and sluiced down with water. A solution of permanganate of potash as cyanicide is supplied from the dissolving box in the left foreground.

Samples of the issuing solutions are taken for assay during treatment, and samples of the residue are taken for assay from the ore-trucks, usually by means of a long iron semi-circular probe, shaped like a cheese-taster,

---

[1] *Eng. Min. Journ.*, 1934, **135**, 126.

[2] Caldecott and Power, *J. Chem. Met. Mng. Soc. S.A.*, 1913, **14**, 119.

which is thrust to the bottom of the vat, then revolved by means of the handle, and withdrawn with the tailing adhering to it.

Fig. 158.—Disposal of Sand Tailing in Underground Workings by Truck Transport. Robinson Deep Mine.

Thurlow[1] has investigated the location and nature of residual gold in sand tailings by intensive cyanide treatment (to remove free gold), followed by aqua regia (to expose and dissolve the gold in the pyrite). The residue

[1] *J. Chem. Met. Mng. Soc. S.A.*, 1932, **32**, 311.

would then be the gold enclosed in the quartz. He found that a theoretical recovery of approximately 50 per cent. of residual gold in the tails would be made if the + 200 mesh material could be ground to − 200 mesh. Heating in hydrogen or direct roasting, followed by cyaniding, reduced the values in the sand from 1·0 dwt. to 0·05 dwt.—the latter being assumed to be encased in quartz.

Fig. 159 [1] shows a simple form of a cyanide plant for the treatment of sand.

Wilfley Sand Pump.—This machine, shown in Fig. 160, is commonly used for pumping all classes of pulp. The chamber is a case A fitted with a follower plate D. A combined runner B and expeller G form a single casting which is a taper fit on the driving shaft C and fixed by a runner bolt.

Fig. 160.—Wilfley Centrifugal Sand Pump.

The runner has an inlet on the side nearer the drive, and the intake chamber is thus subjected to a static pressure as well as that of the head in the feed box. Only a narrow space is left between the shaft and the circular opening in the head E of the short cylinder F. The expeller G acts as a centrifugal seal and prevents pulp leaking through during running. This seal takes the place of the ordinary stuffing box, and its action results from a number of short vanes H on the face of the expeller, radiating from the centre to the periphery, which act as an auxiliary open runner. A stationary member, having a projecting groove, is set close to the revolving portion. Leakage is prevented by the centrifugal action of the expeller wings. Any material tending to creep back from the discharge to the intake is caught by the projecting groove and returned to the wings. The expeller builds up a head which counteracts the static pressure of the intake chamber and feed

[1] Gowland, "*Non-Ferrous Metals,*" p. 281.

box, preventing thereby the pulp getting past the die ring and out through the annular space round the shaft. The intake chamber is exposed to the atmosphere. The pump can therefore have no suction, but must be fed by gravity. Two feed openings are provided, one on each side of the casing. When the pump is idle the annular opening round the shaft is sealed by the packing ring of the check valve K, which is actuated by weights that are thrown outwards by centrifugal force, and are therefore out of action when running.

3. The Cyaniding of Slime.

Several definitions of slime have been suggested. A good definition given by H. A. White[1] describes slime as "that portion of crushed ore which, owing to its minutely subdivided condition and the presence of colloidal substances, settles very slowly in water and cannot be leached without extra pressure." The product of all-sliming, however, usually contains from 5 to 35 per cent. of sand that will not pass a 200 mesh sieve. The pulp contains 5 to 10 parts of liquid and one of solids. The slime separated from sand by cones, bowls or sand collectors is suspended in water, and to promote its settlement a solution of an electrolyte, in practice always lime, is added. The alkalinity is maintained at 0·002 to 0·025 per cent. CaO. This causes agglomeration or flocculation of the particles, "clouds of large and indefinite diameter" are formed, and these quickly subside in the liquid. The largest flocks, according to White, are produced with an alkalinity of about 0·03 per cent. In practice, with an alkalinity of not more than 0·005 per cent., Rand slime settles at the rate of 2 to 4 feet per hour with a clear overflow. If no pyrite be present, 1 to 2½ lbs. of lime per ton of ore is usually added to the mill bins for the purpose. The higher the alkalinity the more moisture is retained by the slime. The lime also neutralises acidity in the ore and is a protection against the loss of cyanide. Heat assists settlement.

There are two principal methods of slime treatment, viz. :—

A. The decantation method, used in South Africa in the early days, but in course of displacement by B. A modified method, counter-current decantation, was used in America.

B. Agitation and slime filtration, used widely in many forms.

A.—The Decantation Method.

Lime is added to the stream of slime on entering the slime plant. Among the devices for adding the lime may be mentioned the use of a grinding pan, into which lime is fed, and out of which it is carried by a small stream of water as milk of lime. Ordinary Rand pyritic banket requires about 6 lbs. of lime per ton of slime, in addition to that fed into the stamp mill. The *collecting vat* is 50 to 70 feet in diameter, and 12 feet in depth, with a conical bottom 7½ feet deep. Usually 8 to 15 square feet of settlement area per ton of dry slime are allowed. The slime enters through a vertical pipe leading from above in the centre of the vat. The stream is delivered below the surface, and strikes a baffle plate, which spreads it out horizontally. Air bubbles pass up through a second larger pipe or sleeve, which surrounds

[1] "*Rand Metallurgical Practice*," vol. i., p. 189, where a full discussion on slime is given.

the feed pipe. The slime settles and clear water overflows the rim of the vat all round, into a launder, at the same time that thin pulp is entering through the feed pipe.

When the collecting vat in the intermittent system has received its charge of settled slime, the supernatant water is drawn off through a *decanter* (see Fig. 161),[1] which consists of a hinged pipe so arranged that the height of its intake can be adjusted to a point near the surface of the settled slime. The slime (in which the percentage of moisture is about 35 to 45) is then sluiced out through the apex of the conical bottom by means of a jet of cyanide solution. It passes into a pump, by which it is delivered tangentially

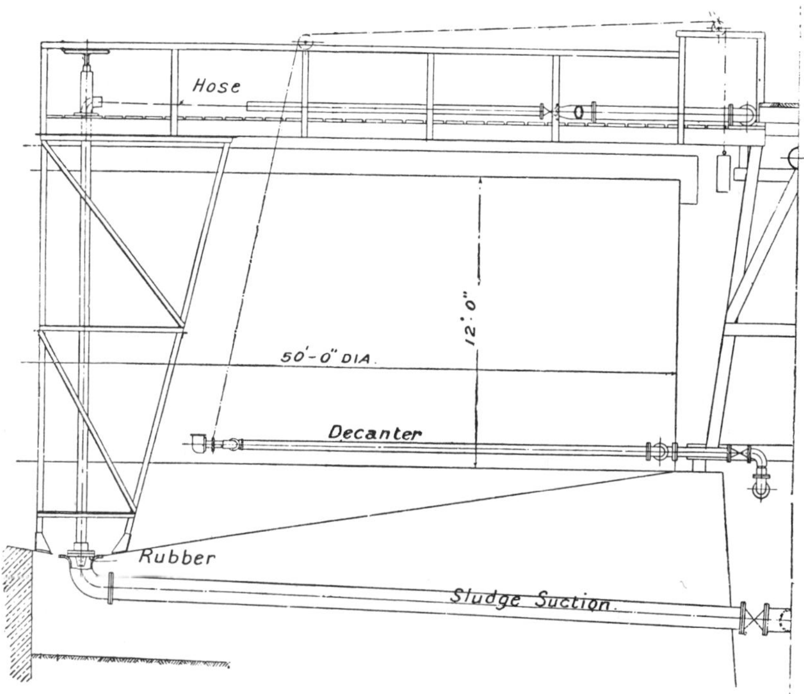

Fig. 161.—Vat with Decanter.

into other vats, giving the charge a rotary motion, so as to mix it thoroughly and keep it in suspension. The ratio of solution to solid is about 3½ or 4 to 1. The strength of the solution is 0·01 or 0·02 per cent. KCN and 0·004 to 0·02 per cent. free alkali. About 80 per cent. of the gold is dissolved in the passage through the pumps, but agitation is continued for a few hours by withdrawing the solution at the bottom and discharging it in oblique jets at the top. As an alternative, the charge may be transferred to an *intermediate vat*, and thence to the *first settlement vat*. Lead acetate solution is added to the charge in the collecting vat in order to precipitate soluble sulphides. Aeration is effected by air drawn into the pumps and by air bubbles carried down by the sludge entering the vats. The solution also becomes aerated in passing through pipes and pumps.

[1] Dowling, "*Rand Metallurgical Practice*," vol. i., p. 217.

After the gold has been dissolved, the slime is allowed to settle, the solution removed by decantation, and the pulp sluiced out with "precipitated solution" which has been through the zinc boxes and still retains some gold and cyanide. A centrifugal pump transfers it to the *second settlement vat*, where it is again allowed to settle and the liquor then decanted. This second wash is used again for the first treatment of the next charge. The settled slime, still containing about one-sixteenth of the dissolved gold, is then diluted and discharged to the dam. Settlement of slime is facilitated by the use of warmed working solutions, for which purpose waste steam is commonly employed (Caldecott). At Noranda, where the solids were in such a dispersed condition that they would not thicken or settle, zinc sulphate was added to facilitate settlement.[1]

The first solution is clarified by being passed through a sand filter vat or filter press (*q.v.*), and then goes to the zinc boxes or to the Merrill Crowe precipitation plant. The solutions are generally heated (say to 86° F.) by waste steam. Centrifugal pumps are in general use.

The time required for dissolving the gold in slime varies from 4 to 24 hours, depending on the nature of the ore, the degree of grinding and the amount of sand mixed with the slime.

Although the gold in slime fresh from the battery is very readily soluble in cyanide solution, it is quite otherwise with slime which has accumulated in dams and settling pits and has been exposed to the weather for some time. The presence in these materials of finely divided ferrous sulphide, ferrous hydroxide, and other ferrous salts and decomposing organic matter gives rise to the rapid abstraction of oxygen from the cyanide solutions used in their treatment, and the solution of the gold is in this way prevented. Even in fresh slime some ferrous sulphide, produced during the crushing in the battery, may be present. Caldecott has shown that this substance is produced by triturating pyrite in a mortar. The remedy for this difficulty is to supply more oxygen, usually in the form of air delivered into the agitation pump, or through perforated pipes in the bottom of a vat. The latter method is also in use with fresh slime.

The decantation method described above has several disadvantages. It requires a number of large vats and much space, and the slime is discharged still containing 6 or 7 per cent. of the dissolved gold. In 1927 the average extraction of gold from slime in the plants of the Rand Mines group was only 89·28 per cent. The residue is, therefore, unduly rich, and the filtration processes described in the next section are preferable in this respect, as they separate 98 per cent. or more of the soluble gold. Besides this, the amount of gold solution yielded by the decantation method requiring precipitation is about 3 tons per ton of slime, whereas by filtration 2 tons or less of gold solution are produced. Vacuum filtration, moreover, has the advantage of being independent of fluctuations in the rate of settlement due to variations in temperature between summer and winter.

A method of continuous decantation has been developed in North America. It is usually called *counter-current decantation*. It has been introduced on some modern plants. The pulp is agitated with cyanide and then goes to a series of thickeners, being diluted after thickening and transferred to another thickener, lastly going to a vacuum filter, which, however, has been omitted in some mills where the solution itself is low in

---

[1] *Min. Ind.*, 1929, 38, 718.

cyanide and contains small metal values.[1] There is a tendency where capital will allow, however, to extend filtration, in this connection, at the expense of thickening, because of the larger percentage of solids in the discharge. Occasionally primary and secondary agitation, separated by a filtration unit, is employed, thus permitting the use of a smaller quantity of solution for washing.

A simple typical flow sheet[2] for counter-current decantation is shown in Fig. 162. It will be noticed that generally the pulp at each step goes towards the discharge end while the solution goes towards the feed end.

Crushing is done in solution, using the overflow from tank T_2. The overflow from T_1 goes to precipitation and the underflow to the agitator A. The barren solution from precipitation dilutes the underflow of T_3 as it enters T_4 together with the overflow of T_5. The overflow of T_4 mixes with the underflow of T_2 to form the feed to T_3, etc. At each succeeding mixture the solution meets a pulp containing more dissolved gold than itself and is enriched, while the pulp is correspondingly impoverished.

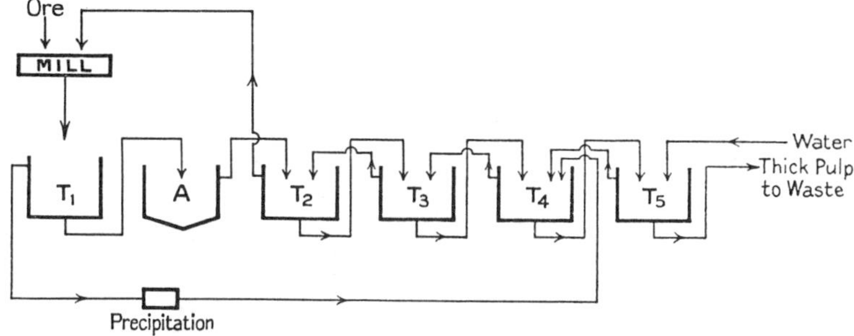

Fig. 162.—System of Counter-current Decantation.

The advantages and limitations of counter-current decantation are discussed by H. St. J. Brooks,[3] who considers it inapplicable to ores in which the greater part of the gold and silver is locked up in sulphides, so that long-continued agitation with cyanide is required to dissolve the metals. High-grade ores requiring strong solutions are also unsuitable. He would apply the system to low-grade siliceous ores requiring solutions of only moderate strength, and finish with filtration.

Owing to the increased price obtainable for gold in 1934 the tailing dumps at Manning Mercur were re-treated by counter-current decantation. After mixing in a Hardinge ball mill, the discharge was classified in Akins machines and then traversed the counter-current system.[4]

B.—**Treatment of Slime by Agitation with Cyanide and Filtration.**

The stages in this process are (1) Thickening, (2) Agitation, (3) Filtration.

Thickeners.—The dissolution of the gold is usually effected easily and rapidly by agitation with cyanide, although some exceptions are noted below. It is necessary that slime should be de-watered or thickened, however, before it is treated in agitators. One method of doing this is by settlement

[1] H. A. Megraw, *Eng. and Mng. J.*, 1914, 97, 1061; also 1914, 98, 683.
[2] Games, *Trans. Amer. Inst. Min. Eng.*, 1917, 57, 144.
[3] *Mng. and Sci. Press*, 1913, 106, 624.
[4] *Eng. and Mng. Journ.*, 1934, 135, 54.

in large vats, as described on p. 342. Early attempts at continuous collection of slime were made on the Rand in 1899 and 1903 and these methods may be considered as the forerunners of the Dorr Thickener, first installed at New Modderfontein in 1918 and later adopted at several other mines.[1] A serious disadvantage of continuous collection is that a record of daily tonnage can rarely be obtained. The appliances for de-watering sand (cones, filter table, etc.), are obviously not suitable for slime. One successful slime de-waterer in wide use is the *Dorr Continuous Thickener* (see Fig. 163). This consists of a vat from 20 to 35 feet in diameter and from 8 to 12 feet deep, in which a central vertical shaft, reaching nearly to the bottom, and carrying four radial arms, revolves slowly. The thin pulp flows into the tank at the centre, just below the surface. For settlement

Fig. 163.—Dorr Continuous Thickener.

it is usual to have a minimum of 8 square feet per ton of slime per 24 hours. Pieces of angle iron attached to the arms are so placed that they move the settled thickened pulp to the centre of the bottom of the vat, where it is discharged through a pipe. The clear liquor overflows continuously at the periphery over a lip or through perforations. The shaft can be raised and lowered while it is running. In some recent Rand plants, however, the Dorr thickener has not been incorporated.[2]

A later type of Dorr thickener delivers the feed from near the circumference to a central perforated vertical cylinder by means of an inverted U-tube. This arrangement saves mill-head, is more compact, and gives a better distribution of feed.[3]

[1] Prentice, *Trans. Inst. Min. Met.*, 1935, 44, 514.
[2] Thurlow and Prentice, *J. Chem. Met. Mng. Soc. S.A.*, 1928, 28, 251.
[3] *Min. Ind.*, 1932, 41, 586.

The Dorr Tray Thickener consists of a tank divided horizontally into a series of compartments by steel trays or diaphragms attached to the sides. Radial arms carrying plough blades are attached to a central revolving shaft and revolve above each tray. The blades cause the thickened material to move towards a discharge opening at or near the centre. By this arrangement floor space is saved, and there is a decreased capital cost for tanks, pipes, and transmission gear. Fig. 164 shows a tray thickener with two compartments.

Fig. 164.—Dorr Two-Compartment Tray Thickener.

The Golden Cycle Thickener[1] was designed by Coe and Blomfield, who realised that zones of increasing density from top to bottom of an ordinary thickener tend to retard the free settling rate of the solids and limit the capacity of the thickener. The level of the thickening zones is drawn down and the bottom layer of settled solids is made more compact, allowing a very thick underflow. To effect this, the bottom of the thickener is made porous only to the liquid and is kept clean by suitable mechanism. The 30 × 12 foot tank has the usual 6-inch false bottom, supporting a bed of clean fine sand. All settled material and a thin layer of sand are removed at each revolution of the scrapers, which are suspended from a superstructure over the tank. When treating roasted Cripple Creek ores, containing 95-98

[1] N. Cunningham, *Eng. Min. Journ. Press*, 1925, 119, 905.

per cent. of material of − 200 mesh, the underflow can be maintained at 64 to 70 per cent. solids. The machine effects a certain amount of filtration.

The Genter Continuous Vacuum Thickener (Salt Lake City) consists, in its simplest form, of tapered canvas socks drawn on to perforated wooden or metal tubes which are held in pipe frames and are totally submerged. The interior of the tubes is connected with an automatic valve which governs the suction cycle. In a later form the tubes are suspended in a circular concrete tank, at the bottom of which are some rakes. When the thickened material is discharged from the socks, the rakes gather it towards the centre, where it is discharged into a pump for transfer to the next stage.

Agitators.—Agitation in the decantation method already described, p. 343, consists in withdrawal from the bottom of the vat and delivery at the top by centrifugal pumps. In Western Australia, revolving-arm or paddle agitators were used. "In these it is not possible to use a thick or very sandy pulp, and in the majority of them air is introduced into the pulp, generally by an air jet at the bottom of the vat, but in some cases in small jets along the agitator paddles, it having been delivered through the hollow shafting of the agitator gear. The smaller sizes of these mechanical agitators are often arranged so that the stirring gear can be raised out of the pulp, when a stoppage is imperative."[1]

Agitators in general use are all air-lift machines, such as the Brown or Pachuca tank, the Dorr Agitator or the Wallace Turbo-Mixer, in order to combine agitation with aeration. In air-lift vats the agitation is effected by reducing the specific gravity of a central column of pulp, by the introduction of air at a pressure which will just overcome the pressure of the column at the point of introduction.

The Brown Agitator or Pachuca Tank.—This agitator was introduced in New Zealand in 1902, and afterwards adopted at Pachuca, in Mexico. In Fig. 165 it is shown in vertical section. It consists of a tall cylinder of steel, conical at the bottom. Inside the cylinder there is a central pipe, B, extending to a point near the bottom. Air is introduced into the central pipe through the internal pipe, D, which discharges at G through a rubber valve. C is a second air pipe outside the tube. It serves to keep the pulp in motion during the filling and discharge of the vat. The pipe, E, is used for supplying solution, water or air to the distributor, F, which discharges through the pipes, I, to prevent packing before agitation is started and when it is stopped. Air is supplied to pipes C, D and E from pipe O, and water or solution from pipe N. An adjustable splash plate, Q, is fitted near the top of the central tube. The slime pulp is fed through the pipe P. The discharge after agitation takes place through a pipe with a stopcock shown close to the apex of the cone. The vats are of various sizes, such as 45 to 55 feet in height and 13 to 33 feet in diameter.

The method of working is as follows:—As soon as the tank is filled with slime and solution, air is turned on in the pipe D and emerges at G, overcoming the pressure of the column at that point. As it bubbles up through the central tube it lightens the column of pulp inside and rapidly lifts it, causing it to overflow. Fresh pulp is drawn in at the bottom, and a perfect circulation is given. The W/S ratio is usually 1·2 : 1 (specific gravity 1·337). On the Rand the time of agitation varies from 3 to 12

[1] R. Allen, *J. Chem. Met. Mng. Soc. S.A.*, 1911, II, 428.

hours. The cyanide strength is 0·01 per cent. KCN and the lime content 0·005 per cent. CaO.

In starting up old charges, or those which have been allowed to pack,

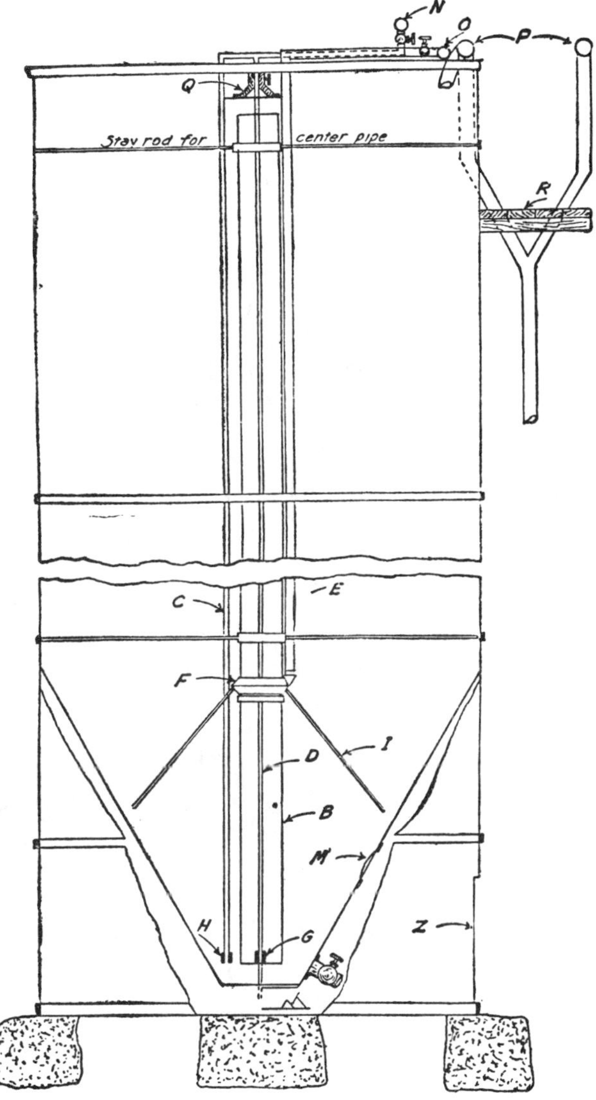

Fig. 165.—Brown Agitator.

the material at the bottom of the cone is softened by the introduction of water or solution through the wash-ring F, so that it may be readily lifted through the tube B or upcast column ; or a suspended pipe with compressed air jet may be left to bubble in the packed centre tube until it works its way down to the cone, when agitation quickly becomes general. The use at the Treadwell Mine of a "Spider" or adjustable hollow annular casting with radiating fingers is described by W. P. Lass.[1]

In beginning agitation, the air pressure required is considerable, but when circulation is fully established, the quantity of air is kept moderate

Fig. 166.—Brown Agitators at the East Rand Proprietary Mines, Johannesburg.

in amount in order to avoid excessive circulation. As originally used, the Brown agitator was intermittent in its action, but it is now common for four to six vats to be placed side by side, and worked continuously as a series. In this case pulp is continuously fed into the first one near the entrance to the air-lift, and is drawn off into a pipe placed about midway

[1] *Mng. and Sci. Press*, 1911, 103, 517.

between the air lift and the periphery of the vat, conveyed to the second vat, and thence in succession to the others. As a safety measure the tops of the vats are connected by an 8 inch pipe to prevent overflows. After passing through the series of vats, the pulp is discharged to the filter plant.

The vats are somewhat costly for their capacity, owing to their great height, which is necessary in order to utilise to the full the expensive compressed air.

Where prolonged treatment of 6 to 8 hours is not required in the air-lift vats, they are used intermittently. The pulp is then sent to storage vats, where it is kept in movement until sent to the filters.

Fig. 166 is from a photograph of the Brown agitators at the East Rand Proprietary Mines, Johannesburg.[1] These four agitators were stated to be continuously agitating 1,600 tons of slime a day with an extraction of 97 per cent.

The advantages of the continuous system are :—(1) The time of emptying and filling is saved and is more usefully employed in agitation. (2) Less attention is required. (3) The cost of individual discharge is saved. (4) The loss of head is almost all saved, the level of the discharge being only about 3 feet below the top of the last vat. The treated pulp thus gravitates to the filter plant.[2]

Usually a longer treatment is given in the continuous system than in the intermittent. The agitator should be emptied at frequent intervals to clean out the coarser particles, which are liable to settle.

Dowsett describes a specific gravity indicator employed at Dome Mines for use in cyanide slime vats. It consists of two tubes, one 5 feet longer than the other, and is compensating with regard to variation in pulp level. It has a fixed position in the Pachuca tank. A piece of cotton duck between the two flanges at the bottom end of the tubes acts as a filtering medium. The longer tube is submerged exactly 5 feet more than the shorter tube, and when both are immersed, the clear solution rises into the glass tubes above the steel ones in the pulp, to establish the hydraulic balance between the outer column of slime pulp and the column of solution inside. The difference in level between the two columns of solution is the distance that the solution would rise above the pulp level in a tube that had a submergence of exactly 5 feet. Each 6 inches on a scale behind the tubes represents 0·1 in specific gravity.

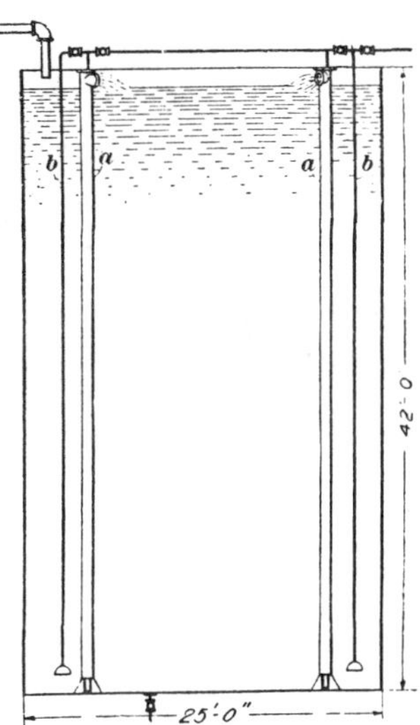

Fig. 167.—Parral Vat.

Hills describes an intermittent agitation and decantation system for small cyanide plants, using a *Tuolumme agitator*, which is of the Pachuca type. It picks up and restores the thickened pulp to suspension. The agitation tank also serves as a thickener.

---

[1] By courtesy of the Cyanide Plant Supply Co.

[2] Thurlow and Prentice, *J. Chem. Met. Mng. Soc. S.A.*, 1928, **28**, 258.

The Parral Vat (Fig. 167) [1] has a flat bottom and two to four air-lifts near the periphery of the tank instead of one at the centre, as in the Pachuca tank. In this way vats of greater diameter and capacity are made possible. An elbow or turn is fixed to the top of the air-lift pipe, and the pulp is delivered circumferentially so that the contents of the vat acquire a rotary motion, which is designed to avoid the settling of solids, and assists agitation. The Parral vat is generally made of less height than the Pachuca tank.

The Dorr Agitator (Fig. 168) is seldom used for the intermittent system

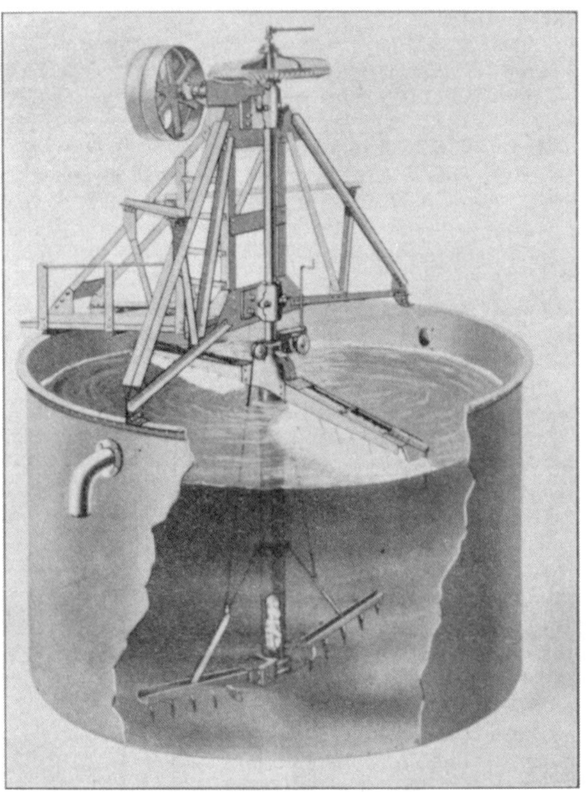

Fig. 168.—Dorr Agitator.

of treatment, but is suitable for the continuous method. It can be affixed to any flat-bottomed vat and is a development of the Dorr thickener. It consists of a central vertical cylinder or pipe carried by a shaft supported from the top of the tank and equipped with two arms carrying ploughs, as in the Dorr thickener, which travel round near the bottom of the vat and draw the pulp to the centre. The pulp (W/S ratio 1·2 : 1) flows in continuously at a point in the periphery and gradually falls to the bottom. It is then raised through the cylinder by an air-lift and distributed evenly over the surface of the charge in the vat by revolving launders. The air

[1] Gowland, "*Non-Ferrous Metals,*" p. 312. See also *Eng. and Mng. J.*, 1914, 97, 422.

pipe for the lift may be inserted through the bottom of the vat. There is no agitation of the charge as a whole, and the solids therefore tend to settle. They are scraped from the bottom towards the centre and so to the air-lift.

Fig. 169.—Denver (Wallace-Type) Super-Agitators, one in operation and another being filled.

There is thus a vertical circulation of particles through a comparatively still solution. This gives them a more pronounced movement relative to the solution than in the Brown tank, and hence tends to give better contact with the solvent.

The arms are hinged so that they can be raised and stand close to the cylinder during a shut-down. On beginning again, they are lowered by degrees until the settled pulp has been brought into suspension, so that there is no danger of the arms being broken.

A Dorr agitator has been combined with a filter-thickener for flotation concentrates in a Canadian gold mill by submerging twelve 5 × 3·5 feet suction filter leaves in the pulp in the agitator.[1] Where it is desired to give coarser particles a longer contact with the liquor, the dilution of the pulp and the discharge have to be regulated. The tanks used (*e.g.* 30 feet diameter by 12 feet deep) are of small depth, so that the air pressure required is small. The arms revolve at from 1 to 4 revolutions per minute.

Ham [2] has developed a formula for calculating the percentage extraction in continuous agitation.

Pachuca tank and Dorr agitators are confined to the treatment of —150 mesh or finer material if building-up at the bottom is to be avoided. Where more sandy material or flotation concentrates have to be treated, a Wallace agitator may be used. This can deal with sand of 40 mesh.

The Wallace Agitator (Fig. 169) consists of a circular tank, 12 × 12 feet, with a hollow 22 inch stand-pipe in the centre. Within this a shaft, to which a 27 inch impeller is attached at a point below the bottom of the pipe, rotates at 200 r.p.m. The stand-pipe is telescopic to allow for adjustment to the height of the charge. Usually this is about 2 feet from the bottom of the tank. Four small air-lifts are situated towards the periphery. In action, the pulp is drawn down the stand-pipe through a vortex which sucks in air. The pulp then rises in the outer space assisted by the air-lifts.[3] The agitator has also been applied in the treatment of flotation concentrates with cyanide solution. Here, owing to the presence of oil, there is a tendency to form a froth, which by ordinary agitation is not completely drawn within the solution. In the Wallace agitator the suction down the stand-pipe caused by the impeller is said to overcome this difficulty.

A *Turbo-mixer*, which may also be used, has a mixing unit consisting of an impeller, with stationary curved deflecting blades, rotating in a horizontal plane. The pulp enters near the centre of the impeller and is expelled tangentially. It is then deflected by the stationary blades surrounding the impeller. Thus an intensive agitation is produced. Fig. 170 shows a turbo-mixer with two series of blades. The lower series keeps the lower portion of the pulp agitated and the upper does likewise for the upper portion, thus maintaining violent agitation throughout the tank.

As has been seen, most modern agitators employ some kind of air-lifting device which also increases the aeration. All types of ore, and especially a flotation concentrate containing gold, need the maximum supply of air. Knapp deprecates the use of artificial oxidisers such as ozone or hydrogen peroxide, which have been suggested, and prefers direct aeration.

Brown, in experimental work, found that by bubbling air through the vats during agitation he could get an average of 25 per cent. of dissolved oxygen just prior to pressing. If the air supply was stopped the content of oxygen rapidly decreased.

---

[1] *Min. Ind.*, 1932, **41**, 586.
[2] *Chem. Met. Eng.*, 1918, **19**, 663.
[3] S. G. Turrell, *Chem. Eng. Min. Rev.*, 1934, **26**, 312.

It has become the practice on some plants to aerate the slime in large vats by ploughing or harrowing. Ore pulp has been sprayed to effect the same result.

Prentice [1] found that solutions from which both nitrogen and oxygen had been removed picked up oxygen more rapidly than similar solutions still containing nitrogen. He also showed that the dissolving power of cyanide solutions might be increased by pumping air into them, and main-

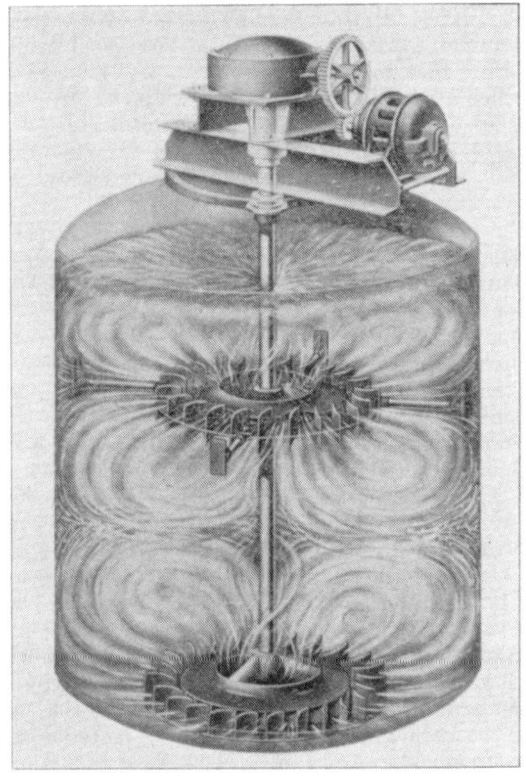

Fig. 170.—Turbo-mixer (*Dorr Oliver Co.*).

taining the air pressure at 65 lbs. per square inch.[2] In this manner he raised the oxygen content from $3\frac{1}{2}$ to $6\frac{1}{2}$ mg. per litre and retained 4 mg. after standing in the open for 22 hours. The process has been applied in practice and made continuous. There is no increase in the rate of dissolution, and the decrease in the value of the tails is not very great, but there is about a 34 per cent. drop in the cyanide consumption and a greater solution efficiency.

Filtration of Ore Slime.—This is effected by a difference in pressure on two sides of a canvas or other permeable medium immersed in a pulp carrying solids in suspension. The liquid passes through, leaving the solids as a cake adhering to the canvas. The weight and thickness of the cake depend on (1) the time of operation, (2) the specific gravity of the solid material and

[1] *J. Chem. Met. Mng. Soc. S.A.*, 1934, 34, 244.
[2] *Eng. Min. Journ.*, 1934, 135, 298.

that of the liquid-solid mixture, and (3) the difference in pressure on the two sides of the filtering medium.

A uniform flow of the solution ensures uniformity in the cake. A flow of a clear solution, under similar conditions, causes the displacement of the solution already in the cake. A smaller difference in pressure, and therefore a longer time required for filtration, corresponds to a greater weight of wash solution necessary for displacement.

The difficulty in the filtration of slime is that even a thin layer of slime packs down and offers great resistance to the passage of liquids. A pressure much greater than that merely of gravity is required. In practice layers of ½ inch to 1 inch are used with vacuum leaching or layers of 2 to 3 inches with pressures of 40 to 100 lbs. per square inch given by pumps. An enormous area of filtering surface is required for operations on a large scale, and this is obtained by using a number of parallel plates or leaves placed side by side, or a continuously revolving drum.

When slime is forced against a filtering surface either by the pressure of the atmosphere (vacuum filtration) or by direct and higher pressure, the slime forms a coherent and approximately homogeneous layer or cake, through which liquids pass almost evenly. There is little tendency for the formation of channels, and so washing is complete and satisfactory with little water.

Warwick discusses the matter as follows [1] :—" In removing the soluble values from slime cakes, the principle of displacement is used rather than the laws of continuous or repeated dilution." In cakes of uniform permeability the wash-water would pass through like a wall pushing the gold-bearing or " pregnant " solution before it. In practice a wash equal in volume to the moisture in the cake and filter leaves will remove 80 to 85 per cent. of the dissolved gold from the slime cake. In order to recover 98 per cent. of the dissolved gold, a volume of wash equal to one and a half times or twice that of the retained solution must be used. To avoid increasing the volume of mill solution unduly, the larger part of the wash consists not of water but of barren solution, low in cyanide. This results in a residue low in gold but containing some cyanide, which is thrown away with the residue.

The filter may be vertical or on the surface of a cylinder. The slime cake forms and gradually thickens or builds up on the canvas, adhering well so long as the vacuum is maintained. Some material builds up a slime cake readily and quickly—*e.g.* 1 inch thick in a few minutes—but with argillaceous or talcose slime the building-up is slower. If cracks form in the cakes, washing becomes impossible, as the liquid passes through the cracks, following the line of least resistance.

In filter-pressing, the pulp to be filtered is thickened, in order to make settlement inside the presses less likely to occur. Such settlement would cause the formation of non-homogeneous cakes, with the result that the washing would not be uniform.

Slime Filters may be classified as follows :—

I. Vacuum or suction filters.

A. Appliances using a thick slime-cake and intermittent in their action (Moore and Butters filters).

B. Appliances using a thin slime-cake and practically continuous in their action (Oliver and American filters).

[1] *Mng. Eng. World,* 1913, 38, 665, 797, 1135 ; *Mineral Industry,* 1913, 22, 347.

II. Pressure filters, in which the liquor is forced through the filtering medium by external pressure.

C. Ordinary filter presses (Johnson and Dehne presses).
D. Sluicing filter presses (the Merrill filter press).

After agitation, and before filtration, it may be necessary to have storage capacity, keeping the pulp still agitated. For this purpose Pachuca tanks or large circular tanks with revolving arms are employed. Air-lift vats are in common use with vacuum filters on the Rand. In one instance a 2 foot 6 inch propeller revolving at 450 r.p.m. has been fixed through the side of a 70 foot tank at an angle and is said to give good results.[1]

The Moore Vacuum Filter.—This is the oldest of the vacuum slime filters. It consists of a series of parallel plates or leaves. Each leaf is simply a light framework with canvas on both sides, and is of large dimensions,

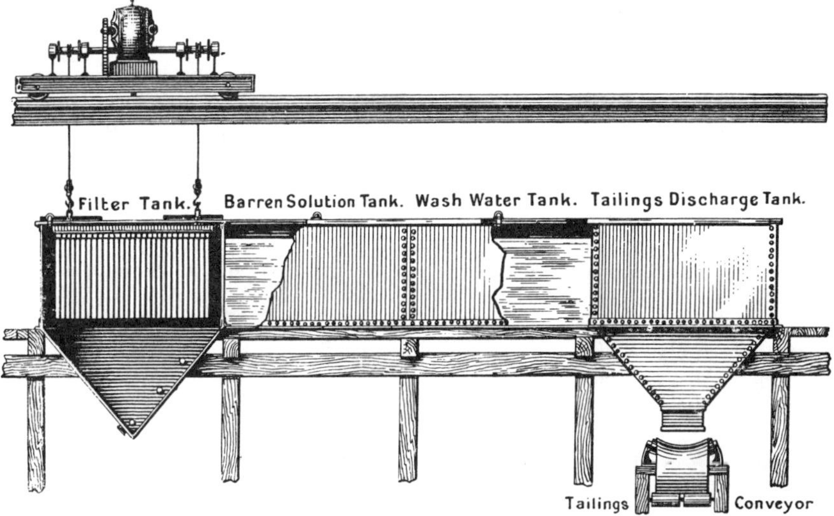

Fig. 171.—Moore Filter Vats.

20 foot by 4 feet, or later 16 feet by 5 feet. A suction pipe and also two pressure pipes communicate with the interior. When a vacuum is formed the canvas sides are prevented from collapsing by wooden strips and wire netting. A number of these leaves (49 in later practice) are hung (4 inches apart) from a steel frame, forming a " basket " or unit. The basket hangs by cables from an overhead traveller, and can be raised or lowered. The suction pipes communicate with one main pipe.

In operation the filter-basket is lowered into the vat containing the thin slime, so as to be submerged, and the vacuum pump started. The slime is agitated to prevent settling. After one or two hours a cake of slime of $\frac{3}{4}$ inch to 1 inch thick is formed on the outer surfaces of each leaf, and during this time the pump is continuously discharging clear gold solution. The basket is then lifted out of the vat and transferred to a vat containing weak cyanide solution, which is drawn through the cakes for washing purposes

[1] Thurlow and Prentice, *J. Chem. Met. Mng. Soc. S.A.*, 1928, **28**, 258.

for twenty minutes. The washing is finished by ten minutes' immersion in a vat of wash-water with continuous pumping, after which the basket is withdrawn and run over the discharge hopper. The cakes are dried by air drawn through them. Then the suction is for the first time discontinued and a blast of air forced into the leaves, by which the waste slime-cakes are dislodged and fall into ore trucks. The arrangement of the vats is shown in Fig. 171.[1]

The Butters Filter [2] has been incorporated in many Rand plants. It is an intermittent filter and consists of a number of parallel plates or leaves, about 30 to each section. These remain stationary in a steel or concrete filtering vat. The leaves are made of a $\frac{3}{4}$ inch perforated iron rectangular framework, 10 × 5 feet, covered with sheets of 10 or 12 oz. canvas having coco-matting or wooden strips within to keep them apart when under a vacuum. They are placed $4\frac{1}{2}$ inches apart and each is attached to a manifold column connected to vacuum and solution pumps. The capacity

TABLE XXXVIII.

TYPICAL CYCLES IN BUTTERS FILTERS (THURLOW AND PRENTICE).

	Crown Mines, Mins.	City Deep, Mins.	New Modder, Mins.	Modder East, Mins.
Filling the boxes from the storage pulp vat,	15	12	8	5
Forming the cakes, vacuum 20 ins., . .	20	26	25	25
Emptying the excess pulp, increasing vacuum to 5 ins., ...	13	10	8	7
Filling with wash solution, ...	10	9	7	6
Washing (vacuum 20 ins.)	60	35	65	60
Sampling and dropping cakes (vacuum shut off), ...	12	10	10	10
Returning the excess wash solution, . .	15	8	7	7
Discharging pulp, ...	5	5	5	5
Total, ...	150	115	135	125
Tons slime per cycle, ...	216	150	150	250

TABLE XXXIX.

RATE OF WASHING (THURLOW AND PRENTICE).

	Modder East.	City Deep.
Filtered solution at start of washing, .	2·09 dwts./ton	2·10 dwts./ton
,, ,, after 10 mins. washing,	1·90 ,,	1·98 ,,
,, ,, ,, 20 ,, .	1·40 ,,	1·66 ,,
,, ,, ,, 30 ,, .	0·37 ,,	0·30 ,,
,, ,, ,, 40 ,, .	0·22 ,,	0·19 ,,
,, ,, ,, 50 ,, .	0·18 ,,	0·19 ,,
,, ,, ,, 60 ,, .	0·14 ,,	...
Value of initial washing solution, . .	0·11 ,,	0·11 ,,

[1] Gowland, "*Non-Ferrous Metals*," p. 317.

[2] *Mineral Industry*, 1906, **15**, 411 ; *Mng. and Sci. Press*, 1907, **94**, 785, 818.

is roughly 85 lbs. per square foot of canvas per 24 hours. Thin pulp, after agitation, is run into the vat, and as liquid is withdrawn by the vacuum

Fig. 172.—" Butters " Slime Filter at Brakpan Mines.

pump more pulp is added in order to keep the vat full. A jet of air from the bottom maintains agitation. When cakes of slime $\frac{3}{4}$ to $1\frac{1}{2}$ inches thick have been formed on each side of the canvas leaves, the pulp is run off either direct into the same pulper, or by gravity into another tank whence it is pumped back into the pulper. The latter arrangement is preferable as it saves time. The box is then filled with barren solution or water for washing as in the Moore filter. When washing is complete, the cakes are detached by water or barren solution under pressure from within and, disintegrating, are carried away. The remaining wash solution is returned by decanter pipes to the stock wash-tanks and the pulp is discharged to the dam. The time for each cycle is about 2 hours.

The canvas filtering surfaces in these filters become clogged with calcium salts in course of time, and these are removed by dilute (2 per cent.) hydrochloric or sulphurous acid. Sometimes a lower vacuum gives better results than a higher one. The solution is deaerated to a considerable extent during its passage under vacuum.

The W/S ratio is about 1·25 : 1 and in addition to the 1·25 tons of gold-bearing solution, about 0·9 ton of wash solution per ton of dry slime passes through the leaves, 0·5 ton remaining in the cake.

The Butters filter installation at Brakpan Mines, Transvaal, is shown in Fig. 172.[1] This plant contained 336 filter leaves, and had a capacity of 1,200 tons of dry slime per twenty-four hours. The slime assayed 2·6 dwts. per ton, and the average extraction was 95·5 per cent.

The time for cake formation varies roughly as the square of the thickness, and as the tonnage treated per cycle.

White gives a full description of the operation of a modern Butters filter plant.[2]

The Ridgway Filter.[3]—This machine consists of 12 or 14 cast-iron horizontal filtering frames, which are in the form of sectors of a circle. They are suspended from hollow arms which radiate from a central hollow revolving column, and also run on rollers at the periphery. The frames travel in a shallow annular trough divided by partitions into three compartments, one filled with slime pulp and the second with wash-water, whilst the third is the discharge chamber. The under side of the frames is the filter surface, and is covered with filter cloth. The frames revolve with suction on, dip into the slime pulp and take on a filter cake of $\frac{1}{8}$ to $\frac{3}{8}$ inch in thickness, rise over the elevated portion and pass to the solution part of the trough, the valve being automatically changed ; here they suck solution through the slime cakes, and then pass over the second elevated portion (the valve changing automatically to compressed air). Finally they discharge the slime cakes into the third part of the trough before again making the cycle as described. The total cycle requires 60 seconds, of which about 13 are in pulp, 30 in wash solution, and 17 on the raised portions of the track and in discharging. Each frame has a filter surface of 4 square feet. The machine has a capacity of 30 to 60 tons per day ; it is shown in Fig. 173.[4] In practice its use is limited, by the small filtering surface, to the treatment of slime which can be rapidly filtered, as, if longer than fifteen seconds is taken for cake formation, the output of the machine is too small. There is also no opportunity of increasing the wash period with the grade of material to be treated. In these respects it is inferior to the Moore filter and to the reciprocating Ridgway filter, an improved form described below.

---

[1] From a photograph kindly supplied by Mr. F. L. Bosqui.
[2] "*Rand Metallurgical Practice*," vol. i., p. 504.
[3] *Mineral Industry*, 1907, 16, 540 ; Gowland, "*Non-Ferrous Metals*," p. 320.
[4] By courtesy of the Cyanide Plant Supply Co.

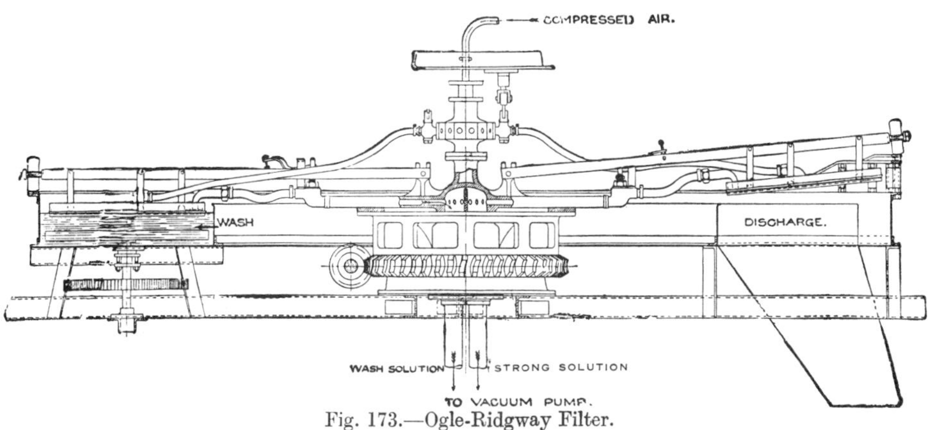

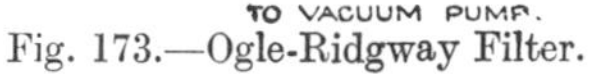
Fig. 173.—Ogle-Ridgway Filter.

Fig. 174.—Reciprocating Ridgway Filter at the Great Boulder Mine lifting Cakes from Pulp to Wash.

The Reciprocating Ridgway Filter.[1]—In this machine a set of twenty vertical filtering frames suspended from an arm is lowered into a slime pulp vat and kept there until the frames have received a cake of about 1 inch in thickness by suction. The frames are then lifted out and rapidly immersed in a washing vat for the necessary time, after which the cakes are discharged. This machine is shown in Fig. 174,[2] as installed at the Great Boulder Mine, West Australia. The cakes are being lifted from the pulp vat to the washing tank.

It is claimed that it is the rapid means of transferring the cake to the solution that constitutes the advantage of this filter, "as it is practically impossible to hold a heavy roasted sulphide cake on for more than sixty seconds without the appearance of cracks."

The Oliver Filter (see Fig. 175), consists of a drum mounted on trunnions

Fig. 175.—Oliver Continuous Filter.

with the filtering area on the outside surface of the cylinder. This surface is divided into 12 to 24 shallow compartments, with a watertight backing of wooden staves from flange to flange and supported on interior spokes. A wooden grating, supporting the filter cloth, is attached to the staves. The cloth itself is held in place by fine wire and should in all cases be sufficiently strong to withstand the applied pressure. It is of fine mesh to retain the precipitate without sacrificing rapidity. A suction pipe is carried from each compartment through a trunnion to an automatic valve controlling the vacuum for forming the cake and washing it. The discharge of the cake is by means of compressed air. The drum is revolved once every three to six minutes, and is partly submerged in pulp, the level of which is such that the filter surface is submerged for one-third to three-fifths

[1] *Mng. and Sci. Press*, 1913, 106, 993.
[2] By courtesy of the Cyanide Plant Supply Co.

of a revolution. Under the influence of the vacuum a slime cake of $\frac{1}{4}$ to $\frac{1}{2}$ inch thick accumulates on the canvas. On emerging from the tank the cake is partly dried and then washed by sprays, which displace the gold-bearing solution. Finally, the vacuum is cut off and air pressure applied inside the filter, the cake is detached, and, assisted by a water spray, slides down over a scraper into a launder, whence it is pumped to the dam. The filter surface then immediately enters the slime pulp again. From the automatic valve the solution passes to clarifiers and precipitators in turn.

The capacity of the Oliver filter depends on the pulp density, the percentage submergence of the drum, the vacuum maintained, the speed of the drum, and the physical condition of the solids. Increase in capacity is not directly proportional to increase in speed, but the latter results in a thinner cake which can be more effectively washed. The best conditions for a 5 foot 4 inch drum were found to be, on the average, one revolution in 3·20 minutes with one-third of the circumference submerged.

The Oliver filter has displaced the Butters in many instances on account of its greater rate of treatment, lower consumption of water, and lower capital cost. The size generally used is 14 feet diameter, and 16 feet long, with a filtering area of 700 square feet. About 0·8 ton of wash solution is used per ton of slime, of which 0·4 ton remains in the slime cake, the remainder going with the pregnant solution. The capacity of the Rand filters is 1,000 to 1,500 lbs. per square foot per 24 hours.

The Dorrco Filter is of the continuous, vacuum, rotary drum type, but the filter cloth is inside the drum, supported by a heavy screen. The drum itself is closed at one end, while the other is open for the discharge of the filter cake and is provided with a flange which assists in keeping the pulp within the drum. Mounting is by means of a trunnion at the closed end and friction rollers at the open end. Rotation is by means of worm gearing. Liquid drawn through the filter cloth passes into a space between it and the drum. The cloth is fastened into grooves at the end of the drum and also along the channels separating the parallel panels into which the surface is divided. Connection to pressure and vacuum pumps is made in sequence through a main valve. The cake is continuously discharged by means of a hopper feeding on to a conveyor which is mounted within the drum and is supported on girders. These girders extend through each end and rest on the foundations.

The pulp feed may enter through the open end of the drum or through a pipe in the closed end. When a panel is under its surface, the application of a vacuum between the shell and the filter cloth causes a thin cake to form on the latter, which then goes through a cycle of operations (dewatering, washing and discharge) similar to that in the case of the external drum filter already described. Discharge is assisted by air pulsations applied to the filter cloth. It is usual to keep the cloth in good condition by a fine water spray, exhaust steam or air, applied on each revolution.

In the operation of this filter, gravity assists cake formation and also assists in discharge when a panel is at the top of a revolution. Any panel can be put out of operation to await a convenient time for repair. Lagging may be applied to the outside of the drum where heat radiation has to be avoided.

The Genter Continuous Vacuum Filter has a series of cylindrical filter units arranged in the form of a squirrel cage and attached round the periphery of a vertical circular disc mounted on a horizontal rotating shaft. The lower portion of the disc is immersed in the pulp. As each unit emerges it has taken on a cake, and at the highest point the suction is replaced by air pressure which blows the cake off into a discharge chute.

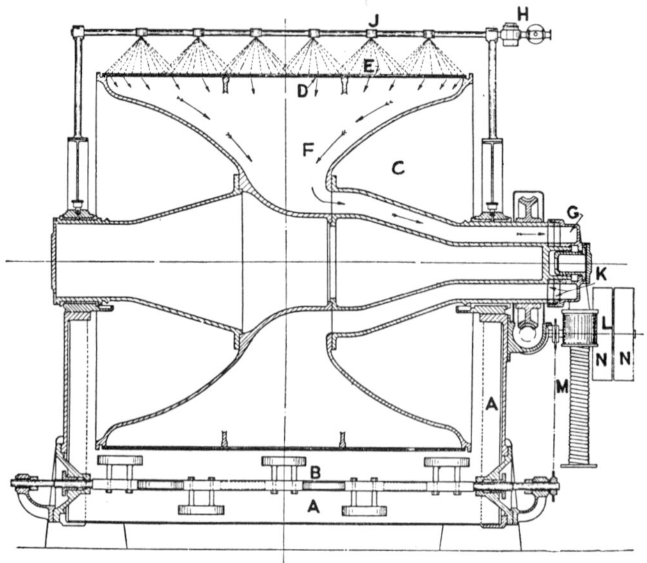

Fig. 176.—Section through " Rovac " Filter.

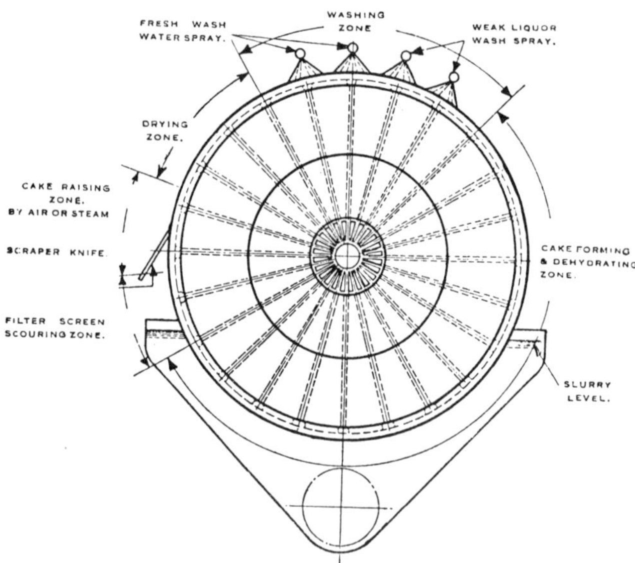

Fig. 177.—Principle of operation of " Rovac " Filter.

The " Rovac " Vacuum Filter (Figs. 176 and 177) is of steel construction, and consists of 18 self-contained, stream-lined vacuum cells built up to form the rotating drum, the surface of which is highly finished. Perforated grooved plates are secured to this surface and act as a support to the filter cloth. The filter is mounted on cast-iron trunnions at each end. On the face of the drum there is complete cell-isolation, which prevents a loss of vacuum at the point of cake discharge due to the blow-back air finding its way through joints, etc., to the cells still under vacuum.

Strong filtrate can be separated from weak liquor or wash by means of a cast-iron valve head, air being admitted beneath the cloth from above and below the point of cake discharge.

Inside the steel trough below the drum an oscillating rake agitator keeps the solids from settling.

The American Filter consists of one or more filter discs up to 8 feet in diameter spaced some 12 inches apart on a horizontal shaft. Each disc has 8 to 10 wooden sectors, held in position by radial rods. The discs are corrugated on both sides, and attached to a light cast-iron draining cap. Cloth bags, which fit over each sector like a drum, are the filtering media. The shaft on which the discs rotate is covered with 8 or 10 filtrate channels, and the drainage cap of each sector fits into one of these channels. All the channels connect with an automatic valve which controls the application of the vacuum for forming and washing the cake and also the air pressure for its discharge. The tank containing the pulp is semicircular and on the discharge side is cut away between the discs. This allows the cake to drop on to a conveyor belt below. No mechanical agitators are used.

The advantages claimed for this filter are as follows :—(1) There is less space required per ton of capacity ; (2) an automatic valve of stronger design and greater bearing surface is used ; (3) no wiring and no mechanical agitation are required. On the other hand the disadvantages are :—(1) Attention to partly choked spray is needed ; (2) the washing zone is covered ; (3) about 15 per cent. of the wash solution runs off the cake and either drains away or dilutes the feed pulp. Capacity cannot be controlled by altering the submergence as in the Oliver filter, as there is a danger of bare canvas being exposed to the atmosphere.[1]

Discussion of Vacuum Filters.—Continuous filters such as the Oliver, have an advantage over the discontinuous types (*e.g.* the Butters) in a somewhat lower initial cost, and a larger capacity. They also require less wash solution to give similar displacement. Both kinds are much superior to washing by decantation. For instance, Prentice states that the loss of dissolved gold in the residue from a vacuum filtration plant was 0·022 dwt. per ton of slime and that from a decantation plant 0·071 dwt.

Kelly [2] advocates the operation of continuous filters in series, and good results have been obtained from such an arrangement in mines in the Porcupine area. At the same time he states that counter-current decantation is simpler in operation and design, and costs less to run than continuous filters in series.

In one plant, the feed to the first filter (W/S = 1 : 1) is taken direct from the agitators. Two-thirds of the solution are recovered during the formation of the cake, which is then spray-washed with 50 tons of barren solution. This displaces about 30 tons of the solution which had at first remained in the cake. After being washed, the cake is discharged with 25 per cent. moisture and about one-tenth of the values of the original feed. The cake is

[1] Thurlow and Prentice, *loc. cit.*
[2] *Eng. Min. Journ. Press*, 1924, 118, 569.

repulped with 67 tons of barren solution, the W/S ratio of 1 : 1 being thus restored. It is then filtered on a second continuous machine and a combined barren solution and water-wash is used to get the maximum efficiency of displacement.

The flow of wash through a filter cake approximates in most cases to capillary action. Under ideal conditions the rate of flow varies inversely as the cake thickness and is nearly proportional to the pressure.[1] The physical condition of the cake, however, materially disturbs any theoretical reasoning. The ideal to be aimed at in washing filter cakes is that water consumption should equal the volume of voids in the cake. Though this ideal is not achieved, new designs have done much to reduce the former inefficiency. Perfection in washing is rarely attained owing to the influence of adsorption, capillary diffusion, the formation of chemical compounds and the re-occurrence of colloids when electrolytes are withdrawn.

Alliott [2] gives an account of the various ways in which cakes may be washed and discusses their relative efficiencies.

Although suction filtration is dependent, like the pressure method, on a difference of pressure, it does not follow that a nearly perfect vacuum is equivalent in efficiency to an extra atmospheric pressure of 15 lbs. per square inch in a filter press. The advantage of vacuum filtration lies in the uniformity of the cake, giving equal resistance to flow at all parts of the filtering medium. In general it may be taken that a pulp that has a high settling rate has also a high filtering rate.

Boreham [3] has proved that a more rapid displacement of the dissolved gold from the pulp is obtained and " bridging " is minimised when the vacuum in a filter plant is balanced, or uniformly distributed. He discusses methods for obtaining this. Everitt [4] suggests the use of baffles in the pipe connections to the vacuum mains for the same purpose. Thomas [5] expresses the opinion that differences in pump speeds, piston displacements and main pipe areas would cause uneven distribution of vacuum.

Ordinary Filter Presses.—An example of one of these is shown in Fig. 178, in use on sulpho-telluride slimed ore. In these presses the action is intermittent. A number of parallel filter leaves are clamped together, and the slime pulp is forced by pressure into the spaces between the filter cloths, when the liquid passes through the cloths and the slime forms a layer on their surface. The slime cakes accumulate until they are 2 or 3 inches thick. When washing is complete, the cakes are partly dried by compressed air, and the press is then opened and the cakes discharged. Pressures of 45 to 100 lbs. per square inch have been used. The output per press is about 8 or 10 charges in 24 hours. A press of standard size has 50 chambers, each 40 inches square, and will hold 9,000 lbs. of dry slime. Two types are known as the " Chamber " and " Frame " presses.[6] In the former the plates are kept apart by flanges, and in the latter by separate frames. The *Dehne press*, belonging to the frame type, has been largely used in Western Australia. In this press (Fig. 179) [7] there is a number of cast-iron plates, with a hollow frame between each two plates. Filter cloths are hung over

---

[1] Alliott, *Journ. Ind. Eng. Chem.*, 1921, 13, 976.
[2] *Loc. cit.*
[3] *Kolar Gold Fields Min. and Met. Soc.*, 1933, 7, *Bull.* 34, 3.
[4] *Ibid.*, p. 7. [5] *Ibid.*, p. 7.
[6] Julian and Smart, "*Cyaniding Gold and Silver Ores,*" p. 358.
[7] Julian and Smart, *loc. cit.*

the corrugated surface of each plate. Two kinds of plates are used, marked A and B respectively in Fig. 179. The pulp is passed into the spaces within

Fig. 178.—Filter Press, Kalgoorlie, Western Australia. From a Photograph kindly supplied by Mr. A. G. Charleton.

the frames, *f*, between the two filter cloths, through which the solution passes. When the spaces are filled with pulp, wash liquor enters from the channel *b*, and is forced horizontally through both filter cloths and the pulp between them and passes out through *d* (see arrows).

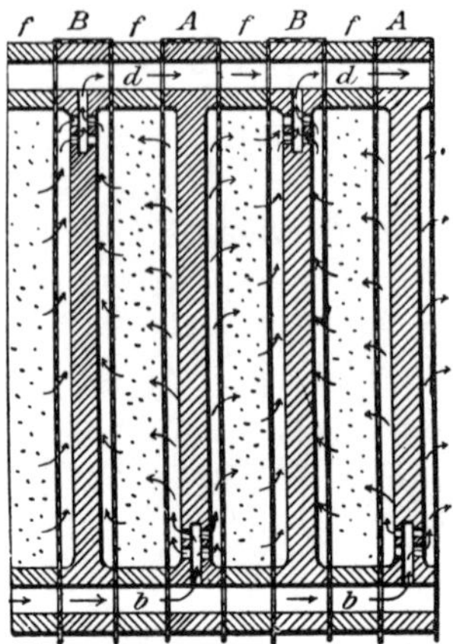

Fig. 179.—Dehne Filter Press (Section).

The Merrill Press has corrugated plates and hollow frames, placed alternately as in the Dehne press. The slime pulp enters each frame through a feed channel at the top, B (Fig. 180). The solution or wash-water is drawn off through the channels, C. The distinguishing feature of the Merrill press is the removal of the slime cake by sluicing instead of by opening the press. The sluicing water, under a pressure of 60 to 90 lbs. per square inch, is introduced through the pipe H, which has a sluicing nozzle, I, opposite the central plane of each frame. In Fig. 180, D is the partially sluiced slime cake, E is the filter cloth, partially removed to show the corrugated filter plate F, and G is a horse-shoe clamp for holding the filter cloth against the filter plate. The pipe H slowly rotates round its axis through an arc of nearly 180°, backwards and forwards, so as to direct the jet of water in all directions inside the pulp chamber. As the cake is washed away by the jet, the mixture of slime-residue and water flows into the channel underneath the sluicing pipe and through the outlet cocks to the waste conduit below.

There are two varieties of Merrill filters—(1) the "solid filling" press, in which the compartment is filled with slime cake and the wash-water passes through the whole cake, as in the Dehne press, and (2) the "partial filling" or "centre washing" press, in which filling is stopped when the cake consists of two equal portions; one of these adheres to the filter cloth on each side, with a vertical space between them extending from the top to the bottom of the frames. The wash solution is then forced into the central space, and passes through the cakes from the centre outwards.

The Merrill press is sometimes used as a treatment press, see pp. 373, 374.

Prentice gives the following results of filter-pressing for 9 months, 1934, at the Central Mining Rand Mines :—

Excluding the all-sliming plant :—

Original value of slime, .	1·764 dwt. per ton.
Value of discarded residue,	0·141 ,, ,,
Extraction, . . .	92·0 per cent.
Grading of slime, . .	86·9 per cent. of −200 mesh.

In the all-sliming plant the results were :—

Original value of slime, .	3·850 dwt. per ton.
Value of residue, . .	0·323 ,, ,,
Extraction, . . .	91·6 per cent.
Grading of slime, . .	65·8 per cent. of −200 mesh.

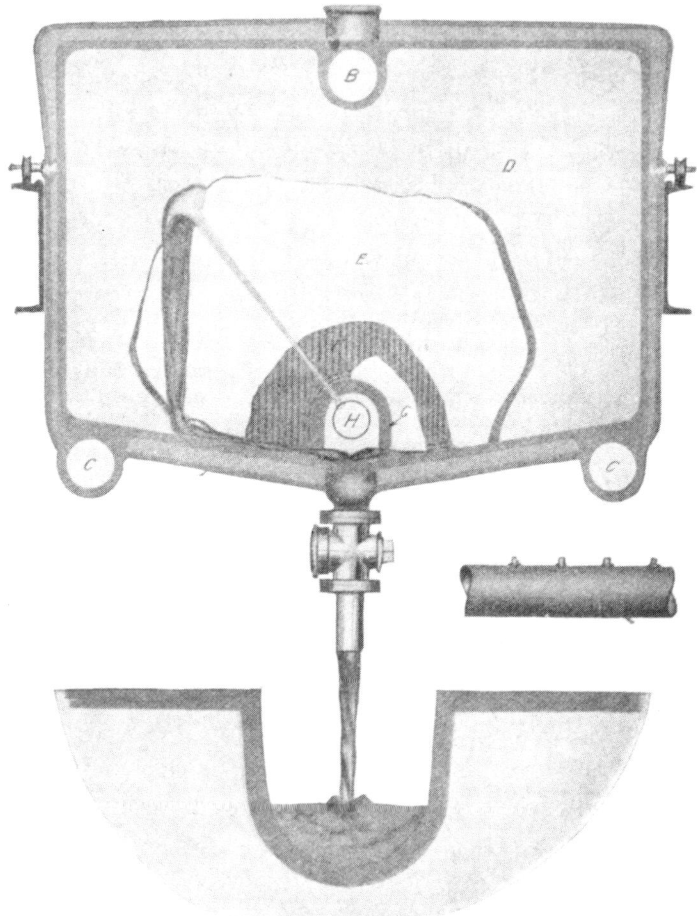

Fig. 180.—Merrill Press.

Factors in Filtration.[1]—Clarke, Ure and Hinchley,[2] following the work of Poiseuille, Hatschek,[3] Lewis and Alny,[4] and others, found that (1) in

[1] For mathematical formulae governing the influence of the variables in filtration, see Sperry, *Met. Chem. Eng.*, 1916, 15, 198 ; *ibid.*, 1917, 17, 161 ; *ibid.*, 1918, 18, 362, 520 ; *ibid.*, 1918, 19, 680 ; *Journ. Ind. Eng. Chem.*, 1921, 13, 986 ; *ibid.*, 1926, 18, 276 ; *ibid.*, 1928, 20, 892.

[2] *Chem. Tr. Journ.*, 1925, 76, 131.

[3] *Journ. Soc. Chem. Ind.*, 1908, 27, 538.

[4] *Journ. Ind. Eng. Chem.*, 1912, 4, 528.

a leaf vacuum filter the cake does not consolidate under increasing pressure, (2) the structure of the cake is not influenced by either its thickness, the composition of the liquid to be filtered, or that of the primary pulp. In a plate-and-frame filter press, however, there was a progressive increase in the percentage of solids in the cake as the pressure increased, and also as the percentage of solids in the primary pulp decreased. The following formula given by the investigators is said to show reasonable agreement with works' practice :—

$$T = \frac{r_m}{P} . W + \frac{rW^2}{4PN}$$

where T = time in hours
r_m = resistance of filter cloth, *i.e.* reciprocal of lbs. of filtrate passing per hour through 1 square inch of cloth per lb. per square inch pressure.
P = pressure of filtration.
W = weight in lbs. of filtrate passing through 1 square inch of cake.
r = resistance of cake to flow of filtrate, *i.e.* reciprocal of lbs. of filtrate flowing through 1 inch cube of cake per lb. per square inch pressure per hour.
N = lbs. of filtrate passing during the deposition of 1 cubic inch of cake on a filtering surface of 1 square inch.

$$N = \frac{a}{2b(1 + c)}$$

where a = weight of 1 cubic inch of dry cake.
b = weight of solids per lb. of filtrate.
c = weight of water associated with each lb. of dry cake.

When a cake t inches thick has to be formed, Nt lbs. of filtrate must be used.

Then $$W = Nt$$

and $$T = N\left(\frac{r_m t}{P} + \tfrac{1}{2} \cdot \frac{r}{2P}t^2\right)$$

In other words, the time to form the cake is proportional to N.

It was also shown that

$$\frac{dW}{dT} = \frac{P}{(rt + r_m)}$$

It was demonstrated that if a cake is thicker than $\frac{3}{4}$ inch, the rate of flow is directly proportional to the pressure, but for cakes under $\frac{3}{4}$ inch the flow depends on some power of the pressure.

In 1926 Underwood [1] put forward the following formulae based on data obtained by other workers :—

T = time of filtration.
V = total volume of liquid per unit area filtered in time T ($V = 0$ when $T = 0$).
$F = \frac{dV}{dT}$, the rate of flow per unit area at any moment.

---

[1] *Trans. Inst. Chem. Eng.*, 1926, 4, 19; Forster, *Trans. Aust. Inst. Min. Met.*, 1932, No. 88, 407.

R = "specific resistance" of filter cloth, *i.e.* resistance of cloth plus the initial layer of solid particles.
r = "specific resistance" of filter cake, *i.e.* resistance of unit thickness of cake.
For an incompressible precipitate r is independent of the pressure, but where a precipitate is compressible, r is a function of the pressure.
t = thickness of cake formed by the filtration of volume V of liquid.
c = a constant defined by the relation $t = cV$. It can be calculated when the percentage of solids in the primary pulp and in the cake and also the specific gravity of the cake or of the particles composing it are given.
p_1 = pressure fall through cloth.
p_2 = ,, ,, ,, cake.
$P = p_1 + p_2$ = total pressure of filtration.
N = flow ratio, *i.e.* the ratio of the rate of flow at the end to that at the commencement of a filtration at constant pressure.
$k = 3 - \dfrac{6}{3 + 2\sqrt{N} + N}$, the flow ratio coefficient.

Underwood arrived at the conclusion that the rate of flow through the filter cloth varies as the mth power of the pressure, where $m = \frac{1}{2}$, 1 or 2. $m = \frac{1}{2}$ is applicable only in the case of wire filter cloth or of fabric cloths at very high pressures. For fabric cloth at usual pressures, $m = 2$.

$$\text{Rate of flow through cloth} = F = p_1^2/R$$

From a consideration of capillary flow, Underwood assumes that

$$\text{Rate of flow through a cake} = F = p_2/rt \text{ and since } t = cV$$

$$P = p_1 + p_2 = \sqrt{RF} + rcVF \quad . \quad . \quad . \quad (1)$$

This equation does not lead to a simple relation between T and V, but the following relations provide a sufficiently close approximation for all practical purposes :—

$$T = \frac{RV}{P^2} + \frac{krcV^2}{2P} \quad . \quad . \quad . \quad . \quad . \quad (2)$$

i.e.
$$\frac{1}{F} = \frac{R}{P^2} + \frac{krcV}{2P} \quad . \quad . \quad . \quad . \quad . \quad . \quad (3)$$

The value of c depends primarily on the thickness of the pulp and should be calculated for each filtration unless the same stock of pulp is used. Various factors may cause variation in the resistance values,[1] viz. :—

(1) Cloth resistance :—

(*a*) Design of feed and drainage channels.
(*b*) Conditions of depositing initial layer of solids.
(*c*) Deterioration of cloth condition by shrinkage, deposition in pores, etc.

[1] Forster, *Trans. Aust. Inst. Min. Met.*, 1932, No. 88, 407.

(2) Cake resistance :—

(a) Compressibility of precipitate.
(b) Unprotected drainage channels.
(c) Intermittent flow.

A lower resistance is obtained if the filtration is started and run for a few minutes at a pressure which is just sufficient to induce filtration over the whole filtering area than if the initial pressure is high enough to produce the normal operating flow. This effect is due to the primary deposition of an almost infinitesimal layer of solids.

When wire screening of suitable mesh is used to support the filter cloth, it prevents the latter from being forced back into the channels, which would hinder filtration.

Constant-pressure filtration gives the quickest filling cycle, but is applicable only in the case of thick pulps or slow-filtering solids. Constant-rate filtration is advantageous in continuous work. It eliminates the necessity of a large storage capacity for the feed to the filters. In practice neither one nor the other is usually fully obtained.

If two or more presses are used simultaneously, there should be separate means of pressure-control on each unit. Otherwise when a clean press is put into operation, considerable variation in pressure will ensue.

In design—[1]

(1) The maximum pressure must not exceed that which can be conveniently held at the press gaskets or economically produced by the pump for filling.

(2) The maximum gasket-pressure will depend on whether the press is operated manually or hydraulically.

(3) Press frames are limited in thickness by the inconvenience of handling heavy members.

(4) Frames should be as thick as other conditions permit in order to keep the cleaning costs low.

(5) The length of a press is limited by conditions of size and weight. The longer a press the more difficult it is to keep the gaskets tight.

(6) With free-filtering materials, the rate of flow should not be high enough to call for undue thickening of the press-plates in order to gain adequate drainage space.

Examples of the use in practice of some of the agitators and filters described are given in Chap. XVI.

4. Clarification.

In precipitation it is necessary to have a perfectly clear solution as otherwise the suspended slime tends to increase the pressure in the filters and to retard the easy flow of solution. It may be deposited on the zinc shavings and thus reduce their efficiency. Leachings are usually sufficiently clear, but it is found essential on the Rand to pass slimy solution through

[1] Forster, *loc. cit.*

beds of sand acting as clarifiers. The suspended matter collects as a thin film on the sand, and when percolation becomes slow, the tank is drained dry and the top skimmed. Prentice advises an area of two square feet of clarifier per ton of solution per 24 hours.

At the Mysore Mine the filter beds are made of a mixture of primary cone underflow sand and river sand to a depth of 12 inches, resting on coir matting and filter cloth.[1] 20 per cent. river sand and 80 per cent. dump retreatment sand also gave no ill-effects.

When a clarifier is charged with new sand, a considerable absorption of cyanide (up to 10 lbs. per 100 tons) takes place immediately, but afterwards only an insignificant amount is absorbed (White).

Johnson[2] states that calcium carbonate and silica sometimes exist in such a fine or colloidal form that they pass through the clarifier and interfere with precipitation and filtration. He also records instances of "white precipitate" containing about 38 per cent. silica.

Prentice[3] recommends two clarifiers in series, using as sand the underflow from the cones which feed the tube mills. The second is at a higher level than the first, from which the solution is pumped. The solution then gravitates to the extractor boxes. Prentice states that the double reducing effect of the sand is as effective as the Crowe process. The results were as follows :—

	Oxygen Content.
Solution from the primary clarifiers, .	3 mg./litre.
,, ,, secondary ,, .	2 ,,
,, entering the extractors, . .	0·5 ,,

The second clarifier completely removed all the slime particles.

At the same time, if the clarifiers were allowed to drain and their contents exposed to the sun for any considerable length of time, re-oxygenation of the solution would occur and a high cyanide consumption result.

Whitehead[4] states that owing to the presence of pyrrhotite the mill sand used for the bed of the clarifiers at the Balaghat Mine absorbed a third of the total cyanide consumption. Much of this loss occurred whilst the solution stood over the filter bed. River sand consumed practically no cyanide.

In Mexico gravity-fed clarifying tanks having two or three curtain filters of jute or coir are used.

Merrill presses and vacuum leaf-filters of the Butters type are used extensively as clarifiers.

The quality of the filtrate and a maximum period of filtration are important. In automatic filtration the separate collection and re-treatment of cloudy filtrates may be necessary.

[1] *Kolar Gold Fields Min. and Met. Soc.*, 1928, *Bull.* 23, 199.
[2] *J. Chem. Met. Mng. Soc. S.A.*, 1920, **21**, 58, 191.
[3] T. K. Prentice, *ibid.*, 1932, **32**, 256.
[4] *Kolar Gold Fields Min. and Met. Soc.*, 1929, *Bull.* 25.

The Merrill Pressure Clarifier (Figs. 181 and 182) is of the flush-plate distance-frame type, having individual discharge cocks A (Fig. 182) to the

Fig. 181.—20-Frame Merrill Clarifying Filter. (Capacity 1,000 tons Solution Daily).

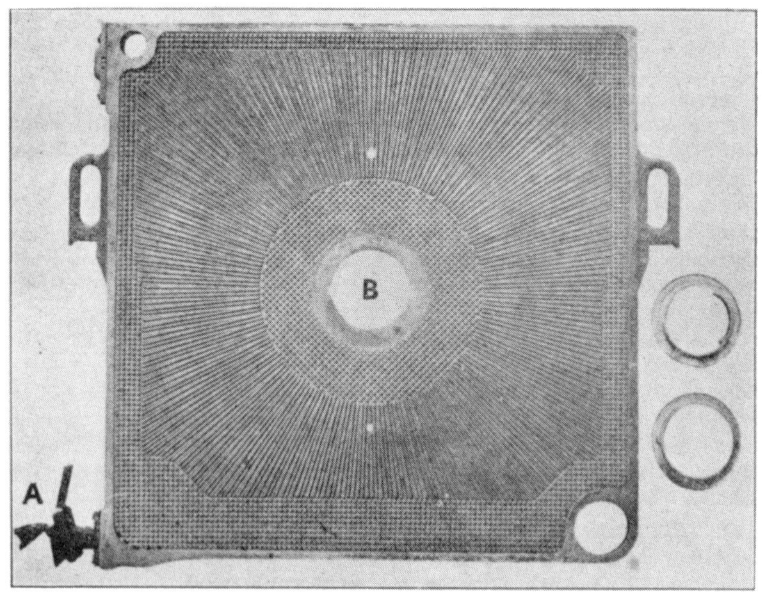

Fig. 182.—Merrill Sluicing Clarifying Filter. Details of Filter Plate.

frames. It has a central rotary sluice pipe (inside B) by which the interior of the frames may be cleaned with water or weak cyanide solution without dismantling the press. The stream of wash water is short and passes along the radiating ribs. Double nozzles give slightly diverging streams of water through $\frac{1}{8}$ inch holes, at 50 to 75 lbs. pressure. Ordinary duck is used on the frames and these are sluiced about three times per day. About every fortnight the duck is treated with 3 per cent. hydrochloric acid to remove the precipitated calcium carbonate, and then washed with water before being again put into use. In starting up a new or freshly washed press, the first runnings should always be returned to the storage of unclarified liquor.

5. Precipitation of the Gold.

This is effected by means of zinc either in the form of shavings or dust. Electrical precipitation, the charcoal method, and other methods are little used, but will be briefly described.

Precipitation by Zinc Shavings.[1]—The shavings are freshly turned on a lathe just before use, and kept dry to avoid oxidation. The material is sometimes called "filiform zinc" or "zinc thread." The shavings form a light spongy mass easily traversed by the solution and presenting a large surface for precipitation. They are $\frac{1}{500}$ to $\frac{1}{1500}$ inch in thickness and of no great width, but are sometimes 2 or 3 feet long. They weigh from 6 to 15 lbs. per cubic foot in the boxes, and are so fine that they can be ignited with a match, burning to zinc oxide. One pound of zinc $\frac{1}{500}$ inch thick exposes an area of 28 square feet (E. H. Johnson).

At Morro Velho, zinc ribbon is used, produced by allowing molten zinc to run in fine streams through a finely perforated dozzle on to a water cooled bronze mandril rotating at 1,000 r.p.m. The ribbon which flies off is 0·1 inch wide and about 0·004 inch thick.[2]

The new zinc going to the extractor boxes is immersed in a solution containing about 10 per cent. of acetate or nitrate of lead (this solution is also dripped into the head box of the series). Lead nitrate is superseding the acetate owing to its lower cost. The zinc soon acquires a dark-coloured hue, due to a coating of precipitated lead, and it is then transferred to the extractor boxes and at once covered with solution to prevent surface oxidation which occurs very rapidly on exposing the couple to the air. The *lead-zinc couple* is more active in precipitating gold than zinc alone, and is especially required for weak and low-grade solutions. It also rapidly removes copper from the solution, and so prevents the "plating" action of copper on zinc shavings, which is a source of trouble in weak solutions if ordinary zinc is used. Any excess of the lead salt added to sand and slime charges to precipitate soluble sulphides tends to keep the zinc active by forming fresh coatings. Lead is also sometimes contained in zinc dust, and lead nitrate solution is added to the mixing device in zinc dust precipitation. An objection to the use of lead is that it contaminates the bullion and requires removal.

The shavings are placed in steel troughs, the "zinc-boxes," which are divided into compartments by partitions. They are supplied with baffle

---

[1] This method is employed only on the older plants on the Rand and is being replaced there and elsewhere by the Merrill zinc dust process.

[2] *Trans. Inst. Min. Met.*, 1933, **42**, 221.

boards, which cause the solution to flow upward through the shavings. Argall used boxes in which the solution flowed alternately downward and upward through the zinc, entering the first compartment at the top. Fig. 183 shows a steel extractor box with twelve compartments.[1]

Fig. 184, from a photograph kindly supplied by Dr. W. A. Caldecott, gives a general view of the interior of the extractor house at the Knight's Deep and Simmer East Joint Plant, Transvaal.

The shavings are supported on wire-screen trays, formed of $\frac{1}{8}$-inch or $\frac{1}{4}$-inch iron wire gauze, so that the finely-divided, precipitated gold falls through to the bottom of the trough, while the unaltered zinc remains on the sieve. The trays are removable and furnished with handles so that they can be lifted out. The bottom of each compartment slopes towards one corner, where there is a plug hole through which the gold slime is washed in cleaning up. The zinc boxes are fitted with lids and kept locked.

There are usually five to twelve compartments, of which the first and last are left empty. The use of iron in a box in contact with the zinc increases the loss of zinc by dissolution, but iron wire trays form the most convenient support. Wooden supports have been used.

The dimensions of the boxes depend on the amount of work to be done. Each compartment is usually about 16 cubic feet in capacity, and holds about 200 lbs. of freshly cut zinc. It is usual to allow from 1 to $1\frac{1}{2}$ cubic feet of zinc per ton of solution to be precipitated in 24 hours. This gives contact for 30 to 45 minutes. The coarser the zinc, and hence the less surface exposed, the more time is required. On the Rand about $2\frac{1}{2}$ tons of solution are precipitated per ton of ore treated.

It is usual on the Rand to have three sets of boxes in a cyanide works, one for the "strong" solution, one for the "weak" solution, and one for very dilute washings. In the last two series of boxes less zinc is consumed, but the gold slime is poorer, less gold being contained in the solution, and greater attention to the control of the precipitation is necessary.

When the solution comes in contact with bright zinc, the latter turns black at once, owing to the deposition on it of finely divided gold. The zinc is gradually dissolved, and the shavings fall to pieces, those in the first compartment being consumed most rapidly. As the precipitation proceeds, the zinc is transferred from the lower compartments to the upper ones, and fresh zinc is added at the foot of the box. In thus "dressing" the zinc boxes, it is essential to distribute the shavings evenly to avoid too close packing in some places and the formation of free channels in others. The corners are carefully filled in. Short and rotten zinc is placed on the top. The boxes are dressed at the clean-up (say once in 7 to 10 days), or in some mills more frequently.

The consumption of zinc is generally from 4 to 20 ozs. of zinc for 1 oz. of gold recovered. The consumption is partly chemical (by dissolution) and partly mechanical (short zinc removed with the gold in cleaning up). The average loss on the Rand is about 0·12 lb. zinc per ton of ore milled. For the chemical reactions occurring during precipitation, see p. 316, *et seq.*

Spent solution from the zinc boxes should contain no more than 0·02 dwt. of gold per ton (Caldecott).

Kasey[2] suggests that improvement in precipitation when using zinc

---

[1] Schmitt, "*Rand Metallurgical Practice*," vol. ii., p. 283.
[2] *Min. Journ.* (*Ariz.*), 1934, **18**, 5. 26.

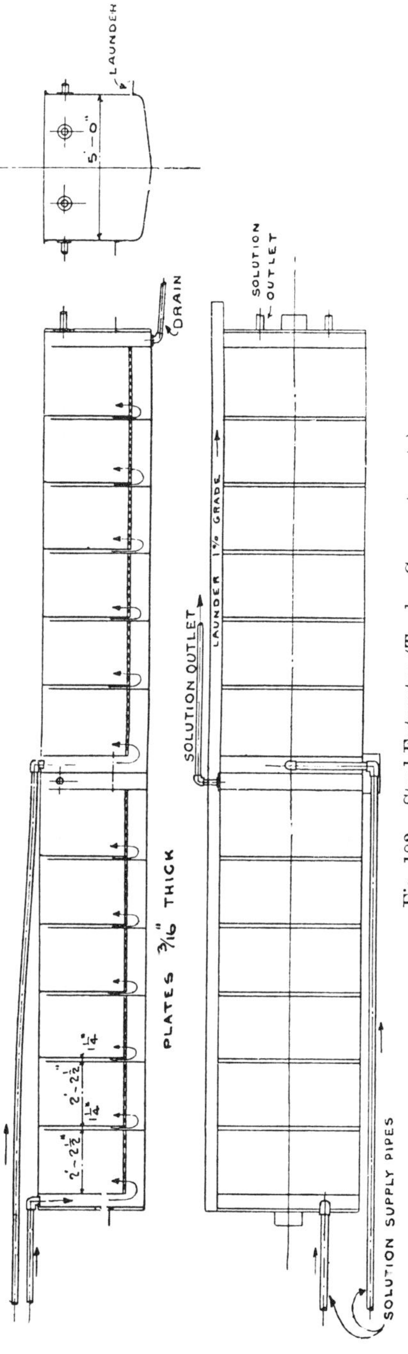

Fig. 183.—Steel Extractor (Twelve Compartments).

Fig. 184.—Interior of Extractor House, Knight's Deep and Simmer East Joint Plant.

shavings can be attained by removing the inevitable oxide on the latter with cold sodium hydroxide and cyanide solutions.

Cooke[1] gives the following assays of solutions passing through the zinc boxes :—

TABLE XL.

Solution Value at Head of each Compt.	1st Compt., dwt.	2nd Compt., dwt.	3rd Compt., dwt.	4th Compt., dwt.	5th Compt., dwt.
I, . .	3·00	0·60	0·06	0·02	0·01
II, . .	3·00	0·30	0·03	0·00	Tr.
III, .	3·00	1·00	0·50	0·04	0·02
IV, .	3·00	2·00	0·75	0·40	0·02
V, . .	3·00	2·50	1·00	0·75	0·40

I. Normal working, 80 per cent. precipitated in 1st compartment.
II. Extra efficient working, 90 per cent. precipitated in 1st compartment.
III. Flow too fast.
IV. } Flow still faster.
V. }

Clean-up of Zinc Boxes.—The clean-up takes place as frequently as is found desirable or convenient, and is often coincident with dressing. As a rule, only the first three compartments are cleaned up. The solution is displaced by clean water, and the latter drawn off down to the level of the wire gratings. The screens containing the undissolved shavings are lifted out and the zinc-gold slime, remaining at the bottom of the boxes, allowed to run out by withdrawing a plug in the bottom of the box, and drained through a 20 or 30 mesh screen, which retains a small part only. After the residue has been gently rubbed on the screen and well washed, it is put back again into the first compartment of the box, on the top of fresh shavings, as it consists mainly of small pieces of unconsumed zinc. The shavings proper are thoroughly washed and rinsed on a sieve to separate the gold slime as much as possible, and are returned to the zinc boxes. Short rotten zinc containing gold is not usually returned to the zinc boxes.

In large plants, the zinc shavings may be washed in a cylinder with screening which can be fitted in a compartment of the zinc box. Zinc shavings are put into this trommel, which is immersed in water to a depth of one-third of its diameter and rotated.

The gold slime is run into a small filtering vat, or more usually into a filter press of the Johnson type, where it is washed and sent to the acid tanks for dissolving out the zinc, or dried for fluxing. The recovered zinc sulphate is often sold for use as a timber preservative.

The value of the dried precipitate depends on the richness of the

[1] *J. Chem. Met. Mng. Soc. S.A.*, 1917, 18, 33.

cyanide solution, and on the time of contact as well as on the quantities of base metals which are present in the solution. By the prolonged action of the cyanide, the zinc shavings become partly corroded and disintegrated, so that the precipitated gold is mixed with zinc debris. Ordinary commercial zinc contains a considerable proportion, generally over 1 per cent., of lead, and a small quantity of carbon, besides other impurities, such as arsenic and antimony. After the introduction of the lead-zinc couple, the proportion of lead rose to about 5 per cent. All these impurities accumulate and are collected with the gold.

In cleaning-up the precipitate from gold solutions it is profitable to separate as much gold from the zinc as possible, to avoid locking up values in the zinc boxes. Zinc containing gold is accordingly not returned to the boxes. It is otherwise with silver solutions.

In the clean-up at Mysore the coarse zinc from the top compartments is screened on either $\frac{1}{16}$, $\frac{1}{10}$ or $\frac{1}{8}$ inch mesh screens, according to the character of the precipitate. An accumulation of short zinc in the boxes beyond half the capacity of the top compartment is prevented. Fresh shavings are added to the 5th compartment and the 6th is left empty.[1]

Brooks[2] gives evidence that the precipitation on zinc of silica taken up originally by water or cyanide solution from an ore may interfere considerably with the efficient recovery of gold.

Precipitation by Zinc Dust.—In this process zinc dust (" zinc fume " or " blue powder ") is agitated with gold solution and the precipitate separated from the impoverished solution by filter pressing. It was introduced by H. L. Sulman at Deloro, Canada, in 1894,[3] and at the Mercur Mine, Utah, in 1896.[4]

Zinc dust used in gold precipitation contains usually from 95 to 97 per cent. of metallic zinc. It is rapidly oxidised in moist air. Owing to this oxidation many samples contain only 85 per cent. of zinc, together with zinc oxide, which is harmless but inoperative, and usually 1 or 2 per cent. of lead, which is advantageous. Cadmium has but little effect. The efficiency of the dust as a precipitant is related to its fineness and therefore to the extended surface of its particles. A large surface of zinc is the most important factor in precipitation. Specifications in 1925 called for 97 per cent. to pass a 350 mesh screen.[5] Distilled dust consists of almost perfect spherules. 1 lb. of zinc dust, assumed to be composed of spheres 0·0001 inch in diameter, has a surface of 1,650 square feet. Atomised dust has a coarse, coke-like appearance under the microscope. Full details of testing zinc dust for use in the cyanide process are given by Sharwood[6] (see also p. 328.).

In some isolated plants zinc dust is run into the gold-bearing cyanide solution, passing directly, without agitation, to the presses. The benefits of the entire removal of oxygen from the solution are, however, so marked, that the Merrill Crowe vacuum precipitation process has superseded other methods in nearly all plants.

---

[1] *Kolar Gold Fields Min. and Met. Soc.*, 1928, *Bull.* 23, 199.

[2] *Trans. Inst. Min. Met.*, 1927, 36, 3.

[3] *J. Soc. Chem. Ind.*, 1897, 16, 961 ; *Trans. Fed. Inst. Mng. Eng.*, 1897, 15, 417.

[4] G. A. Packard, *J. Chem. Met. Mng. Soc. S.A.*, 1899, 2, 720.

[5] Lepsoe, *Min. and Met.*, 1925, 6, 536 ; *Trans. Amer. Inst. Min. Met. Eng.*, 1925, 71, 1061.

[6] *Mng. and Sci. Press*, 1912, 104, 659 ; von Bernewitz, "*Cyanide Practice*, 1910-1913," p. 403.

A continuous feeder is shown in Fig. 185.[1] In this, the zinc dust is conveyed from a hopper, A, by means of a revolving feeder to a mixing cylinder, E, where it is carried by a small stream of solution, entering at C, to the pump-suction, no air being used to assist in the mixing. It is generally the practice to maintain a drip of a solution of lead nitrate or acetate to

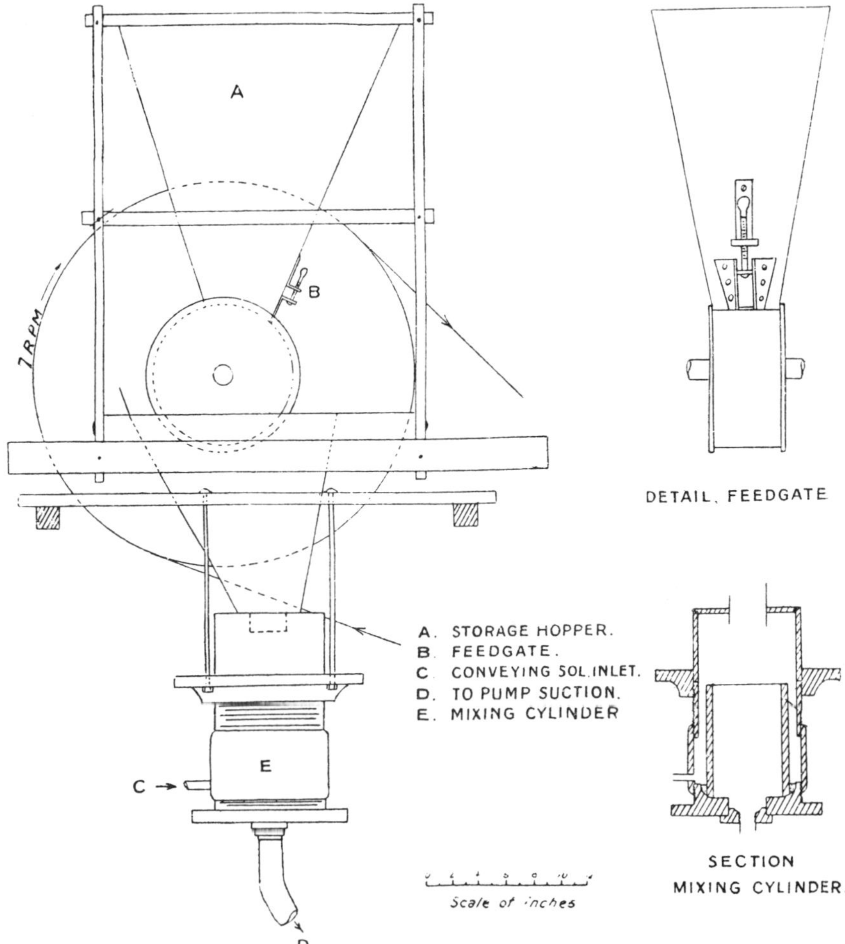

Fig. 185.—Todd Zinc Dust Feeder, Homestake.

the feed pipe carrying the zinc emulsion, to obtain the advantage of the lead-zinc couple.

In *Merrill Precipitate Filter Presses* of triangular section (see Fig. 186)[2], the solution enters at the lowest point or apex of the triangle. The mixing takes place partly in the upcast pipe and partly in the press. The filter cloth

[1] A. J. Clark, *Mng. Mag.*, 1911, **4**, 289; *J. Chem. Met. Mng. Soc. S.A.*, 1911, **12**, 103.
[2] A. J. Clark, *Mng. Mag.*, 1911, **4**, 291.

consists of two thicknesses of twilled cotton. The canvas leaves are cleared of gold slime about every fortnight. The slime is then pumped directly from the bottom of the vat to the acid vat, prior to being sent to the smelting department.

In starting-up a zinc press 50 per cent. excess zinc is usually added to the first charge and the zinc then gradually diminished in amount until normal working is assured. Sharwood [1] gives the minimum quantity of zinc dust for efficient precipitation as 50 to 100 parts per million of solution.

Woodworth [2] mentions the use of diatomaceous earth to the extent of 8 per cent. of the weight of precipitate under treatment. This addition facilitates settlement. The sheeting cover of the filter is also given a dressing of the earth to prevent solids passing through (*cf.* the pulp filter used for many years in assay practice).

Fig. 186.—Triangular Merrill Filters for Gold Precipitate.

The most recent zinc dust precipitation installations on the Rand have utilised a submerged vacuum leaf system instead of the triangle frame press.

Clemes [3] states that an irregular addition of lead salt too near the intake of the presses will result in speedy " blinding " of the filter. He prefers to admit the lead solution directly into the pump which draws the solution away from the slime filters.

The lead feed should be increased when a new clarifier bed is put into use if soluble sulphides are present in the sand (as in sand residue). Otherwise the issuing solution will be barren of lead.

A colorimetric test for lead, by the addition of standard sodium sulphide solution, is used.

---

[1] *Trans. Amer. Inst. Min. Eng.*, 1917-18, 58, 215.
[2] *Eng. and Min. Journ.*, 1934, 135, 159.
[3] *Journ. Chem. Met. Mng. Soc. S.A.*, 1932, 32, 216.

White says there is as yet no evidence to show that lime up to 0·03 per cent. alkalinity (NaOH) has any appreciable effect where gold is the main metal to be deposited. At higher strengths the lead portion of the couple is attacked and the electrolytic effect much reduced.

Wartenweiler[1] doubts the efficacy of adding lead salt prior to clarification, as a large quantity of the metal is precipitated as lead cyanide and remains undissolved on the sand filter.

Macdonald and Adam found that the lead addition for the formation of the zinc-lead couple should be made after the clarifiers. The only reason for adding it before clarification is the possibility that, in the precipitation of the lead, traces of colloidal matter might be brought down and these might interfere with subsequent filtration. At the New State Areas Mill the lead is added in a concentrated form by means of a belt feeder. This carries the crystals to small grinding mills containing quartzite pebbles, to which cyanide solution is added from the main stream. The discharge from the mill goes to a regulating cone and thence is pumped to the top of the deaerating tower. Zinc dust is then added to the solution as it passes to the precipitation plant.[2]

In some instances half the lead salt is added before clarification and half afterwards.

The use of lead dust with zinc dust in precipitation has been tried.

Comparison of Zinc Shavings and Zinc Dust as Precipitant.—The precipitation is about equally efficient in the two processes, but better results and a higher grade precipitate can be obtained with dust than with shavings. The clearest advantage of the use of zinc dust is that all the precipitated gold is cleaned-up and brought to account, whereas with shavings part always remains in the boxes for a future clean-up. With zinc dust less zinc is used (Prentice gives a figure of 0·04 lb. zinc dust consumed per ton of ore milled on the Rand, against about 0·12 lb. of zinc shavings), and less floor space is required. Also there is greater security against theft, less loss in handling, and a saving in the cost of acid treatment and smelting. The difficulty of the "white precipitate" is avoided, and copper in the solution is not prejudicial when zinc dust is used. By the Merrill system, too, the zinc is always kept from the air when wet and surface oxidation is thereby prevented. In addition, in the absence of oxygen, there is greater stability of precipitation.

Some zinc may be saved by by-passing round the boxes or presses a portion of any solutions which are low in gold and high in cyanide, and using it for dissolving more gold.

Merrill-Crowe Precipitation Process.—Reference has already been made to the harmful effect of oxygen in precipitation, resulting in a high consumption of zinc and cyanide. The Merrill-Crowe process, based on Crowe's original method to overcome the influence of the oxygen, is widely used, and is applied mainly to zinc dust precipitation. It can also be used, however, with zinc shavings.

The solution after clarification flows by gravity into the steady head tank A (Fig. 187) or is drawn into it by the vacuum receiver. A float mechanism B within the tank operates a control valve C in the vertical supply pipe and regulates the amount of solution within the receiver. Inside

[1] *J. Chem. Met. Mng. Soc. S.A.*, 1932, 32, 334.
[2] *Ibid.*, 1935, 35, 315.

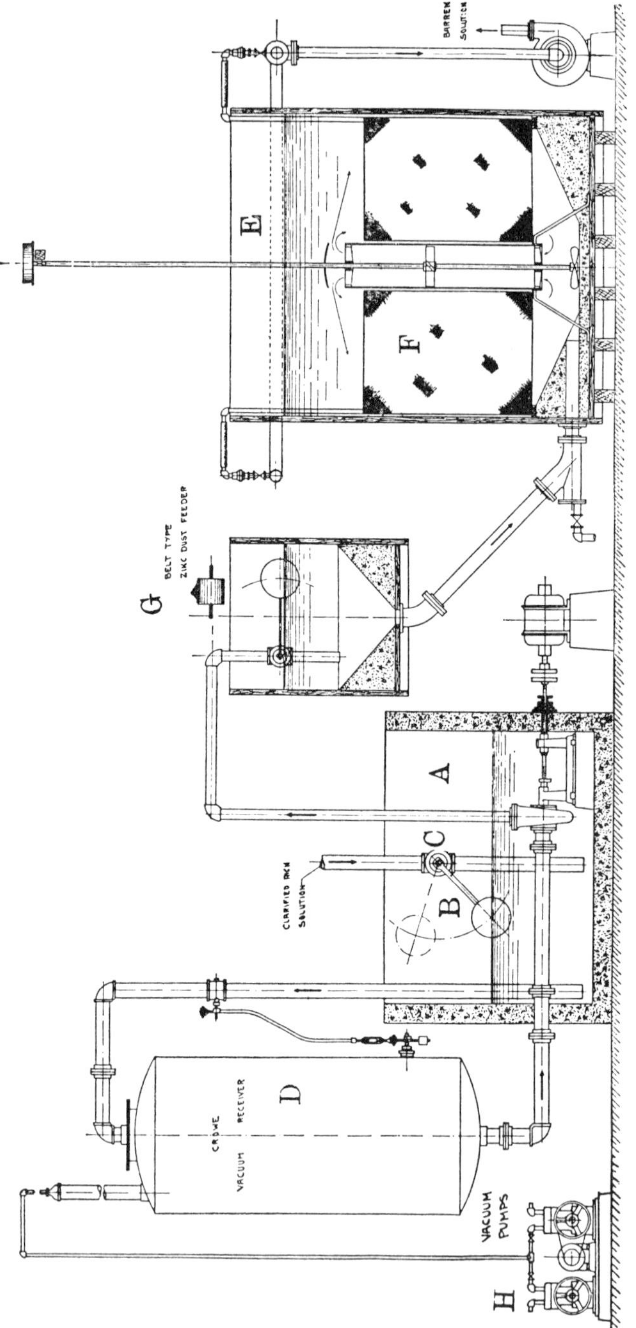

Fig. 187.—Merrill-Crowe Precipitation Process. Low Level Vacuum Tank—Vacuum Leaf Type for capacities of 600 tons of Solution or over.

this chamber D is a number of grids or baffles over which the solution flows in small streams and is thereby exposed to the highly reduced pressure created by vacuum pumps H. This treatment removes most of the absorbed oxygen.

From the receiver the deoxygenated solution is transferred by means of a pump sealed by the liquor in the steady head tank A, into the precipitation tank E containing vacuum filter leaves F. The amount delivered is regulated by a float and valve in G. Zinc dust is fed into the steady head tank G by a spiral belt, or other type of feeder, at a constant rate and is continuously circulated with the solution over the surface of the filter leaves by means of an air-lift. When the cakes on the leaves are fully formed, the filter leaf tank is drained and the precipitate washed from the leaves with a jet of water. The sludge is collected and partly dried in a vacuum clean-up tank. The precipitation tank is provided with a locked cover which is opened only at the clean-up. A directly-connected centrifugal pump draws the solution through the filtering medium.

Wartenweiler [1] states that the vacuum and filling cylinders have a large capacity, amounting to about 100 tons per day per square foot of filling cylinder area. One of the largest plants has a cylinder 6 feet in diameter × 10 feet deep and deals with 6,000 tons of solution per 24 hours, or 212 tons per square foot of area.

Bag filters and plate-and-frame pressure filters have also been used for collecting the gold precipitate. Fig. 188 [2] shows diagrammatically the sequence of operations where a filter press or, alternatively, a zinc extractor box, is used in the precipitation of the gold. The principal difference from the suction plant described above is the placing of the vacuum cylinder at such a height as to be able to feed the press or zinc boxes regularly.

A simultaneous clarification-precipitation system has been devised in which a single liquid-sealed centrifugal pump causes the liquor to pass successively through the clarification, deaeration and precipitation stages. The inflow to the clarifying tank is controlled by an automatic float valve and the level of unclarified liquor in the tank is kept constant. The outlets from the leaves in the tank are connected with the top of the vertical vacuum-tower through a manifold, and here the solution is deaerated. The inlet to, and the level of the solution within, the tower are also controlled by an automatic valve. Zinc dust and oxygen-free precipitated solution are introduced as a steady flow of emulsion into the pump suction midway between the pump and the outlet from the tower. The solution containing the zinc dust is next forced by the centrifugal pump into cylindrical pressure-bags suspended in the steel precipitation tank. The bags are of double thickness —an outer of canvas and an inner of sheeting. The barren solution overflows through a measuring weir into a float-compartment and thence is pumped to a storage tank. At the clean-up the barren solution is pumped out, the bags are blown with air, and the inner sheeting-cases, containing the bullion, are removed. They are turned inside out, cleaned and washed, or burned and the ashes added to the smelting charge. A continuous drip of strong cyanide solution is added to the zinc-emulsion feeder, and one of lead acetate or nitrate, amounting to about 10 per cent. by weight of the zinc dust, is added to the solution entering the clarifying tank.

[1] *Journ. Chem. Met. Mng. Soc. S.A.*, 1932, **32**, 139.
[2] Wartenweiler, *Journ. Chem. Met. Mng. Soc. S.A.*, 1932, **32**, 143.

When lead nitrate or acetate is to be added to the rich solution, it is best done by allowing a continuous drip of a solution of one of these salts into the liquor passing to the unclarified solution storage tank. The lead should not be added with the zinc dust.

The chief advantages of the Crowe zinc dust process as compared with the use of zinc shavings are stated to be :—(1) There is more uniform and efficient precipitation owing to constant fresh exposure of the surface of the

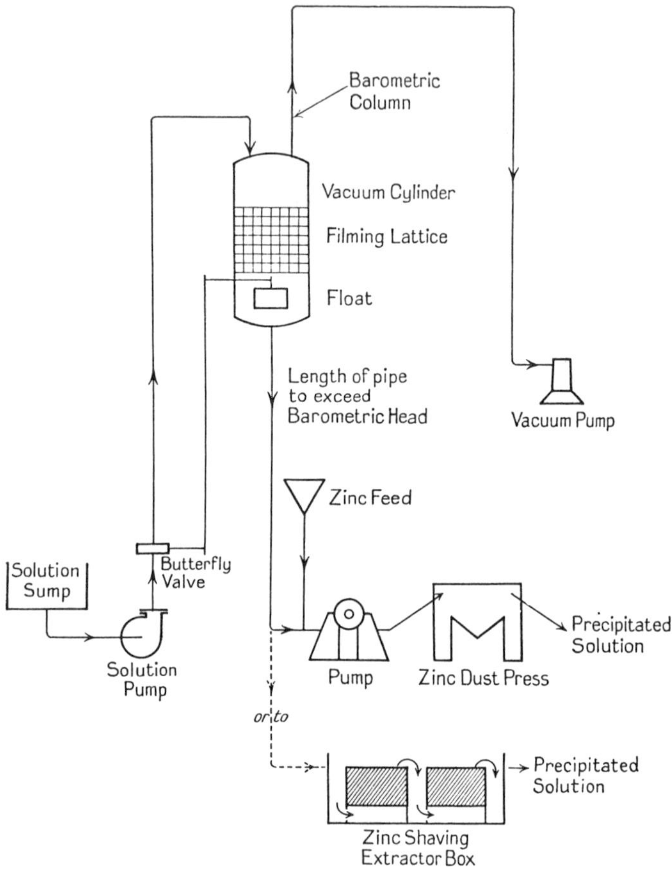

Fig. 188.—Deaeration in Crowe Process. Precipitation with zinc dust using pressure press, or by zinc shavings in extractor box.

precipitant, whereas zinc shavings rapidly become coated with lime and base metals. (2) As the time of contact is short, any coating with lime is reduced to a minimum. (3) There is no formation of "white precipitate" as no oxygen is present. (4) The filter cloths are always coated with a layer of fine precipitant and precipitate ; hence the solution must, even at the last stage, come into contact with some zinc. (5) The effective presentation of surface by zinc dust is much greater than that by shavings, weight for weight.

(6) The interval of time between the contact of the solution with the zinc and its removal therefrom is short. (7) Nascent hydrogen is created throughout the layer of precipitant, giving a reducing condition, but polarisation is minimised by the rapid flow of the solution carrying the hydrogen with it. (8) There is a lower consumption of zinc and cyanide. (9) There is less unconsumed zinc in the precipitate owing to the fineness of the dust and the smaller quantity used. An average figure for unconsumed zinc is 10 to 45 per cent. (10) The labour charges are lower. (11) There is less handling of precipitates, giving greater security from theft. (12) There is a quicker return of the bullion in the precipitation plant and lower interest charges. (13) The value of the tailing is lower.

An average result obtained shows that solution containing $8 or $10 in gold per ton is deprived of gold until only a few cents per ton remain. The zinc consumption depends on the conditions, but is usually about 0·45 oz. per ton of solution. About 0·002 oz. of lead nitrate per ton of solution is needed.

The following oxygen contents of cyanide solutions are given :—[1]

	Entering Crowe System.	Leaving Crowe System.
Modder B.	5·5 mg. per litre	0·5 mg. per litre
City Deep,	4·5 ,,	0·5 ,,

The operation of the Merrill-Crowe precipitation process at Modderfontein B is described by Newton and Fewster as follows :—[2]

When commencing a fresh cycle of precipitation, 500 lbs. of zinc dust are fed into the press on the first day, 350 lbs. on the second, 200 lbs. on the third, and about 150 lbs. each day afterwards. The quality of the zinc dust is important. It is passed through a 90 mesh sieve to remove foreign material. Precipitation in too highly reducing solutions is not complete. This reducing power is due to the mill circuit carrying underground water, and is tested by N/10 permanganate solution. Chloride of lime or sodium hypochlorite is then added in the requisite quantity. Kieselguhr added to the precipitation presses appeared to give a more porous filtering medium. The precipitate is smelted in a Tavener furnace.

When the Crowe process is used, savings have been noted of 0·85d. per ton of ore treated with zinc shavings, and 1·6d. per ton with zinc dust. The items showing decreased cost are zinc, acid, cyanide and fluxes. The plant works almost automatically.

Chemical means of deoxygenation have also been tried [3] as follows :—

(*a*) A highly pyritic sand, *e.g.* the tube mill dewatering-cone underflow or the underflow of the primary cones, is used in the sand clarifiers. To this may be added either fine iron or battery sands. White has suggested the

---

[1] Thurlow and Prentice, *Journ. Chem. Met. Mng. Soc. S.A.*, 1928, **28**, 267.
[2] *Journ. Chem. Met. Mng. Soc. S.A.*, 1922, **22**, 249.
[3] Thurlow and Prentice, *loc. cit.*

use of sodium sulphite. All act as reducing agents. 2 square feet of clarifying area per ton of solution are allowed. Some results are given below.

	Before Clarification, mg. O_2 per litre.	After Clarification, mg. O_2 per litre.
Village Deep sand solution, .	4·0	1·75
,, slime ,, .	2·5	Trace
Durban Deep slime ,, .	4·0	1·5
New Modder sand and slime solution, ...	5·0	0·5

The decrease in zinc consumption was 0·03 lb. per ton milled.

(*b*) Battery chips or wire are put in the head compartments of the extractor boxes. A reduction in oxygen content of the solution of some 30 per cent. was obtained. This process is not as efficient or as economical as (*a*).

Precipitation by Aluminium.—The first practical application of aluminium (originally proposed by Moldenhauer in 1893, see p. 318) was in the form of dust, and was made at Deloro in 1908, and by Kirkpatrick [1] and E. M. Hamilton,[2] working on silver solutions in 1913. At Nipissing, Ontario, Hamilton used Merrill's zinc dust machinery with modifications necessitated by the fact that it is difficult to wet aluminium dust, and that even after it has been wetted it tends to rise to the surface and float as a thick scum. The emulsifier is omitted, and the aluminium is agitated with the solution in tanks by means of paddles or screw propellers. The precipitation occupies from 10 to 15 minutes, and the solution then passes to filter presses. The consumption of aluminium is about 0·02 lb. per oz. of silver (or about 1 part aluminium to 3 parts of silver), and cyanide is regenerated. The presence of caustic alkali is necessary. Some special constituents in the solutions at Cobalt (possibly due to the presence of arsenic) prevented the use of zinc there. A difficulty in the use of aluminium shavings or dust, caused by the formation of quantities of insoluble alumina, is at least in part due to the action of the air.

Electrical Precipitation has been used occasionally, but has met with no substantial success. It requires a heavy current density of 0·3 to 0·8 ampere per square foot. The gold is deposited as a black slime, which can be wiped off the cathodes. The latter are usually of tinned iron. The anodes are peroxidised lead plates or graphite. Perfect precipitation is not obtained, and Lamb suggests that the method should be supplemented by zinc precipitation. The advantages are the avoidance of foul solutions (because base metals are precipitated by the current) and the regeneration of cyanide.

An exhaustive survey of tests on the electrolytic precipitation of gold from cyanide solutions is given by Clevenger.[3]

Lay [4] says that electrical precipitation of gold from clarified mill solutions has the advantages of simplicity, low cost and ease of clean-up. He

---

[1] *Eng. and Mng. J.*, 1913, 95, 1277.

[2] *Eng. and Mng. J.*, 1913, 95, 935.

[3] *Met. and Chem. Eng.*, 1915, 13, 803, 852; *Trans. Amer. Electrochem. Soc.*, 1915, **28**, 263.

[4] *Eng. and Mng. J.*, 1920, 110, 58.

advocates a current density of 0·02 to 0·04 ampere per square foot of cathode and a voltage of 3 to 5. Lead anodes are most desirable, especially as they can finally be sent to a smelter for the recovery of the lead. Thin lead sheets are also recommended for the cathodes. The process is not attractive, however.

Dunstan [1] made unsuccessful experiments on precipitating gold electrolytically from cyanide solutions.

Precipitation by Charcoal.—Charcoal is capable of extracting 7 per cent. of its weight of gold from cyanide solution without showing any change in appearance.

The softer woods appear to produce the most suitable charcoal for precipitating gold.[2] Efficiency in precipitation depends on the total time of heat-treatment during or subsequent to charring. A temperature of 700° to 900° C. appears to be the best. Age, up to one month, does not diminish the efficiency of pine charcoal. Quenching does not increase the efficiency, but may wash out impurities ; the apparent activation by quenching is probably due to the heating prior to quenching. Quenching in certain chemical solutions, *e.g.* nitric acid or zinc chloride, appears to give promising results, but correct heat-treatment is just as beneficial. Excessively fine grinding is detrimental, and also increases the difficulty of filtering. Dry and wet pulverisation are equally effective. Industrial activated charcoals, of which there are now a good many on the market, are good precipitants, but not superior to pine charcoal.

McKee and Horton [3] recommend black ash residue, a waste product of the soda pulp industry, as a good precipitant after treating it with sodium carbonate, heating to 900° C. and quenching in water. It is stated to have a capacity of 2,000 ozs. of gold per ton of the charcoal.

Gold precipitated on charcoal either as adsorbed salts or after their conversion to the metallic state, cannot be amalgamated. All the gold but no silver can be removed from charcoal by sodium sulphide. The dissolving power of the latter, however, decreases rapidly with successive treatments and intermediate precipitations. Mercurous nitrate dissolves the silver salts only. Hydroxides are usually more effective than acids in dissolving power, while oxidisers and reducers react equally well. The power of the dissolving solution normally increases with longer contact, stronger concentration and increase in temperature. Boiling water is also a solvent, but after the first treatment only a small portion of the remainder of the gold is soluble. Hot cyanide solutions of all strengths are good solvents. A disadvantage in dissolving the adsorbed salts in order to use the charcoal again is that they have to be reprecipitated.

Attempts have been made to treat the loaded charcoal so that the adsorbed material may change its condition, remain within the charcoal, and leave the latter free to take on a new load. This is termed " fixing." It was found that heating has no such effect, but that nitric acid improved the capacity of the charcoal by 50 per cent. Attempts to remove the gold from the charcoal other than by smelting have only been partially successful. Sulphides, polysulphides and alcohol were used by Williams.

When charcoal is used as a precipitant a loss of cyanide occurs, probably due to adsorption. The recovery of cyanide which has been adsorbed is

[1] *Journ. Chem. Met. Mng. Soc. S.A.*, 1930, 31, 118.
[2] Gross and Scott, *U.S. Bur. Mines*, 1927, *Tech. Pap.* 378.
[3] *Chem. and Met. Eng.*, 1925, 32, 164.

difficult, but when sodium sulphide is used as a "fixer," cyanide is regenerated and does not appear to be re-adsorbed.

When charcoal is used it is probable that it acts not only as a precipitant but also as a filter. At Yuanmi salts of lime preponderated in the ash derived from the charcoal, even up to 20 per cent. Sodium salts, on the other hand, were present in only small amounts, and but little silver was in the bullion. In the first box the silver was very low, but increased in the third box—presumably when all the trivalent and some of the monovalent and divalent materials had been adsorbed. Summing up, the true adsorptive capacity of the charcoal for gold was greater than was apparent in mill practice owing to part of it being employed in taking up other elements.

Few substances have a bad effect on the precipitation of gold on charcoal. Sodium sulphide, owing to its solvent power, decreases the amount of precipitation, as does also free cyanide. Hydroxides to the extent of less than 0·5 lb. per ton increase precipitation, but in greater amounts diminish it.

McKee and Horton [1] put forward the following advantages of charcoal over zinc :—

(1) Gold can be more completely recovered from foul solutions.
(2) Higher-grade bullion is produced.
(3) The refining cost is less.
(4) No fouling agent is introduced into the solution.
(5) The cyanide consumption is less.

Williams [2] says that as to the efficiency of the process there is no doubt. The gold is precipitated in a few seconds. With high lime or heavy ferrocyanide content, however, stage precipitation might be better. Precipitation in stages involves less time of contact and less charcoal is needed, although the throwing-down of the gold is much faster than that of the silver. The filtration through charcoal should not be too fast. Smelting of ash containing more lime than iron is not without difficulty. The gold produced is of high fineness, but when iron contaminates it toughening is difficult.

Gross and Scott assert that the objections to charcoal as a precipitant are (1) its bulk, (2) the loss of cyanide due to adsorption, (3) the trouble of burning the charcoal in a clean-up. However, it does not contaminate the solution with harmful substances, and is of use in helping to keep a mill solution free from copper. It is comparatively cheap and its preparation is not difficult. It is said to be an excellent precipitant for small amounts of gold in waste solutions containing little or no free cyanide.

Carbon precipitation was formerly used [3] in a number of small plants in Victoria. The charcoal was placed in small tubs, or in boxes resembling ordinary zinc boxes, and the solution passed in succession through a number of these. The precipitation of gold bore no relation to the strength of cyanide, but no mention was made of the precipitation of gold from solutions containing less than 0·04 per cent. of KCN. The sump liquors contained as little as 3½ grains of gold per ton. The cost was supposed to be about double that of precipitation by zinc. When the charcoal became ineffective with use, it was re-burned (see pp. 322 *et seq.*).

In 1916 charcoal replaced zinc as a precipitant at the Yuanmi mine in Western Australia (Moore and Edmands).[4] The charcoal was quenched from

[1] *Loc. cit.* [2] *Min. Mag.*, 1923, **28**, 139.
[3] J. I. Lowles, *Trans. Inst. Min. and Met.*, 1899, **7**, 190.
[4] Aust. Patent, 566, 1917; *Trans. Inst. Min. Met.*, 1917, **27**, 277; U.S. Patent, 1368520, 1921.

a red heat, crushed very fine and then used in successive filters. The sequence of these was changed, as the efficiency of any one diminished, and fresh material was brought into service. The cost as compared with zinc precipitation was roughly 3 : 2.

In a later modification a carbon-water emulsion was mixed with aurocyanide solution and subjected to vacuum filtration in a Butters plant. The cakes were finally discharged, repulped and refiltered. The filters were operated in series, the strongest solution going to the twice-formed cake. The final cake was smelted.[1]

Edquist [2] has suggested the solution of gold and silver by cyanide solution of standard strength, precipitation of the gold by the addition of a precipitant such as finely ground charcoal, and then the removal of the precipitate in a flotation cell, thus avoiding filtration or decantation.

"The use of charcoal may be reasonably recommended in cases where a limited quantity of tailings which contain base metals, such as copper, zinc, etc., has to be treated."[3] There is little hope, however, of producing a charcoal which will generally replace zinc for precipitation, but where special impurities are present the use of charcoal may be effective.

6. Treatment of Barren Solution.

Barren solution may be dealt with in the following ways :—(1) Directly as a wash in the general mill circuit in order to become enriched. (2) For the recovery of the cyanide content (regeneration of cyanide). (3) For the recovery of precious and base metal values.

(1) **As Wash in Mill Circuit.**—Usually the solution will require aeration in order to render it an effective solvent.

Knapp [4] recommends that the effluent from zinc dust precipitation, which contains much hydrogen, should not be sent to the agitators direct, where it operates adversely in reducing the active aeration, but that the hydrogen should be removed by an intermediate oxidation process.

At Kirkland Lake the aeration of barren cyanide solution has been effected by spraying it into the main storage tank. Kieselguhr diffusers, fed with compressed air at 5 lbs. pressure, are also effective for aeration.

At the New England Mill [5] barren solution is revivified by discharging it under pressure, together with air drawn in through an auxiliary opening in the pump, through fine nozzles attached to rotating arms in a steel tank. The jets strike a layer of mercury on the bottom of the tank. It is suggested that the mercury prevents the solutions from becoming foul.

(2) **Regeneration of Cyanide.**—MacArthur, in the original patent for the cyanide process, contemplated the regeneration of cyanide from the solutions after the precipitation stage. He stated that "the metals are recovered by any suitable process, and the cyanogen, cyanide or substance containing or yielding cyanogen may be regenerated."[6] In many fine grinding plants the mechanical loss of dissolved cyanide, in terms of NaCN, in waste solutions, varies from $\frac{1}{3}$ to 1 lb. per ton of solution. The loss of solution used to convey,

---

[1] *Min. and Sci. Press*, 1921, 113, 493. [2] *Chem. Eng. and Min. Rev.*, 1934, 26, 384.
[3] H. A. Megraw, *Eng. and Mng. J.*, 1914, 97, 1234. See also M. Green, *Trans. Inst. Min. Met.*, 1913, 23, 65.
[4] *Min. Mag.*, 1934, 51, 213. [5] *Eng. and Mng. Journ.*, 1928, 126, 213.
[6] Merrett, *Trans. Inst. Min. Met.*, 1932, 42, 243.

tailings to the pond may amount to one ton per ton of dry tailings. Loss also occurs in filtration where there is an excess accumulation of wash effluent, which builds up in cyanide strength. The recovery of this cyanide represents a considerable monetary gain in the process.

(*a*) The *Halvorsen process* [1] combines acid cyanide regeneration and fractional distillation in a vacuum. It has three steps: (1) Conversion of combined cyanogen into hydrocyanic acid by acid treatment; (2) heating and distillation, under high vacuum, of a small fraction of the total solution, the distilling vapours taking with them out of the remaining bulk of solution all the HCN gas; (3) condensation of water and HCN by absorption in lime solution. (Hydrocyanic acid itself is a colourless volatile liquid, boiling at 26·1° C. and solidifying at − 14° C. It is a deadly poison, soluble in water in all proportions.)

The barren solution, after precipitation, is circulated through two acidifying towers in series, where it meets gaseous SO_2 from a sulphur burner. A small amount of steam admitted at the bottom of the acidifier assists in condensing the SO_2 and in heating the solution before it enters the disperser. The acidifying towers are filled with 1 inch earthenware Raschig rings. The lime deposit which forms is washed out once a month. The solution passes from the acidifying towers to a storage tank in which there is a feed-box partially immersed. This maintains a continuous flow of solution to the dispersing tower, which is a fractionating column, 4 × 13 feet. The solution in the storage tank is heated by steam to 100°-120° F., according to the degree of vacuum that is maintained in the disperser. The solution enters at the top of the disperser and trickles down through Raschig rings. Hydrocyanic acid gas diffuses from the solution into the water vapours which continually ascend and pass out at the top to the absorber. At Tonopah Extension Mill, 6,000 feet above sea level, a 22 inch vacuum is maintained.

The absorber is smaller than that used in the Mills-Crowe process. The alkaline solution passing through it contains 1·3 lb. of lime per ton and is used in the proportion of 1·6 − 2 parts to 1 part of decyanided liquor. Any excess HCN which passes forward is trapped in a condensing tower in which a high vacuum is also maintained.

Any gold and silver in the decyanided solution is separated by settling and then drying in a filter press.

It is suggested that steel would be preferable to wood in the construction of the towers in order to stop inward leaks which might be caused by the high vacuum.

Burk and Pettis give the following consumption figures:—

Sulphur, 0·88 lb. per lb. of cyanide transferred.
Lime, 0·78 lb. per lb. of cyanide transferred.

The theoretical consumption of lime is 0·57 lb. for fixing 1 lb. of sodium cyanide, but this figure is increased if an excess of SO_2 is used. The solution for decyaniding contains 1·5 lb. of available cyanide, and 0·20 lb. of lime per ton. The fixing solution shows a transfer of 1·4 lb. cyanide and a loss of 0·85 lb. lime per ton of decyanided solution. 0·333 gallon of oil (150,000 B.Th.U.) is consumed for steam-raising per ton of solution decyanided. The process is applicable to the recovery of cyanogen combined with copper.

[1] Burk and Pettis, *Min. and Met.*, 1925, 6, 136.

Some reactions which probably occur in the Halvorsen process may be represented by the following equations :—

$$2NaCN + SO_2 + H_2O = Na_2SO_3 + 2HCN$$
$$Na_2Zn(CN)_4 + 2SO_2 + 2H_2O = ZnSO_3 + Na_2SO_3 + 4HCN$$
$$Ca(CN)_2 + SO_2 + H_2O = CaSO_3 + 2HCN$$

(*b*) *The Mills-Crowe Process.*—This is a method for recovering cyanogen from spent mill solutions in which it exists as alkali cyanide, double cyanides of zinc and copper, sulphocyanides and ferrocyanides. From alkali cyanide and $Na_2Zn(CN)_4$ nearly complete recovery may be made. The cyanide combined with copper needs special treatment for its liberation. Recovery in this manner reduces cyanide consumption and obviates the pollution of streams. The disadvantage of excess water due to crushing in cyanide solution is lessened if the cyanide can be easily recovered. Stronger solutions may also be used in treating complex ores without fear of the consequences.

The process consists in acidifying the cyanide solution (weak wash-solution or foul mill-solution) by means of sulphur dioxide and then transferring the acidified solution to a closed tank or disperser where agitation takes place. The air leaving the tank is charged with hydrocyanic acid and is passed to a second tank, where it is mixed with an alkaline solution which absorbs the hydrocyanic acid. The air is rendered sufficiently pure to be used over again.

The cyanide solution from the extraction plant is alkaline and the sulphur dioxide neutralises the alkalinity before freeing hydrocyanic acid. While air alone will cause the expulsion of free HCN, a certain amount of SO_2 is necessary in the free state or as bisulphites, which are easily broken down. Sulphuric acid has given place to SO_2 for acidification in many cases.[1]

The solutions may be acidified (1) by passing gas and solution in opposite directions through a scrubbing tower filled with packing or wooden grids, (2) by spraying the solution into the gas in a revolving chamber.

The percentage of the HCN removed from the acidified solution depends upon (1) the acidity, (2) the amount of air used, (3) the amount of residual HCN in the air after the latter has passed the absorption plant.

Lawr[2] states that the amount of cyanide removed from a given volume of solution increases with an increase of air. The impoverished solution may be wasted or used as wash water on the pulp filters.

The results are not affected by the concentration, the temperature or the purity of the acid used. The necessary degree of acidification varies with the nature of the solution treated. Simple mixtures may be left just alkaline to phenolphthalein, but complex ones, *e.g.* those containing copper, must be made more strongly acid.

A typical sulphur burner is cylindrical, 8 feet long by 30 inches diameter, rotating at $\frac{1}{6}$ r.p.m. Liquid sulphur is continuously fed in at the top and the gas issues at the lower end. The burner should be placed near to the discharge of the last disperser and its working should be regulated by the composition of the final spent solution.

---

[1] Lawr, *Amer. Inst. Min. Met. Eng.*, 1929, *Tech. Paper* 208 ; *Eng. and Min. Journ.*, 1929, **128**, 688.

[2] *Loc. cit.*

The theoretical quantities of sulphur required are estimated from the following equation—

$$2KCN + H_2SO_3 = K_2SO_3 + 2HCN$$

from which it may be calculated that 1 part of KCN is equivalent to 0·246 part of S. Similarly 1 part of lime is equivalent to 0·57 part of sulphur.

The main requirements for efficient operation of the sulphur burner are (1) a constant volume of solution-feed to the acidifiers, containing uniform amounts of cyanide and lime, (2) a positive and unchanging flow of air through the burner, and (3) as little manipulation of the dampers as possible.

At Santa Gertrudis two steel towers, 6 × 31 feet, act as acidifiers. They have conical bottoms and are lined with concrete for the lowest 9 feet to protect the steel from corrosion. Inside are wooden grids at 15 inch intervals in the space between 4 feet from the bottom and 5 feet from the top, in order to ensure the intimate contact of gas and solution. The gas from the sulphur burners enters near the bottom, while the solution is spread over the top grid from a launder and passes downward in fine streams. The emerging solution flows over a 40 inch weir, which also acts as a seal, and then goes to the first of two dispersers, where hydrocyanic acid is removed by air agitation.

The dispersers are tall tanks, 8 × 23 feet, filled with wooden grids 15 inches apart. The rate of dispersion increases with increase in acidity and temperature of the solution, and reaches a maximum when the air volume is constant. The air is pumped in at the bottom of the tower while the solution trickles down.

In closed circuit with the dispersion towers are two absorption towers, 8 × 21 feet, each containing nine sets of 4 inch grids. Alkaline solution, consisting of ordinary mill cyanide solution and sufficient milk of lime or sodium carbonate solution to neutralise and absorb the acid gas, passes down these towers in thin streams and removes the hydrocyanic acid from the rising gas.

Considerable trouble may arise owing to lime salts which are precipitated and fill the pipes and grids. This was met at Santa Gertrudis by inserting an inverted truncated cone at the top of the acidifier with a distributing overflow well suspended above it. Four ¾ inch nozzles were placed in the same plane near the bottom of the tower and spray from them impinged, at a pressure of 29 lbs. per square inch, against a 12 inch pipe suspended from the cone. The lime was thus deposited on the side of the towers.

A full scheme for the methods of control and testing at Santa Gertrudis is given by Lawr,[1] who makes the following suggestions for the erection and operating of plants for cyanide recovery :—

(1) The discharge from the acidifiers should flow by gravity to the dispersers and then to the clarifying filter.

(2) Corrosion is best met by dipping all parts in hot asphalt.

(3) Acidifiers, dispersers and absorbers should be made of wood, square in section, with double walls having heavy tarred roofing-paper between them.

(4) The economic balance between the amounts of sulphur burned and of cyanide dispersed must be determined for any given dispersion and absorption capacity.

---

[1] *Loc. cit.*

(5) The exhaust fan connections should be of wood and the SO_2 entering wooden acidifiers should first be cooled.

(6) Elemental sulphur is a costly source of SO_2. It is better to obtain the SO_2 by roasting sulphides.

(7) Dispersion is adversely affected by low temperatures.

(8) To obtain the best results, the discharge from the last disperser should be distinctly acid to methyl orange.

(9) In treating barren solution for the removal of fouling agents, such as copper and zinc, in order to reduce the volume of mill solutions, and to recover the copper, silver and gold from gold-silver ores containing too much copper for ordinary cyanidation and too little for profitable smelting, the following scheme is suggested :—

Precipitate the pregnant solution with zinc dust, and treat the barren solution from this in a recovery plant. Remove CuNCS from the tails in a Butters filter ; make the filtered solution alkaline with lime, and then thicken, using the overflow as a filter-wash.

Bruhl [1] found at Rosario that if the head solution contained more than 2·75 lbs. potassium cyanide per ton, the efficiency of transference nearly always fell, even if the amount of solution to be decyanided was reduced. He found it inadvisable to use the effluent as a barren wash and doubted the efficacy of employing it for sluicing slime presses. The only advantage in this use of it was that, being acid, it might assist in keeping the filter-cloths clean.

At Fresnillo,[2] where barren solution is treated for the recovery of cyanide by the Mills-Crowe process, the final slightly acid solution has a precipitating action on the silver in the filter cake. Some SO_2 is also absorbed with the HCN by the mill solution in the absorber towers. The SO_2 takes oxygen from the solution, whether it remains free or forms calcium sulphite. Measures have to be taken to maintain the oxygen content.

Wright [3] describes the following method for the regeneration of cyanide solutions :—4 lb. of lime were added to each ton of mill sand to neutralise any latent acidity, and the sand was then treated with 2 tons of solution during 14 days. Sump solution, containing 0·06 per cent. NaCN and 0·04 per cent. CaO, was run on first. The run-off was sent to the regenerating plant. Each 50 tons of solution were agitated with 380 to 700 lb. of an 80 per cent. sulphuric acid solution to throw down all the cuprous cyanide. After settling, the clear liquor, containing hydrocyanic acid, was decanted and milk of lime stirred in vigorously until the solution showed 0·09 to 0·13 per cent. free CaO and 0·14 per cent. free NaCN. The solution was then pumped back to the mill circuit. The cuprous cyanide precipitate was washed, dried and sent to the smelter.

(3) **The Recovery of Metals from Barren Solutions.**—If the effluent from the filters contains *silver*, it is precipitated as silver sulphocyanide, mixed possibly with copper sulphocyanide. The precipitate may be settled or filtered and then shipped to a smelter. The main requirements in *silver* and *copper* precipitation are a large volume of air per ton of solution (15 to 20 cubic feet per ton per 24 hours) and the presence of the necessary amount of sulphur dioxide. The precipitation of copper appears to require more air than that of silver.

---

[1] *Eng. and Min. Journ.*, 1928, **125**, 658.
[2] *Trans. Amer. Inst. Min. Met. Eng.*, 1934, **112**, 742.
[3] *Eng. and Min. Journ. Press*, 1923, **115**, 806.

Hamilton[1] describes a process for the recovery of *gold* and *silver* from solution which is withdrawn from the mill circuit in order to control the volume, to keep the precipitation active and to remove deleterious impurities. The *Geco method* consists in adding zinc sulphate equivalent in amount to the cyanide and lime in the spent solutions, and thickening. The overflow contains the precious metals in solution and is returned to the mill circuit if this can be done without increasing the volume too much. Otherwise it is acidified and treated with zinc dust to throw down the gold. The sludge from the thickener is passed to a small lead-lined regenerator and acidified with sulphuric acid *in vacuo*. The HCN gas driven off is collected in lime water. The flow sheet of the Geco process is shown in Fig. 189.[2]

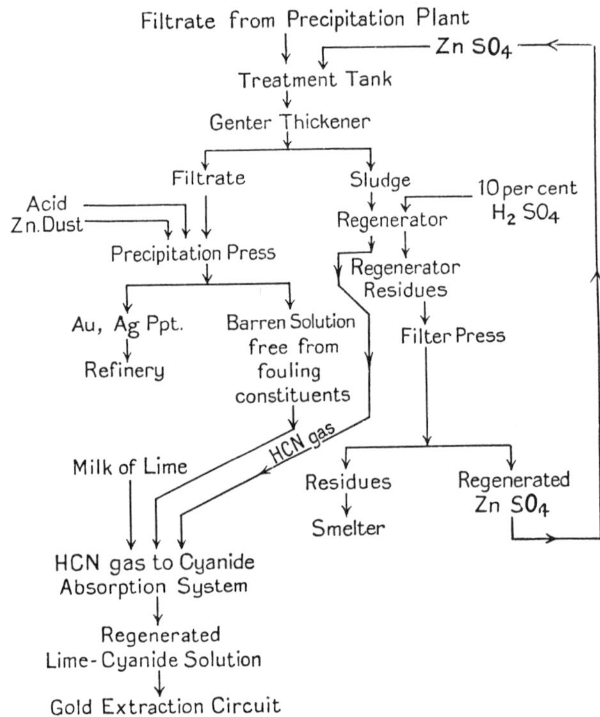

Fig. 189.—Flow Sheet of Geco Cyanide Recovery Process.

Kominsky[3] suggested a process for the recovery of *zinc* from waste gold solutions and its conversion into pigments. The solutions contain about 5 per cent. zinc, together with some ferrous salts and acid. The acidity is reduced by the addition of lime, and sodium chromate is added to convert the iron into the ferric condition. To a small proportion of the original solution sufficient sodium hydroxide is added to throw down the zinc as hydrate, which is filtered off and then made into a thin paste with water and added in excess to the main bulk of the solution. The remaining acid is

[1] *Can. Min. Journ.*, 1931, 52, 809; *J. Chem. Met. Mng. Soc. S.A.*, 1933, 34, 75.
[2] *Eng. and Min. Journ.*, 1932, 133, 53.
[3] *Sth. Afr. Journ. of Industries*, 1921, 4, 894.

thereby neutralised and a quantity of zinc sulphate produced. Ferric and chromic hydroxides are formed. These separate and an equivalent amount of zinc sulphate again goes into solution. The fluid is filtered. The clear solution contains mainly zinc sulphate and sodium sulphate. The sludge in the filter press is principally zinc hydroxide with hydroxides of iron and chromium. The zinc sulphate solution is precipitated with sodium carbonate or chromate to produce the corresponding zinc salts, which are filtered off and washed. The sludge is then dried. The chromate may be used direct for yellow pigments, but the carbonate has to be calcined in order to form zinc white.

7. Production of Bullion from the Precipitate.

Various methods have been suggested for effecting the elimination of the zinc and other base metals. The chief ones are—

(*a*) Direct fusion with fluxes.
(*b*) Roasting, followed by fusion.
(*c*) Treatment with acid, followed by fusion with or without roasting.
(*d*) Reverberatory furnace lead fusion and cupellation (Tavener process), often preceded by acid treatment.
(*e*) Blast furnace lead fusion and cupellation.

Williams[1] gives a summary of trials as to efficiency and costs in the treatment of precipitate, after the clean-up at Mysore, by various methods—(1) smelting and refining; (2) smelting in lined pots; (3) treatment with sulphuric acid; (4) treatment with hydrochloric acid.

(*a*) **Direct Fusion.**—In this method the slime is dried in iron pans or merely by air-blowing in a filter press. If the zinc has been imperfectly separated from the slime, nitre must be added. Other fluxes, such as borax, carbonate of soda, sand and fluorspar are always added.

The slag, which consists of silicates of zinc, soda, etc., corrodes the pots rapidly. Large quantities of zinc oxide are given off as fumes, forming thick crusts in the flues, and evil-smelling products of decomposition of the cyanides are also evolved. The bullion produced by this method is of low fineness and varies in colour from iron grey to pale yellow. It cannot be obtained uniform in composition, so that accurate assays are difficult to obtain.

The slags obtained in this way are always rich in gold, part of which is sometimes in the form of shots, and may be recovered by crushing and panning. The crushed slag is then fused again with the addition of granulated lead, or of litharge, when all the gold is concentrated in the lead. If the lead thus obtained is granulated, it can be used again until rich enough to be worth cupelling.

When, by the careful use of a 40 mesh screen, followed by the squeezing of the slime in a filter bag, the zinc has been almost all eliminated by mechanical means, very little nitre is required. The zinc can be reduced to about 1 per cent., and bullion then obtained 960 fine in gold and silver without the use of nitre, roasting, or acid. The dry slime is fused in graphite pots with about half its weight of borax, a little sodium carbonate, and sand or fluorspar, if necessary. The proportion of the fluxes must, however, be determined in each case by experiment. It is necessary to form a fluid slag,

---

[1] *Kolar Gold Fields Min. and Met. Soc.*, 1928, *Bull.* 23, 213.

but it is difficult to avoid pastiness in the presence of much zinc, and hence acid treatment is preferred.

Where the grade of precipitate varied over a wide range, fusion was found to be the only satisfactory method. In an oxidising fusion zinc tends to oxidise first, and there should be more borax than silica present in order to flux the zinc oxide and silicate completely. Sulphur oxidises next, forming a sulphate "cover" which, as it rarely carries values may, if desired, be skimmed off. It appears that Mo, As, Sb, Te, Se, and V are often concentrated in this sulphate "cover." Thirdly, lead oxidises and enters the slag. Copper is almost the last metal to enter the slag, which is coloured red or green by it according to the state of oxidation. If copper and lead are present together, Weinig found that it was difficult to slag the precipitate in a graphite crucible owing to the reducing action of the latter on the lead. Clay-lined crucibles obviate this difficulty. Griffith Jones[1] recommends that clay liners in graphite pots should be packed with finely crushed graphite crucible, a layer of ganister or quartz being laid on the top.

Oxidising Fluxes.—Manganese dioxide is costly because of its comparatively low percentage of available oxygen. Sodium nitrate is preferable to potassium nitrate owing to the lower melting point of the sulphate. Sodium carbonate or soda-ash is better than the bicarbonate as it goes farther, weight for weight, and some cost of freight is saved.

The ratio of borax glass to silica in the slag should not be less than 2 to 1 where much zinc is present, or less than 1 to 1 where much lead has to be slagged.

Weinig recommends a bisilicate slag. It is easily fusible and corrodes the crucible less than other silicates. He also suggests that the precipitate and fluxes should be mixed while the former is wet from the press, and that the mixed charge should be sintered in an oven furnace to cause the reactions to start. Too much "boiling" in the crucible is thus prevented.

In the Kolar Gold Field[2] the precipitate averages 1 to $1\frac{1}{2}$ oz. of fine gold per lb. and is not given an acid treatment. It is dried and fluxed with 25 per cent. sand and 15 per cent. borax. The slags are finely ground in a Krupp ball mill to —90 mesh, and then amalgamated in a 3 foot pan. Afterwards, when the amalgam thus formed has been separated, about $3\frac{1}{2}$ tons are slowly fed, over a period of 10 hours, to a Wheeler pan with a Spitzkasten discharge. Additions of 5 lb. of cyanide and 25 lb. of mercury are made gradually. The pulp overflows to an air-lift leading to an agitator. When the charging is completed, 15 lb. more of cyanide are added, and then another 10 lb. 2 hours before discharge. After agitation for 16 hours, the charge is first run slowly over an amalgamating table and then over blankets before being passed to a filter tank. The recovery by cyaniding is about equal to that by amalgamation.

The following flux is used at Shasta, California,[3] with 100 lb. of precipitate :—Na_2CO_3 24 lb., sand 32 lb., borax 40 lb., MnO_2 9 lb. After fusion, the gold is separated and remelted into bars assaying 455 parts per 1,000 of gold and 180 of silver. The slag is finely crushed in a ball mill and concentrated on a Deister table. A high-grade concentrate results, which is sold to a smelter, and also fine shot and a middling which is reground, re-tabled and then re-treated.

---

[1] *Kolar Gold Fields Min. and Met. Soc.*, 1928, *Bull.* 23, 222.
[2] *Ibid.*, p. 199.
[3] *Eng. Min. World*, 1931, **2**, 500.

The precipitate at the Globe and Phoenix Mine [1] is granulated, mixed with sulphur and heated slowly in a graphite pot. Most of the copper and other metals form a matte, but antimony is not entirely removed. After fusion, the charge is poured, the matte removed and the bullion re-melted in a clay-lined pot with an oxidising flux. It is then skimmed and toughened by compressed air. The matte is crushed, mixed with an equal weight of borax and one-quarter of its weight of crushed cyanide and heated in a graphite pot until all action ceases, when it is poured. The bullion so obtained is added to that already purified by the sulphur treatment.

In smelting the bullion obtained from the sand plant in the clean-up, objectionable antimony fumes arise. Such bullion is melted with an oxidising flux, and a blast of dry compressed air is blown on to its surface for about 2 hours before the fusion with sulphur.

(*b*) **Roasting followed by Fusion.**—The dried slime is spread in a thin layer on iron trays, and heated to a barely-perceptible red heat, about 500 to 550° C., in the flue of a furnace, or above a special grate under a hood. The carbon, zinc, arsenic, etc., ignite readily, being in a very fine state of division, and roasting proceeds regularly without much stirring, which would tend to cause loss through dusting. Dense fumes of zinc oxide are given off, and the residue consists chiefly of the oxides of lead and zinc with gold and silver and a variable quantity of sand. A little nitre is sometimes mixed with the wet slime to aid in the oxidation of the base metals. Fusion with fluxes is carried out as in the previous case in clay-lined pots. The resulting buttons or culôts are remelted in plumbago pots and cast into bars. The loss by dusting has never been determined.

(*c*) **Acid Treatment.**—In this process, after the mud has been filtered and washed, the zinc is dissolved in sulphuric acid and the residue dried and smelted. Hydrochloric acid is occasionally used, on account of its solvent power on lime and lead. The wet slime from the filter press is charged into a large vat and a solution containing 10 to 15 per cent. sulphuric acid is then added a little at a time. On the Rand the slime is added little by little to the acid. Violent action ensues and the vat must be deep to prevent frothing-over. The fumes are highly poisonous, sometimes containing arseniuretted hydrogen, and must be carried off by a good draught with the aid of a hood. Cold water may be added to moderate excessive action.

After treatment with acid, the slime is settled, washed with hot water, and separated by decantation, by a vacuum filter or a filter press. It is then dried, calcined and fused in the usual way.

Where white precipitate (*q.v.*) is present in the boxes in considerable quantity, special precautions are taken on the Rand,[2] including (*a*) emulsification of the white product and avoidance of the formation of greasy lumps, and (*b*) long contact, say 12 hours, with hot acid in excess. Some of the coarser zinc is then reserved to neutralise the acid. This treatment improves the value of the calcined slime. Care is taken that the solution shall be sufficiently dilute to prevent the separation and crystallisation of $ZnSO_4.7H_2O$. To aid in this the charge is sometimes heated by steam, and must be occasionally stirred. For this purpose a two- or a four-armed stirrer is provided. The vat is made of wood, which may be lined with lead, and is usually from 6 to 10 feet in diameter and 5 or 6 feet deep on the

[1] *J. Chem. Met. Mng. Soc. S.A.*, 1921, **21**, 117.
[2] H. A. White, *J. Chem. Met. Mng. Soc. S.A.*, 1914, **15**, 50.

Rand. The vat is closed with a lid in which is a charging hole and a pipe for the escape of the gases.

J. E. Thomas and G. W. Williams showed[1] that sodium bisulphate is a cheaper and more convenient solvent than sulphuric acid for zinc. The use of bisulphate of soda is general on the Rand. There is less tendency to boil over, but a larger plant and an extra vat for dissolving the bisulphate are required.

Durant[2] advises the use of potassium or sodium dichromate with sulphuric acid in the ratio of 1 to $2\frac{1}{2}$ as an additional treatment after that by sulphuric acid. Agitation, even to the point of attrition of the precipitate, is necessary, and no hydrochloric acid should be present. The action should continue until chromic acid can definitely be detected in the liquor. If it is carried too far, silver will be redissolved.

The following analysis of slime, after treatment with acid at the Brodie Works, Cripple Creek, Colorado, was given by W. R. Ingalls :—[3]

TABLE XLI.

	Per cent.
Gold,	49·85
Silver,	9·54
Silica,	15·60
Lead,	3·00
Copper oxide,	5·58
Zinc oxide,	0·14
Lime,	0·51
Ferric oxide, Alumina,	4·00
Sulphur trioxide,	11·32
	99·54

Ingalls states that in fusing this slime about 50 per cent. of sodium carbonate and 25 per cent. of borax would be required.

The analyses in Table XLII show the composition of acid-treated and calcined precipitate obtained on the Rand.

In No. 1, made by A. Whitby,[4] the organic matter, including insoluble cyanide compounds, was not determined. The high percentage of silica was said to be due to blown sand. In No. 2,[5] given by W. A. Caldecott, some lead existed as sulphate. The slime was from the Robinson Deep Mill. Nos. 3 and 4, made by G. W. Williams,[6] are of slimes from the East Rand Proprietary. No. 3 is before and No. 4 is after calcination. No. 3 also contained from 7 to 10 per cent. of ferrocyanides ($K_4Fe(CN)_6$). About half the sulphur was present as PbS, which was oxidised in the ensuing calcination.

[1] *J. Chem. Met. Mng. Soc. S.A.*, 1905, 5, 334.
[2] *Eng. and Min. Journ.*, 1915, 100, 523.
[3] *Mineral Industry*, 1895, 4, 336.
[4] Johnson and Caldecott, *J. Chem. Met. Mng. Soc. S.A.*, 1902, 3, 47.
[5] T. K. Rose, *Trans. Inst. Mng. and Met.*, 1905, 14, 400.
[6] *Loc. cit.*

TABLE XLII.

	1	2	3	4
	Per cent.	Per cent.	Per cent.	Per cent.
Au,	34·50	34·0	36·07	37·3
Ag,	4·75	4·3	3·5	3·7
Zn,	...	...	7·50	6·16
ZnO,	7·00	22·5	...	...
Pb,	12·50	...	20·81	20·00
PbO,	...	15·6	...	...
Cu,	2·55	...	2·46	2·5
CuO,	...	0·7	...	...
Fe,	...	...	0·02	Trace
Fe_2O_3,	3·65	2·9	...	...
Ni,	...	...	Trace	Trace
NiO,	1·00	0·7	...	...
SO_3,	6·95	6·8	6·3	13·7
SiO_2,	21·00	7·5	12·6	16·2
	93·90	95·0	89·26	99·56

The prepared gold precipitate is melted in graphite crucibles provided with clay liners, and then poured into conical moulds. Even after roasting, the precipitate appears to contain reducing substances, and an oxidiser is added, such as nitre. This does not readily oxidise lead. Manganese dioxide oxidises lead, but carries more silver into the slag than nitre does, and an excess is therefore avoided. If the slag passes to the Tavener furnace the silver is saved there. Other fluxes are sand, borax and sodium carbonate, the latter being especially useful if much zinc is present. Fluorspar is sometimes added. On some mines, fluxes and manganese dioxide are not necessary, but a cover of borax and sand is added.[1] The precipitate and fluxes are mixed, and charged into the crucible by degrees, more being added as the charge melts and sinks down.

A flux used on the Rand is given by E. E. Meyer as follows :—[2]

Acid-treated dried slime,	100 parts.
Borax,	45 ,,
Sodium carbonate,	1 ,,
Fluorspar,	7 ,,
Manganese dioxide,	2 ,,
Sand,	15 ,,

This gave bullion from 875 to 900 fine.

Flux charges given by MacFarren are as follows :—[3]

Precipitate,	100 parts.
Borax glass,	12 to 30 ,,
Sodium carbonate,	6 to 15 ,,
Silica,	3 to 8 ,,

If the fluxes are in suitable proportions the slag is fluid and glassy and contains, after panning, about 20 ozs. of gold per ton, which is recovered by

[1] H. A. White, *J. Chem. Met. Mng. Soc. S.A.*, 1902, 3, 46; 1914, 15, 51.
[2] *J. Chem. Met. Mng. Soc. S.A.*, 1905, 5, 169.
[3] "*Cyanide Practice,*" p. 187.

fusion with litharge. The buttons from the conical moulds, which are sometimes as much as 100 ozs. in weight, are remelted in plumbago crucibles and cast into ingot moulds. The precipitate is sometimes melted in Faber du Four tilting furnaces, fired with coke, oil or gas, and remelted in crucibles. The bullion prepared as above may be as high as 985 fine in gold and silver, and should not be lower than 850 fine. It is sometimes brittle from the

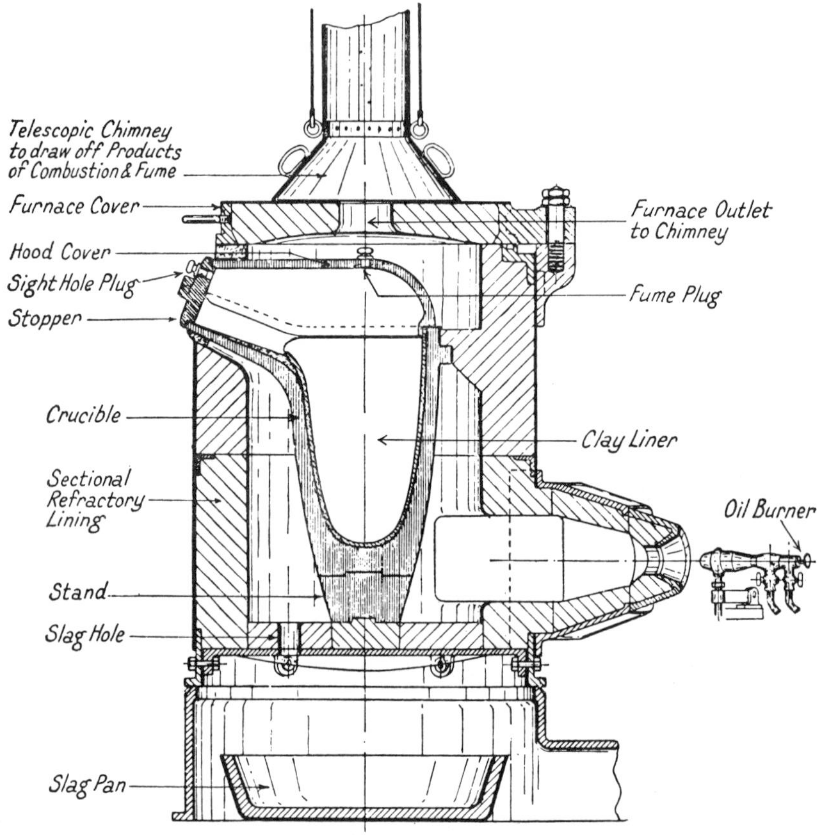

Fig. 190.—Sectional View of Oil-Fired Furnace for Smelting Cyanide Precipitate (*Morgan*).

presence of zinc, etc. The loss in roasting and subsequent handling is estimated by Bettel to be from 0·01 to 0·025 per cent., or less than ½d. per oz.

In cleaning-up on the Rand by modern methods the gold contained in the by-products, such as pots, liners and slag, should not exceed 0·5 per cent. of the gold produced, and in some cases is less than 0·25 per cent.[1]

Crucible tilting furnaces are also in use for melting precipitate. They have the advantage that the crucible remains in the furnace during pouring. The melting is also done more quickly. Oil or gas is used as fuel. On

[1] H. A. White, *J. Chem. Met. Mng. Soc. S.A.*, 1914, **15**, 51.

the other hand, tilting furnaces are more expensive to install. In the stationary furnace the crucible is usually lifted out for pouring with the result that it is cooled down before being replaced in the furnace, and does not last so long as in the tilting furnace. The metal could, however, be ladled out.

Fig. 190 shows an oil-fired tilting furnace used for the smelting of precipitate and also for the melting of crude bullion. The flame enters the base tangentially and the hot gases circulate round the crucible until finally they issue from the top into the telescopic stack. The specially-shaped hood prevents contact between the charge and the gases. A clay lining is provided on the internal surface of the crucible so that the metal may not be contaminated by impurities from the graphite of the main crucible body.

Oil-fired tilting reverberatory furnaces without crucibles were used at the Tonopah-Belmont Mill [1] for melting precipitate without previous acid treatment. The precipitate contained 74 per cent. of gold and silver.[2] The cost of treatment was very low, and for a large plant such furnaces seem to be more suitable than crucible tilting furnaces, the cost of furnace liners being less than that of crucibles, and the labour costs lower.

Electric furnaces are also sometimes used for smelting precipitate.[3]

Acid Treatment with HCl and H_2SO_4 at Hollinger (see Fig. 191).—The composition of precipitate at Hollinger is given by Scott [4] as follows :—

Original Slime.		After HCl treatment.
	Per cent.	Per cent.
Gold in dry material, .	35·0	75·85
Silver, ...	7·2	15·37
Copper, ...	1·4	2·27
Lead, ...	9·3	0·15
Zinc, ...	14·6	1·54
Iron, ...	0·3	0·14 (Iron and Alumina)
Alumina, . . .	0·5	
Lime, ...	11·7	0·32
Silica, ...	0·7	0·46
Sulphur, ...	4·4	0·76

The precipitate is added, 45 lb. at a time, to 35 per cent. hydrochloric acid equal in amount to the weight of precipitate to be treated. Hydrogen and sulphuretted hydrogen are given off and are scrubbed in a water tower on their way to the atmosphere. Steam and air are then admitted and agitation at 160° F. continued for 75 minutes. The precipitate is washed by decantation six times with hot water to remove $PbCl_2$. The residue is discharged to a filter press dressed with heavy twill and filter paper. It is washed there with hot water until no trace of lead can be found in the effluent. Finally the filter cake is blown with air, discharged, and dried.

This material is next treated for 1 hour with sulphuric acid in steel pans heated electrically. To 650 lb. of acid 400 lb. of residue are added. Sulphur dioxide is evolved and silver dissolved. The temperature rises to

[1] A. H. Jones, *Eng. and Mng. J.*, 1913, **95**, 1197.
[2] *Ibid.*, 1914, **97**, 967.
[3] H. R. Conklin, *ibid.*, 1912, **93**, 1191.
[4] *Trans. Inst. Min. Met.*, 1930-31, **40**, 321.

about 520° F. Copper sulphate partially dissolves and partially settles. Lead forms the sulphate and is precipitated. The acid is syphoned off into a tank containing warm water. Fresh acid is added to the pans, the mud stirred, the temperature again raised to 520° F. for 1 hour and the acid again syphoned off. Cooling-off and fume removal are effected by adding cold

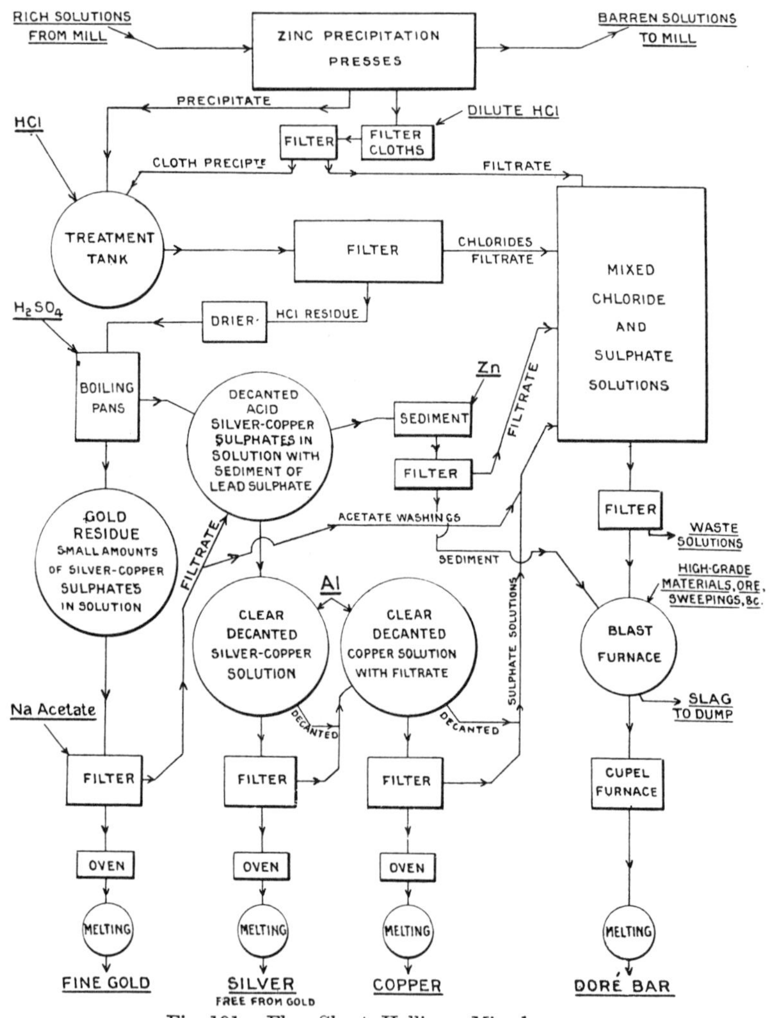

Fig. 191.—Flow Sheet, Hollinger Mine.[1]

acid to the pan. The mud is dropped into hot water and then filtered under suction. The filtrate is sent to the warm water tank. The residue in the filter is washed twice with sodium acetate and then with water to remove lead. It is finally dried, melted in a graphite pot with a flux of silica, borax and manganese dioxide, and cast into bars 994 to 997 fine.

[1] *Trans. Inst. Min. Met.*, 1930-1, 40, 329.

The sulphate liquors containing silver and copper are treated with aluminium powder, of which 1 lb. will throw down 11·6 lbs. of silver or 3·4 lbs. of copper. The silver is deposited first, removed by filtration, dried and melted under soda and borax. The bars are 985 fine. The copper is precipitated from the remaining solutions periodically.

The chloride liquors and the sulphate liquors from the copper tank are mixed. The lead is thrown down as sulphate. The clear liquor is wasted and the precipitate is filter-pressed.

A small water-jacketed cupola is started-up twice a month and takes the lead sulphate slags, floor sweepings, etc., to give a lead bullion which is then cupelled.

Weinig [1] records that at the Liberty Bell Mine wet treatment of precipitate was rejected because owing to lead and lime being present, the action of sulphuric acid following hydrochloric acid was to form sulphates, thus lowering the grade of the precipitate. Subsequent boiling with caustic soda was necessary in order to raise the grade again.

(*d*) **Tavener Process.**—In this process the precipitate is smelted with litharge, fluxes and coal in a small reverberatory (or "pan") furnace, and the auriferous lead cupelled. It was originally introduced [2] with the intention of avoiding acid treatment, but is now generally used for acid-treated slime.

The charge for the pan furnace is approximately as follows :—[3]

Gold slime,	100 parts	
Litharge,	100 ,,	(Varied to produce bullion containing 7 to 10 per cent. gold).
Assay slag,	55 ,,	
Coal dust,	10 ,,	
Silica,	25 ,,	
Iron (any scrap),	13 ,,	

The damp charge is put into the warm furnace and covered with some of the litharge. The fire is then lighted and the charge slowly melted, and the iron scrap afterwards added to clean the slag. Washes of litharge and coal are also used. When action ceases, the slag is run off through the slag door into slag pots and the lead tapped into moulds.

The lead pigs are cupelled in an English cupellation furnace until a cake of gold-silver is obtained on the cupel. The cake is broken up while hot, and the pieces melted in crucibles and cast into ingots. The fineness of the bullion is 960 to 980 in gold and silver, and the losses small.

The slags from the pan furnace are smelted for lead with other by-products, such as assay slag, old cupellation "tests," old crucibles, filter cloths, extractor house sweepings, etc. This is done either in the pan furnace itself or in a small blast furnace,[4] and the lead bullion so produced is cupelled.

(*e*) **Blast Furnace and Cupellation.**—Instead of a reverberatory furnace a small blast furnace is used for smelting gold slime in most of the plants installed by C. W. Merrill in America.[5] The method at the Homestake Mill is typical of this practice.[6] Here the precipitate is partly dried, mixed with

---

[1] *Trans. Amer. Inst. Mng. Eng.*, 1916, 55, 385.
[2] P. S. Tavener, *J. Chem. Met. Mng. Soc. S.A.*, 1902, 3, 112.
[3] L. A. E. Swinney, *Trans. Inst. Mng. and Met.*, 1907, 16, 115.
[4] E. H. Johnson, "*Rand Metallurgical Practice*," vol. i., pp. 277-285.
[5] H. A. Megraw, *Eng. and Mng. J.*, 1914, 97, 606.
[6] Clark and Sharwood, *Trans. Inst. Mng. and Met.*, 1912, 22, 137.

fluxes and briquetted. The briquettes are dried and the rich ones added to the lead bath in the cupel furnace. The lower grade material is smelted with by-products in an ordinary water-jacketed lead blast furnace with three tuyeres, and the lead passed to the cupel. At the Goldfield Consolidated Mill, direct blast-furnace smelting of briquetted precipitate is practised.

Chauvenet [1] recommended the use of a small blast furnace for the smelting of cyanide precipitate containing copper or lead. He calculated appropriate fluxes.

Richards [2] describes the refining practice at Buckhorn, Nevada. Each month the precipitate from the Merrill presses is weighed, sampled and assayed. Flue dust from a previous run is added to the precipitate, and the whole then mixed with twice its weight of litharge, 15 per cent. of soda ash and 5 to 6 per cent. of borax glass and briquetted. The briquettes are charged into a small lead blast furnace containing 300 to 400 lbs. of pig lead. The charge consists of 36 briquettes (4 × 2 inches), 15 to 20 per cent. of coke, 40-50 lbs. of lead blast furnace slag, cupel bottoms, etc. When under full heat the slag is tapped continuously from the furnace. Lead bullion is tapped from the well and cast into small bars, of which 300 to 400 lbs. are retained to start the next run, the remainder being cupelled. From the cupel the bullion is cast into moulds, cooled and then remelted with nitre before final shipping (890 to 950 fine).

The advantages of this method are said to be as follows :—There is no dust loss in charging, no drying of precipitates, no broken crucibles, quick conversion into bullion, and removal of arsenic and antimony by volatilisation during cupellation. The principal disadvantage is the cost of installation. Practically no gold is observed in the flue gases from the smelting or cupellation operations.

---

[1] *Met. and Chem. Eng.*, 1916, **14**, 96.
[2] *Eng. and Mng. Journ.*, 1922, **113**, 865.

CHAPTER XV.

THE CYANIDE PROCESS. SPECIAL METHODS.

Treatment of Concentrate.—When gold and silver in an ore are largely contained in constituents such as sulphides, it is often expedient to separate them by gravity concentration or by flotation and to treat the products separately. The removal of the concentrate enables the residue to be cyanided with comparative ease and simplicity. The concentrate is a fraction of the original ore from which it came and will contain not only gold but also many sulphides and " cyanicides." Special treatment of the concentrate may be necessary, which treatment would otherwise have to be applied at considerable cost and probably not so efficiently to the whole bulk of the material. The methods include " roasting, leaching, agitation, filtration, decantation, oxidation and fine grinding as with ordinary ores " (MacFarren). Free gold dissolves slowly, being as a rule comparatively coarse-grained, and can be recovered in part by amalgamation. " No standard method for the treatment of concentrate has yet been or can be decided on, as it must vary with the nature of the concentrate and the mode of occurrence of gold and silver in it. Hence the practice differs at almost every mill " (Gowland).

Roasting has been generally discontinued except for sulpho-telluride ores (*q.v.*) and certain flotation concentrates. *Percolation* may require as much as three or four weeks to obtain a good extraction. The concentrate tends to pack down and become impermeable, and is sometimes mixed with sand to facilitate leaching. Shallow charges are usual. Strong solutions, containing as much as 0·5 per cent. KCN or more, are used. The work is begun and ended with weak solutions and water washes. It is necessary to apply special methods of aeration, such as the forcing of air through the charge, or turning the material over by shovels, or transferring it to another vat. Acidity and dissolved sulphides are dealt with as usual. Cyanicides, especially iron and copper salts, are sometimes removed by means of dilute sulphuric or hydrochloric acid followed by water washes (MacFarren).

Fine grinding and agitation save much time. Air agitation is preferred to the use of mechanical agitators, owing to the tendency of the charge to pack, on account of its rapid settlement in water. Concentrate is often ground in cyanide solution if the removal of cyanicides is not required. The addition of fresh cyanide solution to the charge during treatment is sometimes found to be necessary.

Decantation is facilitated in the treatment of slimed concentrate by the fact that the latter settles rapidly. For the same reason *filtration* by some of the leaf filters is difficult. The Kelly filter press[1] is mentioned by MacFarren as giving satisfactory results, on the ground that in this

---

[1] For description see MacFarren, "*Cyanide Practice,*" pp. 148 and 201.

machine cake formation by pressure is rapid. With such filters as the Moore or the Butters the concentrate settles in the vat before the cake can be formed. With such a filter press as the Dehne, settlement takes place inside the press, and channels are formed, preventing efficient washing.

Fig. 192 shows a flow sheet for the treatment of concentrates by grinding in a tube mill, agitating with cyanide solution, filtering and then recovering the gold from the liquor by the Merrill-Crowe process. The wash solutions from the filters are used again in the preliminary operations and the pulp may be treated in a smelter for the recovery of base metals.

See also "Cyaniding Flotation Concentrates" (p. 417).

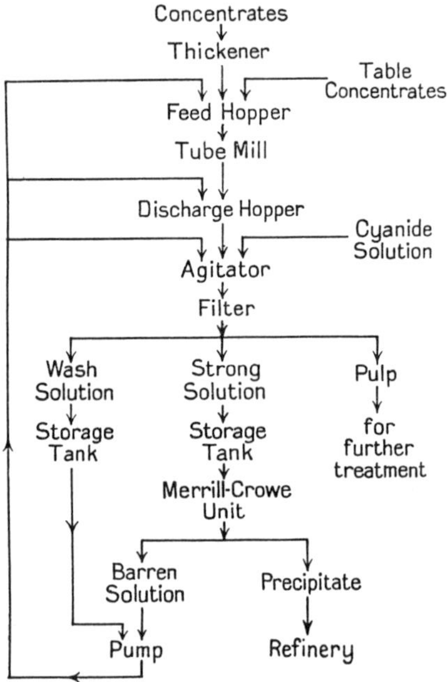

Fig. 192.—Cyaniding of Concentrates.

The Treatment of Sulpho-telluride Ores.—These ores occur in large quantities at Kalgoorlie, West Australia, and at Cripple Creek, Colorado. The gold is contained almost entirely in the sulpho-tellurides, and probably exists mainly in the form of telluride of gold (*q.v.*). The gold cannot readily be extracted from these by cyanide solutions, owing to the slight solubility of telluride in cyanide. The successful methods of treatment have been (1) dead roasting with the expulsion of tellurium, sulphur, etc., followed by cyaniding: used at Kalgoorlie and at Cripple Creek; (2) wet crushing and concentration, followed by treatment of concentrate and various classes of the tailing with cyanide to which cyanogen bromide is added to assist in the dissolution of the gold tellurides: used at Kalgoorlie, and with modifications at Cripple Creek; (3) flotation, tending to supersede the other processes; (4) straight cyaniding. These methods are described in succession in the following.

(1) *Roasting Processes.*—The moisture in the ore is kept at 2 per cent. or lower. The process usually consists in (1) dry crushing in rock breakers, followed by ball mills or roller mills; (2) roasting dead; (3) grinding wet in pans with a large quantity of mercury and hot alkaline cyanide solutions; (4) agitating by means of paddles in large vats with cyanide solutions containing from 0·01 to 0·08 per cent. KCN; (5) filter-pressing; and (6) precipitation of the gold by zinc shavings or dust.

In roasting, a large amount of soluble sulphate is formed, and the losses by dusting must be carefully attended to. Edwards or Merton furnaces are used. They are fired by Salmon gum tree wood. The loss of gold in roasting is 0·5 per cent. A low temperature (600°) is necessary in the early stages of roasting to prevent sintering of the sulphides and melting of the tellurides. The roasted ore is carried by a stream of weak cyanide solution to spitzkasten or cones, where it is separated into sand and slime, the sand going to the Wheeler pans and the slime to the agitation vats. The pans are used primarily to grind the ore to slime, which is necessary to enable the fine gold to be freed so that the cyanide may dissolve it. Any coarse gold is extracted in them by amalgamation. The continuous overflow from the pans is again passed through spitzkasten and the sand returned to the pan. The agitation vats are 15 to 22 feet in diameter and 4½ to 9 feet deep, and are provided with radial arms or stirrers revolved by bevel gearing. The pulp is agitated for about eight hours. It is then drawn off and forced by pumps into filter presses of the Johnson type, where the gold solution is separated and the residue washed. In later practice, vacuum filters have been used. They are worked at a lower cost.

At the Golden Cycle Mill,[1] which is a customs mill, average-grade Cripple Creek ores are treated by roasting, followed by cyanidation. Ore from the storage bins is sent over a magnetic pulley for the removal of tramp iron, and discharged to a stationary screen having ⅝ inch holes. From this the oversize passes to a Symons crusher set at ⅜ inch, while the undersize and the crushed product go to Hum-mer screens, equipped with 4½ mesh Rek-Tang screen cloth. The oversize from here is further reduced by Schmidt dry grinding ball mills operating in closed circuit with the screens.

Roasting is done in Edwards duplex furnaces, 115 feet long and 13 feet wide. A cooling hearth, 44 × 13 feet, adjoins each roaster.

The fire-box is a semi-gas-producer and uses live steam. The consumption of coal is 230 lb. per ton of ore roasted. The temperature of roasting is 800°-900° C., and is controlled by indicating pyrometers. Sizing of the ore on screens before roasting has reduced the coal consumption by 75 to 100 lb. per ton of ore.

The stack losses have been reduced to $0·04 per ton, mainly by lowering the feed discharge point into the roasters, by providing better draught control, and by choking the discharge of the calcine. Calcium sulphide and sulphate produced in the roasting furnace sometimes prove troublesome in cyaniding and precipitation.

The calcine is ground in Chilian mills having screens of 0·0496 inch opening. The pulp goes over blanket tables, the concentrates from which are amalgamated in a small grinding pan. The tailings from the blanket table and the overflow of the grinding pan are treated in a bowl classifier. The sands are leached during 6 to 8 days, with frequent intermediate aeration.

[1] *U.S. Bur. Mines Inf. Circ.*, 1933, 6739.

The fines are thickened to 42 per cent. solids and then subjected to counter-current cyaniding treatment. The overflow from the final thickeners is passed through a sand clarifier and then precipitated with zinc dust.

The precipitate while still wet is put into iron pans with $\frac{1}{8}$ its weight of sodium nitrate, placed in a muffle furnace and heated to redness. The sintered charge is then mixed with ordinary fluxes and smelted.

(2) *Treatment with Cyanogen Bromide* (*Bromocyanogen, Bromocyanide*), BrCN.—In the *Diehl process*,[1] which was introduced at the Hannan's Star Mine, and subsequently used at the Lake View Consols and at Hannan's Brownhill, Kalgoorlie, the ore is wet crushed, amalgamated and then concentrated. The concentrate, which contains 30 to 40 per cent. of the gold, is roasted and sent back to the wet crushing mill. The tailings are separated into sand and slime, the sand reground in tube mills, and the whole agitated with cyanide (to which cyanogen bromide is added) and filter-pressed.

At the Hannan's Star Mill, Kalgoorlie,[2] the ore from the rock-breaker was dry-crushed in Krupp ball mills, mixed with water, passed over amalgamated plates and then ground fine in tube mills. The fine product was agitated in vats with cyanide and cyanogen bromide. Nardin observes [3]—"It is useless to add bromocyanogen at the beginning of the agitation, as it is comparatively rapidly destroyed. The proper time to add it is when no further extraction can be gained by plain cyanide solution. Before adding the bromocyanogen it is essential to neutralise some of the alkalinity of the pulp solution by the addition of H_2SO_4, in order to prevent the rapid destruction of bromocyanogen." The alkalinity was reduced from 0·02 or 0·03 per cent. lime to 0·01 per cent. The method of preparation of cyanogen bromide from bromide and bromate of potassium, sulphuric acid and cyanide is given by Nardin.[4] It was found that finely divided iron and also fine pyrite are destructive of cyanogen bromide.

At Golden Horseshoe, strong cyanide solution is added to the agitated slime pulp, and the protective alkalinity is reduced by the latent acidity developed by the pyritic matter in the slime. An excess of alkalinity would destroy the cyanogen bromide, which is next added through a lead pipe to the extent of 1 lb. per ton of slime. Before the slime pulp is sent to the filter press the protective alkalinity is again increased by the addition of lime—an important point.[5]

At Boulder Perseverance mine the ore is treated by three-stage crushing to $\frac{3}{16}$ inch, closed circuit grinding, strake concentration, cyanidation in ordinary cyanide solution, followed by the addition of bromo-salts in the form of sodium bromide and bromate, at the same time maintaining the alkalinity at 0·0014 per cent. CaO.[6]

Blackett [7] found at Boulder Perseverance that treatment with bromo-cyanide according to the following scheme gave a recovery of 90 per cent. of the gold (the ore contains 6·8 per cent. FeS_2, 20·3 per cent. $MgCO_3$ and $CaCO_3$, 63 per cent. SiO_2):—

---

[1] H. Knutsen, *Trans. Inst. Mng. and Met.*, 1902, **12**, 1.
[2] E. W. Nardin, *Mineral Industry*, 1908, **17**, 444; G. W. Williams, *ibid.*, 1907, **16**, 537.
[3] *Loc. cit.*
[4] *Loc. cit.* See also p. 411.
[5] *Chem. Eng. Min. Rev.*, 1928-9, **21**, 46.
[6] O'Malley, *Chem. Eng. Min. Rev.*, 1933, **26**, 115.
[7] *Eng. and Min. Journ.*, 1932, **133**, 11; *Trans. Aust. Inst. Mng. Met.*, 1933, No. 91, 99.

Grinding to 150 mesh; amalgamation to remove coarse gold (up to 20 per cent. of gold removed); agitation of 1 : 1 pulp in 0·15 per cent. KCN solution for two hours; agitation with 1 lb. bromocyanogen per ton of ore for one hour after addition of 1 lb. of lime per ton of ore; addition of 2 lbs. of lime per ton of ore to neutralise acidity and regenerate metallic cyanides. The cost is roughly 10s. per ton against 12s. to 15s. per ton when using roasting methods.

Only one Kalgoorlie mill[1] continues to use stamps, the remainder employing cone crushers of the Symons type; grinding in two or three stages with one or two of the stages in closed circuit with classifiers is common. The bowl classifier is in use so far only on one mill. The long tube-mill is still retained. Corduroy strakes are used instead of plate amalgamation in all mills. The proportion of gold caught on the strakes may be improved by crushing in clear water instead of in mill solution.

Cripple Creek Procedure.—Ore treatment at Cripple Creek has always been influenced by the fact that neighbouring smelters are available to which rich ore or concentrate can be sold. The usual procedure is to crush the poorer ores in cyanide solution, concentrate, and treat the tailing by cyaniding with the aid of cyanogen bromide. Concentrate is shipped to a smelter or roasted and cyanided.

(3) *Flotation.*—All the ores of the Kalgoorlie district appear to be amenable to flotation and there is little doubt that future production will be bound up with this process. In spite of the small amount of auriferous minerals in the ore, a stable bubble-column, to restrict the frothing area, has been attained by the provision of hoods in the upper part of the flotation cell. The introduction of unit flotation cells between mills and classifiers, as at McIntyre Porcupine, is becoming common.

Bromocyanide treatment is applied to the concentrates obtained from the flotation machines at the Wright-Hargreaves Mine after reducing the alkalinity to about 0·1 per cent., adding cyanide to bring the strength up to 1 lb. per ton and then agitating. 20 lb. of bromocyanide are added to 30 tons of concentrate and water (1 : 1), agitating for 24 hours. More reagent is added if the gold is still high after that period. About 70 lb. in all is found usually to be sufficient. When treating high-grade material each pound of solvent is equivalent to $1\frac{1}{4}$ ozs. gold. The bromocyanide is made by mixing together sodium bromide and bromate, 520 grms., sodium cyanide, 205 grms., sulphuric acid, 485 grms., and water, 7,500 c.c. The method of preparation is described by Willey.[2]

Low-grade ores of the Golden Cycle Mill are first treated by flotation, the concentrate sent to the roasters and the tailings subjected to cyaniding.

(4) *Straight Cyaniding.*—At the Pioneer Mine, B.C.,[3] the ore consists of quartz carrying free gold and associated with pyrite, arsenopyrite and pyrrhotite in small quantities. There is also some telluride. Straight cyaniding has been adopted. The ore is subjected to single-stage crushing in a gyratory machine and two-stage grinding in cyanide solution in ball mills, using forged steel balls. Both grinding stages are in closed circuit with classifiers. Counter-current methods of decantation are employed. The pulp is thickened and the pregnant solution decanted for precipitation by zinc dust. The solutions are heated prior to precipitation by the Merrill-Crowe

---

[1] O'Malley, *Chem. Eng. Min. Rev.*, 1933-4, **26**, 202.
[2] *Eng. and Min. Journ.*, 1928, **126**, 16.
[3] *Min. Mag.*, 1934, **51**, 175.

process. Crushing 375 to 400 tons per day of original ore containing about 1 oz. gold per ton, the overall recovery is 97 per cent. Cyanide consumption is 0·816 lb. NaCN per ton. The consumption of lime, which is added in the mine skips, is about 4 lb. CaO per ton. Trials are afoot to remove the sulphides by flotation between the primary and secondary grinding stages and to treat the concentrates separately.

Sodium peroxide was employed at the Teck Hughes plant in treating telluride ore. About ½ to 1 lb. of the reagent per ton was added to the agitators through a special feeder. Increased extraction and an absence of tellurium fumes and of the so-called " matte " during smelting were noticed. If too much peroxide is added, trouble in filtration appears to arise due to colloidal silica.[1]

Antimonial and Arsenical Ores.—Gold ores containing stibnite are difficult to treat with cyanide. " Antimony sulphide is very soluble in caustic alkali and decomposes cyanide, combining with the alkali and forming antimonite and thio-antimonite ; also some KCNS is formed and HCN is evolved. The antimony compounds act as strong deoxidisers " (Julian and Smart). Hence much cyanide is destroyed and the gold is not dissolved.

The removal of antimony by dissolution in 2 to 4 per cent. caustic soda solution, or by roasting alternately in an oxidising atmosphere and with coal, followed in each case by amalgamating and cyaniding, has not been satisfactory.

Blyth[2] states that the roasting of antimonial ores is not satisfactory. They are best treated, as far as the sand is concerned, by atmospheric oxidation of the crushed ores direct from the mill for a few months in large dams in the absence of lime. But with no lime present the finely divided iron in the slime is converted to ferrous compounds, and these interfere considerably with cyanidation in two ways :—(1) The precipitation of gold by ferrous hydrate ; and (2) the excess consumption of cyanide due to its conversion into ferrocyanide. About ½ lb. of lead nitrate per ton of ore is therefore used and the sand is slimed in acid solution.

At the Globe and Phoenix Mill, Rhodesia,[3] the presence of 1 per cent. of stibnite reduced the extraction of gold by ordinary cyaniding to about 20 per cent. It was found that by exposing the slime (containing up to 6 dwts. gold per ton) to oxidation by weathering in the dam for six months, the extraction by cyanide became satisfactory. The sand was not equally amenable, but by fine grinding most of the stibnite, owing to its brittleness, passed into the slime. In the method adopted, the stibnite was partly removed by hand picking and the remainder crushed to 200 mesh in the stamp battery. The tailing from the stamps after amalgamation was further amalgamated in grinding pans. The discharge flowed over canvas strakes and then through various mercury traps. A small amount of NaCN was added to assist amalgamation by desulphurising. The sand was concentrated and then ground and amalgamated in pans, and again concentrated. The sand was cyanided by percolation, with fair results if the stibnite in it did not exceed 0·2 per cent., and the slime was passed to the dam for weathering. The concentrate was roasted in Merton furnaces, by which the antimony was reduced from 5 per cent. (as stibnite) to 3 per cent. (chiefly in the form of antimonate), and the roasted product ground, amalgamated and then treated

[1] *Eng. Min. Journ. Press*, 1923, **15**, 440.
[2] *Min. Mag.*, 1919, **20**, 224.
[3] *J. Chem. Met. Mng. Soc. S.A.*, 1921, **21**, 117.

with cyanide in agitation vats and filter-pressed. The extraction was from 85 to 90 per cent., but the tailing was rich. A further amount was extracted after the residue had been weathered for some time.

At Hillgrove, New South Wales,[1] antimonial tailing which had been subjected to long weathering was treated by percolation. Care was taken to make the strongly acid tailing almost exactly neutral or very slightly alkaline, to avoid dissolution of antimony, which was thrown down in the zinc boxes and gave trouble in smelting the precipitate. Ore containing stibnite was also cyanided at Bidi, in Upper Sarawak, by direct treatment of coarsely crushed ore with 0·05 per cent. KCN solution. About 75 per cent. of the gold was extracted from 5 dwt. ore.[2]

Leaver and Woolf[3] conclude from their experiments that arsenical and antimonial gold ores are best treated by roasting at 450° C. for about 30 minutes. The foreign minerals are thereby dissociated, leaving the gold fully amenable to cyanide treatment. Roasting at higher temperatures (say above 600° C.) results in the formation of arsenates and antimonates that lock up the gold and protect it against cyanide attack. It is necessary to add lime to the calcine and allow the mixture to stand for about three days before starting cyanide extraction.

Blyth[4] considers that *arsenic* does not influence the straight cyanide extraction of gold, but that, as it is frequently associated with pyrrhotite, antimony and graphite, it shares their bad name. The first two cause premature precipitation of the gold, and graphite varies in its behaviour according to its age and its content of hydrocarbons. Semi-oxidised ores containing arsenic are difficult to treat due to their acidity. A roasted arsenical ore does not, as a rule, yield such a good extraction as roasted pyritic ores with arsenic absent.

Cupriferous Ores.—Unoxidised chalcopyrite is little acted on by cyanide, and does not occasion trouble, but "in some cases, as with oxides and carbonates, a few pounds of copper per ton may prohibit cyaniding by ordinary methods" (MacFarren), the soluble compounds acting as cyanicides. Sometimes the copper is removed by a dilute sulphuric acid wash before cyaniding.[5] A solution of ammonia has also been proposed[6] for the same purpose.

Where chalcopyrite causes trouble in cyaniding, a grading with non-cupriferous ores, so that the copper content of the whole does not exceed 0·4 per cent., gives an increase of 20 to 40 per cent. in cyanide consumption over that for a straight ore. Even then the cost is said to be less than when the copper sulphide is removed by flotation and shipped to a smelter.[7]

Apart from the action of ammonia on copper compounds, the addition of ammonia to cyanide solutions increases their power of dissolving gold and also their selective action on gold. Acid solutions of cyanide have also been used on cupriferous and antimonial tailing (Gitsham process).[8] In this

---

[1] W. A. Longbottom, *Mng. and Sci. Press*, 1912, **104**, 884.
[2] R. Pawle, *Trans. Inst. Mng. and Met.*, 1905-6, **15**, 79.
[3] *U.S. Bur. Mines*, 1928, *Tech. Paper* 423.
[4] *Min. Mag.*, 1919, **20**, 224.
[5] W. S. Brown, *Trans. Inst. Mng. and Met.*, 1905-6, **15**, 445.
[6] A. Jarman and E. Le Gay Brereton, *ibid.*, 1904-5, **14**, 288; H. L. Sulman, *ibid.*, **14**, 363; *Electrochem. and Met. Ind.*, 1908, **6**, 128; *Pacific Miner*, Mar. 1910.
[7] *Eng. and Min. Journ.*, 1925, **119**, 430.
[8] Von Bernewitz, "*Cyanide Practice*, 1910-13," p. 102.

process a little sulphuric acid is added to the cyanide solution, with which the ores are treated. Gitsham suggests the following equations :—

$$2KCN + H_2SO_4 = K_2SO_4 + 2HCN$$
$$4HCN + 2Au = 2HAu(CN)_2 + H_2$$

Sulman and Picard point out that the dissolving action of the HCN depends on the presence in solution of $CuSO_4$ or some similarly acting substance, by which nascent cyanogen is produced. It is stated that very little copper is dissolved by acid cyanide solutions and consequently that ores containing copper carbonate can be treated. The solution is regenerated by alkali before precipitation. Wheelock points out [1] that cupriferous cyanide solutions can be regenerated by the addition of sulphuric acid—

$$K_2Cu_2(CN)_4 + H_2SO_4 = K_2SO_4 + Cu_2(CN)_2 + 2HCN$$

and the insoluble cuprous cyanide may be treated with hydrogen sulphide [2]—

$$Cu_2(CN)_2 + H_2S = Cu_2S + 2HCN$$

Cyanide solutions containing copper give trouble in precipitation on zinc shavings, the copper plating the zinc and preventing its further action. This difficulty is not experienced in zinc dust precipitation. Copper is not precipitated by zinc in solutions containing much KCN, but it is readily brought down by the lead-zinc couple (see p. 375).

Sodium sulphide is a precipitant for copper from acid but not from alkaline solutions. The HCN evolved may be trapped in closed vessels. The amount of sodium sulphide consumed varies as the amount of copper present, averaging, in experiments, 0·7 lb. per lb. of copper precipitated. Acid consumption, under similar conditions, is 4·7 to 6·5 lb. per lb. of copper precipitated. Any cyanate or thiocyanate formed is not regenerated as free cyanide.

Leaver and Woolf [3] say that all common copper minerals, excepting chrysocolla and chalcopyrite, are sufficiently soluble in cyanide solution under usual conditions to cause excessive cyanide loss. The latter may be reduced by regeneration of the cyanide which is combined with the copper. Those minerals containing arsenic and antimony will cause additional loss of cyanide and fouling of solution. From 1·85 to 2·75 lb. of cyanide are used up per lb. of copper dissolved. These investigators [4] also state that gold and silver ores containing 0·5 per cent. of copper which is soluble in cyanide may be treated with sodium sulphide in acid solution without any excessive loss of cyanide; over 80 per cent. of the cyanide used in the dissolution of the copper can be regenerated. The copper precipitate must be removed from the solution while the latter is acid. Only a small part of the gold is thrown down in the copper sulphide and may be recovered by zinc dust at a later stage. "Protective alkalinity" and free cyanide strength should be kept low. The amount of sodium sulphide consumed in the process is proportional to the copper content, while the amount of sulphuric acid used varies as the total cyanide and lime or free alkali in the treated solution. Copper in cyanide solution up to 10 lb. per ton does not diminish the activity of the solution as a solvent for gold and silver if the solution contains free cyanide equal in

---

[1] *Mng. and Sci. Press*, 1909, 99, 814.
[2] Von Bernewitz, "*Cyanide Practice*, 1910-13," *loc. cit.*
[3] *U.S. Bur. of Mines Tech. Paper* 250, 1929.
[4] *U.S. Bur. Mines Tech. Paper* 494, 1931.

quantity to that needed in fresh solvent for maximum extraction. Sodium-copper cyanide solution, in the absence of free cyanide, is only a weak solvent for gold.

At Shasta County, California,[1] where the ore is a low-grade auriferous copper gossan deposit, with the gold very finely distributed, leaching is carried out, followed by the Merrill-Crowe precipitation process. Precipitation is exceptionally good, which is credited to the presence of a small quantity of mercury.

The sulphide-acid method has also been applied to *zinciferous* ores. For these the free cyanide strength of the pregnant solution should be kept higher than in the case of copper.

The precipitated sulphides filter cleanly and contain no combined cyanide. The regenerated solution has extracting power equal to that of fresh solution.

Crushing in Cyanide Solution.—The use of cyanide solution instead of plain water in the stamp battery was tried as early as 1891 in the United States, and in 1892 at the May Consolidated, Transvaal.[2] It was successfully developed at the Crown Mines, New Zealand, in 1897,[3] and applied in South Dakota in 1903.[4] It has not been much practised in conjunction with amalgamation where coarse gold exists, because, although the gold is cleaned and brightened by the cyanide, and thus kept in excellent condition for amalgamation, the corrosive effect on the copper plates and the hardening of the amalgam by cyanide more than neutralises this advantage. Crushing in cyanide is used in many mills, however, in connection with "all sliming" methods and the use of slime filters, amalgamation being omitted. On the Rand, for example, with the adoption of all-sliming methods from 1923 onwards, all the new plants, with the exception of East Geduld, installed cyanide milling circuits.[5]

Crushing in cyanide solution in tube mills was tried at Mysore mill, but gave much trouble and a great deal of fouling of the solution, probably owing to the heat generated by grinding and the presence of finely divided iron.

The residue after milling a gold ore in cyanide solution, followed by agitation and filtration by displacement, is usually higher in value than if the ore had been milled in lime water, thickened and then cyanided, assuming identical mechanical procedure in each instance.[6] One reason for this may be that if a colloidal clay is treated with lime water and milled in water, it adsorbs the calcium ion of the electrolyte. If milled in gold-bearing cyanide solution it may adsorb gold or an aurocyanide compound.

It is probable that the practice of crushing ores in cyanide in stamp batteries will seldom be adopted, although it was found to be satisfactory at the Hollinger mill.[7] With soft ores other machines, such as the Huntington mill, requiring less solution (say 4 tons per ton of ore), can be used in milling, and the difficulty caused by the large amount of solution in the mill circuit is diminished.

---

[1] *Eng. Min. World*, 1931, 2, 530.

[2] Butters and Clennell, *Eng. and Mng. J.*, 1892, 54, 342.

[3] J. McConnell, *Trans. Inst. Mng. and Met.*, 1898-9, 7, 26; Park, *Trans. Amer. Inst. Mng. Eng.*, 1899, 29, 666.

[4] See Papers by C. H. Fulton and J. Gross, *Trans. Amer. Inst. Mng. Eng.*, 1904, 35, 587, 616.

[5] Prentice, *Trans. Inst. Mng. and Met.*, 1935, 44, 492.

[6] Allen, *Trans. Amer. Electrochem. Soc.*, 1931, 60, 67.

[7] N. Cunningham, *Min. and Sci. Press*, 1900, 81, 19.

Among the advantages of crushing in cyanide as an alternative to amalgamation, it is pointed out that—

(1) Dissolution of gold is promoted by the close mixing of ore and solution involved in crushing.

(2) Float gold is rapidly dissolved during crushing without a chance to escape.

(3) If water is used, the de-watered pulp carries some water into the cyanide plant. This amounts only to from 25 to 40 per cent. of water in the case of sand, but in slime it is considerably more. With all-sliming, followed by de-watering in a Dorr thickener, the ratio of water to dry slime is 1 or 1½ to 1. The addition of this quantity of water to the cyanide solution in agitation vats seriously dilutes it. More cyanide must be added, and to avoid a continual increase in the quantity of solution in the cyanide plant, it is necessary to throw some solution away. When ore is crushed in cyanide, an amount equivalent to the solution in the thickened pulp can be returned to the mill circuit, and so the consumption of cyanide is *ipso facto* reduced.

(4) There is no risk of the theft of amalgam.

A. J. Clark considers that crushing in solution is something of a sacrifice of efficiency to expediency.[1]

Carbonaceous Ores.—Carbon being a precipitant for gold, its presence in ores creates much difficulty in obtaining full recovery during gold ore treatment. Of the methods for the removal of this harmful constituent, roasting or a semi-flotation process appears to have been most successful, though each ore has to be treated with special precautions.

Wartenweiler [2] adapted Feldtmann's [3] observation that gold precipitated by charcoal is soluble in a solution of sodium sulphide, to the ores at Prestia, where a carbonaceous schist interferes. The weak sulphide solution (0·20 per cent.) followed the normal cyanide extraction, and was succeeded by a water wash. The gold was precipitated from the sulphide solution by passing the latter through boxes filled with copper shavings. The copper was freed periodically from adhering copper sulphide, as this diminished its efficiency. The use of aluminium or of an alloy of aluminium and zinc was also advocated as a precipitant.

In Ontario, where gold is found associated with a carbonaceous schist (containing up to 5 per cent. of carbon), the graphitic constituent acted as a powerful reprecipitant of gold already dissolved in the cyanide solution. The device was adopted of adding about 3 quarts of kerosene per ton of ore in the ball mill. The pulp was then treated in the normal way. No frothing occurred and the cyanide consumption was not greater than usual, but the value of the tailing was very much less. The kerosene appeared to coat the particles of graphite so that they could not come into contact with the gold cyanide compound.[4]

At the Belledar Goudrean Mine, and at Porcupine,[5] Ontario, graphite interfered with cyanidation. About 80 per cent. of the gold was recoverable by amalgamation, and subsequent concentration raised the extraction to

---

[1] *Min. and Met.*, 1934, **15**, 26.

[2] *Trans. Inst. Min. Met.*, 1917-18, **27**, 372.

[3] *Ibid.*, pp. 302, 311.

[4] *Eng. and Mng. Journ.*, 1925, **119**, 930; Spearman, *Journ. Soc. Chem. Ind.*, 1925, **44**, 179.

[5] *Can. Dept. Mines, Mines Branch*, 1931, *Rep.* 720, 71-81.

99 per cent. In continuous working, however, the graphite coated the flakes of gold. Grinding with lime and kerosene, followed by flotation with pine oil, removed the graphite in concentrates containing 23 ozs. of gold and 25 ozs. of silver per ton, which were smelted. The graphite-free tailings were then cyanided, with a recovery of about 99 per cent.[1]

Dorfman,[2] in treating carbonaceous schist, used crude fuel oil and gas oil as pre-treatment agents, added directly to the ball and tube mills. The pulp was then agitated with cyanide and lime. Excellent gold extraction was obtained, with no increase in cyanide consumption.

The slime portion of the mill tails on the Mother Lode contains the greater proportion of the gold lost, and also has 1·5 to 3 per cent. of carbon. The treatment proposed is ordinary amalgamation followed by flotation concentration of the carbon and sulphides. Cyanidation may be satisfactory if the sand portion has enough gold to warrant additional treatment. The flotation tails are dumped. The concentrates should be given a low-temperature roast (not above 600° C.) and the calcine a treatment by cyanide.[3]

Cyaniding Flotation Concentrates.—With many gold ores the introduction of flotation necessitates the cyanidation of a small quantity of concentrates, which may contain " cyanicides " also in a concentrated form, apart from the low-grade flotation tailings. This has resulted in modifications of the original process and the introduction of new machines. As flotation is mostly operative on quartz sands crushed to little over 60 mesh, the finer crushing is avoided.[4]

In the treatment of these concentrates grinding in cyanide solution is mostly practised. The quantity of solution is small and there is not much justification for a large thickening plant. Lime consumption may be high—up to 100 lbs. per ton—but if the cyanide solution is well aerated before adding it to the mill, a good recovery of gold is normally obtained (Knapp).

Owing to the high specific gravity and fineness of the sulphides in the concentrates there is difficulty in washing and filtering them and the so-called " batch " system of agitation, decantation and final agitation has been revived in one or two instances. There is no necessity to allow for vessels taking more than a 24-hour production, which is comparatively small. Knapp suggests the following cycle :—

First day, . .	No. 1 tank filled.
Second day, .	No. 2 tank filled. No. 1 tank.—Impeller stopped and pulp settled. Pregnant solution decanted. Fresh solution added. Cyanide strength made up. Agitation resumed.
Third day, . .	No. 3 tank filled. No. 2 tank.—Decanted as in the case of No. 1 tank on second day.
Fourth day, .	No. 3 tank.—Decanted as in case of No. 1 and No. 2 tanks.

[1] *Trans. Inst. Min. Met.*, 1923-4, 33, 464.
[2] *Trans. Can. Inst. Min. Met.*, 1922, 25, 109.
[3] Leaver and Woolf, *U.S. Bur. Mines*, 1930, *Tech. Pap. No.* 481.
[4] Knapp, *Min. Mag.*, 1934, 51, 209.

Small amount of barren solution added after final agitation in each case and agitation continued. Allowed to settle. Small quantity of water added and agitated for $\frac{1}{2}$ hour. This cycle repeated with small washes until pulp reasonably free from cyanides. Sludge then sent to waste.

The wash is built up to working solution strength and used with barren solution for the supply to the primary agitation. By using a filter the water wash on it may be used instead of extended washes in the agitator.

Where the slimes are treated in agitators in series, a secondary filtration may be employed by thickening the pulp from the agitators, pumping the underflow to the first filter, using only barren solution as the wash. The filter cake is repulped with barren solution and sent to a secondary filter to be given a water wash.

CHAPTER XVI.

EXAMPLES OF PRACTICE.

Besides the following examples, others have been already given, notably in the section on the treatment of sulpho-tellurides.

Witwatersrand.

The main features of the Rand are the large number of producing mines in close proximity to one another; the large scale of operations; the absence of chemical difficulties in the simple low-grade ore and the use of as few and as large units as possible for each stage of treatment; simplicity of plant arrangement and operation.

If the increase in the price of gold persists, the retreatment of old dumps may come under review.

The older treatment is fully described in "*Rand Metallurgical Practice,*" Vols. I. and II. (Griffin).[1] Briefly, it consisted in sorting out waste rock; breaking the ore in rock-breakers; crushing in alkaline water through coarse screens by stamps; de-watering in cones; crushing in tube mills; diluting and passing the pulp over amalgamated plates; classifying the pulp in cones into sand and slime, with return of a classified underflow to the tube mill; cyaniding of collected sand by percolation; coagulation of slime with lime, and treatment with cyanide by decantation (p. 342); precipitation of gold by lead-zinc shavings; dissolving the zinc in bisulphate of sodium, and fusing the residue with fluxes or with litharge.

There are two main systems of treatment in use on the Rand, viz.:—

(1) Corduroy concentration with barrel amalgamation of the concentrate to recover the coarser gold, and subsequent separate treatment of sand and slime by cyanide to dissolve the fine gold.

(2) All-sliming with a short cyanide treatment by agitation only. There is a tendency to introduce corduroy concentration in this system also.

It is expected that flotation of the pyrite, with cyanide treatment of the residue, will soon pass into use as a third method.

In general, it is not economical to crush banket so fine as to obtain a residue containing, on the average, much less than 0·25 dwt. gold per ton after cyaniding. This necessitates a final tailing pulp containing from about 75 per cent. of — 90 product (*i.e.* passing 90 mesh sieve, with 0·006 inch aperture) with 4-dwt. ore, to 85 per cent. of — 90 product with 7-dwt. ore. About 70 per cent. of the — 90 grade will pass a 200 mesh (0·003 inch aperture) screen. The original free gold in Rand banket after crushing in a stamp mill with a 0·27 inch mesh screen, is in a comparatively fine state of division. At least one-half will pass a screen of 260 apertures to the linear inch (width of opening 0·05 mm. or 0·002 inch). After tube milling practically all the free gold varies from 0·008 mm. to 0·01 mm. mean diameter. The gold should pass a 260 mesh screen for a short treatment by agitation with

[1] See also T. K. Prentice, *Trans. Inst. Min. Met.*, 1935, 44, 179.

cyanide. For the longer leaching treatment, it should all pass a 60 mesh screen.[1]

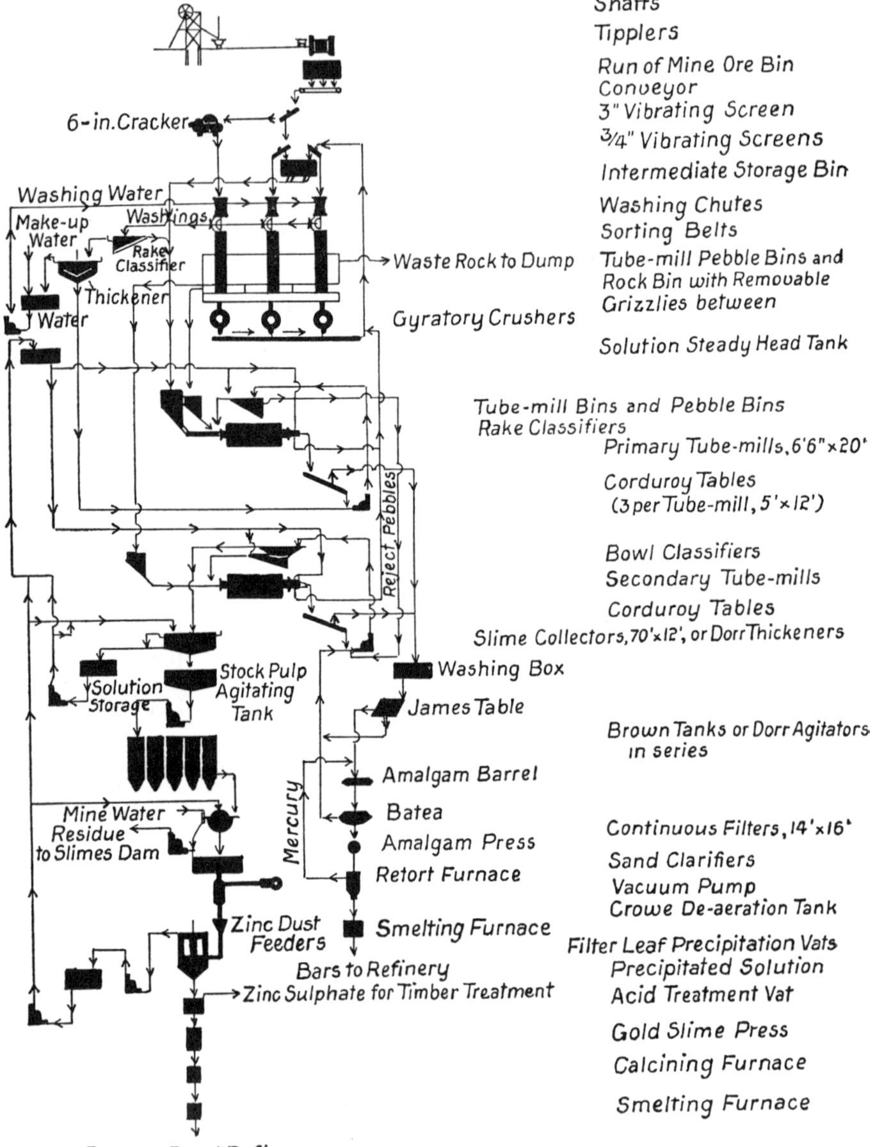

[*Reproduced by permission of the Institution of Mining and Metallurgy.*

Fig. 193.—Metallurgical Practice on the Rand, 1934 (*Prentice*).

[1] Graham and Wartenweiler, *Trans. Inst. Min. Met.*, 1924, 34, [i], 68; Wartenweiler, *J. Chem. Met. Mng. Soc. S.A.*, 1920-1, 21, 217. See also Graham and Wartenweiler, *J. Chem. Met. Mng. Soc. S.A.*, 1924, 24, 290.

A laboratory investigation [1] of ore from a Central district shows the following distribution of free and encased gold in banket :—

(1) Ground to 92 per cent. — 200 mesh :— 47 per cent. in the free state, 9 per cent. encased in gangue, 44 per cent. associated with pyrite minerals.

(2) Ground to 50 per cent. — 200 mesh :— 33 per cent. recoverable by blanket concentration.

Fig. 193 is a flow sheet showing the general scheme of treatment of Rand ores by the all-sliming method. It embraces washing of the ore from the screens, sorting on belts, two-stage crushing in tube mills, followed at each stage by concentration on corduroy tables. The tube mills work in closed circuit with classifiers. The concentrate from the tables is re-treated on James tables. The tailings from these return to the secondary tube mill circuit, while the concentrate is amalgamated, washed in a mechanical batea, the amalgam distilled off and the residual gold smelted. The slime from the classifiers is thickened, agitated in cyanide solution and filtered. The filtrate passes through sand clarifiers and then to the Crowe vacuum precipitation unit. These various steps are dealt with more fully below.

Breaking.—Underground breakers, where used, are set to make a product of 6 to 9 inches, and surface breakers one of 1½ or 2 inches. Grizzlies screen off about 50 per cent. of the ore, which does not pass through the breakers. Both jaw and gyratory breakers are used. The ore is washed in trommels with holes 2½ inches and 1 inch diameter to take out fine material. The washing is done in the upper portion, the water and fines passing to smaller trommels of 9 holes per square inch, while the oversize is delivered to the sorting belts. In the most recent plants unwashed fines are by-passed direct to the milling plant. Vibrating screens are gradually replacing trommels and grizzlies, owing to their high capacity and efficiency in sizing.[2]

Sorting is almost entirely on rubber belts, 36 inches wide, which have a capacity of 80 tons per hour. Removable grizzlies or trommels are situated at the end of the sorting belt to separate tube mill pebbles.

Stamp Milling.—In 1934 there were still more than 5,600 stamps, 88 of them Nissen, in operation on the Rand. They were in use in 31 plants, but in the other 7 plants, all new ones, they have never been installed. Heavy stamps up to 2,000 lbs. in weight are used. The Nissen stamps are especially adapted for small installations. The stamps make about 100 drops per minute, with 8 inches height of drop.

Stamp data in 1934 at one of the groups of mills are given by Prentice as follows :—

Weight of stamps, . .	1,200-1,910 lbs.
Average, ...	1,345 lbs.
Screens, ...	3 to 155 holes per square inch.
Duty, ...	40·6 to 7·4 tons per 24 hours.
Power, ...	15·76 to 26·41 kWh per ton of — 100 mesh.
Cost, ...	3·5d. to 11·36d. per ton stamped.
	20·57d. to 33·13d. per ton of — 100 mesh produced.

[1] Wartenweiler, *Trans. Amer. Inst. Min. Met.*, 1934, 112, 761.

[2] *Min. Mag.*, 1932, 46, 134.

In new plants on the Rand high-speed gyratory crushers of the Symons type are installed instead of stamps to prepare the feed for the tube mills. The principal advantage thus gained in new plants is a decrease in capital cost. Prentice[1] gives the following figures for a plant having a capacity of 50,000 tons per month :—

	Capital Cost.
Stamps and sand treatment, ...	£400,000
Modern all-sliming plant, ...	£300,000

The inclusion of stamps does not much increase the grinding cost. The following figures are also by Prentice :—

Cost of producing 1 ton of . . .	— 100	— 200
(*a*) Crusher-stamp-tube mill, . .	20·6d.	29·7d.
(*b*) Crusher-tube mill, . . .	19·9d.	28·3d.

It is estimated that the electrical energy consumed by either method to produce 1 ton of —200 mesh material is approximately 27 kWh. Thus the retention of stamps where they already exist can be justified, and their replacement by gyratory crushers would involve capital expenditure with but a poor return. In new plants, however, the inclusion of stamps is not attractive.

The ratio of water to solids in the battery pulp is about 6 to 1 ; tube mill pulp contains about 30 per cent. of water. Lime is added during milling and also during cyanide treatment, the total amount being 2 lbs. per ton of ore. In general, launders in the crushing plant are steep, 10 per cent. being a frequent grade, and the proportion of water in the pulp in various stages is correspondingly reduced. A reinforced rubber lining in the launders is common.

For pulp elevation centrifugal pumps are used, with steel or iron liners, but rubber liners are being tried.

Tube Milling—" In the case of plants still incorporating stamp mills the pulp gravitates along slightly inclined rubber-lined launders to pumps which elevate it, together with the outflow from the tube mills, to classifiers. Here the portion which is sufficiently ground is separated in the overflow, while the remainder is re-distributed to the tube mills. The classifiers consist usually of a nest of cones about 4 feet in diameter by 5 feet deep, the overflow passing on to be separated into sand and slime for treatment. The underflow is divided into as many portions as there are tube mills, each portion gravitating to a dewatering cone about 5 feet 6 inches by 7 feet 6 inches, the underflow from which constitutes the feed to the tube mill. (See Fig. 194, which is from a photograph of the Robinson Deep installation taken in January 1914.) With the addition of a little water the pulp passes through the mill with a moisture content of about 30 per cent. The tube mill outflow joins the overflow from the dewatering cone and passes over corduroy tables, each 5 by 12 feet and usually five in number, and then flows to the elevating pumps previously mentioned." (Prentice.) The ratio of stamps to tube mills is approximately 8 : 1.

Crushing in cyanide solution is practised at the all-sliming plants, and one or two other plants, but King[2] states that at the Sub Nigel plant the

[1] *Trans. Inst. Min. Met.*, 1935, 44, 479.
[2] *Trans. Inst. Min. Met.*, 1935, 44, 542.

Fig. 194.—Tube-mill Installation. Robinson Deep Mine.

resulting cyanide consumption was 0·075 lb. KCN per ton, and that less than 5 per cent. of the gold was dissolved in these sections of the plant. Grinding in water is now established there, with a decrease in cyanide consumption and no increase in residue value.

Hopper feeders that can pass pebbles up to 10 inches diameter are in general use. The use of an internal scoop discharge, lowering the level of the liquid pulp in the mill and equivalent to a large outlet trunnion, has increased the crushing capacity, with a corresponding increase in pebble consumption and liner wear. A discharge grid, 2 to 3 inches thick, is fitted with sufficient slots to allow egress of the product without any of the grinding load. Where stamping is not followed, the slots are large enough to discharge $1\frac{1}{2}$ inch pebbles. The rejects are usually screened, the larger pieces being used as pebbles in the secondary or tertiary circuit and the smaller crushed to become part of the pulp at an earlier stage. Mechanical pebble feeders driven from the tube mill are employed in order to maintain continuously adequate pebble loads in each mill. The longitudinal hard steel bar liner, and the block liner, have come into general use. The Osborne is the best-known bar liner. The block liner consists of white iron or steel blocks, 21 × 8 inches × 4 inches thick, made solid or honeycombed to hold pebbles. The blocks are wedged in position.

"In the modern all-tube milling plants (on the Rand), the crusher station product is conveyed by belt to a storage bin built of steel or reinforced concrete, from which it is carried by a belt feeder to each primary tube mill. The outflow from each mill may or may not pass over corduroy prior to being pumped to a straight drag classifier of the Dorr type. The raked product from the classifier is returned to the tube mill with the necessary amount of liquid to give the required dilution. The overflow from the primary classifiers is elevated to bowl classifiers which divide the feed into finished product, which gravitates to the slime plant, and the coarser raked product, which forms the feed to the secondary tube mills."[1] The discharge from the latter is either returned to the bowl classifiers direct or to a straight classifier in closed circuit with the tube mill. In the latter case the sands are either returned to the bowls or diverted to a third stage of tube milling prior to being returned to the bowls.

At East Geduld[2] three-stage crushing in tube mills is employed (see Fig. 195). (1) *Primary.*—The feed is the underflow from a collector tank taking washed ore. The mills are 16 × 8 feet, rotating at 24 r.p.m., and driven by 250 h.p. motors. Pebbles are fed in with the ore. The product is classified in Dorr machines (8 × $18\frac{1}{3}$ feet) from which the coarse product, together with pebbles, goes to the secondary tube mills. (2) *Secondary.*—These are 20 feet × 6 feet 6 inches, driven by 250 h.p. motors, and running at 25·9 r.p.m. The outflow is returned by air lifts to Dorr classifiers in closed circuit, whence the overflow is re-classified and then sent to the slime plant. The coarse raked product passes to the tertiary mills. (3) *Tertiary.*—These are similar to the primary mills, but run at only 18·75 r.p.m. The product is then classified in bowl machines.

Table XLIII gives data as to size and capacity of types of tube mills commonly used.

---

[1] T. K. Prentice, *Trans. Inst. Min. Met.*, 1935, 44, 479.
[2] *Min. Mag.*, 1932, 46, 79.

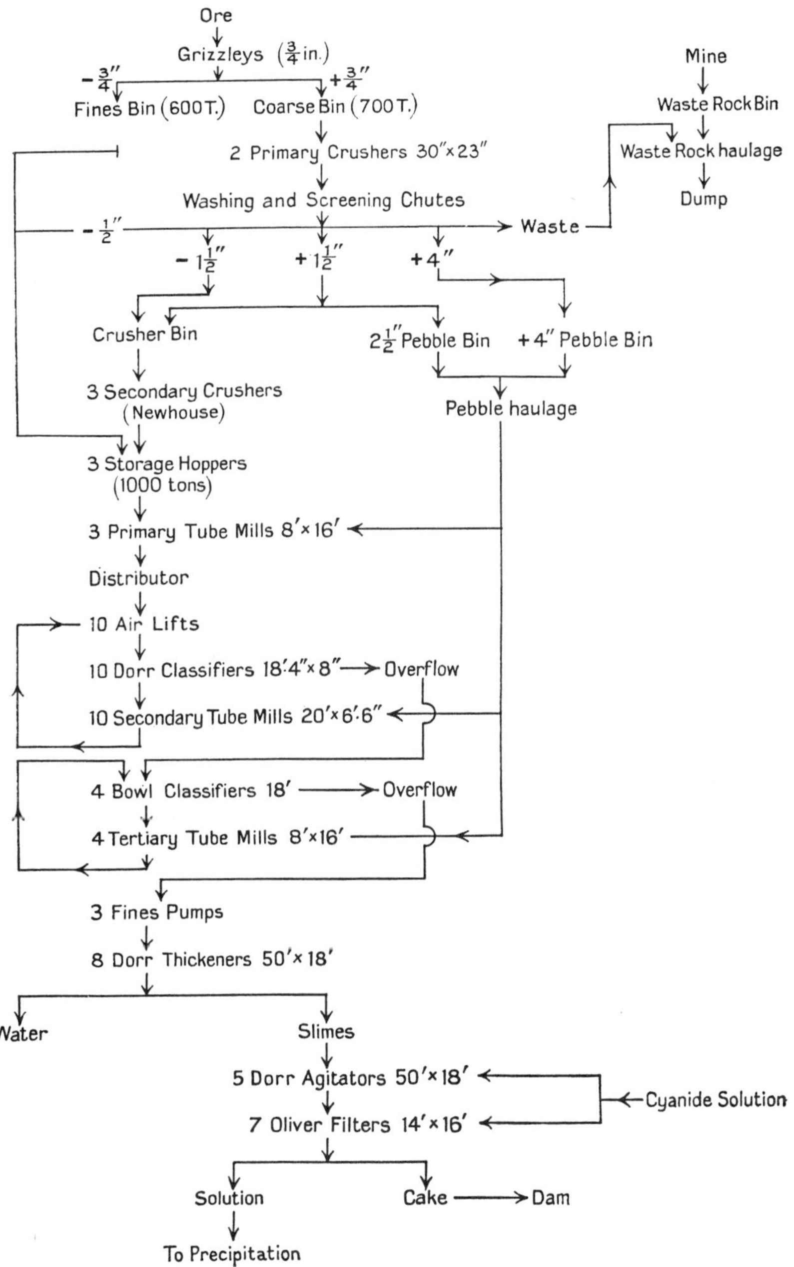

Fig. 195.—East Geduld Flow Sheet, 1927.

TABLE XLIII.

TUBE MILL CAPACITY.

Size, feet.	Feed per 24 hours, tons.	Speed, r.p.m.	Product tons of – 100 mesh per 24 hours.	
			Ordinary load.	Composite load.
22 × 5½	450	29	150	172
20 × 6½	900	25	230	265
16 × 8	1,300	21	...	...

The cost of tube milling depends on the degree of grinding. It is as follows :—

Per ton milled, . 3·7 to 15·0d.
Per ton of – 100 mesh product :
(*a*) All-sliming, . 15·5d.
(*b*) Other plants, . 11·3d.

Amalgamation.—The use of battery screening as coarse as 64 holes per square inch or coarser prevents plate amalgamation owing to the scouring of amalgam, and hence the use of amalgamated plates was discontinued. They were retained in some instances for a time in the tube mill circuit in the form of three stationary plates of the usual dimensions (4½ by 12 feet) per tube mill, but with a grade of 18 per cent. to prevent any banking of the thick pulp which flows over them. Concentration by corduroy followed by barrel amalgamation instead of amalgamation on plates has been described on p. 236.

Concentration.—The pyritic content of the ore is about 1·72 per cent. Concentration of this material has been effected by means of the Johnson concentrator.

In 1933, at the West Rand Consolidated mine, a large jig with a conical bottom was placed at the end of each tube mill to recover the pyritic concentrate from the mill effluent. This concentrate flows over corduroy tables to separate free gold and is then re-dressed on James' tables. The concentrate from these is ground in ball mills to 325 mesh and cyanided. The residue is sold for sulphuric acid manufacture, but it is suggested it might be used for the production of elemental sulphur.

Progress is being made with the flotation of the pyritic portion of banket ore and a few machines have been installed. Experimental results hitherto have been disappointing.

Classification.—The separation of sand and slime has reached a high degree of efficiency, with the result that the sand can be treated in the vat where it is collected, without the cost of transfer and with a much reduced plant, whilst the clean sand residue is fitted for the sand-filling of old mine workings. Incidentally, the classification permits of lime being added to the ore entering the mill bins, instead of being crushed separately in a ball mill for subsequent addition to the slime-pulp.

Safety or return sand cones receiving pulp from the common overflow launder of the main classifiers are found to be useful.

[*To face p.* 427.

[*By courtesy of the Institution of Mining and Metallurgy.*

General View of Modderfontein B Cyanide Plant.

Automatic regulators are in satisfactory operation on some mines. They regulate the ratio of water to sand in the underflow of diaphragm cones which are engaged in classifying sand and slime. The Caldecott diaphragm cone, however, has been generally replaced by Dorr classifiers in de-watering finely-ground pulp for tube milling.

Cyaniding.—The sand formerly constituted not less than 50 per cent. by weight of the ore milled. Prentice states[1] that at the Central Mines Group the amount of sand is about 48 per cent. of the total ore milled. Wartenweiler[2] gives figures as low as 10 per cent. and 20 per cent. for some of the older mines where fine milling has been very much extended. In leaching, the W/S ratio is 1 : 1 or 2 : 1, and there are usually 3 to 7 days' treatment, but in one case up to 20 days. The maximum solution strength is 0·06 per cent. KCN, and the alkalinity 0·005 per cent. CaO. The slime agitation is for 6 to 30 hours in 0·01 to 0·025 per cent. KCN, and the W/S ratio is 1·20 : 1.

Slime is collected continuously in thickeners. Its treatment by decantation is still practised, but it is dealt with in some plants by vacuum filtration subsequent to cyanidation. The last six mills designed on the Rand did not include sand treatment units, but were confined to all-sliming. Cyanide consumption is from 0·15 to 0·50 lb. per ton, the higher figure being for plants grinding in solution. Continuous filtration on drum filters is usual. In filtration, the duty in the Butters filter is 4·5 tons per leaf per day, and in the Oliver 0·65 ton per square foot per day. In both settlement of slime and precipitation, working solutions are warmed by the waste steam heat from the mill engine power plant.

A general view of the Modderfontein B Cyanide Plant, one of the new mills on the Far East Rand, is shown in the accompanying Plate. The tube mills are shown in the foreground.

Precipitation.—Lead-coated zinc shavings are still in use in older mills for precipitation. Zinc dust precipitation, using the Crowe vacuum process for de-aeration, is used in new mills and has also replaced the zinc shaving process in some of the older ones. The chemical method of de-aeration by means of fine iron or pyritic sand, followed by zinc precipitation, is also used. About 5 lbs. of lead acetate or nitrate are added per 1,000 tons of solution.

Table XLIV shows comparative operating data when using zinc shavings and zinc dust.

TABLE XLIV.

PRECIPITATION DATA (Wartenweiler).

	Precipitated Solution Assay, dwt. per ton.	Solution to Precipitation.		Zinc Consumption, lb. per ton milled.	Solution Precipitated per ton treated, tons.
		Per cent. KCN.	Per cent. CaO.		
With shavings—					
Sand solution, . .	0·02	0·028	0·005	} 0·14	1·3
Slime, . . .	0·015	0·011	0·005		2·0
With dust—					
Mixed, . . .	0·015	0·016	0·02	0·05	1·5
From ore-product pulp,	0·020	0·014	0·018	0·06	1·8

[1] *Trans. Inst. Min. Met.*, 1935, **44**, 511.
[2] *Trans. Amer. Inst. Min. Met. Eng.* 1932, **112**, 761.

Zinc-gold precipitate is treated with acid, calcined and smelted in large (No. 100) clay-lined graphite pots which are placed, 25 to 30 at a time, in a reverberatory furnace.

Wartenweiler says that short-cuts by the elimination of acid treatment, or by direct smelting, have not proved very satisfactory.

The overall recovery of gold on the Rand varies from 92 to 98 per cent. The 1934 figure for the Central Mining Rand Mines group was 95·2 per cent. Details of extraction by cyaniding are as follows :—

TABLE XLV.

	All-sliming.	Sand and Slime Treatment.		
		Sand.	Slime.	Combined.
Percentage of ore treated, .	100·0	47·4	52·6	100·0
Original value, dwt. per ton,	3·854	2·884	1·764	2·295
Residue value, ,,	0·323	0·324	0·141	0·228
Percentage extraction, .	91·6	88·8	92·0	90·1
Grading, + 48, . .	0·0	5·0	0·0	2·4
— 48 + 100, .	6·1	46·2	0·8	22·3
— 100 + 200, .	28·1	35·4	12·3	23·2
— 200, . .	65·8	13·4	86·9	52·1

In the all-sliming plants the original value of the pulp for cyaniding is relatively high, as only 23·4 per cent. of the gold is recovered on corduroy, due to refractory ore and limited corduroy area. In the sand and slime plants the corduroy plant is more extensive and recovery rises to 52 per cent.

In Table XLVI the average costs for 1934 in the mills attached to one of the larger Rand groups of mines are given.

TABLE XLVI.

CENTRAL MINING RAND GROUP, COSTS FOR 1934.

	Pence, per ton milled.	
	New All-sliming. Crushers, Tube Mills and Ore-pulp Treatment.	*Older Sand and Slime.* Crushers, Stamps, Tube Mills, Sand and Slime Treatment.
Sorting and transport of waste, .	1·4	1·4
Crushing, ...	4·0	1·8
Transport from crusher to mill, .	0·6	1·4
Stamp milling, ...	...	7·4
Tube milling, ...	14·6	6·3
Concentration and amalgamation, ...	0·9	1·4
Cyaniding (includes smelting and sand disposal), . .	12·3	13·6
Assaying, ...	0·4	0·5
Total, . .	34·2	33·8

Homestake Mills, South Dakota. South Mill, 1933[1] (see Fig. 196).

The ore at this mine is of uniformly low grade, containing about 4 dwt. gold per ton. When derived from deep levels it consists of chlorite or ferruginous hornblende, quartz, pyrite and pyrrhotite. Carbonates of calcium, magnesium and iron cause reactions with lime, and interfere with the use of an alkaline circuit. In some parts arsenopyrite, garnet and mica are present.

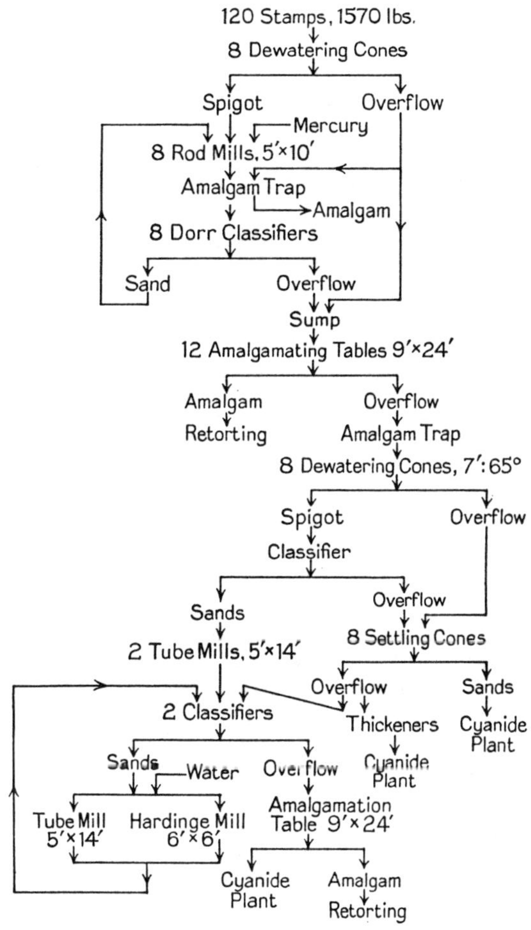

Fig. 196.—Homestake Mill Flow Sheet.

The sulphides and the coarser oxidised ore yield their values at all sizes more readily than either the gangue or the unoxidised material.

The ore is first crushed underground in jaw breakers and passes to Symons cone crushers set to a 2 inch opening. Stamp mills are used because the ore is hard and tends to break into large lumps, but they are retained merely as a primary machine to prepare the ore for the rod mill grinders. There are

[1] *U.S. Bureau Mines, Inf. Circ. No. 6408*, 1931; *Can. Mng. Journ.*, 1933, 54, 299.

120 stamps weighing 1,570 lb. each. Each set of 5 is driven by a separate 25 h.p. motor. The duty per stamp is 15 to 18 tons per day through steel wire screens with apertures of 0·295 to 0·838 inch.

Grinding.—The stamp mill product passes to a number of grinding units each consisting of (1) a dewatering cone, (2) a rod mill, working at 20·3 r.p.m., (3) a Dorr classifier in closed circuit with the rod mill. 30 per cent. of the stamp mill product overflows the dewatering cone, by-passing the rod mills and going direct either to the fine grinding plant via the amalgamation tables or to the discharge end of the rod mills to ensure proper dilution of the pulp there. The latter practice obviates the necessity of the addition of more water. The underflow from the cones is delivered to the rod mills. The initial rod-load is 23,500 lbs. Each rod is of 3 inches diameter and 10 feet long. The rods are made of 1·00 per cent. carbon steel and are hot-rolled and machine-straightened. The rod mill capacity is 190 to 260 tons per day.

Fine Grinding.—The overflows from the classifiers and the cones pass to amalgamation plates and thence to a second series of dewatering cones the underflow of which is treated by a classifier. The sand product of this is recrushed in rubber-lined tube mills and a conical mill. French pebbles are used as grinding agents. More amalgam plates follow the mills. The overflow from the classifier proceeds to cyanidation.

Amalgamation.—Mercury is added to the rod mills in measured quantities at 30 minute intervals and any amalgam is recovered from traps between the mills and the classifiers. The pulp from the traps travels along a launder (the bottom of which is covered by a silvered copper plate 7 feet × 12 inches) to the classifiers. The plate is replaced every four hours by a freshly cleaned and dressed plate. The classifier overflow is carried over amalgamated plates, 9 × 24 feet, covered with 3 ounces of silver per square foot. In the 24 foot length a small drop occurs at the 12 foot mark. The pulp finally runs to a large trap. A final amalgamation plate follows the tube mills in the fine grinding section.

Cyanidation.—Owing to the presence of ferrous compounds it is essential to aerate the solutions regularly and uniformly before cyaniding by forcing air through the drained charge. Sand and slime are separated in cones. The *sand* is sent for leaching through Butters-Mein distributors. The redwood vats for leaching are 44 feet in diameter and 10 feet 5 inches deep. They hold 670 tons of sand. The filter bottoms are of coconut matting, resting on 2 × 4 inch pine strips, covered by 10 oz. canvas duck and caulked on the outer rim. There are four sluicing gates at the level of the filter cloth and a central circular discharge aperture. There are also valved openings below the filter which deliver to the various solution tanks. Leaching is by downward displacement. Solutions are delivered by pipes at four points on the surface of the charge. Air is admitted through the bottom at a pressure of 8 lbs. per square inch, which is just sufficient to overcome the sand pressure. For successful aeration there should be no slime pockets, thorough draining and carefully controlled air-pressure. Lime is fed to the sand as it enters the vat. " Protective alkalinity " is maintained at 0·1 lb. CaO per ton—a very low figure.

Strong cyanide solution contains 1·2 lbs. NaCN per ton and weak solution 0·4 to 0·8 lb. per ton. The solution is passed through two sand charges before precipitation.

A treatment cycle consists of three periods, each divided into the three operations of draining, aerating and leaching. Finally the charge is washed and discharged by sluicing. Barren and newly precipitated solutions are used in the later stages of treatment of a charge, then strengthened in cyanide and used again, without precipitation, in the earlier stages of treatment of a subsequent charge. A typical sand charge will leach at the rate of $1\frac{1}{2}$ to 2 inches per hour.

CYCLE OF OPERATIONS IN LEACHING.

Filling,	8 hours.
First draining,	16 ,,
First aerating,	14 ,,
Second draining,	12 ,,
Second aerating,	11 ,,
Strong solution leaching,	14 ,,
Third draining,	10 ,,
Third aerating,	10 ,,
Weak solution leaching,	12 ,,
Washing,	$29\frac{1}{2}$,,
Sluicing,	$1\frac{1}{2}$,,
Total,	138 ,,

The cost is about 16 cents per ton of sand.

In the treatment of *slime*, air is supplied to the charge, solutions are recirculated before precipitation and "protective alkalinity" is extremely low. Slaked lime mixed with water is added to the supply tanks. The slime pulp contains 99 per cent. of — 200 mesh.

The slime is treated in 34 flush-plate and distance-frame Merrill presses, each having 90 frames, 6 feet × 4 feet 4 inches, operating at 25 to 30 lbs. pressure and treating 24 tons of slime per charge. Each plate is covered by a light muslin cloth underneath a layer of cotton duck, which lasts about two years. These are treated with 0·7 per cent. HCl solution every three weeks. A complete cycle on the same lines as the sand treatment occupies about 9 to 10 hours. Internal channels in the presses have been adopted. Sluicing is automatic and is by a 3 inch pipe through a 6 inch passage opening into each frame at the centre of the bottom. For sluicing, water at 70 lbs. pressure is admitted into the space between the canvas-covered plates.

Permeability of the pulp rather than strength of solution determines the time of treatment. Extraction efficiency depends on the passage of sufficient solution strong enough to give good precipitation. About 0·10 to 0·20 per cent. thiocyanates are formed, which represents a loss of cyanide.

Occasionally trouble at all stages throughout the process is experienced owing to the presence of small quantities of organic material.

Precipitation and Refining.—The Merrill-Crowe process using zinc dust is employed. Sand plant solutions are not clarified, but those from the slime plant are. A small quantity of "Celite" and a drop of strong $Pb(NO_3)_2$ solution are added to the zinc dust, which is admitted on a belt feeder.

At the sand plant, "weak" and "low" solutions are pumped alternately through the same presses. The "low" solutions are wasted after precipitation. In the slime plant this is not practicable.

Barren solutions are tested hourly by the Dowsett colorimetric method, which shows as little as 0·02 dwt. of gold per ton of solution.

When full, the presses are blown with air for 2 hours and the cakes sent to the refinery.

The precipitate may be treated with acid, but this is not the usual practice. Normally it is partially dried, mixed with litharge, borax and coke, and briquetted. The briquettes are dried, charged on to a bath of lead in an oil-fired cupellation furnace of the English type, having a test 4 feet 1 inch × 3 feet 6 inches, made of 75 per cent. cement and 25 per cent. crushed limestone. A slag is drawn into slag pots, and the lead is then oxidised to litharge. A Doré bar, about 960 fine in gold and silver, remains and is afterwards melted in a graphite crucible.

Morro Velho, Brazil.

At Morro Velho (see Fig. 197, which indicates the replacement of blanket strakes by James tables) the ore is complex and contains pyrrhotite, which

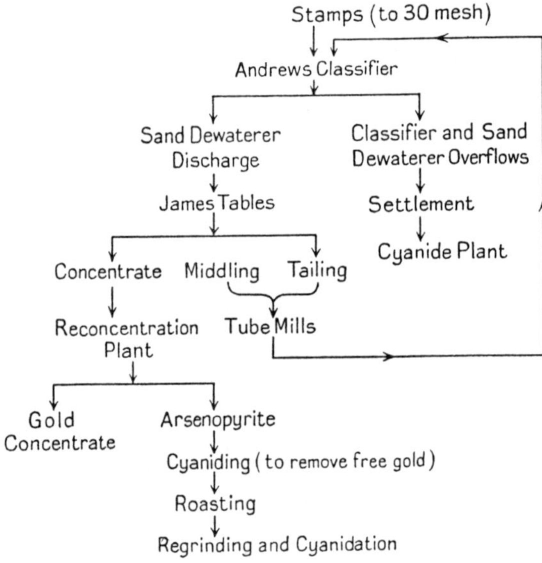

Fig. 197.—Morro Velho Flow Sheet.

is a vigorous "cyanicide." The ore contains an average of 12 dwt. gold per ton. The gold and silver occur as a natural alloy, containing 77 per cent. gold and 23 per cent. silver. The alloy is very fine and is generally associated with arsenopyrite as well as pyrrhotite. This is somewhat similar to the occurrence at Homestake.[1] Owing to the large amount of arsenopyrite at Morro Velho, the ore is not amenable to amalgamation, and as it consists mainly of readily floatable sulphides, flotation is not indicated.[2]

The reduction practice consists of concentration followed by cyanidation, with roasting of some of the concentrates. The ore is crushed in jaw breakers, whence it passes to 140 Californian stamps (850 to 1,000 lbs. each)

[1] *Bull. U.S. Geol. Surv.*, 1924, No. 765, p. 43.
[2] *Trans. Inst. Min. Met.*, 1933, 42, 189, 262.

where it is reduced to 30 to 60 mesh. The duty is 5 tons per head per 24 hours.

A small Andrews kinetic elutriator removes finely divided sulphides, which are dewatered and then concentrated. Until 1935, concentration on canvas tables was used (see p. 235). James tables have now been adopted in place of the strakes in order to get a better separation of arsenopyrite.

The concentrate is smelted direct in graphite pots, using the following charge :—

	Parts by Weight.
Concentrate,	100
Sand,	50
Borax glass,	20
MnO_2,	40
Saltpetre,	45

The bullion assays gold 765, silver 225 parts per thousand.

The mixed tailing is classified in spitzkasten. The overflow from the V-boxes goes direct to the cyanide plant, and the underflow is thickened in diaphragm cones, from which the underflow is sent to the tube mills. The dewaterer overflow joins the elutriator overflow. The tube mill discharge is concentrated on strakes to remove fine gold.

The pulp going to the cyanide plant is subjected to settlement and divided into sands and slimes. The settled slimes are agitated with air forced in at 15 lbs. pressure through perforated pipes lying on the bottom of the vats.

The sand ($W/S = 1:3$) is stirred by mullers for 12 hours in 0·25 per cent. NaCN solution, while compressed air at 15 lbs. pressure is admitted through a perforated peripheral pipe at the bottom. Longer periods result in reprecipitation of the gold by the pyrrhotite. Lime is added at the rate of $\frac{1}{2}$ lb. per ton to the tube mill feed. Higher alkalinity increases the value in the tails.

Moore leaf-filters are used for the filtration of the cyanide pulps. Sand and slime pulps are sometimes mixed to facilitate filtration. Only the original solution is clarified by settlement and then in a plate and frame press.

Precipitation is in zinc boxes after de aeration. The zinc is brittle owing to impurities, and cannot be dressed by ordinary methods. Zinc consumption is 0·25 lb. per ton of ore treated. At the clean-up all zinc is treated with acid after being roughly washed. Treatment with caustic soda was given when acid was expensive.

The slimes are melted in the following charge :—

	Parts by Weight.
Slime,	100
Soda-ash,	60
Borax glass,	35
Sand,	50
Saltpetre,	33

The bullion assays gold 700 to 850, silver 130 to 300 parts per thousand. It is refined by the chlorine process.

Flin Flon.

Cyanidation (Low Oxygen Content in Solutions).

The cyanide plant at the Flin Flon Mine,[1] which treats flotation concentrates, is unique in three respects: (1) Agitation is at a pulp dilution of 65 per cent. solids and lasts only 2½ to 4½ hours. (2) The oxygen concentration in the solution is very small. (3) All solution which has been in contact with the ore is discarded after regeneration, and there is no circulation. The low W/S ratio results in few agitators being necessary and removes the necessity for thickening before filtration. The short time of contact with cyanide solution is essential because of progressive cyanide consumption, and the low oxygen content results from the highly reducing nature of the ore. A higher concentration of oxygen causes increased cyanide consumption.

The barren solution from precipitation is acidified with sulphuric acid and passed through two dispersing towers in series. The HCN gas liberated is removed by a current of air and absorbed in lime solution, which is then used in the agitators for treating more ore. The residue is treated for the recovery of copper.

A full description of the Flin Flon concentrator and cyanide plant is given by Lowe.[2]

Alaska Juneau (1927).

Jaw Breakers—Sorting—Rolls—Ball Mills—Sizing—Sand and Slime Treatment—Table Concentration.

The ore consists of quartz carrying galena, chalcopyrite and other sulphides. These are friable, whereas the free quartz is very hard. The ore is first crushed in jaw crushers set to 8 inches, and the broken material passes over a 3½ inch grizzley on to picking belts. The picked material goes to gyratory crushers and the undersize from the grizzlies joins the crushed product on 2½ inch impact screens. The oversize on the grizzley is recrushed. The oversize from the screens goes to a series of rolls and then to ball mills, whence the product is screened in short trommels attached to the end of the mills. The oversize then goes to waste, as it is barren quartz which has resisted crushing. The undersize from the trommels, about ¼ inch material, is ground in tube mills, the daily throughput here being only about half the original feed. The tube mill product is concentrated and classified on two sets of tables into concentrates, sands and slimes. The sands and slimes are cyanided. The concentrates go through two succeeding concentrations, and finally three products are obtained: (1) Heads (coarse gold), which fall into an amalgamating pot. (2) Fine gold and galena, which go to an amalgamating barrel. (These two products (1) and (2) contain 80 per cent. of the gold.) (3) Galena carrying 20 ozs. gold per ton sent to a smelter. The tails from the last tables are returned to the circuit.[3]

---

[1] *Min. Mag.*, 1935, 52, 376.
[2] *Can. Inst. Min. Met.*, 1935, 38, 163.
[3] Spalding, *Kolar Gold Fields Min. and Met. Soc.*, 1931, 6, 4.

Lake Shore Plant, 1932.

(a) *All-sliming.* (b) *Flotation of Cyanide Tailing.* (c) *Cyanidation of Flotation Concentrate.*

The treatment is as follows :—(*a*) All-sliming, cyanidation of the bowl classifier overflow, with primary and secondary agitation and thickening, followed by filtration of the thickened pulp. (*b*) Flotation of the cyanide tailing and preparation of a concentrate. (*c*) Very fine grinding in tube mills, intense agitation and cyanidation of the flotation concentrate in an independent cyanide unit of the mill. (See Fig. 198).

To free the gold in the flotation concentrate, extremely fine grinding is necessary, 60 per cent. of the pulp being brought down to 10 microns. Fortunately from an economic point of view this is done only on some 0·75 per cent. of the mill feed. About 100 lbs. of lime per ton are added to the tube mill and a cyanide strength of 0·075 per cent. KCN is used.

Waihi Mine, New Zealand, 1925.[1]

The ore is refractory, containing 2½ to 7 per cent. of sulphides, chiefly pyrite, associated with small quantities of lead, zinc, copper and silver sulphides. The gold is finely divided. The ratio of silver to gold is about 7 : 1.

The whole of the crushing is done in water. The ore is first broken down in gyratory machines, then passes to stamp mills, and afterwards to classifiers. The duty per stamp is 5·8 tons of −10 mesh material, and the coarse product from 15 stamps goes through one tube mill. After regrinding, 82 per cent. of the ore passes a 200 mesh sieve, and proceeds to concentrating tables. The sulphide thus recovered is reground in a tube mill and then agitated with 0·1 to 0·15 per cent. KCN solution. The time of agitation and the solution strength are dependent on the silver content, which may reach 50 ozs. per ton of concentrate. About 98 per cent. of the gold and 96 per cent. of the silver are thus extracted. The tailing from the concentrators is classified ; the sands are treated by percolation with cyanide solution, and the slime by agitation and vacuum filtration.

The gold slime from precipitation is smelted in oil-fired furnaces. The bullion is then cupelled and finally refined electrolytically.

Porcupine District.

The accompanying flow sheets indicate three schemes which have been adopted for the treatment of gold ores in the Porcupine district. The ores in all three cases are similar in many respects and arise in like geological formations.

It will be noticed that at *Hollinger* (Fig. 199) the general scheme is crushing, grinding in cyanide solution, classification, table concentration of the sulphides for additional grinding and more intensive cyanidation, agitation and two-stage filtration. At *Dome* (Fig. 200) where the ore is much richer in gold, blanket concentration is followed by cyaniding, and at *McIntyre-Porcupine* (Fig. 201) flotation is followed by cyaniding of all products.

---

[1] *Trans. Inst. Min. Met.*, 1925, 34, 469.

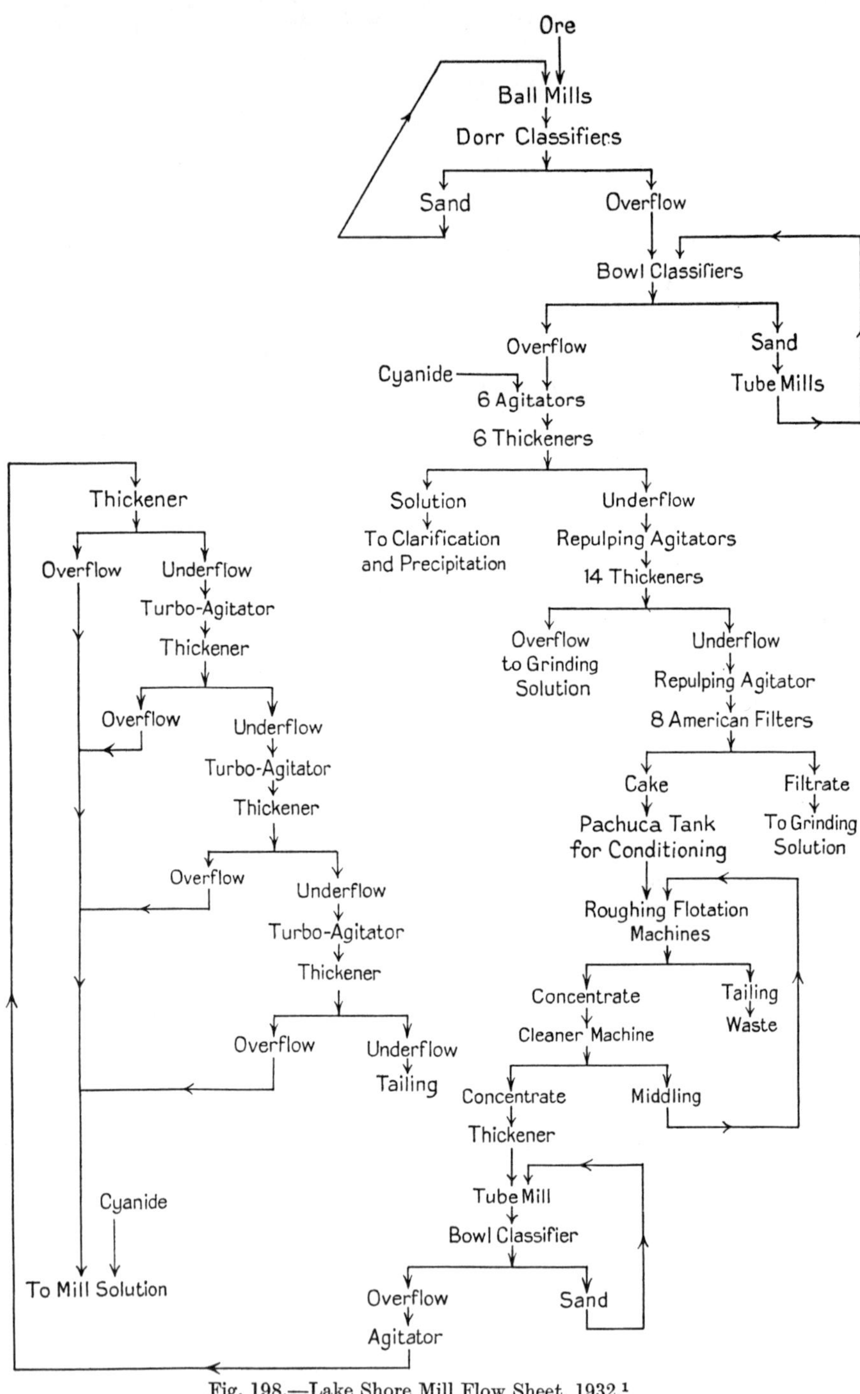

Fig. 198.—Lake Shore Mill Flow Sheet, 1932.[1]

[1] *Trans. Amer. Inst. Min. Met. Eng.*, 1934, **112**, 649.

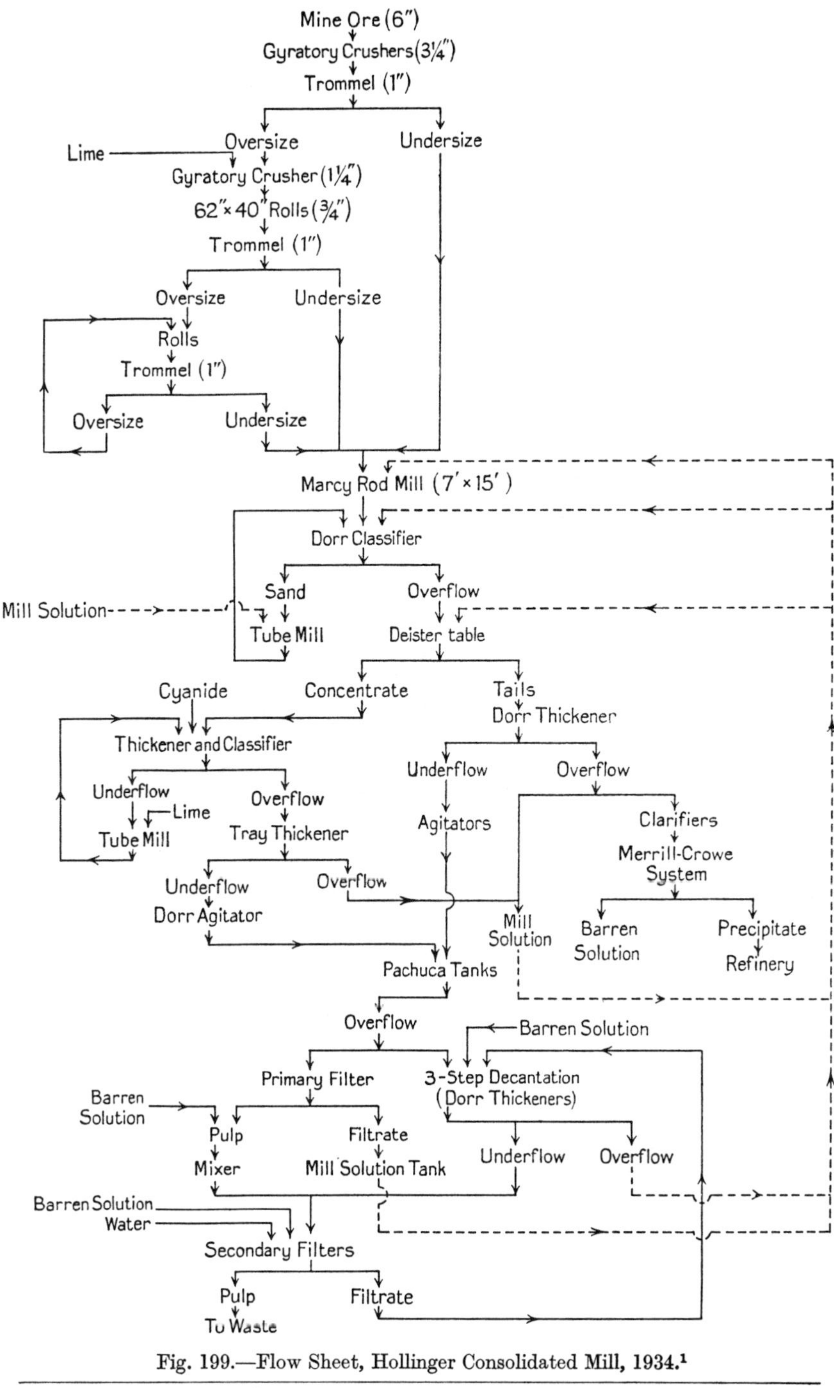

Fig. 199.—Flow Sheet, Hollinger Consolidated Mill, 1934.[1]

---

[1] *Trans. Amer. Inst. Min. Met. Eng.*, 1934, **112**, 524.

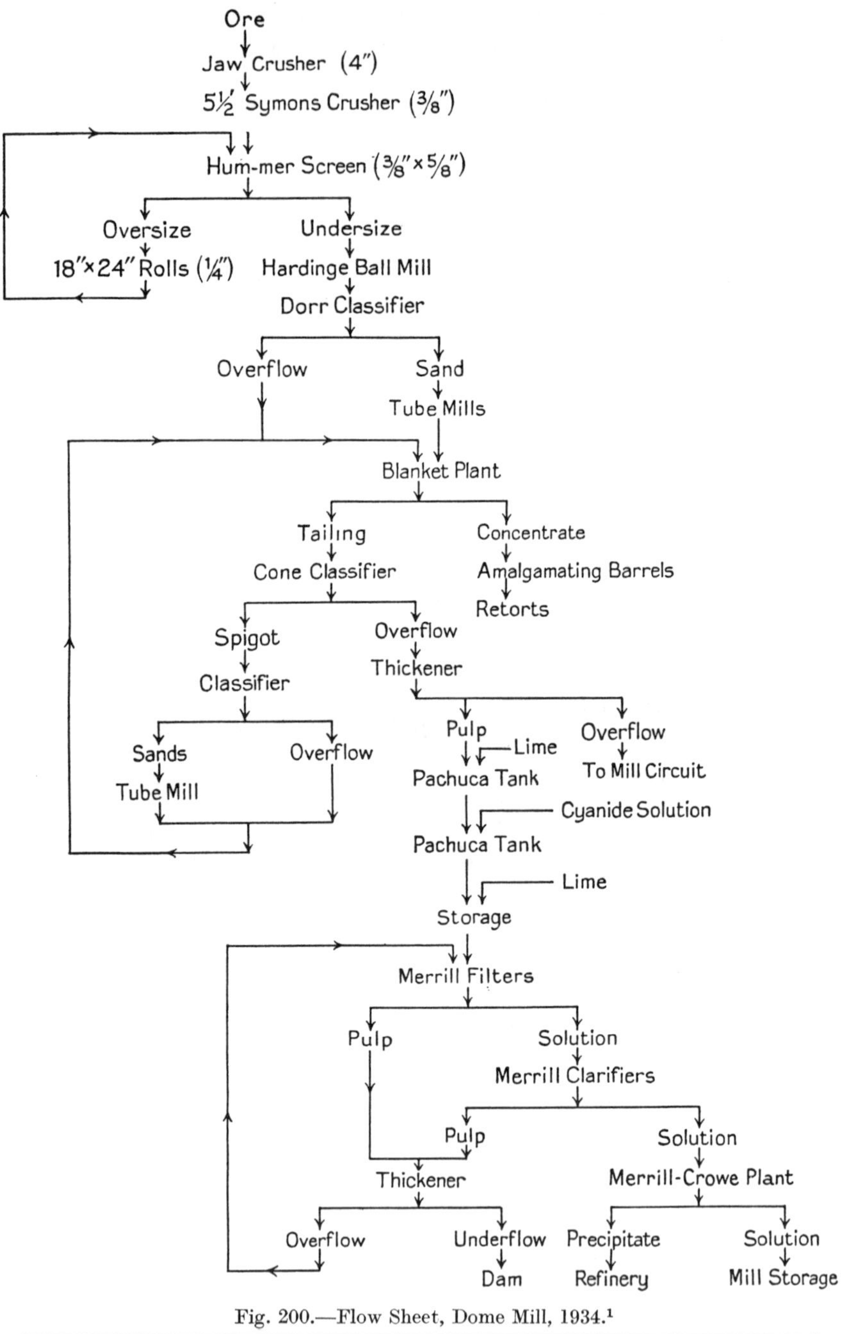

Fig. 200.—Flow Sheet, Dome Mill, 1934.[1]

[1] *Trans. Amer. Inst. Min. Met. Eng.*, 1934, **112**, 621.

The treatment at *Dome* (see Fig. 200) consists of crushing in jaw crushers and then in Symons crushers, screening, grinding, classification, concentration on blanket tables, amalgamation of concentrate and classification of tailing, reclassification, agitation, precipitation. The blanket plant is newly built.

The feed to the blanket tables amounts to 3,500 tons daily. It is unclassified product from the primary and secondary grinding circuits (60 per cent. — 200) (W/S = 5/1). There are 28 tables, each $4\frac{1}{2} \times 6$ feet, with an inclination of $1\frac{3}{8}$ inches per foot. The pulp is fed to each table by a flexible pipe. The blankets themselves are 28 inches × 5 feet to allow for shrinkage during use. The high cords of the corduroy are towards the flow of the stream. The corduroy is washed 15 times in 24 hours.

There are about three tons of concentrate per day. One ton is charged to each amalgam barrel (5 feet × 31 inches). Cast-iron shells without liners or wooden shells with white iron liners are used. Each barrel carries 500 lb. of $1\frac{1}{2}$ inch balls and rotates at 19 r.p.m. Lime is added and rotation started. After about 18 hours mercury is added, rotation continued for $1\frac{1}{2}$ hours, and then the barrels are emptied. Mercury and amalgam are pressed and taken to the refinery for retorting. The blanket recovery is 76·1 per cent. of the total.

When carbonaceous material is present in the ore, an abnormal accumulation of froth is obtained in the agitators and thickeners, giving an increase in the value of the residue. The washed residue is diverted to two six-cell sub-aeration flotation units before it is discharged to waste. On account of the effect of cyanide remaining from the prior treatment, 95 per cent. of the material is depressed, and the values are recovered as a concentrate which is returned to the cyanide circuit.

Plant Details. McIntyre Porcupine.

(1) Total Operating Costs.

	Cents per Ton.
Crushing and Conveying,	10·70
Flotation,	36·33
Cyanidation,	27·34
Refining,	2·22
Assaying,	1·52
Mill Alterations,	0·41
Total,	78·52

(2) Steel Consumption.

	Flotation Circuit.	Cyanide Circuit.
Balls,	1·791 lb. per ton.	0·585 lb. per ton.
Liners,	0·034 ,,	Rubber-lined.

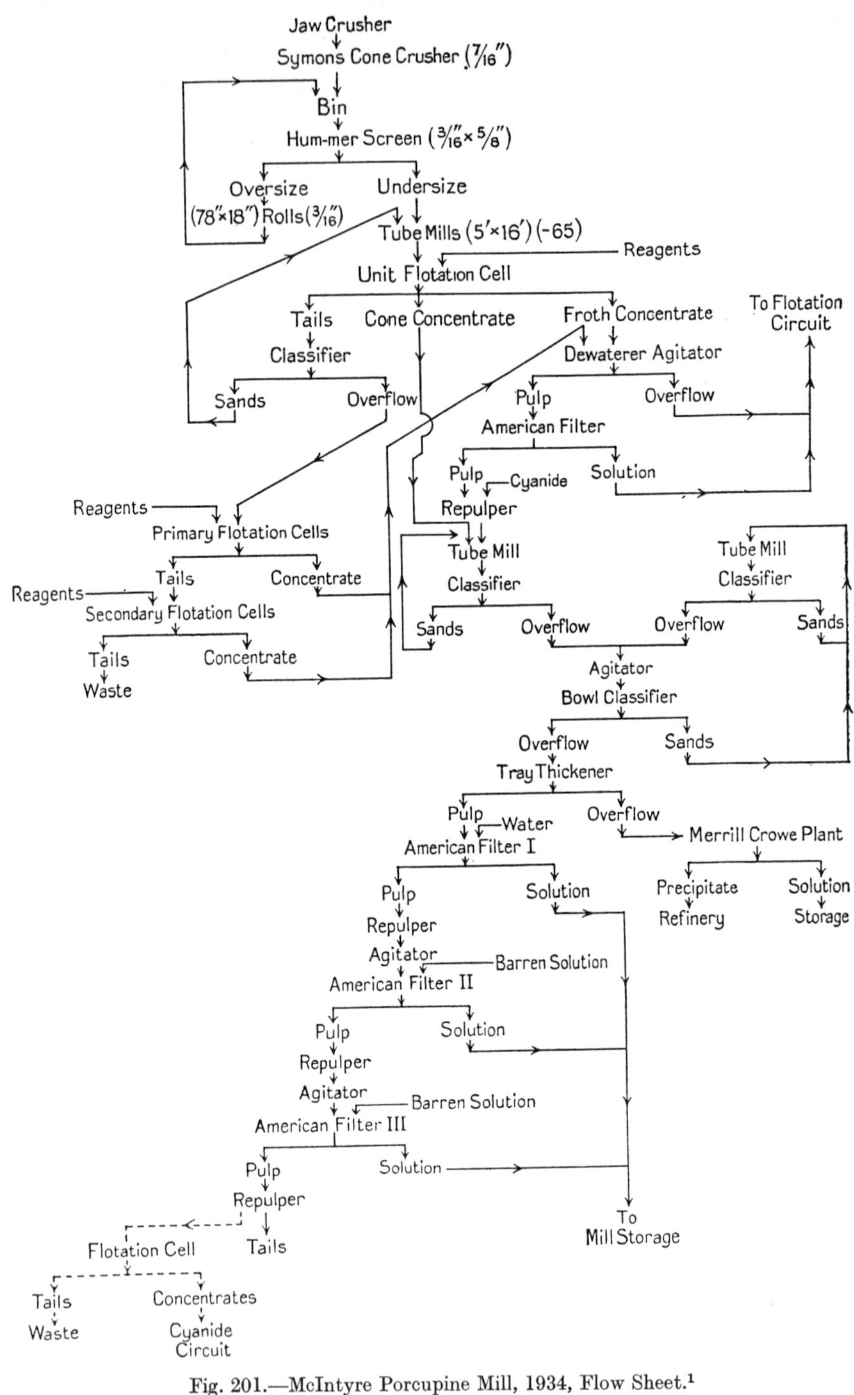

Fig. 201.—McIntyre Porcupine Mill, 1934, Flow Sheet.[1]

[1] *Trans. Amer. Inst. Min. Met. Eng.*, 1934, 112, 630.

(3) Flotation Concentrate.

	Per cent.		Per cent.
Pyrite, . .	64·89	Copper (chalcopyrite), .	0·61
Silica, . . .	15·41	Zinc (sphalerite), .	0·71
Fe_2O_3, . .	3·17	Lead (galena), . .	0·02
Al_2O_3, . . .	8·61	Gold, . 62·21 dollars per ton.	
CaO, . . .	3·19	Silver, . 0·01 oz. per ton.	
MgO, . . .	2·50		

(4) Raw Precipitate.

	Per cent.		Per cent.
Au, . . .	33·29	SiO_2, . . .	1·48
Ag, . . .	5·76	Al_2O_3, . . .	1·55
Cu, . . .	11·41	CaO, . . .	1·90
Pb, . . .	19·41	MgO, . . .	0·62
Zn, . . .	14·43	Na_2O, . . .	1·31
Fe, . . .	1·46		——
S, . . .	6·88		

(5) Consumption of Chemicals (Cyaniding).

Cyanide (NaCN), . . .	0·628 lb. per ton.
Lime, ...	1·106 ,,
Zinc dust, ...	0·084 ,,
Lead acetate, . . .	0·022 ,,

(6) Solution Strengths.

Cyanide, ...	3·5 lb. per ton.
Lime, ...	0·75 ,,

(7) Values in Residues.

Heads, ...	7·61 dollars.
Flotation tails, . . .	0·253 ,,
Cyanide residue, . . .	0·849 ,,
Soluble loss, ...	0·010 ,,
Combined tails, . . .	0·319 ,,

(8) Recoveries.

Flotation.	Per cent.	Cyanidation.	Per cent.
Unit cells, . .	76·00	Primary grinding, .	65·00
Flotation, . .	21·06	Agitation regrind, .	31·50
		Filters, etc., . .	2·20

Total recovery, 95·8 per cent.

Alkalinity of Flotation Solutions.[1]

	pH
Lake water,	7·80
Tube mill feed,	8·10
,, discharge,	8·90
Classifier overflow,	8·70
Primary cell feed,	8·75
,, discharge,	8·60
Secondary cell feed,	8·80
,, discharge,	8·40
Filtrate from concentrate,	8·30

Howey Mill, 1931.

Symons Crusher—Ball Mills—Classification—Merrill-Crowe Precipitation.

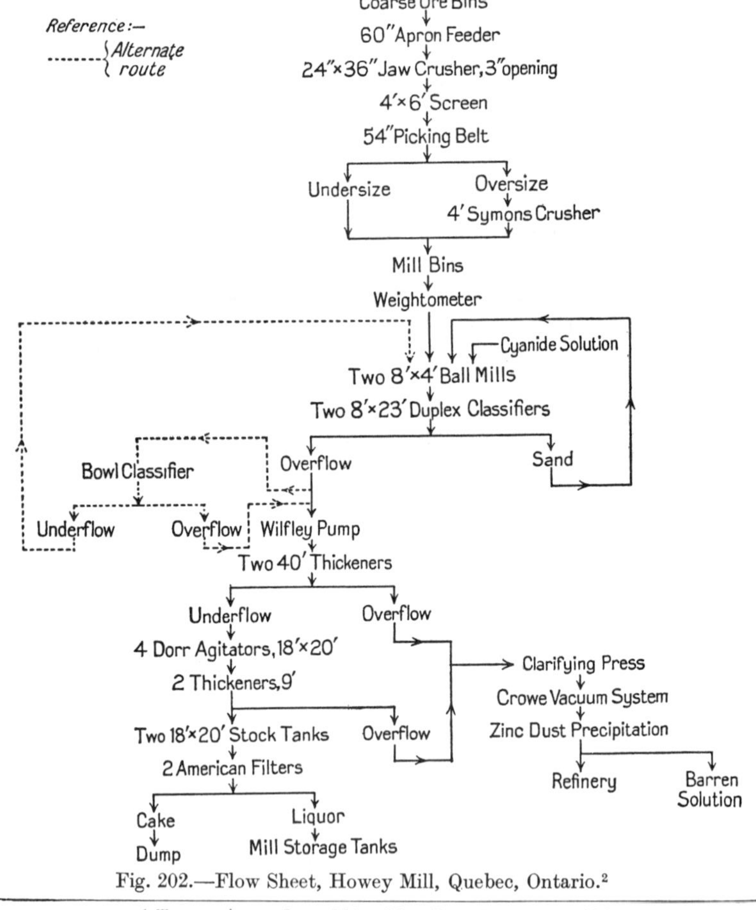

Fig. 202.—Flow Sheet, Howey Mill, Quebec, Ontario.[2]

[1] *Trans. Amer. Inst. Min. Met. Eng.*, 1934, **112**, 636.
[2] *Eng. Min. World*, 1931, **2**, 545.

The Howey Mine, Ontario, was only discovered in 1925, but now has a 500 ton plant fully working. The ore consists of quartz and pyrite, with other sulphides. The gold is occasionally coarse, but generally is fine, and evenly distributed. 95 per cent. of it is recovered by straight cyaniding a 40 mesh product. The flow sheet (Fig. 202) shows the simplicity of the plant. The classifiers are placed above the ball mills and not below, as is customory, thus allowing the mills to be on an unobstructed floor at a low level. Flotation may ultimately be employed to reduce the cost, which even now is only about 12/- per ton.

Cariboo, B.C.

Jaw Crusher—Ball Mill—All-sliming—Thickening—Agitation—Merrill-Crowe.

The ore consists of quartz containing free gold, and also pyrite having gold on its surface. The mill feed contains 15 to 20 per cent. of sulphides. Lime is added to the ore from the mine and the mixture crushed in a Traylor jaw crusher to 20 per cent. + 1 and 50 per cent. — 4 mesh. The product then goes to a Marcy ball mill operating in closed circuit with a Dorr classifier. Cyanide is fed into the classifier circuit. The overflow (23 per cent. solids) contains 2 per cent. + 65 mesh and 55 per cent. — 200 mesh material, while 40 per cent. is — 325 mesh. This is elevated to the primary thickener, the underflow from which, containing 50 per cent. solids, is sent to three agitators for 24 hours' treatment. The washing of the residue is carried out counter-current fashion, and the sludge is repulped between the thickener steps by means of air lifts. The underflow from the last thickener of the washing series is given a final wash with water on a Dorrco filter, the cake from which is repulped in an Oliver repulper and then discharged to waste. Gold solution overflowing the primary thickener is clarified and then precipitated by the Merrill-Crowe process. (See fig. 203).

Ore Treatment in Small Mills.

The mill for a small gold mine is an important and pressing problem, particularly in these days of intensive search for gold. The question is discussed from the point of view both of profit and production by J. A. Baker.[1] He says that, generally, gold ore containing values of 12/- to 14/- per ton can rarely be mined and treated profitably by underground methods. Ore assaying 20/- to 24/- per ton may pay on a scale of 100 tons per day under favourable circumstances and good management. It is more profitable when worked at the rate of 200 to 300 tons per day. In general 48/- to 60/- ore is required if the daily output is only 25 to 50 tons. There are exceptions to these figures, and they do not apply to tailing dumps, where grinding has already been done.

The quick-dropping Californian stamp mill with outside amalgamation plates, crushing to, say, 30 mesh, is still good practice for small mills. The amalgamation process gives easily marketed bullion and affords the simplest and cheapest method of recovering the free gold content of an ore. Not al

---

[1] *Min. and Met.*, 1932, 13, 209; see also Gardner and Johnson, *U.S. Bur. of Mines Inf. Circ.* 6800, 1934.

the gold, however, can be extracted by amalgamation. As an intermediate crushing stage the ball mill-classifier closed circuit is common, and a product of the desired mesh can thereby be obtained without undue ore grinding. At the Porcupine United Gold Mines, for example, the ore is treated in this manner. Amalgamation extracts 75 per cent. of the gold, of which 40 per

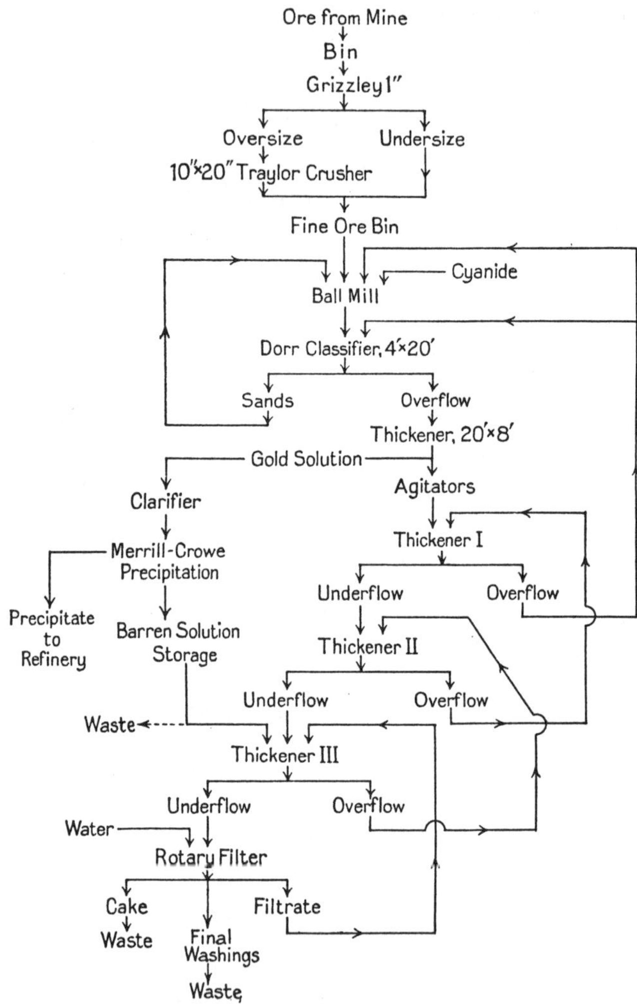

Fig. 203.—Flow Sheet, Cariboo Mill.[1]

cent. is caught on the plates and 35 per cent. won by barrel treatment of blanket concentrates. Grinding is in ball mills to 48 mesh. The overall recovery is 83·6 per cent. and on a 44/- ore the profit is estimated to be 5/- to 6/- per ton. Operating costs are shown below.

[1] *Min. Mag.*, 1935, 53, 57.

TABLE XLVII.

Operating Costs.

Mill heads, ... $11·00.
Amalgamation tailings, . . . $2·80.
Table tailings, ... $1·80.
Ball and liner consumption, . . 2 lb. per ton.
Power consumption, . . . 40 kWh. per ton.
Mercury consumption, . . . 0·5 oz. per ton.

	Labour, $	Power, $	Supplies, $	Total, $
Crushing, . . .	0·214	0·135	0·035	0·384
Grinding, . . .	0·429	0·180	0·162	0·771
Classifying, Screening, Refining, etc., . .	0·429	0·090	0·107	0·626
Miscellaneous, . .	...	...	0·070	0·070
Total, .	1·072	0·405	0·374	1·851

Cyanidation works most effectively and quickly on fine gold, although hours of contact are required in the dilute solutions used. Amalgamation and cyanidation are therefore supplementary. When all-cyaniding is used, the ore is usually crushed in cyanide solution and the use of the grinding mill-classifier combination retains the coarse gold automatically in the circuit until it is ground fine enough to be readily dissolved. Sand and slime can then be separated, the former to be treated by leaching and the latter by agitation.

All-sliming too is often practised, as at the Elko Prince Mine, where a hard brittle quartz, containing 2 per cent. sulphides, is treated. The ore contains 48/- to 100/- of gold and 10 oz. of silver per ton. It is worked in a 40 ton per day plant. A ball mill and tube mill in closed circuit with a classifier are used. The pulp contains 5 per cent. of + 150 mesh and 82 per cent. — 200 mesh material. After agitation it is washed by decantation in thickeners and then passed to a filter. The extraction is 97 per cent. of the gold and 87 per cent. of the silver.

Given an ore that responds well to cyanidation when coarsely ground, and carries 32/- to 36/- of gold per ton, the economics of a 100 ton plant appear attractive from the point of view of first cost and operating expenses.

The cyanide process can be employed effectively on low-grade surface ores mined by open cut or "glory hole" methods. Oxidised gossans are porous and often permit the leaching of ore crushed through a relatively coarse mesh.

Gayford [1] gives a summary of average figures for water and power consumption and for general plant on small mines.

The advent of chemical promotors, such as xanthates and dithiophosphates, has greatly helped in the development of the flotation of gold ores. Flotation is attractive to the small mill owner, on account of its simplicity and cheapness. With modern reagents, high ratios of concentration and high recoveries can be obtained. Floating auriferous sulphides is, in general, not widely

[1] *Trans. Amer. Inst. Min. Met. Eng.*, 1934, **112**, 558.

different from floating the same sulphides without the gold. Metallic gold, if not too coarse, is easy to float (see Chap. X).

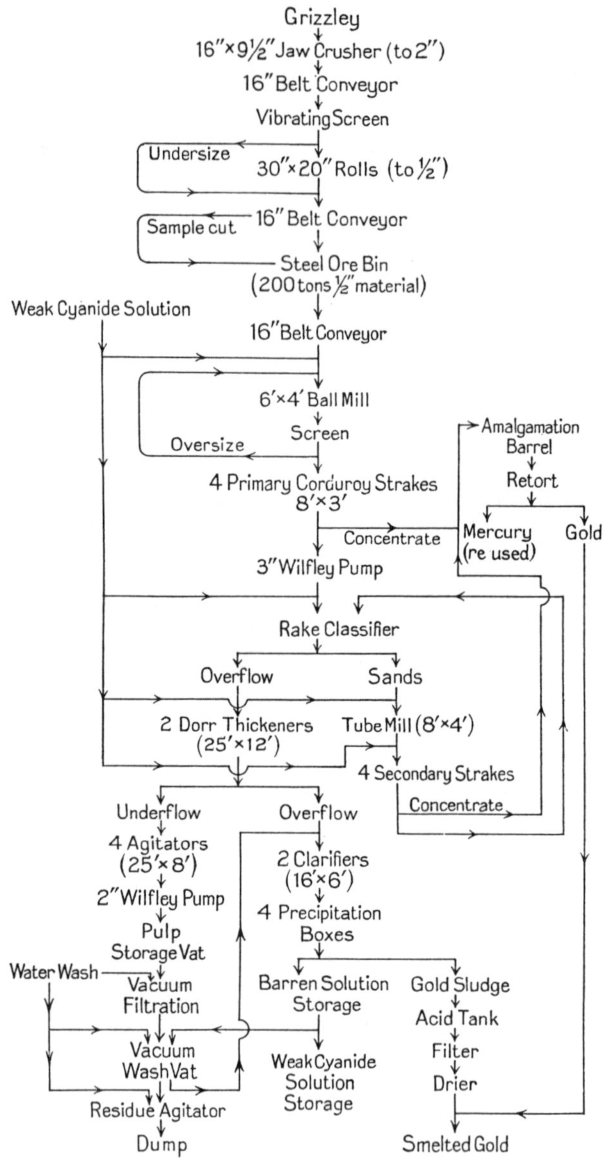

Fig. 204.—Flow Sheet, Golden Plateau Mill, 1933.[1]

While amalgamation and cyanidation ultimately produce a doré bullion that can usually be sold for almost its full value in gold and silver, flotation

[1] *Chem. Eng. Min. Rev.*, 1933, 25, 321.

produces a concentrate which has to be treated further, either by amalgamation, cyanidation or smelting. It thus incurs an extra treatment charge of 10 to 15 per cent. of the entire milling cost.

Some of the more common combinations of flotation and other methods of treatment are mentioned on p. 279.

Generally speaking, the smaller operator would do well to employ amalgamation where 60 per cent. of the gold can be so recovered, and then treat the tailing by flotation. Where direct flotation is indicated, the treatment of the concentrate needs careful study, unless it can be sent to a near-by smelter or a customs cyanide plant.

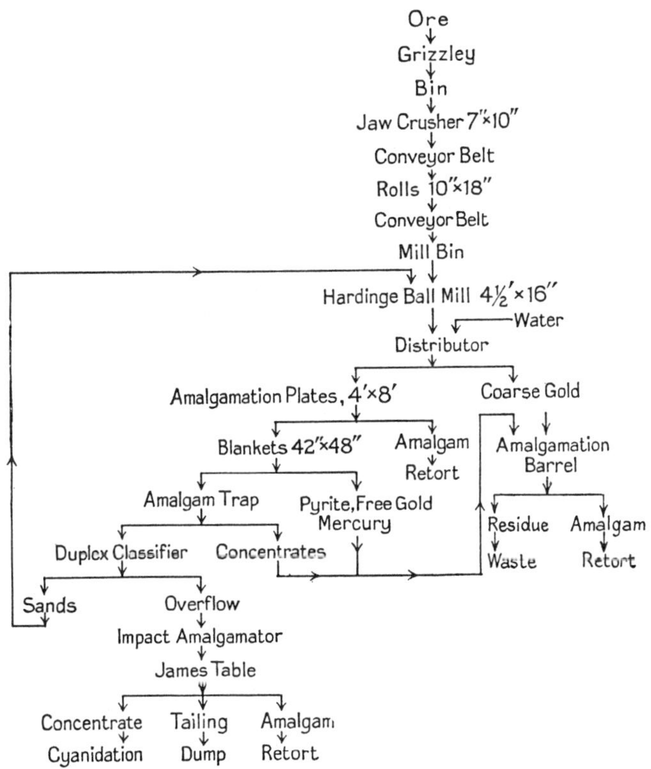

Fig. 205.—Flow Sheet, Porcupine United Mill.[1]

The Department of Mines at Ottawa, in a general warning against the indiscriminate application of flotation to gold ores, points out that the size of the mill may be a determining factor. Thus, while a large concern may be able to operate profitably on the dual process flotation-cyanidation, a small mill cannot do so and may reasonably be expected to run better on an all-cyanide treatment.[2]

The flow sheets appended (Figs. 204-207) indicate typical processes

[1] *U.S. Bur. Mines Inf. Circ.* 6800, 1934.
[2] *Min. Mag.*, 1931, 45, 369.

for different classes of ore available in relatively small quantities. Fig. 204 is the flow sheet for the new mill at Golden Plateau (1933).

Typical Small Amalgamation-Concentration Mill.—For small plants of about 25 tons per day capacity, two-stage crushing is unnecessary and uneconomical. Gyratory crushers are also unsuitable. The flow sheet of Porcupine United 25 ton plant, 1929, is given in Fig. 205.

The flow sheets of a 35 ton flotation plant (Fig. 206) and of the 50 ton

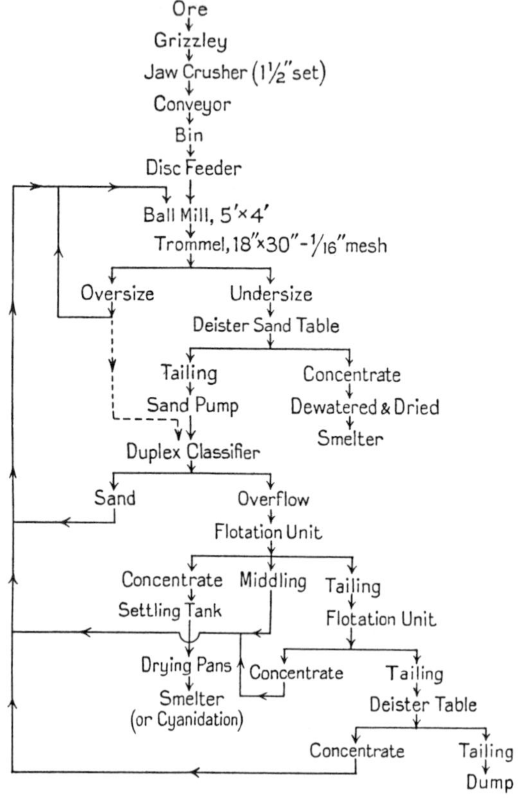

Fig. 206.—Flow Sheet of 35 Ton Gold Flotation Plant.[1]

Telluride Mill, Oatman, Arizona (Fig. 207) are appended.

Fig. 208 is a diagram of a plant to handle 150 tons of gold ore per day. It embodies two Blake crushers (20 × 10 inch) which reduce the ore to 3 inch size. The ore then passes to a battery of 2,200 lb. stamps, followed by amalgam tables and blanket strakes. The tailings from the latter are ground in two tube mills to 100 mesh. Parral tanks are used for agitation, which occupies 10 hours. The strake concentrates are agitated in Dorr agitators which are equipped for decantation. The pregnant solution is treated in a Merrill-Crowe precipitation plant.

---

[1] *U.S. Bur. Mines Inf. Circ.* 6800, 1934.

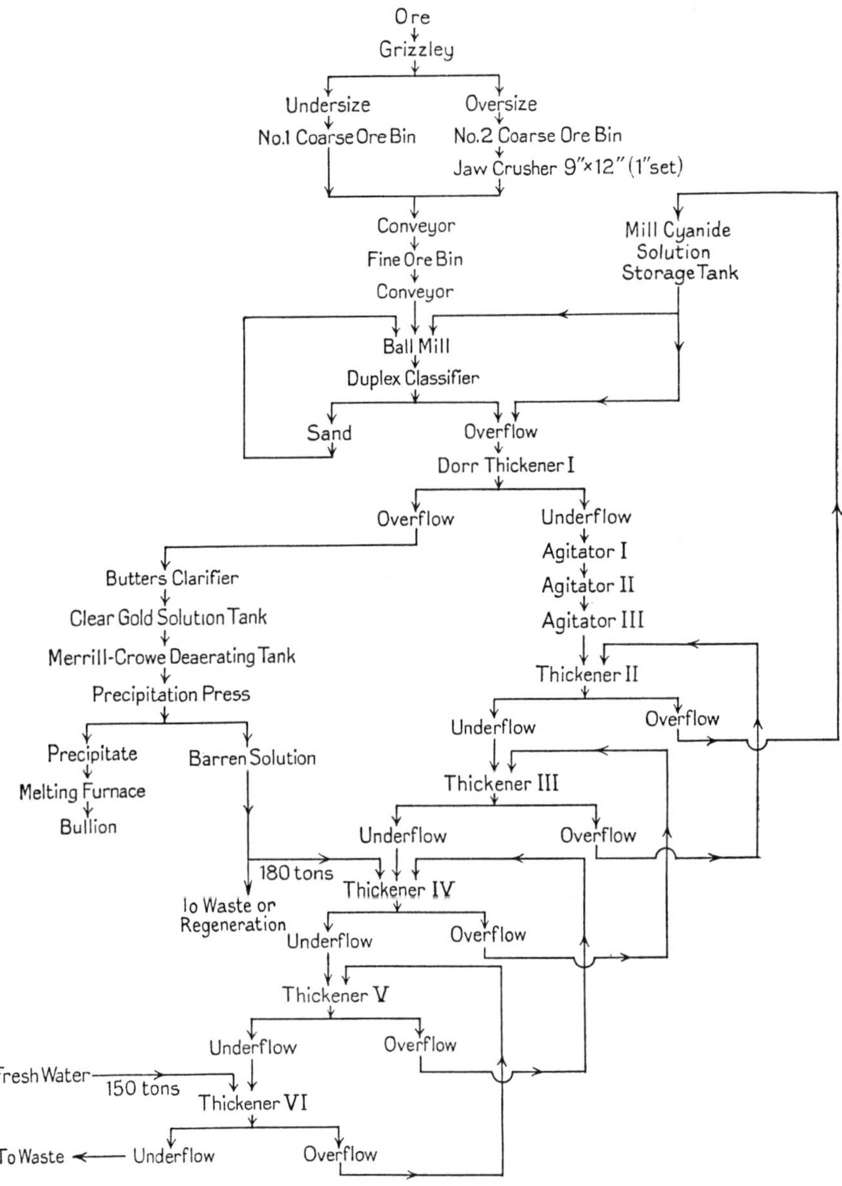

Fig. 207.—Flow Sheet of 50 Ton Telluride Mill, Oatman (Ariz.).

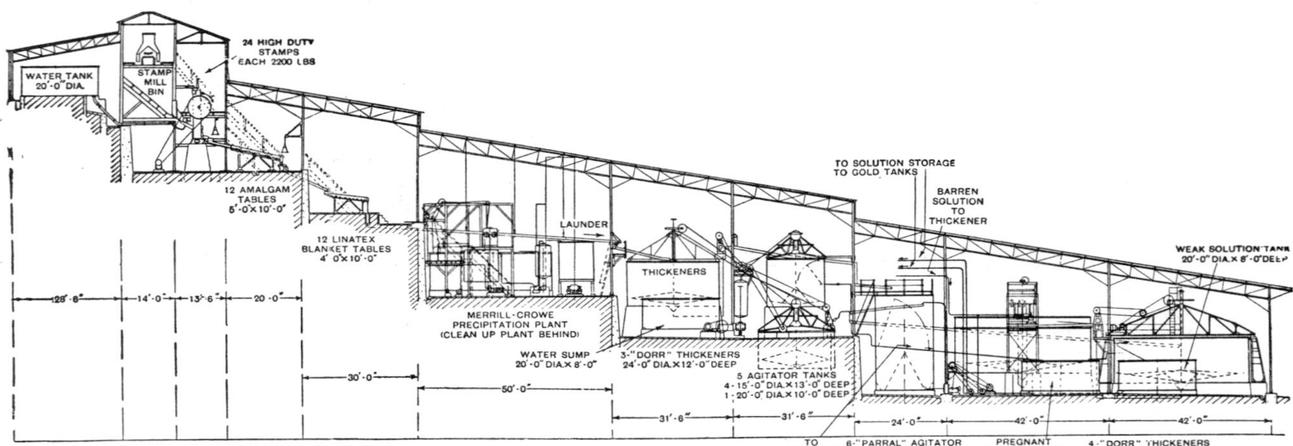

Fig. 208.—150 Ton Cyanide Plant, with Amalgamating Plates to catch Free Gold, and Blanket Strakes to remove Cyanicides in the form of Sulphide Minerals, followed by Agitation and Counter-current Decantation. (*Fraser and Chalmers.*)

Fig. 209 shows a plant for dealing with 100 tons of gold ore per 24 hours. It embodies coarse, intermediate and fine crushing in Blake crushers, Symons crushers and tube mills, respectively. Agitation is carried out in Pachuca

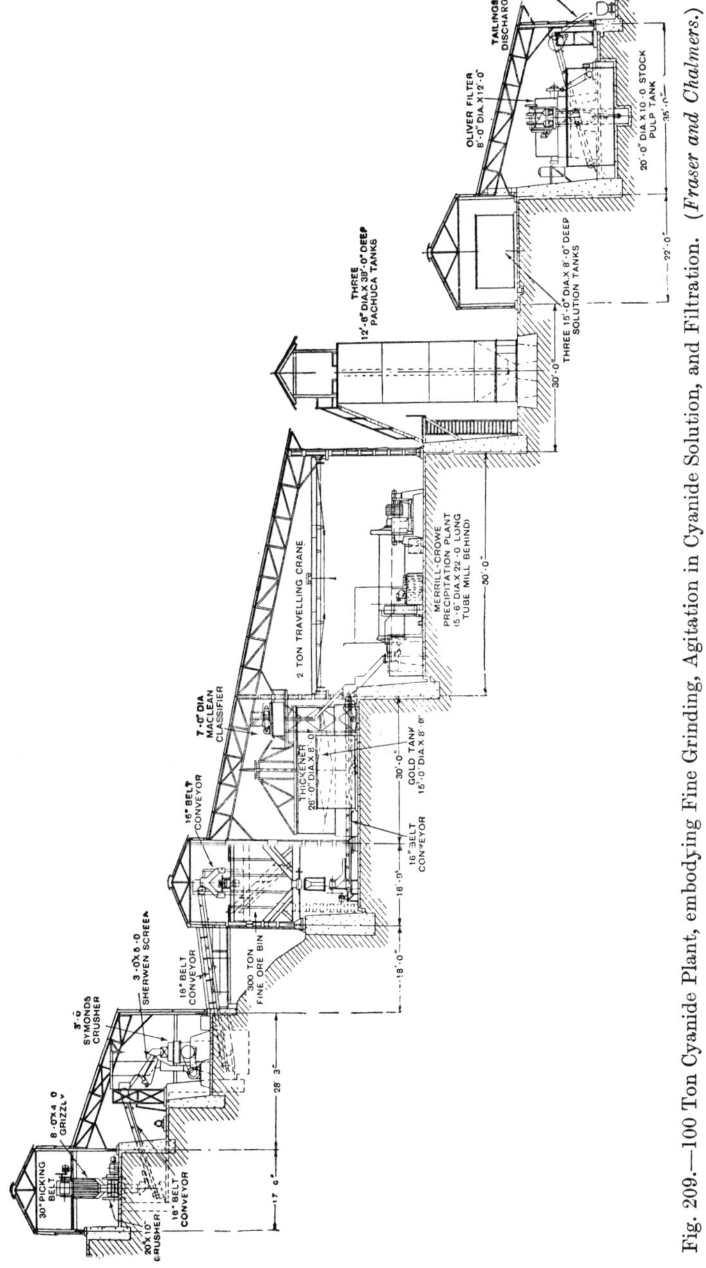

Fig. 209.—100 Ton Cyanide Plant, embodying Fine Grinding, Agitation in Cyanide Solution, and Filtration. *(Fraser and Chalmers.)*

tanks working in series; filtration is on Oliver filters and precipitation by zinc dust in a Merrill-Crowe plant.

Fig. 210 is the flow sheet for a small mill for the treatment of ore containing appreciable quantities of fine gold. Gold jigs are introduced after the stamp battery and after the tube mill in order to recover metallic material as early as possible. At each stage of classification the sands are returned to the tube mill and the overflow passed to blanket strakes. The concentrate collected on the latter joins that from the jigs and is amalgamated in a Brittain pan.

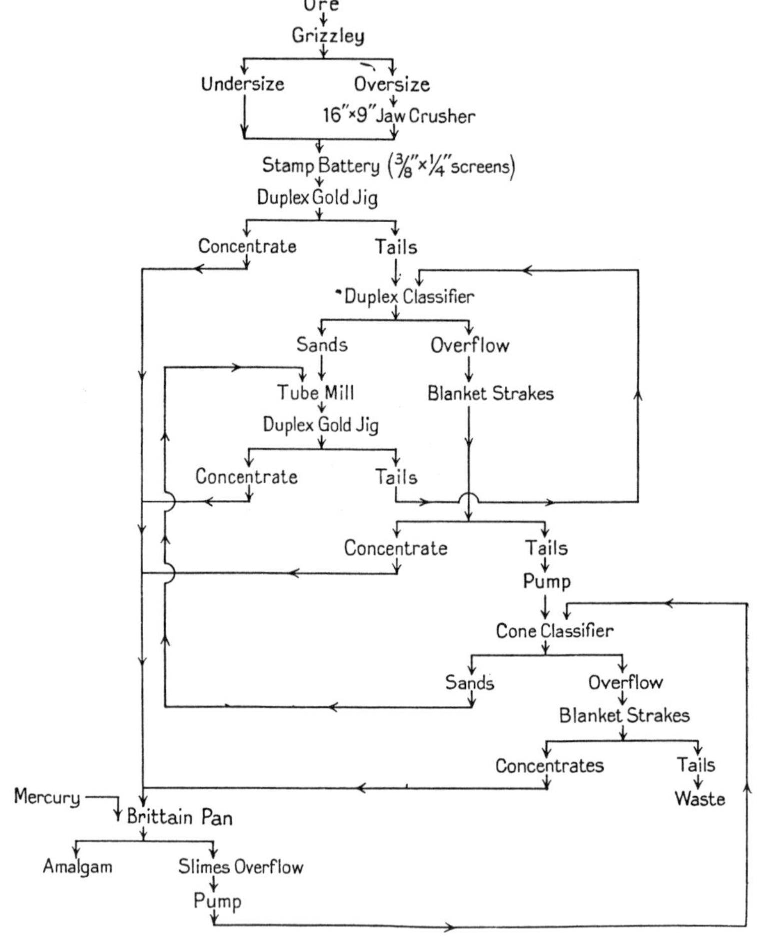

Fig. 210.—Suggested Flow Sheet for the Extraction of Gold, using Jigs and Rubber Blanket Strakes (*E. A. Knapp*).

The following costs are given by Jackson and Knaebel[1] for gold ore treatment at various mills. Allowances should be made for locality, scarcity of labour, nature of ore, etc.

[1] *U.S. Bur. of Mines*, 1932, *Bull.* 363, p. 137.

TABLE XLVIII

Mill and Year.	Tons Milled per Period.	Average Tons per Day.	Cost per Ton Milled.					Remarks.
			Labour $	Supplies $	Power $	Misc. $	Total $	
Porcupine United, 1929-1930,	...	25	1·072	0·374	0·405	...	1·851	Amalgamation. Table concentration.
Argonaut, 1929,	89,684	246	0·23	0·10	0·19	0·44	0·96	Amalgamation. Gravity concentration.
Homestake, 1929,	1,437,935	4,000	...	...	...	...	0·503	Amalgamation. Sand and slime cyanidation.
Alaska Juneau, 1929,	2,904,578	10,718	0·1081	0·0994	0·0288	...	0·2363	Sorting. Gravity concentration. Barrel amalgamation.
Conianrum, 1930,	122,972	337	...	...	...	...	0·939	All-slime continuous agitation with cyanide.
McIntyre, 1931,	558,115	1,530	0·2108	0·3953	0·1302	0·0147	0·9274	Continuous cyaniding with countercurrent decantation +200 ton pilot flotation plant
Vipond, 1930,	113,281	310	...	...	...	...	1·19	All-slime agitation with cyanide
Hollinger, 1930,	1,625,868	4,479	...	...	...	...	0·6355	All-slime cyanide agitation. Countercurrent decantation.
Lake Shore, 1931,	698,624	1,914	...	...	...	...	0·990	All-slime cyanide agitation.
Teck Hughes, 1931,	396,200	1,085	...	...	...	...	1·14	All-slime cyanide agitation.
Kirkland Lake, 1930,	52,768	145	0·4289	0·5715	0·3677	0·0209	1·3890	All-slime cyanide agitation. Continuous countercurrent decantation.
Spring Hill, 1929-1931,	40,930	153	0·426	0·374	0·207	...	1·007	Bulk flotation, concentrates shipped.
Mountain Copper,	16,847	543	0·182	0·176	0·031	0·043	0·432	Bulk leaching with cyanide.

CHAPTER XVII.

THE MELTING AND REFINING OF GOLD BULLION.

Introduction.—Crude bullion from the mills is melted and cast into bars so that its value may be ascertained, and that it may be put into a form convenient for transportation and sale. The appellation "base bullion," or "pig-copper," is given to the metal which has been obtained in smelting operations, or as the result of melting worn-out amalgamated copper plates. Such materials may contain a few parts per thousand of gold and silver, the main portion consisting of base metals. The treatment of base bullion, however, belongs to the metallurgy of argentiferous lead and copper, and the descriptions given here apply only to bullion which is valuable almost entirely on account of the gold and silver contained in it.

The bars shipped from gold mills are very different from one another in quality. Retorted metal is generally higher in gold than cyanide bars, but may contain considerable amounts of iron as well as copper. From large mills, the quality of the bars is usually good, and at some mills the gold is refined on the spot and requires no further treatment. Most gold, however, is sent to refineries situated at convenient centres for treatment. Cyanide gold is often shipped as rough, discoloured, pitted bars which require to be toughened before they can be valued.

Melting Furnaces.—All classes of fuel have been employed for melting, viz.: wood, coal, coke, gas and oil. Electric furnaces have also been tried. The choice is determined principally by the cost and availability of the different varieties of fuel. Coke and gas-fired furnaces are most common, while oil furnaces are becoming more frequent. The electric furnace has been the least used. It is the only one, however, which does not need a flue and stack and in which the temperature of melting can be automatically controlled, while the pressure and atmosphere within the chamber can also be varied if necessary. An induction furnace, installed in the United States Mint in 1931, is in use for the melting of deposits.[1] It is lined with silicon carbide and gives little radiation of heat and a low melting loss.

Coke furnaces may be round or square, with walls consisting of an outer layer of ordinary brick and an inner layer, at least 4 inches thick, of the best firebrick, preferably separated by an air space of 1 to 2 inches. There is often a complete outer casing of iron, which is useful in keeping the furnace from falling to pieces, but radiates more heat than the bricks. The chamber in a small coke furnace may be about 1 foot square and about 2 feet deep. Below is an ashpit, usually lined with a cast-iron tray and provided with an iron working door, through which the air-supply of the furnace passes, and by which it is regulated. The fire-bars are movable, and their ends rest loosely on iron supports. The top of the furnace may be made flat or sloping up towards the back at an angle of about 30°, in which case a wide flat ledge

---

[1] *Report of the Director of the U.S. Mint*, 1931 . 4.

is provided at the front, on which crucibles and moulds can rest. The top is always made of a cast-iron flanged plate, with an opening of the same area as the chamber. This opening is closed by a cast-iron sliding door made in one or two pieces, and preferably lined with firebrick and running on rollers. Furnaces may be placed with their mouths flush with the floor of the melting house, to facilitate charging and withdrawing by hand. The flue is placed at the back of the chamber near the top ; in a small 12-inch square furnace the cross-section of the flue should have an area of about 16 or 18 square inches—*e.g.* 4 inches square—varying, however, with the height of the

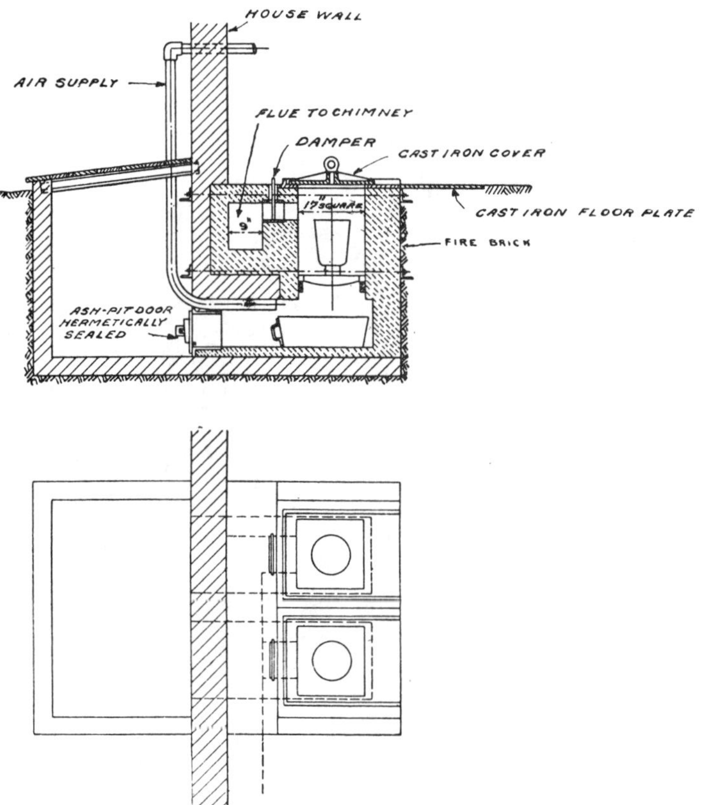

Fig. 211.—Bullion or Cornish Furnace, Two Pots.

stack, a higher stack going with a smaller flue. The flue communicates with a stack, which must be of firebrick for a distance of 2 or 3 feet from the furnace, but may be of wrought-iron tubing in its upper part. The height of the stack will depend on the position and size of the furnace. No mortar is used in construction, fireclay, mixed with an equal bulk of sand, being substituted for it. A sliding damper in the flue regulates the draught. The coke should be of high quality.

A melting furnace in use on many mines on the Rand is shown in plan and elevation in Fig. 211.[1] Air admitted through the "air supply"

[1] Schmitt, "*Rand Metallurgical Practice*," vol. ii., p. 197.

pipe is found to be preferable to a supply drawn through an open ash-pit, and accelerates the rate of melting. The fuel is coke. The stack is in this case about 50 feet high.

Crucible tilting furnaces fired by coke, oil or gas are also used for melting, solid fuel being less used than gas or oil. They work more quickly than the stationary furnaces, but are more expensive to install. One of their advantages is that the pot remains in the furnace during pouring and recharging, and is less cooled down than pots which are lifted out before being poured. The reduction in the amount of alternate heating and cooling gives a longer

Fig. 212.—Tilting Gas Furnace.

life to the crucible. A gas-fired tilting furnace is shown in Fig. 212. The crucible is held in place by three projecting bricks, and the gas nozzles are seen on the left.

An installation embodying a tilting furnace used in Rhodesia is shown in Fig. 213. The shape of the pots is clearly indicated. The mouth of the crucible needs careful attention and should be packed uniformly to give supporting strength. During the first few days of work it will be necessary to readjust the setting to allow for contraction of the pot.

Crucibles usually consist of fireclay, or of graphite mixed with fireclay, or graphite lined with clay. Detachable clay liners are sometimes used, or a clay pot may be fitted inside a graphite guard pot, as in chlorine refining. The size of the crucibles and the weight of the charges of bullion vary greatly, but in extraction mills, as a general rule, a gold-charge does

Fig. 213.—Morgan Tilting Furnace.

not exceed 600 ozs. In mints and refineries much larger crucibles are employed, holding amounts up to 6,000 ozs. of metal.

Melting the Bullion.—All crucibles must be thoroughly annealed before being used; otherwise, the contained moisture being suddenly converted into steam when the crucible is heated rapidly, the pots crack. The crucible is kept on a shelf near the flue or in a specially heated chamber, for as many days or weeks as convenient, before being used. It is then placed on the top of the furnace or in the ashpit for a few hours. It is preferable to place a newly annealed crucible in a cool furnace or in one that is being heated from cold. With Salamander crucibles, a less degree of care in annealing will suffice, as they are well annealed before being sold and in some instances have been heated to full working temperature by the makers. The crucible in a coke-fired furnace rests on a firebrick about 3 inches thick, which is laid on the bars of the grate. If the firebrick were omitted, the bottom of the pot, resting directly on the fire-bars, would be too cold, while a layer of fuel, if placed below the pot, would soon be burnt out, and could not readily be replaced, so that the pot would sink down to the bars. The fuel is built up round the pot until it reaches to its rim, or the top of the muffle, and the fire urged until the whole pot is at a full red heat. Borax is then thrown into the crucible by means of a scoop to slag off metallic oxides and so assist the metal to melt. A pure borax slag is usually too thin when molten and tends to follow the metal into the moulds. It can be thickened later with

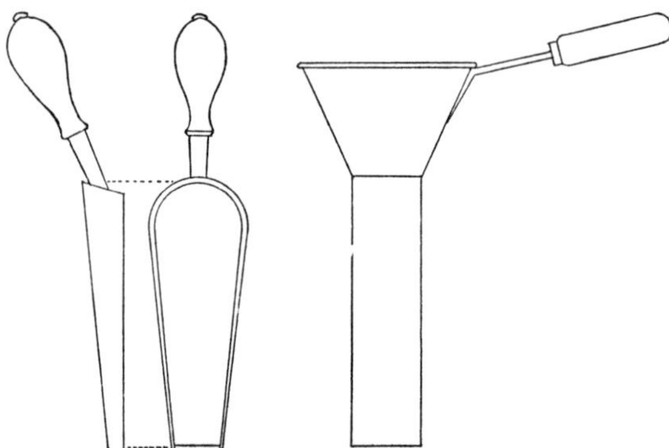

Fig 214.—Charging-scoop and Shoot.

bone-ash or a mixture of soda and sand so that it can be rolled back and remain behind. As soon as the borax is melted, the introduction of the bullion is commenced. The safest way to do this is to use the shoot shown in Fig. 214. In this way the melter avoids all danger of loss which might be encountered if the metal scrap were wrapped in paper and added by the tongs. Large pieces of metal are added with the crucible tongs. The cover, which must also have been previously well annealed, is kept on the crucible during melting as much as possible.

Loss of Gold in Flue Gases.[1]—During melting, a small proportion of gold is invariably lost by volatilisation and the presence of volatile impurities may increase this considerably. In removing zinc from gold-zinc precipitate by volatilisation, Sulman [2] observed a serious loss of gold, which he ascribed to its volatility in metallic zinc vapour. On the other hand, Rose [3] found that tellurium does not cause volatilisation of gold at temperatures below

---

[1] For a discussion of losses incurred during the melting of industrial gold alloys, see p. 66.

[2] *Trans. Inst. Min. and Met.*, 1919, 28, 153.

[3] *Brit. Assoc. Rept.*, 1897, p. 623.

1,100° C. in a reducing atmosphere, and very little under oxidising conditions.

Gold carried into the flues may be trapped in specially constructed large chambers through which the flue gases have to pass on their way to the stack. The essentials are (1) reduction in the flue gas velocity, (2) reduction in temperature, (3) a tortuous course provided by baffles behind which the particles are deposited. The reduced gas velocity is obtained by enlarging the flue section at specified points, and the reduction in temperature by introducing water sprays, damp blanketing, etc.—none of which, however, is very effective.

In one or two instances the Lodge-Cottrell process has been introduced for the recovery of gold carried in flue gases. Briefly the process is as follows :—The fumes, preferably at a temperature of 90° C., are conducted into a chamber in which are suspended a number of tubes, down the centres of which are a corresponding number of wires with cross-shaped sections or roughened surfaces, and weighted at their ends to keep them taut. Plates may replace the tubes, in which case two wires are placed opposite to each 4 or 6 inch plate. A high-voltage unidirectional current (produced by the rectification of an alternating current) passes across the gap between the wires (positive electrodes) and the tubes or plates (negative electrodes). The particles of gold and other materials in the intervening space are thereby given a positive charge of electricity. They then flow to the negative electrode where their charge is neutralised and they may accumulate until shaken or washed off periodically. The voltage used is usually between 50,000 and 100,000, and the intensity is augmented by the peculiarly roughened shapes of the anodes. It is stated that 97 to 98 per cent. of the gold and 98 to 99 per cent. of the silver passing away as fume is thus recoverable. Fig. 215 shows the installation at a London refinery. Two sides of the chamber have been removed so that the tubes may be seen.

Toughening.—Certain base metals such as zinc, antimony, lead and tellurium cause segregation on solidification in gold bullion, so that the metal is not uniform in composition, and sampling becomes difficult or impossible. The bullion is then, also, often brittle. Moreover, the metals which cause brittleness and segregation (especially lead) are troublesome in the refining operations and are better removed beforehand. The object of toughening is to remove these metals in a preliminary operation. Traces of these impurities may remain in the refined gold, however, making it still brittle, at any rate after it has been alloyed with copper for use in coinage, or in the arts. Toughening is therefore sometimes again necessary as the final operation in refining.

The method of toughening to be used depends partly on the composition of the bullion, and partly on the means at the disposal of the operator. Cupellation belongs properly to the metallurgy of silver and lead, and need not be dealt with here. (See, however, Tavener's process, p. 405.) The other operations are usually carried out in crucibles, although reverberatory or tilting furnaces are sometimes used for work on a large scale.

Toughening by Air or Oxygen.—The use of a blast of air on the surface of molten gold or silver is probably the oldest of the refining processes. The method appears to be described in the Book of Ezekiel,[1] to which the date B.C. 593 is assigned, and the principle is the same as in cupellation, which

[1] Ezekiel, chap. xxii., 18-22 ; see also Jeremiah, chap. vi., 28-30.

was practised by the ancients before 500 B.C., and perhaps as early as B.C. 2500.[1] In cupellation, the oxides of the base metals are dissolved and slagged off by litharge, but if a blast of air is used when lead is not present in large quantities, it is necessary to add other fluxes, as many oxides are almost infusible by themselves. In 1580, Ercker recommended the metallurgist to

Fig. 215.—Lodge-Cottrell Plant for Recovery of Gold from Fume.

melt brittle gold "with good Venetian borax and drive it before the bellows till it endureth the blowing."[2] The method is in use at some mills to raise the fineness of low-grade bullion. If the oxides of iron, etc., are not slagged off,

[1] See Hoover's "*Agricola*," p. 465, "Historical Note on Cupellation."
[2] Pettus, "*On Metals*" (translation of Ercker's book. London, 1686), p. 218.

the dross collects much gold which is difficult to separate by heat alone. If large quantities of base metals are present, the slag soon covers the surface of the metal, and prevents the access of the air. It is, therefore, necessary to skim off slag at frequent intervals, keeping the middle of the charge free from slag to enable oxidation to proceed.

In certain cases, the same object may be attained by granulating the bullion, roasting the granulations spread out on trays at a red heat, with frequent stirring, and melting the product with borax and sand. After two or three repetitions of such treatment, the bullion may be refined sufficiently to be sold.[1]

A stream of air or oxygen may be passed through molten bullion by means of pipes as in the chlorine process.[2] The base metals are oxidised successively in the order zinc, iron, antimony, arsenic, lead, bismuth, nickel, tellurium, copper. The oxidation of these metals, however, proceeds to some extent simultaneously, some copper being oxidised before the last traces of zinc are eliminated from the bullion. The oxides rise to the surface of the metal, and are slagged off with borax or a mixture of borax and sand. Tin oxide may be slagged off by pearl ash. The end of the operation is difficult to determine, except by dipping out part of the metal, casting it, and bending the ingot. If it is tough, only gold, silver and copper remain unoxidised. A little silver oxidises simultaneously with copper and passes into the slag. The losses of gold in the slag are insignificant. The slag prevents loss by projection or volatilisation. It is skimmed off before the metal is poured.

The losses by volatilisation in air-refining of bullion were investigated at Pachuca.[3] The refining at the Loreto and Guerrero mills was necessary as it was proposed to separate the gold and silver on the spot by electrolysis, instead of shipping the raw bullion. Cyanide precipitate containing 75 to 85 per cent. of gold and silver was charged into an oil-fired reverberatory furnace with borax and finely-ground bottle glass. The slag was tapped and skimmed and the air-pipe then immersed in the bullion and moved about. The average percentage losses by dusting and volatilisation were as follows :—

TABLE XLIX.

	Guerrero Bullion.		Loreto Bullion.	
	Au	Ag	Au	Ag
In melting precipitate, .	0·0150	0·0625	0·0280	0·1074
In air-refining, . .	0·0039	0·0156	0·0133	0·0497

The air was passed for 10 hours. Some copper was left in the toughened bullion, but no other impurities. The assay of the skimmings after the air-refining was Ag 125·336 kg., Au 0·091 kg. per metric ton. The dross was

[1] J. S. MacArthur, *Trans. Inst. Min. and Met.*, 1905, 14, 429.
[2] T. K. Rose, *Trans. Inst. Min. and Met.*, 1905, 14, 378.
[3] G. H. Clevenger, F. S. Mulock and G. W. Harris, *Min. and Met.*, 1922, 3, No. 181, 11.

skimmed ten times at intervals of an hour, and the base metals in the different skimmings were as follows :—

Cu,	first skimming,	2 per cent., then higher (9-12 per cent.).
Pb,	,, ,,	57 per cent., last skimming, 22 per cent.
Zn,	,, ,,	12·4 ,, ,, ,, 2 ,,
Fe_2O_3, Al_2O_3,	} first skimming, 17·4 per cent., last skimming, 20·4 per cent.	

At the 5th skimming the Fe_2O_3 and Al_2O_3 together were 26 per cent.

The doré bullion contained silver 981 and gold 6 parts per thousand after air-refining.

Oxidation by Nitre.—This ancient method was described by Ercker in 1580,[1] who explained that the nitre must be "projected upon the gold just before it melts, as it has little effect on molten gold." The method is still in wide use. The bullion is melted in clay crucibles, and a little nitre (potassium nitrate) or sodium nitrate is thrown on to the surface of the metal. Violent bubbling at once ensues, as heat converts the nitrate into nitrite with evolution of oxygen. The nitre is pressed down with a stirrer, and the nitrite and undecomposed nitrate oxidise some of the base metals. The nitrates and oxides corrode the pots, and it is better to have a ring of bone ash next the pot, and to throw the nitre into the "eye" of metal in the centre. The oxides are absorbed by the bone ash, which protects the crucibles from attack. After a minute or two, the action of the nitre moderates, and it is then, together with the bone ash, skimmed off with a ladle, and the operation repeated as often as necessary. If the metal is allowed to become pasty, so that it can be mixed with the nitre, the action is much more rapid and effective. If too much nitre is added at one time the charge boils over.

Iron and zinc can be removed in this way, but the oxidation of lead is more tedious, and bismuth and tellurium are very troublesome. The losses by spirting are heavy, and large quantities of both silver and gold are entangled in the dross skimmed off. Potassium permanganate has been used as a substitute for nitre.[2]

Oxidation by Black Oxide of Copper, CuO, was formerly used in certain cases. The oxide is stirred into the molten metal, and the whole then allowed to remain in the furnace for a short time before pouring. All base metals are oxidised, the cupric oxide being reduced first to cuprous oxide, and then to metallic copper. The cuprous oxide is dissolved in the metal, and so carries oxygen to all parts of the molten mass. The process is efficacious, but the gold is, of course, contaminated with the reduced copper.

Chlorination.—Chlorine gas is often used for toughening in the refineries which use the chlorine process (*q.v.*). Sal-ammoniac, NH_4Cl, is sometimes used to remove lead from gold bullion. When much lead is present, alternate additions of nitre and sal-ammoniac have been recommended. It is probable that the sal-ammoniac acts by decomposing basic compounds of lead which resist the action of nitre. Cupric chloride acts like ch orine gas, chloridising base metals and being reduced to cuprous chloride which is volatilised.

Use of Sulphur.—When retorted metal is infusible from the presence of

---

[1] *Op. cit.*, p. 216.

[2] W. R. Dowling, *J. Chem. Met. Mng. Soc. S.A.*, 1905, 5, 224.

large quantities of iron free from carbon, it may be refined by melting with sulphur or sulphides, according to W. McDermott.[1]

Osmiridium in Gold Bars.—If gold contains osmium and iridium in perceptible quantities, a fact which is not usually detected until after it has been refined, it is remelted in a clean crucible and kept fused at a high temperature for about half an hour, when the osmiridium will settle to the bottom of the crucible. Osmiridium does not alloy with gold, but being of high density and very infusible, settles through the liquid gold. The crucible is then gently lifted out of the furnace, and the greater portion poured into a mould; the remainder, which contains almost all the osmiridium, is allowed to cool in the crucible until it has solidified. An alternative plan is to allow the whole charge to solidify in the crucible, and then to cut off the lowest portion, which is set aside. The osmiridium settles better from an alloy chiefly consisting of silver than from pure gold, and the rich bottoms are consequently melted several times with silver, the lowest part being cut off each time. The gold is thus gradually replaced by silver, which eventually forms by far the greater part of the mass. It is then granulated and parted, and the resulting powder of gold and osmiridium is treated with aqua regia, by which the gold is dissolved, and the osmiridium separated as a black powder.

Refining.

Cementation,[2] melting with sulphide of antimony [3] or with sulphur,[4] boiling in nitric acid,[5] and other old processes are now obsolete. The only processes of refining in use in the Western World are the sulphuric acid process, the chlorine process and the electrolytic processes. These will now be described.

Refining by Sulphuric Acid.—The earliest reference to the use of sulphuric acid in parting gold from silver was made by Scheffer in 1753, but the process was not used on the large scale until the year 1802, when it was introduced into France by C. D'Arcet, and worked in a refinery built in Paris for the purpose. It was established in London at the Royal Mint Refinery in 1829 by Mathison, and is still used on the Continent of Europe. It has been much used in the United States, but has been superseded there in the mint refineries by the electrolytic process.

The gold is toughened if necessary by melting with nitre or by one of the other methods already described. It is then melted with silver in the proportion of $2\frac{1}{8}$ to 4 parts of silver to 1 part of gold. The silver is of course in the form of doré silver or other unrefined silver if available. Silver containing at least 2 grains of gold per pound troy or 0·35 part per 1,000 is usually regarded as "doré silver." If less than this quantity of gold is present, it is not profitable to refine it.

The amount of base metals present in the alloy is carefully regulated, as their sulphates are little soluble in concentrated sulphuric acid, and consequently are precipitated and interfere with the progress of the operation.

---

[1] *Trans. Inst. Min. and Met.*, 1905, 14, 433.
[2] J. Percy, "*Metallurgy of Silver and Gold*," 1881, p. 384.
[3] *Op. cit.*, p. 367.
[4] *Op. cit.*, p. 356.
[5] *Op. cit.*, p. 437; F. Cunningham-Hughes, *Trans. Mng. and Geol. Inst. India*, 1918, 13, 15, states that nitric acid is still used in India by local refiners.

Bars of auriferous copper, such as those formed from worn-out amalgamated plates, are added to the parting alloy, as a small amount of copper facilitates the solution of the silver. The proportion of copper is always less than 10 per cent., and is usually much less. A small quantity of lead is said to assist in the solution of copper, and a maximum amount of 5 per cent. of lead is said not to interfere with the operation.

The parting alloy is usually granulated by pouring it into water, but is sometimes cast into slabs, on which the action of the acid is less violent.

Dissolution of the Silver.—This is usually effected in kettles, made of a fine-grained white cast-iron. The kettle is slowly dissolved by the strong acid, and freely if dilute acid comes into contact with it, ferrous sulphate being formed. The perfect exclusion of air from the interior increases the length of life. The vessels are rectangular or cylindrical, with flat or hemispherical bottoms. They are covered with cast-iron lids, which are bolted tightly to the vessels, and have bent leaden pipes fitted to them for carrying off the fumes, which consist largely of SO_2. This is sometimes reconverted into sulphuric acid in leaden chambers arranged for the purpose. The cover has also an opening (supplied with a lid made air-tight by a water joint) through which the alloys and acids are added and the operation watched. Heat is supplied by a wood or coal fire.

The charge for the pots varies from 3,000 to 15,000 oz. of alloy, and the amount of acid required varies from 2 to $2\frac{1}{2}$ times the weight of the alloy, depending on its composition. About one-half of the acid, which is strong commercial acid of 66° B. (sp. gr. 1·85) is added at first, and the temperature raised to boiling point, when the pot is closely watched, and, if the ebullition becomes too violent, the temperature is lowered by regulating the fire and by adding cold acid a little at a time. The charge is stirred occasionally with an iron tool, particularly towards the end of the operation, when the undissolved granules of metal must be freed from the surrounding sediment, consisting of sulphates of the base metals, and exposed to the action of the acid. The ebullition gradually subsides and action ceases in about five or six hours, the presence of a greater proportion of base metals increasing the length of time required. The reactions are as follows—

$$(1)\ 2H_2SO_4 + 2Ag = Ag_2SO_4 + SO_2 + 2H_2O$$
$$(2)\ 2H_2SO_4 + Cu = CuSO_4 + SO_2 + 2H_2O$$

and similar reactions with tin and lead. It is obvious that 63 parts of copper decompose as much sulphuric acid as 216 parts of silver, so that an increase in the percentage of copper present necessitates an increase in the amount of sulphuric acid required.

One part of sulphate of silver is soluble in $\frac{1}{4}$ part of boiling concentrated sulphuric acid, but the solubility rapidly falls off as the temperature and concentration diminish, so that 180 parts of cold acid of specific gravity 1·08 are required for the same purpose. Sulphate of copper dissolves slightly in the boiling concentrated acid, but is almost all precipitated in the form of the white anhydrous salt on cooling. Tin and zinc behave similarly, and lead makes the solution turbid and milky.

When the dissolution is complete, the solution is siphoned into a large lead-lined covered vessel containing hot water (the settling tank); a gold siphon may be used. If enough copper is present, cooling of the acid precipitates some sulphate of copper, which falls to the bottom and clarifies the liquid. The clear silver solution is then siphoned off or ladled out with iron ladles into lead-lined rectangular wooden vats already partly filled with hot water, in which the precipitation is subsequently effected.

Washing and Melting the Gold Residue.—The residue in the dissolving pot, if the amount of base metals present is not large, is then boiled at least twice more with fresh concentrated sulphuric acid. There are sometimes as many as seven re-boilings. The gold is dipped out with an iron-strainer and transferred to a lead-lined filter-box, where it is thoroughly washed with boiling water, after which it is pressed, dried and melted with nitre and bone ash. It is sometimes brittle, from the occurrence in it of traces of lead, but is usually tough and about 996 or 998 fine.

If the amount of base metals present is very large, the gold residues are ladled into a vessel containing hot dilute sulphuric acid and boiled with it by means of steam. In this way most of the sulphate of silver and the whole of the copper, zinc, iron, etc., remaining with the gold are rapidly dissolved. Care is taken to add the residues a little at a time, as otherwise the anhydrous sulphate of copper will form lumps, which are only slowly dissolved. The gold may also be purified by heating in a furnace in small iron pots with about half its weight of bisulphate of potash, by which some additional silver is converted into sulphate. The temperature is not raised much above the fusion point of the salt. The fused mass is then boiled in sulphuric acid, and again washed.

Precipitation of the Silver.—On pouring the sulphuric acid solution into cold water, most of the silver sulphate is precipitated at once in the form of small crystals, and the liquid must then be raised to boiling, by means of steam, in order to redissolve them. When the original alloys contain much lead this is not redissolved, and it is, therefore, necessary in this case to let the solution settle and transfer the clear liquid to another vessel. Some particles of gold are usually found in the precipitate thus formed.

The reduction and precipitation of the silver may be effected by means of copper, which takes its place in solution. The copper is usually added in the form of scrap while the liquid is being heated up by steam. The precipitation is assisted by constant stirring by means of wooden paddles. The solution should be of about 24° B.; if it is much more concentrated than this, the precipitation of the silver is imperfect. The end of the reaction is detected by testing with salt solution, and when complete, the stirring is stopped, the solution allowed to settle, and the clear liquid drawn off. The precipitated silver is washed with boiling water in wooden filters lined with lead. The metal is then pressed, dried and melted, and is usually from 998 to 999 fine. The copper is recovered by electro-deposition or by alternate evaporation and crystallisation of the copper sulphate in lead-lined wooden tanks.

The silver may also be recovered direct from the crystals of silver sulphate formed on cooling the sulphuric acid. This was done in the Gutzkow process. The crystals were placed on a filter and a hot solution of ferrous sulphate poured on them. The silver was reduced by the following reaction:—

$$2FeSO_4 + Ag_2SO_4 = Fe_2O_3.3SO_3 + 2Ag$$

The silver may also be reduced by sheet iron.

The Chlorine Process.—The use of chlorine gas for the purification of molten gold was proposed by L. Thompson in 1838.[1] In 1867, F. B. Miller[2] installed the process at the Sydney Mint, and it is now in wide use. The process is a simple one. Chlorine is passed into molten gold covered with borax, and combines with all metals present except, perhaps, some of those belonging to the platinum group. Gold itself is attacked very slightly at first. The chlorine is initially completely absorbed. The chlorides rise to the surface and are baled out, together with the borax, when they have become inconveniently bulky. Some chlorides volatilise and pass out of the furnace. When the gold approaches a fineness of 990, gold chloride is formed in rapidly increasing amounts and begins to appear in appreciable quantities in the fumes which pass through the slag. The stream of chlorine, which has already been gradually reduced to a mere trickle, is stopped soon afterwards. The end-point is marked by a peculiar stain (caused by gold chloride) on a cold clay-pipe stem held in the issuing fume. The remainder of the chloride slag is then removed and the gold cast into ingots. It is usually about 996 fine, the residue being mainly silver, with 0·5 or 0·6 per 1,000 of base metal, partly copper.

A clay pot is used to contain the molten gold in the refining operation, as a graphite pot reduces silver chloride and is therefore unsuitable.

The chloride slag is found to contain about 2 per cent. of the gold, mainly in the form of spangles and minute crystals of gold reduced from chloride of gold which has entered the slag. The gold is collected by the addition of carbonate of soda to the molten slag. Part of the silver chloride is reduced, the reaction being

$$4AgCl + 2Na_2CO_3 = 4Ag + 4NaCl + 2CO_2 + O_2$$

The metallic silver settles to the bottom, carrying the gold with it. This portion again passes through the process. The remainder of the silver chloride is freed from the base chlorides and reduced with iron plates.

A modification of the process consists in the addition of air to the chlorine passed into the molten gold, so that part of the base metal is oxidised instead of being chloridised. This modification was introduced by R. R. Kahan at the Perth Mint in 1917.[3]

The chlorine process is especially suitable for refining bullion over 700 fine in gold, with the remainder mainly silver. Gold of lower fineness can be treated, but the time of treatment becomes longer and the cost greater. For example, metal from New Guinea containing gold 570, silver 425, is thus refined at Melbourne.[4] Bullion of suitable composition is obtained when possible by mixing. If the silver is below 50 per 1,000 there is some difficulty in removing base metals. The process is obviously unsuitable for doré bullion or platiniferous gold.

Base metals differ considerably in their behaviour. Owing to mass action, all the metals present, including gold, are attacked by the chlorine simultaneously from the beginning of the operation to the end. Because, however, of the continual increase in the proportion of gold in the molten

---

[1] *J. Soc. Arts*, 1840-1, [i.], 53, 16.

[2] *J. Chem. Soc.*, 1868, 21, 506; *Trans. Roy. Soc. of New South Wales*, 1869, 3, Dec.

[3] 47*th Ann. Rept. Royal Mint*, 1916, p. 91; 48*th Ann. Rept. Royal Mint*, 1917, p. 83; *Trans. Inst. Min. Met.*, 1918, 28, 35.

[4] O. G. Reynolds, 64*th Ann. Rept. Royal Mint*, 1933, p. 110.

metal, the amount of gold chloride formed is always increasing. The chlorine attacks the different metals and metalloids preferentially in an order believed to be something like the following :—Zinc, iron, antimony, tin, arsenic, copper, lead, bismuth, silver, tellurium, selenium, gold, with the platinum metals somewhere near gold.

It is certain that dense fumes of the volatile chlorides of zinc, iron, antimony, etc., are given off in the early stages. Lead appears to be erratic in its behaviour. Under the conditions at the Perth refinery, where air is used in addition to chlorine, it was found [1] that lead was removed at an even rate, regardless of the amount of chlorine passing through. The time of treatment was the main factor. The last traces of lead are difficult to remove and occasionally remain in the finished product of refined gold, making it brittle when alloyed with copper, so that it has to be toughened or refined again.[2] If much lead is present, it is better that the proportion of silver should be high, so that the time of treatment is prolonged. It is possible that air or oxygen is more efficacious than chlorine in removing lead.

Tellurium is the most difficult of all elements to eliminate. It remains to the end and has to be removed by means of repeated additions of nitre.

Copper is chloridised with the silver and if there is much present, the fumes of cuprous chloride escaping into the atmosphere have distressing effects on the workmen.

The platinum metals remain with the gold. At the Ottawa Mint [3] a charge containing Yukon and British Columbian gold with some jewellery and dental scrap was given $2\frac{1}{2}$ hours extra treatment. The refinage was then found to be of the following composition :—

Gold, ...	991·0
Silver, ...	2·0
Palladium, ...	5·75
Platinum, ...	0·72
Copper and nickel, . .	Traces.

After two more hours' chlorination the gold was raised to 992·0 fine.

The passage of gold into the chloride slag is of especial interest. At Melbourne, where no air is used with the chlorine, it was found [4] that gold was perceptible in the slag after 30 minutes and increased rapidly in the final stages, reaching 16 per cent. or more in the last baling. The gold was mainly metallic, in the form of minute crystalline particles, which were over 999 fine, although the refined gold charge was only 995·7 fine. Globules of gold were scarce and globules of low fineness were very rare, but doubtless more globules of this kind would be found if air were used. There was some water-soluble gold observable towards the end of refining (presumably $AuCl_3$ or $AgAuCl_4$). Law concluded that mechanical spirting played little part in the transfer of gold to the chloride slag (in the absence of air), but that $AuCl_3$ was formed and volatilised, and then caught in the slag and nearly all dissociated into metallic gold and chlorine.

---

[1] H. R. Hillman, *62nd Ann. Rept. Royal Mint*, 1931, p. 104.
[2] C. Bowyer, Private Communication.
[3] A. L. Entwistle, *57th Ann. Rept. Royal Mint*, 1926, p. 131.
[4] R. Law, *54th Ann. Rept. Royal Mint*, 1923, p. 97.

The exact figures in one experimental run were as follows :—

Original Assay of Bullion.

	Per cent.
Gold,	92·24
Silver,	4·21
Lead,	2·51
Iron,	0·725
Copper,	0·300
Undetermined,	0·015
	100·000

Assay of Metal Dips taken at Half-hourly Intervals.

	Gold.	Silver.	Base.
Original,	92·24	4·21	3·54
After ½ hour,	94·88	4·30	0·82
,, 1 hour,	96·08	3·70	0·18
,, 1½ hours,	99·36	0·40	0·13
,, 1 hour 55 mins.,	99·55	0·20	0·18
Finish,	99·57	...	...

Gold and Silver Content of Silver Chloride Slags.

	½ hour.	1 hour.	1½ hours.	1 hour 55 mins.
TOTAL GOLD AND SILVER :—				
Silver,	4·46	47·20	56·77	50·33
Gold,	0·12	1·57	10·32	16·285
METALLICS :—				
Silver,	0·095	0·38	0·65	0·33
Gold,	Trace	1·53	8·61	13·03
WATER-SOLUBLE :—				
Silver,	0·02	0·015	0·01	0·02
Gold,	Trace	Trace	1·26	3·03

The amount of gold carried into the slag is stated by Kahan [1] to be less when air is used in conjunction with chlorine than when chlorine is used alone. In the absence of exact figures to prove this statement, however, it appears likely that owing to mass action the loss of gold in the slag is small in raising the fineness to about 990 and is always large in raising it from 990 to 996. It seems generally to be in the neighbourhood of 2 per cent. of the fine gold produced. The percentage of gold in the chlorides naturally varies with the fineness of the rough gold. It is often from 5 to 10 per cent.

The recovery of the silver from the chloride slag presents some difficulties. If reduced directly, the silver is contaminated with much copper. If the slag is leached, the cuprous chloride tends to form insoluble basic compounds. Moreover the chlorides, solidified from fusion, are tough and difficult to grind

[1] *Trans. Inst. Min. Met.*, 1918, 28, 37.

finely owing to dusting. The usual practice is to break the crusts into about 1 inch cubes and to boil them in a saturated solution of common salt acidified with sulphuric acid. The copper chloride (together with some silver chloride) is dissolved in the course of a few days and the washed silver chloride is reduced with iron plates in very dilute hot sulphuric acid. The sponge silver, after being washed, is dried, melted, toughened if necessary with oxygen or nitre, and cast into bars, which are from 996 to 999 fine.

Any silver dissolved by the brine is recovered with the copper by running the liquor over scrap iron.

The condensing chambers in use for treating the gases rising from the pots are of different kinds. At Ottawa water sprays and baffles, and at the other refineries expanded metal screens, baffle walls and other devices, are used to clean the fumes. The Cottrell process (see p. 459) is used at the Royal Mint Refinery, London, and appears to be far more effective than other condensing plants. At Ottawa in 1925,[1] where air is not used, the fine gold contained in the chamber sweep was 173 ozs. or 1·4 per 1,000 of the fine gold treated. It is usually less than this, however.

The loss of unrecovered gold in working the process seems to be about 0·6 per 1,000 of the fine gold produced when the Cottrell process is not used.

One great advantage of the chlorine process is its rapidity, as it brings 98 per cent. of the gold to account in a marketable form in a few hours. The plant costs little and can be expanded readily to meet any requirements. The running costs are low and the loss of gold is small.

The chlorine is usually supplied from cylinders when these can be obtained. As an alternative it may be generated by warming a mixture of common salt, manganese dioxide and sulphuric acid, or a mixture of manganese dioxide and hydrochloric acid. The proportions normally used are 56 lbs. crushed manganese dioxide (73 per cent. MnO_2) and about 150 lbs. of hydrochloric acid or, instead of the hydrochloric acid, salt and sulphuric acid in the ratio of 1 : 2.

The chlorine generators for small-scale operations are domed stoneware vessels with three necks, closed with indiarubber plugs, $1\frac{3}{8}$ to $1\frac{3}{4}$ inches in diameter. The first is used for charging in the manganese dioxide and salt, the second for the acid pipe, which is fitted with a glass stop-cock, and the third for the glass tube delivering the chlorine. The bottom of the generator is covered with quartz pebbles to prevent choking of the acid pipe, which reaches to within 1 inch of the bottom of the vessel. The generator is warmed by a water-bath or steam-jacket. Connections consist of rubber tubing with $\frac{3}{8}$ inch wall. These can be closed by a screw as an alternative to the use of glass stop-cocks. A safety valve is fitted either to a side tube in the acid pipe or to the chlorine-delivery tube, or in a separate neck. It consists of stout combustion tubing about $\frac{3}{4}$ to 1 inch in diameter. Eight such generators, 2 feet high and about 15 inches in diameter, were enough at Melbourne for the treatment of 1,000,000 ozs. per annum. In 1901 they were superseded by the Edwards generator, which has a lead lining to an iron case (see Fig. 216). It is bowl-shaped and is oscillated three times per minute by a crankshaft. The capacity is 90 gallons, and with 100 lbs. manganese ore, 125 lbs. salt, 12 gallons of water and 265 lbs. of acid, 1,560 ozs. of base metal and silver have been removed in one working day.

In small isolated plants, where chlorine in cylinders cannot be bought

[1] P. W. Bond, *56th Ann. Rept. Royal Mint*, 1925, p. 136.

and where the amount of gold to be refined does not warrant the installation of large generators, a smaller type, diagrammatically represented in

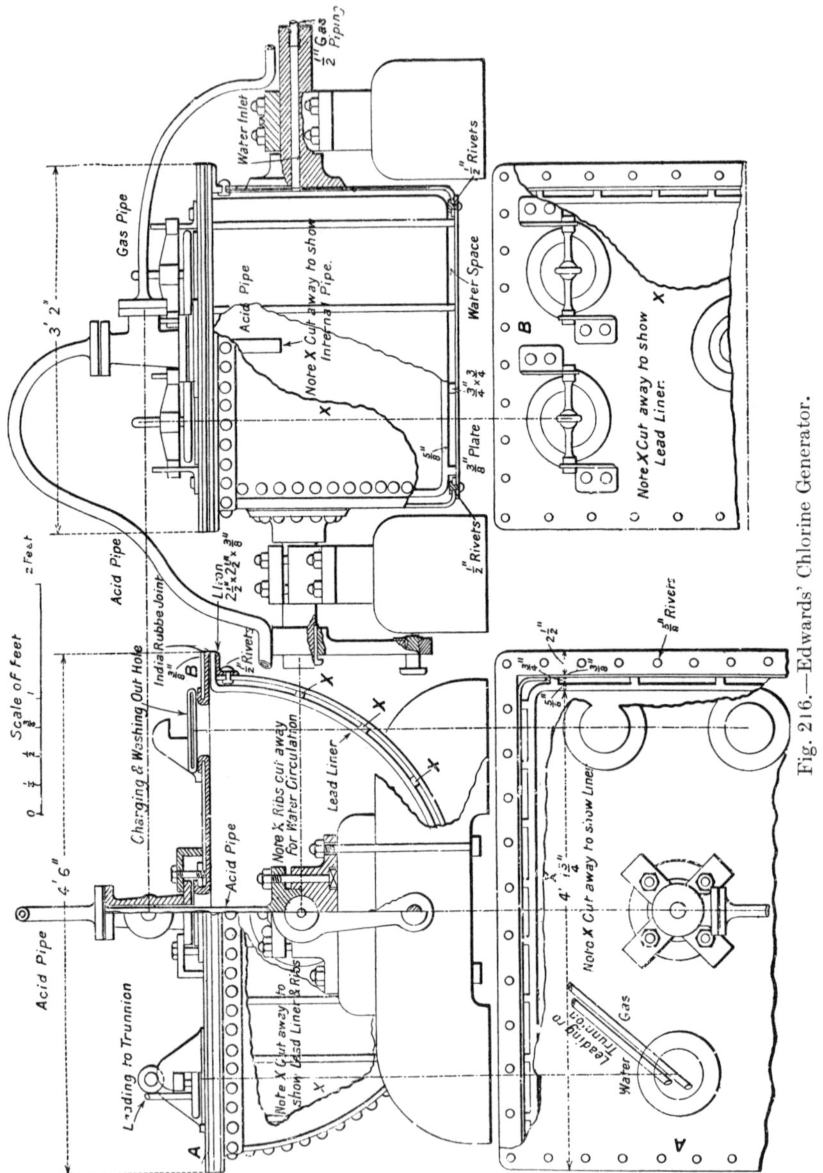

Fig. 216.—Edwards' Chlorine Generator.

Fig. 217, may be set up. Manganese dioxide and salt are added to the earthenware or lead-lined steel container in weighed quantities according to the equation

$$MnO_2 + 2NaCl + 2H_2SO_4 = Na_2SO_4 + MnSO_4 + Cl_2 + 2H_2O$$

The sulphuric acid, mixed with an equal volume of water, is added gradually in order to generate a steady stream of chlorine. Hydrochloric acid alone may be used instead of the salt-sulphuric acid combination. The contents of the generator may be warmed by surrounding the vessel with water heated by steam coils. The pipes in contact with chlorine should be made of lead. The section between the delivery pipe and the chlorine tube entering the crucible should, however, be made of rubber, so that the pulsations of the gas entering the metal may be felt and the flow thereby controlled.

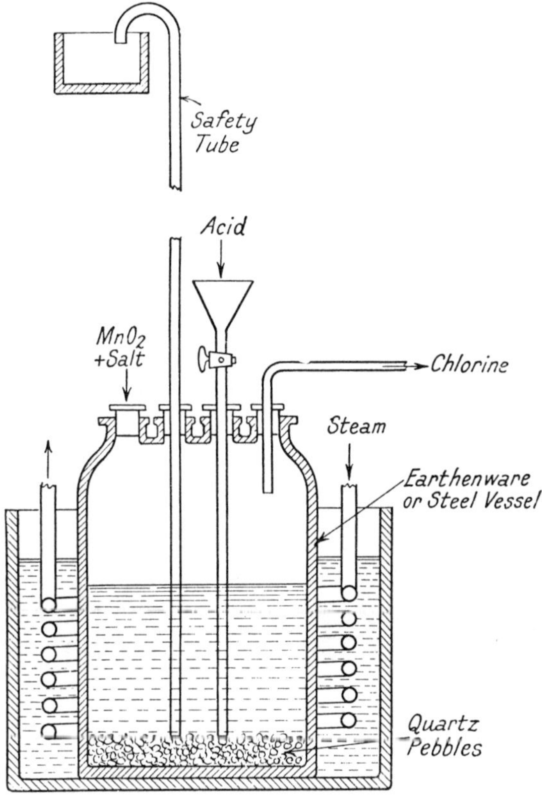

Fig. 217.—A Chlorine Generator for Small Plants.

In order to avoid a sudden increase in pressure within the vessel, and as a gauge to the actual pressure attained, it is essential to provide a safety tube about 10 to 12 feet long if a depth of 7 inches of molten metal is contained in the refining crucible, but proportionately less if less metal is being treated. This tube should be made of glass, and should open at its upper end into a container so that in case of accident acid, etc., may not be spilled on to operators below. The stoppers through which the tubes pass may be screwed in, and in any case should be held down by clamps.

Practice at Perth, W.A.[1]—There are eight refining furnaces (see Fig.

[1] Private Communication from C. Bowyer, by permission of the Deputy Master of the Perth Mint.

218), with fire-boxes 12 inches square and 22 inches deep, built of firebrick with cast-iron top and front. The furnaces have tile covers strengthened with iron rods. They are connected by a horizontal main flue with condensing chambers and the stack. The ash-pit is lined with cast-iron. The fuel is coke. The air blast for combustion is supplied through a 3 inch main at a pressure of 4 to 5 ozs. per square inch, with a $1\frac{1}{2}$ inch pipe for each furnace, furnished with a regulating tap. It is used only for a short time, while heating-up.

The refining crucible is of clay, and is about 11 inches in height and $5\frac{3}{4}$ inches in diameter at the top. It fits into a plumbago guard pot which stands on a half-brick placed on the firebars. A dished clay cover, which

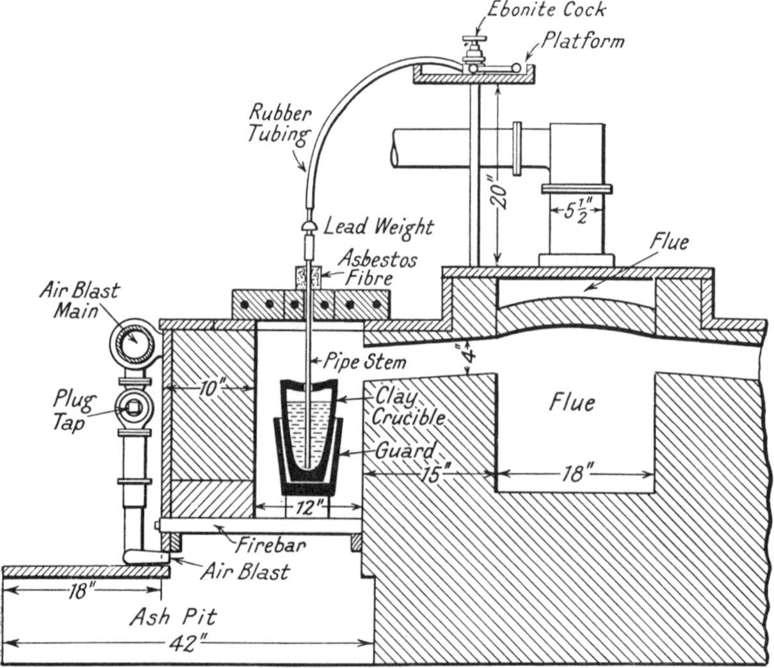

Fig. 218.—Section of Chlorine Refining Furnace (Perth).

fits the top of the clay crucible, has two holes for the pipestems. These are 24 inches long and $\frac{3}{8}$ inch in diameter, with $\frac{1}{8}$ inch bore. They are wedge-shaped at the bottom to facilitate the escape of the chlorine when they are resting on the bottom of the pot. Rubber tubing connects the pipestem with the chlorine cock and is protected from heat by asbestos. A lead weight keeps the end of the pipestem on the bottom of the pot. The chlorine cock is of ebonite, with a brass valve and rubber washer.

The chlorine is supplied from cylinders each containing 150 lbs. of liquid chlorine. The gas passes through a needle valve by copper connections to lead piping. A combined pressure gauge and safety valve is connected with the piping to regulate the supply of gas (see Fig. 219). The pressure gauge, about 11 feet in vertical height, contains a solution of cupric chloride, which is easily seen. The working pressure is about 5 or 6 lbs. per square inch, enough to overcome the resistance of about 7 inches of molten metal. If the pressure rises too high, the chlorine escapes up the stack. Compressed air for refining is supplied by a rotary blower.

In refining operations, the crucibles are heated slowly to dull redness and two ingots of gold, slipper-shaped to fit the crucible, are charged in. The charge is 650 to 700 oz. of bullion. A ladleful comprised of a few ozs. of borax is added to glaze the crucible and act as a slag cover, and more is added during the refining. After about 30 minutes' further heating, the charge begins to melt and two pipestems, previously heated to redness at the lower end, are gradually inserted, with gentle streams of chlorine and air passing through them. The chlorine pipe is put in first and the heat of formation of the chlorides helps the metal to reach complete fusion.

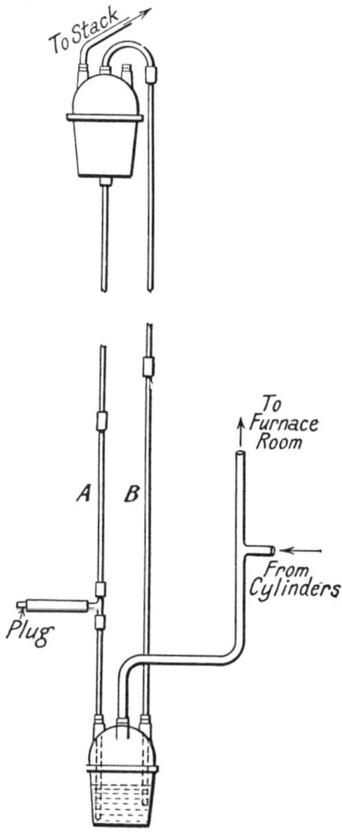

Fig. 219.—Pressure Gauge and Safety Valve attached to Chlorine Cylinders (Perth).

The chlorine is entirely absorbed, but the air bubbles rise through the charge, so that only a small quantity can be used, as projection of metal out of the pot must be avoided. The volatile chlorides are formed first and given off, fuming strongly, and as they pass off the stream of chlorine is increased more and more until it reaches full pressure. The chlorides of copper and silver are now being formed. The pot soon fills, and the chlorides are baled out. For this, a small triangular clay pot is used, the streams of gas being turned off. The chlorides are poured into a large mould, 12 inches long, and the viscous mass of borax and oxides, floating on the chlorides, is returned to the pot. After a second baling the air pipe is removed and the stream of chlorine gradually reduced. The passage of the gas into the molten metal can always be detected by the throbbing of the rubber tubing between the finger and thumb. Towards the finish, chlorine can be observed leaving the pot, and the appearance of a peculiar yellow stain on a cold pipestem held in the fumes signalises the end of the operation. The stain is due to escaping chloride of gold. Copper in excess has a tendency to mask the end-point, making it darker. The chlorine supply is now turned off, and the remaining chlorides baled out. The last remnants of the chlorides are thickened with bone ash and removed with a clay stirrer. Nitre is then stirred in, the slag removed, and the operation repeated until the slag is light green in colour. (Apparently this toughening is to remove tellurium.) The metal is cast in slipper-shaped moulds and the ingots dipped in 5 per cent. sulphuric acid, washed and dried. They are remelted for export. The average assay is

Fig. 220.—Furnaces for Chlorine Process, Melbourne Mint.

Fig. 220A.—Chlorine Generators, Melbourne Mint, prior to 1901.

996·0, and 13,000 to 16,000 ozs. of rough gold are refined in about six hours, two rounds being worked in each furnace.

The fumes arising from the chloride moulds and from the pot during the baling pass into a large galvanised iron box, 3 feet cube, which covers the furnace. This box contains hessian screens for the collection of precious metal. It is connected with a fan which draws the fumes into the box and discharges them into the flues.

The box has replaced a cowl similar to that used at Melbourne, which is shown in Fig. 220. The Melbourne furnaces resemble those at Perth, except that the fireboxes are circular, 12 inches in diameter and 21 inches deep. The earlier Melbourne chlorine generators are shown in Fig. 220A.

The mixed chlorides are weighed into lots of 300 ozs. and melted in graphite pots, together with the borax from the day's work and some common salt. 10 ozs. of bicarbonate of soda are added without stirring, and a second similar addition is made 10 minutes later. When all action has ceased, the charge is stirred with an iron rod to assist settlement, and the pot is allowed to cool. As soon as the metal has solidified, the viscous borax is removed and the molten chlorides are poured into flat moulds, 12 × 12 inches, where they form cakes, 1 inch thick. A button containing silver and gold is obtained from each pot and is subsequently refined. The borax is remelted with a little soda and also yields a button of silver and gold. The borax is used again in the refining pot.

The chloride cakes are broken into 1 inch cubes in a Blake crusher and treated once a month. The analysis of this material in Aug.-Nov., 1930, was as follows :—

	Per cent.
Ag, calculated as $AgCl$,	67·11
Cu, ,, Cu_2Cl_2,	19·47
Pb, ,, $PbCl_2$,	5·13
Te,	0·03
Au,	0·01
NaCl (including traces of Fe and Zn),	8·66
	100·41

The copper chloride is removed by boiling with dilute sulphuric acid for several days and the washed silver chloride is reduced with iron plates. After being toughened (to remove tellurium) by passing oxygen through it and by the addition of nitre, the silver is 996 to 999 fine.

Bowyer observes that very little air is used in the refining process owing to the danger of projection of metal from the pot by the issuing nitrogen. The air, however, is an advantage in the elimination of lead, which is the most troublesome metal. Lead is sometimes left in the finished gold, which would in that case make brittle alloys, and has to be refined again. Lead is more readily removed if much silver is present.

Another difficulty has occurred with copper in bullion containing very little silver. Gold from the Wiluna gold mine, as received at the Perth Refinery, contains 89-93 per cent. gold and ½ per cent. silver. The base metal is principally copper, with a small percentage of lead. According to H. R. Hillman,[1] the lack of silver causes considerable trouble in refining, as the end-point of the reaction is not given distinctly. This may lead to large losses of gold chloride by volatilisation. The trouble has been overcome

[1] *Royal Mint, Perth,* Private Communication.

by adding 50 ozs. of molten silver to each refinage of 700 ozs. after the first amount of chloride has been baled off, and necessitates a longer period of treatment to the extent of about 12 minutes. The total time of refining is about 1 hour 50 minutes.

The average composition of the bullion refined at the Perth Mint is approximately gold 800, silver 140, base metal 60. The large amount of base metal is perhaps the reason for extending the time of the air refining beyond that used at the Rand Refinery. At the Ottawa Mint, where no air is used, the amount of base metal is still higher, averaging about 80 per 1,000; in 1924 the base metal was 231 per 1,000, and the gold only 656.[1]

At the Perth Mint, the crucibles, pipestems, etc., and the rest of the "sweep," are crushed in a Chilian mill to pass a 1 inch screen and are then passed to a ball mill. In this the screens are 40 and 60 mesh, and many metallics are retained. The crushed material is washed by water sprays over a set of brushes in a box, where more metallics are recovered. The residue is treated with mercury in a Wheeler amalgamating pan and the tailing collected in a settling pit, dried and sold to a smelter.

Practice at the Rand Refinery.—All the gold produced on the Rand, amounting to over 10,000,000 ozs. per annum, is treated at this refinery. The average assay of the rough gold in 1934 was :—

Gold, ...	889·4
Silver, ...	84·5
Base metal, ...	26·1

The rough gold is of fairly even composition, the silver ranging from 68 per 1,000 in the gold received from the Consolidated Main Reef Mines to 106 per 1,000 in that from Rose Deep.[2] Some bullion produced from by-products contains more silver.

The general practice [3] is similar to that at the Perth Mint. There are 64 coke-fired refining furnaces, the charge being two ingots of 450 ozs. each, prepared by melting in another type of furnace (see Fig. 221). If the bullion contains too much base metal it is toughened before being refined. The toughening is done by passing air or chlorine through the molten metal, or occasionally by means of nitre or cupric chloride. In refining, the pipestems are $\frac{7}{16}$ inch in diameter, with $\frac{3}{16}$ inch bore. The air and chlorine together are passed in slowly until the silver and base metals are forming chlorides freely, when large volumes of chlorine are introduced and are absorbed without any "spirting" of the metal taking place. The air pipe is then removed and the refining carried on by chlorine alone. Kahan observes—"The introduction of air, simultaneously with the chlorine, has been found to be of decided advantage in bringing about the rapid oxidation of certain base metals which appear to react but slowly with the chlorine alone, and their removal as fusible borates and borosilicates appears to 'open up' the charge in such a way as to render it more easily acted on by chlorine alone." (As has been already stated, one of these base metals is lead.) Later in the operation, a peculiar brownish-red fume makes its appearance at the hole in the crucible cover and the volume of chlorine passing through is then reduced to a mere trickle to avoid "spirting" of the metal. A cold clay rod held near the

[1] J. H. Campbell, *57th Ann. Rept. Royal Mint*, 1926, p. 128.
[2] R. R. Kahan, *J. Chem. Met. Mng. Soc. S.A.*, 1930, **31**, 18.
[3] Private Communication from R. R. Kahan, The Manager, 1935.

fume acquires a clear yellow stain which smells of chlorine. The metal at this stage is about 985 fine, and after 15 minutes' further slow chlorination, the gold reaches 995 fine, at which point refining ceases.

The gold is freed from chloride and transferred to a tilting furnace of a capacity of 7,000 ozs. When fully melted, the metal is stirred, samples are taken for assay and the gold is cast into ingots of 400 ozs. each. The mould is placed on scales and the metal poured in from a pot, acting as a ladle, until the beam rises, so that an almost exact weight is obtained.

Fig. 221.—Pouring of Crude Gold in Rand Refinery.

Immediately the pouring is finished, an acetylene flame is played on the surface of the metal, retarding the surface crystallisation of the gold, and filling up the cavity which tends to form, so that a bar with an even specular surface is produced.

The chlorine is generated in Edwards' generators (see Fig. 216) from common salt, sulphuric acid and manganese dioxide, using an excess of acid.

The cakes of mixed chlorides are crushed in a jaw-breaker and washed first in hot water and afterwards in hot strong brine. The purified silver chloride is reduced with iron plates in very dilute sulphuric acid heated by a steam jet. The silver sponge is washed, dried and melted down, without toughening. (Rand gold contains no tellurium.) The bars are 996 to 999 fine. All sump liquors are passed over scrap iron.

The total cost of treatment in 1934, according to the published accounts, was about 1·16 pence per fine ounce of gold.

Practice at Morro Velho, Brazil.[1]—The bullion is of different qualities. The part from the blanket strakes contains about 765 gold, 225 silver and 10 base metal; the cyanide gold, after fluxing, contains rather more base metal, with silver ranging from 50 to 300 per 1,000. Bullion containing about 450 ozs. gold is melted in a No. 60 graphite pot with clay liner. Chlorine from cylinders of liquid "gas" is passed in through fireclay tubes for 2½ to 3 hours, according to the fineness. The pot is then removed from the fire, the gold allowed to solidify and the chloride poured off into a large flat mould. The gold cake is turned out by inverting the pot, and after cleaning from adhering chlorides, is remelted with a cake from another charge and given another chlorination for half-an-hour before casting into a 900-oz. bar. The primary chlorination yields gold 993-994 fine and the secondary treatment raises the grade to 996-997 fine.

The chlorides are put in alternate layers with iron plates in acidulated water in a porcelain tank, and next day the reduced silver is dried and melted with fluxes. Doré bullion averaging 100 gold, 880 silver and 20 base metal, is produced and cast into anode plates. These are electrolysed and pure silver, 999 fine, and anode sludge assaying 920-940 gold, result. The sludge is dried and returned for re-chlorination.

All the melting is carried out in oil-fired furnaces. The exhaust gases pass into large brickwork chambers, and a system of condensing flues follows.

The refinery costs are given as 0·404 pence per ton of ore milled in 1932, the gold content of the ore averaging about 12 dwt. per long ton.

Electrolysis in Silver Nitrate Solution.—The method presents some advantages over acid parting, among them being the absence of noxious fumes, except during acid treatment of the gold residues, the small amount of labour and chemicals required, the cleanliness and rapidity of working, and the absence of by-products, so that the loss of metal is reduced to a minimum. In consequence of these advantages, doré silver is generally treated in this way, if it cannot be combined with material richer in gold. The method is also in general use to prepare gold bullion for electrolysis in gold chloride solution.

Observations on the conductivity of electrolytes used in the process are as follows:—[2]

An electrolyte containing approximately 60 grms. copper and 60 grms. silver per litre shows good conductivity and gives heavy, dense silver crystals. The potential drop per cell is about 3 to 5 volts. Generally speaking, copper is more effective than silver in increasing the conductivity, but for *equivalent* concentrations silver nitrate has a higher conductivity than copper nitrate. Old electrolytes invariably behave better than new ones, this being attributed to the presence of ammonium nitrate, formed during the solution of excess silver and copper in nitric acid, or by cathodic reduction. Ammonium nitrate equivalent to up to 35 grms. of ammonia per litre is said to increase the electrical conductivity of the electrolyte used in parting. Its effect is greatest when the copper and silver contents are lowest. The silver crystals become shorter and more compact. Increase in silver content of the electrolytes causes the production of coarser crystals, while with additions of copper the aggregates are more compact.

---

[1] J. H. French and H. Jones, *Trans. Inst. Min. and Met.*, 1933, **42**, 227.
[2] Colcord, Kern and Mulligan, *Trans. Amer. Inst. Min. Eng.*, 1926, **73**, 108.

Sodium and potassium nitrates have also been used to increase conductivity.[1]

The Moebius Process.—The apparatus required consists of a number of earthenware vessels or wooden vats coated with asphalt or graphite paint. The electrolyte contains about 3 per cent. of $AgNO_3$ and 2 per cent. of nitric acid at the U.S. Mint refineries, but rather less at some other establishments. This is soon converted into silver and copper nitrate, in which a small percentage of free nitric acid is kept. At the mints of the United States and Canada the anodes usually contain from 30 to 35 per cent. of gold, and not more than 15 per cent. of base metals. They consist of plates of bullion about ½ inch thick, 18 inches long, and 10 inches wide, which are hung in muslin or cotton bags tied at the top round the suspending hooks. The bags retain the insoluble anode residue, consisting of gold, lead (as peroxide), platinum metals, antimony, etc., after the silver, copper, etc., have been dissolved. The anodes are usually entirely immersed in the electrolyte and are suspended by gold hooks. The cathodes consist of thin rolled plates of pure silver, slightly oiled to prevent adhesion of the deposited metal. Sometimes these plates are continually scrubbed by a mechanical arrangement of wooden brushes, by which their surfaces are kept free from loose crystals of electro-deposited silver. The loose silver falls on to trays placed below, which are removed at intervals, and the silver collected from them. At the U.S. mints and Ottawa, the silver accumulates on the cathodes, which are lifted out and scraped about every 8 hours. The washing filter used at Ottawa for the silver crystals is shown in Fig. 222. When melted, the cathode silver is 999 to 999·5 fine. It is necessary to agitate the electrolyte.

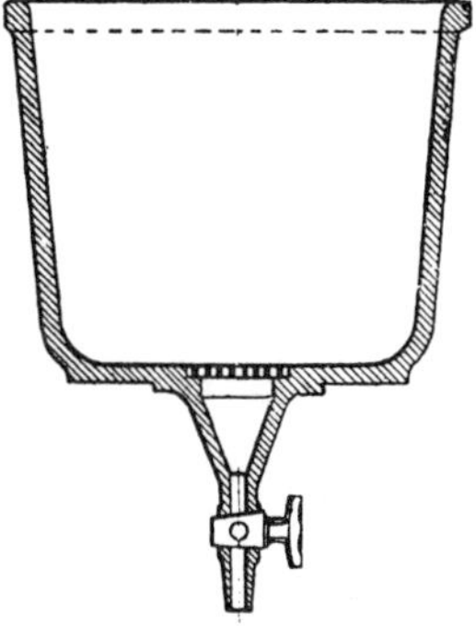

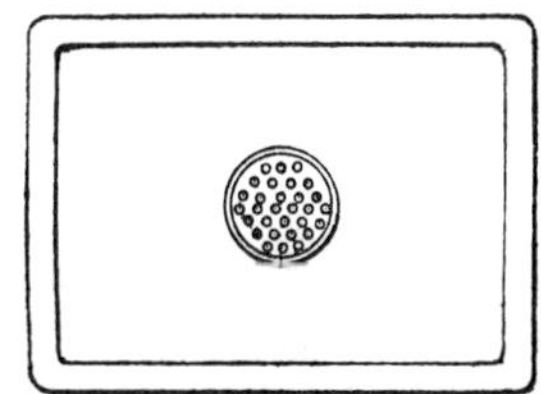

Fig. 222.—Earthenware and Porcelain Filters (Section and Plan).

The current requires an E.M.F. of 1·0 to 1·5 volts for each vat. The current density is usually 20 or 30 ampères per square foot of cathode. The copper is not deposited unless the solution becomes too weak in silver or too rich in copper. As copper accumulates in the solution, fresh acid and some $AgNO_3$ are added, and the current density is diminished. When the proportion of copper has reached about 5 per cent., part of the impure electrolyte is drawn off and replaced by silver nitrate and water. The silver is recovered from the waste liquor by precipitation with scrap copper, or, as at Ottawa, by electrodeposition using iron anodes and silver cathodes. Any silver still left in the waste liquor is precipitated with common salt.

At the U.S. mints the anode butts and cores are transferred to Balbach-Thum cells.

The process was installed at Perth Amboy, N.J., in 1897.[2] The anodes weighed 100 ozs. and the cathodes were sheets of silver about $\frac{1}{32}$ inch thick.

---

[1] Pyne, *ibid.*, p. 120.
[2] G. G. Griswold, *Eng. and Min. Journ.*, 1919, **107**, 789.

There were 4 bags (each containing 4 anode plates) and 5 cathodes in each cell. Wooden sticks with a reciprocating motion continually brushed the silver from the cathodes and at the same time stirred the electrolyte. The latter was a neutral nitrate containing 15-20 grms. silver and 30-40 grms. copper per litre. The current was 19½ ampères per square foot, raised to 40 ampères later. The bags of gold-slime (anodic) were boiled in sulphuric acid of 66° B. and then washed and cast into anodes for the Wohlwill cells.

It was found at Ottawa [1] that if the passage of the electric current is continued after the anode reaches 940 in gold, the silver almost ceases to dissolve, and oxygen is freely evolved at the anode. Accordingly anodes of this composition were melted down for transfer to the Wohlwill cells at that mint.

A rotating cathode was installed at the Ottawa Mint refinery in 1922 by A. H. W. Cleave.[2] The cell is round and the anodes are grouped round the cathode, which is mechanically scraped as it revolves. The detached silver falls into trays. A current density of 75 ampères per square foot, at 1·5 volts, was used in practice, but 150 ampères, with higher voltage, seemed possible. Anodes containing 85 per cent. silver, 5 per cent. gold and 10 per cent. base metal, mostly copper, were among those treated. The efficiency was about 73 per cent.

At Lithgow, in New South Wales,[3] the mud from the electrolytic refining of blister copper, with electrolyte of copper sulphate and sulphuric acid, is sifted, washed, heated to a dull red heat and boiled in concentrated sulphuric acid. The residue is then washed, dried and fused with sodium carbonate in a small cupel furnace with the production of bullion which usually contains 12 to 16 per cent. gold and 82 to 86 per cent. silver. This bullion is parted electrolytically in silver nitrate solution, a silver plate forming the cathode. The gold is caught as a mud in calico bags surrounding the anodes, and after being heated with nitric acid, is fused, giving a product usually from 990 to 998 fine. The silver is removed from the cathodes by moving wooden arms as fast as it is deposited. The detached crystals are collected and fused, the silver being not less than 996 fine.

The Balbach or Balbach-Thum Process.[4]—The chief characteristic of this process is that the anodes and cathodes are laid horizontally in the cells, instead of being hung vertically as in the Moebius process. The flat cells in the U.S. mints are 39 × 19 inches and 12 inches deep. They take the anode cores from the Moebius cells. The bullion is put in baskets or trays of wood or earthenware. The cathodes are carbon plates with silver connections. One of these is placed at the bottom. The anode connections consist of an alloy of gold 50 per cent., silver 50 per cent., which resists the action of the current and on which the crude bullion rests. The density of the current is 14 ampères per square foot and the E.M.F. is 5 volts per cell. The electrolyte is similar to that in the Moebius process, but there is no circulation. The black anodic gold is melted into anodes for the Wohlwill cells.

At the parting plant of the U.S.S. Lead Refinery [5] the cells are of acid-proof stoneware, 52 × 24 × 9 inches, and are in series electrically. The current density is 40 ampères per square foot and the E.M.F. per tank about

---

[1] A. L. Entwistle, Private Communication, 1914.
[2] *53rd Ann. Rept. Royal Mint*, 1922, p. 110; *Eng. and Min. Journ.*, 1923, 116, 21.
[3] *Trans. Australasian Inst. Mng. Eng.*, 1911, 15, 36; *J. Soc. Chem. Ind.*, 1914, 33, 202.
[4] *Elec. Chem. Ind.*, 1904, 2, 306.
[5] *Trans. Amer. Electrochem. Soc.*, 1926, 49, 351.

3 volts. The electrolyte contains 50 grms. silver and 60 grms. copper as their respective nitrates, and a small quantity of free nitric acid, per litre. The gold slime collecting on the duck lining of the anode baskets is gathered about every fifth day, washed with water, boiled twice with sulphuric acid to remove the last of the silver, which is afterwards cemented out from its solution by metallic copper. The gold is again washed with hot water and then melted down in a graphite crucible.

Mulligan [1] states that an increase in the copper content of the electrolyte from 55 to 75 grammes per litre resulted in a better production of silver crystals.

Electrolysis in Gold Chloride Solution.—In this process a solution of gold chloride containing hydrochloric acid is subjected to electrolysis. The anode consists of impure gold. Gold and most other metals are dissolved at the anode, and nearly pure gold is deposited at the cathode, which consists of a thin sheet of pure gold, sometimes corrugated to prevent buckling.

The process was described in 1863 by Charles Watt, of Sydney, in a communication addressed to the Master of the Mint, London, and now in the Mint Library.[2] Watt used a solution to which he added " from one-fourth to one-fifth its bulk of hydrochloric acid." This he " carefully added " until bubbles of chlorine ceased to be evolved from the anode. He does not mention the heating of the solution nor the addition of chloride of gold. The original solution was to contain 0·5 oz. of gold per pint, or 27 grammes per litre. The voltage used was 1·7, and anodes $\frac{1}{10}$ inch thick were dissolved in 24 hours. This would correspond to a current density of about 30 ampères per square foot. The silver chloride was brushed from the anodes if necessary. The process was not adopted at the Sydney Mint, on the ground that no parting process was necessary there. Watt also described in 1863 the electrolytic parting of gold-silver alloys, " containing not less than three parts of silver to one part of gold," in nitric acid solution. Watt was negotiating " for the introduction of his processes in Europe."

The process was installed at Hamburg in 1878 by E. Wohlwill [3] and spread through Germany to Russia. It was worked out independently in America and installed at Philadelphia in 1902.[4] It is now in wide use, especially for platiniferous gold.

The reactions at the anode, as given by Wohlwill, are as follows :—

$$(1)\ HCl + Au + 3Cl^- + 3 \oplus = HAuCl_4$$
$$(2)\ HCl + Au + Cl^- + \oplus = HAuCl_2$$

where 3 $\oplus$ denotes 3 units of positive electricity and $\oplus$ one unit.

At the cathodes the reactions are :—

$$(3)\ HAuCl_4 + 3H^+ + 3 \ominus = Au + 4HCl$$
$$(4)\ HAuCl_2 + H^+ + \ominus = Au + 2HCl$$

According to these equations the electrolyte would remain constant but for the impurities in the anode.

[1] *Trans. Amer. Electrochem. Soc.*, 1926, 49, 360.

[2] T. K. Rose, *Trans. Inst. Mng. and Met.*, 1915, 24, Presidential address.

[3] *Electrochemical and Metallurgical Industry*, 1904, 2, 221 and 261 ; *Zeitsch. Elektrochem.*, 1898, 4, 379, 402, 421 ; *J. Chem. Soc.*, 1899, 76, [ii.], 105.

[4] D. K. Tuttle, *Report of the Director of the U.S. Mint*, 1901 2, p. 123.

On the assumption that only trivalent gold is formed and exists in the solution, 2·45 grammes of gold should be dissolved and deposited per ampère-hour, but with monovalent gold the amount should be 7·35 grammes. Wohlwill's experiments, and the results obtained in working on a large scale, show that from 2·5 to 3 grammes per ampère-hour are usually deposited, and that the loss at the anode is greater than the amount deposited. It is, therefore, clear that a mixture of trivalent and monovalent gold exists in the solution.

Part of the *anodic loss* is accounted for by the separation of fine particles of gold, which are found for the most part in the silver chloride mud, into which the silver alloyed with the gold of the anode is converted. In ordinary working with the direct current, the particles of gold in the mud are equal to about one-tenth of the amount of gold deposited on the cathode. They are not merely mechanically detached from the anode, but are due to the formation of aurous chloride at its surface, which subsequently decomposes into metallic gold and auric chloride. Thus—

$$3AuCl = 2Au + AuCl_3$$
$$3HAuCl_2 = 2Au + 2HCl + HAuCl_4$$

In support of this view, Wohlwill has shown that the particles are much purer than the anode itself, that aurous chloride can be detected in the solution, and that under certain conditions minute crystals of gold separate out in all parts of the solution.

A series of experiments, with varying anodic current densities, made by Wohlwill, shows that the formation of monovalent gold ions diminishes as the current density increases. In experiments with very small current densities, the non-electrolytic dissolution of gold in the hot acid solution of $HAuCl_4$ (with formation of $HAuCl_2$) must be allowed for. The equation is as follows—

$$HAuCl_4 + 2Au + 2HCl = 3HAuCl_2$$

This appears to be reversible, a sheet of gold losing weight when the temperature is raised and gaining when it is lowered again. After allowing for this chemical attack, a current density of 1 ampère per square metre was found to correspond to a cathode deposit of 4·33 grammes and an anode loss of 6·01 grammes per ampère-hour. It is evident in this case that 72·9 per cent. of the gold had dissolved in the aurous state, but that only 38·6 per cent. of the deposited gold was from solution in the aurous state. This difference corresponds to a large amount of gold thrown down in the anode mud.

With 1,500 ampères per square metre (about 140 ampères per square foot), 2·48 grammes of gold per ampère-hour were deposited (as against 2·45 grammes, if pure $HAuCl_4$ had been present), and in two experiments the losses and gains at the anode and cathode were nearly equal, thus :—

	Loss at Anode.	Gain at Cathode.
(1)	105·2	104·5
(2)	107·7	105·0

Here, therefore, trivalent gold was present almost exclusively. Such results are obtained only when the liquid is well stirred. If the current density is too high, however, chlorine is given off at the anode and the electrolyte becomes impoverished in gold. A higher current density may be used if more HCl is present, and also if the temperature is higher. A temperature of 65° to 70° C. with 5 per cent. of HCl or a temperature of 20° C. with 10 per cent. of HCl represents average practice. It may be added that unusually high current densities may be used with unusually strong solutions of $AuCl_3$, but only if the cathode is rotated to prevent " treeing "[1] (see p. 480).

The existence of $AuCl_4$ ions in the solutions may be regarded as proved, one piece of evidence in support of this being the reddish-yellow unstable precipitates of $AgAuCl_4$ which are formed on adding silver nitrate. The ions of $HAuCl_4$ are H^+ and $AuCl_4^-$. On electrolysis hydrogen is released at the cathode, but is not evolved. It displaces gold in the compounds $HAuCl_2$ and $HAuCl_4$, so that the gold is precipitated and hydrochloric acid formed. At the anode the ions $AuCl_2$ and $AuCl_4$ are released and the nascent chlorine formed by the decomposition of these ions combines with the gold at the anode. Owing to the cathode deposition and the migration of the ions containing gold away from the cathode, the impoverishment in gold of the electrolyte near the cathode is very rapid, and it is nĕcessary to stir the solution. The circulation is effected by air-blowing or by a propeller, not by a pump. It can also be effected by a rotating cathode. In any case the anode slime is disturbed and may adhere to the cathode.

As the presence of monovalent gold causes its deposition in the anode sludge, it is desirable to avoid the formation of such gold, and to favour the dissolution of trivalent gold. This is a strong point in favour of using a high current density, and the need for rapid dissolution and reprecipitation of the gold in order to reduce the charges for interest on the value of the gold in the refinery is an even more cogent reason.

The influence exerted by the *composition of the anode* is of great importance. *Silver* is converted into chloride, but is not dissolved, except in small quantities, and remains adhering to the anode, reducing its effective surface and consequently increasing the anode current density and tending to cause chlorine to be given off. It has been found in practice with an electrolyte containing 5 per cent. of hydrochloric acid, and with the direct current used alone, that when the proportion of silver in the material to be refined exceeds 6 per cent., it is necessary periodically to brush off the deposit of silver chloride adhering to the anodes, in order to avoid the evolution of gaseous chlorine and to expose more anode to the electrolyte. With the pulsating current (see p. 486), this necessity is greatly modified, and gold with 15 to 17 per cent. silver can be refined.

The dissolved silver is in part precipitated with the gold on the cathode, and particles of silver chloride floating in the solution also adhere to the cathode and reduce the fineness of the gold. R. Pearson [2] obtained a gold deposit only 996·8 fine, containing 2·5 of silver, in some experimental work.

Pure *platinum* is not dissolved if it forms the anode, but platinum alloyed with gold is dissolved, and is not reprecipitated with the gold on the cathode. In a cold solution containing 2 per cent. of platinum and 5 per cent. of gold, some platinum is deposited with the gold, but if the solution is heated no platinum is deposited.[3] At 70° C. the platinum in the solution should not

---

[1] T. K. Rose, *loc. cit.*
[2] *44th Annual Report of the Royal Mint*, 1913, p. 182.
[3] T. K. Rose, *loc. cit.*

exceed 4 per cent.[1] In America small amounts of platinum are recovered from the solution, but the amount, if any, electro-deposited with the gold is unknown. R. Pearson [2] found some cathodic gold to contain a distinct trace of platinum, but does not give the analysis of the electrolyte.

In the year 1911-12 gold equivalent to 5,682,461 fine ozs. was treated in the refineries at San Francisco, Denver and New York. An amount of 362·25 ozs. of platinum sponge, or 0·06 per 1,000, was recovered, and in addition 27·77 ozs. of palladium and 1·18 ozs. of osmiridium.[3] The amounts recovered are very variable, and, according to D. K. Tuttle,[4] are derived only from dental and jeweller's scrap. Entwistle finds [5] that about 10 ozs. of platinum were recovered per 100,000 ozs. of the gold refined at Ottawa, which was received in about equal quantities from Porcupine and the Yukon.

Palladium goes into solution with the platinum, but is not allowed to reach a concentration of more than 0·5 per cent. in order to avoid reprecipitation. *Iridium*, *osmiridium* and *rhodium* do not dissolve, but accumulate in the anode sludge. *Lead* is converted into peroxide and causes passivity at the anode ; it also reaches the cathode and makes the gold brittle. Lead is accordingly removed beforehand. *Copper* accumulates in the solution, and if more than 5 per cent. is present in the anode the results are bad. Cupric chloride may crystallise out on the cathodes. Cuprous chloride seems to form in the solution and precipitate gold in the slime. An anode containing gold 85, silver 5 and copper 10 gave a slime containing 50 per cent. of the gold.[6] Selenium, tellurium, arsenic, antimony and bismuth, like lead, should be removed before the electrolysis, to prevent the deposited gold from being brittle. When the electrolyte becomes foul, owing to an accumulation of copper or other impurities, it is drawn off, and the copper and other metals are precipitated and recovered. The electrolyte is renewed most conveniently by withdrawing a part of it each day, and making good the deficiency with a solution of pure gold chloride.

Composition of the Electrolyte.—Under ordinary conditions with hot solutions the electrolyte contains from 25 to 40 grammes of gold and 50 grammes of hydrochloric acid per litre. If the anode becomes passive by the presence of too much silver or of an impurity such as lead the acid tends to increase by the liberation of HCl at the cathode. Too much acid causes strong fuming at 70° C. and is bad for the workmen. It is often necessary to reduce the acid by withdrawing part of the electrolyte and replacing with water containing $AuCl_3$. In cold solutions 10 to 12 per cent. HCl is used, with a lower current density. In laboratory tests [7] the following results were obtained :—

(1) When the amount of free hydrochloric acid in the bath is raised to about 30 per cent., a current of 500 ampères per square foot of anode surface can be used at 62° C. without causing the evolution of chlorine. Anodes of $\frac{3}{8}$ inch thickness are dissolved in about seven hours.

(2) Under these conditions the amount of gold passing into the anode mud is less than 0·1 per cent. of the amount dissolved.

---

[1] E. Downs, *Met. Ind.*, 1930, 36, 141.
[2] *Loc. cit.*
[3] *Report of the Director of the U.S. Mint*, 1912, p. 17.
[4] *Ibid.*, 1903, p. 63.
[5] Private communication, 1914.
[6] E. Downs, *loc. cit.*
[7] T. K. Rose, *loc. cit.*

(3) A further result is that the proportion of silver in the anode is of little importance, as the silver chloride flakes off under the action of the heavy current, and the pulsating current described below (p. 486) is unnecessary. Anodes containing 20 per cent. of silver are readily treated.

(4) The formation of a satisfactory *deposit of gold at the cathode* depends on the proportion of gold in solution as chloride and on the temperature. If the current density is not too great, the deposit of gold is yellow and coherent. If the density exceeds a certain maximum, the deposit is dark coloured and pulverulent. With 3 per cent. of gold in solution, a current of 30 ampères per square foot at 20° C. gives a black or dark brown powdery deposit. With 8 per cent. of gold in solution a current of 40 ampéres at 20° gives a coherent sheet of gold. With 20 per cent. of gold in solution, a yellow coherent sheet is formed at 25° with 150 ampères per square foot, and an almost equally good deposit at 62° with 500 ampères per square foot. This deposit is about 998 fine and of perfect quality, as regards its malleability after being melted, even when the anodes are very impure—*e.g.*, gold 780, silver 195, copper 25. With these high current densities and rapid deposition of gold it is necessary to use a rotating cathode to avoid "treeing."

In practice it is preferred that the gold should be deposited on the cathode in the form of particles which are easily detached, but which are sufficiently adherent to permit of thorough washing without loss. The crystals are not usually detached from the plate, but are melted with it. An increase in the concentration of the electrolyte gives a more closely adherent deposit. The presence of platinum in the electrolyte is also said to improve the density of the gold deposit,[1] but palladium has the opposite effect.

The deposited gold is fine-grained, the crystals being too small for measurement. It is of a high degree of purity, and the process offers a cheap and ready method of preparing proof gold, 1,000 fine, for use in bullion assaying. The deposited gold is usually from 999·5 to 999·9 fine, and is seldom of less fineness than 999.

The electrolyte gradually becomes weaker in gold chloride, owing to the impurities in the anode, even if no evolution of chlorine at the anode is allowed to take place. It is, therefore, necessary to add gold chloride to the bath. This is usually prepared by dissolving gold in aqua regia, but the electrolytic method worked out by R. L. Whitehead[2] is used in the United States mints. The process consists in passing an electric current through a solution of strong hydrochloric acid with anodes of gold 990 fine, and cathodes of gold suspended in a porous cup or cell. A strong current is used, "200 ampéres at 5 volts," and much heat is generated. At Ottawa only 17 to 20 ampéres per square foot at 2·4 volts are used. Acid is added occasionally to compensate for loss by evaporation, and for the conversion of HCl and Au into $HAuCl_4$. It is obvious that stirring is unnecessary, and that as the ions $AuCl_4^-$ migrate away from the cathodes, there will be little or no tendency for gold to appear inside the porous cups and to be precipitated on the cathodes.

The amount of gold chloride solution to be added to the electrolyte depends on the composition of the anodes. Thus, Wohlwill calculates[3] that if the anode gold contains 15 per cent. of silver, then 11·5 kilogrammes of the alloy must be dissolved "chemically" for every 100 kilogrammes deposited electrically, or about 10 per cent. If the composition is gold 90,

[1] J. B. C. Kershaw, *Electrician*, 1898, 41, 187.
[2] *Electrochemical and Metallurgical Industry*, 1908, 6, 357.
[3] *Loc. cit.*, 1904, 2, 261.

copper 10, then 20 per cent. of the alloy must be dissolved chemically. It is usual to add more gold chloride to the solution than is called for by theoretical calculations.

The introduction of the *pulsating current* by Wohlwill in 1908 [1] marked a distinct advance in the electrolytic process. The pulsating current is obtained by superimposing an alternating current on the direct current. The a.c. has a higher voltage than the d.c., so that the combined current is alternating. The advantages are :—(1) A stronger d.c. can be used without the evolution of chlorine at the anode ; (2) the proportion of silver in the anode may be as much as 15 per cent. without scraping of the electrode being necessary ; (3) less gold accumulates in the anode slime. Downs [2] adds that with an anode containing little silver, less HCl and a lower temperature may be used and the voltage may be reduced owing to the absence of passivity phenomena.

The ratio of a.c. to d.c. is usually 110 to 100 with about 10 per cent. silver in the anode, but with more silver a higher a.c. is required. A hot-wire ammeter is used to measure the heating effect, C_c, of the combined current, a.c. and d.c., and an ordinary ammeter gives C_d, the current reading for the d.c. Then

$$C_c = \sqrt{C_d^2 + C_a^2}$$

Hence C_a, the measurement of the a.c., is determined.[3] The frequency of the a.c. does not usually exceed 50 cycles. A lower frequency gives almost equally good results.

The effect of the a.c. on the AgCl layer on the anode is to make it softer and induce it to fall off. The anode is momentarily charged with hydrogen at each change of current, and the AgCl is alternately reduced and re-formed. This may help to make it more porous.

Allmand and Puri [4] have found that with d.c. alone solution of gold in hydrochloric acid begins at a certain minimum potential, depending on the acidity, and ceases at a certain maximum current density owing to the anode becoming passive. Further increase in potential then causes evolution of chlorine. With a superimposed alternating current the percentage increase in activity is greater the higher the acidity, the higher the temperature and the ratio $\frac{\text{a.c.}}{\text{d.c.}}$, and the lower the frequency of the a.c. The solution potential of gold decreases with increase in the $\frac{\text{a.c.}}{\text{d.c.}}$ ratio and with decrease in the frequency. The potential at initial evolution of chlorine is lower the higher the $\frac{\text{a.c.}}{\text{d.c.}}$ ratio and the lower the acidity.

In a study of electrolytic refining Kimata [5] found that the range of current density over which an impure gold anode is active is relatively small when d.c. alone is used, but is somewhat enlarged when the full-wave or half-wave from a mercury rectifier is applied. He gives various formulae for the ratios of Au^+ and Au^{+++} which take part in the reaction.

As an example of the effect of the pulsating current, if gold bullion containing 10 per cent. of silver is refined by means of the direct current

[1] *Met. and Chem. Eng.*, 1910, 8, 82.
[2] E. Downs, *loc. cit.*
[3] E. Downs, *loc. cit.*
[4] *Trans. Far. Soc.*, 1925, **21**, 1.
[5] *J. Min. Inst., Japan*, 1930, 46, 315.

only, an anodic current density of not more than 75 ampères per square foot must be used, and the silver chloride must be scraped off the anode every 45 minutes. If an alternating current is used the strength of which is about 1·1 times that of the direct current, a direct current of 125 ampères per square foot of anode can be used, without scraping the anodes. Gold containing 20 per cent. of silver can be refined by using the asymmetrical current, with $\frac{\text{a.c.}}{\text{d.c.}} = \frac{1{\cdot}7}{1{\cdot}0}$, and a direct current density of 120 ampères to the square foot. In practice, about 80 or 90 ampères per square foot are used.

The amount of gold passing into the anode mud is diminished by the use of the pulsating current. According to Wohlwill, with the pulsating current minute particles of gold form in the slime only at the beginning of treatment, when the anode has not yet been covered with silver chloride. At the end of the operation, when the anode finally breaks up, small pieces of gold are mechanically detached. Between these periods the accumulating slime consists almost entirely of chloride of silver. The result is that the pulsating current is an advantage even if the anodes contain very little silver.

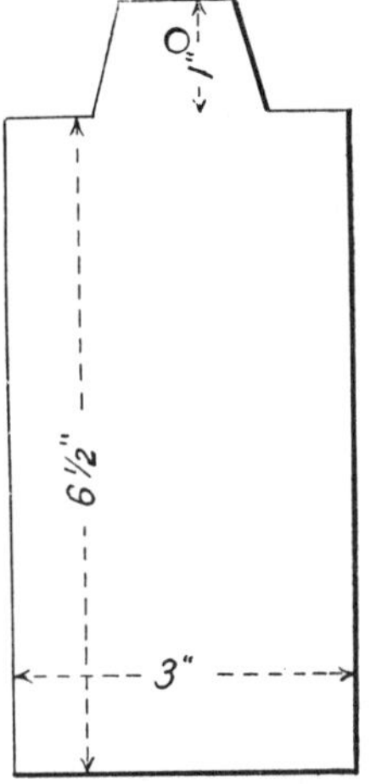

Fig. 223.—Anode Plate.

When using the pulsating current the amount of hydrochloric acid in the bath can be reduced from about 3 per cent. HCl (or, say, 6 or 7 per cent. of acid of sp. gr. 1·19) to a quarter of that amount if the solution is heated to 65° C. and the direct current density is about 100 ampères to the square foot.

Practice at the American Mint Refineries.—The anode slime from the Moebius and Balbach cells (see p. 479) is melted and cast into anodes containing over 85 per cent. gold, for the Wohlwill cells. The anodes are about 8 inches long, 3 inches wide and ½ inch thick, and weigh about 75 ozs. They taper downwards, since the metal dissolves more quickly at the top. At Ottawa (see Fig. 223), the anodes are ¼ inch thicker at the top than at the bottom. They are suspended by gold hooks, and the top is shaped so as to leave little metal suspended at the end of the operation. The anode is not completely immersed. The porcelain cells at Ottawa are shown in Fig. 224.

The cells at New York[1] are 19 by 15 inches and 12 inches deep. The solution contains 50 or 60 grammes of gold per litre and 5 to 7 per cent. of HCl. The current density (d.c.) is about 50-70 ampères per square foot and the pulsating current is used. The anodes dissolve in about 36 to 48 hours. The solution is assayed for gold by adding an excess of ferrous ammonium sulphate and titrating back with potassium permanganate.[2]

The cathodes are washed, drained, dried and melted into fine gold bars. The anode slime is treated in a porcelain tank with granulated zinc, by which the silver chloride is reduced. After filtering and washing, the slime, containing from 25 to 50 per cent. of gold, 3 to 4 per cent. of base metal and

[1] B. P. Wirth, *Report of the Director of the U.S. Mint*, 1912, p. 50.
[2] D. M. Liddell, "*Non-Ferrous Metallurgy*," 1926, 2, 991.

the rest silver, is melted with other material to form anodes for the silver cells. The slime contains about 5 per cent. of the gold in the original Wohlwill anodes.

The impure electrolyte withdrawn from the cells at intervals is treated by precipitating most of the gold with sulphate of iron or in a "plating" cell and subsequently recovering the platinum by precipitation with ammonium chloride or iron plates.

Fig. 224.—Gold Cells, Ottawa Mint.

At Denver [1] the introduction of the pulsating current enabled the minimum fineness of the anodes to be reduced from 900 to 800, so that half the silver cells were thrown out of use. The finely divided and scrap gold in the slime fell from 17·19 per cent. of the fine gold produced to 11·37 per cent. The "cost of operating the refinery" fell from 3·2 cents per oz. of fine gold to 2·0 cents.

At the *American Smelting and Refining Co.*,[2] Perth Amboy, N.J., the anode slime from the Moebius cells was melted, cast into anodes, and subjected to a current of 150 ampéres per square foot in 5 stoneware cells standing in a lead-lined box filled with hot water. The heat was supplied by steam. The cathodes were of sheet gold. The electrolyte contained 30

[1] *Report of the Director of the U.S. Mint*, 1913, p. 41.
[2] G. G. Griswold, *Eng. and Min. Journ.*, 1919, 107, 789.

per cent. of free HCl and 80-85 grammes of gold per litre. Some platinum and palladium were recovered.

Treatment of Anode Slime from the Electrolytic Refining of Copper.—At the Hitachi works in Japan[1] the slime from the electrolytic vats contains gold, silver and platinum. The impure mud is melted into anodes and used as such in a Wohlwill cell, employing a temperature of 70° C. and a maximum hydrochloric acid content of 5 per cent. Alternating current is superimposed on direct current. The platinum elements, with the exception of iridium, pass into solution. The iridium goes into the slime and is easily separated.

Electrolysis in Sulphuric Acid.—The method is similar to that used in the electrolytic refining of copper. It was applied in Nicaragua[2] to low-grade bullion containing gold 146, silver 100, zinc 20, copper 734. The bullion was produced by cyaniding the tailing from amalgamated plates. The precipitate was melted and cast into slabs measuring 20 by 12 inches and $\frac{3}{4}$ inch thick. These were suspended in canvas sacks in a wooden cell 36 × 24 inches and 15 inches deep. The cathodes were of sheet lead. The electrolyte contained 17·6 per cent. of $CuSO_4.5H_2O$ and 5 per cent. of H_2SO_4. It was heated to 150° F. and circulated. The electric current used was 50 ampères at 2 volts. The anode slime assayed 920 to 970 in gold, and nearly all the silver was contained in the copper which had been deposited on the lead.

---

[1] *Japan J. Min.*, 1928, 44, 785.
[2] T. W. Bouchelli, *Eng. and Min. Journ.*, 1913, 95, 238.

CHAPTER XVIII.

THE ASSAY OF GOLD ORES.

Introduction.—The assay of gold ores was probably practised by the Romans,[1] and was elaborated in the Middle Ages. The German assayers had already reached a fair degree of proficiency in the year 1500.[2]

The assay of gold ores is generally conducted in the *dry* way—*i.e.* by furnace methods. The plan of operation is to concentrate the precious metal in a button of lead either (1) by fusion in a crucible ; or, more rarely, (2) by scorification. The button of lead obtained by either method is then subjected to *cupellation*, by which the lead is oxidised and removed, and the resulting bead of precious metal is weighed. In these operations silver and the metals of the platinum group remain with the gold, and are subsequently separated by *inquartation* and *parting*, and, in the case of platinum and its allies, by further special methods.

The assay by scorification was preferred to pot-fusion by the German assayers of the 16th century. When fusion in a crucible was recommended, it was only as a preliminary to scorification.

Sampling the Ore and Preparation of the Sample for Assay.—The value of an assay depends largely on the care with which a sample of the ore is selected. Sampling should be, as far as possible, automatic, and independent of the will or judgment of the assayer. The sampling of ore in place in mines is described in books on mining.

The size of the sample depends on the size of the particles. A larger sample is required for larger particles. Argall [3] gives the following figures for gold ore :—

Average size of ore cubes, .	1·0″	0·25″	8 mesh (0·062″)	30 mesh (0·017″)
Sample, percentage of bulk,	20	1·25	0·078	0·005

Heaps of ore, placer deposits and impounded tailing are sometimes sampled in the same way as vat or bin charges by driving-in at regular intervals iron tubes, 1 or 2 inches in diameter and open at one side. The samples withdrawn in the tubes are then mixed. A form of sampling iron suitable for very dry material is described by Richards.[4] It consists of two

[1] See Hoover's translation of "*Agricola*," 1912, p. 219, note 1.

[2] For an account of the "Probierbüchlein," published in Germany between 1510 and 1556, see Hoover's "*Agricola*," appendix B, p. 612; they are the earliest treatises on assaying known to exist. See also Ercker, "*Allerfurnemisten Mineralischen Eerzt u. Bergwerks Arten*," Frankfort, 1580, Book i., chap. 10 ; and Book ii., chap. 8. "*Laws of Nature in Assaying Metals*," by Sir John Pettus, London, 1686, is a translation of Ercker's work.

[3] W. R. D. Jones, *Met. Ind.*, 1928, **33**, 125, 199.

[4] "*Ore Dressing*," vol. ii. (1903), p. 845.

iron tubes, each with nearly half its circumference cut away. One tube fits inside the other, and is fixed to a wooden handle by a T-piece, so that it can be rotated inside the outer tube. The lower or entering part of the outer tube is not cut away, and is sharpened at the end or drawn out to a point. The tubes are driven into the ore with their openings arranged as shown in the cross-section *a* (Fig. 225), and the inner tube is then rotated, passing through the position *b* (Fig. 225), until it is filled with ore. It is then returned to its original position, and both tubes withdrawn. The outer tube must also be clamped to a cross-handle, to facilitate its withdrawal.

When the ore to be sampled can be moved, it is reduced in bulk either by hand or by automatic machines, and is generally crushed between each successive reduction of bulk in accordance with requirements based on theoretical considerations.[1]

If the ore is in course of removal by shovelling, every second, fifth, or tenth shovelful may be set aside as a sample, or the whole heap may be piled up to form a perfect cone, each shovelful being thrown as nearly as possible on the exact apex of the cone, so that the ore runs down on all sides. The heap may be made into a new cone to ensure thorough mixing, or may

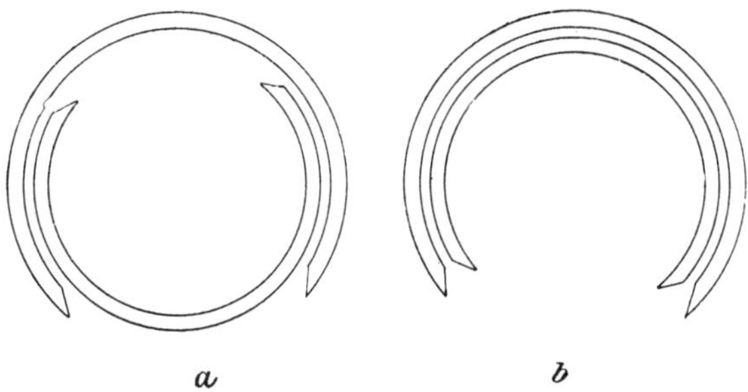

Fig. 225.—Sampling Tube for Dry Ore.

be at once flattened into a circular cake, and divided into four quarters, along two diameters at right angles to each other. Two opposite quarters are removed and mixed, and the operation repeated if the ore is fine enough. "Coning and quartering" is usually preferred to the method of setting aside particular shovelfuls.

Automatic samplers give better results than hand sampling, and their working costs are smaller. They are always used in customs work, for the sale of ore. They are divisible into two classes—(1) those which continuously take a part of a moving stream of ore, and (2) those which momentarily take the whole stream of ore at regular intervals of time. The former class of machine is the cheaper, but is not much used because the values are never evenly distributed across the stream. In many machines a narrow scoop passes steadily across the stream of ore at regular intervals of time, and this is considered to be equivalent to the method used in the machines of class (2). General principles to be observed in accurate ore sampling are given by Warner.[2]

The *Vezin sampler*[3] is well known in all parts of the world. It consists

[1] Smith, "*The Sampling and Assay of the Precious Metals,*" p. 90.

[2] *Eng. and Min. Journ. Press*, 1925, **120**, 175.

[3] Argall, *Trans. Inst. Min. and Met.*, 1902, **10**, 240, 241, with working drawings.

of two sheet-steel truncated cones, *b*, *c* (Fig. 226), with their bases bolted together. The ore falling from a shoot, *a*, into a hopper, *d*, is carried outside the cone *c*, and is delivered from the hopper through another shoot, *e*. A scoop, *f*, made of sheet-steel, with a sector-shaped opening above, is rivetted or hooked on to the cone *b*. The angle of the sector may subtend any desired portion of the circumference of a circle, such as one-tenth (36°), or one sixteenth (22½°). Both cones and the scoop are rotated at about 25 or 30 revolutions per minute by bevelled gearing, and when the scoop comes below the shoot *a*, the whole stream of ore falls into it, and is led into the interior of the cone *c*, and thence to a separate truck or bucket. In this way a sample is taken about every two seconds. A Vezin sampler of about 3 feet in diameter, and requiring a fall of about 6 feet, is trifling in cost, and treats 30 or 40 tons an hour. The Snyder type is also useful.[1]

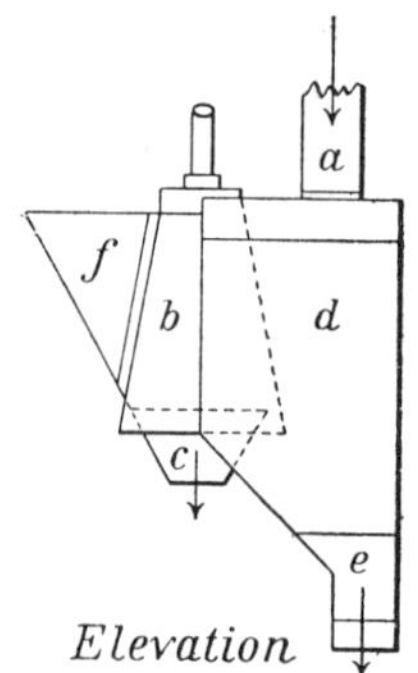

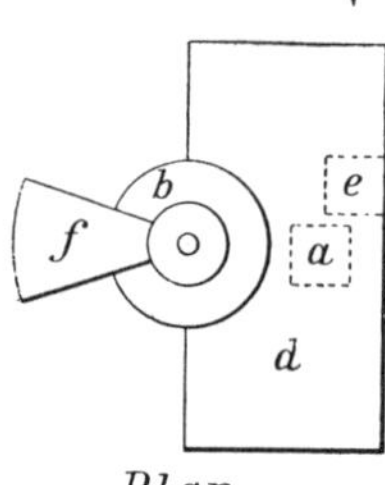

Fig. 226.—Vezin Sampler.

Argall recommends[2] the following course of procedure in sampling telluride ores containing 10 or 15 ozs. per ton:—From 100 short tons crushed to 1 inch cubes, take 20 tons. Crush this to ¼ inch, and take 2 tons. Crush this to 8 mesh (= $\frac{1}{16}$ inch), and reduce by "riffling" to 250 lbs. Dry and crush the sample to 30 mesh (= $\frac{1}{60}$ inch), and riffle down to 15 lbs. Crush by a sample-grinder to 90-100 mesh, and riffle down to 1 lb. This is crushed on the buck board to pass 120 mesh (= $\frac{1}{250}$ inch), and divided for assay. Poorer and less "spotty" ores may be sampled somewhat more simply, and without such fine preliminary crushing.

A modern plant for commercial ore sampling in Utah is described by Hofstrand.[3] The units handle 100 tons or more per hour. The general practice is to crush to ⅜ inch maximum size before sampling begins and to crush finer as sampling proceeds. The sample cutters move at a uniform velocity across the stream of ore and make from 39 to 47 cuts per minute. Finally a sample of 4 lbs. per ton is obtained, after passing through 10 mesh screens. This is further crushed and divided in the Assay Office.

In the Assay Office the sample, however obtained, may be further reduced in bulk by an automatic machine, similar in principle to the large samplers, or by the *riffle* or "*splitter*" (Fig. 227).[4] An even stream of ore being let fall from a shovel on to this sampler, half is retained in the troughs, while half passes through. Repetition of the process reduces the sample to any required extent. A sampler with troughs 1 inch wide is suitable for treating materials which include lumps of not more than about ⅕ inch in diameter. For finely ground materials a convenient width for the troughs is ⅜ inch. The *split shovel* (see Fig. 228)[5] is similar in principle to the riffle, and in the *Jones*

---

[1] W. R. D. Jones, *loc. cit.*
[2] *Op. cit.*, p. 238.
[3] O. B. Hofstrand, *Eng. and Min. World*, 1930, I, 136.
[4] E. A. Smith, "*The Sampling and Assay of the Precious Metals*," p. 122.
[5] Smith, *op. cit.*, p. 99.

sampler,[1] instead of troughs and spaces, two sets of troughs are substituted sloping at a high angle in opposite directions. The troughs clear themselves, discharging half the ore at each end of the machine.

Mechanical appliances for sampling in the Assay Office, however, are objected to by Johnston[2] because it is difficult to clean them, so that "salting" results, and also on account of dusting.

When the sample has been reduced to from 5 to 20 lbs. in weight it is crushed through a 20 mesh sieve, and from 1 to 5 lbs. selected for the assay sample.[3] This sample is used for the estimation of moisture, and is then dried and finally crushed through sieves of 100, 120 or finer mesh.

For pulverising the sample, disc machines are generally used, with a rotating disc working against a stationary disc. These grinders take $\frac{1}{4}$ inch ore and give a product which will pass through a 100 mesh, or even finer, screen. Thorough cleaning of the machines is necessary to avoid "salting." Bugbee[4] recommends jaw-breakers and rolls, as being easily cleaned. Small samples may be ground on the bucking-board.

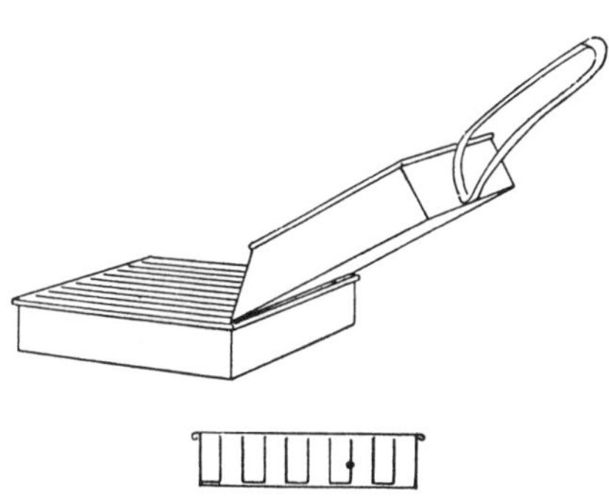

Fig. 227.—Sampling Riffle.

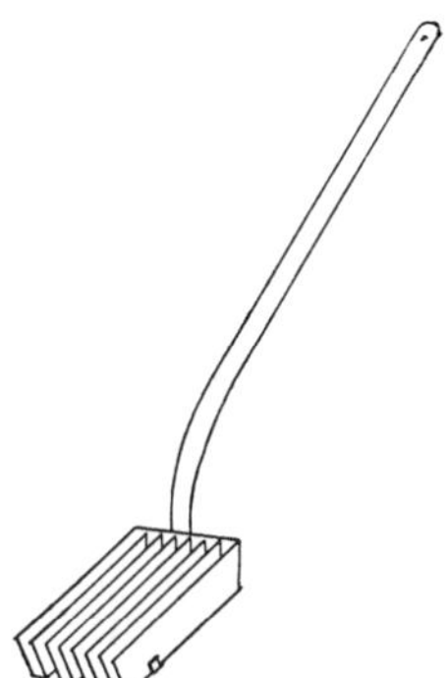

Fig. 228.—Split Shovel for Sampling.

H. A. White[5] recommends that all pulverising machines should be carefully housed and connected with an exhaust to keep dust out of the Assay Office.

The prepared sample is mixed on glazed paper or sheet rubber by "rolling," *i.e.* lifting up the corners in succession, each time shifting the whole sample. The ore is then spread out evenly and portions for assay taken with a spatula, each dip representing a complete section of the stuff from top to bottom. The application of Poisson's formula to ore sampling is discussed by Beringer[6] and applied by Edmands,[7] who shows that the error

---

[1] Described by P. Argall, *Trans. Inst. Mng. and Met.*, 1902, **10**, 271; *Eng. and Mng. J.*, 1903, **76**, 729.

[2] "*Rand Metallurgical Practice,*" 1926, vol. i., p. 291.

[3] The views of various authorities as to weights of samples and fineness of preliminary crushings in the sampling of large lots of ore are given in Richards, "*Ore Dressing*" (1st ed., 1903), p. 848. Richards gives a rule that for any given ore, the weight taken for a sample should be proportional to the square of the diameter of the largest particles.

[4] E. E. Bugbee, "*Text-book of Fire Assaying,*" 1933, p. 48.

[5] "*Rand Metallurgical Practice,*" vol. i., p. 528.

[6] "*Text-book of Assaying,*" 11th Ed., 1908, p. 444.

[7] *J. Chem. Met. Mng. Soc. S.A.*, 1920, **20**, 180; and 1921, **21**, 45.

in a 2 A.T. sample may be 1·65 dwt. per ton if gold particles of $\frac{1}{100}$ inch diameter are present.

Metallics.—Coarse gold in ores is retained on the fine screens after grinding. The particles which have resisted abrasion are collected carefully, fused or scorified and then cupelled separately. The value per ton is calculated and added to that obtained by ordinary assay. It is not good practice to pass metallics through the screen.

Assay by means of the Blowpipe.[1]—This method, though less exact than the furnace methods, is sometimes useful. The amount of powdered ore taken is usually 100 milligrammes, and this is mixed with borax and about 1 gramme of granulated lead. The whole is wrapped in paper and heated on charcoal in the reducing flame of a blowpipe until the fusion is complete, and then for a short time in the oxidising flame. The lead is separated from the slag and heated on a bone-ash cupel until it is all converted into litharge. The nearly spherical bead of silver and gold thus obtained is weighed or its diameter measured.[2]

Fusion or Crucible Process of Assay.

This process is divided into four parts, viz. :—(1) Fusion ; (2) Cupellation ; (3) Parting ; (4) Weighing the gold.

(1) **Fusion.**—The object of this operation is to concentrate the precious metals in a button of lead, while the whole of the remainder of the ore forms, with suitable fluxes, a fusible slag in which lead sinks. The fusion is made in a fireclay crucible, which is placed in a muffle furnace, the usual practice in America, or on the floor of a reverberatory furnace, as in South African and Australian practice. The crucible may also be embedded in coke in a pot furnace, an old-fashioned practice not yet obsolete.

Muffles usually consist of fireclay, but carborundum [3] muffles, though costly to install, last longer and soon save their cost. The muffle is a closed oven, heated from outside, with a domed roof and an open front or mouth which can be closed by a firebrick door, hinged or otherwise. An opening in the upper part of the door lets in air and permits observation, but it can be partly or entirely closed with sliding sheets of transparent mica or with firebrick. If the muffle is being used for cupellation, the products of combustion of the fuel are not allowed to enter it, but a current of air passes in at the mouth and out through a vent controlled by a sliding damper at the back. For precise work, the current of air may be drawn through the muffle by a regulated pump. When used for pot fusion, the muffle does not need the current of air through it.

The fuel used is of any available kind. Bituminous coal is most favoured for heating both muffles and reverberatory furnaces. The bed of coal does not touch the bottom of the muffle, which is heated by the flames. Two or more muffles may be heated in one furnace. In West Australia wood is used, as it is cheaper than coal.[4] The flames from the wood strike the roof of the reverberatory furnace and are reflected on to the pots, which last for many fusions.

---

[1] For a full description of this method, see Plattner's "*Blowpipe Analysis,*" translated by Prof. Cornwall, 8th ed., New York, 1902, pp. 350-384.

[2] L. Wagoner, *Trans. Amer. Inst. Min. Eng.*, 1901, **31**, 798.

[3] *62nd Ann. Rept. Royal Mint*, 1931, p. 121.

[4] M. W. von Bernewitz, *Chem. Eng. and Min. Rev. (Aust.)*, 1934, **27**, 47.

Muffles are also heated by gas, either natural or manufactured, with or without an air blast, or by kerosene, fuel oil or petrol, with air blast. Electric resistance muffle furnaces are also used, notably at Hollinger,[1] where they replaced oil furnaces with a saving of cost. (See also Chapter XIX.) Coke pot furnaces usually take only two or three pots, but as many as 20 " 20 gramme " pots can be placed in a large muffle, and more still in a reverberatory furnace. One of the last-named described by Johnston [2] as typical of Rand practice has space for 40 " F " Battersea pots or 16 to 20 " J " pots. It is heated by coal.

Slag overflowing from the pots may corrode the muffle or the floor of the reverberatory furnaces, but a lining of tiles will take up the wear. A covering of bone-ash is also useful.

In weighing the materials for a crucible charge the use of a set of *assay-ton* weights saves much labour in calculation. The assay-ton is a weight which contains as many milligrammes as a ton contains ounces. Thus an English or long ton of 2,240 lbs. contains 32,666 Troy ounces, so that the corresponding assay-ton must weigh 32,666 mg. or 32·666 grammes. If the weight of the resulting bead of gold (or silver) from an assay-ton of ore is 1·5 milligrammes, then the ore contains 1·5 ozs. of gold per statute ton. If the value per short ton of 2,000 lbs. is required, the weight of the assay-ton is 29·166 grammes. Other assay-tons are sometimes used. For example, at Morro Velho [3] an assay-ton of 1,568 grains is used, so that 0·001 grain of gold in the assay represents 10 grains of gold per long ton of ore. In the following pages 1 A.T. is equal to 29·166 grammes.

The weight of ore taken for assay varies according to circumstances. 1 A.T. is a suitable quantity if the amount of gold is not less than 0·5 oz. per ton. With mill samples, 5 to 10 A.T. may be taken, and with poor residues, 20 A.T. If several A.T. of ore are taken, the charge is divided between two or more pots, and the resulting lead buttons scorified down and cupelled together.

The charge is made up of litharge or red lead,[4] charcoal, and suitable fluxes. The amount of lead reduced is usually from 25 to 30 grammes per A.T. of ore.

The amount of the *charcoal* added varies with the reducing power (percentage of ash, etc.) of the particular sample which is employed, as well as with the degree of oxidation of the ore. If too little charcoal is used, the amount of reduced lead will be too small, and if too much charcoal is added, the charge remains pasty and emits bubbles of combustible gas (CO). If ores contain much sulphur, no charcoal is used, and nitre may even be added to burn off the excess of sulphur. Flour, mealie meal or argol (potassium bitartrate) are preferred to charcoal by most assayers as reducing agents. 1 gramme of charcoal is capable of reducing about 25 grammes of lead from litharge, and its place may be taken by 2½ grammes of flour or 3 grammes of argol. These equivalents, however, are approximate, and vary with the sample of the reducing agent. The reducing power of each sample is tested

---

[1] W. R. Dodge, *Can. Min. and Met. Bull. No.* 225, 1931, p. 115.

[2] A. M'A. Johnston, "*Rand Metallurgical Practice*," vol. i., p. 297.

[3] J. H. French and H. Jones, *Trans. Inst. Min. Met.*, 1933, 42, 228.

[4] Litharge, PbO, is used by most assayers, but red lead (Pb_3O_4) is preferred by some on account of its greater purity. Litharge is of course objectionable if it is seriously impure. It always contains silver, the amount of which is determined. The gold in litharge is generally insignificant in quantity.

beforehand. One advantage of argol is that it increases fusibility and avoids the foaming and boiling caused by charcoal or flour.[1]

Sodium carbonate is used to flux silica and to take up sulphur. Bicarbonate of soda is less convenient than the normal carbonate. If it is used, the water and excess carbon dioxide must be driven off by slow and cautious heating.[2] *Borax* is valuable to flux metallic oxides, and render the slag more liquid. The relative amounts required are judged from the appearance of the ore and from the results of panning in the first place, and afterwards modified according to the success of the fusion. Even when the ore is entirely siliceous, some borax is added, but too much borax causes the slag to adhere to the lead button. The most convenient form is anhydrous borax.

Silica is used for ores full of lime, baryta, compounds of the base metals, etc., or generally whenever the ore does not contain much quartz. It aids fusion in these cases, and protects the crucible from corrosion. From $\frac{1}{4}$ to $\frac{1}{2}$ A.T. of silica to 1 A.T. of ore is generally enough. About two parts of glass may be used instead of one part of silica. *Fluorspar* is added to the charge when the ore contains sulphate of barium or calcium, and in the fusion of cupels. Like borax, it increases the fluidity of almost any charge, but it attacks the crucible, and care must be taken to avoid deficiency of silica when it is used. It is not good practice to use it instead of borax. In general it may be noted that for basic impurities an acid flux is used and for an acid gangue a basic flux. Flux, if too acid, sends values into the slag ; if too basic, it corrodes the pot.[3] A cover of *common salt* to protect the charge from gases is falling into disuse. A cover comprised of part of the flux is sometimes used. Iron is sometimes added in the shape of large nails or hoop iron. Sulphur, arsenic, etc., are thus kept out of the lead button.

The charge for any given ore is made up by the assayer according to his judgment and experience.

For an average Rand ore Johnston [4] recommends—

Ore, ...	1 A.T.
Litharge, ...	1·5 ,,
Sodium carbonate, . .	1·5 ,,
Borax, ...	0·5 ,,

with reducing agent to give 25-30 grms. lead, and if the ore is heavily pyritic, 0·5 A.T. litharge extra. More flux and litharge are advised by other Rand assayers.[5] Bugbee [6] suggests 70 grms. litharge for 1 A.T. of siliceous ore and less soda and borax than that given above.

Methods of Operation.—The ore is weighed accurately. The charcoal is always weighed with great care ; the litharge is best measured by a ladle or shot measurer ; the fluxes may also be measured out by a ladle more rapidly than they can be weighed. The various ingredients are thoroughly mixed together on rubber cloth or in the crucible in which the fusion is to take place. Part of the borax is often kept separate and used as a cover, being put on the top of the rest of the charge after transference to the crucible. If a large number of assays is to be made on an ore, as at a mine,

[1] W. R. Dodge, *Can. Min. and Met. Bull. No.* 225, 1931, p. 115.
[2] J. Bettel, *J. Chem. Met. Mng. Soc. S.A.*, 1899, 2, 467.
[3] W. R. Dodge, *Can. Min. and Met. Bull. No.* 225, 1931, p. 115.
[4] "*Rand Metallurgical Practice*," vol. i., p. 302.
[5] J. Moir and G. H. Stanley, "*Rand Assay Practice*," 1923, p. 181.
[6] E. E. Bugbee, "*Fire Assaying*," 1933, p. 163.

a standard flux is made up, the ingredients being mixed in a barrel. The flux is then measured out as required. The fluxed sample may be put into a paper bag and dropped into the pot.

The crucible is annealed before using. The fire should be at a low red heat on charging in—that is, at a temperature of 600° or 700° C.—and should not be urged at first. The objects to be attained are as follows :—

(1) The gold is to be brought into such a condition that it will be readily taken up by the molten lead before the subsidence of the latter to the bottom of the pot.

(2) The mixture is to be brought finally to a state of quiet fusion, with low viscosity, so that the reduced lead may subside completely and collect into one button.

(3) There must be no mechanical losses.

Reduction of the lead begins at 500°C. and is rapid at 600°. Sodium carbonate melts at 852° and litharge at 883°, but they begin to sinter with silica at about 700°. Borax melts at 742° and begins sintering at 500°C. The lead begins to collect into globules visible to the naked eye as soon as the charge begins to work and effervesce, but does not sink until the effervescence has proceeded for several minutes. The slower the melting (*i.e.* the more the "fritting" stage is prolonged) the more chance is afforded of bringing the lead into intimate contact with all parts of the pasty mass and of collecting the gold from the ore. Also the slower the melting, the less chance there is that part of the charge will be projected from the pot. With charges containing nitre, however, it is recommended to fuse quickly in a hot fire. The final temperature need not, as a rule, be above 1,100° C., and may be lower in certain cases. In the electric muffles at Hollinger, the pyrometer readings during fusion range from 870° to 1,010°C.[1]

The effervescence is due chiefly to the escape of carbon dioxide from the sodium carbonate as it unites with silica, and is of value in moving about the lead globules.

After a lapse of 30 to 50 minutes the charge is in a state of tranquil fusion. A high temperature for finishing is recommended. To clean the slag, it is "washed" by adding some litharge and a pinch of charcoal. Then after a brief pause, the pot is taken out of the furnace, allowed to cool, and broken by a hammer to extract the lead button, or the charge may be at once poured into an iron mould. The mould must be cleaned, blackleaded, and warmed before being used. The slag is detached from the lead button with a hammer on an anvil.

If the charge does not fuse completely, so that the slag is pasty or has lumps in it, it is advisable to recommence the assay, making such alterations in the charge as experience suggests. The slag should be glassy and homogeneous ; if it is streaked it is probable that the fusion has not been perfect. It is green and transparent if the ore is nearly pure quartz, but black and opaque if much iron is present.

Washing the slag, as described above, is not always enough. If slags are rich, they are re-fused in the same pot with the addition of litharge, flux and a reducing agent.

The lead button should be soft and malleable. If it is hard or brittle it may contain sulphur, arsenic, antimony, copper, etc. Sulphur and arsenic are kept from entering the lead by the addition of iron, and may form with

[1] W. R. Dodge, *Can. Met. and Min. Bull. No.* 225, 1931, p. 115.

the iron a matte or speiss, which separates as hard blackish-grey or white layers found just above the lead. They are richer than the slag, and may often yield appreciable quantities of gold on further treatment by scorification or fusion. If the quantity of sodium carbonate is large enough, the sulphur is retained in the slag. Antimony makes the lead hard, white, brittle and sonorous. The lead must be completely freed from the slag, very small quantities of the latter interfering seriously with the cupellation, forming a scoria and occasioning loss.

(2) **Cupellation.**—This operation is conducted in a muffle furnace.[1] Bone-ash is sprinkled on the floor of the muffle to prevent its corrosion by litharge in case of the upsetting of a cupel. The cupels are cleaned by gentle rubbing or blowing before being charged in, and are again cleaned with bellows before the lead is added. They are placed in the furnace one by one, or, better, charged in together on a graphite or nichrome tray.

Cupels [2] are made of bone-ash. In their manufacture the bone-ash is finely powdered so that it will pass a 40 mesh sieve, then slightly moistened with water, put into a mould and compressed by hand or in a press. Cupels must be dried carefully and slowly, but completely, as otherwise cracks appear when they are heated. Cupels made several months before being used and dried slowly are often said to be better than those only a few days old, but J. T. King [3] found no difference in the results. If too much water is used in mixing the bone-ash, the cupels lose part of their porosity; if too little is used they are too soft and crumble readily. About 6 or 7 per cent. of water answers very well.[4] Care must be taken to preserve cupels from the access of nitrous or other acid fumes. These are absorbed by the bone-ash and given out in the furnace so that they are liable to cause spitting. Lengthened heating of an hour or more in the muffle before charging in the lead buttons will prevent loss from this cause. It is always advisable to heat cupels for 15 minutes after they have attained the temperature of the muffle before charging in the lead buttons.

Magnesite (mabor, morganite), or rather a mixture of magnesite, lime and some silica, is also in use as a substitute for bone-ash. It is strongly calcined, and does not absorb acid fumes. The cupels require to be made under high pressure, and should be baked at a high temperature before use. They are then as hard and strong as fireclay. Magnesite absorbs and holds about two-thirds of its weight of litharge, and bone-ash about its own weight. Good cupels are made of mixtures of bone-ash and magnesite, with or without an addition of cement, and also of mixtures of 25 to 50 per cent. of bone-ash with Portland cement. "Cupel clay" also makes good cupels.

Portland cement is cheaper than other materials for making cupels, but gives inferior results. One objection is that cement sticks firmly to the button of gold and silver [5] and cannot be removed. Unlike bone-ash, it is not soluble in nitric acid, leaving gelatinous silica which may cause

---

[1] H. R. Edmands exceptionally uses his reverberatory fusion furnace for cupellation without a muffle. The atmosphere is reducing at first, then oxidising. *J. Chem. Met. Mng. Soc. S.A.*, 1920, **20**, 180.

[2] For an exhaustive discussion of the manufacture and use of cupels, with many experiments on cupel loss, see J. T. King, *Univ. of Toronto Eng. Res. Bull. No.* 147, 1934.

[3] *Op. cit.*, p. 40.

[4] Bugbee, in his "*Text-book of Fire Assaying,*" recommends about 12 per cent. water. The standard practice in America is apparently about 10 per cent.

[5] J. W. Merritt, *Min. and Sci. Press*, 1910, **100**, 649.

embarrassment or error. The cupels are made in the usual way with 6 to 10 per cent. water.[1]

The cupels having been sufficiently heated in the muffle, the lead buttons are charged in by tongs. The buttons collapse and lose their shape in a few seconds if the temperature is sufficiently high, but the molten mass formed is covered by a black crust for some time later. The crust then breaks up, and the brilliant surface of a liquid bath is seen. The muffle door should be closed immediately after the charging in is completed, and kept closed until all the assays have thus " uncovered." The door is then opened. A current of air oxidises the lead, and the litharge, floating to the edge of the bath, is absorbed by the cupel, together with the oxides of other base metals. Uncombined metals are not absorbed by the cupel, and only traces of gold and silver are carried into it by the litharge (see p. 500).

As the cupellation advances, the lead bath becomes more convex and brighter, owing to local heating by the oxidation of the lead, which becomes hotter than the cupel, especially if this is made of bone-ash. Magnesia has a higher conductivity than bone-ash and conducts away more heat from the lead. It follows that a higher muffle temperature and a high finishing temperature are required with magnesia. Finally the bead becomes suddenly much duller in appearance, thus indicating that the cupellation is at an end. The cupels may then be removed from the muffle. As silver absorbs oxygen when molten and gives it off suddenly when solidifying, " sprouting " or " spitting " may take place in argentiferous-gold beads unless a trace of copper is present. Where spitting is to be feared, therefore, the cupel containing the silver bead is covered with a red-hot empty cupel, and is withdrawn gradually from the muffle so that the bead may cool slowly. Sprouting is avoided in this way. When the beads are very large (0·2 gramme or more) the muffle is closed and luted up and the fire allowed to die down as in the assay of silver bullion. Very small beads do not spit.

When cooled the beads often " flash "—*i.e.* brighten suddenly at the moment of solidification. This is due to the fact that the latent heat of fusion being released raises the temperature of the bead enormously.

Temperature of Cupellation.—Vauquelin, perhaps the earliest writer on the temperature of cupellation, observes [2] that the lead is to be charged in when the cupels are seen to be " rouge légèrement blanc," and that during cupellation the fumes of litharge rise and wind about in the interior of the muffle. This appears to correspond to a temperature of the muffle walls of about 900°C. He proceeds to state that the heat is too great if the colour of the cupels is white, the fused metal is seen with difficulty, and the fumes, scarcely visible, rise rapidly to the vault of the muffle without winding about (say 1,000°C.). If the fumes appear heavy and dark, moving sluggishly in a direction almost parallel to the bottom of the muffle, the temperature is too low (say 800°C.). These correct observations have been repeated by many other writers since Vauquelin. A pool of litharge may form on the cupel, followed by freezing, if the temperature is too low.

The lowest temperature at which litharge is absorbed by a cupel is about 840°C.[3,4] or, according to Bradford, 906°C.[5] The temperature of

[1] E. E. Bugbee, "*Fire Assaying,*" 1933, p. 116.
[2] "*Manuel Complet de l'Essayeur,*" Paris, l'an vii. (1800). New edition, Paris, 1836, p. 45.
[3] Fulton, *Western Chemist and Metallurgist,* 1908, 4, 47.
[4] T. K. Rose, *Trans. Inst. Min. and Met.,* 1909, 18, 463.
[5] *J. Ind. Eng. Chem.,* 1909, 1, 181.

uncovering is about 850°C., and the finishing temperature usually recommended well over 900°C. The temperature of the air in the muffle is always lower than that of the muffle walls and the cupel. At Hollinger,[1] the pyrometer readings range from 760° to 870°C. The cupels are probably hotter. Many assayers prefer to finish the cupellation at a temperature low enough to enable feather-like crystals of litharge to form in a ring round the cupel. Feathers are more readily formed if the current of air is checked. Other assayers prefer the litharge to be completely absorbed.

If the assays are long in uncovering they may sometimes be started by dropping on them a little charcoal powder wrapped in tissue paper, or, still better, by placing a piece of charcoal near them. If one freezes before completion it is restarted in the same way, or fresh lead is added or the temperature is raised, but the results are not good. Spirting or "spitting" of the lead may occur, due to the use of raw cupels, or of cupels which have been made with too much pressure, or to the presence of sulphur, arsenic, etc., in the lead.

The bead thus obtained should be well rounded and bright, loosely adherent to the cupel and slightly crystalline although malleable.[2] If it contains lead it is more globular and brittle and its surface is very brilliant, while it does not adhere at all to the bone-ash. If copper is present in large quantities, the bead adheres firmly to the cupel, and in extreme cases its surface is blackened.

Loss in Cupellation.—The absorption of gold by the cupel is not serious in the absence of impurities, amounting usually to 1 or 2 per cent. in small beads. An empiric rule [3] is that the apparent loss of weight of precious metal is directly proportional to the surface of the button of fine metal remaining, if other conditions remain the same. It follows that the percentage loss of weight varies inversely as the diameter of the button or inversely as the cube root of the weight. This is proved by a large number of experimental results by different observers.

The loss is somewhat greater with a larger amount of lead and with an increase of temperature, but these variations of loss are not important in ordinary work—for example,[4] a loss of 0·9 per cent. gold with 10 grammes of lead compared with one of 1·2 per cent. with 25 grammes of lead, in cupelling 1 milligramme of gold and 4 milligrammes of silver; and an increase of temperature of 200° raised the loss of gold from 0·4 to 1·2 per cent.[5] An increase in the proportion of silver reduces the absorption of gold.

Cupel absorption is slightly less in magnesite cupels than in bone-ash. Portland cement absorbs most of all. The quality of the cupel may be important, as T. L. Carter [6] found losses up to 10 per cent. of gold with a particular set of cupels. The shape of the cupel has also some influence on the loss, which is greater in shallow saucer-shaped cupels than in deep ones.

Effect of Base Metals in Cupellation.—The loss is greater in the presence of certain impurities. Copper carries gold into the cupel; 1 gramme of

---

[1] W. R. Dodge, *loc. cit.*

[2] For the effect of small quantities of the platinum metals on the microscopic structure of the bead, see C. O. Bannister and G. Patchin, *Trans. Inst. Min. and Met.*, 1914, **23**, 163.

[3] W. J. Sharwood, *Trans. Amer. Inst. Min. Eng.*, 1915, **52**, 179.

[4] T. K. Rose, *J. Chem. Met. Mng. Soc. S.A.*, 1905, **5**, 165.

[5] See also Hillebrand and Allen, *U.S. Geol. Surv.* 1905, *Bull. No.* 253, p. 20.

[6] *Eng. and Min. Journ.*, 1902, **73**, 728.

copper in 25 grammes of lead causes a loss of 10 per cent. of a bead of gold 1 milligramme in weight, and even 0·05 gramme of copper doubles the ordinary loss.[1] Tellurium and selenium are still more dangerous in large amounts. S. W. Smith[2] has shown that tellurium is slowly removed into the cupel during cupellation, and if there is not enough lead to carry it off, so that towards the end the tellurium becomes equal in amount to the gold plus silver, then the surface tension of the globule breaks down completely and the alloy spreads over a wider area, " wets " the cupel and is completely absorbed. The lead should be 80 to 120 times as much as the tellurium to obtain good results in cupellation. The loss of gold is then small.

The elements copper, bismuth, sulphur and especially tellurium are oxidised with difficulty and tend to remain in the button, increasing its weight. Nickel[3] is intractable in cupellation because its oxide is insoluble in litharge, and if copper is also present, the difficulty of removing nickel is greatly increased.

The formation of scoria on the cupel owing to the presence of certain base metals is usually accompanied by low results. Minute beads of fine metal may be entangled in the scoria.

Indications of the presence of some base metals in cupellation are as follows :—Copper stains the cupel green if a small quantity is present and dark brown or black if the amount is large. If the amount is large enough, the button spreads out and adheres to the cupel. The green stain is used by assayers, who add copper to some of the assay pieces in a large muffle charge so that the stains produced will indicate the correct order of the assays. Tellurium sometimes gives a pink stain and sometimes merely a brown colour in a ring round the litharge stain. Bismuth gives an orange-yellow ring. Nickel and cobalt form black or dark green scoriæ with a green stain.[4]

Rare Metals in Cupellation.[5]—Traces of iridium, rhodium, osmium, ruthenium and iridosmine prevent " flashing." Palladium does not, and platinum only if 6 per cent. is present. In a silver-gold bead, 1·6 per cent. of platinum causes a frosted appearance and 8 per cent. makes the surface of the bead rugose. Rhodium (0·04 per 1,000) causes crystallisation, and 0·3 per 1,000 gives silver a blue-grey colour. As little as 0·04 per 1,000 ruthenium gives a black crystalline deposit attached to the bead, usually at the bottom edge. Iridium also gives black spots on the bottom of the bead. Palladium is distinguished by giving a yellow colour to the parting acid.

For the separation of these metals from the gold, see Chap. XIX.

(3) **Inquartation and Parting.**—The bead of silver and gold obtained by cupellation is squeezed between pliers, or flattened with a hammer on a clean anvil, to loosen the bone-ash adhering to its lower surface, and is then cleaned with a brush of wires or stiff bristles. It is then weighed, the silver removed by solution in nitric acid, and the weight of the residual gold taken, when the difference between the two weighings represents the silver. If the bead contains more than one-fourth its weight of gold, more silver is added

[1] R. W. Lodge, "*Notes on Assaying*," 1905, p. 142.
[2] *Trans. Inst. Min. and Met.*, 1908, **17**, 468.
[3] E. E. Bugbee, "*Fire Assaying*," 1933, p. 130.
[4] For indications of other metals, see T. K. Rose, *J. Chem. Met. Mng. Soc. S.A.*, 1905, **5**, 165.
[5] J. van Riemsdijk, *Chem. News*, **41**, 126, 266 ; C. O. Bannister and G. Patchin, *Trans. Inst. Min. and Met.*, 1914, **23**, 163.

to it, as otherwise some of the silver will remain undissolved, being protected from the action of the acid by the outer layers of gold. A pale yellow bead always contains more than 60 per cent. of gold, but a perfectly white bead may not "part" completely. The addition of the silver is sometimes effected in the case of small beads by fusion on charcoal by the blowpipe, but it is better to cupel the bead with the additional silver, wrapped in as small a piece of lead foil as possible. This is called "*inquartation.*" In routine work, the silver may be mixed with the stock flux in the form of AgCl, as at Hollinger,[1] or metallic silver or drops of silver nitrate solution from a burette, as on the Rand, may be dropped into the pot before fusion. The resulting bead is cleaned, flattened by a hammer, and dropped into nitric acid of about specific gravity 1·10, which is at or near its boiling point.

When the amount of gold is small, it is convenient to use a large proportion of silver in parting. Suitable proportions of silver for different weights of gold are as follows :—

Weight of Gold.	Ratio of Silver to Gold.
Less than 0·1 mg., ...	20 or 30 to 1
About 0·2 mg., ...	10 to 1
,, 1·0 ,, ...	6 to 1
,, 10 mg., ...	4 to 1
More than 50 mg., ...	2½ or 3 to 1

With these proportions parting is very rapid and the gold does not break up if boiling acid is used. Most assayers use less silver to avoid breaking-up, which is more likely to occur if the acid is stronger or is much below boiling point. Acid of specific gravity 1·20 may be used safely with a ratio of 4 or 5 silver to 1 gold in beads of medium size.

The *parting* is effected in test tubes or porcelain cups arranged on a sheet iron tray divided into compartments. Porcelain trays with hollows for 12 beads, instead of separate cups, are also used.[2] The freedom of the nitric acid from chlorides is ensured by adding nitrate of silver and settling. The acid attacks the bead instantly and violently, turning it black and giving off nitrous fumes. With little beads, containing only a small proportion of gold, the silver is dissolved in a few seconds and decanting may be at once proceeded with, but the boiling is usually continued for a minute or two. If the amount of gold is large, the boiling is continued for some minutes. The acid is then decanted, with the aid of a suction pump if convenient (the fine point of a glass tube is inserted in the cup and the liquid sucked away), and the residue is washed twice with hot water by decantation. At Hollinger the parted gold is washed three times with 2½ per cent. ammonia. If the bead is very large fresh acid is added of specific gravity 1·20, but with the beads from almost all gold ores no second treatment with acid is required. A small amount of silver, weighing about 0·1 or 0·2 per cent. of the gold, obstinately resists the action of the acid, and remains as a surcharge which may be neglected in most cases The gold usually remains as a single piece if it weighs less than 0·1 mg., even if the proportion is only one part of gold to 40 to 50 parts of silver. If the ore is rich, the gold sometimes breaks up if hot acid is used, and invariably breaks up if the parting is begun with cold acid. The finer particles may float and be lost in decantation

[1] W. R. Dodge, *Can. Min. and Met. Bull. No.* 225, 1931, p. 115.
[2] "*Rand Metallurgical Practice,*" vol. i., p. 530

Particles floating on the surface may be sunk by touching with a glass rod or by a drop of water let fall on them. Continued heating of the gold in acid makes it agglomerate to some extent, so that it is easier to wash. Certain base metals form compounds or eutectic mixtures with the silver or gold, and these appear to disrupt the continuity of texture of the alloy. The gold then breaks up into powder, no matter what the temperature and strength of acid may be.[1] This may happen if the fine metal falls to 990, but cupelled beads are usually 996 to 998 fine in gold and silver.

After washing, the porcelain cups and their contents are dried at a gentle heat, and then annealed at a low red heat. The tray carrying the cups is placed in a muffle or on a red-hot cast-iron block for annealing. The gold which was previously black and soft, being in a fine state of division, now resumes its usual yellow colour, and hardens so that it can be removed to the pan of a balance and weighed. At a bright red heat the glaze in many porcelain cups softens and the gold sticks to it.

If a test-tube is used for the boiling, the parted gold is transferred to an unglazed Wedgwood crucible. To effect this the tube is filled with water, the crucible placed over its mouth and the whole inverted.

When large pieces of silver or copper containing very little gold are parted, whatever method of parting is used, very finely divided gold may remain suspended in the liquids, and may thus be lost in the course of decantation. Loss by decantation after parting may be reduced by adding a globule of mercury free from gold, and stirring with a glass rod until all the black particles of gold have been absorbed. The spent acid is then poured off and fresh nitric acid of density 1·2 added. On gently warming, the mercury is slowly dissolved and the gold remains as a coherent spongy mass. It is washed with nitric acid and then with water, annealed, re-heated with concentrated nitric acid, washed, re-annealed, and weighed. A little mercury remains with the gold.

By the operation of parting, silver, palladium and some platinum are removed in solution, but part of the platinum, and all the rhodium, iridium, etc., remain with the gold. If the presence of these metals is suspected they must be looked for and removed by special methods (*vide infra*, p. 529).

(4) **Weighing the Gold.**—Great care is exercised in placing the balance, to avoid vibration from machinery, traffic, etc. In certain cases the foundation is laid at a depth of many feet and a column built for the support of the balance. The column is not allowed to touch any part of the superstructure.

Assay balances for mine and mill assays are usually sensitive to 0·01 or 0·005 milligramme (value per scale division). In special balances the sensitivity is 0·0025 milligramme. Fig. 229 shows a balance of this latter kind. The scale division can be divided by eye with the aid of a fixed lens, or by a vernier, whichever is provided; more accurate readings, at the expense of time, are obtained by the method of swings. Weighing by substitution has the advantages of eliminating the bias of the balance and also any error due to unequal length of the arms of the beam, which may be caused by unequal heating. A large shield in front of the balance reduces this inequality, which is often caused by the breath of the operator. The principle of the method is to counterpoise the gold, then remove it from the pan, and compensate for the loss of weight by a rider. If direct weighing is preferred, the bias is determined from time to time. In reading the

---

[1] E. Keller, *Trans. Amer. Inst. Min. Met. Eng.*, 1919, 60, 706.

indications of the balance, parallax is avoided by applying the eye to a pinhole in a card fixed at about 12 inches from the ivory scale.

In the method of swings, the balance is set swinging, passing zero each time, and the positions of arrest, l_1, $-l_2$, l_3, observed, disregarding the first swings. The true position of equilibrium is then given by $\frac{l_1 - 2l_2 + l_3}{4}$. The value of a change of one scale division in the position of equilibrium must be determined. It does not vary from day to day.

Weights are often made with less precision than balances. For this reason riders are preferable to small weights, but the difficulty is to determine their exact position on the beam. Riders should be of platinum or aluminium.

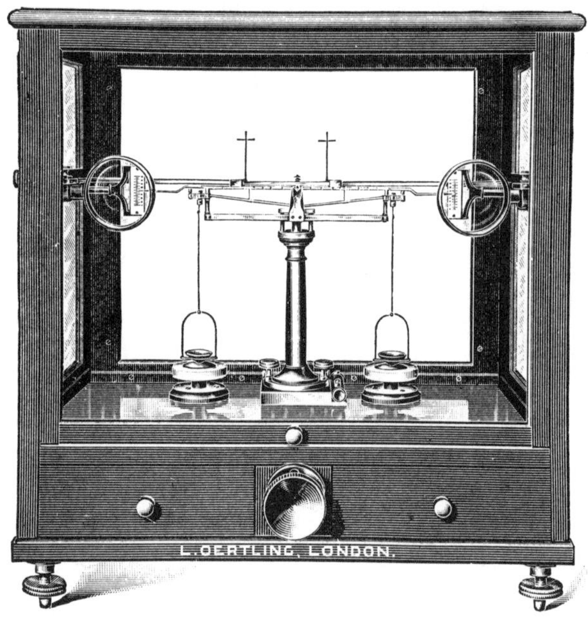

Fig. 229.—Assay Balance.

Brass riders increase in weight, especially if gilded. The precise weight of a rider, if not guaranteed by a competent authority, may be determined in reference to any guaranteed weight by "building-up." A rider may be reduced in weight by light rubbing on a sheet of ground glass, or by other mechanical means. A platinum rider, if too light, may be gilded in a cyanide bath, but light aluminium riders are rejected. When used for both mill heading and tailing assays, an inexact rider, if too heavy, would give low recovery results, and so the bullion recovered should exceed expectations.[1]

All weights are covered with a film of air, moisture, etc., which is removed, together with a part of the metal, by wiping with silk.[2] Wiping is therefore

[1] It is pointed out by W. J. Kemnitzer (*Eng. and Mng. Journ. Press*, 1926, **21**, 807) that checking ore assays against bullion recovered is very difficult, but should always be attempted. Accurate weights are necessary if the attempt is to be made.

[2] J. J. Manley, *Phil. Mag.*, Ser. 7, 1933, **16**, 489.

avoided. The film is soon renewed. Platin-iridium weights acquire less film and keep better than brass weights. Cupro-nickel and chromium-plated brass weights also seem to keep well.

The storage of weights is of importance, especially in hot climates. Glue in weight-boxes gives off acid fumes which attack and corrode weights, making spots on brass and causing it to increase in weight. Gilded brass is far worse. Weights should be stored in ivory, brass or African blackwood boxes, not in mahogany, which corrodes weights, especially if there is a felt lining. If boxes are lined, white woollen felt in a metal box, without glue, is best.

Assays are sometimes reported in ozs. and decimals, but more often in dwts. and decimals. Mine assays are reported to 0·1 dwt. per ton and mill samples and residues to 0·01 dwt. Poor residues are occasionally reported tò 0·001 dwt., which is equivalent to 0·001 mg. gold in 20 A.T.

It may be mentioned that very minute amounts of gold, too small for measurement on an assay balance or even on a micro-chemical balance, may be measured under a high-power microscope (2 mm. lens) and their weight calculated, as proposed by L. Wagoner.[1] It is necessary to fuse the gold, or cupel it, when it becomes spherical, with or without a slight flattening. The error in reading is about 0·001 mm., or about 6 per cent. on the weight of a gold bead weighing 0·001 mg. A bead of 0·05 mm. diameter would weigh about 0·0012 mg.

Fusion of Used Cupels.—All clean bone-ash is detached and thrown away, the remainder being crushed through an 80 mesh sieve, and the charge made up as follows :—[2]

Cupel, ...	45 grammes.
Sodium carbonate, ...	25 ,,
Borax, ...	45 ,,
Litharge, ...	100 ,,
Argol, ...	2½ ,,

Salt cover.

Other assayers recommend the addition of fluorspar and silica, with more soda and less borax.

E. A. Smith gives the following charge for fusing magnesia cupels :—[3]

Cupel (magnesite), ...	40 to 60 grammes.
Sodium carbonate, ...	20 ,,
Borax, ...	20 ,,
Litharge, ...	40 ,,
Silica, ...	15 ,,
Argol, ...	2·5 ,,

Portland cement cupels require at least twice this proportion of silica, according to E. E. Bugbee, and about twice the proportion of sodium carbonate.

Assay by Scorification.—This time-honoured process is especially applicable to (*a*) complex ores, (*b*) very rich ores, (*c*) ores which are mainly valuable for their silver contents. The losses are small and the operations

[1] *Trans. Amer. Inst. Min. Eng.*, 1901, 31, 798.

[2] A. M'A. Johnston, "*Rand Metallurgical Practice*," vol. i., p. 310.

[3] "*The Sampling and Assay of the Precious Metals*," p. 180.

are easy to conduct, and need not be varied much for different classes of ore. For these reasons the process is occasionally used; several assays are made and the lead buttons scorified together. The chief disadvantage of the process lies in the small quantity of ore that can be treated.

Scorification is conducted in a muffle at a higher temperature than that required for cupellation. It must be high enough to melt litharge when contaminated by silica and oxides of copper, iron, manganese, etc. A temperature of 1,050° to 1,100° C. is usually enough. The charge is placed in a *scorifier*, a shallow circular fireclay dish 2 to 3 inches in diameter. The charge consists of about $\frac{1}{10}$ A.T. of ore with 1 or 2 A.T. of granulated lead, and a few grains of borax glass.[1] The ore is mixed with half the granulated lead, the mixture put in the scorifier, the rest of the lead spread over evenly and the borax put on the top. An addition of some litharge to the cover is often made. After charging in, the door of the muffle is closed until fusion takes place. As soon as the lead is melted the door is opened, and a current of air allowed to pass over the bath of metal. Some of the ore is now seen to be floating on the surface of the lead, and is rapidly oxidised, partly by the air and partly by the litharge which immediately begins to form. The sulphur, arsenic, antimony, etc., are thus soon eliminated, while copper, iron, and other bases oxidise and slag off with the borax, and the silica and other acids form fusible compounds with the litharge. Effervescence and spirting may occur, especially if the scorifier has not been well dried by warming before it is used. The slag soon forms a ring completely encircling the bath of metal. As oxidation of the lead proceeds, the litharge flows to the sides and increases the quantity of slag until at length the ring closes completely over the metal, leaving a flat uniform surface. This usually happens after from thirty to forty minutes. The slag should be " cleaned " by means of a pinch of charcoal powder before withdrawal. The fusion being quiet again, the charge is poured into an iron mould, and the lead button cleaned from slag with a hammer. If the button is too large to cupel at once it is re-scorified in the same dish, fresh lead being added if it is not soft and malleable.

The slag from the scorifier seldom contains much gold.

Assay of Complex Ores.—*Auriferous Tinstone.*[2]—For alluvial tin ore from Lower Burma, Bannister found that fusion gave the best results. The charge is :—

Ore,	25 grammes.
Litharge,	60 ,,
Sodium carbonate,	40 ,,
Borax,	10 ,,
Charcoal,	1·5 ,,

Only half the lead is reduced, so that the reduction of tin is prevented. Slag cleaning is desirable for ores containing over 1 oz. gold per ton.

An amalgamation assay is also recommended.

Antimonial Gold Ores.[3]—These ores may be scorified if they are

[1] By the use of a flatter scorifier than is usually made, or even by using a roasting dish, $\frac{1}{2}$ A.T. of ore may be taken with a charge of 75 grammes of lead. E. H. Simonds, "*Cal. Mines and Minerals*," 1899, p. 226.

[2] C. O. Bannister, *Trans. Inst. Min. and Met.*, 1906, **15**, 513.

[3] W. Kitto, *Trans. Inst. Min. and Met.*, 1907, **16**, 89; Holloway, *op. cit.*, p. 96; E. A. Smith, *Trans. Inst. Min. and Met.*, 1901, **9**, 334; H. R. Edmands, *J. Chem. Met. Mng. Soc. S.A.*, 1920, **20**, 180, and **21**, 45.

rich enough, but a better general method appears to be partial oxidation with nitre, followed by fusion. The charge given by Kitto for nearly pure stibnite is as follows :—

Ore,	1 A.T. (32·6 grammes).
Litharge,	200 grammes.
Nitre,	40 ,,
Sodium carbonate,	20 ,,
Borax,	10 ,,
Silica,	20 ,,

Melt at a very low temperature and cupel direct, or scorify if the lead is brittle.

Holloway prefers a large excess of soda in order to keep the antimony in the slag as sodium antimonate. His charge with ore containing 70 per cent. stibnite and 20 per cent. silica is :—

Ore,	20 grammes.
Sodium carbonate,	60 ,,
Litharge,	80 ,,
Nitre,	14 ,,

A large button of lead, 50 grammes, is produced, and is scorified. To save the pot from corrosion, if that is necessary, he prefers fireclay to sand.

For 2 per cent. stibnite ore at the Globe and Phoenix Mine, S. Rhodesia, Edmands uses 1 A.T. ore and 130 grammes of a mixture of PbO 12 parts, sodium carbonate 20 parts, borax 8 parts, mealie meal 0 to 1 part.

Telluride Ores.[1]—There is no loss by volatilisation when the ore is roasted, but it is better to fuse with excess of litharge at a moderate temperature, getting a large lead button. Tindall points out that fine crushing is essential —120 mesh sieves for poor ores and 200 mesh sieves for rich ores.

Holloway and Pearse use the following charges for West Australian ores :—

	Rich Ore.	Average Ore.
Ore,	10 grammes.	50 grammes.
Litharge,	120 ,,	120 ,,
Sodium carbonate,	30 ,,	60 ,,
Borax,	20 ,,	20 ,,
Flour,	7 ,,	7 ,,

The lead button of 50 grammes or more is scorified.

Hillebrand and Allen found that the crucible assay is satisfactory for Cripple Creek ores. For ores with 20 oz. gold the best charge is :—

Ore,	1 A.T.
Sodium carbonate,	1 ,,
Litharge,	6 ,,
Borax,	10 grammes.
Salt,	Cover.

and a reducing agent to give a large button of lead.

[1] G. T. Holloway and L. E. B. Pearse, *Trans. Inst. Min. and Met.*, 1907, 17, 171; Tindall, *Trans. Inst. Min. and Met.*, 1901, 9, 354; Hillebrand and Allen, *U.S. Geol. Surv., Bull.* 253, 1905; *Chem. News*, 1906, 93, 100, 109; E. A. Smith, "*Sampling and Assay of the Precious Metals,*" 1913 p. 225.

Scorification often gives low results, owing to large quantities of tellurium entering the lead. Litharge is added to check this. S. W. Smith explains that, at 700° to 900° C., litharge is reduced by tellurium, which passes into the slag, thus :—

$$2PbO + Te = Pb_2O + TeO$$

If the lead button is brittle it is re-scorified with litharge. The cupellation of lead buttons from telluride ores has already been discussed (see "Loss in Cupellation," pp. 500, 501).

Arsenical Ores.—These may be roasted at a low temperature to expel part of the arsenic. The product may then be amenable to fusion, the lead button being scorified if brittle. The unroasted ore may also be fused direct, with the addition of iron. A speiss is formed and collected and then scorified with the lead button. Scorification of the original ore sometimes gives satisfactory results.

Nickel-Cobalt Ores.[1]—The arsenides and sulphides of nickel and cobalt which occur at Cobalt, Ontario, may be treated direct by the crucible method, if the ore is low-grade. If much nickel and cobalt are present, a combination method is used, attacking the ore with nitric acid and fusing the residue.

Copper Ores.—These ores may be treated in three ways : [2] (*a*) Fusion with much litharge ; the lead button becomes cupriferous, and is scorified together with the matte. (*b*) Roasting, followed by fusion and scorification. (*c*) Treatment with nitric acid, by which the sulphur and copper are removed. The silver is precipitated from the solution by common salt and the insoluble residue fused and cupelled for gold.

For ore containing such quantities as 30-45 per cent. copper, 30-40 per cent. iron and 24-27 per cent. sulphur, the crucible method is now preferred in America.[3] With 1 A.T. of ore as much as 6 or 7 A.T. of litharge, 4 A.T. of silica and 18 grammes of nitre are used, with a flux of soda and borax. The charge is covered with soda, borax and silica.

Pyrrhotite Ore.[4]—The Morro Velho ore contains quartz, calcite, dolomite, siderite, ankerite, pyrrhotite, pyrite and arsenopyrite, with small amounts of chalcopyrite and chlorite. Pyrrhotite is the predominating sulphide. A typical analysis gave SiO_2 19·6, Fe 32·15, S 14·66, As 3·45, CaO 2·4, MgO 7·56, CO_2 15·7 per cent., besides other constituents. Pot fusion is now in use. The charge is :—

Ore, ...	1,568 grains.
Red lead, ...	800 ,,
Soda, ...	800 ,,
Borax, ...	800 ,,
Nitre, ...	200 ,,
Glass, ...	75 ,,
Fuba (maize meal), ...	125 ,,

With hoop iron and a cover of some of the flux.

[1] *Eng. and Min. Journ.*, 1910, **90**, 809 ; A. M. Smoot, *Trans. Can. Min. Inst.*, 1914, **17**, 244.

[2] Percy, "*Metallurgy of Silver and Gold,*" 1881, p. 247.

[3] E. E. Bugbee, "*Fire Assaying,*" 1933, p. 201.

[4] J. H. French and H. Jones, *Trans. Inst. Min. and Met.*, 1933, **42**, 228.

The regulus and slag are re-fused in the same pot with red lead 200 grains and flux 400 grains. Attempts to dispense with the re-fusion of the slag were not successful. The charge is fritted slowly to allow the nitre to oxidise the sulphur and later, at a higher temperature, the maize meal reduces the iron to the ferrous state.

Zinc Ores.[1]—The blende may be desulphurised with metallic iron in the fusion, when the zinc is volatilised. Fusion with nitre is recommended as an alternative. The amount of sodium carbonate should be four or five times that of the zinciferous ore.

Bismuth Ores[2] are fused with comparatively large quantities of sodium carbonate and borax, using a low temperature. Metallic bismuth containing gold[3] may be cupelled for approximate work, but for more precise work, boil the metal in concentrated sulphuric acid, cool, dilute with sulphuric acid of half strength and heat gently to dissolve the bismuth oxysulphate. Precipitate the silver as AgCl. This carries down the gold. Assay the AgCl and gold by fusion.

Assay of Mill Products.—*Cyanide Solutions.*[4]—The solution may be evaporated in a lead basin, avoiding spirting, and the lead cupelled, but the method is not suitable for solutions containing less than 5 dwt. gold. Usually 10 or 20 A.T. of solution are required, with results reported to 0·01 dwt. This quantity is too large for lead basins, but may be evaporated in an oiled porcelain basin with litharge and charcoal. The method though accurate is tedious. The residue is fused with a little sand and cupelled.

A more convenient method (the Chiddey method) for 20 A.T. is to add cyanide to bring the strength to 0·05 per cent., and then boil for half an hour with 25 c.c. of lead acetate saturated solution and 3 grammes of zinc dust or shavings rolled into a ball. Then add 25 c.c. of dilute HCl and boil again slowly to dissolve excess zinc. The lead precipitate is made into a ball by gentle pressing with the fingers, washed, dried and scorified or cupelled direct. Silver nitrate for parting may be added to the original solution. Edmands states that heating the original solution is unnecessary.

A rapid and easy, if sometimes inexact, method is to precipitate the gold in cyanide solution with a sheet of aluminium and a few c.c. of dilute H_2SO_4. The precipitate is fused and cupelled.

At Hollinger[5] the Chiddey method is used, but with aluminium plates instead of zinc for precipitation.

When $Ca(CN)_2$ solutions were introduced in place of NaCN at Kirkland, it was found that the Chiddey method was inexact for higher-grade solutions if they were foul. A new method[6] found to be satisfactory in these conditions was as follows:—Take a sample of 20 A.T. Add 1 c.c. of lead nitrate saturated solution, 15 c.c. of NaCN saturated solution, and

---

[1] E. A. Smith, "*Sampling and Assay of the Precious Metals,*" p. 222.

[2] E. A. Smith, *Trans. Inst. Min. and Met.*, 1901, 9, 346.

[3] *Eng. and Min. Journ.*, 1923, 114, 636.

[4] *Eng. and Min. Journ.*, 1903, 75, 473; C. T. Creed and C. F. Caxton-Boxall, *J. Chem. Met. Mng. Soc. S.A.*, 1932, 33, 190; J. Moir and G. H. Stanley, "*Rand Assay Practice,*" 1923, p 39; A. M'A. Johnston and H. A. White, "*Rand Metallurgical Practice,*" vol. i., pp. 321, 529; H. R. Edmands, *J. Chem. Met. Mng. Soc. S.A.*, 1920, 20, 180.

[5] W. R. Dodge, *Can. Min. and Met. Bull. No.* 225, 1931, p. 115.

[6] J. H. Dickson, *Rhodesian Min. J.*, 1930, 4, 293.

2 grammes of zinc dust. Shake until the precipitate of lead settles. Transfer the whole to a porcelain dish lined with lead foil, decant the solution and melt the lead and the precipitate with litharge, flour, borax and soda. Cupel.

The precipitation of gold by cuprous chloride is much favoured, especially with the addition of potassium ferrocyanide. To 20 A.T. of solution, 10 c.c. of well-reduced Cu_2Cl_2 are added and the liquid allowed to stand, with occasional agitation, for 30 minutes. Then a few drops (not more than 1 c.c.) of 5 per cent. potassium ferrocyanide solution are added and the liquid shaken for a few minutes. Excess of ferrocyanide leads to loss of gold by redissolving. The ferrocyanide gives a brown precipitate of copper ferrocyanide, which aids in the collection of the gold. The gold is filtered off, fused and cupelled. The same result is given with $CuSO_4$, Na_2SO_3 and H_2SO_4, instead of Cu_2Cl_2. If much sulphocyanide is present, Cu_2Cl_2 gives low results.

Colour Test for Gold in Cyanide.—This is useful for rapidly testing barren or very dilute solutions. To 1,000 c.c. of solution in a stoppered bottle add NaCN to make up to 0·10 per cent. strength, two drops of lead acetate saturated solution, and 2 grammes of zinc dust, shaking well. Allow to settle, and decant the clear solution. Add 10 c.c. of aqua regia to the residue and evaporate nearly to dryness. Take up with 2 c.c. of concentrated HCl and transfer to a test-tube. Cool and add a few drops of stannous chloride solution. A purple ring is obtained with as little as 0·02 dwt. of gold per ton of solution.

Assay of Slime Residues.—These may contain some gold in solution. For the complete gold value of the dry solids, add a few drops of a cuprous chloride saturated solution to the charge in the pot.[1] Creed and Caxton-Boxall,[2] however, recommend that 1,500-2,000 c.c. of slime pulp be thinned down to specific gravity 1·40 in a Winchester quart bottle or enamel bucket. Then add 15 c.c. of well-reduced Cu_2Cl_2 solution and agitate for at least 10 mins. Press off the liquor, dry the residue, pass through a 30 mesh screen to break up lumps, mix and sample for assay. The addition of potassium ferrocyanide is not necessary as the slime is an efficient catching medium for the precipitated gold. Filtering before drying prevents re-solution of gold, but White[3] adds excess of permanganate to destroy the cyanide and so prevent further dissolving of gold. If much sulphocyanide is present the results with Cu_2Cl_2 may be low, and Edmands[4] recommends in such cases the use of an emulsion of finely powdered carbon instead of, or with, Cu_2Cl_2.

If both the dissolved and the undissolved gold in slime residues are required separately, then (*a*) part of the sample may be dried and assayed, and (*b*) part of the sample is filtered, and the residue washed and assayed. The errors in (*a*) are due to loss of gold giving a low result, which is avoided for the most part by adding cuprous chloride. In (*b*) the assay almost certainly gives a high result. It seems better to check the result by assaying the filtrate and washings.[5]

---

[1] H. A. White, "*Rand Metallurgical Practice,*" vol. i., p. 531.
[2] *J. Chem. Met. Mng. Soc. S.A.*, 1932, 33, 190.
[3] *J. Chem. Met. Mng. Soc. S.A.*, 1911, 12, 89.
[4] *Loc. cit.*
[5] L. R. Benjamin, *J. Cham. of Mines of W. Australia.*, 1914, 13, 235.

Assay of Battery Chips and Screens.[1]—The charges for chips, according to Wilmoth and Jolly, respectively, are as follows :—

	Wilmoth.	Jolly.
Chips,	1·0 A.T.	1·0 A.T.
Sodium carbonate,	1·5 ,,	1·5 ,,
Borax,	1·5 ,,	1·0 ,,
Litharge,	2·0 ,,	2·0 ,,
Sulphur,	0·5 ,,	...
Silica,	0·5 ,,	1·5 ,,
Sodium bisulphate,	...	0·75 ,,
Charcoal,	1·0 gramme	4·0 grammes

and an iron nail.

A high temperature is used and the pot is left in the furnace for 20 minutes after quiet fusion is attained. Wash with litharge and charcoal. Wilmoth's charge gives a matte which contains silver but no gold.

Screens (iron), according to Wilmoth,[2] are weighed and then roasted to oxide in a roasting dish, pulverised and mixed. The charge is :—

Iron oxide,	1·0 A.T.
Sodium carbonate,	1·5 ,,
Borax,	1·0 ,,
Litharge,	2·0 ,,
Silica,	0·5 ,,
Charcoal,	1·5 grammes

Heat in furnace longer than usual and wash with litharge and charcoal just before pouring.

Assay of Metallic Copper.—For *disused copper plates* Johnston gives a fusion charge as follows :—[3]

Drillings,	0·5 A.T.
Sulphur,	0·25 ,,
Borax,	0·4 ,,
Glass,	0·4 ,,
Sodium carbonate,	1·75 ,,
Litharge,	1·0 ,,
Charcoal,	1·5 grammes

Assay of Copper Precipitate containing a Little Gold.—Dissolve[4] in nitric acid. The gold is mainly in the silica residue. Evaporate. Dilute and precipitate the silver as AgCl by $CuCl_2$, followed by H_2SO_4 and lead acetate to carry down the gold and silver. Filter and wash. Pass H_2S to precipitate any remaining gold and silver—they come down before the bulk of the

---

[1] A. M'A. Johnston, "*Rand Metallurgical Practice,*" vol. i., p. 318; L. J. Wilmoth, *J. Chem. Met. Mng. Soc. S.A.*, 1908, **8**, 230, 379; H. R. Jolly, *op. cit.*, 1908, **8**, 343, 378.

[2] *Loc. cit.*

[3] "*Rand Metallurgical Practice,*" vol. i., p. 317.

[4] *Chem. Trade J.*, 1925, **76**, 608.

copper. Wash and dry the precipitate and fuse with litharge, fluxes and iron. Cupel and part.

A crucible charge for *copper bullion* [1] given by the United States Metals Refining Co. is :—Mix 0·25 A.T. drillings with 1·2 grammes sulphur. Add 15 grammes soda, 240 grammes litharge and 8 grammes silica. Heat quickly. The lead is soft and can be cupelled direct.

For *copper matte* the usual fusion charge, according to Chase,[2] is :—

Matte,	0·25 A.T.
Sodium carbonate,	12 grammes
Litharge,	80 ,,
Silica,	6 ,,
Nitre,	5 ,,

Stir and cover with litharge 40 grammes, borax 15 grammes, sodium carbonate 12 grammes. The slag losses and cupel absorption are high with rich mattes, and accordingly Chase scorifies the Tacoma matte (containing 10 to 50 ozs. gold, 20 to 200 ozs. silver and 20 to 50 per cent. copper), mixing ⅛ A.T. matte, 35 grammes lead and 2 grammes silica, and covering with 35 grammes lead and 2 grammes borax. The lead buttons from two charges are put together, made up to 65 grammes lead and re-scorified. Two of the final buttons are then cupelled and the beads parted in the same cup, which now contains the gold from ½ A.T.

Assay of Graphite Crucibles.[3]—The pulverised material is washed through a 200 mesh screen to separate metallics, and each product assayed separately to reduce the difficulty of sampling. The sieved portion is mixed and may be roasted and scorified, which however gives low results for gold. Fusion is better, using much litharge, and to avoid roasting Johnston gives the following charge :—

Ground product, unroasted,	0·25 A.T.
Sodium carbonate,	0·50 ,,
Litharge,	4·00 ,,
Borax,	0·25 ,,
Silica,	0·50 ,,

Furnace ash may be treated in the same way, but if, as usual, there is less carbon, a larger quantity may be taken for assay.

H. R. Hillman [4] used the following charge in the assay of carborundum crucibles :—

Ground crucible,	½ A.T. (163 grains)
Litharge,	2,000 grains
Sodium carbonate,	50 ,,
Borax,	150 ,,

[1] E. E. Bugbee, "*Fire Assaying,*" 1933, p. 223.
[2] R. E. Chase, *Eng. and Min. Journ.*, 1916, 102, 1139.
[3] A. M'A., Johnston, "*Rand Metallurgical Practice,*" vol. i., p. 317 ; L. J. Wilmoth, *Mex. Min. Journ.*, 1912, 14, 23 ; J. Loewy, *J. Chem. Met. Mng. Soc. S.A.*, 1898, 2, 205 ; J. Watson, *Eng. and Min. Journ.*, 1922, 114, 199.
[4] 61*st Ann. Rep. Royal Mint*, 1930, p. 110.

The melt did not quieten down readily, continual sparking occurring. This was remedied by the addition of an excess of soda. A lead button about 1,650 grains in weight was obtained.

Assay of White Precipitate.—This product from zinc extractor boxes is of variable composition, sometimes containing much magnesium carbonate, calcium carbonate or silica in addition to the usual zinc hydrate. A crucible charge is as follows :—[1]

Dry precipitate, ...	0·5 A.T.
Sodium carbonate, ...	0·5 ,,
Borax, anhydrous, ...	0·5 ,,
Litharge, ...	1·5 ,,
Maize-meal, ...	0·1 ,,

Fuse, cupel and part.

Zinc-Gold Slimes may be scorified or alternatively fused with 20-30 parts of litharge, and enough flour to give the usual lead button. Some soda, borax and silica are also added.

Gold in Solution as Chloride.—It is sometimes necessary to estimate traces of gold in a chloride solution in laboratory work. A few drops of a dilute solution of ferrous sulphate may be added, using Nessler tubes, and comparing the colour with that from a known solution of $AuCl_3$. The purple colour produced by a mixture of stannic and stannous chlorides is still more delicate, showing the presence of 0·0001 per cent. of gold. With benzidine acetate, gold can be detected by the appearance of a green or blue colour.[2]

Gold may be determined potentiometrically in solutions of $AuCl_3$ with $SnCl_2$ in HCl at 20° C. The reduction proceeds to complete deposition of the gold.[3]

Microscopic quantities of gold in solution may be estimated by boiling a piece of clean Japanese silk in the feebly acid solution. The gold deposits on the silk and colours it purple. By collecting the silk on a glass filter and then burning it in a platinum crucible the increase in weight of the latter gives the weight of the original gold.

Arrangement of Assay Office.—The Hollinger Assay Office [4] may be taken as typical of a large modern office attached to a mine. About 950 fire assays for silver and gold are made per day. The grinder room contains a Sturtevant roll jaw crusher giving a $\frac{1}{2}$-inch product, crushing rolls with a 4 mesh product and McColl pulverisers grinding to − 100 mesh. Samples of 250 grammes are obtained by a Jones splitter. The machines are covered by housing and connected with an exhaust. They are cleaned with an air jet at a pressure of 100 lbs. The grinder room is outside the main building.

The fluxing room contains the pulp balances and the charges are mixed on rubber cloth. There is also a flux-mixing and storage room. The stock flux contains silver chloride for parting. For mill samples the charges are from 5 to 10 A.T. with 20 A.T. for the final residues.

The furnace room contains seven electric muffle furnaces with pyrometers, the muffle size being 8 × 13 × 22 inches. The temperature is 1600°-1850° F.

---

[1] J. Watson, *Eng. and Min. Journ.*, 1922, **114**, 199.
[2] E. Wichers, *J. Ind. Eng. Chem.*, 1927, **19**, 96.
[3] See also E. Müller and R. Bernewitz, *Z. anorg. Chem.*, 1929, **179**, 113.
[4] W. R. Dodge, *Can. Min. and Met. Bull.* No. 225, 1931, p. 115.

for fusion and 1400°-1600° for cupellation. The time for both fusion and cupellation is 30 minutes.

The parting room contains a revolving button brush, an electric annealing oven, electric hot-plates and porcelain ware parting trays for 24 cups.

The weighing room contains three button balances, sensitivity 1/300 milligramme, for mine assays, three button balances, sensitivity 1/500 milligramme, for fine mill assays, and one balance, sensitivity 1/400 milligramme, for bullion assays. The balances are on a table supported by posts set in concrete blocks in the solid rock foundation.

The electric furnaces replaced oil and cost $7,600 to install, but saved more in a year in cost of running.

CHAPTER XIX.

THE ASSAY OF GOLD BULLION.

Introduction.—The "parting assay" was first mentioned in a decree of King Philippe of Valois, published in the year 1343.[1] The methods of procedure in the 16th century have been described by Agricola [2] and by Ercker,[3] and those in the 17th century by Savot [4] and by J. Reynolds.[5] In 1666 Pepys saw the parting assay being practised at the Mint in the Tower of London, and from his description it is clear that the method then employed bears a surprisingly strong resemblance to that of the present day.

Selection of the Sample.—Alloys of gold with either silver or copper or with both are practically uniform in composition if they have been melted and well mixed, and the ingot has not been pickled after casting. In these cases a single outside cut is representative of the composition of the whole of the ingot. The cut must, of course, be clean, and is usually taken in the middle of one of the ends of the ingots at the bottom. A gouging tool is used, worked either by hand or power. Chisel cuts from diagonally opposite corners are also sometimes taken.

When other metals are present, the solidified ingot is not uniform in composition. In particular bars produced by cyaniding undergo segregation, as was first pointed out by E. Matthey.[6] Any brittle bar is not uniform. The bars are sampled by (*a*) drilling or (*b*) dip-sampling as the bar is being poured. The latter is the better general practice.[7]

(*a*) *Drilling.*—The drills are taken with an $\frac{1}{8}$ or $\frac{3}{16}$ inch bit, using no oil, in the positions shown in Fig. 230. The depth of the drills is about 1 inch, the surface often being rejected. The two top drillings are mixed and also the two bottom drillings, so as to give two samples. Templets are used to find the right places for drills or saw cuts.[8]

(*b*) *Dips.*—One sample may be taken from the top just before pouring and a second sample from the pour towards the end of pouring, or a single dip taken from about the middle of the pot before pouring. The metal is melted under a cover of borax in a graphite pot and well stirred with a graphite stirrer. Iron would absorb sulphur, etc., but is itself attacked by molten gold. The sampling tool is a graphite rod with a cavity in it to hold

---

1 *First Annual Report of the Royal Mint*, 1870, p. 103.

2 Hoover's "*Agricola*," p. 247.

3 Pettus' "*Ercker*" ("The Laws of Art and Nature in Assaying, etc., Metals"), bk. ii., chap. xv.

4 "*Discours sur les Medalles Antiques*" (Paris, 1627), p. 72.

5 "*A New Touchstone for Gold and Silver Wares*" (London, 1679), p. 362.

6 Proc. Roy. Soc., 1896, 60, 21.

7 F. P. Dewey, *Trans. Amer. Inst. Min. Eng.*, 1912, 44, 853; J. H. Hance, *ibid.*, 1916, 54, 536; C. O. A. Thomas, *Proc. Austr. Inst. Min. Met.*, 1931, No. 81, p. 17.

8 W. R. D. Jones, *Met. Ind.*, 1928, 33, 201.

about 5 grammes of metal with its cover of borax. The sample is poured into a mould under borax or granulated by pouring into water.

At the Rand Refinery iron dipping rods, heavily coated with a graphite mixture, and having two small cups, 6 to 8 inches apart, are used. It is stated that thereby samples from two different points in the molten metal are obtained.

A summary of the results given by Hance is as follows :—

(1) Drillings near an edge of the bar are very variable but are generally higher than dips. They may be 1 or 2 per 1,000 higher.

(2) Drills near the centre are generally lower than dips.

(3) In general, top drillings and top dips are higher than the bottom ones (gravity, however, may also act in enriching the bottom).

(4) Variations in the drillings (often 2 or 3 per 1,000, occasionally 5 per 1,000) are greater than in dips (sometimes 1 or 2 per 1,000).

(5) The difficulties in drill determinations vanish if some silver is present, as lead, zinc, etc., are soluble in the Au-Ag alloy but not in gold free from

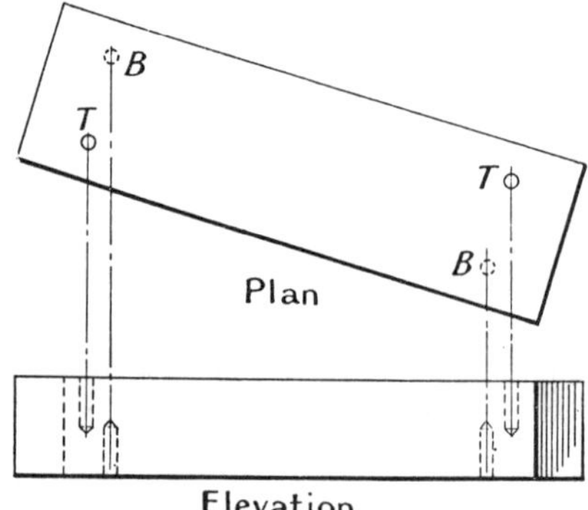

Fig. 230.—Plan and Elevation of an Ingot with Drill Holes for Sampling. Two in Top (T) of Ingot, Two in Bottom (B).

silver. 90 to 100 of silver per 1,000 is ample and prevents discrepancy in drill assays due to segregation.

Dewey found [1] that bars from the Mercur mine containing cadmium and nickel gave discrepant assays, even a single dip ranging in assay from 818·3 to 821·0. He observes [2] that drill samples are not good on brittle bars generally, because the coarse particles are 2 to 4 per 1,000 richer than the fine particles.

The difficulties in valuing base brittle bars may be insuperable. Mints and refining offices are often compelled to refine the bars partially before sampling them. E. Matthey [3] found that one ingot weighing 400 ozs. and containing gold 614, silver 75·8, lead 164, zinc 95, other metals (by diff.) 51·2, as ascertained by separating the whole of the constituents, could not be assayed correctly. Seven dip assays varied from 562 to 622 in gold. The average of 14 other assays of the ingot was gold 576, silver 90. In another

[1] *Eng. and Min. Journ.*, 1912, 93, 733.
[2] *Trans. Amer. Inst. Min. Eng.*, 1912, 44, 853.
[3] *Proc. Roy. Soc.*, 1896, 60, 21.

bar the gold assays were, top 657, centre 785, bottom 790, gravity thus playing a part. The addition of silver to prevent segregation must be equal to not less than two-thirds of the quantity of zinc and lead taken together. Thus an alloy of gold 55, silver 20, zinc 7, lead 18 was practically homogeneous.

The sampling of gold wares is difficult, because the outside is usually finer than the interior, in consequence of the pickling of the wares after manufacture. "It is the usual practice to remove the 'colour' from gold wares by a preliminary scraping or by 'buffing' before scraping to obtain the sample proper."[1] The sampling of "base bullion" belongs to the metallurgy of lead or copper.[2]

Taumann and Loebich[3] have compared the following methods as means of detecting base metal impurities in gold: (1) Tarnishing by annealing. (2) Action on bacteria. (3) Action of dry iodine vapour. As little as 0·1 per cent. of copper, iron, and nickel can be detected by the first method, while nobler metals show no response. The second method is useful for discovering poisonous impurities if they are not actually dissolved in the gold. Dry iodine vapour reveals the characteristic presence of any base impurity on the surface of gold sheet.

Preparation of the Assay Piece for Cupellation.—The assay piece is "flatted" on a clean anvil by means of a machine or a hammer with a rounded face, weighing about 11 lbs., and a portion, weighing 0·5 gramme, is obtained by cutting with shears and filing. Other weights of sample are also in use, such as 5 grains.

The bead balances used in ore assays may be used for bullion assays, but more rapid balances with a sensitivity of 0·05 milligramme per scale division are more convenient if large numbers of assays are made. The weighed piece is wrapped in pure lead foil together with the silver necessary for parting, and some copper unless it is present in the assay piece. The corners of the lead packets are squeezed down so as to fit the cupels by pliers with concave rounded faces and the assays are then ready to be charged into the furnace.

The usual proportion of gold to silver is 1 to 2½, on the ground that less silver is then retained by the cornet than in any other case. If much more than three parts of silver are present the gold breaks up in the acid. The test silver should be assayed for gold, but the presence of a small quantity does not matter if pieces of the same silver are used for "proof" and ordinary assays.

The amount of lead used varies with the proportion of base metals present. For gold 900 fine and upwards, eight times its weight of lead is generally used, but the copper is not completely removed in the course of cupellation. More lead is required, up to about 30 times the weight of the assay piece, if the bullion contains much base metal.

Instead of attempting to remove all the copper present in an alloy of low standard by one operation, using large quantities of lead, it saves time and gives more uniform results if a smaller amount of lead is used in two successive cupellations. For one part of gold bullion 400 fine, 16 parts of lead are enough if added in this way. The object in view is not to remove all the copper by cupellation, but to obtain a well-formed, clean and bright button suitable for parting.

---

[1] E. A. Smith, "*The Sampling and Assay of the Precious Metals*," p. 324.
[2] For description of methods, see E. A. Smith, *op. cit.*, chap. xx.
[3] *Z. anorg. Chem.*, 1927, 168, 255.

In the cupellation of triple alloys of gold, silver and copper, E. A. Smith recommends the quantities given in the following table :—[1]

TABLE L.

Standard.		Amount of Lead for One Part of Alloy.
Gold in Parts per 1,000.	Carats.	
916·6	22	8
750	18	15
625	15	18
500	12	20
375	9	24

In this case, the 22 carat contains a little silver, and the other standards usually from 100 to 125 parts of silver per 1,000. The amounts of lead would remain the same if part of the copper were replaced by other base metals. Gold-tellurium alloys require more lead.

Cupellation.—Muffle furnaces for cupellation have been described in the previous chapter. The main requirements in bullion assaying are uniformity

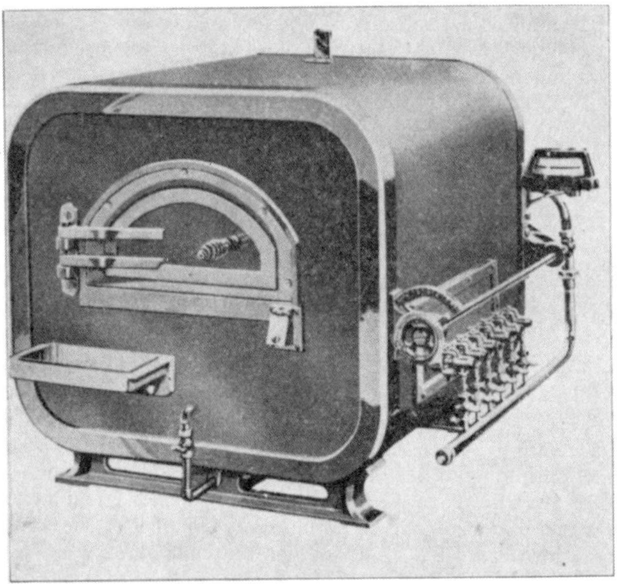

[*By permission of the Gas Light & Coke Co.*

Fig. 231.—G.L.C. Assay Furnace.

of temperature and air supply throughout the muffle in order that the cupel absorption of gold may be regular.

[1] E. A. Smith, "*Sampling and Assay of the Precious Metals,*" p. 326.

A gas furnace in use at H.M. Mint, Bombay, and designed to economise gas, is shown in Figs. 231 and 232.

The brick lining has a low thermal capacity and conductivity, and is

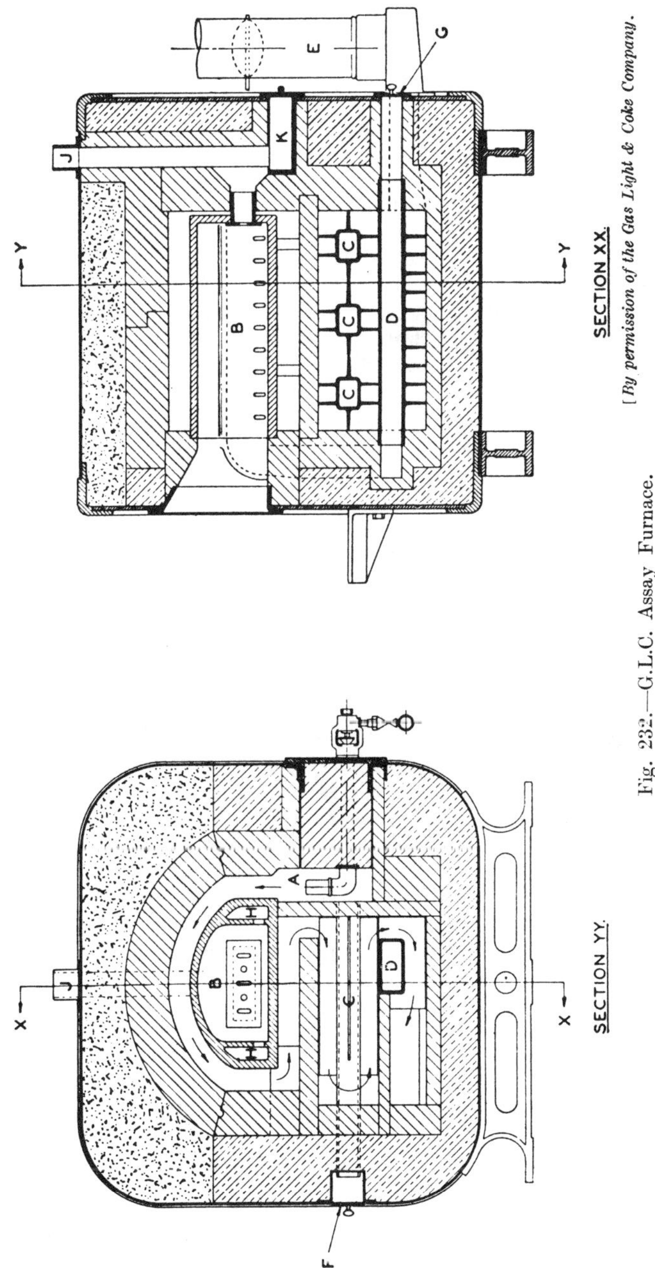

Fig. 232.—G.L.C. Assay Furnace.

[By permission of the Gas Light & Coke Company.

enclosed in a polished aluminium casing to reduce loss by radiation. There are six gas burners, A. The secondary air supply for them enters at F and is pre-heated by passing through the steel recuperators C, on its way to the burners. The flames from the burners pass round the muffle B, and the recuperators, as shown by the arrows, to the flue E. The air supply to the muffle enters at G and is pre-heated by passing through D to the passages H, at the side of the muffle, which it enters through a number of slots. The air leaves the muffle by an opening at the back leading to the flue J, which can be closed by a sliding damper.

Starting from cold, the muffle reaches a temperature of 1,000° C. in $1\frac{1}{4}$ hours. This temperature is maintained by a gas consumption of 65 cubic feet per hour. The size of the muffle is $6 \times 11\frac{1}{2} \times 15\frac{1}{4}$ inches.

In most muffles, the air enters the muffle through its mouth or front, a simple arrangement which works very well.

At the Ottawa Mint fireclay muffles in the gas furnaces have been replaced by carborundum ones, which have greater conductivity and are cheaper to run, though their first cost is greater.

Electrically-heated cupellation furnaces with nichrome rod resistances are in use at the Mints in Ottawa and Pretoria.[1] The one at Pretoria consists of two rectangular muffles one inside the other with a $\frac{1}{2}$ inch air space between them. The whole is enclosed in an iron case. Air passes into the air space through a hole in the outer muffle and then through 10 smaller holes ($\frac{3}{16}$ inch diameter) into the inner muffle, being thereby pre-heated in its travel and at the same time acting as an insulating layer. Lead oxide fumes leave the muffle by the usual separately controlled flue. The inner muffle is heated by a grid of $\frac{5}{16}$ inch nichrome rods suspended from the roof. A current of 300 ampères at 21 to 25 volts is required. Starting from cold, the muffle is ready for use in $\frac{3}{4}$ hour. This muffle has been found to be much cheaper than the coke-fired furnaces previously employed.[2]

In the Ottawa Mint the electric muffle is fitted with nichrome resistance wires in the sides. Alternating current at 550 volts is supplied to the coils, and if one fails an indicator lamp lights up. Pyrometer and lead oxide fume tubes are provided as in older types of furnaces. The energy consumption is 9 kWh.[3]

When large numbers of bullion assays are to be made, it is convenient to range the cupels on a graphite, nichrome,[4] or nickel-chrome or manganese steel tray which fits the muffle loosely and is charged-in and withdrawn by a peel. Large muffles take from 72 to 96 cupels. The lead packets may be charged-in one by one with tongs or placed in a nickel charging tray. This consists of a perforated plate with a non-perforated sliding plate underneath. The lead packets are put into the compartments made for them and the tray placed in the muffle above the cupels. The sliding plate is then withdrawn and the lead packets fall through the holes into the cupels.

After charging, the muffle door is closed to allow time for the lead to uncover, which occurs in about two minutes. Air is then admitted and the draught started by opening the damper. Cupellation is completed, as far as can be judged by the eye, in about 15 minutes, but it is usual to leave the "prills" or cupelled buttons in the furnace for 10 to 20 minutes longer, so that the last traces of lead may be oxidised and absorbed.

---

[1] *58th Ann. Rep. Royal Mint*, 1927, p. 127.
[2] *59th Ann. Rep. Royal Mint*, 1928, p. 128.
[3] *62nd Ann. Rep. Royal Mint*, 1931, p. 121.
[4] *55th Ann. Rep. Royal Mint*, 1924, p. 114; Do., **62**, 1931, p. 121.

The usual temperature in the muffle is between 1,000° and 1,100° C., but it is of more consequence that the muffle should be equally hot throughout than that any particular temperature should be maintained, as the use of checks eliminates errors due to uniform losses of gold by cupel absorption. If checks are not used, the temperature of the muffle walls and cupels may be kept at about 1,050° C. with good results. Pyrometers are often used.

The cupels are usually withdrawn on the tray while the "prills" are still fluid, but they may be allowed to solidify in the furnace. "Flashing" (see p. 499) occurs in the absence of the platinum metals, but there is no spitting if at least 50 parts of copper per 1,000 of gold were originally present, so that some remains in the "prills." If spitting should occur, the assay is rejected.

The buttons (which are of the form represented at *a*, Fig. 233) are removed from the cupels, after cooling, by a pair of pliers, cleaned by means of a stiff brush (sometimes revolving), and placed in a tray. If the bone-ash is not completely removed from their lower surface it is of little moment, since bone-ash is readily dissolved by nitric acid on parting. The surface of the

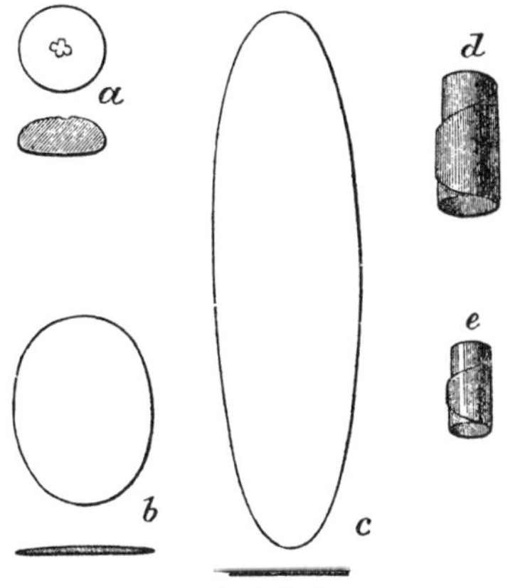

Scale, full size.
Fig. 233.—Stages in Working a Gold Bullion Assay Piece.

cupels must be carefully examined for minute beads of metal due to spirting of the lead bath. If any such beads are found in a cupel the fact is noted and the assay repeated.

If traces of lead remain in the button it is more globular, separates more easily from the bone-ash of the cupel, and has a brilliant steely surface. The effect of the presence of other metals is discussed on p. 501.

Preparation of the Assay Buttons for Parting.—The buttons are flattened (*b*, Fig. 233) by a hammer with a convex face on a clean anvil used for this purpose only.

After being annealed in the muffle or by a blowpipe, the flattened buttons are passed in succession through a pair of special rolls which are kept clean and bright. The oil is removed as completely as possible from the rolls before they are used, as otherwise the first fillets come out thinner than the remainder. The rolls are adjusted so that one passage through them reduces the buttons to the required thickness, which is about 0·25 millimetre, or 0·01 inch. The "fillets" (*c*, Fig. 233) thus obtained should all be of uniform size and thickness, with "wire edges," as ragged edges (due generally to traces of lead) expose them to loss during the boiling. After being rolled they are annealed at a dull red heat. They must not be made too hot, as that entails a loss of gold in the parting acid. The object of the annealing is to enable the fillets to be rolled into " cornets " or spirals, *d*, between the finger and thumb, or round a glass rod, and also to put the metal into a suitable physical condition for parting. Unannealed fillets tend to break up in nitric acid.

According to Law[1] annealing of the fillets has a definite effect on the surcharge :—

	Surcharge
No annealing,	$20\frac{1}{2}$
Normal annealing,	$7\frac{1}{4}$
Annealing at 330° C.	$3\frac{1}{2}$

A thick fillet (0·033 inch) has a higher surcharge than one rolled to the usual thickness (0·011 inch).

Parting.—This was formerly effected by boiling with nitric acid in glass "parting flasks." Platinum or fused silica boiling trays save time, and are used when possible. The silver is dissolved by the acid, which should be free from chlorine in any form, sulphuric and sulphurous acids, or sulphides from which sulphuric acid may be formed. These substances dissolve gold in the presence of boiling nitric acid. It is sometimes stated that the acid must be free from nitrous fumes, but this is not necessary, as silver protects gold from the action of nitric and nitrous acids, and as soon as parting begins great quantities of nitrous fumes are generated.

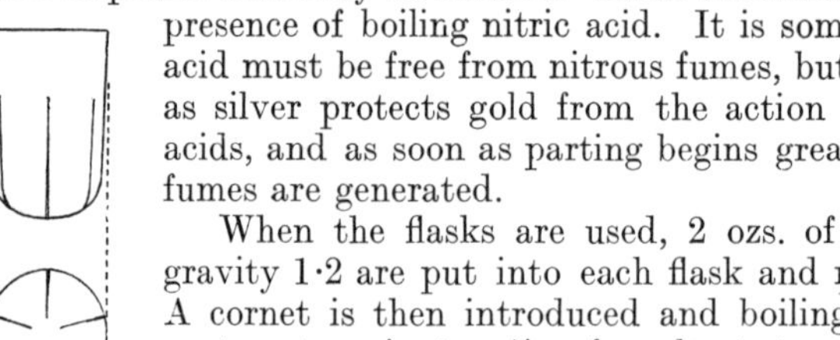

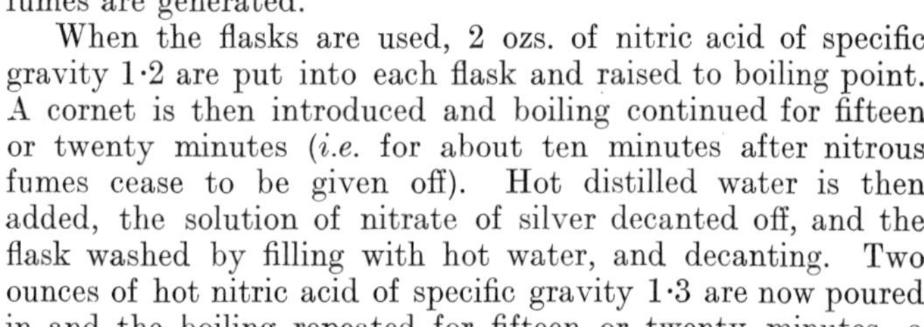

Fig. 234.—Platinum Cup for Parting.

When the flasks are used, 2 ozs. of nitric acid of specific gravity 1·2 are put into each flask and raised to boiling point. A cornet is then introduced and boiling continued for fifteen or twenty minutes (*i.e.* for about ten minutes after nitrous fumes cease to be given off). Hot distilled water is then added, the solution of nitrate of silver decanted off, and the flask washed by filling with hot water, and decanting. Two ounces of hot nitric acid of specific gravity 1·3 are now poured in and the boiling repeated for fifteen or twenty minutes, a piece of fireclay or capillary glass tube being added to prevent bumping; decantation and washing are then twice performed. Another boiling with acid of specific gravity 1·3 is usual, with the object of making the surcharge (*q.v.*) about zero. If any small particles of gold have become detached from the cornet, time must be allowed for them to settle before each decantation. After the last decantation the flask is filled with hot water, the top covered by a small porous crucible, and the whole carefully inverted; the pure gold, which is of a dark brown colour and exceedingly fragile, falls through the liquid and rests in the crucible, the water which enters with it being afterwards poured off. The crucible and cornet are dried and then annealed at a red heat over gas or in the muffle, when the gold shrinks greatly, though still preserving its shape, hardens and regains its ordinary pale yellow colour. It can now be weighed. (*e*, Fig. 233.)

When a platinum boiler is used the cornets are put on platinum pins, or into silica or platinum cups, one of the latter being shown in Fig. 234. The silica cups are of the same general shape, but have five small perforations

[1] *59th Ann. Rep. Royal Mint*, 1928, p. 89.

in the bottom. These cups are supported in a platinum or silica tray and the whole lowered by a platinum hook into a platinum vessel containing hot nitric acid of specific gravity 1·2. When the ratio of 2·5 parts of silver to 1 part of gold is used, some care is necessary in putting the assay pieces into the acid. A temperature of 90°C. should not be departed from widely; if the acid is colder than this the cornets tend to break up. Cornets containing 2 parts of silver to 1 of gold are less delicate.

Gentle boiling is kept up for about thirty minutes, and the tray is then withdrawn, washed by dipping vertically in and out of a vessel of hot distilled water, and boiled once or twice more in nitric acid of specific gravity 1·2 to 1·3 free from silver. The platinum tray of cornets is then washed, dried and annealed.

Weighing the Cornets.—The balance may be the same as that used for weighing-in. For more rapid work on cornets of widely different weights, balances are made to register the weight, on coming to rest, by means of a pointer moving over a dial or curved scale. Refined bar gold and mint assays are reported to 0·1 per 1,000, and unrefined gold bars usually to 0·25 or 0·5 per 1,000.

The checks or proofs, consisting of assay pieces of known composition, are weighed first and their "surcharge" (*vide infra*) applied as a correction to all the cornets worked with them. This may be done by means of a light rider. Fine gold is used for check assays.

In refined gold and mint assays the weighing is usually by substitution, the proof cornets being followed in the same pan by the ordinary cornets, so that the weight in the other pan is a mere counterpoise, and its error, if any, is immaterial.

The weights, if not certified by a competent authority, should be compared with a certified weight and corrected by rubbing down, if high, or by gilding, if low. Riders are used instead of small weights. For the care of weights, see p. 505.

Surcharge.—Gold is lost by (*a*) volatilisation; (*b*) absorption by the cupel; (*c*) solution in the acid. On the other hand, the cornet always retains some silver. The algebraical sum of these losses and gains is called the "surcharge," since the cornet usually weighs more than the gold originally present in the assay piece; if the reverse is the case, the work is less accurate. When checks are used, satisfactory results are obtained with surcharges of + 0·2 to + 1·0, or even more, per 1,000. The various losses and gains are discussed in detail in the following.

Losses of Gold.—It was found at the Royal Mint in 1910 that 0·0001 per cent. of the gold worked in a gas-fired muffle was contained in the litharge condensed in the flue. The loss of gold by volatilisation of one part in a million thus indicated is of course inappreciable in the assay.

The gold absorbed by the cupel in the assay of fine gold with the addition of 50 parts per 1,000 of copper and 4 grammes of lead is usually 0·4 or 0·5 per 1,000. More copper gives greater losses of gold. If the proportion of silver is increased, the absorption of gold is diminished. Doubling the amount of lead gives twice the cupel absorption. The loss also depends on the temperature, higher temperature causing higher losses. The presence of other base metals in addition to copper increases the absorption, but the effect is insignificant if the quantities of base metals do not exceed one or two per cent. (See below.)

The amount of gold dissolved by the acid increases with the strength of the acid and with the proportion of silver present. Under ordinary conditions it does not seem to exceed 0·005 per 1,000 and so is negligible.[1]

Silver Retained in the Cornet.—After boiling in the first acid the amount of silver retained is about 2·5 to 3·0 parts per 1,000. The amount left undissolved by the second acid varies with the length of time of its action. Under normal conditions with a surcharge of 0·6 to 0·8, the silver left in the cornets is from 1·0 to 1·2.

By continuing to boil in the second acid kept at about specific gravity 1·25 by occasional additions of water, the fineness of the gold may be raised to about 999·7 in five or six hours. The surcharge is then about − 0·3 per 1,000.

The platinum metals are also in great part retained in the cornets and require special treatment.

Assay of Bullion containing Base Metals.—Cyanide bullion is often brittle, pitted and discoloured. In such bars Hance [2] found, as impurities, Hg, Pb, Bi, As, Sb, Zn and Cr. Dewey also gives Cd, Ni, Fe and Cu. Some of these do not affect cupellation, but in general if a scoria is formed there is naturally danger of loss of entangled gold. If necessary the sample may be scorified with lead before cupellation, or the gold may be recovered from the cupel. Otherwise the results are low. Dewey [3] found that bars containing nickel and cadmium gave very discrepant results. He also found [4] that zinc in small amounts has no effect on cupellation, but in large amounts, such as 10 per cent., the Zn-Au-Ag alloy comes to the surface and carries gold to the cupel. Cadmium, if present with the zinc, tends to protect the gold from loss. If 1 or 2 per cent. of tellurium or selenium is present, a larger amount of lead, 15 grammes, is used in cupellation, but if much more tellurium is present it must be removed before cupellation. Another method for bullion containing base metals is to use the cadmium parting method, see p. 525.[5]

Preparation of Fine Gold.—Fine gold for use as standards or check pieces in assay may be purchased. It is prepared by precipitation. The finest gold available (or cornets, which contain about 1 per 1,000 of silver) is dissolved in nitric acid, one part, and hydrochloric acid, four parts, and the excess of acid driven off on a water-bath. The blackish-red liquid remaining consists mainly of $HAuCl_4$ and solidifies at 70° C. When cool it is dissolved in distilled water and diluted to contain about 30 grammes of gold per litre. Silver chloride is precipitated and allowed to settle. The decanted solution is further diluted and allowed to stand until it becomes clear by settlement. Finally the clear solution of gold chloride is siphoned into a saturated solution of SO_2 in distilled water. The gold is precipitated in a fine state of division and is thoroughly washed by shaking and boiling with distilled water. Finally the precipitate is dried and melted in a clay pot which has previously been washed out with molten borax. After casting, the bar is cleaned, rolled, and again cleaned by scrubbing with fine sand and water.

Oxalic acid is often used as a precipitant instead of sulphurous acid and is generally regarded as the best precipitant if platinum is present. The solution is kept warm and allowed to stand until precipitation is nearly

---

[1] F. P. Dewey, *J. Amer. Chem. Soc.*, 1910, 31, 318.
[2] *Trans. Amer. Inst. Min. Eng.*, 1916, 54, 536.
[3] *Eng. and Min. Journ.*, 1912, 93, 733.
[4] *Trans. Amer. Inst. Min. Eng.*, 1917, 58, 850.
[5] F. P. Dewey, *Eng. and Min. Journ.*, 1915, 99, 355.

complete. J. W. Pack [1] uses aluminium to precipitate the gold, and Krüss [2] adopted H_2O_2 and KOH for the purpose. In preparing fine gold at the Melbourne Mint,[3] chlorine gas is passed through molten gold previously purified by precipitation and washing. Fine gold is also prepared by electro-deposition from the chloride solution, as in the Wohlwill process used in the United States Mints.

Fine gold is made a little purer by scraping it just before it is assayed, and also by heating it to redness, as suggested by S. W. Smith. This is probably due to the removal from the surface of impurities such as grease, moisture, dust, condensed vapour, etc.[4]

Assay of Gold by means of Cadmium.—Balling has shown [5] that cadmium may be substituted for silver in the operation of parting. The ½ gramme of gold alloy is placed in a porcelain crucible in which a little fragment of potassium cyanide has been previously fused in order to protect the metal from the air. Cadmium is then added in the proportion of 2½ to 1 of gold. If silver is present in addition, the combined weight of cadmium and silver must be 2½ times that of the gold. The whole is fused and then cooled and plunged into hot water to clean the button, which is then parted in nitric acid (specific gravity 1·2), boiled in water for some minutes, dried and weighed. The silver, if any, can be estimated by precipitation as chloride from the acid solution, or, better, by titration with sulphocyanide. By this method the losses of gold and silver incidental to cupellation are entirely avoided. A similar method, employing zinc in place of cadmium, had previously been recommended by von Jüptner.[6]

Alloys of Gold, Silver and Copper.—These may be assayed by the method just given, the copper being estimated as difference. The method of double cupellation, by which the button of silver and gold is weighed and then subjected to inquartation and parting, is less accurate than cadmium parting and sulphocyanide titration. The temperature of cupellation must be lower than for gold, approximating more to that used for silver. Proofs of similar composition must be used and the operations require much practice before the necessary skill is acquired. The difficulty is that part of the silver is lost by cupel absorption and part of the copper is retained in the prill. It is almost impossible to arrange that these two amounts shall be equal.

Assay by Parting without Cupellation.—In this method [7] the gold alloy is melted with copper in an atmosphere of steam and is then parted in nitric acid and annealed in the usual way. The amount of copper to be added, together with the copper originally present in the alloy, should be from 2 to 2¼ times the weight of the gold present. Care is taken to prevent the elimination, before parting, of the constituents other than zinc contained in the sample, but when zinc is present it is volatilised and oxidised in the steam. Other metals are left in the bead and may be recognised by its appearance, but do not in general interfere with the accuracy of the assay. A convenient weight for the sample is 5 grains (or say 0·33 gramme). The weighed assays

---

[1] *Min. and Sci. Press*, 1908, 96, 324.
[2] *Liebig's Annalen*, 1887, 43, 237; Vanino and Seemann *Ber.*, 1899, 32. 1968.
[3] F. R. Power and R. Law, *44th Ann. Rep. Royal Mint*, 1913, pp. 44, 138.
[4] T. K. Rose, *J. Inst. of Metals*, 1913, 10, 160.
[5] *Oestr. Zeitsch. für Berg. und Httnwesn.*, 1879, 27, 597.
[6] *Zeitsch. anal. Chem.*, 1879, 18, 104.
[7] A. Westwood, *J. Inst. Metals*, 1922, 27, 307.

are each put into a separate cup of clay, with the added copper. The cups are deep, steep-sided depressions in a clay stick, 9 inches long, which is slid into a silica tube. There are 10 cups in each stick. The arrangement is shown in Fig. 235. A gas muffle furnace with the muffle removed has 5 silica tubes, each $1\frac{1}{8}$ inches bore and 3 feet long, passing through it and projecting at each end. At the back these are connected with a copper boiler, so that steam passes freely through the tubes. Five minutes' heating in the furnace is usually enough, and clean, bright, malleable beads are obtained which can be hammered and rolled out into fillets without annealing. The fillet is not coiled into a spiral, but is merely doubled on itself and is then parted, annealed and weighed. The method has been in use at the Assay Office, Birmingham, for hall-marking purposes since 1919, with satisfactory results, the errors apparently being not more than $\pm$ 0·1 or 0·2 per 1,000.

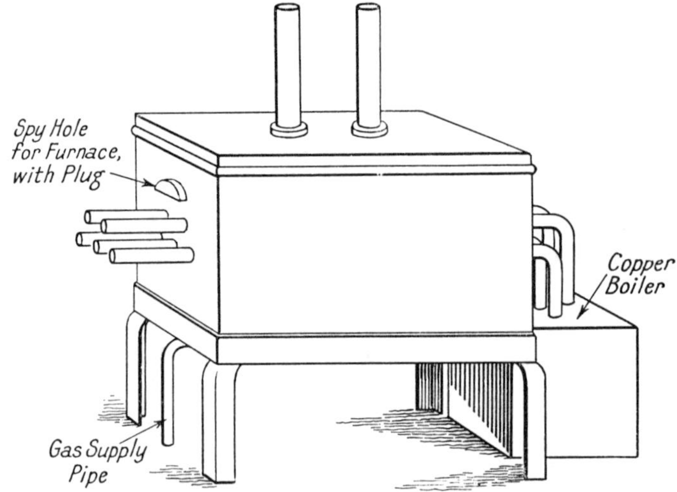

Fig. 235.—Furnace for Fusing Gold Alloy with Copper in Westwood's Method.

Volumetric Assay of Gold Bullion.—A method was devised in 1916 by W. B. Pollard for assaying gold wares and was introduced at the Cairo and Alexandria Assay Offices in Egypt.[1] The method consisted in dissolving a known weight of the alloy in aqua regia and precipitating the gold as metal by a standard solution of mercurous nitrate. Precipitation was known to be complete when a further addition of the reagent no longer produced a purple colour but only a white cloud of mercurous chloride.

Although the accuracy of the original process did not exceed $\pm$ 1 part per 1,000, it displaced the furnace assay in Egypt owing to the greater speed of working, and lower costs. The sensitivity of the method was improved by the use of *o*-tolidine dissolved in HCl or acetic acid.[2] This indicator gives an intense yellow colour when mixed with a solution containing minute traces of gold in the form of chloroauric acid. The end-point in precipitating the gold was thus made more precise. Other defects, however,

[1] W. B. Pollard, *Trans. Inst. Min. Met.*, 1932, 41, 434.
[2] W. B. Pollard, *Trans. Inst. Min. Met.*, 1923, 32, 242 ; *Analyst*, 1919, 44, 94.

were discovered [1] and were overcome by giving up mercurous nitrate and using instead a solution of an arsenite, potassium iodide and mercuric chloride, the end-point being determined by *o*-tolidine or *o*-dianisidine.[2]

For making the assays, the following solutions are prepared :—

Sodium arsenite reagent.—400 grammes of sodium arsenite are dissolved in 2 litres of water, filtered, diluted to 30 litres, and made acid with HCl. 60 grammes of mercuric chloride dissolved in 1 litre of water and 0·3 gramme of KI dissolved in 25 c.c. water are added to the arsenite solution. 50 c.c. of the reagent thus prepared, with further dilution as required, are equivalent to 0·5 gramme of gold.

Urea-potassium bromide solution, containing 10 grammes urea, 5 grammes KBr and 2 c.c. strong hydrochloric acid per litre.

Decimal hydroquinone solution, containing 0·8372 gramme hydroquinone and 10 c.c. strong hydrochloric acid per litre.

Indicator solution, containing *o*-tolidine or *o*-dianisidine dissolved in acetic acid, with sodium fluoride added.

In the assay, 0·5 gramme of gold is dissolved by heating in 3 c.c. of HCl and 1 c.c. of HNO_3 and air passed through the solution to remove gaseous by-products. 50 c.c of the urea—KBr solution are then added and the temperature raised to 40° C. The arsenite solution is then run in from a burette until no more brown colour (due to the conversion of chloroauric acid to bromoauric acid) is observed. The gold is precipitated and re-solution is prevented by the urea, which destroys all traces of nitrous acid. 20 c.c. of the *o*-tolidine indicator are then added and the yellow colour titrated out with decimal hydroquinone. The number of c.c. of decimal hydroquinone used gives the number of milligrammes of gold remaining in solution at the end of the arsenite titration, so that the weight of gold precipitated by the volume of arsenite used is obtained. The arsenite and hydroquinone solutions are standardised by the use of samples of pure gold, and more exact determinations are obtained by the use of a 50 c.c. Stas pipette. Full details, including the effects of interfering metals, are given in Pollard's papers. Palladium and cadmium are the most troublesome impurities. Alloys with high silver content are insoluble in aqua regia and cannot be handled.

Assay of various Gold Alloys.

It is often impracticable to apply the ordinary parting assay to the examination of low-standard alloys of gold with other metals. These are tested by various other methods, of which a summary is given below, the alloys being grouped in four series for convenience :—

A. Alloys requiring scorification.
B. Amalgams.
C. Alloys containing members of the platinum group.
D. Tellurium and selenium alloys.

A. Scorification of Alloys.—Alloys of gold containing *arsenic* or *antimony* are reduced to a fine powder and scorified with thirty parts of lead and a half part of borax. If the slag becomes pasty towards the end of the operation more borax is added, a little at a time. If the lead button

[1] Pollard and Ridge, *J. Chem. Soc.*, 1926, p. 529.
[2] W. B. Pollard, *Trans. Inst. Min. Met.*, 1932, 41, 434.

obtained is hard, a second scorification is necessary, with the addition of more lead. There is some loss of gold in the slag.

Iron or Manganese Alloys.—The operation is tedious and difficult with these alloys, as they are difficult to fuse, having higher melting points than pure gold, and the oxides of iron do not form easily fusible compounds with the litharge. Ten parts of lead, one of borax, and one of silica usually suffice.

Cobalt and Nickel.—Twenty parts of lead are used, but no borax at first, so that the oxidation of the nickel may not be hindered. A very high temperature and the subsequent addition of two parts of borax are necessary. Several successive scorifications are required as nickel and cobalt are difficult to oxidise.

Zinc.—Oxide of zinc does not form a fusible mixture with litharge, and the slag is only rendered pasty by borax, unless it is added in large quantities. Gold is lost in the slag, but the loss is minimised by slagging off the zinc as rapidly as possible. Use 15 to 20 parts of lead and two to three parts of silica, with a little borax. Instead of silica, caustic soda may be used.

Tin.—Twenty parts of lead are required ; oxide of tin is rapidly formed but the slag is not easily fusible. Large amounts of borax are necessary, or still better, borax mixed with caustic potash or soda which forms a fusible stannate with SnO_2.

Aluminium.—Alloys containing this metal cannot be assayed by scorification and cupellation. As soon as fusion takes place in the muffle, the aluminium is rapidly oxidised, producing alumina which forms an exceedingly infusible scoria not easily removed by litharge. The production of the latter, moreover, is checked by the scum. Caustic soda would form a fusible aluminate.

Edward Matthey observes [1] that the removal of aluminium by digestion in hydrochloric acid, and the collection of the residual gold, does not yield satisfactory results. The process he recommends is as follows :—Accurately weighed portions of 50 grains each of the alloys are fused with litharge, under a flux of potassium carbonate and borax with a small proportion of powdered charcoal, and the resulting slag re-fused with a further small quantity of litharge and powdered charcoal. The lead buttons containing all the gold (the aluminium having combined with the fluxes employed) are cupelled, and the resulting gold cupelled with silver and parted with nitric acid in the usual manner. The assays must be worked with checks or standards of fine gold and pure aluminium.

B. Amalgams.—The alloy is placed in a weighed porcelain crucible and gradually heated so as to drive off the mercury. After the greater part of the mercury has been driven off the temperature is raised to a full red heat which is maintained for half an hour. About 0·1 per cent. of mercury still remains in the gold after this treatment and can only be completely removed by cupellation and parting. The loss of gold in the operation may amount to 1 per 1,000.

A better method is to dissolve the mercury in nitric acid mixed with an equal volume of water. A porcelain crucible or basin should be used and gentle heat applied. If the action is not too violent the gold does not break up, but is left as a spongy coherent mass. It is washed, first with nitric acid, and finally with water, and ignited. After ignition, further treatment

---

[1] *Phil. Trans. Roy. Soc.*, 1892, 183, **A,** 647.

with concentrated nitric acid for a few moments is necessary. The gold still contains some mercury, which is removed by cupellation and parting.

Cupellation and parting give more accurate results than nitric acid when the proportion of mercury is small.

Mercury may be used to collect very finely divided precipitated or parted gold into a coherent mass, the mercury being removed with nitric acid.

C. Platinum Group.—Cupellation is effected at a higher temperature than usual (see p. 43 for the melting points of platinum alloys).

Platinum-Gold Alloys.—The button obtained by cupellation is dull and crystalline. If the alloy contains as much as 7 or 8 per cent. of platinum the cupellation proceeds slowly and brightening is only obtained at a very high temperature; the button appears flattened, and has a rough crystalline surface and a grey colour. If more than 10 per cent. is present, brightening does not occur at all, and the other features just mentioned are more strikingly exhibited. On parting, the platinum is partly dissolved with the silver, but the assay-piece must be boiled in acid for a long time, and the parting is incomplete. When the ordinary parting assay is used the results are not satisfactory if more than 1 or 2 parts of platinum are present per 1,000 of alloy. If the cornets are not of the normal gold colour owing to retention of platinum, they may be cupelled and parted again.

The most satisfactory method is to dissolve the sample in aqua regia, evaporate nearly to dryness to drive off the free acid, add water and precipitate the gold with oxalic acid, warming the solution until precipitation is complete. The platinum is estimated in the solution. J. Phelps finds[1] that the best way to do this is to evaporate to dryness and heat until copious oxalic acid fumes come off. All the platinum metals, except possibly palladium, are thus reduced.

Alloys of Gold, Platinum, Silver and Copper.[2]—If too much silver is present for the alloy to be dissolved in aqua regia, even with the addition of common salt, the sample is inquarted with silver (at least 75 per cent. Ag is needed) and parted in nearly pure sulphuric acid in a silica or platinum vessel until action ceases. If glass is used it is protected from the air with asbestos cloth to avoid cracking. Bumping is reduced by capillary tubes. The boiling may take several hours. The acid is allowed to cool before decantation. The residue of gold and platinum is treated as already described. An alternative method is by precipitating platinum from solution as chloride with ammonia and ammonium chloride.[3]

Palladium-Gold Alloys.—The palladium is dissolved in parting if the weight of silver is at least three times that of the gold, yielding an orange coloured solution. Matthey recommends double parting. Separation may also be effected by fusion with six to eight parts of potassium bisulphate, and dissolving out the dark brown palladious sulphate by boiling water.[4] A second fusion is usually necessary to render the gold residue quite pure.

As in the case of platinum, palladium may be separated from gold by

---

[1] Private Communication.

[2] Forest, 6*me Rapport par l'Administration des Monnaies et Médailles*, 1901, p. xxix. See also E. A. Smith, "*Sampling and Assay of the Precious Metals*," pp. 410-429.

[3] *Eng. and Min. Journ.*, 1911, **92**, 259. See also A. F. Crosse, *J. Chem. Met. Mng. Soc. S.A.*, 1914, **14**, 373.

[4] H. Rose, quoted by E. Cumenge and E. Fuchs, "*L'or dans le laboratoire*" (Frémy, *Ency. Chim.*, p. 177.)

dissolving the alloy in aqua regia, evaporating to dryness, taking up with water and precipitating the gold by oxalic acid from a very dilute solution: the palladium remains in solution.

Rhodium- and Iridium-Gold Alloys.[1]—Iridium, if present, always sinks to the bottom of the cupelled button as it is very dense (specific gravity = 22·5), and is not usually fused at the temperature of the muffle, but occurs in the state of fine black crystalline particles. Hence, when the button is rolled into a cornet with the lower face outwards iridium occurs as black sooty spots or streaks which are seen by a lens to fill up depressions in the surface of the gold.

Both rhodium and iridium are almost insoluble in aqua regia. If gold alloys containing both of them are parted in the ordinary way with nitric acid, only a small quantity of rhodium goes into solution with the silver. The residue consisting of the gold, the iridium and most of the rhodium may then be attacked by 10 per cent. aqua regia, when the gold is dissolved together with only traces of the other metals. These may be separated by evaporating the solution to dryness and heating to dull redness, when the reduced metals, being no longer alloyed, may be completely separated by dissolving the gold in aqua regia.

The estimation of osmium, iridium and ruthenium in gold bullion has also been studied by Riche, Leidié, and Quennessen,[2] assayers at the French Mint.

D. Tellurium and Selenium Alloys.—Tellurium alloys may be dissolved in nitrohydrochloric acid, and the solution containing both gold and tellurium evaporated with a large excess of hydrochloric acid until no more chlorine is given off, when both gold and tellurium are readily precipitated by a current of sulphur dioxide gas. On attacking the precipitate with dilute nitric acid the tellurium is dissolved in the state of tellurous acid, and the gold residue may be dried and weighed, and its purity ascertained by inquartation and parting. The greater part of the tellurium may be removed from gold-tellurium alloys by boiling in nitric acid, and the residue can be cupelled and parted with very little loss of gold.

Clennell finds[3] that gold bullion containing 1 to 2 per cent. selenium gives little trouble. He cupels 0·5 gramme with 15 grammes lead, when the selenium is eliminated.

Wet Methods of Assay of Gold Alloys, Compounds, etc.—Assays or complete analyses of gold bullion, natural minerals, etc., can be made by the ordinary chemical methods. From 1 to 5 grammes of bullion are usually enough, but a much larger amount is necessary if the alloy is nearly pure gold. M. Forest[4] takes 300 grammes of gold bullion when examining it for small quantities of impurities. In general the residue left after prolonged action of nitric or sulphuric acid is not sufficiently pure to weigh as gold, and complete solution in aqua regia is usually necessary. From the solution the gold may be precipitated by (*a*) ferrous sulphate, (*b*) sulphurous acid, (*c*) oxalic acid, (*d*) sulphuretted hydrogen, (*e*) ammonium sulphide, followed by the addition of hydrochloric acid. The following remarks may be of value in aiding the chemist in his choice of a precipitant in any particular case.

---

[1] H. A. White, "*Rand Metallurgical Practice,*" vol. i., p. 531; C. Toombs, *J. Chem. Met. Mng. Soc. S.A.*, 1914, 14, 4; J. Gray, *op. cit.*, 14, 2.

[2] "*Huitième Rapport par l'Administration des Monnaies et Médailles,*" 1903, p. xxix.

[3] J. E. Clennell, *Eng. and Min. Journ.*, 1906, **82**, 1,057.

[4] "*Neuvième Rapport par l'Administration des Monnaies et Médailles,*" 1904, p. xxxi.

Nitric acid must always be expelled from the solution by warming with successive additions of hydrochloric acid. The acid solution must not be heated too strongly or loss of gold chloride by volatilisation occurs. Some other chlorides escape more freely. Ferrous sulphate and sulphurous acid act well in strongly acid (HCl) solutions; oxalic acid, sulphuretted hydrogen, and ammonium sulphide act best in presence of small quantities of HCl. The solution should be dilute (say 1 part of gold in 300 of water), so that other metals may not be carried down by the gold. Sulphate of iron gives a very finely divided precipitate which is difficult to wash by decantation without loss; precipitation is slow in cold solutions. Oxalic acid causes plates and scales to form which are readily washed and are very pure; it acts best in boiling liquids, but a temperature of 80° C. for forty-eight hours suffices; in the cold or in the presence of much hydrochloric acid or alkali chlorides the action is very slow and partial; a large excess of the precipitant must be present. Oxalic acid is used for solutions containing metals of the platinum group, which are not precipitated by it. Alkali oxalates may be used instead of the free acid.

Sulphurous acid is an excellent precipitant for most solutions. It acts rapidly and completely in the cold, and does not readily precipitate other metals, except tellurium. Sulphuretted hydrogen is used in the absence of all other metals whose sulphides are insoluble in hydrochloric acid.

In all cases careful consideration must be given to the nature of the base metals present, and the precipitant which will not render any of them insoluble must be selected.

Other Methods of Bullion Assay.—Other methods have been proposed at various times, and may still be of service occasionally in particular cases.

1. Trial by the Touchstone.—This is the oldest method of assay. It was described by Theophrastus, about 300 B.C.,[1] and the methods in use in Germany in the 16th century are fully detailed in "Agricola."[2] The assay consists in rubbing the gold bullion to be tested on a hard dark-coloured smooth stone, and comparing the appearance and colour of the streak with those made by carefully prepared touch needles of known composition. The effect of the action of nitric acid and dilute aqua regia on these streaks is also noted. Touchstones usually consist of Lydianstone or of silicified wood, and black or dark green stones are best. Only alloys of gold and copper or of gold and silver can be thus tested. The trial is more sensitive for alloys below 750 fine than for higher standards. The amount of gold in alloys between 700 and 800 fine can be determined correct to 5 parts per 1,000.

2. Colour and Hardness of Alloys.—These properties form a guide to the composition of copper-gold alloys, an increase of copper corresponding to a heightening of the colour and an increase of the hardness as tested with shears or a knife. On heating the alloy to redness in air, the degree of blackening of the surface is a further indication of the percentage composition if compared with plates of known fineness.

3. Density of Gold-Copper and Gold-Silver Alloys.—The method is used for medals, coins and similar articles in order to avoid mutilation, and is satisfactory because neither expansion nor contraction occurs to any noticeable extent when silver or copper is mixed with gold. The samples are

---

[1] See Hoover's "*Agricola,*" p. 252, note 37.

[2] *Op. cit.*, pp. 252-260.

weighed in air and water successively and the density of the alloy calculated by means of the formula—[1]

$$D = \frac{W}{W_1} \cdot \frac{1 + K t}{1 + \delta_t} - \frac{W - W_1}{W_1} (1 + Kt)\alpha$$

Where D = the density
W = weight in air
W_1 = loss of weight in water
K = coefficient of cubic expansion of the alloy
t = temperature (°C.)
$\frac{1}{1 + \delta_t}$ = the density of water at the temperature t
$\delta_t = 1$ — density of water at $t°$ C., a very small number
$\alpha = 0{\cdot}001293$, the weight of one c.c. of air at N.T.P.

The composition of the alloy is calculated by means of the formula—[2]

$$\frac{W}{D} = \frac{w_1}{d_1} + \frac{w_2}{d_2}$$

Where W = the weight of the sample
D = the density of the sample
w_1 = the weight of the gold
w_2 = the weight of the other metal
d_1 = the density of gold
d_2 = the density of the other metal.

4. Assay by means of the Spectroscope.—The spectroscope is not of much service in determining the composition of gold alloys.[3] J. Phelps observes:[4] "By the use of Hilger's Quartz Spectrograph small amounts of other metals in fine gold can be detected, as for example one part of silver or copper in a million. The lines of silver, copper, silicon and calcium are quite strong in the arc spectra of gold which assays 1,000 fine against other gold which is nearly free, spectroscopically, from these elements." Such small quantities of these elements cannot be determined in any other way. Phelps continues: "The persistent lines of gold seem to vanish from the spectrum when there is still enough present in other metals to be detected by assay." Similar results have been obtained in attempts to detect traces of gold in minerals,[5] and it is doubtful whether such methods are useful in this direction. In a study of the detection of lead in gold, it was found that the intensity of the lead lines in the spark spectrum fell until, after a period of 1-2 minutes, a constant intensity was reached.[6] (See also p. 6.)

---

[1] Kohlrausch, "*Physical Measurements*."
[2] *7th Ann. Rept. Royal Mint*, 1876, p. 43.
[3] Lockyer and Roberts-Austen, *Phil. Trans. Roy. Soc.*, 1874, 164, [ii.], 495.
[4] Private Communication, 1935.
[5] Frémy, *Ency. Chim.*, vol. iii., "L'or," p. 134.
[6] Gerlach and Schweitzer, *Z. anorg. Chem.*, 1928, 173, 92.

5. Microchemical Analysis.—Microchemical analysis of the normal streak of gold alloys made on a roughened microscope slide will indicate the nature of the alloying elements. A whitish streak, soluble in nitric acid, indicates a silver alloy. By precipitating the silver with a drop of hydrochloric acid, and treating the clear liquor with a drop of $K_2Hg(CNS)_4$, the appearance of green crystals indicates the presence of copper, and brown crystals copper and cadmium together. The gold streak may be dissolved in aqua regia, the solution treated with H_2S, the precipitate digested with ammonium sulphide to remove gold and the remainder dissolved in nitric acid. Minute drops of this solution are tested for lead with potassium nitrite and acetate (precipitating the triple potassium-lead-copper nitrite), for bismuth with potassium sulphate, for palladium with dimethylglyoxime in acid solution, for nickel with the same reagent in alkaline solution, for cobalt and zinc with $K_2Hg(CNS)_4$, and for aluminium with alizarinsulphonic acid.[1]

---

[1] Strebinger and Holzer, *Mikrochemie*, 1930, **8**, 264; *ibid.*, 1931, **10**, 306.

CHAPTER XX.

ORE TESTING.

THE steps to be taken in the testing of an ore body preparatory to its development and the erection of a mill for its treatment may be summarised as follows :—

(1) **Sampling** of the ore body. With this we are not directly concerned here, as it is more properly the function of the pioneers. Generally speaking, however, the method of sampling will depend on the nature of the deposit, whether it be alluvial or whether it be in vein formation. In the former instance drilling usually suffices, while in the latter, drilling, cross-cutting or the sinking of small pits may be severally or collectively employed. During the early investigations the extent and direction of the formation will have been roughly determined from visual observation, possibly after the overburden has been partially removed. Geological indications, such as the general nature of the ore-body, the dip and strike, and the position and characteristics of outcrops, are all essential features in arriving at a preliminary estimate of the value of the deposit, and will in most instances be considered before sampling is attempted.

When the general trend of the valuable ground has been noted, sampling may proceed on broad lines, and the assays from the various samples will give some confirmation as to the position of those areas carrying most gold. The sampling should be done in regular fashion according to a predetermined plan, marking off the ground into equal areas, which may in the first instance be rather large. When the assays reveal the portions which are likely to yield values, sampling at closer intervals—but still according to a plan—will afford more precise information. In all cases, by whatever means the sample is obtained, the selection should follow general rules :—

(*a*) The sample should embrace portions from all sections of the drilling or cut.

(*b*) It should be sufficiently large to be reasonably representative of the ground the drill has traversed or through which an adit or cross-cut has been made.

(*c*) The sampler should be as unbiassed as possible in the selection of the ore for test purposes.

(*d*) Sufficient ore should be ground initially for all the tests, as it is often difficult to reproduce exactly the same conditions.

Too much emphasis cannot be laid on the necessity for correct and orderly sampling, for on it depends the subsequent development of the ore body and the estimates of possible profits.

(2) **Hand Examination** of specimens from geological and petrographical viewpoints will enable the sampler to determine the general characteristics of the ore body and the nature of the principal minerals, so that an estimate of the difficulties to be met in recovering the gold can be obtained.

(3) **Microscopical Examination** of the ore samples. Attention is being directed to this to a much greater extent than formerly. By it a true idea of the associated minerals, their frequency and amount, as well as of the nature, characteristics and mode of occurrence of the gold, can be obtained. In some measure this may determine the general lines of treatment for the ore, according as the gold is free, encased in pyrite or other mineral, associated with cyanicidal minerals, *e.g.* containing copper or tellurium, or is merely a valuable secondary constituent of a base metal deposit.

(4) **Chemical Analysis.**—Mineralogical examination should be followed by chemical analysis and a determination of the physical properties, both of which are complementary to a knowledge of the minerals in the ore. They do not suffice in themselves to determine a line of treatment, and should not be given undue weight in the summing up; they are useful, however, in assessing the importance, and the results, of preliminary treatment such as roasting.

(5) **Acidity of Ore.**—Take 150 to 250 grammes of finely crushed ore and shake with an equal weight of distilled water. Filter. (*a*) Titrate the filtrate with standard alkali. (*b*) Add standard lime solution to the filtrate until it gives a slightly alkaline reaction with phenolphthalein. Calculate the amount of lime thus necessary to add. Repeat with mill water to see if this affects the result. Mill water should be thoroughly tested for organic matter, sewage, acid, etc.

(6) **Crushing and Grinding Tests.**—By these are determined: (*a*) the hardness of the ore, which may be confirmed by ordinary hardness tests, (*b*) the friability of the ore, which will determine the general nature of the coarse crushing plant that will be required, (*c*) the ease with which the gold is separated from the gangue. This last can be most effectively gauged by carrying out sizing tests as follows:—

A weighed quantity of the ore is crushed in a laboratory crusher to about 80 to 100 mesh and is then passed through a nest of standard sieves which may be shaken by hand or mechanically (British Standards Institution). The portion remaining on each sieve is weighed and its percentage of the whole sample calculated; it is then assayed. The relative values obtained enable one to say approximately to what degree of fineness it is necessary to crush the ore in order to liberate the gold.

A somewhat similar classification of the ground product may be obtained in a wet way by using an elutriator or small hydraulic classifier. In this the ore is fed into a column of water flowing upwards under a constant head. The lighter particles are removed at the top, while the concentrates and heavier particles which are not lifted by the water sink towards the bottom.

Further microscopical examination of the graded samples at this stage will confirm, or otherwise, the previous conclusions, and particularly will determine whether free gold is obtained at an early stage (and therefore recoverable after comparatively cheap treatment), or whether it is so encased as to necessitate fine crushing in order to expose the metal to attack by the various reagents. It should also be remembered that the introduction of flotation processes has vitally modified the probable conclusions to be drawn from such examination, for if the microscope reveals, say, pyrite encasing the gold, or even if the gold itself is exposed say at 60 mesh, then it is possible to obtain a rich concentrate which is but a fraction of the total tonnage and can be treated separately.

To test for encased gold, the ore may be digested gently with diluted aqua regia for several hours. The solution is then poured off, the residue washed well with water, dried and assayed.

(7) **Concentration.**—The object of this test is to determine whether a concentrate containing high values in gold may be recovered at an early stage in order (*a*) to obtain a quick return of profits, and (*b*) to relieve the subsequent cyanide section of the burden of dissolving comparatively coarse gold, which would take a long time and lock up valuable material. Concentration tests are usually carried out on large samples (1 or 2 cwts.) in laboratory machines which are small replicas of plant units of similar type, *e.g.* Wilfley tables, blankets. A simple batea may also be used. Assays of the concentrate and of the tailings will provide the following particulars: (*a*) Ratio of concentration (*i.e.* of weight of concentrate to that of original ore); (*b*) grade of concentrate to be expected; (*c*) the loss in the tailing. Microscopical examination of the concentrate and the tailing will give information as to the condition of the gold—whether it is free or enclosed. The variables are (1) fineness and grading analysis of the ore, (2) pulp density, (3) percentage of concentrate.

(8) **Flotation Tests.**—Miniature laboratory flotation cells of various types are now available for tests of this nature. The tests are carried out on large quantities of ore (1 or 2 cwts.) and the variables to be noted are :—

(*a*) Minimum degree of fineness to which ore should be crushed.
(*b*) Pulp density.
(*c*) Acidity or alkalinity of the pulp, *i.e.*—*p*H value.
(*d*) Amounts and nature of reagents :—

(*a*) Frothers.	(*c*) Dispersers.
(*b*) Stabilisers.	(*d*) Promoters.

(*e*) Amount of conditioning required.
(*f*) Where best to add the reagents.
(*g*) Degree of agitation.
(*h*) Percentage and grade of froth concentrate.
(*i*) Whether the froth needs to be treated in a cleaner cell.
(*j*) Effect of temperature.
(*k*) Kind of process and apparatus that are most appropriate.

It should be remembered that testing for flotation treatment is neither so easy nor so satisfactory as the ordinary tests, principally because of the extremely small proportion of reagents which is required to effect the recovery of a suitable concentrate.

(9) **Amalgamation Tests.**—Though amalgamation is not widely practised at the present time, such tests add to the evidence of the presence of free gold. They are usually performed by—

(*a*) Stirring a weighed quantity of the ore (10-15 A.T.) made into a thin pulp with water (in definite proportions) with a weighed quantity of mercury for a known period.
(*b*) By rotating or shaking a similar pulp with mercury for a definite time.

If the ore contains harmful impurities, small quantities of appropriate chemicals, *e.g.* caustic soda or hydrochloric acid, may be added to keep the mercury active.

Afterwards, the mercury, together with any amalgam that may have been formed, is recovered by washing away the pulp into a separate container. The amalgam may be recovered by squeezing the mercury through chamois leather. It is weighed and then treated with dilute nitric acid to remove the mercury and leave the gold. Alternatively the whole of the mercury and amalgam may be so treated. In either case the residual gold is thoroughly washed with cold and hot water, dried and finally heated to redness to anneal it. Its weight will enable the percentage so recovered to be calculated, and by assay the fineness can be determined. If the tailing from the separate container is now sampled and assayed, the sum of the two results should amount to the original value of the ore.

An amalgamated copper pan may also be used for the amalgamation tests.

By performing amalgamation tests of this nature on each of the sieve samples a clearer conception of the range over which gold is susceptible to amalgamation may be obtained.

(10) **Cyanidation Tests.**—These resolve themselves into the determination of the correct values for—

(1) Strength of cyanide solution.
(2) Total alkalinity.
(3) Protective alkalinity.
(4) Time of agitation.
(5) Percentage extraction.
(6) Effect of heat.
(7) Cyanide consumption.
(8) Leaching of sands.

They may be carried out on :—(*a*) Each sieve sample. (*b*) Each sieve sample after amalgamation. (*c*) On the coarse and fine products of classification after grinding to the most suitable mesh, previously determined, for the liberation and exposure of the gold. (*d*) On the whole ore crushed to the consistency of slime, *i.e.* to – 200 mesh. (*e*) On the tailings from any flotation tests. (*f*) On the tailings from concentration tests.

Cyanide Extraction after Amalgamation.—A weighed charge (say 300 grammes) of the tailing from amalgamation is mixed with 0·3 gramme of lime and 600 c.c. of cyanide solution of average strength (0·25 per cent.). The charge may then be treated in one of three ways :—[1]

(*a*) Placed in a bomb, and air or oxygen admitted to give a pressure of 100 lbs. per square inch. The bomb is then rotated for 5 or 6 hours, when the solution is filtered from the residue. The operation is repeated, the residue washed with water and both portions assayed.

(*b*) Allowed to remain in contact with 0·25 per cent. of barium peroxide for 24 hours with occasional shaking. After decantation, more barium peroxide and cyanide solution are added and contact maintained for a further 24 hours. The solution is again decanted, the residue washed and then both solution and residue assayed.

(*c*) Agitated in a miniature Brown tank.

Cyanide Extraction without Amalgamation.—A weighed sample of the ore in the form of pulp of known fineness and water-solid ratio may be rotated in a bottle with a definite quantity of standard cyanide solution and alkali for a stated time. At the end of this period the mixture is allowed to settle, the clear liquor then siphoned off through a filter (to retain stray particles of solid material) and its gold content determined. A check may be obtained by drying and assaying the solid residue.

Further tests may be carried out on separated sand and slime. Each is

[1] "*Rand Metallurgical Practice*," vol. i, p. 350.

treated with solutions containing known amounts of cyanide and lime and during treatment sufficient air should be present to provide the necessary oxygen.

Leaching of Sands.—The effect of percolation or leaching of sands may be investigated by taking a weighed sample, washed free from slime, of known mesh and sufficient to fill a Buchner funnel, having filter paper or cloth at the bottom, to half its depth. To this 0·1 per cent. of lime and 0·05 per cent. of lead acetate are added. A known volume of cyanide solution, whose strength is known (0·02 to 0·5 per cent.), is then poured on and allowed to pass through the sands. The filtrate is returned to the funnel and the process repeated several times, after which weak cyanide and water washes are given in similar manner. Finally the filtrate and the residue are assayed for gold and will give the percentage which may be expected by leaching. The sum of the results should equal the original assay of the ore.

An alternative method is to leave the cyanide solution in contact for one to two hours, and then to open the leaching tap and draw off 100 c.c. The tap is then closed and the remainder allowed to stand for 12 hours. The tap is again opened, and the solution allowed to drip quickly through, so that drainage is complete in 24 hours. A weaker cyanide solution is then poured on in measured quantity and the drip maintained. Similar procedure is adopted with two still weaker washes, and finally with water. To assist aeration, an interval should be allowed between successive washes, or the solids may be transferred to another vessel.

Agitation of Slimes.—The slime, free from sand, is mixed in weighed quantities with water to bring the water-solid ratios to about 3 : 1. A known volume of cyanide solution of correct strength (0·01 to 0·05 per cent.), and containing lime, is then added to each sample. The samples are then agitated for definite periods up to, say, 10 hours, when settlement is allowed to take place. The liquors are decanted or filtered, replaced by a weaker solution of cyanide, and the operations repeated. A similar process is undertaken with water. The solutions and the residues are assayed for gold.

The variables here are : (1) Pulp density, (2) strength of cyanide solution, (3) quantity of cyanide solution, (4) amount of lime, (5) time of agitation, (6) number of decantations.

Settlement of Slime.—This is of importance in indicating the possibility of interferences with decantation. The slime should be the natural material and should not have been treated or dried, thereby altering its physical condition. Moreover, slime which has stood by and become weathered tends to form colloids, containing especially silicic acid, which hinder settlement. Small-scale tests in inverted Winchester bottles from which the bottoms have been removed may be employed, but are often misleading. It is preferable to use a tank into which the pulp (W/S = 3/1) is poured and agitated. The time and temperature are noted, and also the time when complete settlement has occurred—usually indicated by bubbles appearing on the surface. The rate of settlement may be followed by withdrawing samples at definite periods and estimating the amount of solids they contain.

Filtration of Slime.—The filtration test may be carried out on a laboratory-type machine, or, as a rough guide, on a Buchner funnel using the suction provided by a pump attached to a water tap. Weighed quantities of pulp of known water-solid ratio should be employed, and a strong cloth or other filtering medium used. The variables to be noted are :—(*a*) Density of the pulp, (*b*) vacuum employed, (*c*) thickness of pulp layer on filter cloth, (*d*) quantities of solution employed for washing, (*e*) weight of dry slime produced, (*f*) assays.

Cyanide Consumption.—A weighed quantity of the pulp is placed in an inverted Winchester quart bottle, which has had the bottom removed and a tube with tap inserted through the cork. Half the pulp weight of solution of a strength found best by previous tests is added, together with 0·1 per cent. of lime, the mixture agitated for a short while and allowed to stand overnight. The liquor is decanted or drained, and then measured and assayed for its cyanide content. Weaker cyanide solution equal in quantity to the original solution is then added to the pulp and the process repeated. Finally water is used in a similar way. By deducting the total cyanide of the used solutions from that of the unused ones the loss may be found.

Temperature Test.—The temperature of the solutions may influence the results of the foregoing tests. They should therefore be repeated under different temperature conditions up to 100° F. and the variations in extraction noted. It should be remembered, however, that an increase in temperature may result in greater evil effects from organic material.

In all tests on the ore it is essential that not more than one variable be altered during any one series of tests, and that in each succeeding type of test those conditions which were found best in previous tests should be adopted and maintained constant. Eventually, when all tests have been completed and a perspective of possible treatment has been obtained, it may be necessary to repeat some of the experiments in slightly different combinations and over somewhat narrower ranges of proportions and composition, in order to obtain data for maximum extraction.

A knowledge of the primary characteristics of the ore from the previous tests—its response to amalgamation, cyanidation, flotation, etc.—should be applied in extending the treatment to combinations of those processes in order (*a*) to recover the gold as quickly as possible, (*b*) to obtain a product which is reasonably pure, (*c*) to attain the maximum extraction. Some help may be obtained in this connection from a knowledge of the processes adopted in the case of similar ores elsewhere. A summary of combinations which have been successful is given on p. 279.

While ore testing is a necessary feature of all development work, it should be remembered that it is merely a guide to the probable treatment for a particular ore. Each deposit is a problem in itself and the information derived from a series of tests should bring into prominence the characteristics to which most attention must be paid in plant operation. It is not safe, however, to proceed direct from the results of laboratory testing to the erection of a large plant for bulk treatment, which almost invariably necessitates modifications of the laboratory procedure, especially as the latter often yields the higher results. It is better, and more customary, to erect a pilot plant of, say, 10 tons per day capacity, following this with another of, say, 100 tons per day capacity, before finally embarking on the full production. Alternatively a unit of an already existing plant, if available, may be set apart for the trials.

Tests on an Operating Plant.

It may be necessary to investigate or check the process work of a mill already in operation. To a certain extent the tests already described under Ore Testing will be useful.

Among details to be examined the following may be mentioned :—

(1) *General*:
(*a*) Examine current records and investigate inconsistencies.
(*b*) Before altering a flow sheet in a major respect, endeavour to run the plant at the maximum efficiency under current conditions.
(*c*) In making alterations, change but one factor at a time and note the effect.

(2) *Ore*:
Note any change in general characteristics from the general run hitherto supplied.

(3) *Stamp Duty*:
(*a*) Rigidity of foundations and general structure.
(*b*) Order of drop of stamps and distribution of ore in mortar box.
(*c*) Height of discharge.
(*d*) Speed of stamps.
(*e*) Regularity and amount of feed.
(*f*) Nature and quantity of mill water.
(*g*) Nature and grading of ore.

(4) *Amalgamation*:
(*a*) Quality of tailing.
(*b*) Quality of amalgam.
(*c*) Periods between dressings.
(*d*) Presence or otherwise of mercury in cyanide plant.
(*e*) Relative times allowed for amalgamation.

(5) *Cyanide Plant*:
(*a*) Examine grading analysis to see if crushing is correct.
(*b*) Note degree of separation of sand and slime.

(6) *Precipitation*:
(*a*) Area of zinc exposed to the solution.
(*b*) Zinc consumption per ton of solution.
(*c*) Flow of solution.
(*d*) Amount of oxygen or air passing into the precipitation section.

(7) *Assaying*:
The gold content of all products should be determined and related to the steps in the process.

The following methods are suggested for testing the tonnages of sand and slime :—[1]

Sand.—Drop into the vats boxes (without tops) of known capacity and having sides perforated with small holes through which the solutions may pass. When the vat has drained, dig out the boxes and level the pulp with a spatula. Tip the material out, dry, weigh and thus estimate the weight of the cubic contents. A simple proportion will give the content of the vat.

A small proportion of the pulp should be reserved for specific gravity determination.

Slime.—The entering and discharge streams should be sampled for definite periods and the times noted to fill and discharge the vats.

[1] "*Rand Metallurgical Practice,*" vol. i., p. 540.

INDEX

BIBLIOGRAPHY.

It has been found impracticable to enumerate the articles and paragraphs relating to the metallury of gold, which have from time to time appeared in the various periodicals, as very little search results in the accumulation of thousands of such references. In general, therefore, only the names of some of the publications which contain important or interesting matter on the subject are given; but a few exact references on special points have been added, and many others occur in the footnotes to the text.

PERIODICAL LITERATURE.

Anales de la Mineria Mexicana, ó sea; Revista de Minas. Mexico. From 1861.
Anales de Minas. Madrid. From 1841.
Annalen der Berg- und Hüttenkunde. Salzburg, 1802-5.
Annales des Mines. Paris. From 1816.
Annuaire du Journal des Mines de Russie. St. Petersburg. First published in 1840.
Annuaire des Mines et de la Métallurgie Françaises. Paris. First published in 1876.
Annual Reports of the Californian State Mineralogist. Sacramento. From 1881 to 1890. First Biennial in 1892.
Annual Reports of the Deputy-Master of the Royal Mint. London. From 1870.
Annual Reports of the Director of the United States Mint. Washington.
Annual Reports on Gold Mining. Victoria, British Columbia. From 1875.
Australian Mining Standard. Sydney and Melbourne. From 1872.
Berg- und Hüttenmannisches Jahrbuch. Vienna. From 1866.
Berg- und Hüttenmannische Zeitung. Freiberg. From 1842.
Biennial Reports of the Nevada State Mineralogist. From 1871.
Boletin oficial de minas. Madrid, 1844-5.
Bulletin de l'Association amicale des anciens élèves de l'École des Mines. Paris. From 1869.
Bulletin de la Société de l'Industrie Minerale. Quarterly. St. Etienne. From 1855.
Dingler's Polytechnisches Journal. From 1815.
Engineering and Mining Journal. New York. From 1866.
Jahrbuch für Berg- und Hüttenwesen. Freiberg. From 1827.
Journal des Mines de Freiberg. Koehler. Freiberg, 1788-1793.
Journal des Mines de Russie. 1832 to 1835.
La Mineria. Mexico, 1843.
Mineral Industry. New York. Annually from 1892.
Mines and Minerals, formerly *Colliery Engineer.* Monthly. Scranton, Pa., U.S.A. From 1897.
Mining and Scientific Press. San Francisco. From 1865.
Mining and Smelting Magazine. London, 1862-65.
Mining Journal. London. From 1836.
Mining Magazine, formerly *Pacific Coast Miner.* Monthly. New York. From 1904.

BIBLIOGRAPHY.

Mining Review. Denver, Colorado, 1873-76.
Mining World. London. From 1871.
Nouveau Journal des Mines de Freiberg. Koeler & Hoffmann. Freiberg, 1795-1804.
New Zealand Government Mining Journal. Wellington. From 1897.
Oesterreich Zeitschrift für Berg- und Hüttenwesen. Otto Freihern. Vienna. From 1853.
Precious Metals of the United States, Annual Reports on. Washington. From 1880.
Proceedings of the California Academy of Sciences. San Francisco.
Proceedings of the Chemical and Metallurgical and Mining Society of South Africa. Johannesburg. From 1894.
Proceedings of the Colorado Scientific Association.
Reports of the Mining Commissioner of New Zealand. Wellington, New Zealand. From 1871.
Reports of the South African Association for the Advancement of Science. Johannesburg. From 1903.
Revista Minera y Metalurgica. Madrid. From 1850.
Revue Universelle des Mines. Paris. From 1857.
School of Mines Quarterly. Columbia, U.S. From 1879.
Scientific American. New York. From 1846.
Silliman's American Journal of Science and the Arts. New Haven and New York. From 1816.
South African Mines, Commerce, and Industries. Johannesburg.
Transactions of the American Institute of Mining Engineers. Philadelphia. From 1871.
Transactions of the Australasian Institute of Mining Engineers. Sydney. From 1893.
Transactions of the Institution of Mining Engineers. Newcastle-on-Tyne. From 1852.
Transactions of the South African Association of Engineers. Johannesburg. From 1902.
Transactions of the Institution of Mining and Metallurgy. London. From 1892.

GENERAL METALLURGY OF GOLD.

Geber, the Works of. Translated by R. Russell. London, 1686.
Biringuccio. De la Pirotechnia. Venice, 1540. French translation, Rouen, 1627.
Agricola (Georgius). De re metallica. Bale, 1556.
Michaelis (Johannis). De Oro. Leipzig, 1630.
Barba (Alphonzo). Arte de los metales. Madrid, 1639. French translations, Paris, 1751, and La Haye, 1782.
Schluter. Principles of Metallurgy and Assaying. Brunswick, 1738.
Vargas (Perez de). Traité singulier de metallurgie. Translated from the Spanish. Paris, 1743.
Lewis (Wm.). Commercium Philosophico-Technicum. London, 1763.
Valerius. Grundriss der Metallurgie. Ulm, 1768.
Cramer. Principes de metallurgie et docimasie. Blankenburg, 1774.
Jars. Voyages metallurgiques. Paris, 1774-81.
Karsten. System der Metallurgie. Breslau, 1818.
Kiessling. Die Metallurgie. Dresden, 1841.
Ansted (D. T.) The Gold-Seekers' Manual. London, 1849.
Landrin (H.) Traité de l'or. Paris, 1850 and 1863.
Rammelsberg. Lehrbuch der chemischen Metallurgie. 1850.
Phillips (J. A.) Gold Mining and Assaying. London, 1852.
Phillips (J. A.) *Encyclopædia Metropolitana.* Article on "Metallurgy." London, 1854.

BIBLIOGRAPHY.

Rivot (L.) Principes generaux du traitement des minerais metalliques. Paris, 1859, nouv. ed , 1871-73.

Kerl (B.) Die Rammelsberger Hüttenprozesse. Clausthal, 1861.

Kustel (G.) Processes of Gold and Silver Extraction. San Francisco, 1863.

Kerl (B.) Handbuch der metallurgischen Hüttenkunde. Clausthal, 1865.

Phillips (J. A.) The Mining and Metallurgy of Gold and Silver. London, 1867.

Overman (F.) A Treatise on Metallurgy. New York, 1868.

Crookes (Wm.) and Röhrig (E.) Treatise on Practical Metallurgy. Translated from the German. London, 1868-70.

Blake (W. P.) Report on the Precious Metals. Washington, 1869.

Raymond (R. W.) Mineral Resources West of the Rocky Monntains. 7 vols. Washington, 1869-74.

Raymond (R. W.) Mines, Mills, and Furnaces of the Pacific States. New York, 1871.

Schiern (F.) Sur l'origine de la tradition de fourmis qui ramassent l'or. Copenhagen, 1873.

Makins (G. H.) Manual of Metallurgy, particularly Precious Metals. 2nd edition. London, 1873.

Greenwood (W. H.) Manual of Metallurgy, vol. ii. London, 1875.

Kerl (B.) Grundriss der Metallhüttenkunde. Leipzig, 1880.

Simonin (L.) L'Or et l'Argent. Paris, 1880.

Industrial Progress in Gold Mining. Philadelphia, 1880.

Percy (John). Metallurgy of Silver and Gold, vol. i. London, 1880.

Ryan (J.) Gold Mining in India. London, 1881.

Lock (A. G.) Gold : Its Occurrence and Extraction. With a Bibliography. London, 1882.

Egleston (T.) The Progress of the Metallurgy of Gold and Silver in the United States. New York, 1882.

Balch (W. R.) Mines, Miners, and Mining Interests of the United States in 1882. Philadelphia, 1882.

Restrepo. Estudio sobre las minas de oro y Plata de Colombia. Bogota, 1884.

Gore (G.) Art of Electro-Metallurgy. New York, 1884.

Zoppeti. L'electrolisi in metallurgica. Milan, 1885.

Egleston (Thos.) Metallurgy of Silver, Gold, and Mercury in the United States. 2 vols. London, 1887-90.

Balling (C.) Grundriss der electrometallurgie. Stuttgard, 1888.

Frémy. L'Or dans le laboratoire. *Encyclopædie Chimique*, vol. iii. Cahier 16e. Paris, 1888.

Watt (A.) Electro-Metallurgy Practically Considered. London, 1889.

Lock (G. W.) Practical Gold Mining. London, 1889.

Juptner (H.) Traité pratique de chimie metallurgique. Paris, 1891.

Frémy. L'Or dans les centres de travail et de l'industrie. *Encyclo. Chim.*, vol. v. Cumenge and Fuchs. Paris, 1891.

Phillips (J. A.) and Bauerman. The Elements of Metallurgy. London, 1891.

Raymond (R. W.) Gold and Silver: Report of the 11th Census of the U.S. New York, 1892.

Hatch and Chalmers. Gold Mines of the Rand. London, 1895.

De la Coux (H.) L'or. Gites auriferus—Extraction de l'or. Paris, 1895.

Launay (L. de). Les Mines d'Or du Transvaal. Paris, 1896.

Becker (H.) L'Or, Minerais auriféres et auro-argentiferes. Paris, 1896.

Schnabel (C.) Translated by H. Louis. Handbook of Metallurgy. 2 vols. London, 1898.

Coignet (F.) Traitement des Quartz Auriferes. *Bulletin de la Société L'Industrie Minerale.* Tome xii.-xiii. 1898-9.

BIBLIOGRAPHY.

Begeer (B. W.) Metallurgy of Gold on the Rand. Freiberg, 1898.
Eissler (M.) The Metallurgy of Gold. 5th edition. London, 1900.
Wade (E. M. and M. L.) A Compendium of Gold Metallurgy (pocket book). Los Angelos, 1901.
Hatch (F. H.) The Kolar Goldfield. *Memoirs of the Geological Survey of India.* Calcutta, 1901.
Roberts-Austen (Sir W. C.) Introduction to the Study of Metallurgy. 5th Edition. London, 1902.
Charleton (A. G.) Gold Mining and Milling in Western Australia. London, 1903.
Encyclopædia Britannica. 10th edition. Article on "Gold." London, 1903.
Clark (D.) Australian Mining and Metallurgy. Sydney, 1904.

CHAPTERS I. AND II.—PROPERTIES OF GOLD, ITS ALLOYS AND COMPOUNDS.

Budelius (R.) De monetis. Coloniæ Agrip, 1591.
Savot. Discours sur les Medalles Antiques. Paris, 1627.
Potier (M.) Philosophica Chemica. Francfort, 1648.
Borrichius. Hermetes Ægyptiorum et Chemicorum Sapientia. Copenhagen, 1674.
Gobet. Les anciens mineralogistes du royaume de France. Paris, 1679.
Gellert (C. E.) Metallurgic Chemistry. Translation. London, 1796.
Hatchett (J.) Wear of Coins. *Phil. Trans. Roy. Soc.*, 1803, p. 43.
Hatchett. Experiences et observations sur l'or. Paris, 1804.
D'Arcet. L'art de dorer le bronze. Paris, 1818.
Schmieder. Geschichte der Alchemie. Halle, 1832.
Boue (P.) Traité d'orfèvrerie, bijouterie, et jouaillerie. Paris, 1832.
Gmelin (L.) Translated by H. Watt. Handbook of Chemistry. Chapter on Gold. Vol. vi., pp. 200-251. London, 1851.
Ansel (G. F.) A Treatise on Coining. London, 1862.
Rossignol (J. P.) Les metaux dans l'antiquité. Paris, 1863.
Ronchaud (L. de). Dictionnaire des antiquités grecques et romaines. Article, "aurum."
Mommsen. Histoire de la monnaie romaine. Paris, 1868.
Lepsius. Les metaux chez les Egyptiens. Paris, 1877.
Wright (C. R. A.) Metals. London, 1878.
Cripps. Old English Plate. London, 1878, 1891.
Lenormant. La Monnaie dans l'antiquité. 3 vols. Paris, 1878.
Gee (G. E.) The Goldsmith's Handbook. London, 1879.
Pollen (J. H.) Ancient and Modern Gold and Silversmith's Work. London, 1879.
Noback (Fr.) Münz-, Maass- und Gewichtsbuch. Leipzig, 1879.
Douan. Inoxydation, dorure, et platinage des métaux. Paris, 1880.
Brandis. Das Münz-mass und Gewichtwesen in Vorder-Asien bis auf Alexander den Grossen. Berlin.
Wagner (A.) Gold, Silber, und Edelsteine. Vienna, Leipzig, and Pesth. 1881.
Wheatley and Delamotte. Art Work in Gold and Silver. London, 1881.
Bloxam (C. L.) Metals: their Properties and Treatment. London, 1882.
Achiardi (T.) Metalli. Milan, 1883.
Kenyon (R. L.) The Gold Coins of England. London, 1884.
Roberts-Austen (W. C.) *Cantor Lectures, Soc. of Arts.* London, 1884, 1888, 1892, 1897, and 1901.
Berthelot (M.) Origines de l'alchemie. Paris, 1885.
Streeter (E. W.) Gold: the Standards of all Countries. London, 1885.
Kopp (H.) Die Alchemie in älterer und neuerer Zeit. Heidelberg, 1886.

Schaefer (H. W.) Die Alchemie, &c. Flensborg, 1887.
Riche (A.) Monnaie, Medailles et Bijoux. Paris, 1889.
Brannt (W. T.) Metallic Alloys. London, 1889.
Guettier (A.) Practical Guide for the Manufacture of Metallic Alloys, 1865. Translated from the French by A. A. Fesquet. New York. 1890.
Wagner (A.) Gold, Silber, und Edelsteine. Handbuch für Gold-, Silber-, Bronze-Arbeiter und Juweliere. Leipzig, 1895.
Smith (E. A.) Dental Alloys. London, 1897.
Hiorns (A. H.) Mixed Metals. London, 1901.
Griffiths (A. B.) Dental Metallurgy. London, 1903.

CHAPTER III.—MODE OF OCCURRENCE AND DISTRIBUTION OF GOLD.

Holzchul. Remarques sur l'or des mines de Saxe. Penig, 1805.
Atkinson (S.) Discoverie and Historie of the Gold Mines in Scotland. Edinburgh, 1825.
Miers (J.) Travels in Chile and La Plata. London, 1828.
Dupont (S. C.) De la Production des Metaux precieux au Mexique: considerée dans ses rapports avec la metallurgie, &c. Paris, 1843.
Tchihatchef (Pierre de). Voyage Scientifique dans l'Altai Oriental. Paris, 1845.
Papers relating to the Discovery of Gold in Australia. 2 vols. London, 1852-57.
Calvert (J.) The Gold Rocks of Great Britain and Ireland. London, 1853.
Clarke (W. B.) Researches in the Southern Gold Fields of New South Wales. Sydney, 1860.
Davison (S.) Geognosy of Gold Deposits in Australia. London, 1861.
Rosales (H.) Essay on the Origin and Distribution of Gold in Quartz Veins. Melbourne, 1861.
Brown (J. R.) Mineral Resources of the United States. Washington, 1867.
Lovell (J.) Gold Fields of Nova Scotia. Montreal, 1868.
Smyth (R. Brough). Gold Fields of Victoria. Melbourne, 1869.
Mackay (J.) Report on the Thames Gold Fields. Wellington, N.Z., 1869.
Cotta (B. von). Treatise on Ore Deposits. Translated. New York, 1870.
Henwood (W. J.) On the Gold Mines of Minas Geraes, Brazil, Gold Fields of Kumaon and Garhwall, &c. *Transactions of the Royal Geological Society of Cornwall.* Vol. viii., 1871.
Bateman (A. W.) South African Gold Fields. *The Times*, Sept. 28, 1874.
Domeyko (J.) Ensayo sobre los Depositos Metaliferos de Chile. Santiago, 1876.
Bain (A. G.) Gold Regions of S.E. Africa. London and Cape Colony, 1877.
Church (J. A.) The Comstock Lode: its Formation and History. New York, 1879.
Jenney (W. P.) Mineral Resources of the Black Hills of Dakota. Washington, 1880.
Ball (Prof. B.) Diamonds, Coal, and Gold in India: their Occurrence and Distribution. London, 1881.
Blake (W. P.) Geology and Mineralogy of California. Sacramento, 1881.
Jervis (G.) Dell'Oro in Natura. Rome, 1881.
Emmons (S. F.) and **Becker (G. F.)** Precious Metals: being vol. xiii. of U.S. Census Reports of 1880. Washington, 1885.
Handbook of New Zealand Mines. Wellington, 1887.
Ferguson (A. M.) All about Gold in Ceylon and S. India. Colombo, 1888
Liversidge (J.) Minerals of New South Wales. London, 1888.
Posewitz (T.) Borneo: its Geology and Mineral Resources. London, 1892.

BIBLIOGRAPHY.

Lock (C. G. W.) Economic Mining. London, 1895.
Phillips (J. A.) Edited by H. Louis. Treatise on Ore Deposits. 2nd ed. London, 1896.
Schmeisser (K.) and **Vogelsang** (K.) Translated by H. Louis. The Gold Fields of Australasia. London, 1898.
Steuart (D. S. S.) Mineral Wealth of Zoutpansberg. *Transactions Institution of Mining Engineers*, vol. xvii.
Anderson (J. W.) The Prospector's Handbook: a Guide for the Prospector and Traveller in search of metal-bearing and other valuable minerals. 9th edition. London, 1902.
Truscott (S. J.) The Witwatersrand Gold Fields. 2nd ed. London, 1902.
Posepny (F.) The Genesis of Ore Deposits. New York, 1902.
MacLaren (J. M.) The Occurrence of Gold in Great Britain. *Transactions of the Institution of Mining Engineers*, vol. xxv., pp. 435-508, 1902.
Fuchs (E.) and **Launay** (L. de). Traite des gîtés mineraux et metalliferés. 2 vols. Paris, 1903.
Launay (L. de). Les richesses minérales de S. Afrique et de Madagascar. Paris, 1903.

CHAPTERS IV. AND V.—PLACER MINING.

Moneeram. Native Account of Washing for Gold in Assam. *Journ. Asiat. Soc. Bengal*, vol. vii., p. 621, 1838.
Abbott (Capt. J.) Account of the process employed for obtaining gold from the sand of the River Beyass: with a short account of the gold mines of Siberia. *Journ. Roy. Asiat. Soc. Bengal*, vol. xvi., pp. 266-272, 1847.
Delesse. Gisement et exploitation de l'or en Australie. Paris, 1853.
Report of the Royal Commission appointed to inquire into the best methods of removing sludge from the gold fields. Melbourne, 1859.
Radde (Gustav). Reisen im Süden von Ost-Sibirien in den Jahren, 1855-59. St. Petersburg, 1863.
Debombourg (G.) Gallia aurifera. Études sur les alluvions auriféres de la France. Lyons, 1868.
Christy (S. B.) Ocean Placers of San Francisco. *Proc. Cal. Acad. Sci.*, August, 1878.
Egleston (T.) Hydraulic Mining in California. London, 1878.
Whitney (J. D.) Auriferous Gravels of the Sierra Nevada of California. Cambridge, U.S., 1880.
Hammond (J. H.) Auriferous Gravels of California and the Methods of Drift Mining. *Prod. of Gold and Silver in U.S. for 1881.* Washington, 1882.
Bowie, Jr. (A. J.) Practical Treatise on Hydraulic Mining in California. New York, 1885.
Bowie, Jr. (A. J.) Mining Debris in Californian Rivers. San Francisco, 1887.
Hammond (J. H.) Auriferous Gravels of California. *Ninth Annual Report California State Mining Bureau*, pp. 105-138, 1889.
Gould (E. S.) Practical Hydraulic Formulæ for the Distribution of Water through long Pipes. New York, 1891.
Kirkpatrick (T. S. G.) Hydraulic Gold Miner's Manual. New York, 1891.
Wagenen (T. F. van). Manual of Hydraulic Gold Mining. New York, 1891.
Levat (E. D.) Memoire sur l'Exploitation de l'Or en Siberie Orientale. Paris, 1896.
Johnson (J. C. F.) Getting Gold. A Practical Treatise for Prospectors. London, 1897.
M'Kay (A.) Older Auriferous Drifts of Central Otago, New Zealand Mines Dep. Wellington, 1897.

BIBLIOGRAPHY.

Batz (Baron René de). Les Gisements Aurifères de Siberie. Paris, 1898.
Wilson (E. B.) Hydraulic and Placer Mining. London and New York, 1898.
Grothe (A.) Gold Dredging in the United States. *Mineral Industry*, vol viii., pp. 326-336, 1899.
Radford (G. K.) Mining for Gold in the Auriferous Gravels of California. *Transactions Institution of Mining Engineers*, vol. xvii., pp. 452-481, 1899.
Barbour (T. J.) Gold Dredging in California. *Proceedings of the California State Miners' Association.* San Francisco, 1900.
Schrader (F. C.) and **Brooks** (A. H.) Preliminary Report of the Cape Nome Gold Region. *Professional Papers U.S. Geological Survey.* Washington, 1900.
Winslow (G.) Notes on Gold Dredging. *Proceedings of the Institution of Civil Engineers*, vol. cliii., 1902-3.
Longridge (C. C.) Hydraulic Mining, in four parts. London, 1903.
Longridge (C. C.) Gold Dredging. London, 1905.

CHAPTERS VI., VII., AND VIII.—QUARTZ CRUSHING AND AMALGAMATION.

De Born. Amalgamation des minerais d'or et d'argent. Vienna, 1786. English translation by R. E. Raspe, 1791.
Sonneschmied. L'Amalgame espagnol. Leipzig, 1811.
Sonneschmied. Traité sur l'amalgamation. Ronneburg, 1811.
Rivot. Nouveau procédé de traitement des minerais d'or et d'argent (a manuscript in the archives of the École des Mines de Paris). Paris, 1818.
Ortmann. Kurze Geschichte der Amalgamation in Sachsen. Freiberg, n.d.
Lawson (G.) Improvements in Amalgamation. *Trans. Nova Scotia Inst.* 1866.
Hague (J. D.) Gold Mining in Colorado. *Report on the Fortieth Parallel*, vol. iii. Washington, 1870.
Keith (N. S.) Amalgamated Copper Plates. *Eng. and Mng. Journ.*, vol. xi., p. 270, 1871.
Blake (W. P.) Mining Machinery. New Haven, 1871.
Fonseca. Memoire sur l'amalgamation Chilienne. Paris, 1872.
Bergmann (E. von). Die Anfänge des Geldes in Ægypten. Vienna, 1872.
Sonneschmidt (F) Tratado de la Amalgamacion de Mexico. Mexico, 1876.
Egleston (T.) Californian Stamp Mills. London, 1880.
Egleston (T.) Treatment of Gold Quartz in California. London, 1881.
Randall (P. M.) Quartz Operators' Handbook. New York, 1888.
M'Dermott & Duffield. Gold Amalgamation and Concentration. London and New York, 1890.
Curtis (A. Harper). Gold Quartz Reduction. *Proc. Inst. C.E.*, vol. cviii. (1892), part ii.
Charleton (A. G.) The Choice of Coarse and Fine Crushing Machinery and Processes of Ore Treatment. *Trans. Fed. Inst. Mng. Eng.*, 1892-94. Seven Papers.
Louis (H.) Handbook of Gold-Milling. London, 1894.
Rickard (T. A.) Variations in Gold Milling. New York, 1895.
Rickard (T. A.) The Stamp Milling of Gold Ores. New York and London, 3rd edition, 1901.
Lock (C. G. W.) Gold Milling: Principles and Practice. London, 1901.
Adams (W. J.) Hints on Amalgamation and the General Care of Gold Mills. Chicago, 1901.
Richards (R. H.) Ore Dressing. 2 vols. New York, 1903.

BIBLIOGRAPHY.

CHAPTER IX.—CONCENTRATION IN GOLD MILLS.

Gaetzschmann (M. F.) Die Aufbereitung. (Mechanical Concentration of Ores.) Leipzig, vol. i., 1864; vol. ii., 1872.

Rittinger (P. von). Lehrbuch der Aufbereitungskunde. Berlin, 1867, 1870, and 1873.

Smyth (Sir W. W.) Dressing, or the Mechanical Preparation of Gold Ores. Lectures on Gold. *Mng. Journ.*, 1873.

Schmidt (A. W.) Der Schlamfänger auch Kornfänger genannt. Dillenburg, 1877.

Habermann (J.) Comparison of the Salzburg Table, the Rittinger Table, and the Hand Buddle. *Oesterr. Zeitschr. für Berg- und Hüttenwesen*, 1879, No. 8.

Cazin (F. M. F.) Dynamical Metallurgy or Mechanical Ore Concentration. *Mining Record*, 1881-2.

Callon (J.) Lectures on Mining. Translated by Le Neve Foster and W. Galloway, vol. iii. London, 1886.

The Settling of Solid Particles in Liquids. Bulletin No. 36. *United States Geological Survey*. Washington, 1886.

Ore Dressing in California. *Sixth Report of the Cal. State Min.*, 1886. This is a full account of the machines actually at work.

Lock (C. G. W.) Mining and Ore Dressing Machinery. London, 1891.

Kunhardt (W. B.) The Practice of Ore Dressing in Europe. New York, 1891.

Commans. Concentration and Sizing of Crushed Ore. *Proc. Inst. Civil Eng.*, 1894.

Rosales (H.) Report on the Loss of Gold in the reduction of Auriferous Veinstone in Victoria. Melbourne, 1895.

De La Goupilliere (H.) Cours a'exploitation des Mines (chapter on Ore Dressing). Paris, 1896-7.

Lock (C. G. W.) Gold Milling. London, 1901.

MacLaren (J. M.) Queensland Mining and Milling Practice. *Queensland Geological Survey*. Brisbane, 1901.

Davies (E. H.) Machinery for Metalliferous Mines. London, 1902.

Richards (R. H.) Ore Dressing. New York, 1903.

Foster (C. Le Neve.) Ore and Stone Mining. 6th edition. London, 1905.

CHAPTER XII.—ROASTING OF GOLD ORES.

Plattner (C. F.) Die metallurgische Röstprozesse. Freiberg, 1856.

Kustel (G.) Roasting of Gold and Silver Ores. San Francisco, 1880.

Stetefeldt (C. A.) On Salt Roasting. *Trans. Am. Inst. Mng. Eng.*, vol. xiii. New Haven, 1885.

Christy (S. B.) Losses of Gold in Roasting. *Trans. Am. Inst. Mng. Eng.*, 1888.

Adams (W. H.) Pyrites: Practical Methods for Extraction of Gold, &c. New York, 1892.

CHAPTERS XIII, XIV., AND XV.—CHLORINATION.

The earliest researches on the chlorination of gold were made by—

(1) **Duflos (Dr.)** *Die schles. gesell. Uebersicht.* Breslau, 1848.

(2) **Richter (Theo.)** *Journ. für Prak. Chem.*, vol. li. (1849), p. 151.

(3) **Lange (Herr).** *Karsten's Archiv*, vol xxiv., pp. 396-429. 1852.

(4) **Plattner (C. F.)** Probirkunst, p. 570. 1853.

(5) **Percy (J.)** *Phil. Mag.*, vol. xxxvi. (1853), pp. 1-8.

Whelpley and **Storer**. Method of separating Metals from Sulphurets. Boston, 1866.
Küstel (G.) Concentration and Chlorination. San Francisco, 1868.
Pyrites: Report of the Board appointed to report on the methods of treating Pyrites and Pyritous Veinstuff, as practised on the Gold Fields. Melbourne, 1874.
Egleston (T.) Leaching Gold and Silver Ores in the West. New York, 1883.
Egleston (T.) Leaching Gold Ores containing Silver. London, 1886.
Stetefeldt (C. A.) Lixiviation of Silver Ores. New York, 1888. This work gives many details applicable to all wet processes.
Aaron (C. H.) Notes on the Hydro-Metallurgy of Gold (Loss of Gold by Smelting). *Report of the California State Mining Bureau.* San Francisco, 1888.
O'Driscoll (F.) Notes on the Treatment of Gold Ores. London, 1889.
Rothwell (J. E.) Recent Improvements in Chlorination. *Mineral Industry for 1892, for 1896*, and *for 1901.*
France (Ch. D.) Extraction par voie humide du cuivre, de l'argent, et de l'or. Paris, 1897.
Wilson (E. B.) The Chlorination Process. London and New York, 1897.
Thomson (F. A.) and **Goodall (S. L.)** The Portland Mill. (Description of the Chlorination Plant.) *Mines and Minerals*, Oct. and Nov. 1904.

CHAPTERS XVI. AND XVII.—CYANIDE PROCESS.

Articles in *Mineral Industry* from 1892.
Scheidel (A.) The Cyanide Process. Sacramento, 1894.
Reunert (T.) Diamonds and Gold in South Africa. London, 1894.
Gaze (W.) Handbook of Practical Cyanide Operations. London, 1898.
Eissler (M.) The Cyanide Process. 3rd edition. London, 1902.
Fulton (C. H.) The Cyanide Process. *Bulletin No. 5, South Dakota School of Mines.* Rapid City, Dakota, 1902.
Wilson (E. B.) The Cyanide Process. 3rd ed. New York and London, 1902.
James (A.) Cyanide Practice. 3rd edition. London, 1903.
Uslar (M.) and **Erlwein (G.)** Cyanid Prozesse zur Goldgewinnung. Halle, Germany, 1903.
Julian (H. F.) and **Smart (E.)** Cyaniding Gold and Silver Ores. London, 1904.
Clennell (J. E.) Chemistry of Cyanide Solutions resulting from the Treatment of Ores. New York, 1904.
Bosqui (F. L.) Practical Notes on the Cyanide Process. 3rd edition. New York, 1904.
Park (Jas.) The Cyanide Process of Gold Extraction. Auckland and Melbourne, 1896. London, 1904.

CHAPTER XVIII.—REFINING AND PARTING OF GOLD BULLION.

Goddard (Jonathan). Experiments on Refining Gold with Antimony. *Phil. Trans. Roy. Soc.*, 1676.
Egleston (T.) Parting Gold and Silver in California. New York, 1877.
Egleston (T.) Parting Gold and Silver by means of Iron at Lautenthal. New York, 1885.
Egleston (T.) The Separation of Silver and Gold from Copper at Oker. Washington, 1885.
Hugon. Etude sur le raffinage electrolytique du cuivre noir. Paris, 1885.

BIBLIOGRAPHY.

Egleston (T.) Treatment of Gold and Silver at the United States Mint. London, 1886.
Gumbinner (Sven). Parting Gold and Silver in the United States. *Ninth Report of the Cal. State Min.*, 1889, pp. 62-104. This is a complete account of the methods in use in the United States.
Ulke (Titus). Parting and Refining Gold and Silver. *Mineral Industry for 1895*, pp. 343-366.
Iles (M. W.) Notes on the Moebius Process for Parting Gold and Silver Ores. *Mineral Industry*, vol. viii., 1899; also vols. ii., iv., and v.
M'Millan (W. G.) Treatise on Electro-Metallurgy. London, 1899.
Watt (A.) and **Philip** (A.) Electro-Plating and Electro-Refining of Metals. London, 1902.
Langbein (G.) Translated by W. T. Brannt. Complete Treatise on the Electro Deposition of Metals. Philadelphia, 1902.
Borchers (W.) Translated by W. G. M'Millan. Electric Smelting and Refining. 2nd edition. London, 1904.

CHAPTERS XIX. AND XX.—ASSAYING.

Ercker (L.) Allerfurnemisten Mineralischen Eerzt und Bergwerks Arten. Frankfort, 1580. Another edition, Frankfort, 1629.
Carranza (A.) El Ainstamieto i Proporcion de las Monedas de Oro, Plata i Cobre i la reduccion distos Metales a su Debida estimaccion. Madrid, 1629.
Bedrock (Wm.) A New Touchstone for Gold and Silver Wares. London, 1651. 2nd edition, 1679.
Le Febure (N. R.) Compleat body of chymistry. 1670.
Pettus (Sir John). Laws of Nature in Assaying Metals. Translated from the German of L. Ercker. London, 1686.
Cramer (J. A.) Elementa artis docimasticæ. Lugduni Batavorum (Leyden), 1741.
Symonds (W.) Essay on the Weighing of Gold, &c. London, 1756.
Cramer, M. D. (J. A.) Elements of the Art of Assaying Metals. Trans. from Latin by C. Mortimer, M.D. 2nd edition. London, 1764. An account of this book by Prof. Austin of Michigan University is given in the *Eng. and Mng. Journ.*, July 28, 1904, p. 144.
Pouchet. Le nouveau titre des matières d'or et d'argent. Rouen, 1789.
Aldridge (W. J.) The Goldsmith's Repository, containing Treatise on Assaying. London, 1789.
Citoyen Tauquelin. Manuel de l'Essayeur. Paris, An. vii. (1800).
Becquerel. Gold and Electricity. *Ann. de Chimie et de Physique*, vol. xxiv. (1823); also Article by **Oersted**, d°, vol. xxxix. (1828), p. 274.
Chaudet. L'Art de l'essayeur. Paris, 1835.
Bodemann (Th.) Anleitung zur Berg- und Hüttenmannischen Probirkunst. Clausthal, 1845.
Berthier. Traité des essais par la voie séche. Paris, 1847.
Pettenkofer. *Bergwerksfreund*, vol. xii. (1849). Article on Gold Bullion Assaying.
Watherston (J. H.) The Gold Valuer. London, 1852.
Plattner (C. F.) Probirkunst. Freiberg, 1853. 6th ed., Leipzig, 1897.
Terrell (A.) Atlas de Chemie analytique minerale. Paris, 1861.
Bodemann and **Kerl**. Treatise on Assaying. Translated by W. A. Goodyear. New York, 1868.
Domeyko (J.) Tratado de Ensayes, tanto por la via seca como por la via humeda. Chile, 1873. 5th edition. Mexico and Paris, 1889.
Foord (G.) Mechanical Assay of Quartz. *Trans. Roy. Soc. Victoria*, vol. x., pp. 139-147. 1874.

BIBLIOGRAPHY.

Broch (**Dr. O.**) Assay of Gold by means of its Density. [*Norwegian*] *Nyt. Mag. für Naturvsk.* Christiania, 1876.
Phillips (**J. S.**) Explorers and Assayers' Companion. San Francisco, 1879.
Attwood (**G.**) Practical Blowpipe Assaying. London, 1880.
Chapman (**E. J.**) Assay Notes. Practical instructions for the determination by furnace assay of Gold and Silver in rocks and ores. Toronto, 1881.
Balling (**C.**) Probirkunde. Pibram, 1879.
Balling (**C.**) L'Art de l'essayeur. Paris, 1881.
Kerl (**B.**) Metallurgische Probirkunst. Translated by W. T. Brannt (*Assayers' Manual*). London, 1883. Translated by Garrison. 2nd ed. Philadelphia, 1889.
Jagnaux (**Raoul**). Traité pratique de analyses chimiques et d'essais industriels. Paris, 1884.
Rössler (**H.**) Article on Gold Bullion Assaying. *Dingler's Polyt. Journ.*, vol. ccvi. (1884).
Black (**J. G.**) Chemistry of the Gold Fields. Dunedin, N.Z., 1885.
Rivot (**L. E.**) Docimasie Traité d'Analyse des substances minerales. Tome v. Paris, 1886. [Containing useful notes on assaying complex ores].
Mitchell (**W.**) Manual of Practical Assaying. Edited by Wm. Crookes. 6th edition. London, 1888.
Hoirns (**A. H.**) Practical Metallurgy and Assaying. London, 1888.
Frémy. L'or dans le Laboratoire. Cumenge and Fuchs. *Ency. Chim.*, vol. iii., c. 16e. Paris, 1888.
Ross (**W. A.**) Blowpipe Analysis. London, 1889.
Brown and **Griffiths**. Manual of Assaying of Gold, Silver, &c. London, 1890.
Lieber (**O. M.**) Assayer's Guide. New York.
Plattner. Blowpipe Analysis. Enlarged by **Richter** (**Th.**) Translated by H. B. Cornwall. New York, 1890.
Riche (**A.**) L'Art de l'essayeur. Paris, 1892.
Furman. Practical Assaying. New York, 1894. 5th edition, 1900.
Ledoux (**A. R.**) Assay of Copper and of Copper Materials for Gold and Silver. *Am. Inst. Mng. Eng.*, October, 1896.
Campredon (**L.**) Guide pratique du chimiste, Metallurgiste et de l'essayeur. Paris, 1898.
Aaron (**C. H.**) Manual of Assaying. 3rd edition. San Francisco, 1900.
Brown (**W. L.**) Manual of Assaying. Gold, Silver, &c. 9th edition. Chicago, 1900.
Simonds (**E. H.**) Practical Course in the Fine Assaying for Gold, Silver, &c. San Francisco, 1900.
Ricketts and **Miller**. Notes on Assaying. 3rd edition. 1900.
Merritt (**W. H.**) Field Testing for Gold and Silver Ores. London, 1900.
Merritt (**W. H.**) Gold and Silver Ores, what is their value? New York, 1901.
Miller (**A. S.**) Manual of Assaying. 1901.
Smith (**E. A.**) Assaying of Complex Gold Ores. *Trans. Inst. of Mining and Metallurgy*, vol. ix., pp. 315-361. London, 1901.
Rhead (**E. L.**) and **Sexton** (**A. H.**) Assaying and Metallurgical Analysis. London, 1902.
Lindgren (**W.**) Tests for Gold and Silver Ores in Shales from Western Kansas. *Bulletin No. 202, U.S. Geol. Survey.* Washington, 1902.
Beringer (**J. J. & C.**) Manual of Assaying. 9th edition. London, 1904.
Macleod (**W. A.**) and **Walker** (**C.**) Metallurgical Analysis and Assaying. London, 1904.
Lodge (**R. W.**) Notes on Assaying and Metallurgical Analysis. New York, 1904.

BIBLIOGRAPHY.

CHAPTER XXI.—STATISTICS OF GOLD PRODUCTION.

Jacob (Wm.) A History of the Precious Metals. 2 vols. London, 1831.
Humboldt (A. von). Fluctuations in the Supplies of Gold. London, 1839.
Blake (W. P.) The Production of the Precious Metals. London and New York, 1869.
Soetbeer (A.) Edelmetall Production. Gotha, 1879.
Del Mar (A.) A History of the Precious Metals (containing a Bibliography). London, 1880. New York, 1902.
O'Brien. Treatise on Gold and Silver. London, 1884.
Roswag (C.) L'Argent et L'Or, Production, consommation et circulation des métaux précieux. 2 vols. Paris, 1889-90.
Rothwell R. P.) Gold and Silver: their Production, Uses, and Logical Ratio. *Eng. and Mng. Journal*, vol. lx., pp. 76-8, 100-102. New York, 1895.
Welton (W. S.) Practical Gold Mining: its commercial aspects. London, 1902.

For later statistics see—

Reports of the Director of the Mint (United States).
Annual Reports of the Royal Mint (London).
Mineral Statistics (*Home Office Reports*).
Reports of Chamber of Mines of Transvaal.
Mining Statistics of Western Australia (*Government Gazette*).